A SOURCE BOOK IN THE FUNDAMENTALS OF CLASSICAL AND STATISTICAL THERMODYNAMICS

A SOURCE BOOK IN THE FUNDAMENTALS OF CLASSICAL AND STATISTICAL THERMODYNAMICS

Hanna A. Rizk

Professor of Chemistry,
Faculty of Science, University of Cairo

Library of Congress Control Number:		2012900658
ISBN:	Hardcover	978-1-4653-0926-6
	Softcover	978-1-4653-7318-2
	Ebook	978-1-4653-7319-9

This book was printed in the United States of America.

To order additional copies of this book, contact:
Xlibris Corporation
0-800-644-6988
www.xlibrispublishing.co.uk
Orders@xlibrispublishing.co.uk
302897

ABOUT THE AUTHOR

Professor Hanna Abdel- Messih Rizk, PhD, is a well-travelled , dedicated man, lecturer of over 4 decades. He completed his Bachelor's, Master's and Doctorate at the Cairo University (formerly known as Fouad El-Awal University).

After a brief stint working for the British Oil Company Shell, in Suez, Dr.Rizk pursued his calling and love of education, starting as a demonstrator in the Chemistry Departement, Faculty of Science for the University of Cairo sand then quickly promoted to lecturer for the same department to teach thermodynamics at various levels.

In November 1958, he joined Baghdad University, Iraq, as a lecturer in the Chemistry Department, Faculty of Sciences, for a period that had been assumed to be one year. But for political reasons following the revolution in Iraq, this period was not completed and resumed his post at Cairo University by March of 1959.

In June 1960, Dr. Rizk was seconded to Khartoum University in Sudan to work as a senior lecturer of chemistry until June 1962.

On his return to Cairo, the head of the National Research Centre in Egypt appointed Dr. Rizk to supervise an established research unit for carrying out research in the field of dielectrics, at the same time, working as an assistant professor in the Chemistry Department of Cairo University.

In January 1964, Dr. Rizk was offered a research fellowship at the Royal Institute of Technology, Stockholm, Sweden, for one year, concentrating on the nuclear magnetic resonance spectra of cellulose. An extension of the fellowship was then made in January 1956 for six months, ending in June 1965.

In August 1969, he was appointed by UNESCO, Pairs, as an export in chemistry for a period of two years. This period was extended twice, for another two years and, one more year thereafter.

In August 1974, Dr. Rizk resumed his scientific research activities in Cairo University with the rank of Professor. In the course of five years, the scientific activities he carried out took him to various places. In August 1970, he was recruited to Haile Selassie University, Faculty of Science, Addis-Ababa, Ethiopia for two years, which was then extended by additional one year. During the three years in Addis-Ababa, the Professor made use of his accumulated holiday leave of about three months and spent it at Massachusetts Institute of Technology, Boston MA from July 1971 to September 1971. The laboratory which he selected to work in was that of Professor J.Wilson's, where spectroscopic measurements involving Infra Red and Raman Spectra were properly carried out. At this Institute, he was also appointed as a visiting scientist without salary for the period from July to September 1971.

UNESCO awarded Prof. Rizk a study leave of three months starting from July 1972 to September 1972. He spent this leave at Harvard University, Cambridge, MA. He joined the research and laboratory team of Professor L. Nash who is regarded as one of the most brilliant teachers in the field of thermodynamics. At that time, Prof. Nash was the head of the Chemistry Department. Besides providing intellectual input in the many problems in the field of thermodynamics, Professor Nash was extremely kind to give Prof. Rizk access to one of the laboratories where Prof. Rizk was able to carry out experimental measurements tracing the possibility of complex formation in the Water-Pyridine system. He was appointed as an Honorary Research Associate for the period from July to September 1972, at the Harvard University.

On his return to Cairo University in August 1974, and culminating from a period of over thirty years, he was able to establish in steps laboratories in the Chemistry Department, Faculty of Sciences, to carry out research in the field of dielectrics. These laboratories were equipped with the most recent types of apparatus at that time. The required equipments were purchased from United Kingdom and Germany through tenders or invitations for biddings with the help of the Ministry of Foreign Affairs and the Ministry of International Cooperation in Egypt. Thus he was able to obtain the financial American Aid required for equipping the laboratories stepwise. These laboratories are called, Thermal Analyses Laboratories. Technicians from the selling companies in England and Germany were invited for few days to install the various types of equipment, set them into action, and teach the research students how to use them properly. More than 10 research students were able to obtain their master degree or Doctorate degree under the supervision of Prof. Rizk by using these equipments to carry out their experimental measurements.

In 1981 and 1982, Prof., Rizk taught thermodynamics in the Science Division of the American University in Cairo, as a part-time professor, for graduate and the undergraduate students for five semesters.

From 2000 onwards, Prof. Rizk dedicated himself to editing his text entitled. A Source Book in the Fundamentals of Classical and Statistical Thermodynamics. Writing of this book ended in August 2010.

A 10-year project, this book is his legacy to the Math, Physics and Sciences enthusiasts and learners around the globe.

CONTENTS

PART I

CHAPTER 1

A Brief Review of Classical Thermodynamics

Symbols for the important thermodynamic functions

In the table below, symbols for the important thermodynamic functions are listed.

Year	Author(s)	Entropy	Energy	Helmholtz free energy	Gibbs free energy	Enthalpy	Partial potential
1876	Gibbs	η	$\in$	ψ	ζ	χ	μ
1923	Lewis and Randall	S	E	A	F	H	$\overline{F}$
1924	Partington	S	U	F	Z	H	μ
1929	Schottky	S	U	F	G	W	μ
1933	Guggenheim	S	E	F	G	H	μ
1936	Fowler	S	E	F	G	H	μ
1936	Brönsted	S	E	F	G	H	μ
1937	Zamansky	S	U	F	G	H	μ
1939	MacDougall	S	E	A	F	H	μ
1939	Slater	S	U	A	G	H	–
1940	Mayer	S	E	A	F	H	–
1946	De Boer	S	U	F	G	W	μ
1948	Fowler and Guggenheim	S	E	F	G	H	μ
2010	This book	S	E and U	F	G	H	μ

1.1. Introduction

Thermodynamics deals with the relation of heat and work. Like mechanics, it is a science of great power and universality because it employs a small number of primary principles or postulates, from which may be drawn a variety of far-reaching deductions. The science is based on two very general principles, which are known as the *first* and the *second laws of thermodynamics*. The first of these two principles is the *law of conservation of energy*, which we are already familiar with. Thermodynamics differs from the kinetic-molecular interpretation of heat phenomena in that it is independent of any special theory of the mechanism of heat transfer and of the structure of the substances involved. It deals with the *equilibrium conditions* that exist at the beginning and at the end of the process and not with the forces and the conditions that intervene, and for this reason requires *no model of the structure of the matter.* Thermodynamics achieved its greatest usefulness, however, when it was properly correlated with a kinetic theory point of view, for it then furnished a general theory which is far in advance of any that could have been reached by either point of view taken alone.

The *first* law is described as the law of conservation of energy. This law was defined accurately by *Helmholtz, in 1847,* in the following words.

> *In all processes occurring in an isolated system, the energy of the system remains constant.*

The *second* law expresses the limitations of the *first* law by showing the relation between heat and work and the direction of energy changes. It was stated in different ways, one of which is

> *It is impossible for a self-acting machine, unaided by an external agency, to convey heat from a body at a low temperature to one at a higher temperature. Or heat cannot by itself (i.e. without the performance of work by some external agency) pass from a cold to a warmer body (Clausius).*

The analytical formulation of the first law for an infinitesimal change is

$$d'q = du + d'W \tag{1}$$

Here u is the *molar specific* internal energy, d'W represents an infinitesimal amount of work done by the system per mole, unlike du since it is not a state (or thermodynamic) function of the system being dependent upon the path followed in doing the work by the system. The same can be said with respect to d'q so that d'w and d'q should be expressed as inexact differentials. For a finite change, Eq. (1) is generalised into

$$q = \Delta u + W \tag{2}$$

1.2. Enthalpy, heat of reaction, and specific heat

1. Enthalpy. When the volume of a system is not kept constant, the pressure–volume work done by the system must be taken into account. This type of work may be of great importance, as it is for *heat engines.* Also, it may be just incidental, and this is the usual situation in chemistry. Chemical reactions are usually performed in vessels open to the atmosphere so that the pressure under which a reaction takes place remains essentially constant. If gases are evolved or used up or the volume of the system changes in any other way during the reaction, the resulting pressure–volume work done reversibly by the system at constant pressure is

$$\int_1^2 P dV = P\Delta V. \tag{1}$$

This work, as done on the surroundings, is in fact of no practical interest. Nevertheless, it must be taken into account, owing to the fact that the energy required for it must come from somewhere. It would be handy to deal with this incidental work *implicitly.* For this purpose, let us separate the work done by the system into two terms: the *pressure–volume work under constant pressure* $P\,\Delta V$ and the remainder which we call as *net work* and denote by w'; that is,

$$W = P\,\Delta V + w' \tag{2}$$

Subsequently, what is needed is a formulation in which the net work w' performed by the system appears *explicitly*. This can be accomplished for isobaric processes by introducing a *new function* called *enthalpy,* denoted by *H* and defined by the equation

$$H = U + PV \tag{3}$$

Since *U*, *P*, and *V* are functions of state, the enthalpy *H* is one also.

The enthalpy concept was introduced into thermodynamics by the American physicist and physical chemist *J. Willard Gibbs*, who was one of the founders of modern chemical thermodynamics.

It is noteworthy that the term $\Delta(PV)$ differs from the term $d(PV)$, as can be seen from the following:

$$\Delta(PV) = (PV)_2 - (PV)_1 = P_2 V_2 - P_1 V_1 \text{ by definition.}$$

Therefore,

$$= (P_1 + \Delta P)(V_1 + \Delta V) - P_1 V_1,$$
$$= P_1 V_1 + P_1 \Delta V + \Delta P V_1 + \Delta P \Delta V - P_1 V_1.$$

Hence,

$$\Delta(PV) = V_1 \Delta P + P_1 \Delta V + \Delta P \Delta V, \tag{4}$$

whereas

$$d(PV) = V dP + P dV. \tag{5}$$

The last term in Eq. (4) must not be overlooked, even though $d(PV)$ as given by Eq. (5) may tempt one to do so. On the other hand, if P = constant, Eq. (4) reduces to

$$\Delta(PV) = P \Delta V. \tag{6}$$

Based on this, we can derive the enthalpy change ΔH for a constant pressure process. Thus according to Eq. (3), we have

$$\Delta H_P = \Delta U_P + P\,\Delta V. \tag{7}$$

At constant volume, this equation reduces to

$$\Delta H_V = \Delta U_V \tag{8}$$

that is, the enthalpy change is equal to the internal energy change at constant volume.

It is therefore evident that by introducing the concept of enthalpy or heat content *H* defined by Eq. (3), the pressure–volume work is taken into account *implicitly*.

A modification of the analytical formula of the first law given by Eq. (1) using this new concept *H* is thus required.

Equation (1) for non-specific thermodynamic functions such as internal energy may be rewritten as

$$d'q = dU + d'W \tag{9}$$

So for a finite change, we have

$$q = \Delta U + W. \tag{10}$$

Also, for a *constant pressure change*, the external work *W* done by the system against the surroundings is equal to the sum of the pressure–volume work and the net work which is given by Eq. (2), so substitution of this equation in Eq. (10) yields

$$q_P = \Delta U_P + P\,\Delta V + w' . \qquad (11)$$

But the enthalpy change ΔH is related to the internal energy change by Eq. (7), hence

$$q_P = \Delta H_P + w'. \qquad (12)$$

In the absence of net work, Eq. (12) reduces to

$$q_P = \Delta H_P , \qquad (13)$$

and in the absence of both net work and volume change, Eq. (1) takes the form

$$q_v = \Delta H_v \qquad (14)$$

2. Heat of reaction. The heat of a reaction might be simply defined as *the heat absorbed by the reaction during its occurrence at a given temperature at a constant pressure or at a constant volume.* The symbols used are q_P and q_v, respectively. According to the thermodynamic conventions, when heat *q* is absorbed by a system, it is given a positive sign, and when it is evolved from or rejected by a system, it is given a negative sign. On this basis and in the light of Eqs. (13) and (14), for *constant pressure* reactions, we have either

$$q_P = \Delta H_P = \text{positive quantity}$$

for endothermic reactions, in the sense that enthalpy, heat content of the system, increases at constant pressure, or

$$q_P = \Delta H_P = \text{negative quantity}$$

for exothermic reactions, implying that there is a decrease in the heat content of the system.

In like manner, for *constant volume* reactions, we have either

$$q_v = \Delta H_v = \text{positive quantity}$$

in the case of endothermic reactions, or

$$q_v = \Delta H_v = \text{negative quantity}$$

in the case of exothermic reactions.

Examples of representing heat of reaction at constant pressure ΔH_P or q_P for a given temperature are

$$H_2(g) + \frac{1}{2}O_2(g) = H_2O(P),\ \Delta H = -68.32\ \text{kcal/mole at } 25°\text{C}.$$

$$H_2O\ (\text{ice}) = H_2O(\ell),\ \Delta H = 1.44\ \text{k cal/mole at } 0°\text{C}.$$

$$H_2O(\ell) = H_2O(g),\ \Delta H = 9.71\ \text{k cal/mole at } 100°\text{C}.$$

(The constant pressure is the normal atmospheric pressure.)

3. Specific heat. To deal with specific coordinates, we shall use *small* letters for their representation.

If $d'q_V$ be the amount of heat absorbed reversibly by 1 mole of a substance to raise its temperature at a constant volume by d*T*, we *define* its specific heat c_V as

$$d'q_V = c_v \, dT_V . \quad (1)$$

According to the first law, for an infinitesimal change, we have

$$d'q = du + d'W . \quad (2)$$

If d''W represents an infinitesimal amount of mechanical work only done by the system per mole reversibly, it will be given by

$$d'W = P dv . \quad (3)$$

Therefore Eq. (3) becomes

$$d'q = du + P dv .$$

At constant volume,

$$d'q_v = du_v . \quad (4)$$

Putting $U = u(T, v)$, we have

$$du = \left(\frac{\partial u}{\partial T}\right)_V dT + \left(\frac{\partial u}{\partial v}\right) dv \quad (5)$$

At constant volume, this equation reduces to

$$du_V = \left(\frac{\partial u}{\partial T}\right)_V dT_V , \quad (6)$$

so using that and according to Eq. (4), we have

$$d'q_V = \left(\frac{\partial u}{\partial T}\right)_V dT_V .$$

But according to the definition given above by Eq. (1), we have

$$c_v \, dT_V = \left(\frac{\partial u}{\partial T}\right)_V dT_V . \quad (7)$$

Hence

$$c_v = \left(\frac{\partial u}{\partial T}\right)_V \quad (8)$$

This is the way by means of which heat capacities entered thermodynamics.

In like manner, c_P entered thermodynamics as can be seen from the following:
Enthalpy, by the definition of *Gibbs* as shown above, is given by

$$h = u + Pv . \quad (9)$$

At constant pressure

$$dh_P = du + P dv . \quad (10)$$

The first law of thermodynamics as shown above is

$$d'q = du + d'W \tag{11}$$

for an infinitesimal change. If $d'W$ is only mechanical work, that is, pressure–volume work carried out by the system reversibly per mole, it is given by

$$d'w = P\,dv \tag{12}$$

So the first law (Eq. 11) takes the formulation

$$d'q = du + Pdv \tag{13}$$

Comparing Eq. (13) with Eq. (10), at constant pressure, we have

$$d'q_P = dh_P \tag{14}$$

Based on the definition of c_P as the amount of heat absorbed reversibly by 1 mole of a substance to raise its temperature at constant pressure by dT, that is

$$d'q_P = c_P\,dT. \tag{15}$$

Then it follows from Eqs. (14) and (15) that

$$c_P\,dT = dh_P,$$

or

$$c_P = \left(\frac{\partial h}{\partial T}\right)_P \tag{16}$$

This equation can be obtained by another way, by which Eq. (8) for c_v is obtained. Thus putting

$$h = h(T,P),$$

we have

$$dh = \left(\frac{\partial h}{\partial T}\right)_P dT + \left(\frac{\partial h}{\partial P}\right)_P dP.$$

At constant t pressure, this equation reduces to

$$dh_P = \left(\frac{\partial h}{\partial T}\right)_P dT_P.$$

When this equation is compared with Eq. (6), we obtain

$$d'q_P = \left(\frac{\partial h}{\partial T}\right)_P dT_P,$$

so using the definition given above for c_P by Eq. (15), we have

$$c_P\,dT_P = \left(\frac{\partial h}{\partial T}\right)_P dT_P,$$

or

$$c_P = \left(\frac{\partial h}{\partial T}\right)_P. \tag{16}$$

1.3. The pressure-volume-temperature relations for an adiabatic process of an ideal gas

In the case of an ideal gas which is undergoing an adiabatic process, the first law

$$d'q = dU + d'W$$

reduces to

$$dU = - d'W.$$

If the work done by the gas against the surroundings is only reversible mechanical work, then for an infinitesimal amount of work the expression is

$$d'W = P\,dV,$$

and thus

$$dU = - P\,dV = - RT\frac{dV}{V}.$$

But for an ideal gas, the heat capacity at constant volume is given by the expression

$$c_V = \left(\frac{\partial U}{\partial T}\right)_V$$

or

$$c_V\,dT = dU,$$

so that

$$c_V dT = -\frac{RT}{V}dV.$$

Subsequently

$$c_V \ln\frac{T_2}{T_1} = R\ln\frac{V_1}{V_2}.$$

Using this equation together with the equation

$$C_P - C_V = R$$

for any quatity of an ideal gas the heat capacity at constant volume is given by

$$C_V \ln\frac{T_2}{T_1} = \left(C_p - C_V\right)\ln\frac{V_1}{V_2},$$

$$C_V \ln\frac{T_2}{T_1} + C_V \ln\frac{V_1}{V_2} = C_p \ln\frac{V_1}{V_2},$$

Therefore,

$$\ln\frac{T_2}{T_1} + \ln\frac{V_1}{V_2} = \frac{C_P}{C_V}\ln\frac{V_1}{V_2}.$$

Putting $\frac{C_P}{C_V} = \gamma$, we obtain

$$\ln \frac{T_2}{T_1} = (\gamma - 1) \ln \frac{V_1}{V_2},$$

$$= \ln \left(\frac{V_1}{V_2} \right)^{\gamma - 1}.$$

Hence

$$T_1 V_1^{\gamma-1} = T_2 V_2^{\gamma-1}$$

or

$$T V^{\gamma-1} = \text{const.} \qquad (1)$$

Using the ideal gas equation $PV = RT$ per mole as the substitution of *T* in Eq. (1) by $\frac{PV}{R}$, we get

$$\frac{PV}{R} \cdot V^{\gamma-1} = \text{const.},$$

or

$$P V^{\gamma} = \text{const.} \qquad (2)$$

Also, by substituting in Eq. (1) for V, from the ideal gas equation, by $\frac{RT}{P}$, we obtain

$$T \left(\frac{RT}{P} \right)^{\gamma-1} = T^{\gamma} \left(\frac{1}{P} \right)^{\gamma-1} = \text{const.},$$

or,

$$T^{\gamma} P^{1-\gamma} = \text{const} \qquad (3)$$

Hence, for an ideal gas undergoing an adiabatic process, we have the following three equations:

$$T V^{\gamma-1} = \text{const.} \qquad (1)$$

$$P V^{\gamma} = \text{const.} \qquad (2)$$

$$T^{\gamma} P^{1-\gamma} = \text{const.} \qquad (3)$$

Finally, Eq. (3) can be utilised to find an expression for $\gamma = \left(\frac{C_P}{C_V} \right)$ in terms of pressure and temperature for an ideal gas as follows.

Thus according to this equation, we have

$$T_1^{\gamma}\, P_1^{\,1-\gamma} = T_2^{\gamma}\, P_2^{\,1-\gamma} \tag{3}$$

$$\gamma \ln T_1 + (1-\gamma)\ln P_1 = \gamma \ln T_2 + (1-\gamma)\ln P_2 ,$$

or

$$\gamma \ln\left(\frac{T_1}{T_2}\right) - \gamma \ln\left(\frac{P_1}{P_2}\right) = -\ln\frac{P_1}{P_2},$$

Hence,

$$\gamma = \frac{\ln\dfrac{P_2}{P_1}}{\ln\dfrac{P_1}{P_2} - \ln\dfrac{T_1}{T_2}}. \tag{4}$$

1.4. Reversible (a) isothermal and (b) adiabatic expansion of an ideal gas

a. Isothermal expansion.

According to the first law, we have

$$d'q = dU + d'W \tag{1}$$

for an infinitesimal change. Since the process is isothermal and the gas is ideal, the change in the internal energy is zero. This is because this energy is a function of temperature alone. Therefore, for a constant volume process or a constant pressure process at a given temperature, we have

Therefore, that is, for a constant it is of volume or pressure,

$$d'q = d'W$$

If $d'w$ stands only for mechanical, that is, pressure–volume work carried out reversibly by the system against the surroundings, we have

$$d'q = P\,dV$$

According to this equation, if the pressure varies, we have

$$\int d'q = \int_{P_1}^{P_2} P\,dV$$

$$= \int_{V_1}^{V_2} RT\,\frac{dV}{V},$$

so that

$$q = RT\ln\frac{V_2}{V_1},$$

or

$$q = RT\ln\frac{P_1}{P_2}. \tag{2}$$

b. Adiabatic expansion

Since the process is adiabatic, then according to the first law given by Eq. (1), we have

$$0 = dU + d'W.$$

If $d'w$ stands only for pressure–volume work, then we have

$$d'W = P\,dV$$

But since the process is adiabatic, the equation of state may be taken as

$$PV^{\gamma} = \text{const.}$$

so that

$$P = \frac{\text{const}}{V^{\gamma}},$$

and

$$d'W = \text{const.}\,\frac{dV}{V^{\gamma}}.$$

Therefore,

$$W = \text{const.}\int_{V_1}^{V_2}\frac{dV}{V^{\gamma}}$$

$$= \text{const.}\int_{V_1}^{V_2} V^{-\gamma}\,dV$$

$$= \frac{\text{const.}}{1-\gamma}\left[V_2^{1-\gamma} - V_1^{1-\gamma}\right] \qquad (3)$$

Since it is an *expansion* process, V_2 is greater than V_1, and the sign of the bracket will be negative because $V_1^{1-\gamma}$ will be greater then $V_2^{1-\gamma}$. And since this negative sign is preceded by another negative, *w* will be positive, that is, work is done by the system in adiabatic expansion. This results in a lowering of the temperature of the gas.

It will be evident that for an adiabatic reversible *compression*, Eq. (3) will be of negative sign since V_2 will be less than V_1 and hence $V_1^{1-\gamma}$ will be less than $V_2^{1-\gamma}$ with the result that the positive sign of the bracket is preceded by a negative one thus making *w* negative, which implies work is done on the system in adiabatic compression. A rise in the temperature of the gas thus occurs.

1.5. Conversion of heat into work

1. Heat engine. The flow of heat from a hotter to a colder body is an irreversible or spontaneous process and can be made use in doing work by means of appropriate arrangements. The mechanism which obtains work from the flow of heat from a body at a higher temperature to one at a lower temperature is known as a *heat engine*.

According to the second law, no work can be obtained from the passage of heat from one body to another at the same temperature, or from a colder to a hotter body. This is because in neither case does heat flow of its own accord (spontaneously) in that direction.

On the other hand, in the passage of heat in a spontaneous process from a hotter to a colder body, work can *only* be obtained through the agency of some working substance that can perform mechanical work by expansion, etc.

We wish now to find out the maximum work obtainable from the passage of a definite quantity of heat from a body at a higher temperature, T_2, to a body at a lower temperature T_1. It is to be noted that, in order to exclude any work done due to a change in the working substance, it is necessary to bring the latter into the original condition at the end of operations. This means that it is necessary to consider a *cyclic process*.

According to the first law, we have

$$q = \Delta U + W, \qquad (1)$$

for a finite change, as shown before, where q is the heat absorbed by the system, ΔU the increase in the energy of the system, and *w* the work done by the system on the surroundings. The work done may be mechanical (as in an expansion against the pressure of the atmosphere) or electrical (as in the production of a current, which may be used to drive an electric motor). There are other ways in which work can be done, for example, against magnetic forces. For the present, we shall consider mechanical work alone.

According to the first law expressed by Eq. (1), we can obtain a general relation applicable of cyclic processes at once. If a substance is put through a series of operations such that it is finally left in a state identical to its original state, we have

$$\Delta U = 0,$$

and

$$q = W,$$

or

$$\Sigma q = \Sigma W, \qquad (2)$$

that is, the sum of the amounts of heat absorbed in the cycle of operations is equal to the *algebraic sum* of the amounts of work done.

2. Carnot's cycle. In order to obtain the maximum work obtainable in a cycle of operations, it is necessary that every stage should be carried out *reversibly. Carnot's cycle* is a typical reversible cycle of operations. We shall first consider the case in which the working substance is *1 mole of a perfect gas.* We may suppose that the gas is confined in a cylinder fitted with a piston. The stages of expansion and compression are conducted reversibly, that is, during an expansion, the pressure on the piston is always less by an infinitesimal amount than that exerted by the gas and during a compression, it is infinitesimally greater. Thus the maximum work is obtained in every state.

The operations are as follows (Fig. 1):

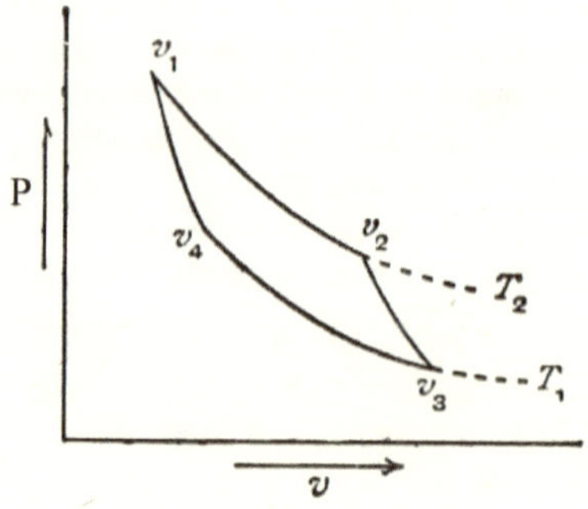

1. The gas is expanded *isothermally* and *reversibly* at temperature T_2 from volume v_1 to volume v_2.

   Work done by the gas, $W_2 = \int_{V_1}^{V_2} P\,dv = \int_{V_1}^{V_2} R T_2 \frac{dv}{v}$

$$= R T_2 \ln \frac{v_2}{v_1}.$$

   Heat absorbed by the gas $= q_2$.
   Therefore,

$$q_2 = W_2$$

2. The gas is *thermally isolated* so that it cannot receive heat from or give heat to its surroundings, and it is expanded further (adiabatically) from v_2 to v_3; the temperature drops to T_1.

   Work done by the gas $= W_2'$.

This must be equal to the decrease in the internal energy of the gas, that is,

$$-du = W_2'.$$

Thus if c_P is the specific heat of the gas, then we have

$$\frac{du}{dT} = c_v$$

or

$$du = c_v\, dT.$$

Therefore,

$$\Delta u = c_v\,(T_1 - T_2)$$

or

$$-\Delta u = c_v\,(T_2 - T_1).$$
$$W_2' = c_v\,(T_2 - T_1).$$

3. The gas is compressed *isothermally* from v_3 to v_4 (which is on the adiabatic curve going through v_1).

Work done on the gas = $W_1 = R\,T_1 \ln \frac{v_3}{v_4}$

(Work done by the gas is negative; it is given by

$$\int_{v_3}^{v_4} R\,T_1 \frac{dv}{v} = R\,T_1 \ln\frac{v_4}{v_3})$$

Heat evolved by the gas = $-\,q_1$

Therefore,

$$-q_1 = W_1$$

(Here q_1 is of negative sign. It represents the heat absorbed by the gas.)

4. The gas is *thermally isolated* and further compressed adiabatically to the original volume v_1.

   The temperature *rises* to T_2.

   Work done on the gas = W_1'

   This is equal to the increase in the internal energy of the gas, $d\,u = W_1'$. Thus if c_v is the same as before, we have

$$W_1' = c_v\,(T_2 - T_1).$$

Therefore, for the whole cycle of operations, by Eq. (2)

$$\Sigma q = \Sigma W \qquad (2)$$

we have

$$\underset{\text{Total heat absorbed}}{q_2 - (-q_1)} = \underset{\text{Total work done by the gas.}}{W_2 + W_2' - W_1 - W_1'}$$

Summing the *work* terms and denoting the summation by *W*, we find

$$W = R\,T_2 \ln\frac{v_2}{v_1} - R\,T_1 \ln\frac{v_3}{v_4}.$$

But since $\frac{v_2}{v_1}$ can be proved, as will be shown below, to be equal to $\frac{v_3}{v_4}$, we have

$$W = R\,(T_2 - T_1) \ln\frac{v_2}{v_1}, \qquad (3)$$

and

$$\frac{W}{q_2} = \frac{R\,(T_2 - T_1)\ln\frac{v_2}{v_1}}{R\,T_2 \ln\frac{v_2}{v_1}} = \frac{T_2 - T_1}{T_2} \tag{4}$$

The ratio W/q_2, that is, the ratio of the work obtained in the cycle to the heat absorbed at the higher temperature is known as the *efficiency* of the process. It may be denoted by η.

When the temperature difference between the two isothermal states of a Carnot cycle is a small amount, dT, we may write Eq. (4) in the form

$$\frac{dW}{q} = \frac{dT}{T}$$

or

$$dW = q\,\frac{dT}{T}, \tag{5}$$

where dW is the work obtained in a cyclic process with temperature difference dT, in which heat *q* is absorbed at a temperature *T*.

Now we want to prove that

$$\frac{v_2}{v_1} = \frac{v_3}{v_4}.$$

Consider the first adiabatic process in which the volume changes from v_2 to v_3 (Fig. 1). According to the first law, we have

$$0 = \Delta U + W.$$

Since ΔU is given by

$$\int_{T_2}^{T_1} c_v\, dT,$$

and

$$W = \int_{V_2}^{V_3} RT\,\frac{dv}{v},$$

where $\frac{RT}{V}$ is substituted for *P* in the integral

$$W = \int_{V_2}^{V_3} P\,dv,$$

we have

$$0 = \int_{T_2}^{T_1} c_v\, dT + \int_{V_2}^{V_3} RT\frac{dv}{v}$$

or

$$\frac{c_v}{R}\int_{T_2}^{T_1}\frac{dT}{T} = -\int_{V_2}^{V_3}\frac{dv}{v}.$$

Therefore,

$$\frac{c_v}{R}\ln\frac{T_1}{T_2} = -\ln\frac{v_3}{v_2}.$$

or

$$\frac{c_v}{R}\ln\frac{T_2}{T_1} = \ln\frac{v_3}{v_2} \qquad (6)$$

In like manner, along the adiabatic path the volume changes from v_4 to v_1.

$$\frac{c_v}{R}\ln\frac{T_2}{T_1} = \ln\frac{v_4}{v_1}. \qquad (7)$$

It is evident from the preceding two equations (6) and (7) that

$$\ln\frac{v_3}{v_2} = \ln\frac{v_4}{v_1}.$$

Hence,

$$\frac{v_3}{v_2} = \frac{v_4}{v_1}$$

or

$$\frac{v_2}{v_1} = \frac{v_3}{v_4}.$$

It may be noted that the efficiency of Carnot's cycle, which is expressed by Eq. (4) is rewritten as

$$\eta = \frac{W}{q_2} = \frac{T_2 - T_1}{T_2}, \qquad (4)$$

or when W is substituted by the total heat absorbed, then the expression

$$\eta = \frac{W}{q_2} = \frac{q_2 + q_1}{q_2} = \frac{T_2 - T_1}{T_2}$$

is expressed in other references by

$$\eta = \frac{q_2 - q_1}{q_2} = \frac{T_2 - T_1}{T_2}, \qquad (8)$$

where q_1 here represents the heat rejected by the gas to the reservoir or the lower temperature bath, that is, the heat evolved by the gas, which is of positive sign.

Also, in other references, the higher temperature is denoted by T_1 and the lower by T_2, and subsequently, the heat absorbed by the gas is denoted by q_1, whereas that rejected is by q_2. So the efficiency of Carnot's cycle is given by

$$\eta = \frac{q_1 - q_2}{q} = \frac{T_1 - T_2}{T_1} \qquad (9)$$

However, all formula developed for the efficiency of Carnot's cycle in various references are equivalent.

3. Carnot's theorem. This result, deduced for an ideal gas, that the efficiency of the cycle is given by

$$\eta = \frac{W}{q_2} = \frac{q_2 + q_1}{q_2} = \frac{T_2 - T_1}{T_2} \tag{4}$$

holds good whatever be the *nature* of the working substance, be it a solid, liquid, or gas. For according to an important theorem, first given by Carnot

Every perfect engine working reversibly between the same temperature limits has the same efficiency, whatever the working substance.

A perfect engine is one which, working without frictional losses, obtains the maximum work obtainable from its cycle of operations. That is, to say, every stage of the process is carried out reversibly, with the system being displaced only infinitesimally from a state of balance and, therefore, infinitely slowly.

This theorem is proved by showing that if a working substance could be found for which the efficiency was greater than that of any other, it would be possible, without the application of outside effort, to transfer heat from a lower to a higher temperature, which contravenes the second law of thermodynamics.

Suppose we have two perfect engines I and II working between the same temperature limits T_2 and T_1. Let their efficiencies be η_I and η_{II}, respectively. Suppose η_{II} is greater than η_I and engine II is arranged to run engine I in the reverse direction.

In the working of engine II, for the absorption of q_2 units of heat at the higher temperature T_2, an amount of work W equal to $\eta_{II} q_2$ is obtained as given by the equation

$$\eta_{II} = \frac{W}{q_2} = \frac{q_2 + q_1}{q_2} = \frac{T_2 - T_1}{T_2}, \tag{4'}$$

and a quantity of heat $(-q_1)$ is given out at the lower temperature T_1. This quantity as derived from this equation of efficiency is rewritten in the form

$$\eta_{II} = 1 + \frac{q_1}{q_2}$$

which becomes

$$-q_1 = (1 - \eta_{II})\, q_2 = q_2 - \eta_{II}\, q_2 .$$

Expressed in units of heat:

The work W may be used to run engine I backwards so that it *absorbs* heat at the lower temperature T_1, and as noted above, at the higher temperature T_2. Since the efficiency of engine I is η_I, the amount of heat given out (or evolved) at the higher temperature T_2 in its reversed action for an amount of work W is W/η_I or $q_2\, \eta_{II}/\eta_I$ since $W = q_2\, \eta_{II}$ as given by Eq. (4′).

The net quantity of heat taken in or absorbed at the lower temperature T_1 by engine I in its reversed action is thus

$$\frac{W}{\eta_I} - W = \frac{\eta_{II}}{\eta_I} q_2 - \eta_{II} q_2 .$$

The above results may be tabulated in the following way:

	Engine II η_{II}	Engine I η_I
(1)	heat absorbed at $T_2 = q_2$,	heat evolved at $T_2 = \frac{\eta_{II}}{\eta_I} q_2$,
(2)	work obtained $W = \eta_{II} q_2$,	work exp ended $W = \eta_{II} q_2$,
(3)	heat evolved at $T_1 = -q_1 = q_2 (1 - \eta_{II})$,	heat taken in or absorbed at $T_1 = \frac{\eta_{II}}{\eta_I} q_2 - \eta_{II} q_2$

We find that the net result of the working of both the engines' processes is the *absorption* of $\frac{\eta_{II}}{\eta_I} q_2 - \eta_{II} q_2 - q_2 + \eta_{II} q_2 = \frac{\eta_{II}}{\eta_I} q_2 - q_2$ units of heat at the *lower* temperature T_1. This is evident from step (3) in the above table just by subtraction.

On the other hand, the *evolution* of heat at the *higher* temperature T_2 is also $\frac{\eta_{II}}{\eta_I} q_2 - q_2$ units of heat as it is evident from step (1) in the above table just by subtraction. Therefore, in the working of both the engines' processes there is an absorption of heat at the lower temperature T_1 and an evolution of the *same amount* at the higher temperature T_2. Accordingly, if η_{II}/η_I is greater than 1, that is, η_{II} is greater than η_I, as we have assumed at the beginning, it would be possible to transfer heat from a lower to a higher temperature without the expenditure of work. We, therefore, conclude that if the second law is true, all perfect engines have the same efficiency, and Eq. (4) which has been deduced for a perfect gas is *universally valid.*

In 1848, *Lord Kelvin* made use of Carnot's theorem suggesting a new temperature scale which was then called the absolute scale of temperature in the sense that it would be independent of any particular property of any thermometric substance. He was able to realise this scale by the result of his famous porous-plug experiment (that he carried out in conjunction with Joule and known as the *Joule-Thomson effect*), in a thermodynamical way, using the entropy concept. In this way, Lord Kelvin proved that the ice-point, which is taken arbitrarily on the centigrade scale to be equal to 0, that is, $t_0 = 0°C$, is on his suggested absolute scale of temperature, which is equal to the reciprocal of the coefficient of thermal expansion α given by the *Charles law* in the case of an ideal gas, that is,

$$T_0 = \frac{1}{\alpha}.$$

Since α for an ideal gas is equal to $1/273.16$, then it follows that the ice-point on the Kelvin scale is 273.16 K, and subsequently, the zero on this scale corresponds to $-273.16°$ C

The treatment of the Joule-Thomson effect that has led to this result will be considered later*.

---

* See Section 5.4.

CHAPTER 2

The Second Law of Thermodynamics

2.1. Heat and work

The convertibility of heat and work as expressed by the first law of thermodynamics places heat energy on the same basis with all forms of mechanical energy. According to this law, a quantity of heat q communicated to any body may expend itself in one or more of several ways. A portion of it, ΔW_t, may be employed in raising the temperature of the body, and another portion, ΔW_a, in increasing the energy inside the molecules. Since the increase in temperature is in general accompanied by an increase in volume, a part of the heat energy q may also be expended in two other ways. For if the body be subjected to external forces, external work of expansion, ΔW_e, will be done against these forces while the volume is changing. Also, internal work of expansion, ΔW_i, will be done against internal forces, such as molecular attractions, while the volume is changing. Then in general,

$$jq = \Delta W_t + \Delta W_a + \Delta W_e + \Delta W_i,$$

where the quantities in the right-hand side are expressed in mechanical units, q is expressed in thermal units, and j is Joule's equivalent. This equation is simply an expression for, in this case, of the first law of thermodynamics. A more general mathematical statement of the first law is

$$jq = \Delta U + \Delta W.$$

This is an equation which states that a quantity of heat q absorbed by a system is in general used up partly to produce an increase of ΔU in the internal energy of the system and partly to cause the system to do the work of amount ΔW.

According to the first law, for example, the heat generated when a rotating flywheel is brought to rest by friction is equal to the loss of kinetic energy of the wheel or again, the heat generated when a falling object hits the ground is accounted for by the loss in the kinetic energy of the object. Now so far as the first law is concerned, the reverse of these processes would also be possible: a flywheel at rest can suddenly start rotating, thus gaining kinetic energy while the bearings lose an equal amount of heat energy and becomes cooler; or an object lying on the ground can suddenly jump up into the air while the ground, at the same time, becomes cooler. Neither of these processes violates the first law. Yet they never have been observed to happen. Consideration of this kind lead to the conclusion that when heat energy is one of the forms of energy involved, *conservation of energy alone is not a sufficient condition for the occurrence of a process in nature, although it is a necessary condition. The first law of thermodynamics is a statement of experience without any theoretical basis; that is, it is an empirical law.* It can be described as a law which equates the energy lost from one body to the energy gained by another body. The fact that when various forms of energy are converted into heat energy, there is always a constant ratio between the amount of energy that disappears and the amount of energy that appears; for example, there are different ways in which work can be converted into heat. But although the methods used to effect the transformations are different, the same ratio of work to heat is always found. *Maxwell* stated this conclusion in the following words:

> *When work is converted into heat or heat into work, the quantity of work is mechanically equivalent to the quantity of heat.*

This important ratio is known as the *mechanical equivalent of heat*, which, when determined accurately, has the value

$$1 \text{ calorie} = 4.182 \times 10^7 \text{ ergs.}$$

The ratio of heat to work is known as *Joule's equivalent.*

This fact indicates that in all the processes, when various forms of energy are converted into heat, energy is not created or destroyed but is merely transformed from one form to another. This is a *partial* expression of the first law of thermodynamics, which was defined more accurately, as noted above, by *Helmholtz,* in 1847, in the following words:

> *In all processes occurring in an isolated system, the energy of the system remains constant.*

The validity of the first law is also shown by the fact that no one has yet succeeded in devising a machine which produces energy continuously without, at the same time and at a definite rate, absorbing a certain amount of energy of another form, in other words, without causing the disappearance of another form of energy. The machine which can produce motion without taking an equivalent amount of energy is *imaginary* and known as *a perpetual motion machine of the first type*. It is called of the 'first type' because it is denied by the first law of thermodynamics. Thus an electric lamp gives radiant energy only as long as it is supplied with electrical energy. The validity of the law is so convincing that whenever a process is found to liberate energy continuously as in the case of a radioactive element, it is at once assumed that other form of energy disappears. In the case of atomic disintegration, it is believed that an enormous amount of energy is locked up in the atoms themselves so that the process is accompanied by the evolution of a large amount of energy.

By means of the first law, we can calculate the change of energy if there is a change, but we cannot tell whether a change would take place in practice or not. Again the law, for example, does not prohibit the transference of heat from a cold to a warmer body since there is no creation of energy. Also, it does not prohibit the conversion of heat into work continuously at one temperature. Thus we can imagine the existence of a machine which would pump water from the sea, take heat energy from it, and return cooler water to the sea, using the heat energy abstracted to drive a ship. This process does not contradict the first law, but it has been proved impossible in practice to construct a machine which will continuously develop mechanical energy at the expense of the heat of the surrounding bodies *at the same temperature*. The machine which can convert heat into work *continuously at the same temperature is imaginary,* and it is known as *a perpetual motion machine of the 'second type'* – being denied by the second law.

Hence, we can say that the first law does not summarise the facts concerning the direction of energy changes in practice, and consequently, being an insufficient law of energy, another law of thermodynamics concerning the conditions under which changes can occur in practice is necessary.

The second law of thermodynamics which expresses the limitation of the first law by showing the relation between heat and work in various processes was stated in different ways. The second law as *a restricting principle* grew out of the work of Carnot. It was first formulated by *Clausius* and *Kelvin* and was called by the former as the second law of thermodynamics. One of its simplest formulations according to *Kelvin* is as follows:

> *There is no natural process the only result of which is to cool a heat reservoir and do work as exemplified by raising a weight.*

The law really is a generalisation of the principle that when two bodies at different temperatures are placed in contact, the flow of heat is always from the hotter to the colder. It does not exclude the transfer of heat from a colder to a hotter body, a process which happens, for example, in mechanical refrigerators. But it does say that this transfer cannot occur, unless some outside agent does work or there is some other additional effect, that is, a compensation for this unnatural result. Based on this restricting principle, *Clausius* gave the following statement of the second law.

> *It is impossible for a self-acting machine, unaided by external agency, to convey heat from a body at a low to one at a higher temperature. Or heat cannot of itself (i.e. without the performance of work by some external agency) pass from a cold to a warmer body.*

Also, *Kelvin* gave the following statement for the second law in addition to the one given above.

It is impossible by an inanimate material agency to derive mechanical effect from any portion of matter by cooling it below the temperature of the coldest of the surrounding bodies.

On the other hand, this law limits the extent to which heat can be converted into other forms of energy. Thus since electrical energy and mechanical work are interconvertible, it is possible to construct an electric motor or a dynamo with an efficiency of perhaps 90 per cent, allowing only 10 per cent for the loss of energy by friction. Heat energy is *unique* among the various forms of energy in the *sense that it can be converted into other forms of energy only partially and temporarily.*

The *quantitative expression* for the second law has to do with the efficiency of any heat engine based on Carnot's theorem. According to this remarkable theorem, which was published in 1824, the maximum efficiency of an engine working through any fixed small range of temperature is *independent of the nature of the working substance used in the engine and is a function of the temperature alone* (as shown before in Section 1.5). The maximum efficiency can simply proved to be equal to $(q_2 + q_1)/q_2$, where q_2 is the heat drawn from the boiler, and q_1 is the heat rejected to the condenser during a given time. This will be clear if one notes that $(q_2 + q_1)/q_2$ is the fraction of the heat received which is transformed into work. Also, it can be proved that $(q_2 + q_1)/q_2$ is equal to $(T_2 - T_1)/T_2$, where T_1 and T_2 are the absolute temperatures of the source of heat and the condenser, respectively. Now the second law, in accordance with Carnot's theorem, states that even if a heat engine is entirely free from friction, its efficiency $(q_2 + q_1)/q_2$ cannot exceed $(T_2 - T_1)/T_2$, that is,

$$\frac{q_2 + q_1}{q_2} < \frac{T_2 - T_1}{T_2} \quad . \tag{1}$$

For example, with the boiler of a steam engine at 160°C and the condenser at 70°C, the efficiency cannot be more than 21 per cent, even if no frictional or other losses are involved. The actual efficiency probably would be about 15 per cent. The point evidently is that *only a part* of the total heat q_2 drawn from the boiler can be converted into work. The remainder, q_1, although not annihilated, is only wasted in the sense that it has descended to the temperature of the surroundings and has become unavailable for doing work.

> *Therefore, the second law in its quantitative form provides a measure of this loss of availability or degradation of energy.*

Also, according to the quantitative expression of the second law, the efficiency of a heat engine approaches unity as necessitated by the first law only when T_2 is very large compared with T_1, or if T_1 approaches absolute zero. But T_1 cannot be lower than the temperature of the surroundings of the heat engine, so the lower limit is *pre-determined.* Therefore, attempts to increase the efficiency must be directed towards increasing T_2.

After all, the second law expresses quantitatively what all of us feel intuitively that although energy is conserved whenever heat is evolved, yet there is something that has been lost; the heat in an object is not so available for doing work as is the kinetic energy of the object when it is in motion as a whole. This implies, as we have noted above, that energy is unique among the various forms of energy; it can be converted into other forms only partially and temporarily.

Furthermore, according to the above quantitative expression of the second law, it is evident that no useful work would be produced if T_2 is equal to T_1, that is, if the cyclic process of the engine takes place at *one temperature only.* The existence of the engine which would produce work *isothermally* and *continuously* is thus denied by the second law and that is why it is called *perpetual motion machine of the second type* as noted above. If all temperature differences in the universe are eventually wiped out, then the gross mechanical motion will be no longer possible, and we must approach what the philosophers of the nineteenth century called the '*Warmetod*' or '*heat** *death*'. This expectation was first predicted by *Clausius* in accordance of his statement of the second law in terms of entropy. This statement is '*The entropy of the world drives towards a maximum*'. Once this maximum has been reached there would be no driving force left to make every thing happen.

2.2. The entropy concept

1. Introduction. In 1854, Clausius gave another interpretation of the second law, in the course of which he introduced a new concept. The inequality in the quantitative expression of this law as derived from Carnot's cycle of any heat engine is given by

$$\frac{q_2 + q_1}{q_2} < \frac{T_2 - T_1}{T_2}. \qquad (1)$$

---

* The state of heat death must not be imagined as a death of flame and fire, but rather a *prosaic* end of completely uniform, uninteresting temperature and *monotonous* randomness in everything else.

This can be rearranged in the form

$$\frac{q_2}{T_2} < \frac{q_1}{T_1}, \tag{2}$$

where q_2/T_2 depends only on the source of heat, and q_2/T_1 only on the condenser. The very important quantity q_2/T_2 is called the entropy of the heat at the temperature T_2 by Clausius. The foregoing inequality, then, states that in any actual heat engine, entropy of amount q_2/T_2 is taken from the source of heat and a larger amount of entropy, q_1/T_1, is given to the condenser. Such considerations lead to another way of stating the second law.

In any isolated system, every change results in an increase in the entropy of the system.

Clausius summed up all of this in his famous statements of the two laws of thermodynamics which are respectively:

(a) the energy of the universe is constant and

(b) the entropy of the universe is always increasing, tending to a maximum.

Today, we should be more cautious about extrapolating our laws into regions where they have not been tested; experiments have not been made in all parts of the universe. Moreover, it must be remembered that thermodynamics ignores the mechanisms of a process and makes no assumptions with regard to the structure of matter. When the behaviour of the molecules and atoms of matter is taken into account by means of the methods of statistical mechanics, the state of maximum entropy predicted for any isolated system by the second law comes to be interpreted as *the most probable state*. This will be shown later.

The second law with its implications of constantly increasing unavailable energy of the universe applies to the average probable condition to be met over exceedingly long periods of time. From this point of view, the 'heat death', noted above, or 'Warmetod' need not be regarded as inevitable.

2. Consequences of Carnot's theorem

According to Eq. (1), the efficiency of a *perfect heat engine* is expressed as

$$\frac{W}{q_2} = \frac{T_2 - T_1}{T_2}, \tag{3}$$

where *W* is the work obtained, which is determined *solely* by the difference between the temperatures of the hotter and colder isothermal stages and the absolute temperature at which heat is absorbed. In chemical thermodynamics, we are concerned with the relations between the properties of substances which must hold if the theorem is to be universally true and the study of various cases of chemical equilibria in connection with these relations. Thus we shall first state below some general results.

a. Isothermal cycles. When all the stages of a reversible cyclic process occur at the same temperature, the work obtained is zero since $T_2 - T_1 = 0$. Therefore, the maximum work obtained by a system in going from state I to state II at the same temperature is same whatever the path or th intermediate stages, so long as the temperature remains constant throughout the change. If this were not the case, it would be possible to go from I to II by one path and return by another and so obtain work by an isothermal cycle. Therefore, we conclude that *the maximum work of an isothermal change is definite and depends only on the initial and final stages of the system.*

Thus we may write

$$W = A_I - A_{II}, \tag{4}$$

where A_I and A_{II} are quantities determined by the initial and final states of the system only. Just as the heat evolved in a change is equal to the decrease in the energy content or heat content (according as the volume or pressure is constant), provided *the temperature remains constant, the maximum work done in an isothermal change is equal to the decrease in the quantity A. A is known as the 'maximum work function' or the 'Helmholtz free energy function' of the system which is also denoted by F.* It can also be regarded as the maximum work '*content*' of the system. When the system performs isothermal work, maximum work content decreases; thus for a given isothermal change

$$\underset{\text{Maximum work}}{W} = \underset{\text{Decrease, in A}}{-\Delta A,} \tag{5}$$

or

$$= -\Delta F.$$

b. Changes at constant temperature and pressure; the net work function or the Gibb free energy function *G*. The net work obtained in an isothermal change $w' = W - P\,\Delta V$ differs from the maximum work *W* by the term $P\,\Delta V$ or $P(V_2 - V_1)$. But the term $P\,\Delta V$ is the same whatever the path or intermediate stages, *provided that pressure remains constant throughout. Therefore, the net work of a change at constant temperature and pressure* w' *is the same for all possible paths and, under these conditions, depends solely on the initial and final states of the system.*

Thus

$$w' = G_I - G_{II}, \tag{7}$$

where G_I and G_{II} are quantities determined only by the initial and final states of the system. Analogous to the maximum work function, *G is the 'net work function' or the 'Gibbs free energy function' of the system as usually known. Thus the net work done by the system in a given change at constant temperature and pressure is equal to the decrease in the free energy G of the system, that is*

$$\underset{\text{Net work}}{w'} = \underset{\text{Free energy decrease}}{-\Delta G} \tag{8}$$

c. Conditions of equilibrium in isothermal changes and in isothermal–isobaric change. These conditions involve, respectively, the preceding two thermodynamic functions: the Helmholtz free energy function (ΔA or ΔF) and the Gibbs free energy function (ΔG). The changes which occur spontaneously are characterised by their ability to perform work. We have seen that for changes which occur at constant temperature, the maximum work obtained is definite and equal to the decrease in the maximum work function or the Helmholtz function F (or A) of the system. Thus if the system undergoes a spontaneous change at constant temperature, then its maximum work function must decrease. Therefore, a system is in equilibrium if there is no change which can spontaneously occur under the given conditions, so the condition of equilibrium for changes at constant temperature is that there is no possible change whereby the maximum work function can decrease, that is, so long as the temperature remains constant, the function F (or A) is a minimum.

This may be expressed by the condition

$$(\delta F)_T \text{ or } (\delta A)_T \geq 0, \tag{9}$$

for all possible changes satisfying the condition of constant temperature.

Similarly, for changes occurring at constant temperature and pressure the net work is definite and equal to the decrease in the net work function or the Gibbs free energy function G of the system. Thus if the system undergoes a spontaneous change at constant temperature and pressure, its net work function or Gibbs free energy function G must decrease. *Therefore, a system is in equilibrium if there is no change which can spontaneously occur under the given conditions of constant temperature and pressure, whereby its free energy function G can decrease, that is, so long as the temperature and pressure remain constant, the function G is a minimum.*

This condition of equilibrium may be expressed by

$$(\delta G)_{T,P} \geq 0, \tag{10}$$

for all possible changes satisfying the condition of constant temperature and pressure.

These conditions of equilibrium, or these statements, are however, not true if the temperature varies during the change from the initial to the final state. This is because when there is a temperature variation, there will be varying amounts of work that can be obtained according to the path actually followed. Thus, if the temperature is kept constant we refer to the maximum work function F (or A) and if the temperature and pressure are simultaneously kept constant we refer to the free energy function G. Consequently, we have only shown that the maximum work function F (or A) and the free energy function G can be used to define the criteria of equilibrium. If the temperature is kept constant we refer to F. If the temperature and pressure are simultaneously kept constant we refer to G defines and that of constant temperature and pressure, respectively, is maintained, respectively.

Now we wish to remove this limitation and to show that there are quantities such as energy U (or E) which have definite values in any state of the system irrespective of how it has been brought into that state, and at the same time, independent of the condition of constant temperature or of constant temperature and pressure that must be maintained as considered above.

The construction of such functions requires the consideration of another quantity, which is also a thermodynamic function, namely, the entropy S. It is also regarded as a consequence of Carnot's theorem as has been first considered by Clausius, as pointed out above.

d. The introduction of the state variable *S*. Consider a simple Carnot cycle in which a quantity of matter, which is supposed to be always in a state of internal equilibrium, is put through a reversible cycle of operations (Fig. 1), consisting of two isothermal stages I and III at temperatures T_2 and T_1 ($T_2 > T_1$) respectively, and two connecting adiabatic stages II and IV.

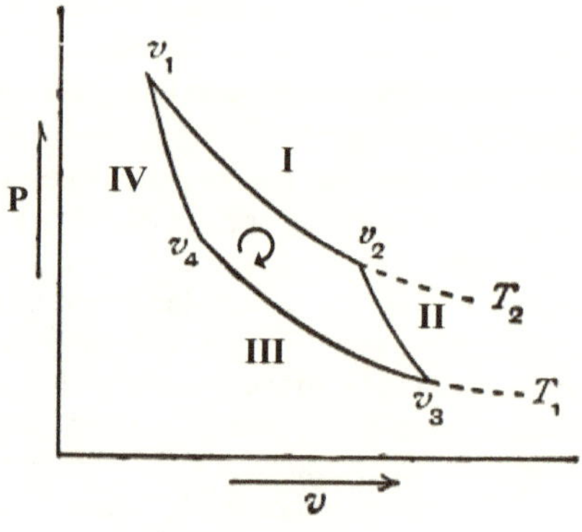

Fig. 1.

Let the quantity of heat absorbed from surrounding bodies in the isothermal stage I be q_2, and that absorbed in the isothermal state III be q_1 (normally heat is evolved in the isothermal stage III, at the lower temperature T_1; q_1 is then a negative quantity). Then *W*, the work obtained in the whole cycle, must be equal to the total amount of heat absorbed, that is

$$W = q_2 + q_1 \tag{11}$$

The ratio of the work obtained to the heat absorbed at the higher temperature, that is W/q_2, is called the efficiency of the engine as given before.

Now Carnot's theorem states that all reversible cyclic processes working between the same two temperatures have the same efficiency. If this were not true, as has been shown above, it would be possible that by the use of two such cyclic processes which have different efficiencies and working – one in the forward and one in the backward direction – to transfer heat from a colder to a hotter body, that is, to reverse a spontaneous change without the application of any outside effort. This is contrary to the second law of thermodynamics.

The ratio W/q_2 is, therefore, the same for all reversible cycles working between the same two temperatures and is independent of the nature of the system employed as the working substance.

It has been pointed out before that this remarkable theorem of Carnot led Lord Kelvin to suggest a new scale of temperature called *absolute scale.* Absolute in the sense of being independent of any particular property of any particular thermometric substance, and hence it is called the *absolute thermodynamic scale of temperature* and has, therefore, great advantages over a scale based on the thermal expansion of any actual substance. Also, the independency of the efficiency η or the ratio W/q_2 of the nature of the system employed as the working substance is expressed in the form, as given in Chapter 1,

$$\eta = \frac{W}{q_2} = \frac{T_2 - T_1}{T_2} \tag{12}$$

This leads to the suggestion that a new absolute scale of temperature may be defined in such a way that the efficiency is equal to the temperature difference between the two isothermal stages divided by the temperature of the first stage.

Therefore, Carnot's theorem can be regarded as the *original source of the absolute thermodynamic scale of temperature* the suggestion of which was first made by Lord Kelvin twenty-four years later after N. L. Sadi Carnot had published his remarkable memoir* on the theory of heat engines, in 1824.

Substituting now Eq. (11) in Eq. (12), we have

$$\frac{q_2 + q_1}{q_2} = \frac{T_2 - T_1}{T_2}$$

or

$$\frac{q_1}{q_2} = -\frac{T_1}{T_2},$$

that is,

$$\frac{q_2}{T_2} + \frac{q_1}{T_1} = 0. \qquad (13)$$

Thus in a simple Carnot cycle, the algebraic sum of the quantities of heat absorbed, each divided by the absolute temperature at which the absorption takes place, is zero.

Now any reversible cycle, whatever may be, can be resolved into a number of elementary Carnot cycles. Consider a cyclic process tracing out the closed path AB on the P–V diagram (Fig. 2). We may resolve this cycle into a large number of simple Carnot cycles, each having two isothermal and two adiabatic stages.

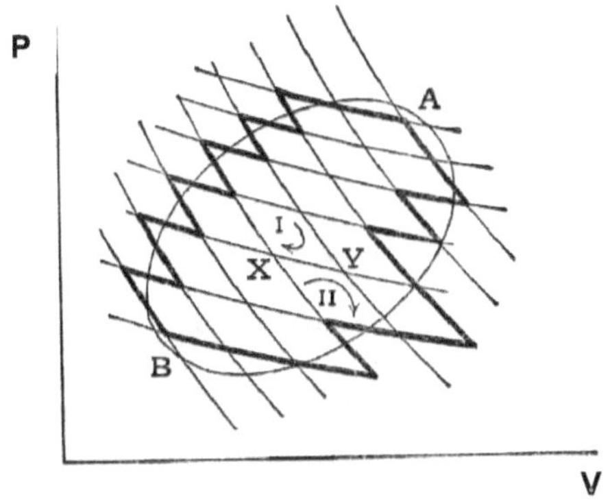

Fig. 2

On Fig. 2 are drawn some isothermal and adiabatic lines for the system which form a large number of Carnot cycles whose outside boundaries approximate roughly with the closed cycle AB. Every section of the isothermals which is not on the outside boundary is shared by two adjacent Carnot cycles. Thus the clement XY is shared by the cycles I and II, and the quantities of work obtained in passing along the isothermal XY in the two cycles are equal and opposite. The heat terms corresponding to the element XY also cancel out. It follows that all the heat and work terms corresponding to the shared sides balance out, and *we are left with only the terms for the outside boundary of the collection of Carnot's cycles.* It is evident that by drawing the isothermal lines and the adiabatic lines very close together, we

* N. L. S. Carnot, Réflectxions sur la puissance du Feu (1824). A translation of this memoir had been published under the title *Reflections on the Motive Power of Heat*, ed. R.H. Thurston (Wiley, 1897).

may make the outside boundaries of the Carnot cycles agree as closely as we wish with the actual boundary of our cycle AB.

For a single Carnot cycle we may write Eq. (13) as

$$\Sigma \frac{q}{T} = 0.$$

The same applies to a collection of Carnot cycles, and therefore, to any reversible cyclic process, which as we have just seen can be resolved into a number of Carnot cycles. Therefore, we may *write for any reversible cyclic process*

$$\Sigma \frac{q}{T} = 0, \qquad (14)$$

the summation being taken right around the cycle.

Now if we have two states A and B of a system at different temperatures, we may make a non-isothermal cycle by proceeding from A to B by path I and returning to the original state A by a different path II as shown in Fig. 3.

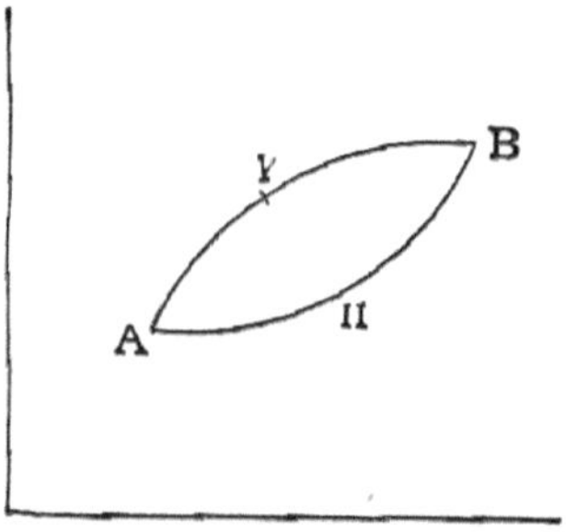

Fig. 3

If every stage is carried out reversibly, we have by Eq. (14)

$$\Sigma \frac{q}{T_{(A\rightarrow B)_I}} + \Sigma \frac{q}{T_{(B\rightarrow A)_{II}}} = 0 , \qquad (15)$$

where the first term represents the sum of the q/T terms for the passage from A to B by path I, and the second term the same quantity for the passage from B to A by path II. Reversing the direction of the second term, we have

$$\Sigma \frac{q}{T_{(A\rightarrow B)_I}} - \Sigma \frac{q}{T_{(A\rightarrow B)_{II}}} = 0$$

or

$$\Sigma \frac{q}{T_{(A\rightarrow B)_I}} = \Sigma \frac{q}{T_{(A\rightarrow B)_{II}}} . \qquad (16)$$

Thus since we have placed no restriction on the path from A to B, except that it shall be *reversible*, the quantity $\Sigma \frac{q}{T_{(A\rightarrow B)}}$ must be constant for all reversible paths from A to B.

This quantity thus depends only on the initial and final states A and B and not on the intermediate stages and may be regarded as the difference between the values of a function of the state of the system in the two given states. This function is called *entropy* and denoted by *S*. If S_A represents the entropy of the system in state A and S_B in state B, then the entropy change in going from A to B is

$$\Delta S = S_B - S_A = \sum \frac{q}{T_{A\to B}}, \tag{17}$$

the summation in the last term being taken over any reversible path between A and B.

It is important to notice that since the entropy change depends only on the initial and final states of the system, it is the same whatever the change is conducted, whether reversibly or irreversibly. But it is only equal to

$$\sum \frac{q}{T},$$

when that latter is evaluated for a reversible path, then *q* here should be q_{rev}.

As a consequence of this, we can define a fundamental distinction between reversible and irreversible processes. Consider a change of a system from a state A to state B. If this change is *carried out reversibly, then the entropy change of the system is*

$$dS = \frac{d'q_{rev.}}{T}, \tag{18}$$

for an infinitesimal change, and

$$\Delta S_{system} = \sum \frac{q_{rev.}}{T}, \tag{18'}$$

for a finite change.

Now all the heat absorbed by the system must come from some surrounding bodies, and it is a condition of reversibility that every element of heat absorbed must be taken from a body which has the same temperature as the system itself in this particular state.[3] Therefore, the entropy change of the surrounding bodies must be

$$\Delta S_{surroundings} = -\sum \frac{q_{rev.}}{T}, \tag{19}$$

since the surroundings give up the quantities of heat which are absorbed by the system and at the same temperatures. Thus for a *reversible process,*

$$\Delta S_{system} + \Delta S_{surroundings} = 0. \tag{20}$$

That is, in a reversible change, the total entropy change of the system and its surroundings is zero.

On the other hand, in an *irreversible change,* the entropy change of the system in passing from the initial state A to the final state B is the same whether the change is irreversible or reversible.

$$\Delta S_{system} = S_B - S_A.$$

But this is no longer equal to $\sum q_{irr}/T$ for the change. This is because the entropy change in only equal to $\sum q/T$ when heat *q* is absorbed in a reversible way as pointed out above. When a system undergoes an *irreversible* change, it performs *less* amount of work

---

[3] The absorption of heat from bodies at other temperatures would necessarily introduce some irreversibility.

than the maximum amount of work that can be obtained when it undergoes a reversible change, and since $q = \Delta U + W$, the system absorbs in the irreversible change a *smaller* quantity of heat *q* from the surrounding bodies than that absorbed in the corresponding reversible change. Thus with respect to the surroundings $\sum q/T$ is less in the irreversible change than in the reversible change so that the *entropy decrease* of the surroundings in this case of irreversible change is less than the *entropy increase* of the system, that is,

$$-\Delta S_{\text{surroundings}} < \Delta S_{\text{system}}$$

or

$$\Delta S_{\text{system}} + \Delta S_{\text{surroundings}} > 0. \tag{21}$$

It is to be noted that the temperature of the bodies from which heat is absorbed by the system cannot be less than the temperature of the system, for heat will not pass from a body at a lower temperature to a body at a higher temperature by itself. Hence, in general, the entropy decrease of the surroundings may be less than $\sum q_{\text{rev.}}/T$ for the system, but it cannot be greater.

It is obvious from above that we have made a fundamental distinction between two types of processes: a reversible process and an irreversible process, and we have assumed that in each of these two processes, the system which undergoes the change from an initial state A to a final state B is in contact with the surrounding bodies.

To sum up, let us consider the system which undergoes changes and the outside bodies with which heat is exchanged as one 'system', which we may suppose to be entirely *isolated* from the action of any other bodies. Thus we may state the following rules:

1. *When a reversible change occurs in any part of an isolated system, the total entropy remains unchanged.*
2. *When an irreversible change occurs in any part of an isolated system, the total entropy increases.*

Examples for the calculation of entropy changes in various processes concerning reversible and irreversible changes and whether these changes are isothermal or adiabatic will be given later.

e. Conditions of equilibrium in terms of entropy *S* and internal energy *U*. Irreversible changes can only occur in a system of bodies which is not in a state of equilibrium. When equilibrium is established, no irreversible changes are possible and consequently, when the absorption of heat from outside bodies is excluded, there are no possible changes whereby the entropy can increase. Thus when an isolated system of bodies, the energy of which is constant, is in equilibrium, its entropy has *a maximum value*. Considering the universe as an isolated system, Clausius, therefore, summed up the generalisation of natural tendencies which is contained in the second law by the statement, '*The entropy of the universe is always increasing, tending to a maximum*'. This statement has been given before, when Clausius introduced the entropy concept in classical thermodynamics, and he considered it as a statement of the second law. It is also expressed as, '*The entropy of the universe tends to be a maximum*' or '*The entropy of the universe is always increasing*'.

We have here a very valuable criterion of chemical equilibrium. A system is in equilibrium if no variations in its state can occur spontaneously. If there is any variation in the state which does not alter its energy but causes an increase of entropy, the system cannot be in equilibrium, for this variation may occur spontaneously. Gibbs, therefore, gave the following proposition as a *general* criterion of equilibrium:

> *For the equilibrium of any isolated system it is necessary and sufficient that in all possible variations in the state of the system which do not alter its energy, the variation of the entropy shall either vanish or be negative.*

In using this criterion, we need only to consider infinitesimally small variations, for an infinitesimally small variation must necessarily precede a finite one, and if the former cannot occur, neither can the latter. The criterion can, therefore, be written in the form:

$$(\delta S)_U \leq 0, \tag{22}$$

$(\delta S)_U$ denotes the variation of the entropy in an infinitesimal change of the state of the system, for which the energy remains constant.

This criterion can be expressed in an alternative form which is more convenient for practical use. Consider some particular state of a system of a certain amount of energy and a certain amount of entropy but which is not necessarily in a state of equilibrium. If there is *another* state which has *less* energy but the same entropy, we can, by adding heat, arrive at a state which has the *same* energy but *more* entropy than the original state. That is, if there is a state with less energy and the same entropy, there is also a state with more entropy and the same energy. But if this is the case, the original state cannot be a state of equilibrium. It is, therefore, a characteristic of a state of equilibrium that the *energy is a minimum for constant entropy*. This is stated in the following second proposition of Gibbs as a general criterion of equilibrium.

> *For the equilibrium of any isolated system, it is necessary and sufficient that in all possible variations in the state of the system which do not alter its entropy, the variation of its energy shall either vanish or be positive.*

For the same reasons as those given in connection with the first criterion, we may express this condition as follows:

$$(\delta U)_S \geq 0, \tag{23}$$

where $(\delta U)_S$ denotes the variation of energy in an infinitesimal variation for which the entropy remains constant.

f. Conclusions. From the foregoing, the following conclusions might be obtained:

(a) The quantity that plays an important role in the second law of thermodynamics is an integral that has no obvious meaning; this integral being the entropy integral is given by

$$\int_1^2 \frac{d'q_{rev.}}{T},$$

where $d'q$ is the heat absorbed by the system from the surroundings, and T is the *source temperature* at which heat $d'q$ is transferred to the system rather than the temperature of the system itself in case they are different. The nature of this integral turns out to depend upon whether the path followed from the starting point 1 to the final point 2 is *reversible* or *irreversible*. This gives the second law its importance and its very special character. This is because it permits the formulation for the criteria of the reversibility of a process in an isolated system, and since reversibility is related to equilibrium, the criteria for the existence of equilibrium can be formulated. On the other hand, based on the first law

$$q = \Delta U + W,$$

the integral

$$\int_1^2 \frac{dq}{T}$$

along an irreversible path in the direction actually traversed is *path-dependent* and always *smaller* than the same integral over a *reversible path.* This is because the work *W* done by the system reversibly is greater than that done irreversibly so that since ΔU is *path-independent*, q_{irr} is less than q_{rev}. (Or $d'q_{irr} < d'q_{rev}$).

For any spontaneous *adiabatic* process in a system between state 1 and state 2, the integral $\int_1^2 d'q/T = 0$, because there is no heat absorbed. But since there is an entropy change ΔS owing to the fact that *S* is a thermodynamic function, this change must be calculated regardless of the zero value of this integral. For this purpose, *a new path other than the actual path between the initial and final states must be invented*, at least in thought, to replace the path practically followed by the system in this change, provided this invented path must be *reversible* so that ΔS can be calculated, but that need not to be adiabatic; reversible *adiabatic* paths need not exist. Evidently, the invented path is just an auxiliary path. Thus following this procedure, ΔS for an adiabatic irreversible process will be always positive, so it may be concluded that there is no process in an adiabatic, that is, thermally insulated system which would result in a decrease of its entropy. The entropy of the system is either constant, which is the case for equilibrium processes, or undergoes increment if the processes are spontaneous, that is, natural. Also, if that is followed for an unnatural adiabatic processes, then the entropy of the system would decrease so that such processes can be considered forbidden by thermodynamics.

(b) Since all processes occurring in nature are irreversible, the entropy of an adiabatic system should increase in this way. Once the entropy has attained a value beyond which it cannot increase, all processes must cease. Hence, equilibrium in an adiabatic system is characterised by the maximum possible value of the entropy. Putting it in another way, when the entropy of an adiabatic system has reached its maximum value, equilibrium exists. Conversely, if an adiabatic system is not in equilibrium, then changes tend to occur in the direction of increasing entropy until the entropy reaches its maximum possible value whereby all changes must cease. The tendency of the system towards increase in entropy represents an important driving force for chemical and physical processes. Consequently, the entropy concept can be used to give thermodynamical criteria for the direction in which a chemical reaction is liable to proceed. But unfortunately, it is not possible to establish, by thermodynamics, how fast a reaction proceeds towards chemical equilibrium. Thermodynamics may thus more properly be called thermostatics; the name thermodynamics to this branch of science was adopted by the early days of steam-engine theory.

(c) The two important concepts derived as consequences of Carnot's theorem, namely, the Gibbs free energy concept *G*, and the Clausius entropy concept *S* indicate that there are two forces that drive a reaction, one of them is in the direction of lowest possible free energy, and the other is in the direction of highest possible entropy (or disorder). Which of the two forces is more important turns out to be dependent upon temperature as will be shown below.

2.3. The driving forces of reactions

It might be expected that the direction in which a thermodynamically stable equilibrium lies is that of lowest energy in the case of reactions at constant volume and temperature, and that of lowest enthalpy in the case of reactions at constant pressure and temperature. *Berthelot and Thomsen* recognised, about a hundred years ago, that the production of heat of reaction (i.e. lowering of enthalpy in our terms) is an important driving principle. It was thus supposed at one time that the heat given out when two substances interact could be used as a measure of their chemical affinity for one another. This supposition is only roughly correct, for there are many physical and chemical changes which occur with absorption of heat. Examples of such changes are the spontaneous evaporation of liquids, the solution of salts in water, and the formation of acetylene from carbon and hydrogen at high temperatures.

Later, it has been realised thermodynamically that on one hand, the driving force of a reaction cannot be determined by the enthalpy change ΔH alone but also by the free energy change ΔG, that is, by the combination

$$H - TS,$$

rather than by just H solely. And on the other hand, the true measure of the affinity of a change is the maximum useful work, that is, the net work w', which is equal to the decrease in the free energy G of the system $(-\Delta G)$ when the change is carried out reversibly at constant temperature and pressure.

The direction in which constant temperature and constant pressure process tends to proceed is towards lower free energy, that is,

$$\Delta G = \Delta H - T\Delta S < 0.$$

Since $\Delta G = \Delta H - T\Delta S$, it follows that while a decrease of H (or a negative value of ΔH) is indeed desirable, an increase of S (or a positive ΔS value) is also desirable. Also that the influence of the entropy change becomes more important as the temperature increases. Thus the temperature can be regarded as a weighting factor of the importance of the entropy change, relative to that of the enthalpy change. Temperature exerts a randomising influence. At low temperatures, ΔH is important; at high temperatures, ΔS.

In detail, depending on the signs of ΔH and ΔS, four different cases exist:

1. **When $\Delta H < 0$ and $\Delta S > 0$,** the reaction may proceed at one temperature. The signs of ΔH and ΔS may of course change with temperature, in which case ΔG may not be negative at all temperatures.
2. **When $\Delta H < 0$ and $\Delta S < 0$,** the reaction may proceed only when

$$|T\Delta S| < |\Delta H|.$$

The reaction must, therefore, be sufficiently exothermic to overcome the handicap of the entropy decrease. At sufficiently large T, the term $T\Delta S$ overcome the ΔH term, and the reaction can no longer proceed.

When ΔH and ΔS are known at a given temperature T', the temperature T above which the reaction becomes prohibited by thermodynamics is given by the equation

$$\Delta G = \Delta H - T\Delta S < 0,$$

which is approximately equal to the ratio $\Delta H(T')/\Delta S(T')$, that is

$$T = \frac{\Delta H(T')}{\Delta S(T')}.$$

This relationship is approximate, because ΔH and ΔS depend on temperature.

3. **When $\Delta H > 0$ and $\Delta S > 0$,** the reaction tends to proceed only so long as $T\Delta S > \Delta H$. Here the advantage of the entropy increase must overcome the handicap or the disadvantage of the endothermic character of the reaction, and for this reason T must be sufficiently large. The heat required by the reaction comes from the surroundings, that is, from the constant temperature bath needed to maintain the temperature of the reaction at a certain constant level. When T is sufficiently small, the reaction is prohibited.
4. **When $\Delta H > 0$ and $\Delta S < 0$,** the reaction is prohibited thermodynamically at any temperature, unless a sign change of ΔH or ΔS with temperature occurs in such a way that the sign of ΔG also changes.

The influence of enthalpy change and entropy change on chemical reactions discussed above is summarised in the following table.

Case	ΔH	ΔS	Reaction Proceeds
1	<0	>0	Always
2	<0	<0	When T is sufficiently low
3	>0	>0	When T is sufficiently high
4	>0	<0	No

2.4. Thermodynamic potentials

1. The potential functions. The functions F, G as well as H are defined by the following equations referring to a system subjected to pressure–volume work alone

Internal energy U

Enthalpy $H = U + PV$ (1)

Helmholtz function $F = U - TS$ (2)

Gibbs function $G = U - TS + PV$

$= H - TS.$ (3)

The significance of these thermodynamic functions, which are also known as *thermodynamic potential functions* because of the role they play together with U in determining the conditions of equilibrium of systems under various constants, becomes a little more apparent from their differential forms which are followed by differentiating the Eqs. (1), (2), and (3) and using the result we already derived for dU in the context of the first law. Thus we have

$$dU = TdS - PdV, \tag{4}$$

$$dH = dU + PdV + VdP = TdS + VdP, \tag{5}$$

$$dF = dU - TdS - SdT = -SdT - PdV, \tag{6}$$

$$dG = dU - TdS - SdT + PdV + VdP$$
$$= -SdT + VdP. \tag{7}$$

Because the thermodynamic potentials are functions of state, as noted above, their differentials are exact. Each has two terms on the right-hand side corresponding to two degrees of freedom of the system. These terms are derived from the *two pairs* of the fundamental variables (or primary variables) which appear in the expression for dU, namely, (T, S) and (P, V). Then Eqs. (4), (5), (6), and (7) show that each potential has a different pair of fundamental variables as its *natural or proper independent variables*:

$$U = U(S, V),$$
$$H = H(S, P),$$
$$F = F(T, V),$$
$$G = G(T, P).$$

The energy of a system may vary in three ways:

1. By the absorption of heat.
2. By performing work on the surroundings.
3. By a change in its amount or composition.

Consider now a change of state of a system from state I to state II, both of which are at the same temperature *T* and pressure *P*. Then the change in *G* is

$$\begin{aligned} \Delta G &= G_{II} - G_I \\ &= (U_{II} - TS_{II} + PV_{II}) - \\ &\quad (U_I - TS_I + PV_I) \\ &= (U_{II} - U_I) - T(\Delta S_{II} - \Delta S_I) + P(V_{II} - V_I) \\ &= \Delta U - T\Delta S + P\Delta V \\ &= \Delta H - T\Delta S. \end{aligned} \tag{8}$$

Or

$$\begin{aligned} \Delta G &= G_{II} - G_I \\ &= (H_{II} - TS_{II}) - (H_I - TS_I) \end{aligned}$$

$$= (H_{II} - H_I) - T(S_{II} - S_I)$$

$$= \Delta H - T\Delta S. \qquad (8)$$

This is the change in the free energy *G* of a system undergoing a process at constant temperature and pressure and exerting pressure–volume work only.

The variation of *G* with temperature at constant pressure for a system, and conversely, the variation of *G* with pressure at constant temperature are given directly from the differential form of *G*.

Thus

$$\left(\frac{\partial G}{\partial T}\right)_P = -S, \qquad (8)$$

and

$$\left(\frac{\partial G}{\partial P}\right)_T = V. \qquad (9)$$

It obviously follows that since $\Delta G = G_{II} - G_I$,

$$\left\{\frac{\partial(\Delta G)}{\partial T}\right\}_P = -(S_{II} - S_I) = -\Delta S \qquad (10)$$

and

$$\left\{\frac{\partial(\Delta G)}{\partial P}\right\}_T = (V_{II} - V_I) = \Delta V \qquad (11)$$

Now by Eq. (8) we have,

$$\Delta G = \Delta H - T\Delta S, \qquad (8)$$

substituting the value of ΔS given by Eq. (10) in this equation, we have

$$\Delta G = \Delta H - T\left\{-\frac{\partial(\Delta G)}{\partial T}\right\}_P$$

or

$$\Delta G - \Delta H = T\left\{\frac{\partial(\Delta G)}{\partial T}\right\}_P. \qquad (12)$$

This equation will be familiar as the *Gibbs-Helmholtz* equation. It can be written in another form as follows.

Differentiating the quotient $\Delta G/T$ with respect to temperature at constant pressure, we have

$$\left\{\frac{\partial(\Delta G/T)}{\partial T}\right\}_P = \frac{1}{T}\left\{\frac{\partial(\Delta G)}{\partial T}\right\}_P - \frac{\Delta G}{T^2}.$$

But according to Eq. (11), we see that

$$\frac{\Delta G}{T^2} - \frac{1}{T}\left\{\frac{\partial(\Delta G)}{\partial T}\right\}_P = \frac{\Delta H}{T^2}.$$

Therefore,

$$\left\{\frac{\partial(\Delta G/T)}{\partial T}\right\}_P = -\frac{\Delta H}{T^2}. \qquad (13)$$

This form is sometimes convenient in practice.

It can easily be seen by similar methods that

$$\left\{\frac{\partial(G/T)}{\partial T}\right\}_P = -\frac{H}{T^2}. \qquad (14)$$

Again from the differential form of the Helmholtz function, that is, from Eq. (6) given above,

$$dF = -SdT - PdV, \qquad (6)$$

we have

$$\left(\frac{\partial F}{\partial T}\right)_V = -S, \qquad (15)$$

and

$$\left(\frac{\partial F}{\partial V}\right)_T = -P. \qquad (16)$$

Substituting these in Eq. (2),

$$F = U - TS, \qquad (2)$$

we have

$$F = U + T\left(\frac{\partial F}{\partial T}\right)_V$$

or

$$F - U = T\left(\frac{\partial F}{\partial T}\right)_V. \qquad (17)$$

Since $\Delta F = F_2 - F_1$ and $\Delta U = U_2 - U_1$, by using Eq. (17), we obtain

$$\Delta F - \Delta U = T\left\{\frac{\partial(\Delta F)}{\partial T}\right\}_V. \qquad (18)$$

Also, from Eq. (17), it is easy to obtain the equation

$$\left\{\frac{\partial(F/T)}{\partial T}\right\}_V = -\frac{U}{T^2} \qquad (19)$$

Further, by differentiating the quotient $\Delta F/T$ with respect to temperature at constant volume and then comparing the result with Eq. (18), we obtain

$$\left[\frac{\partial(\Delta F/T)}{\partial T}\right]_V = -\frac{\Delta U}{T^2}.$$

It seems worthwhile to digress for a moment here, *prior* to the consideration of the important properties of the potential functions for a system, to give a summary of some common contributions to the work done on a system in an infinitesimal reversible change, as can be seen from the following table:

System subjected to work by	Work done on the system (positive sign)	
Tensional force	$f\mathrm{d}L$	f = force
Electric current	$\varepsilon\, \mathrm{d}Z$	ε = potential difference and Z = charge
Hydrostatic pressure	$-P\mathrm{d}V$	P = pressure and V = volume
Change of surface area	$\gamma \mathrm{d}A$	γ = surface tension and A = area
Electric field	$E.\mathrm{d}p$	E = field strength and p = dipole moment
Magnetic field	$B.\mathrm{d}m$	B = magnetic induction and m = magnetic moment
Generalised force	$X_i\, \mathrm{d}x_i$	X = force and x = displacement

The analogous functions to the functions are expressed by Eqs. (1)–(3) as shown above, where the work done on the system is not done by hydrostatic pressure but are obtained by replacing $-P$ and V by the appropriate pair of variables (γ, A); (B, m), etc, as shown by the above table. Also, if the system is subjected to work done by hydrostatic pressure as well as other types of work done on it, use can be made of this table. To avoid confusion, the first law or the differential $\mathrm{d}U$ is expressed as $\mathrm{d}U = T\mathrm{d}S$ + total work done on the system.

Now we can return to the differential forms of the thermodynamic potentials given above referring only to $P-V$ work

If any one of the potentials is known explicitly in terms of its proper variables, which are (S, V) for U; (S, P) for H; (T, V) for F, and (T, P) for G, then we have complete information about the system. This is because any of the parameters of state may be calculated from one function. As an example, we will consider the Helmholtz function F, which is *particularly important* in connection with quantum statistical mechanics, as will be considered later,[4] since the expression for F in terms of statistical parameters is very simple *and forms a link between the microscopic analysis and the macroscopic variables.* Thus the proper variables for F are T and V so that if F is given explicitly in terms of these, we see from the differential form

$$\mathrm{d}F = -S\mathrm{d}T - P\mathrm{d}V \qquad (6)$$

that the other two fundamental variables S and P follow immediately as shown above,

$$\left(\frac{\partial F}{\partial T}\right)_V = -S \qquad (15)$$

[4] See Section 9.6.

and

$$\left(\frac{\partial F}{\partial V}\right)_T = -P . \qquad (16)$$

The expressions for *U*, *H*, and *G* are then constructed from their definitions as follows:

$$U = F + TS = F - T\left(\frac{\partial F}{\partial T}\right)_P = -T^2\left\{\frac{\partial (F/T)}{\partial T}\right\}_V ,$$

$$H = U + PV = F - T\left(\frac{\partial F}{\partial T}\right)_P = -V\left\{\frac{\partial F}{\partial V}\right\}_T ,$$

and

$$G = F + PV = F - \left(\frac{\partial F}{\partial V}\right)_T = -V^2\left\{\frac{\partial (F/V)}{\partial V}\right\}_{T.}$$

It should be pointed out that, as with other functions of state, given suitable information, it is always possible to calculate how a potential function changes when the system goes from one state to another. For instance, if we know *G* at T_0 and P_0 but wish to calculate G at T_0 and P_1, we may write

$$G(T_0 , P_1) - G(T_0 , P_1) = \int_{P_0}^{P_1}\left(\frac{\partial G}{\partial P}\right)_T dP$$

$$= \int_{P_0}^{P_1} VdP.$$

To evaluate this, the only information required is *V* as a function of *P*.

Some of the more important properties of the potential functions for a system subjected to work by *hydrostatic pressure only* may be summarised as follows.

Internal energy *U*. For a thermally insulated system we have

$$-dU = d'W = PdV = \text{work done by the system} .$$

This means that the decrease in the internal energy is equal to the work done by the system. If the change of state takes place quasi-statically, that is reversibly, then the entropy change will be equal to zero, that is, the process is *isentropic*. For an isovolumic, non–adiabatic change, we have

$$dV = 0 \quad \text{and} \quad d'W = 0 ,$$

so that

$$dq_V = dU$$

or

$$C_v dT = dU = TdS .$$

Hence,

$$C_V = \left(\frac{\partial U}{\partial T}\right)_V = T\left(\frac{\partial S}{\partial T}\right)_V .$$

Enthalpy *H*. As in the preceding case, it can be simply proved using the differential form of the first law

$$dU = TdS - PdV,$$

the differential form of *H as*

$$dH = TdS + VdP,$$

and at constant pressure, we have

$$C_P dT = dq_P = dH_P = TdS,$$

so

$$C_P = \left(\frac{\partial H}{\partial T}\right)_P = T\left(\frac{\partial S}{\partial T}\right)_P.$$

Helmholtz free energy function *F*. According to the relation

$$dF = -SdT - PdV,$$

it may be seen that in an isothermal change,

$$dF = -PdV.$$

This means, in an isothermal change, the decrease in the Helmholtz function is the maximum amount of mechanical work that can be done by the system. (Hence that alternative name, *Helmholtz free energy.*) Thus *F* becomes a useful energy function for isothermal processes. In an isovolumic change, the change in *F* is related to the change in temperature is given by

$$\left(\frac{\partial F}{\partial T}\right)_V = -S,$$

Gibbs free energy function *G*. The importance of the Gibbs potential is that it remains constant in reversible processes occurring at constant temperature and at constant pressure, that is, under isothermal isobaric conditions as implied by its differential form,

$$dG = -SdT + VdP.$$

These are the conditions applying to many physical and chemical changes. The constancy of the Gibbs function may then be used to represent the system constraints. We shall later develop its applications to determine the equilibrium states of systems containing several phases (Chapter 6) and several components (Chapters 13 and 14).

2. The Legendre differential transformations. In systems with more than two degrees of freedom, there are correspondingly more thermodynamic potentials, and their differentials contain correspondingly more terms. As in the case of two-parameter system, one first constructs the expression for the differential form of the internal energy and from that derives the other potentials.

For a system with *n* degrees of freedom, the differential form of d*U* contains *TdS* and $n-1$ work-like terms, each of the form $X_i dx_i$. The system, therefore, has 2*n fundamental or primary* variables forming *n conjugate pairs*, the product of the variables in each pair having the dimensions of energy. These pairs are like (*T*, *S*), (*P*, *V*), and (*E*, *p*). The four potential functions for the system with two degrees of freedom, which were discussed above, correspond to all the possible combinations of independent variables when one is taken from each conjugate pair. Thus when we combine *S* and *P*, we obtain the enthalpy *H*, in whose differential the proper or natural variables are *S* and *P*. The combination of *T* and

V corresponds to the Helmholtz function F, and that of T and P corresponds to the Gibbs function G. When these three potentials are added to the internal energy, whose proper independent variables are those when S is combined with V, we obtain the *four* thermodynamic potentials. Accordingly, for a system with n degrees of freedom, there will be 2^n potential functions corresponding to the two-fold choice offered by each pair of the n conjugate pairs. For example, let us choose our system as a wire under tension, that is, a wire subject to a tensional force f, and also for which volume changes are important. Therefore, for an infinitesimal reversible change, the total work done on the wire is

$$d'W = f\,dL - P\,dV,$$

and hence the differential dU or the first law is given by

$$dU = T\,dS + f\,dL - P\,dV.$$

It is evident from this equation that for this system, there are three pairs of fundamental or primary variables:

$$(T,S),\ (P,V),\ (f,L),$$

that is, three conjugate pairs corresponding to the three degrees of freedom, so that there will be $2^3 = 8$ potential functions. These functions correspond to the following sets of independent variables.

$$\begin{array}{llll} T, f, P & T, L, P & S, f, P & S, f, V \\ T, f, V & T, L, V & S, L, P & S, L, V. \end{array}$$

It is clearly a great advantage to have a systematic way of generating these potentials as and when required. The simplest method is the following one.

Firstly, the expression for dU is written down. This consists of TdS plus all the work terms and has as the independent variables the members of the conjugate pairs. To obtain a potential with a different set of independent variables, one picks out the terms in which the wrong member of the pair is the independent variable and adds to or subtracts from dU, the differential of the product of the conjugate pair, so as to remove the unwanted term and replace it by the required one. This produces a new differential expression, still with n terms, but with a different set of independent variables. Obviously, it is an exact differential because it is obtained by adding or subtracting exact differentials, namely dU and terms like $d(PV)$. It also has the dimensions of energy, and is therefore, the differential of a new potential function. This procedure is known as a *Legendre differential transformation.* We illustrate it by returning to the above example of the wire subjected to work by tension and hydrostatic pressure.

For the wire, the differential form of the first law is

$$dU = TdS - PdV$$

referring to a system subject to work by tension and hydrostatic pressure becomes

$$dU = TdS + f\,dL - PdV,$$

which has as natural or proper independent variables S, L, and V. Suppose we wish to construct the potential with proper variables T, L, and P. Then the first and last terms need to be transformed. We may affect this by adding $-d(TS) + d(PV)$. This generates the differential of the new potential

$$\begin{aligned} dG'(T, L, P) &= dU - d(T,S) + d(PV) \\ &= TdS + f\,dL - PdV + PdV \\ &\quad + VdP - TdS - SdT \\ &= -SdT + f\,dL + VdP, \end{aligned}$$

in which T, L, and P are the proper variables as required. Accordingly, the new potential function is

$$G' = U - TS + PV.$$

It is important to point out that although the four thermodynamic potentials we stated before in Section 2.4 (1), namely

internal energy	U
enthalpy	$H = U + PV$
Helmholtz function	$F = U - TS$
Gibbs function	$G = U - TS + PV$

by referring to a system subject to work by hydrostatic pressure only, we could have generated them from the differential form of dU

$$dU = TdS - PdV$$

by applying the Legendre differential transformations (as illustrated above) to obtain all the possible combination of independent variables.

3. The Maxwell relations. For systems with two degrees of freedom, we obtain the following differential forms of the potential functions U, H, F, and G, provided the system is subject to work done on it by hydrostatic pressure *only* which is given by $-PdV$

$$\begin{aligned} dU &= TdS - PdV \\ dH &= TdS + VdP \\ dF &= -SdT - PdV \\ dG &= -SdT + VdP \end{aligned}$$

From these equations four important, useful equations known as *Maxwell's relations* can be derived. They relate the partial derivatives of the fundamental thermodynamic variables, which are T, S, P, and V.

Thus if we form the partial derivative of U with respect to its proper or natural variables S and V, we obtain

$$\left(\frac{\partial U}{\partial S}\right)_V = T \quad \text{and} \quad \left(\frac{\partial U}{\partial V}\right)_S = -P.$$

Differentiating again with respect to the opposite variables, we get

$$\frac{\partial^2 U}{\partial V \partial S} = \left(\frac{\partial T}{\partial V}\right)_S$$

and

$$\frac{\partial^2 U}{\partial S \partial V} = -\left(\frac{\partial P}{\partial S}\right)_V.$$

It is shown in text books on calculus that if a variable *x* is a function of two independent variables *y* and *z*, then

$$\frac{\partial}{\partial z}\left(\frac{\partial x}{\partial y}\right)_z = \frac{\partial}{\partial y}\left(\frac{\partial x}{\partial z}\right)_y$$

or

$$\frac{\partial^2 x}{\partial z \partial y} = \frac{\partial^2 x}{\partial y \partial z},$$

provide the function *x* and its partial derivatives are continuous. Since these conditions are satisfied by *U* and its partial derivatives, we have, in the notation of thermodynamics

$$\frac{\partial^2 U}{\partial S \partial V} = \frac{\partial^2 U}{\partial V \partial S},$$

that is, the so-called mixed second-order partial derivatives are independent of the order of differentiation. Therefore,

$$\left(\frac{\partial T}{\partial V}\right)_S = -\left(\frac{\partial P}{\partial S}\right)_V \qquad (1)$$

The same result can also be obtained immediately by using the condition for d*U* to be an exact differential. Since this condition is fulfilled in the case of the internal energy *U* being a function of state, then it follows that by applying the condition to the coefficients *T* and *P* on the right-hand side of

$$dU = T\,dS - P\,dV,$$

we obtain,

$$\left(\frac{\partial T}{\partial V}\right)_S = -\left(\frac{\partial P}{\partial S}\right)_V. \qquad (1)$$

We already know that if *x* is a continuous function of *y* and *z*, it is always possible to write the infinitesimal change in *x* which results from infinitesimal changes in y and z in the exact differential form

$$dx = \left(\frac{\partial x}{\partial y}\right)_z dy + \left(\frac{\partial x}{\partial z}\right)_y dz$$

or

$$= Y dy + Z dz,$$

where $Y = \left(\frac{\partial x}{\partial y}\right)_z$ and $Z = \left(\frac{\partial x}{\partial z}\right)_y$.

Also, we know that the differential of a function of state must always be exact, since a function of state is, by definition, a single-valued function of the state variables

Again if x is a continuous function and also so its partial derivatives we have, as shown above,

$$\frac{\partial}{\partial z}\left(\frac{\partial x}{\partial y}\right)_z = \frac{\partial}{\partial y}\left(\frac{\partial x}{\partial z}\right)_y,$$

or

$$\frac{\partial^2 x}{\partial z \partial y} = \frac{\partial^2 x}{\partial y \partial z}.$$

Thus applying this result to the exact differential equation of

$$dx = Ydy + Zdz,$$

we obtain

$$\left(\frac{\partial Y}{\partial z}\right)_y = \left(\frac{\partial Z}{\partial y}\right)_z.$$

Hence, this equation can be regarded as a *necessary and sufficient condition* for dx to be exact differential.

Following now the same procedure with H, F, and G, we obtain three more equations of a similar form, as can be seen from the following.

$$dH = TdS + VdP$$

$$\left(\frac{\partial T}{\partial P}\right)_S = \left(\frac{\partial V}{\partial S}\right)_P \qquad (2)$$

$$dF = -SdT - PdV$$

$$\left(\frac{\partial S}{\partial V}\right)_T = \left(\frac{\partial P}{\partial T}\right)_V \qquad (3)$$

$$dG = -SdT + VdP$$

$$\left(\frac{\partial S}{\partial P}\right)_T = -\left(\frac{\partial V}{\partial T}\right)_P. \qquad (4)$$

Eqs. (1), (2), (3), and (4) obtained above are known as the *Maxwell Relations*. It may be instructive to rewrite these relations together so that some useful rules may be extracted.

$$\left(\frac{\partial T}{\partial V}\right)_S = -\left(\frac{\partial P}{\partial S}\right)_V \qquad \text{(from d}U\text{)} \qquad (1)$$

$$\left(\frac{\partial T}{\partial P}\right)_S = \left(\frac{\partial V}{\partial S}\right)_P \qquad \text{(from d}H\text{)} \qquad (2)$$

$$\left(\frac{\partial S}{\partial V}\right)_T = \left(\frac{\partial P}{\partial T}\right)_V \qquad \text{(from d}F\text{)} \qquad (3)$$

$$\left(\frac{\partial S}{\partial P}\right)_T = -\left(\frac{\partial V}{\partial T}\right)_P \qquad \text{(from d}G\text{)} \qquad (4)$$

Apparently, these relations may be recalled easily by remembering the following rules:

1. Cross multiplication of the variables always gives the form
$$TS = PV,$$
with the dimensions of energy.
2. The sign is positive if *T* appears with *P*.
3. Opposite pairs of variables are constant.

On the other hand, if we choose to introduce differential coefficients of the potential functions, we could derive many other equalities between differential coefficients as can be seen from the following example:

In the differential d*U*
$$dU = TdS - PdV,$$
the differential coefficients of *T* and *P* are d*S* and d*V*, respectively, and in the differential d*H*,
$$dH = TdS + VdP,$$
where *T* and *V* are the coefficients of the differentials d*S* and d*P*. Then, since
$$\left(\frac{dU}{\partial S}\right)_V = T$$
and
$$\left(\frac{dH}{\partial S}\right)_P = T,$$
we have
$$\left(\frac{\partial U}{\partial S}\right)_V = \left(\frac{\partial H}{\partial S}\right)_P.$$
But unlike the Maxwell relations, these are rarely useful and it is easier to deduce them when they are needed.

2.5. The derivation of standard forms for some partial derivatives

Since we shall refer in this section to *molal specific values* of the variables for a system, we shall use small letters to represent these variables.

On the other hand, an expression for any quantity is spoken of as a *standard form*, if it contains terms of the state variables as well as the terms of the coefficient of volume expansion (or isobaric-expansivity cubic) denoted by $\beta \left\{ = \frac{1}{v}\left(\frac{\partial v}{\partial T}\right)_P \right\}$, isothermal compressibility denoted by $\kappa \left\{ = -\frac{1}{v}\left(\frac{\partial v}{\partial P}\right)_T \right\}$, and specific heat denoted by c_v or c_p.

i. The standard form of c_v

Put $u = u(T, v)$, therefore,

$$du = \left(\frac{\partial u}{\partial T}\right)_V dT - \left(\frac{\partial u}{\partial v}\right)_T dv. \qquad (1)$$

Since

$$h = u + Pv,$$
$$dh = du + pdv + vdP. \qquad (2)$$

Substituting Eq. (1) for d*u* in Eq. (2), we obtain

$$dh = \left(\frac{\partial u}{\partial T}\right)_V dT + \left[\left(\frac{\partial u}{\partial v}\right)_T + P\right] dv + vdP. \qquad (3)$$

Put $v = v(T, P)$,
therefore,

$$dv = \left(\frac{\partial v}{\partial T}\right)_P dT + \left(\frac{\partial v}{\partial P}\right)_T dP \qquad (4)$$

Substituting Eq. (4) for d*v* in Eq. (3), we have

$$dh = \left(\frac{\partial u}{\partial T}\right)_V dT + \left[\left(\frac{\partial u}{\partial v}\right)_T + P\right]\left[\left(\frac{\partial v}{\partial T}\right)_P dT + \left(\frac{\partial v}{\partial P}\right)_T dP\right] + vdP \qquad (5)$$

At constant pressure, this equation reduces to

$$dh_P = \left(\frac{\partial u}{\partial T}\right)_V dT_P + \left\{\left(\frac{\partial u}{\partial V}\right)_T + P\right\}\left(\frac{\partial V}{\partial T}\right)_P dT_P$$
$$= \left[\left(\frac{\partial u}{\partial T}\right)_v + \left\{\left(\frac{\partial u}{\partial v}\right)_T + P\right\}\left(\frac{\partial v}{\partial T}\right)_P\right] dT_P. \qquad (6)$$

Therefore,

$$\left(\frac{\partial h}{\partial T}\right)_P = \left(\frac{\partial u}{\partial T}\right)_V + \left\{\left(\frac{\partial u}{\partial v}\right)_T + P\right\}\left(\frac{\partial v}{\partial T}\right)_P. \qquad (7)$$

But since $\left(\frac{\partial h}{\partial T}\right)_P = c_P$, and $\left(\frac{\partial u}{\partial T}\right)_V = c_V$

as has been proved before, and $\left(\frac{\partial v}{\partial T}\right)_P = v\beta$, where the isobaric coefficient of thermal expansion of a substance is $\frac{1}{v}\left(\frac{\partial v}{\partial T}\right)_P = \beta$, then Eq. (7), is

$$c_P = c_V + \left[\left(\frac{\partial u}{\partial v}\right)_T + P\right] v\beta. \qquad (8)$$

Now we want to compute first the standard form of the partial derivative $\left(\frac{\partial u}{\partial V}\right)_T$ and then the other partial derivative forms.

The partial derivative $\left(\frac{\partial \mathbf{u}}{\partial \mathbf{v}}\right)_{\mathbf{T}}$ in standard form

$$\text{Put} \quad v = v\,(T, P),$$

therefore,

$$dv = \left(\frac{\partial v}{\partial T}\right)_P dT + \left(\frac{\partial v}{\partial P}\right)_T dP. \tag{1}$$

Also, put $p = p\,(T, v)$

Therefore,

$$dP = \left(\frac{\partial P}{\partial T}\right)_V dT + \left(\frac{\partial P}{\partial v}\right)_T dV. \tag{2}$$

Substituting for d*P* in Eq. (1) from Eq. (2), we obtain

$$dv = \left(\frac{\partial v}{\partial T}\right)_P dT + \left(\frac{\partial v}{\partial P}\right)_T \left[\left(\frac{\partial P}{\partial T}\right)_V dT + \left(\frac{\partial P}{\partial v}\right)_T dV\right],$$

that is,

$$dv - \left(\frac{\partial v}{\partial P}\right)_T \left(\frac{\partial P}{\partial v}\right)_T dv = \left(\frac{\partial v}{\partial T}\right)_P dT + \left(\frac{\partial V}{\partial P}\right)_T \left(\frac{\partial P}{\partial T}\right)_V dT$$

or

$$\left[1 - \left(\frac{\partial v}{\partial P}\right)_T \left(\frac{\partial P}{\partial v}\right)_T\right] dv = \left[\left(\frac{\partial v}{\partial T}\right)_P + \left(\frac{\partial v}{\partial P}\right)_T \left(\frac{\partial P}{\partial T}\right)_v\right] dT. \tag{3}$$

The changes d*v* and d*T* are independent; that is, we can assign any value to d*v* and any other value to d*T*. Suppose we let d*T* = 0 and d*v* any other value but not zero, that is, $dv \neq 0$. Then to satisfy Eq. (3), we must have the coefficient of d*v* = 0, that is,

$$1 - \left(\frac{\partial v}{\partial P}\right)_T \left(\frac{\partial P}{\partial v}\right)_T = 0 \tag{4}$$

or

$$\left(\frac{\partial v}{\partial P}\right)_T = \frac{1}{\left(\frac{\partial P}{\partial v}\right)_T}. \tag{5}$$

Similarly, we can set $dv = 0$, $dT \neq 0$, then to satisfy Eq. (3), we must have the coefficient of $dT = 0$, that is,

$$\left(\frac{\partial v}{\partial T}\right)_P + \left(\frac{\partial v}{\partial P}\right)_T \left(\frac{\partial P}{\partial T}\right)_V = 0, \tag{6}$$

or

$$\left(\frac{\partial P}{\partial T}\right)_V = -\frac{\left(\frac{\partial v}{\partial T}\right)_P}{\left(\frac{\partial v}{\partial P}\right)_T} . \tag{7}$$

But $\frac{1}{v}\left(\frac{\partial v}{\partial T}\right)_P = \beta$ and $-\frac{1}{v}\left(\frac{\partial v}{\partial P}\right)_T = \kappa$,

therefore,
Eq. (7) reduces to

$$\left(\frac{\partial P}{\partial T}\right)_V = -\frac{\beta v}{\kappa v},$$

that is,

$$\left(\frac{\partial P}{\partial T}\right)_V = \frac{\beta}{\kappa}. \tag{8}$$

This standard form of the partial derivative $\left(\frac{\partial P}{\partial T}\right)_V$ will be made use of in deriving the standard form of $\left(\frac{\partial u}{\partial v}\right)_T$, as will be shown below.

Put $u = u(T, V)$,
therefore,

$$du = \left(\frac{\partial u}{\partial T}\right)_V dT + \left(\frac{\partial u}{\partial v}\right)_V dv. \tag{9}$$

According to the first law of thermodynamics,

$$d'q = du + d'W, \tag{10}$$

if $d'W$, which is the work done by the system against the surroundings in an infinitesimal change, is only a pressure–volume work, and is carried out reversibly, then we have

$$d'W = Pdv. \tag{11}$$

Consequently, the first law takes the form

$$d'q = du + Pdv, \tag{12}$$

or

$$du = d'q - Pdv. \tag{13}$$

On the other hand, the second law of thermodynamics states that for an infinitesimal reversible process, the differential element of heat $d'q$ is given by

$$d'q_{rev.} = Tds, \tag{14}$$

where *s* is the specific entropy of the system.
Consequently, for an infinitesimal reversible process in which we consider *Pdv* work only, Eq. (14) may be combined with Eq. (13) to give

$$du = Tds - Pds. \tag{15}$$

This equation is known in thermodynamics as the *combined first and second laws' equation.* Writing this equation in the form

$$Tds = du + Pdv, \tag{16}$$

and putting s as a function of T and v, we obtain

$$ds = \left(\frac{\partial s}{\partial T}\right)_V dT + \left(\frac{\partial s}{\partial v}\right)_T dv. \tag{17}$$

Also from Eq. (16), ds is given by

$$ds = \frac{1}{T}du + \frac{P}{T}dv. \tag{18}$$

Now substituting in this equation for du from Eq. (9), we have

$$ds = \frac{1}{T}\left[\left(\frac{\partial u}{\partial T}\right)_V dT + \left(\frac{\partial u}{\partial v}\right)_T dv\right] + \frac{P}{T}dv$$

or

$$ds = \frac{1}{T}\left[\left(\frac{\partial u}{\partial v}\right)_T + P\right] dv + \frac{1}{T}\left(\frac{\partial u}{\partial T}\right)_V dT. \tag{19}$$

Since the variables T and v in Eqs. (17) and (19) are independent, the coefficients of their exact differentials dT and dv in these two equations must be equal. Thus we obtain

$$\left(\frac{\partial s}{\partial T}\right)_V = \frac{1}{T}\left(\frac{\partial u}{\partial T}\right)_V, \tag{20}$$

and

$$\left(\frac{\partial s}{\partial v}\right)_T = \frac{1}{T}\left[\left(\frac{\partial u}{\partial v}\right)_T + P\right]. \tag{21}$$

It is shown, as noted before, that if a variable w is a function of two independent variables x and y, we have

$$\frac{\partial}{\partial x}\left(\frac{\partial w}{\partial y}\right) = \frac{\partial}{y}\left(\frac{\partial w}{\partial x}\right)$$

or

$$\frac{\partial^2 w}{\partial x\, dy} = \frac{\partial^2 w}{\partial y\, \partial x},$$

provided w and its partial derivatives are continuous. Since these conditions are satisfied by the entropy s and its partial derivatives, we must have from Eqs. (20) and (21)

$$\left[\frac{\partial}{\partial v}\left(\frac{\partial s}{\partial T}\right)_V\right] = \left[\frac{\partial}{\partial T}\left(\frac{\partial s}{\partial v}\right)_T\right]_V,$$

that is,

$$\frac{\partial^2 s}{\partial v\, \partial T} = \frac{\partial^2 s}{\partial T\, \partial v}.$$

In other words, the order of differentiation is immaterial in the so-called 'mixed' second order partial derivatives. Consequently, differentiating Eq. (20) with respect to *v* and Eq. (21) with respect to *T*, respectively, we obtain

$$\frac{\partial}{\partial v}\left[\frac{1}{T}\left(\frac{\partial u}{\partial T}\right)_V\right] = \frac{1}{T}\frac{\partial^2 u}{\partial v\,\partial T}$$

and

$$\frac{\partial}{\partial T}\left[\frac{1}{T}\left\{\left(\frac{\partial u}{\partial v}\right)_T + P\right\}\right] = \frac{1}{T}\left[\left(\frac{\partial^2 u}{\partial T\,\partial v}\right) + \left(\frac{\partial P}{\partial T}\right)_V\right]$$

$$- \frac{1}{T^2}\left[\left(\frac{\partial u}{\partial v}\right)_T + P\right].$$

Thus, equating the two right-hand sides of these two equations, we see that

$$\frac{1}{T}\frac{\partial^2 u}{\partial v\,\partial T} = \frac{1}{T}\left[\left(\frac{\partial^2 u}{\partial T\,\partial v}\right) + \left(\frac{\partial P}{\partial T}\right)_V\right]$$

$$= -\frac{1}{T^2}\left[\left(\frac{\partial u}{\partial v}\right)_T + P\right].$$

Therefore,

$$\frac{1}{T}\left(\frac{\partial u}{\partial v}\right)_T = T\left(\frac{\partial P}{\partial T}\right)_V - P. \tag{22}$$

But $\left(\frac{\partial P}{\partial T}\right)$, as deduced above, is given by

$$\left(\frac{\partial P}{\partial T}\right)_V = \frac{\beta}{\kappa}, \tag{8}$$

hence,

$$\left(\frac{\partial u}{\partial v}\right)_T = \frac{\beta T}{\kappa} - P. \tag{23}$$

This is the standard form of the partial derivative $\left(\frac{\partial w}{\partial v}\right)_T$ which we are looking for.

Substituting now for $\left(\frac{\partial u}{\partial v}\right)_T$ in the equation of c_P, which has been obtained above, namely

$$c_P = c_V + \left[\left(\frac{\partial u}{\partial v}\right)_T + P\right] v\beta,$$

by Eq. (23), we obtain

$$c_V = c_P - \frac{\beta^2 Tv}{\kappa}. \tag{24}$$

This is the standard form of the specific heat at constant volume c_V, which is also the form required.

Accordingly, the general expression for the difference between c_P and c_V in standard form is

$$C_P - C_V = \frac{\beta^2 Tv}{\kappa}. \tag{25}$$

It is evident from this equation that the difference $c_P - c_V$ can be computed for any substance for which β and κ are known. The quantities *T*, *v*, and κ are always positive. Usually β is positive, but it may also be negative or zero. For water at atmospheric pressure, β is zero at about 4°C and is negative between 0°C and 4°C. Nevertheless, β^2 is always positive or zero and consequently c_P is never smaller than c_V. In the case of a perfect gas, it can simply be proved that the term $\frac{\beta^2 Tv}{\kappa}$ is equal to *R*, so that in this case $c_P - c_V = R$.

The partial derivative $\left(\frac{\partial u}{\partial T}\right)_P$ in standard form

Put $u = u\,(T, v)$,

therefore,

$$du = \left(\frac{\partial vu}{\partial T}\right)_V dT + \left(\frac{\partial u}{\partial v}\right)_T dv. \tag{1}$$

Also put $v = v\,(T, P)$,

$$dv = \left(\frac{\partial v}{\partial T}\right)_P dT + \left(\frac{\partial v}{\partial P}\right)_T dP. \tag{2}$$

Substituting Eq. (2) for d*v* in Eq. (1), we obtain

$$dv = \left(\frac{\partial u}{\partial T}\right)_V dT + \left(\frac{\partial u}{\partial v}\right)_T \left[\left(\frac{\partial v}{\partial T}\right)_P dT + \left(\frac{\partial v}{\partial P}\right)_T dP\right],$$

$$= \left[\left(\frac{\partial u}{\partial T}\right)_V + \left(\frac{\partial u}{\partial v}\right)_T \left(\frac{\partial v}{\partial T}\right)_P\right] dT$$

$$+ \left(\frac{\partial u}{\partial v}\right)_T \left(\frac{\partial v}{\partial P}\right)_T dP. \tag{3}$$

At constant pressure, this equation reduces to

$$du_P = \left[\left(\frac{\partial u}{\partial T}\right)_V + \left(\frac{\partial u}{\partial v}\right)_T \left(\frac{\partial v}{\partial T}\right)_P\right] dT_P \ ,$$

so that

$$\left(\frac{\partial u}{\partial T}\right)_P = \left(\frac{\partial u}{\partial T}\right)_V + \left(\frac{\partial u}{\partial v}\right)_T \left(\frac{\partial v}{\partial T}\right)_P . \qquad (4)$$

Substituting in this equation for $\left(\frac{\partial u}{\partial T}\right)_V$ by C_V and for $\left(\frac{\partial u}{\partial v}\right)_T$ by $\left(\frac{\beta T}{\kappa} - P\right)$, which is Eq. (23) computed above and for $\left(\frac{\partial v}{\partial T}\right)_P$ by $v\beta$, we obtain

$$\left(\frac{\partial u}{\partial T}\right)_P = c_V + \left(\frac{\beta T}{\kappa} - P\right) v\beta$$

or

$$\left(\frac{\partial u}{\partial T}\right)_P = \left(c_V + \frac{\beta^2 Tv}{\kappa}\right) - \beta Pv . \qquad (5)$$

But since $c_P = c_V + \frac{\beta^2 TV}{\kappa}$, which is also computed above, we have

$$\left(\frac{\partial u}{\partial T}\right)_P = c_P - \beta Pv , \qquad (6)$$

which is the equation required.

The partial derivative $\left(\frac{du}{\partial P}\right)_T$ in standard form

We consider again Eq. (3), which has been derived in the preceding part. At constant temperature, this equation reduces to

$$du_T = \left(\frac{\partial u}{\partial v}\right)_T \left(\frac{\partial v}{\partial P}\right)_T dP_T$$

so

$$\left(\frac{\partial u}{\partial P}\right)_T = \left(\frac{\partial u}{\partial v}\right)_T \left(\frac{\partial v}{\partial P}\right)_T . \qquad (1)$$

Since

$$\left(\frac{\partial u}{\partial v}\right)_T = \left(\frac{\beta T}{\kappa} - P\right),$$

and

$$\left(\frac{\partial v}{\partial P}\right)_T = -v\kappa ,$$

Eq. (1) takes the form

$$\left(\frac{\partial u}{\partial P}\right)_T = \left(\frac{\beta T}{\kappa} - P\right) \times (-v\kappa),$$

$$= \kappa P v - \beta T v.$$

2.6. The application of partial derivatives to a perfect gas

In a qualitative fashion, an ideal or perfect gas may be defined as that gas between whose molecules there are no attractive or repulsive forces, as a consequence of which its internal energy remains the same at various pressures or volumes when its temperature is kept constant. In other words, the internal energy of a perfect gas is a function of temperature only, independent of its volume or pressure. That is,

$$\left(\frac{\partial u}{\partial v}\right)_T = 0 \text{ , or } \left(\frac{\partial u}{\partial P}\right)_T = 0$$

This again means that the isothermal expansion or compression of a perfect gas does not change its internal energy. The postulate can be proved quantitatively by considering the standard forms we have calculated for $\left(\frac{\partial u}{\partial v}\right)_T$ and $\left(\frac{\partial u}{\partial P}\right)_T$, for any given substance.

These forms are

$$\left(\frac{\partial u}{\partial v}\right)_T = \frac{\beta T}{\kappa} - P \tag{1}$$

and

$$\left(\frac{\partial u}{\partial P}\right)_T = \kappa P v - \beta T v \tag{2}$$

For a perfect gas, the equation of state is

$$Pv = RT, \tag{3}$$

for one mole.
Therefore,

$$Pdv + vdP = RdT$$

At constant pressure,

$$\left(\frac{\partial v}{\partial T}\right)_P = \frac{R}{P},$$

so

$$\beta = \frac{1}{v}\left(\frac{\partial V}{\partial T}v\right)_P = \frac{R}{Pv}.$$

Hence,

$$\beta = \frac{1}{T} \qquad \text{for a perfect gas.}$$

At constant temperature,

$$\left(\frac{\partial v}{\partial P}\right)_T = -\frac{v}{P},$$

but

$$\kappa = -\frac{1}{v}\left(\frac{\partial v}{\partial P}\right)_T,$$

hence,

$$\kappa = \frac{1}{P}. \qquad \text{for a perfect gas.}$$

Now substituting for $\beta\left(=\frac{1}{T}\right)$ and $\kappa\left(=\frac{1}{P}\right)$ in the above two standard forms (Eqs.1 and 2) we see that

$$\left(\frac{\partial u}{\partial v}\right)_T = \frac{\frac{1}{T}\times T}{1/P} - P = P - P = 0,$$

and

$$\left(\frac{\partial u}{\partial P}\right)_T = \frac{1}{P}.Pv - \frac{1}{T}.Tv = v - v = 0.$$

Therefore, with the help of standard forms $\left(\frac{\partial u}{\partial v}\right)_T$ and $\left(\frac{\partial u}{\partial P}\right)_T$ we proved in a quantitative way that the isothermal expansion or compression of a perfect gas does not result in any change in the internal energy of this gas. In other words, the internal energy of a perfect gas is a function of temperature alone. For this reason, an can replace the partial derivatives of u at constant temperature by complete derivatives, that is

$$\left(\frac{\partial u}{\partial v}\right)_T = \left(\frac{du}{dv}\right)_T, \text{ and } \left(\frac{\partial u}{\partial P}\right)_T = \left(\frac{du}{dP}\right)_T.$$

The enthalpy change with pressure or volume at constant temperature for a perfect gas is also equal to zero. This can be proved as follows:
Since

$$h = u + Pv,$$

$$dh = du + Pdv + vdP \qquad (1)$$

Again, since

$$Pv = RT, \qquad \text{for one mole}$$

$$Pdv + vdP = RdT. \qquad (2)$$

At constant temperature, we have from Eq. (2)

$$Pdv = -VdP. \qquad (3)$$

Accordingly, Eq. (1) reduces to

$$dh_T = du_T$$

so

$$\left(\frac{\partial h}{\partial P}\right)_T = \left(\frac{\partial u}{\partial P}\right)_T$$

and

$$\left(\frac{\partial h}{\partial v}\right)_T = \left(\frac{\partial u}{\partial v}\right)_T .$$

Hence,

$$\left(\frac{\partial h}{\partial P}\right)_T = 0$$

and

$$\left(\frac{\partial h}{\partial v}\right)_T = 0 .$$

The same result can be arrived at by another way as follows:

Since

$$h = u + Pv ,$$

we have

$$dh = du + Pdv + vdP ,$$

and

$$\left(\frac{\partial h}{\partial P}\right)_T = \left(\frac{\partial u}{\partial P}\right)_T + P\left(\frac{\partial v}{\partial P}\right)_T + v \qquad (1)$$

For a perfect gas, we have again

$$Pv = RT ,$$

therefore,

$$Pdv + vdP = RdT .$$

At constant *T*, this equation reduces to

$$Pdv + vdP = 0 ,$$

so

$$P\left(\frac{\partial v}{\partial P}\right)_T = -v \qquad (2)$$

Hence, substituting in Eq. (1) for $P\left(\frac{\partial v}{\partial P}\right)_T$ by $-v$, we obtain

$$\left(\frac{\partial h}{\partial P}\right)_T = \left(\frac{\partial u}{\partial P}\right)_T - v + v$$

$$= \left(\frac{\partial u}{\partial P}\right)_T .$$

In like manner,

$$\left(\frac{\partial h}{\partial v}\right)_T = \left(\frac{\partial u}{\partial v}\right)_T + P + v\left(\frac{\partial P}{\partial v}\right)_T . \qquad (3)$$

But $$v\left(\frac{\partial P}{\partial v}\right)_T = -P, \qquad \text{for a perfect gas,}$$

therefore,

$$\left(\frac{\partial h}{\partial v}\right)_T = \left(\frac{\partial u}{\partial v}\right)_T.$$

Hence,

$$\left(\frac{\partial h}{\partial P}\right)_T = 0,$$

and

$$\left(\frac{\partial h}{\partial v}\right)_T = 0.$$

2.7. The reciprocal and reciprocity theorems

Suppose that three variables are related.

$$F(x, y, z) = 0.$$

Then, in principle, this equation may be rearranged to express one of the variables in terms of the other two as *independent variables.*

$$x = x(y, z).$$

Differentiating by parts, we have

$$dx = \left(\frac{\partial x}{\partial y}\right)_z dy + \left(\frac{\partial x}{\partial z}\right)_y dz.$$

We may write an analogous equation for dz

$$dz = \left(\frac{\partial z}{\partial x}\right)_y dx + \left(\frac{\partial z}{\partial y}\right)_x dy.$$

Substituting this in the previous equation, we obtain

$$dx = \left(\frac{\partial x}{\partial y}\right)_z dy + \left(\frac{\partial x}{\partial z}\right)_y \left\{\left(\frac{\partial z}{\partial x}\right)_y dx + \left(\frac{\partial z}{\partial y}\right)_x dy\right\}.$$

Therefore,

$$\left\{1 - \left(\frac{\partial x}{\partial z}\right)_y \left(\frac{\partial z}{\partial x}\right)_y\right\}dx - \left\{\left(\frac{\partial x}{\partial z}\right)_y \left(\frac{\partial z}{\partial y}\right)_x + \left(\frac{\partial x}{\partial y}\right)_z\right\}dy = 0.$$

Since the variables dx and dy are independent in this equation, their coefficients must vanish separately. Thus we have

$$1 - \left(\frac{\partial x}{\partial z}\right)_y \left(\frac{\partial}{\partial x}\right)_y = 0$$

or

$$\left(\frac{\partial x}{\partial z}\right)_y = \frac{1}{\left(\frac{\partial z}{\partial x}\right)_y}.$$

This is the *reciprocal theorem* which allows us to replace any partial derivative by the reciprocal of the inverted derivative *with the same* variable held constant.
Also we have

$$\left(\frac{\partial x}{\partial z}\right)_y \left(\frac{\partial z}{\partial y}\right)_x + \left(\frac{\partial x}{\partial y}\right)_z = 0$$

or

$$\left(\frac{\partial x}{\partial y}\right)_z = -\left(\frac{\partial x}{\partial z}\right)_y \left(\frac{\partial z}{\partial y}\right)_x.$$

Therefore,

$$\left(\frac{\partial x}{\partial y}\right)_z = -\frac{1}{\left(\frac{\partial y}{\partial z}\right)_x \left(\frac{\partial z}{\partial x}\right)_y}$$

or

$$\left(\frac{\partial x}{\partial y}\right)_z \left(\frac{\partial y}{\partial z}\right) \left(\frac{\partial z}{\partial x}\right)_y = -1$$

This is the *reciprocity theorem*. It may be written starting with any derivative then following through the other variables in cyclic order. This gives a dimensionless product or combination. This relation is most often used to split up a derivative into a product of more convenient derivatives.

CHAPTER 3

The Application of Thermodynamics to Changes of State.The Phase Rule

3.1. The application of partial derivatives to simple systems

As an example of this application, we consider here the *isothermal–adiabatic transformation of moduli*. The derivative of an *intensive variable* with respect to its associated *extensive variable* is called a *stiffness coefficient*. The reciprocal derivative is a *compliance coefficient*. These are important physical quantities, and like all thermodynamic coefficients, they are *partial derivatives* since their values depend on the conditions under which they are measured. Two common constraints are that the system is kept isothermal or thermally insulated. We have already used the isothermal compressibility in Section 2.5, and we may similarly define an *adiabatic compressibility*

$$\kappa_S = \frac{1}{V}\left(\frac{\partial V}{\partial P}\right)_S,$$

where we have assumed that the changes are thermodynamically reversible and have replaced the condition of the constancy of temperature by the condition of constancy of entropy.

A very simple relationship exists between the ratio of the isotheramal and adiabatic coefficients and the specific heats. We show this for the case of the compressibilities.

$$\frac{\kappa_T}{\kappa_S} = \frac{\left(\frac{\partial V}{\partial P}\right)_T}{\left(\frac{\partial V}{\partial P}\right)_S}$$

$$= \frac{\left(\frac{\partial V}{\partial T}\right)_P\left(\frac{\partial T}{\partial P}\right)_V}{\left(\frac{\partial V}{\partial S}\right)_P\left(\frac{\partial S}{\partial P}\right)_V} \quad \text{(reciprocity theorem)}^*$$

$$= \frac{\left(\frac{\partial S}{\partial V}\right)_P\left(\frac{\partial V}{\partial T}\right)_P}{\left(\frac{\partial S}{\partial P}\right)_V\left(\frac{\partial P}{\partial T}\right)_V}, \quad \text{(reciprocal theorem)}$$

* For the numerator, we assume that the variables *P*, *V*, *T* are related by the function (*P*, *V*, *T*) and for the denominator we assume that the variables *P*, *V*, *S* are related by the function (*P*, *V*, *S*), each of these functions being equal to zero.

$$= \frac{\left(\frac{\partial S}{\partial T}\right)_P}{\left(\frac{\partial S}{\partial T}\right)_V}.$$

But $$dS = \frac{d'q_{rev.}}{T},$$

therefore, $$dS = C_V \frac{dT}{T},$$

$$\left(\frac{\partial S}{\partial T}\right)_V = C_V/T$$

or $$dS = C_P \frac{dT}{T},$$

$$\left(\frac{\partial S}{\partial T}\right)_P = C_P/T.$$

Accordingly,

$$\frac{\kappa_T}{\kappa_S} = \frac{C_P}{C_V} = \frac{c_P}{c_v} = \gamma.$$

Similar results hold for coefficients formed from other conjugate pairs of variables. For example,

Permittivity $$\epsilon = \left(\frac{\partial D}{\partial E}\right): \quad \frac{\epsilon_T}{\epsilon_S} = \frac{C_E}{C_D} = \frac{c_E}{c_D},$$

where *D* is the electric displacement, *E* electric field strength, $\epsilon\ (= \epsilon_r \epsilon_0)$ permittivity or dielectric constant, ϵ_0 permittivity of a vacuum, and ϵ_r relative permittivity.

Young's moduli, $$E = \frac{L}{A}\left(\frac{\partial f}{\partial L}\right): \frac{E_T}{E_S} = \frac{C_L}{C_f} = \frac{c_L}{c_f},$$

where *f* is the tensional force, *L* length of filament or rod, and *E* modulus.

Magnetic susceptibility $$\chi = \left(\frac{\partial M}{\partial H}\right): \frac{\chi_T}{\chi_S} = \frac{C_H}{C_M} = \frac{c_H}{c_M},$$

where *H* is the magnetic field strength, *M* magnetisation, and χ magnetic susceptibility.

It is to be noted that in the three examples given above, the permittivity and magnetic susceptibility are considered as compliance coefficients, whereas the Young modulus is considered as stiffness coefficient.

3.2. The application of thermodynamics to changes of state

1. The Clausius-Clapeyron equation. It is known that the melting point of a substance depends on the applied pressure. For a given pressure *P*, the solid and liquid are in equilibrium with each other only at a definite temperature *T*, the *melting point.* At this temperature, since the two forms are in equilibrium, the net work of the change solid–liquid is zero so that the maximum work obtainable is the work done through the change in volume. For a given weight of substance, this is equal to $P(V_\ell - V_S)$, if *V* is the volume of the liquid and V_S that of the solid. We can introduce this change of state into a cyclic process in the following way (Fig. 1).

1. Melt a *given weight* of the solid at the melting point *T* corresponding to applied pressure *P*.

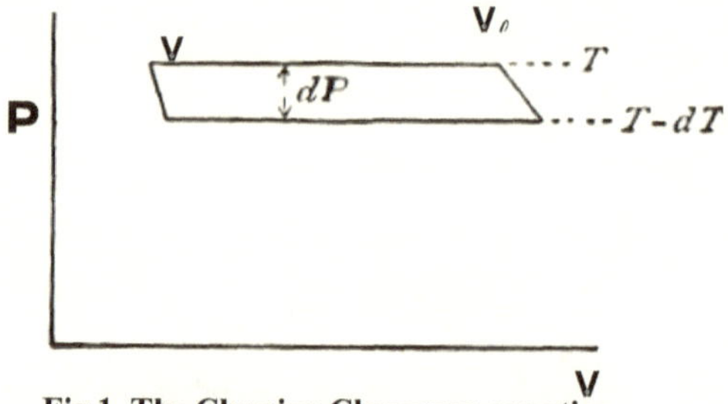

Fig.1. The Clausius-Clapeyron equation.

2. Reduce the pressure on the liquid from *P* to *P*-d*P* and reduce the temperature from *T* to *T*-d*T*, which is the melting point at the new pressure.
3. Cause the liquid to solidify at the temperature *T*-d*T* and pressure *P*-d*P*.
4. Bring the solid back to temperature *T* and pressure *P*.

The work obtained in the whole cycle is equal to the area on the *P–V* diagram, which to the first order of small quantities is

$$dW = dP.\,(V_\ell - V_S). \tag{1}$$

Use can be made here of the efficiency of Carnot's cycle derived before

$$\eta = \frac{W}{q_2} = \frac{T_2 - T_1}{T_2},$$

by assuming that the temperature difference between the two isothermal stages of this cycle is a small amount d*T*, so that the efficiency equation may be written in the form

$$\frac{dW}{q} = \frac{dT}{T}$$

or

$$dW = q\frac{dT}{T}. \tag{2}$$

Therefore, by substituting for dW in this equation by the right-hand side of Eq. (1), we obtain

$$dP\left(V_\ell - V_S\right) = q\,\frac{dT}{T},$$

or

$$\frac{dT}{dP} = \frac{T\left(V_\ell - V_S\right)}{q}, \tag{3}$$

where q is the amount of heat absorbed in the first stage, that is, in melting the quantity of the substance to which V_S and V_ℓ refer. Thus if V_S and V_ℓ are the volumes of 1 gram of solid and liquid, respectively, q is the *latent heat of fusion per gram.* The quotient dT/dP is the rate at which the melting point changes with the applied pressure.

A similar relation holds in all cases of change of state. In general, if V_I is the volume of a given amount of substance in the initial state, V_{II} the volume in the final state and q is the heat absorbed in the change I → II (for the same amount), the relation between the temperature and the equilibrium pressure at which the two forms are in equilibrium with each other is given by

$$\frac{dT}{dP} = \frac{T\left(V_{II} - V_I\right)}{q}. \tag{4}$$

This is the Clausius-Clapeyron equation.

Example. In the case of ice and water, the volume of the solid is greater than that of the same weight of liquid, hence $\left(V_\ell - V_S\right)$ is negative, and since q is a positive quantity, dT/dP is negative, that is, melting point is lowered by the increase of pressure.

Let us calculate the lowering produced by an increased pressure of 1 atmosphere.

At 0°C, $\quad V_\ell = 1.000$ c.c. per gram,

$T = 273$ K

$V_S = 1.091$ c.c. per gram,

therefore,

$$V_\ell - V_S = -0.091 \text{ c.c.}$$

Also $\quad q = 80.0$ calories per gram.

The volume change is expressed in c. c., and we are going to reckon pressure in atmospheres, so that the work done $dP(V_\ell - V_S)$ appears in the equation (Eq. 3) in c. c. atmospheres. The heat absorbed *q* is measured in calories. In order to obtain our result this must be converted into the same units. Since 1 calorie = 41.37 c. c. atmospheres, we have

$$\frac{dT}{dP} = \frac{-0.091 \times 273}{80.0 \times 41.37} \frac{[\text{c.c. degree}]}{[\text{c.c. atmosphere}]}$$

$$= -0.0075^\circ \text{ per atmosphere.}$$

(Note that the numerator is a volume in c. c. multiplied by temperature, and the denominator is a quantity of energy expressed as volume in c. c. multiplied by pressure in atmospheres. The quotient is, therefore, a temperature divided by pressure in atmospheres).

Kelvin determined the effect of pressure on the melting point of ice in 1850 and found $dT/dP = 0.0072$ in very fair agreement with the calculated value.

In like manner, we may calculate the elevation produced by a decreased pressure of 100 atmospheres at 0°C. It is $+0.0075 \times 100 = 0.75°C$. This explains, as will be seen below, why the triple or ternary point of water occurs at a temperature higher than 0°C.

2. The Le Chatalier principle.* Since *q*, the latent heat of fusion, is always positive, we see that if $V_I > V_{II}$, where V_I is the volume of a given amount of the substance in the initial state (the solid state), then dT/dP will be negative, this means that the melting point is lowered with increase of pressure. If $V_{II} > V_I$, that is, the volume of the liquid for a given amount of the substance is greater than that of the solid, then dT/dP will be positive, that is, the melting point is raised with increase of pressure. This is in accordance with the *Le Chatalier principle*, which predicts *qualitatively* the effect of a change of conditions on the equilibrium state of a system. This principle states that

> *If a constraint is applied to a system in equilibrium, the change which occurs is such that it tends to annul the constraint, or if a constraint is applied to a system at equilibrium, the equilibrium is shifted in the direction which partially nullifies this constraint.*

This principle may be stated in other words as follows:

> *If one of the conditions of a system in equilibrium is altered, the system will, if possible, adjust itself in such a direction as to partially neutralise the change in condition.*

This principle applies to physical as well as to chemical changes. An example of a *physical change* is the melting of solids. The solid and liquid phases of a pure substance can exist together in a stable equilibrium only at a definite temperature *T*, the melting point at which, for a given external pressure, the vapour pressure of the solid is equal to the vapour pressure of the liquid. Thus, for example, ice and water saturated with air under a pressure of 1 atmosphere exist in equilibrium at a definite temperature which is taken as the zero

---

* Also associated with the name of *Braun*.

degree centigrade (0°C). When ice melts, it absorbs its latent heat of fusion from the surrounding bodies, and the melting is accompanied by contraction in volume since the specific volume of ice is greater than that of liquid water at the same temperature. In other words, ice is less denser than liquid water at the same temperature.

Suppose that we have an isolated system of ice and water in equilibrium at 0°C and pressure of 1 atmosphere and that a compression is applied on it. Then in order that the system be relieved from compression, a decrease in volume takes place by the melting of some ice. Since the system is isolated, the latent heat of fusion is taken from the kinetic energy of the system, and accordingly, the temperature is lowered until a new state of equilibrium between ice and water is attained at the applied higher pressure when the vapour pressure of liquid becomes equal to the vapour pressure of solid. Hence, the equilibrium between ice and water at a pressure higher than 1 atmosphere will not be at 0°C but will be at a lower temperature. This explains why the melting point of ice or the freezing point of water decreases with increase of pressure as found experimentally.

If the solid expands on fusion, that is, the specific volume of the substance in the solid state is smaller than that in the liquid state at the same temperature, which is the case for most substances, a relief from an applied compression results from the solidification of some of the liquid since the volume of the system, in this case, decreases. The evolved heat of solidification raises the temperature of the isolated system until a new state of equilibrium is established at the higher pressure when the vapour pressures of the two phases become equal.

In general, the freezing point of most substance increases with the increase of pressure. Conversely, the decrease of pressure results in a decrease in the freezing point of most substances. In the case of water, the state is opposite, that is, the freezing point decreases with the increase of pressure and increases with the decrease of pressure.

As pointed out above the Clausius-Clapeyron equation predicts the same behaviour as the Le Chatalier principle. It goes further than the latter however, for it gives quantitatively the change produced by a change of pressure, so it can be regarded as a quantitative expression of the Le Chatalier principle when applied to a change of state.

Further, we have seen above that by using the Clausius-Clapeyron equation to predict the effect of pressure on the melting point of ice or the freezing point of liquid water, we found that the lowering produced by an increased pressure of 1 atmosphere is 0.0075°, that is,

$$\frac{dT}{dP} = 0\,0.0075°\text{C per atmosphere}$$

Conversely, this means that a decrease of pressure of 1 atmosphere causes an increase in the freezing point of water equal to 0.0075°C. This explains in part why water, saturated with air under a pressure of 1 atmosphere, freezes at a definite temperature, taken as the 0°C, the freezing point at which the two phases ice and liquid water are in equilibrium, whereas when ice and water from which dissolved air has been removed are allowed to evaporate into a previously evacuated container, the pressure on ice and water will be 4.57 mm instead of 760 mm (i.e. 1 atmosphere), and at the same time the temperature at which freezing takes place will be slightly higher than 0°C, namely + 0.0099°C. The difference of 0.0099°C is due to two factors:

i. The decrease of pressure from 760 mm to 4.57 mm raises the freezing point by 0.0075°C as has been calculated above, from Clausius equation.

ii. The removal of air at atmospheric pressure is sufficient to increase the freezing point by 0.0024°C, note that the presence of a solute in a solvent lowers the freezing point of the solvent.

The point in a temperature–pressure diagram for water, which represents the only temperature and pressure at which the three states of aggregation (phases) – solid, liquid, and vapour – *can exist together in equilibrium* is called the *triple point or ternary point.* It is automatically fixed at 0.0099°C and 4.59 mm. This means that in this system of three phases, neither temperature nor pressure can be varied, that is, the system has no degrees of freedom, or the number of independent variables is equal to zero in this system.

The significance of the triple or ternary point becomes a little more apparent from the following figure.

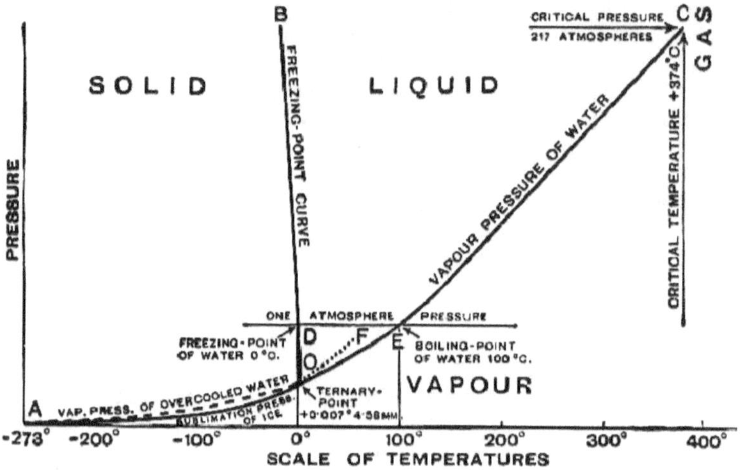

Fig. 1. Schematic triple-point diagram for water or equilibrium diagram for water (not drawn to scale).

Some notes on this diagram. (1) The vapour pressure curve of ice is described as the *sublimation pressure of the solid.* That is, ice has a small vapour pressure, which is proved by the fact that snow can disappear in frosty weather when that temperature is never high enough to melt it. It is found that this curve representing the vapour pressure of ice at different temperatures is not a continuation of the curve OC, which represents the vapour pressure of water, but falls off more rapidly as shown be the full line OA. This line cuts the curve OC at C. The point O, therefore, shows the freezing point of water, not at atmospheric pressure but under the vapour pressure of melting ice. This vapour pressure is about 4.57 mm; and since the freezing point of water is lowered to the extent of about 1° for each additional 100 atmospheres, it will be raised slightly when the pressure is reduced from 760 to 4.57 mm, as noted above.

(2) At the upper limit of temperature, the vapour pressure curve of liquid water terminates at the *critical point* C, since above this temperature, liquid and vapour are *indistinguishable*, and the term vapour pressure has no meaning. The *critical temperature* is 374°C, and the critical pressure is 217 atmospheres.

(3) At the lower limit, the curve would normally terminate at the point O, at which water freezes into ice under a vapour pressure of about 4.57 mm, but if the water could be prevented from freezing, the vapour pressure curve could be continued below the point O, as indicated by the broken line OA which is labelled 'vapour pressure of over-cooled or super-cooled water'.

3.3. Transition points

The considerations applied above to investigate the effect of pressure on the melting point of a substance also apply to the effect of pressure on the change of a substance from one modification to another. At every pressure, the two modifications of a substance are in equilibrium with each other only at a definite temperature, the *transition point.* Thus *rhombic* and *monoclinic* sulphur are in equilibrium at 95.5°C under a pressure of 1 atmosphere. At lower temperatures, the rhombic is the stable form, at higher temperatures the monoclinic. We may calculate the effect of change of pressure on the transition temperature by using the Clausius-Clapeyron equation.

$$\frac{dT}{dP} = \frac{T\left(V_{II} - V_{I}\right)}{q},$$

where V_{II} and V_{I} are the volumes of a given weight of the two forms, and q is the heat absorbed in the change from $I \rightarrow II$ (for the same amount). Thus for the change or transition of rhombic sulphur to monoclinic sulphur, the equation can be written in the form

$$\frac{dT}{dP} = \frac{T\left(V_{m} - V_{r}\right)}{q},$$

where V_m and V_r are the volumes of a given weight of the two forms, and q is the heat absorbed in the change $S_r \rightarrow S_m$ for the same amount. In this case,

$$V_m - V_r = 0.0126 \text{ c. c. per gram}$$

$$q = 2.52 \text{ calories per gram}$$

$$T = 95.5°C + 273$$

$$= 368.5 \text{ K.}$$

Hence, we find

$$\frac{dT}{dP} = \frac{368.5 \times 0.0126}{2.52 \times 41.4}$$

$$= 0.045 \text{ per atmosphere}.$$

The observed value is 0.05°C.

Note: It is important to know that heat is always absorbed in the change from the substance state, which is the stable form below the transition temperature, to that above it. This is because according to the Le Chatalier principle, if we add heat to a system, that is, attempt to raise the temperature, a change will occur in the direction which absorbs heat. Hence q is positive for the change from the form stable below the transition point to that stable above it; the same rules for the effect of pressure apply for the melting point also.

3.4. The phase rule

In order to obtain this rule, we must first consider the conditions of equilibrium in heterogeneous systems. For simplicity, we consider the case of a system of two phases, both containing the components S_1 , S_2 , S_3, ..., S_n. The quantities of these components in the first phase are $n_1', n_2', n_3', ..., n_n'$, and the quantities in the second phase are $n_1'', n_2'', n_3'', ..., n_n''$. Then the variation of the free energy of the first phase when the quantities of its components are varied by small quantities $\delta n_1'$, $d'n_2$,... etc, at constant temperature and pressure is given* by

$$\delta G_1' = \overline{G}_1' \delta n_1' + \overline{G}_2' \delta n_2' + ... + \overline{G}_n' \delta n_n',$$

and similarly, we may write for similar variations of the second phase

$$\delta G_1'' = \overline{G}_1'' \delta n_1'' + \overline{G}_2'' \delta n_2'' + ... + \overline{G}_n'' \delta n_n'',$$

where $\overline{G}_1'$ is the *partial* free energy of the component S_1 in the first phase and $\overline{G}_P''$ is the value in the second phase, etc. We shall assume that the *temperature and pressure are the same in both phases*, since it has already been shown that this is a necessary condition. This being the case, it is necessary according to the criterion of equilibrium, previously considered in Section 2.2, that is, in all possible variations in the state of the system, which do not alter its temperature and pressure, the variation of its free energy *G* shall either vanish or be positive, that is,

$$\left(\overline{\delta G'}_1 + \overline{\delta G''}_1\right)_{T,P} \geq 0,$$

for all such variations

Let us consider what variations are possible. We will limit ourselves at first to the case in which none of the components can be formed out of other components. Accordingly, the possible variations are those in which small quantities of the different components pass from one phase to the other. All such variations must therefore be in accordance with the equations.

$$\left.\begin{aligned} \delta n_1' + \delta n_1'' &= 0 \\ \delta n_2' + \delta n_2'' &= 0 \\ \cdots\cdots\cdots&\cdots\cdots \\ \delta n_n' + \delta n_n'' &= 0 \end{aligned}\right\}, \qquad (1)$$

that is, the total amount of each component is a constant. The change of free energy of the whole system in any variation, in which the temperature and pressure remain constant is given by

$$\begin{aligned} \delta G &= \delta G' + \delta G'' \\ &= \overline{G}_1' \delta n_1' + \overline{G}_2' \delta n_2' + ... + \overline{G}_n' \delta n_n' \\ &+ \overline{G}_1'' \delta n_1'' + \overline{G}_2'' \delta n_2'' + ... + \overline{G}_n'' \delta n_n'' . \end{aligned} \qquad (2)$$

---

* $\overline{G}_1 = \left(\dfrac{dG}{dn_1}\right)_{T,P,n_2,etc}$, $\overline{G}_2 = \left(\dfrac{dG}{dn_2}\right)_{T,P,n_1,etc}$

This quantity must be zero or positive for all possible values of $\delta n_1'$, $\delta n_2'$, ...etc. Since $\delta n_1' = -\delta n_1''$, etc. according to Eq. (1), we may, therefore, write the condition of equilibrium as

$$\left(\overline{G}_1' - \overline{G}_1''\right)\delta n_1' + \left(\overline{G}_2' - \overline{G}_2''\right)\delta n_2' \ldots + \left(\overline{G}_n' - \overline{G}_n''\right)\delta n_n' = 0. \quad (3)$$

Since $\delta n_1'$, $\delta n_2'$, ..., $\delta n_n'$ are independent variables, then Eq. (3) cannot be true except when the coefficients of these variables, $\delta n_1'$, $\delta n_2'$, etc vanish separately, that is, when

$$\overline{G}_1' = \overline{G}_1'', \overline{G}_2' = \overline{G}_2'', \ldots \overline{G}_n' = \overline{G}_n'' \quad (4)$$

It is, therefore, necessary for equilibrium that the partial free energy of each component shall be the same in both phases. This argument can be extended to a system containing any number of distinct phases and to show that it is necessary for equilibrium that the partial free energy of each component shall have the same value in every phase in which it is actually present.

Finally, there is the case in which same components can be formed out of others. Suppose that the components C, D can be formed out of the components A, B according to the equation.

$$aA + bB = cC + dD, \quad (5)$$

where *a*, *b*, *c*, and *d* represent the numbers of formula weights of these substances which enter into the reaction. It can be proved as above by considering variations in which the amount of each component remains constant that

$$\overline{G}_1' = \overline{G}_1'', \overline{G}_2' = \overline{G}_2'', \text{etc.}$$

If Eq. (4) is satisfied for every possible variation, then it must be satisfied for any selection of the possible variations. Thus Eq. (2) can be written as

$$\overline{G}_A \Sigma\delta n_A' + \overline{G}_B \Sigma\delta n_B' + \overline{G}_C \Sigma\delta n_C' + \overline{G}_D \Sigma\delta n_D' = 0, \quad (6)$$

where $\delta n_A'$ is the total change in the amount of *A* throughout the system. Evidently, the quantities $\Sigma\delta n_A'$, $\Sigma\delta n_B'$, $\Sigma\delta n_C'$, $\Sigma\delta n_D'$ must be such that Eq. (5) is satisfied, that is, they must be proportional to $a, b, -c, -d$. We, therefore, have

$$a\overline{G}_A + b\overline{G}_B = c\overline{G}_c + d\overline{G}_D. \quad (7)$$

When we compare Eq. (7) with Eq. (6), we see that the relation between the partial free energies is the same as that between the chemical formulae of the substances involved.

Now we are in a position to introduce *the phase rule of Gibbs*. Every distinct kind of body which is present in a heterogeneous system is a phase, whereas bodies differing only in amount or shape are examples of the same phase.

Consider a *single phase* containing the quantities $n_1, n_2, \dots n_n$ of the independent components $S_1, S_2, \dots, S_n$, and having the entropy *S* and volume *V*. The phase is, therefore, characterised by $n + 2$ intensity factors T, P, $\overline{G}_1, \dots \overline{G}_n$. But these quantities are not all independent, for their variations are related by the equation.

$$+SdT - VdP + n_1 d\overline{G}_1 + n_2 d\overline{G}_2 + \dots$$

$$+ n_n d\overline{G}_n = 0. \qquad (8)$$

This equation can be arrived at as follows. We already know that the variation of *G* with *T* and *P* for a body is given by

$$dG = -SdT + VdP,$$

and the variation of *G* with n_1 and n_2, etc. at constant *T* and *P* is given by

$$dG = \overline{G}_1 dn_1 + \overline{G}_2 dn_2 + \dots + \overline{G}_n dn_n,$$

so when these two equation are combined, we obtain d*G* as

$$dG = -SdT + VdP + \overline{G}_1 dn_1 + \overline{G}_2 dn_2 + \dots$$

$$+ \overline{G}_n dn_n.$$

Integrating this at constant *T* and *P* for a change in which n_1 varies from 0 to n_1, n_2 from 0 to n_2, etc, we have

$$G = \overline{G}_1 n_1 + \overline{G}_2 n_2 + \dots + \overline{G}_n n_n.$$

Differentiating this generally, without any regard for the significance of the quantities, we obtain

$$dG = \overline{G}_1 dn_1 + n_1 d\overline{G}_1 + \overline{G}_2 dn_2 + n_2 d\overline{G}_2 + \dots$$

$$+ \overline{G}_n dn_n + n_n d\overline{G}_n. \quad \text{(constant } T \text{ and } P\text{)}.$$

But if

$$\overline{G}_1 dn_1 + \overline{G}_2 dn_2 + \dots + \overline{G}_n dn_n$$

represents completely an infinitesimal change of *G*, the sum of the remaining terms at constant *T* and *P* must be zero, that is,

$$n_1 d\overline{G}_1 + n_2 d\overline{G}_2 + \dots + n_n d\overline{G}_n = 0.$$

This is the *generalised form of the Duhem–Margules equation.* Subsequently, instead of this equation, in which *T* and *P* are supposed to be constant, we obtain

$$+SdT - VdP + n_1 d\overline{G}_1 + n_2 d\overline{G}_2 + \dots n_n d\overline{G}_n = 0, \qquad (8)$$

which is the required equation.

According to Eq. (8), if $n + 1$ of the quantities *T, P,* $\overline{G}_1, \overline{G}_2, \dots \overline{G}_n$ are varied, the variation of the last is given by or fixed automatically by this equation. A single phase is then capable of only $n + 1$ independent variations, or we may say that it has $n + 1$ *degrees of freedom.*

Now suppose that we have two phases in equilibrium with each other, each containing the same *n* components S_1, ... S_n. It is necessary for equilibrium that the *n* + 2 intensity factors characterising each phase, which are

$$T, P, \overline{G}_1, \ldots \overline{G}_n,$$

shall be the same in the two phases. But there are now two equations like Eq. (8), one for each phase. It follows that only *n* of these quantities can be varied *independently*, the variations of the last two quantities being given by the two equations. Therefore, the two phases have $n + 2 - 2 = n$ *degrees of freedom.*

In general, if there are *r* phases, each containing the same *n* components, there will be *r* relations like Eq. (8) between the *n* + 2 quantities $T, P, \overline{G}_1, \overline{G}_2, \ldots \overline{G}_n$, which are the same throughout the system as a necessary condition for equilibrium. Therefore only *n* + 2 – *r* of the *n* + 2 quantities can be varied independently, and the number of degrees of freedom *F* of *n* components in *r* phases is thus

$$F = n + 2 - r.$$

This is the Phase Rule of Gibbs.

It does not matter if some of the components are absent from certain phases. Taking the system as a whole, we shall have the $n + 2$ quantities $T, P, \overline{G}_1, \ldots \overline{G}_n$ characteristic of the system and *r* relations (equal to the number of phases) like Eq. (8) which limit their variations. It is sometimes convenient to choose components which are *not completely independent of each other.* Let *n* be the number of *independent components*, that is,. the least (or minimum) number of components in terms of which the *composition of the system or every variation of the system can be expressed.*

Let there be *additional h components.* These components can be formed out of the others by reactions similar to that represented in Eq. (5), and for each such relation between the components, there is a corresponding relation between their partial free energies, similar to Eq. (7). There will thus be *h* relations between the partial free energies like Eq. (7). Hence the total number of variable quantities is *n* + *h* + 2, and the total number of relations between them is *h* + *r*, so the number of degrees of freedom is given by

$$\begin{aligned} F &= \text{total number of variable quantities} \\ &\qquad - \text{total number of relations between them} \\ &= (n+h+2)-(h+r) \\ &= n+2-r. \end{aligned}$$

We will briefly survey the application of the phase rule to some typical systems. In the first place, *a system of one component in one phase* according to Eq. (9)

$$\begin{aligned} F &= n+2-r \\ &= 1+2-1=2, \end{aligned}$$

has two degrees of freedom. That is, since there is only one phase, there is a single relation like Eq. (8), viz.

$$+ SdT - VdP + n_1 d\overline{G}_1 = 0,$$

between the three quantities *T*, *P*, and $\overline{G}_1$. Thus *T* and *P* can be varied at will, but for every value of *T* and *P*, there is a corresponding value of $\overline{G}_1$,

Again, a single component in two phases by Eq. (9) has *one degree* of freedom, since there are two relations like Eq. (8) between the three quantities *T*, *P*, and $\overline{G}_1$, these equations are

$$+ S_1' dT - V_1' dP + n_1' d\overline{G}_1' = 0,$$

and

$$+ S_1'' dT - V_1'' dP + n_1'' d\overline{G}_1'' = 0.$$

It is necessary for equilibrium that these three quantities shall be the same in the two phases. Also, as regards the partial molar free energy $\overline{G}_1$, its *complete* infinitesimal change is given by

$$dG_1 = \overline{G'}_1 dn_1' + \overline{G''}_1 dn_1'',$$

and at equilibrium, it reduces to zero, and as we are dealing with a closed system in which the mass is conserved, we have the constraint

$$dn_1' + dn_1'' = 0,$$

so

$$\overline{G_1'} = \overline{G''}_1,$$

that is, $\overline{G}_1$ is the same in the two phases at equilibrium for a constant temperature and pressure.

By means of the above two relations between *T*, *P*, and $\overline{G}_1$, we can get a relation between *T* and *P*, namely

$$\frac{dT}{dP} = \frac{V_1'' - V_1'}{S_1'' - S_1'} = \frac{\Delta V}{\Delta S}.$$

$$= \frac{T\,\Delta V}{q},$$

which is the Clausius-Clapeyron equation obtained above, where *q* being the heat absorbed on passing from volume V_1' to V_1'' and ΔV the volume change in this process.

Finally, when a single component is present in three phases, there are three equation like Eq. (8) between the variables *T*, *P* and $\overline{G}_1$. So according to Eq. (9), no variation is possible. An example of this case is the *triple or ternary point of water*, that is, the point in the pressure–volume diagram for water which represents the only temperature and pressure at which the three phases or states of aggregation can coexist in equilibrium (see Fig. 1, Section 2.)

Condensed systems. Since liquids and solids have only a small compressibility, the effect of pressure on equilibrium of systems containing only liquids and solids is very small. Thus as long as the pressure is maintained higher than the vapour pressure of the system so that no gaseous phase exists, small variation of pressure will not alter appreciably the form of the curves which represent the relationships between the other two variables, namely, the temperature and composition. Such a system is termed as a *condensed system.* For condensed systems, we may write a '*reduced*' phase rule, namely

$$F' = n - r + 1.$$

This will give the remaining degrees of freedom which the system can possess, in addition to the pressure, which, as we have seen, is capable of independent variation as long as it is higher than the vapour pressure. It is convenient to use this rule in the discussion of the solubilities of liquids and solids.

Examples of condensed systems may also be found in the *eutectic mixtures, where two solids and their saturated solution in one another are in equilibrium at one fixed temperature and one fixed, composition.* It must be remembered, however, that although small changes of pressure have only a negligible influence on a condensed system, yet the effect of pressure is a real one and the conditions of equilibrium are altered appreciably under high pressures, just as the melting point of a solid is altered under these conditions as shown above. The 'reduced' phase rule, whilst very convenient, is therefore only an *approximate guide* to the behaviour of condensed systems and can only be used when no large changes in pressure occur.

3.5. The effect of temperature on vapour pressure

We are now concerned with the effect of temperature on the equilibrium vapour pressure of two phases: liquid and vapour or solid *and* vapour. For this purpose use can be made of the Clausius-Clapeyron equation in its inverted form, that is

$$\frac{dP}{dT} = \frac{q}{T\left(V_v - V_\ell\right)} \cdot \qquad (1)$$

In this case, it will be convenient to take as the amount of substance to which the terms apply *1 mole of vapour*. Then q is the latent heat of vaporisation of the substance per mole and v_g and v_ℓ, the volumes of the same amount as vapour and as liquid (or solid). Since v_g is much greater than v_ℓ, we may neglect the latter, and if we assume that the vapour obeys the ideal gas law, we may put

$$v_g = \frac{RT}{P}.$$

Thus Eq. (1) becomes

$$\frac{1}{P} \cdot \frac{dP}{dT} = \frac{q}{RT^2} \qquad (2)$$

or

$$\frac{d \ln P}{dT} = \frac{R}{RT_2}. \qquad (3)$$

In this form, an integration can be made to give the variation of *P* with *T* over a range of temperature. If the range of temperature is comparatively small, so that *q* can be taken constant, we find

$$\ln P = \int \frac{q}{RT^2} dT$$

$$= -\frac{q}{Rt} + K, \qquad (4)$$

where *K* is the integration constant. Thus the relation between $\ln P$ and $\frac{1}{T}$ is a linear one, and if the values of $\ln P$ be plotted against corresponding values of $1/T$, a straight time is obtained (so long as *q* is constant). The slope of the line obtained is equal to $(-q/R)$ so that *q* may be determined by measuring the graph.

The method of evaluating *q* is, perhaps, made clearer by taking two points on the line obtained, say $\ln P_1$ and $\ln P_2$ corresponding to $1/T_1$ and $1/T_2$, respectively, that is,

$$\ln P_1 = -\frac{q}{R} \cdot \frac{1}{T_1} + K,$$

and

$$\ln P_2 = -\frac{q}{R} \cdot \frac{1}{T_2} + K,$$

therefore,

$$\ln P_1 - \ln P_2 = -\frac{q}{R}\left(\frac{1}{T_1} - \frac{1}{T_2}\right) \qquad (5)$$

so that

$$\frac{\ln P_1 - \ln P_2}{\left(\frac{1}{T_1} - \frac{1}{T_2}\right)} = \frac{q}{R},$$

and the left-hand side is obtained directly from the graph.

On the other hand, Eq. (5), in the form

$$\ln (P_1/P_2) = -\frac{q}{R}\left(\frac{1}{T_1} - \frac{1}{T_2}\right) = \frac{q}{R}\left(\frac{T_1 - T_2}{T_1 T_2}\right), \qquad (6)$$

can be *used directly without plotting the data.* Thus if we know the vapour pressures at two temperatures T_1 and T_2 we may find *q*, and conversely, if we know *q* and the vapour pressure *P* at one temperature, then we can calculate the vapour pressure at another temperature within a range of temperature over which *q* can be taken as constant.

Example: The latent heat of vaporisation of water at 100°C is 536 calories per gram, or 9660 calories per mole. The vapour pressure of water at 100°C is 760 mm.
What is the rate of variation of vapour pressure with temperature at this temperature?

Inserting these values in Eq. (2) above, which is

$$\frac{1}{P} \cdot \frac{dP}{dT} = \frac{q}{RT^2},$$

we obtain

$$\frac{dP}{dT} = 760 \times \frac{9660}{1.98 \times (273)^2}$$

$$= 26.5 \text{ mm per degree.}$$

3.6. The change of dissociation pressure with temperature

This case may be treated in the same way as vaporisation. For example, in the reaction

$$CaCO_3 = CaO + CO_2(g)$$

the increase in the volume can be taken without appreciable error, as *the volume of gas produced.* The pressure at which the gas is in equilibrium with the solid phases is the dissociation pressure of Ca CO_3. Thus again the equation

$$\frac{dP}{dT} = \frac{q}{T\left(v_g - v_\ell\right)}, \qquad (1)$$

can be reduced to

$$\frac{dP}{dT} = \frac{q}{Tv_g}, \qquad (2)$$

and then, if we apply the equation $PV = RT$ to the production of 1 mole of CO_2, assuming that this gas obeys the ideal gas law, we may put $v_g = \frac{RT}{P}$ so that Eq. (2) becomes

$$\frac{1}{P} \cdot \frac{dP}{dT} = \frac{q}{RT^2} \qquad (3)$$

or

$$\frac{d \ln P}{dT} = \frac{q}{RT^2}, \qquad (4)$$

where q is the heat of dissociation of Ca CO_3 per 1 mole of CO_2 produced.

Integrating Eq. (4), we have

$$\ln P = -\frac{q}{R} \cdot \frac{1}{T} + K, \qquad (5)$$

so that as long as q can be taken as constant there is a linear relation between $\ln P$ and $1/T$. Similarly, integrating between limits P_1 and P_2 corresponding to temperatures T_1 and T_2, we obtain

$$\ln\left(\frac{P_1}{P_2}\right) = -\frac{q}{R}\left(\frac{1}{T_1} - \frac{1}{T_2}\right)$$

$$= \frac{q}{R}\left(\frac{T_1 - T_2}{T_1 T_2}\right). \qquad (6)$$

Example: The dissociation pressure of calcium carbonate is 34.2 cm of mercury at 840°C, 42.0 cm at 860°C. Calculate the heat of dissociation.

Applying Eq. (6), we obtain

$$q = R\,(\ln P_1/P_2)\,\frac{T_1 T_2}{T_1 - T_2}$$

$$= 1.98 \times 2.303\left(\log\frac{42.0}{34.2}\right).\frac{1113 \times 1133}{20}$$

$$= 31.530 \text{ calories}.$$

It is to be noted that the dissociation of salt hydrates is exactly similar.

CHAPTER 4

Entropy and Free Energy. Engineering, Second-LawAnalysis.

4.1. The calculation of entropy changes in various processes

1. Isothermal changes. In order to find out the entropy change of an isothermal process, it is necessary to carry out the process reversibly. Then we divide the heat absorbed in the process by the absolute temperature of the source of heat.

For example, at the melting point of a solid, the solid and liquid forms of the substance are in equilibrium with each other (i.e. the vapour pressure of the solid is equal to the vapour pressure of the liquid), and the absorption of heat at constant external pressure causes a change of the solid into liquid under reversible conditions. Therefore, the entropy change of fusion of a given mass of the solid is

$$\Delta S = \frac{\Delta H}{T}, \qquad (1)$$

where ΔH is the heat absorbed reversibly, that is, the latent heat of fusion of the given mass of the solid, and T is the absolute temperature at which the fusion takes place. The entropy change in vaporisation is similarly obtained.

Examples:

1. The latent heat of fusion of water at its melting point (273.1 K) under a pressure of 1 atmosphere is 1438 calories per gram molecule. Therefore, the entropy change is

$$\Delta S = \frac{1438}{273.1} = 5.27 \qquad \text{calories/degree.}$$

2. The latent heat of vaporisation (vaporisation) of water at 273.1 K (1 atmosphere pressure) is 9730 calories per gram molecule.
Therefore, the entropy change is

$$\Delta S = \frac{9730}{373.1} = 26.5 \text{ calories/degree.}$$

This is the difference between the entropy of a gram molecule (18 grams) of water vapour at 1 atmosphere pressure and 373.1 K, and that of liquid water at the same temperature and under a pressure of 1 atmosphere, that is,

$$\underset{(273.1\text{K},1\text{ atm.})}{S_{vapour}} - \underset{(273.1\text{K},1\text{ atm.})}{S_{liquid}} = \Delta S = 26.5 \text{ calories / deg ree.}$$

In the isothermal expansion of a perfect gas, the internal energy change ΔU is zero and the heat absorbed *q* is equal to the work *W* performed by the gas as required by the first law.

$$q = \Delta U + W . \tag{2}$$

Therefore, for a reversible isothermal expansion of a gram molecule of the gas from pressure P_1 to pressure P_2 the heat absorbed is

$$q = W = \int_{V_1}^{V_2} PdV = RT\ln\frac{V_2}{V_1}$$

$$= RT\ln\frac{P_1}{P_2}, \tag{3}$$

and the entropy change is

$$\Delta S = \frac{q}{T} = R\ln\frac{P_1}{P_2} . \tag{4}$$

It should be observed that the entropy change of the gas has this value whether the *expansion is conducted reversibly or not*, but it can only be calculated from the *heat absorbed in the reversible process*. In an irreversible expansion, the work done by the gas and, consequently, the heat absorbed by the gas from the *surroundings* is less than what is given above. Accordingly, the entropy *decrease* of the surroundings is less than the entropy *increase* of the gas. This means that when a gas expands *isothermally* and *irreversibly,* then there is *an increase* in the total entropy of the gas and its surroundings.

2. Non-isothermal changes. To find the entropy change of a reversible non-isothermal process, we must sum the quantities of heat absorbed, each divided by the absolute temperature at which the absorption takes place *with these quantities of heat being absorbed reversibly.* It is always possible to add heat *reversibly* to a body, thereby increasing the temperature, if the body is placed in contact with sources of heat which have a temperature only *infinitesimally higher than that of the body itself.* Under these circumstances, the absorption of heat takes place very slowly, and the body remains in a state of *thermal* equilibrium (uniform temperature) throughout. The heat absorbed in raising the temperature of a body from *T* to *T* + d*T*, at a constant pressure is $dH = C_P dT$, and the entropy change is, therefore,

$$dS = \frac{dH}{T} = \frac{C_P}{T} . dT . \tag{5}$$

To find the entropy change in a finite change of temperature from T_1 to T_2, we must integrate this expression between the given temperatures, that is,

$$\int_{T_1}^{T_2} dS = \int_{T_1}^{T_2} \frac{C_P}{T} . dT , \tag{6}$$

thus

$$\Delta S_{T_1}^{T_2} = \int_{T_1}^{T_2} \frac{C_P}{T} \,.\, dT \qquad (7)$$

Examples:

1. The molar heat capacity of helium (a monatomic gas) between $-200°C$ and 0°C is 5.0. The entropy change between these temperatures is therefore,

$$\Delta S_{73}^{273} = \int_{73}^{273} \frac{5}{T} \,.\, dT = \int_{73}^{273} 5 \,.\, d\ln T$$

$$= 5 \ln \frac{273}{73}$$

$$= 6.61 \text{ calories/degree.}$$

2. The molar heat capacity of liquid mercury between – 40°C and + 140°C, is given by the equation

$$C_p = 8.42 - 0.0098\,T + 0.0000132\,T^2.$$

The entropy change between these temperatures is thus

$$\Delta S_{233}^{413} = \int_{233}^{413} \left(\frac{8.42 - 0.0098T + 0.0000132T^2}{T} \right) dT$$

$$= 8.42 \ln \left(\frac{413}{233} \right) - 0.0098 \,(413 - 2.33)$$

$$+ \frac{0.0000132}{2} \left(413^2 - 2.33^2\right)$$

$$= 4.82 - 1.76 + 1.54$$

$$= 4.60 \text{ calories/degree.}$$

3. Irreversible adiabatic changes. Consider a system to be initially at a temperature T_1 and of an entropy S_1. After the spontaneous adiabatic change, it has acquired a temperature T_2 and an entropy S_2.

The initial state is represented by point A and the final by B in the P–V diagram shown in Fig. 1. Since the change is irreversible, it cannot be graphically represented between A and B.

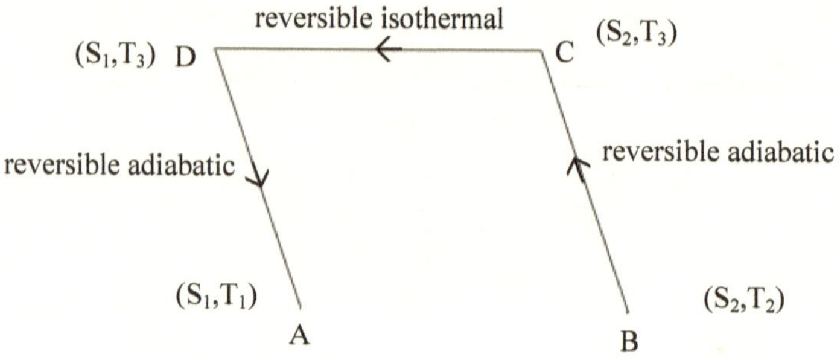

Fig. 1. P–V diagram.

Now we want to bring the system back to the initial state A from the final state B in the following reversible way:

Starting with B, let the system undergo *reversible adiabatic* process BC. At C, the system has the same entropy as B but a different temperature T_3. This is because along BC, $dq_{rev.} = 0$ so that $dS = dq_{rev.}/T = 0$.

At C, the system is conducted *reversibly and isothermally* until the point D, where its entropy becomes equal to the *initial* value S_1, but the temperature remains the same as at C, that is, T_3.

At D, the system is conducted *reversibly* and *adiabatically* along DA. Thus the entropy S_1 remains *unchanged* whereas the temperature T_3 changes to T_1

The temperature T_3 can be chosen arbitrarily higher or lower than either T_1 or T_2 or in between these two values. It is important to point out in this respect that *reversible isothermal processes can be conducted only at certain specified temperatures such as the freezing of water at 0°C and 1 atmosphere. These processes are known as equilibrium processes since their free energy changes* ΔG's *are equal to zero.*

In the above processes, heat is absorbed by the system along CD only, say, *q*.

As the process from A to B and then along BCDA is cyclic, then $\Delta U = 0$, and we have from the first law

$$q = \Delta U + W,$$

$$q = W.$$

Here *W* is the total work done by the system. Suppose that *q* and *W* were positive then it follows that the amount of heat *q* absorbed by the system from the surrounding bodies along the *reversible isothermal path CD was completely converted into work in the cyclic process without a simultaneous change in the system or its surroundings.* This is a *violation* of the second law which in one of its formulations says

> *It is impossible to convert heat completely into work in a cyclic process carried out at one temperature.*
> *Or*
> *Conversion of heat into work completely and continuously in a cyclic process is impossible.*

Consequently, *q* should not be positive, that is, it should be less than zero. Thus applying the relation

$$dS = \frac{d'q_{rev.}}{T}$$

along the reversible isothermal path CD, we get

$$\Delta S = S_{final} - S_{initial} = S_1 - S_2 = \frac{q_{rev.}}{T}.$$

Since *q* must be negative, it follows that S_1 is less than S_2 or $S_2 > S_1$.

Hence in an irreversible (or spontaneous) adiabatic change, there is an *increase* in the entropy of the system.

This principle can now be applied to two simple reversible processes: the first concerns the transfer of heat and the second deals with the isothermal expansion of a perfect gas into a vacuum.

a. Transfer of heat. We consider the transfer of heat $d'q$ from a body of temperature T_1 to another body of temperature T_2, where T_2 is lower than T_1.

The two bodies can first be combined into one isolated system, so *a priori* we expect that the total entropy change in that isolated system due to the transfer of heat must be positive. Then the following two reversible processes are invented in thought, starting and ending with the same corresponding states.

i. An amount of heat $d'q$ from body 1 is transferred reversibly to a constant temperature bath of temperature T_1 so that the process would be a reversible isothermal process. The entropy change of body 1 is, therefore, given by

$$dS_1 = -\frac{d'q_{rev.}}{T_1}. \qquad (1)$$

ii. An equal amount $d'q$ is reversibly transferred from a constant-temperature bath at T_2 to body 2 at temperature T_2. Thus the entropy change of body 2 is

$$dS_2 = \frac{d'q_{rev}}{T_2}. \qquad (2)$$

In Eq. (1), T_1 refers to the temperature of body 1 acting as the source of heat, and in Eq. (2), T_2 refers to the temperature of the bath from which heat is extracted, that is, it is the source of heat.

The total entropy change in this isolated system due to the transfer of heat from body 1 to body 2 is

$$dS_{total} = dS_1 + dS_2 = -\frac{d'q_{rev}}{T_1} + \frac{d'q_{rev}}{T_2}$$

$$= d'q_{rev.}\left(\frac{1}{T_2} - \frac{1}{T_1}\right)$$

$$> 0.$$

This is in agreement with the above principle that an adiabatic irreversible or spontaneous change is accompanied by an entropy increase.

In continuation of this example, we may consider the example of the calculation of entropy change of mixing two liquids of different temperatures as can be seen form the following:

Let us consider an isolated system that consists initially of two quantities of a liquid, say, water at different temperatures. Since the process of mixing the two quantities is definitely an irreversible adiabatic process, then in order to calculate ΔS for this process, a reversible path should be thought of theoretically between the initial state of the system and the final state. This path is what we bring each quantity of water separately to the final temperature *reversibly*, and then the two quantities having the same final temperature are mixed. In this way, the mixing does not correspond to a change of state, and thus, there is no entropy change.

The quantity of water which is of higher temperature than the other has to be cooled *reversibly* to the common, final temperature. Conversely, the quantity of water having a temperature lower than the common final temperature has to be heated *reversibly* to this temperature. *Reversible heating* entails transfer of heat from a source of temperature that is kept adjusted to be *infinitesimally higher* than the temperature of the quantity of water to be heated. In like manner, *reversible cooling* requires transfer of heat from the quantity of water to be cooled to a sink that is kept adjusted to match; that is to be *infinitesimally lower* than the temperature of this quantity of water to be cooled. It is evident that each of these two processes of cooling and heating represents a reversible *non-adiabatic* process due to the exchange of heat energy between water and the surroundings.

In order to know the details of calculation, we solve the following example:

100 grams of water at 0°C are mixed adiabatically with 100 grams of water at 100°C. Both are in liquid form. Calculate ΔS.

We first calculate the final common temperature. Thus we consider the equation

$$200 \times 1 \times (t-0) = 100 \times 1\ (t-0) + 100 \times 1 \times\ (100\ - t)$$

Assuming constant C_P of water between 0°C and 100°C as equal to 1.

Therefore,

$$t = 50^\circ C \quad \text{or} \quad T = 323\ K.$$

Let us denote the 100 grams of water at 100°C as system 1 and the 100 grams of water at 0°C as system 2, so

$$dS_1 = \frac{dq_r}{T} = C_P \frac{dT}{T},$$

or

$$\Delta S_1 = 100 \times 1 \int_{373}^{323} \frac{dT}{T} = 100 \ln \frac{323}{373}$$

and

$$\Delta S_2 = 100 \times 1 \int_{273}^{323} \frac{dT}{T} = 100 \ln \frac{323}{273}$$

Hence,

$$\Delta S_{total} = 100\left(\ln \frac{323}{373} + \ln \frac{323}{273} \right)$$
$$= 2.43 \text{ calories/degree.}$$

Again this means that the entropy change in this adiabatic irreversible or spontaneous process is of positive valve as is expected.

b. Isothermal expansion of a perfect gas into a vacuum. Let V_1 and V_2 be the initial and final volumes, respectively. This expansion being opposed by a pressure equal to zero is an irreversible or spontaneous expansion. Accordingly, in order to calculate the entropy change accompanying this change, a reversible path is invented. According the first law, we have

$$d'q = dU + d'W.$$

Here $d'W$ is equal to zero since no work is done by the gas against the surroundings. Also dU is equal to zero since the process is *isothermal* and the gas is ideal. Hence,

$$d'q = 0.$$

This means that in this special case of *irreversible isothermal expansion* of a perfect gas into a vacuum, the *isothermal* path is *coincident* with the *reversible adiabatic* path which is characterised by $d'q_{rev.} = d'q_{act.}$ = zero. In both cases, although no heat is absorbed by the system, the calculation of the entropy change requires a reversible path which can be invented, at least in thought which being the path between the initial and final states. For an actually reversible adiabatic path (which in fact has no existence), $d'q_{rev.} = 0$, so that $\Delta S = 0$. But if this process is an irreversible adiabatic process, there should be an entropy increase, as pointed out before. Now to calculate the entropy change in an irreversible expansion of perfect gas into a vacuum, applying the first law of thermodynamics,

$$d'q_{rev.} = dU + d'W$$

to the theoretically reversible isothermal path leads to

$$d'q_{rev.} = d'W$$

since the gas is ideal.

Here $d'W$ represents the work done by the gas on expansion which is supposed to be carried out reversibly by reducing the pressure infinitesimally from the initial volume V_1 to the final volume V_2. That is,

$$d'W = PdV.$$

Hence, for *one mole of a perfect gas*, we have[*]

$$d'q_{rev.} = RT\frac{dV}{V},$$

that is,

$$TdS = RT\frac{dV}{V},$$

or

$$dS = R\frac{dV}{V}$$

Therefore,

$$\Delta S = S_2 - S_1 = \int_{V_1}^{V_2} R\frac{dV}{V}$$

$$= R\ln\frac{V_2}{V_1}$$

Since $V_2 > V_1$, ΔS is positive, that is, there is an entropy increase of the gas in this process.

It is thus evident that the entropy change in a process is irrelevant of the actual path followed in that process, and to be calculated, we must consider, at least in thought, a reversible path between the initial and final states.

4.2. The free energies of perfect gas reactions

1. The free energy of a perfect gas. In the isothermal expansion of a perfect gas, the energy change is zero, and therefore, the heat absorbed, *q*, according to the first law

$$q = \Delta U + W \tag{1}$$

is equal to the work performed by the gas, that is,

$$q = W. \tag{2}$$

[*] We shall use, in the subsequent analysis, small letters for extensive specific variables.

For a reversible expansion of one gram molecule of the perfect gas from P_1 to P_2, we have

$$q = W = \int_{P_1}^{P_2} PdV = \int_{V_1}^{V_2} PdV$$

$$= RT\left[\frac{dV}{V}\right]_{V_1}^{V_2} = RT\ln\frac{V_2}{V_1}$$

$$= RT\ln\frac{P_1}{P_2} \qquad (3)$$

The entropy change ΔS is thus

$$\Delta S = q_{rev.}/T$$

$$= R\ln\frac{P_1}{P_2} \qquad (4)$$

Again, it should be observed that the entropy change of the gas has this value whether the expansion is conducted reversibly or not.

By Eq. (4), the entropy change when a gram molecule of a perfect gas is expanded at constant temperature from an unit pressure to pressure *P* is

$$\Delta S = R\ln\frac{1}{P} = -R\ln P$$

If S° is its *entropy at unit pressure* and *S* is its entropy at the same temperature at pressure *P*, we have

$$S - S^\circ = -R\ln P \qquad (5)$$

The Gibbs free energy *G* is given, as shown before, by

$$G = U - TS + PV,$$

so by introducing the value of *S* from Eq. (5), we have

$$G = U - T\left(S^\circ - R\ln P\right) + PV$$

$$= U - TS^\circ + PV + RT\ln P. \qquad (6)$$

Since the energy U and the product $PV\,(=RT)$ for a perfect gas are not changed by change of pressure at constant temperature, $U - TS^\circ + PV$ is the free energy of the gas at the temperature *T* and a unit pressure. If we denote this quantity by G°, that is,

$$G^\circ = U - TS^\circ + PV, \qquad (7)$$

then we have

$$G = G^\circ + RT\ln P, \qquad (8)$$

which represents the free energy per gram molecule of a perfect gas at pressure *P* and temperature *T*.

2. A perfect gas mixture. A perfect gas mixture is *one for which the total pressure is equal to the sum of the pressures which each constituent would exert if present by itself in the same space.*

Or since the total pressure may be regarded as the sum of the partial pressures of the different constituents, we may define a perfect gas mixture *as one for which the partial pressure of each constituent is equal to the pressure it would exert if it occupied the same place only*; that is, the partial pressure of each constituent is unaffected by the pressure of the others.

If follows that the free energy of a perfect gas mixture may be regarded as *the sum of the free energies of the various constituents, each of which is equal to the free energy which this constituent would have if it occupied the same space by itself.*

Suppose that we have a *perfect gas mixture* at temperature *T*, containing the molecules A, B, C, and D, at *partial pressures* P_A , P_B , P_C , P_D. Thus the free energies per gram molecule of these substances in the mixture are

$$\left.\begin{aligned} G_A &= G_A^o + RT\ln P_A, \\ G_B &= G_B^o + RT\ln P_B, \\ G_C &= G_C^o + RT\ln P_C, \\ G_D &= G_D^o + RT\ln P_D, \end{aligned}\right\} \qquad (9)$$

where G_A^o, G_B^o, G_C^o, G_D^o are the free energies of the respective gases at temperature *T* and at unit pressure.

3. Equilibrium in perfect gas mixtures. Suppose that C and D can be formed out of A and B by the reaction

$$A + B = C + D.$$

The free energy change when a gram molecule each of C and D at partial pressures P_C and P_D are formed from a gram molecule each of A and B at partial pressures P_A and P_B is

$$\begin{aligned} \Delta G &= G_C + G_D - G_A - G_B \\ &= \Delta G^o + RT\ln \frac{P_C P_D}{P_A P_B}, \end{aligned} \qquad (10)$$

where ΔG^o stands for

$$G_C^o + G_D^o - G_A^o - G_B^o.$$

That is, the free energy change in the reaction when all the substances taking part are *at unit pressure.* In other words, ΔG^o is the free energy change when a gram molecule of C and a gram molecule of D, each at unit pressure, are formed from a gram molecule of A and a gram molecule of B, each at unit pressure. This becomes more evident if we substitute for each of the partial pressures by unity in Eq. (10), so we get $\Delta G = \Delta G^o$.

Now when the resultants and the reactants are in *chemical equilibrium* with each other, the free energy change ΔG is zero. Writing the *equilibrium pressures* as P_A^e, P_B^e, P_C^e, P_D^e, we have, therefore.

$$0 = \Delta G^o + RT \ln \frac{P_C^e \, P_D^e}{P_A^e \, P_B^e}. \tag{11}$$

The quantity or the quotient

$$\frac{P_C^e \, P_D^e}{P_A^e \, P_B^e}$$

is the *equilibrium constant* of the reaction, usually expressed as K_P, so that

$$\Delta G^o = -RT \ln K_P. \tag{12}$$

This equation is known as *van't Hoff's isotherm.*

4. The change of the equilibrium constant K_P with temperature

The equation $\dfrac{\partial(\Delta G/T)}{dT} = -\dfrac{\Delta H}{T^2}$,

which was obtained before,* can be modified into

$$\frac{\partial\left(\Delta G^o/T\right)}{dT} = -\frac{\Delta H^o}{T^2}, \tag{13}$$

where ΔH^o is the change of heat content corresponding to ΔG^o for the change of the reactions at unit pressure into the products at unit pressure.

According to van't Hoff's isotherm given by Eq. (12), we have

$$\ln K_P = -\Delta G^o/RT = -\frac{\left(\Delta G^o/T\right)}{R},$$

therefore,

$$\left(\frac{\partial \ln K_P}{\partial T}\right)_P = -\frac{1}{R}\,\frac{\left(\Delta G^o/T\right)}{\partial T}.$$

Combining this equation with Eq. (13), we obtain

$$\left(\frac{\partial \ln K_P}{\partial T}\right)_P = \frac{\Delta H^o}{RT^2} \tag{14}$$

this equation is known as *van't Hoff's isochore.*

---

* See Chapter 2. Section 4, Eq. (13).

i. Integration of the van't Hoff isochore. This equation can be integrated over a limited range of temperature over which ΔH^o can be regarded as constant. Thus we obtain

$$\int d \ln K_P = \int \frac{\Delta H^o}{RT^2} . dT, \quad (15)$$

at constant pressure, so

$$\ln K_P = \frac{-\Delta H^o}{RT} + \text{const.} \quad (16)$$

Hence, as long as ΔH^o remains constant, the relation between $\ln k_P$ and $1/T$ is a linear one.

Equation (16) can be rewritten as

$$\log_{10} K_P = \frac{-\Delta H^o}{RT} . \frac{1}{2.303} + \text{const.}, \quad (17)$$

So when the values of $\log_{10} K_P$ are plotted against $1/T$ for a given reaction at constant pressure a line or a nearly linear curve is obtained. The slope of this curve at a given point is equal to $\frac{-\Delta H^o}{2.303\ R}$, and ΔH^o may be found. On the other hand, if we integrate Eq. (15) between two temperatures T_1 and T_2, at which the equilibrium constants are K_{P_1} and K_{P_2}, respectively, we obtain

$$\ln \frac{K_{P_2}}{K_{P_1}} = \frac{\Delta H^o}{R} \left(\frac{1}{T_1} - \frac{1}{T_2} \right) \quad (18)$$

or

$$\ln \frac{K_{P_1}}{K_{P_2}} = - \frac{\Delta H^o}{R} \left(\frac{1}{T_1} - \frac{1}{T_2} \right)$$

It is evident from Eq. (18) that if the equilibrium constants K_{P_1} and K_{P_2} are known at two temperatures T_1 and T_2, then the heat content change ΔH^o of the reaction can be calculated, provided it is constant within this range of temperature.

van't Hoff's isochore in its expression given by Eq. (18) predicts the following experimental facts: (1) If the forward reaction is exothermic, that is ΔH is of negative sign (following the convention in thermodynamics), and temperature T_2 is higher than temperature T_1, then it follows that K_{P_1} will be greater then K_{P_2}, indicating that the equilibrium constant for an exothermic reaction *decreases* with *rise* of temperature. This is in agreement with Le Chatalier's principle as well as with the law of chemical equilibrium. (2) On the other hand, If the forward reaction is endothermic, that is ΔH is of positive sign, and temperature T_1 is less than temperature T_2, then K_{P_1} will be less than K_{P_2}, indicating that the equilibrium constant *increases* with *increase* of temperature for an endothermic reaction.

Knowing that the heat of reaction ΔH is the heat content change when the reaction goes *completely* in the forward direction, it can be expected that, for certain reactions, the heat of reaction might be difficult to measure. In such cases, use is made of van't Hoff's isochore, that is, the equilibrium constant K_P of the reaction is measured at two different temperatures, and then ΔH is directly calculated by Eq. (18), assuming its approximate constancy over the temperature range applied.

Over a *wide* range of temperature, van't Hoff's isochore may be integrated if the variation of ΔH with temperature is known. In this respect, we could first give a brief account of heat of formation and heat capacities.

ii. Heat of formation. Since under ordinary conditions, reactions occur most frequently at constant pressure rather than at constant volume, thermochemical data are usually tabulated as heat content changes. It is easy to calculate the corresponding energy changes. The heat content change ΔH in the formation of a compound is the difference between the heat content of the compound and that of its elements in a specified state at the same temperature. It may be regarded as the heat content of the compound relative to its elements. In a chemical reaction, the amounts of the elements on each side of the equation are necessarily equal so that we may get the heat content changes in a reaction by taking the difference between the heat contents of the resultants and the reactants.

iii. Heat capacities. The heat capacity of a system is the amount of reaction required to raise the temperature 1°. This depends on conditions. If the volume is kept *constant*, all the heat added goes to increase the energy of the system, so that if C_V is the heat capacity under this condition, when the temperature is raised from T_1 to T_2, the heat absorbed $C_V\ (T_2 - T_1)$ is equal to the increase in the energy of the system $U_2 - U_1$, that is

$$C_V\ (T_2 - T_1) = U_2 - U_1$$

or

$$C_V\ \Delta T = \Delta U,$$

so in the limit, for a small change of temperature d*T*, we have

$$C_V = \left(\frac{\partial U}{\partial T}\right)_V . \tag{19}$$

On the other hand, when a system is heated at constant pressure, it may expand, and in doing so, perform work against the applied pressure, that is, the external pressure. The quantity of heat required to produce and increase of temperature of 1° is then greater than that at constant volume by the amount of work done in expanding. If *V* is the volume of the system, the increase in volume for 1° rise of temperature is dV/dT, then the work done by the system is equal to the pressure multiplied by the corresponding increase in volume, that is, it is equal to $P\ dV/dT$. Hence the heat capacity of the system at *constant pressure* is

$$C_P = C_V + P\left(\frac{dV}{dT}\right)_P$$

or

$$C_P = \left(\frac{\partial U}{\partial T}\right)_V + P\left(\frac{dV}{dT}\right)_P . \qquad (20)$$

But since by definition

$$H = U + PV .$$

Therefore,

$$\left(\frac{\partial H}{\partial T}\right)_P = \left(\frac{\partial U}{\partial T}\right)_V + P\left(\frac{\partial V}{\partial T}\right)_P , \qquad (21)$$

so from Eqs. (20) and (21), we have

$$C_P = \left(\frac{\partial H}{\partial T}\right)_P . \qquad (22)$$

It must be emphasised that the heat capacities must refer to the amounts of substances given in the chemical equation defining ΔU or ΔH.

iv. Kirchhoff's equation. The energy change in a reaction $A \rightarrow B$ is given by

$$\Delta U_{A\rightarrow B} = U_B - U_A ,$$

Therefore,

$$\left\{\frac{\partial(\Delta U_{A\rightarrow B})}{\partial T}\right\}_V = \left(\frac{\partial U_B}{\partial T}\right)_V - \left(\frac{\partial U_A}{\partial T}\right)_V$$

$$= (C_V)_B - (C_V)_B . \qquad (23)$$

Thus the rate at which $\Delta U_{A\rightarrow B}$ changes with temperature is equal to the *difference* between the heat capacities at *constant volume* of the final system B (resultants) and the original system A (reactants).

In the same way

$$\Delta H_{A\rightarrow B} = H_B - H_A ,$$

therefore,

$$\left\{\frac{\partial(\Delta H_{A\rightarrow B})}{\partial T}\right\}_P = \left(\frac{\partial H_B}{\partial T}\right)_P - \left(\frac{\partial H_A}{\partial T}\right)_P$$

$$= (C_P)_B - (C_P)_B . \qquad (24)$$

The temperature coefficient of ΔH is, therefore, related in the same way to the heat capacities at *constant pressure*.

Equations (23) and (24) are the two forms of *Kirchhoff's* equation.

Examples

:

1. Knowing the heat capacities of water as liquid and as vapour, we can find the variation of the heat of vaporisation with the temperature.

$$H_2O(\ell) = H_2O(g), \ \Delta H_{373} = 9650 \text{ calories}$$

$$H_2O(\ell), \ C_{P(373)} = 17.82 \text{ cal/mole}$$

$$H_2O(g), \ C_{P(373)} = 8.37 \text{ cal/mole.}$$

Hence,

$$\left(\frac{\partial \Delta H}{\partial T}\right)_P = (C_P)_g - (C_P)_\ell$$

$$= 8.37 - 17.82$$

$$= -9.45 \text{ calories per degree.}$$

This means that the heat of vaporisation decreases 9.45 calories of each *degree* rise of temperature. Therefore, at 120°C, ΔH is given by

$$\Delta H = 9650 - (9.45 \times 20)$$

$$\simeq 9460 \text{ calories .}$$

For an extended range of temperature, it would be necessary to take into account the variation of the heat capacities with temperature.

The relations between heat capacity and temperature are simpler for gases than for solids or liquids, and can in most cases be represented empirically over a considerable range of temperature by equations of the form

$$C_P = a + bT + cT^2 + dT^3 \ldots, \qquad (25)$$

where a, b, c, and d, . . . are constants.

We can obtain ΔC_P for a reaction by subtracting the equations for the reacting gases from those of the products of the reaction, taking note of the amount of each gas which enters into the reaction. Thus we obtain, in general, an equation of the form

$$\Delta C_P = \alpha + \beta T + \gamma T^2 + \delta T^3 + \ldots \qquad (26)$$

Thus substituting Eq. (26) in Eq. (24), we have

$$\left\{\frac{\partial (\Delta H)}{\partial T}\right\}_P = \alpha + \beta T + \gamma T^2 + \delta T^3 \ldots$$

and integrating, we find

$$\Delta H = \alpha T + \frac{\beta}{2} T^2 + \frac{\gamma}{3} T^3 + \ldots + \Delta H_0, \qquad (27)$$

where ΔH_0 is the integration constant.

Evidently, ΔH_0 *is the value of* ΔH *when* $T = 0$, but it cannot be identified with the value of the heat content change at absolute zero. This is because the heat capacity equations on which Eq. (27) is based are never valid at low temperatures in the region of absolute zero.

2. For the reaction

$$H_2 + 1/2\ O_2 = H_2O(g),$$

we have

a. H_2O $\quad C_P = 8.81 - 0.0019T + 0.00000222T^2$

b. H_2, $\quad C_P = 6.5 + 0.0009\,T$,

c. $1/2\ O_2$, $\quad C_P = 3.25 + 0.0005\,T$,

whence

$$\Delta C_P = (C_P)_{H_2O} - \left\{(C_P)_{H_2} + (C_P)_{\frac{1}{2}O_2}\right\}$$

$$= -0.94 - 0.0033T + 0.00000222T^2$$

$$= \left(\frac{\partial \Delta H}{\partial T}\right)_P .$$

Integrating, we have

$$\Delta H = -0.94\,T - \frac{0.0033T^2}{2} + \frac{0.00000222\,T^3}{3}$$

$$+\Delta H_0$$

Thus if we have ΔH at any one of the value of *T*, we can obtain the value of the integration constant ΔH_0 by substitution. When $T = 373$ K, ΔH has been found to be $= -57780$ calories. Substituting these values in the above equation, we find that

$$\Delta H_0 = -57410 \text{ calories}.$$

Therefore,

$$\Delta H = -\,57410 - 0.94\,T - 0.00165\,T^2 + 0.00000074\,T^3.$$

3. We consider the reaction

$$1/2\ N_2 + 3/2H_2(g) = NH_3(g).$$

From the heat capacity equations, using Kirchhoff's equation and then integrating, we can express ΔH in the following equation

$$\Delta H = \Delta H_0 - 4.96T - 0.0006T^2 + 0.0000017T^3.$$

By the use of thermochemical data* it is found that $\Delta H_0 = -9500$ calories.

Now returning to Eq. (27) obtained above, it can be more conveniently rewritten in the form

$$\Delta H = \alpha' T + \beta' T^2 + \gamma' T^3 + \ldots + \Delta H_0 \tag{28}$$

to represent the variation of ΔH with temperature.

It is to be noted that the heat content of a perfect gas does not vary with the pressure, so ΔH^o, the value of ΔH for unit pressure, is the same as that determined for any convenient pressure. For actual gases *at moderate pressures,* the distinction between ΔH, the heat content change for any given pressure, and ΔH^o can be neglected.

Substituting for ΔH^o in Eq. (14) by ΔH, we have

$$\left\{\frac{\partial \ln K_P}{\partial T}\right\}_P = \frac{\Delta H}{RT^2}, \tag{29}$$

and then introducing the value of ΔH given by Eq. (28) in Eq. (29), we obtain

$$\left\{\frac{\partial \ln K_P}{\partial T}\right\}_P = \frac{\Delta H_0}{RT^2} + \frac{\alpha'}{RT} + \frac{\beta'}{R} + \frac{\gamma' T}{R} + \ldots$$

Integrating this equation, we find

$$\ln K_P = -\frac{\Delta H_0}{RT} + \frac{\alpha'}{R}\ln T + \frac{\beta' T}{R} + \frac{\gamma' T^2}{2R} + \ldots + J, \tag{30}$$

where *J* is *an integration constant*; *J* can be evaluated when the value of K_P for one value of *T* is known, provided the integration constant ΔH_0 is first evaluated as shown in the examples given above. On the other hand, since

$$\Delta G^o = -RT \ln K_P, \tag{12}$$

we obtain by multiplying Eq. (30) by $-RT$:

$$\Delta G^o = \Delta H_0 - \alpha' T \ln T - \beta' T^2 - \frac{\gamma'}{2} T^3 \ldots - IT, \tag{31}$$

where $I = JR$.

This equation gives the variation of ΔG^o with the temperature over the same range of temperature as that to which the heat capacity equations apply.

Examples:

Evaluation of *J* and *I*.

We have found above that the heat content change ΔH in the reaction

$$H_2 + \frac{1}{2}O_2 = H_2O(g) \tag{a}$$

* See *Chemical Thermodynamics* by J. A. V. Butler (Macmillan, London, 1960, p. 19).

can be represented for a range of temperature from 0°C to over 1000°C by the equation

$$\Delta H = -57410 - 0.94T \_ 0.00165T^2 + 0.00000074T^3.$$

Inserting this value into

$$\left\{\frac{\partial \ln K_P}{\partial T}\right\}_P = \frac{\Delta H}{RT^2}$$

and integrating, we have

$$\ln K_P = \frac{57410}{RT} - \frac{0.94}{R}\ln T - \frac{0.00165\,T}{R} + \frac{0.00000074T^2}{2R} + \ldots + J, \quad (32)$$

where *J* is the integration constant.

In order to evaluate *J*, we may make use of *Nernst and Wortenberg's* measurements of the dissociation of water vapour; they found that at 1480 K, the percentage dissociation of water vapour is 0.0184.

The equilibrium constant K'_P of the reaction

$$H_2O(g) = H_2 + 1/2\, O_2(g)$$

at 1480 K is thus given by

$$K'_P = \frac{P_{H_2} \cdot P_{O_2}^{1/2}}{P_{H_2O}}.$$

The value of which can be determined as follows.

Commencement	1	0	0
	$H_2O(g)$ =	$H_2(g)$ +	$O_2(g)$
Equilibrium	1 – 0.000184	0.000184	0.000092

This representation means that if one gram molecule of water vapour is heated at 1480 K and allowed to dissociate until equilibrium is attained, at which 0.000184 gram molecules will have dissociated and produced 0.000184 gram molecules of hydrogen and 0.000092 gram molecules of oxygen, then the total number of gram molecules at equilibrium will be

$$1-0.000184 + 0.000184 + 0.000092 = 1 + 0.000092 \simeq 1,$$

It is thus evident that there is an increase in the number of moles as the reaction proceeds towards equilibrium. If the total pressure of the system at equilibrium is 1 atmosphere, then the partial pressures of the constituent gases assuming that they are ideal is

$$P_{H_2O(g)} = \frac{1 - 000184}{1 + 0.000092} \times 1 \simeq 1,$$

$$P_{H_2} = \frac{0.000184}{1 + 0.000092} \times 1 \simeq 0.000184,$$

and

$$P_{1/2\,O_2} = \frac{0.000092}{1 + 0.000092} \times 1 \simeq 0.000092,$$

so the equilibrium constant K'_P of the reaction

$$H_2O(g) = H_2(g) + 1/2\ O_2(g),$$

in terms of partial pressures is given by

$$K'_P = \frac{P_{H_2} \cdot P_{O_2}^{1/2}}{P_{H_2O(g)}} = \frac{0.000184 \times (0.000092)^{1/2}}{1}$$

$$= \frac{1}{5.68} \times 10^{-2} \qquad (T = 1480\ \text{K}).$$

Therefore, the equilibrium constant K_P for our reaction

$$H_2(g) + 1/2\ O_2(g) = H_2O(g), \tag{a}$$

at 1480 K, is the reciprocal of K'_P, that is

$$K_P = 5.66 \times 10^2.$$

Introducing this figure and the value of T in Eq. (32), we find that the value of J is equal to -1.98. Therefore, $\ln K_P$ is expressed as

$$\ln K_P = \frac{57410}{RT} - \frac{0.94}{R}\ln T - \frac{0.00165\,T}{R} + \frac{0.00000074\,T^2}{2R} - 1.98 \tag{33}$$

Multiplying this equation by $-RT$, we have

$$\Delta G^\circ = -57410 + 0.94\,T\ln T + 0.00165\,T^2 - 0.00000037T^3 + 3.94\,T, \tag{34}$$

where $3.94 = -JR$.

Equation (34) may be made use of to evaluate ΔG° at any other temperature. Thus when $T = 298$ K (25°C), we have

$$\Delta G^\circ_{298} - -54590\ \text{calories}.$$

2. It is of interest to calculate from this result that the theoretical electromotive force of the reversible oxygen–hydrogen cell at 25°C as follows.

We first consider the reaction

$$H_2(g) + 1/2\ O_2(g) = H_2O(\ell), \qquad \text{(b)}$$

which we denote by (b) to distinguish it from the above reaction denoted by (a). We see that the product in (a) is water vapour, while in (b), it is liquid water.

$$H_2(g) + 1/2\ O_2(g) = H_2O(g),\ \Delta G^{o}_{298} \qquad \text{(a)}$$

$$H_2(g) + 1/2\ O_2(g) = H_2O(\ell),\ \Delta G_{298} \qquad \text{(b)}$$

ΔG^{o}_{298} for reaction (a) is the free energy change in the reaction when all the substances taking part are at *unit* pressure, or in other words, when the reactants, each at *unit* pressure are converted *completely* into the products each at *unit* pressure. Accordingly, the free energy change in the change of water vapour, at *unit* pressure and at 25°C, to liquid water at the same temperature and at the equilibrium vapour pressure, which is 23.8 mm, that is, $\Delta G_{298}(g \rightarrow \ell)$ for the reaction

$$\underset{760\ \text{mm}}{H_2O(g)} = \underset{23.8\ \text{mm}}{H_2O(\ell)}, \qquad \text{(c)}$$

Water vapour at a pressure of 23.8 mm of mercury at 25°C has the same free energy per mole as that of liquid water, since both phases of water are in *equilibrium* with one another. So we need only to evaluate the free energy change in the isothermal expansion of water vapour from unit atmosphere to a pressure of 23.8 mm. For this purpose, we can use the following equation

$$G = G^{o} + RT\ \ln P,$$

which was obtained above. It represents the variation of free energy of a perfect gas with pressure *P* at constant temperature *T*. Therefore, we have

$$\Delta G_{298}(g \rightarrow \ell)\ \text{for the reaction (c)} = RT \ln \frac{23.8}{760}$$

$$= -2053\ \text{calories}.$$

Also, ΔG_{298} for the reaction (c) can be obtained by using the equation

$$\left(\frac{\partial G}{dP}\right)_T = V,$$

which by integration

$$\int_{P=760}^{P=23.8} dG_{298} = \int_{760}^{23.8} VdP$$

gives

$$\Delta G_{298} = RT \int_{760}^{23.8} \frac{dP}{P}$$

$$= RT \ln \frac{23.8}{760}$$

$$= -\ 2053 \quad \text{calories}$$

Hence, the total free energy change ΔG_{298} for the reaction (b) is

$$\Delta G_{298} = -\ 54590 - 2053$$

$$= -\ 56643 \quad \text{calories} \ .$$

From this value and the relation*

$$\Delta G = -\,n\,EF\,,$$

we can calculate the theoretical (reversible) electromotive force *E* of the oxygen–hydrogen cell in order to bring about the reaction represented by (b). Thus if *E* is the reversible electromotive force corresponding to the passage of two faradays of electricity through this cell to bring about the reaction (b), we have $2\,EF = 56640 \times 4.182$ joules

or

$$E = \frac{56643 \times 4.182}{2 \times 96490}$$

$$= 1.227 \quad \text{volts.}$$

Example:

4. From the heat capacity equations we have derived above the equation, we have

$$\Delta H = \Delta H_0 - 4.96\,T - 0.000575\,T^2 + 0.0000017\,T^3,$$

for the heat content change in the reaction

$$1/2\ N_2(g) + 3/2\ H_2(g) = NH_3(g)\,,$$

and by the use of thermochemical data, it is found that $\Delta H_0 = -9500 \text{calories}$. We want now to express $\ln K_P$ of this reaction and also ΔG° as a function of the temperature *T*.

For this purpose, we use van't Hoff's isochore

$$\left(\frac{\partial \ln K_P}{\partial T}\right)_P = \frac{\Delta H}{RT^2},$$

which was obtained above.

---

* See *Chemical Thermodynamics* by J.A.V. Butler (Macmillan, London 1960, pp. 118–120).

Thus by integrating this equation over a wide range of temperature where the variation of ΔH with temperature is

$$\Delta H = -9500 - 4.69T - 0.000575T^2 + 0.0000017T^3,$$

we have

$$\int d \ln K_P = \int \frac{\left(-9500 - 4.96T - 0.000575T^2 + 0.0000017T^3\right)}{RT^2} dT,$$

$$\ln K_P = \frac{9500}{RT} - \frac{4.96}{R} \ln T - \frac{0.000575}{R} T + \frac{0.0000017\, T^2}{2R}$$

$$+J,$$

and

$$\Delta G^o = -RT \ln K_P$$

$$= -9500 + 4.96T \ln T + 0.000575T^2 - 0.00000085$$

$$+IT$$

From measurements of the equilibrium constant at various temperatures and pressures, Lewis and Randall concluded that the best value of I is -9.61. Thus ΔG^o is given as a function of temperature by

$$\Delta G^o = -9500 + 4.96T \ln T + 0.000575\, T^2$$

$$-0.00000085\, T^3 - 9.61T$$

for the reaction

$$1/2\ N_2(g) + 3/2\ H_2(g) = NH_3(g).$$

A brief account of this reaction which is accompanied by an evolution of heat and a decrease in the number of gram molecules is given below.

An important point in this reaction, which is of great industrial importance, is the investigation of the *percentage of ammonia* in equilibrium mixtures between nitrogen, hydrogen, and ammonia and also *the values of* K_P *at different temperatures for different pressures.*

The representation of the reaction by the above chemical equation implies that the formation of ammonia from its elements is accompanied by a decrease in the number of moles. The forward reaction thus produces either *a diminution in volume if the pressure is kept constant* or *a diminution in pressure if the volume is kept* constant.

According to Le Chatelier's principle, if the pressure on the ammonia system is increased, *lessening* of this pressure results from *the formation of a further amount of ammonia.* That is, the forward reaction is *favoured* with *increase* of external pressure. This conclusion which is obtained in a qualitative way by applying the principle of Le Chatelier to the reaction can also be arrived at in a quantitative way, when the law of chemical equilibrium is applied to the equilibrium between nitrogen, hydrogen, and ammonia at different temperatures for different pressures.

The following table (Table 1) which was obtained experimentally gives the *percentage of ammonia in the equilibrium mixtures* and also *the values of the equilibrium constant,* K_P , referring to the ammonia equilibrium when written as indicated above by the chemical equation,

$$1/2\ N_2(g) + 3/2\ H_2(g) = NH_3(g)$$

that is,

$$K_P = \frac{P_{NH_3}}{P_{N_2}^{1/2} \cdot P_{H_2}^{3/2}}$$

at different temperatures and also for different pressures.

The values of K_P listed in this table are, therefore, equal to the square root of the values which would be calculated by means of the equation

$$K'_P = \frac{P^2_{NH_3}}{P_{N_2} \cdot P^3_{H_2}}$$

referring to the equilibrium written as

$$N_2(g) + 3H_2(g) = 2NH_3(g).$$

On calculating the values of K_P at a given temperature for various values of pressure, as will be seen from the table, the partial pressure of every constituent at equilibrium is taken as proportional to its mole fraction in the mixture; that is, Dalton's law of partial pressures is applied, assuming that the gases are ideal. However, it will be evident that the equilibrium constant, which should remain constant at each temperature regardless of the pressure, varies slightly or does not vary much at low pressures for a given temperature, but varies considerably when the pressure is above 50 atmospheres at the same temperature. This is because the ideal gas laws do not hold good at the higher pressures, especially in the case of ammonia being easily liquefiable gas, and consequently, the partial pressures can no longer be used to represent the concentration of the gases at equilibrium.

Table 1
Equilibrium between nitrogen, hydrogen, and ammonia.

t°C		Pressure in atmospheres		
		10	30	50
300	%NH_3	7.35	17.80	25.11
	K_P	0.0266	0.0273	0.0278
400	%NH_3	3.85	10.09	15.11
	K_P	0.0129	0.0129	0.0130

500	$\%NH_3$	1.20	3.48	5.58
	K_P	0.00381	0.00385	0.00388
t°C			Pressure in atmospheres	
		100	300	400
300	$\%NH_3$	...	...	...
	K_P	...	...	...
400	$\%NH_3$	24.91	...	...
	K_P	0.0137	...	...
500	$\%NH_3$	10.40	26.2	42.1
	K_P	0.00402	0.00498	0.00651

The following is an example illustrating the calculation of K_P from the percentage of ammonia formed under certain applied pressure at a given temperature, and then the use of this value of K_P in calculating the percentage of ammonia formed under another higher pressure at *same temperature* in order to prove in a quantitative way that the increase of pressure increases the proportion of ammonia in the equilibrium mixture.

Example:

A mixture of hydrogen and nitrogen at the ratio 3 moles to 1 mole is heated to 400°C and subjected to a pressure of 10 atmospheres in presence of a suitable catalyst (to hasten attaining equilibrium). It is found that 3.85 per cent (molar per cent = mole fraction in the equilibrium mixture multiplied by 100) of the mixture at equilibrium is ammonia. Calculate the equilibrium constant in terms of partial pressures for the reaction written as

$$N_2(g) + 3H_2(g) = 2NH_3(g).$$

Since the initial concentration of hydrogen is three times that of nitrogen, the ratio of 3 moles of hydrogen to 1 mole of nitrogen is maintained whatever the amount of ammonia is formed. Accordingly, one-quarter of the remaining gas mixture at equilibrium *which is not ammonia*, is nitrogen and three-quarters of it is hydrogen.

Therefore,

$$\text{Percentage of nitrogen} = \frac{1}{4}(100 - 3.85) = \frac{1}{4} \times 96.15$$
$$= 24.03 \text{ (by moles), and}$$

$$\text{Percentage of hydrogen} = \frac{3}{4}(100 - 3.85) = \frac{3}{4} \times 96.15$$
$$= 72.12 \text{ (by moles).}$$

Subsequently, the mole fractions of nitrogen, hydrogen, and ammonia at equilibrium are 0.2403, 0.7212, and 0.0385, respectively. Hence, assuming that the gases at the given temperature (400°C) and pressure (10 atmospheres) behave as perfect gases so that Dalton's law of partial pressures can be applied, we have

$$P_{N_2} = \text{mole fraction} \times 10$$

$$= 0.2403 \times 10 = 2.403 \text{ atm},$$

$$P_{H_2} = 0.7212 \times 10 = 7.212 \text{ atm},$$

$$P_{NH_3} = 0.0385 \times 10 = 0.385 \text{ atm},$$

Therefore, the equilibrium constant expressed in terms of partial pressures for the reaction is

$$N_2(g) + 3H_2(g) = 2NH_3(g),$$

and which we shall denote by K'_P (to distinguish it from K_P which refers to the equilibrium $1/2\ N_2(g) + 3/2\ H_2(g) = NH_3(g)$ and which is shown in Table 1), is given by

$$K'_P = \frac{P^2_{NH_3}}{P_{N_2} \cdot P^3_{H_2}} = \frac{(0.385)^2}{2.403 \times (7.212)^3}$$

$$= 0.000164 \quad \text{at } 400^{\circ\circ}\text{C (and 10 atmospheres)}$$

If we take the square root of this value we find that

$$\sqrt{K'_P} = \sqrt{0.000164},$$

$$K_P = 0.0129$$

This is practically the value shown in Table 1.

Now we shall use this value of $K'_P\ (= 0.000164)$, at 400°C and pressure equal to 10 atmospheres, in calculating the percentage of ammonia formed under a pressure of 50 atmospheres at the same temperature.

Again, whatever the amount of ammonia is formed, the ratio 3:1 of hydrogen to nitrogen moles in the remaining gas mixture at equilibrium, *which is not ammonia,* is maintained. This is because, as noted above, the initial molar concentration of hydrogen is three times that of nitrogen. Accordingly, the partial pressure of hydrogen at equilibrium is *three times* that of nitrogen, assuming that the gases are ideal, so the partial pressure of each gas in the mixture is proportional to its concentration in this mixture. Thus we have

$$P_{H_2} = 3P_{N_2},$$

and since $P_{H_2} + P_{N_2} + P_{NH_3}$ is equal to 50 atmospheres, we obtain

$$3P_{N_2} + P_{N_2} + P_{NH_3} = 50,$$

or

$$4P_{N_2} + P_{NH_3} = 50$$

so

$$P_{NH_3} = 50 - 4P_{N_2}$$

Substituting now for the partial pressures in the equilibrium constant equation

$$K'_P = 0.000164 = \frac{P^2_{NH_3}}{P_{N_2} \cdot P^3_{H_2}},$$

we obtain

$$0.000164 = \frac{(50 - 4P_{N_2})^2}{P_{N_2} \cdot (3P_{N_2})^3}$$

$$= \frac{(50 - 4P_{N_2})}{27 P^4_{N_2}}.$$

Therefore,

$$\frac{50 - 4P_{N_2}}{P^2_{N_2}} = \sqrt{0.000164 \times 27}$$

$$= 6.65 \times 10^{-2}$$

or

$$P_{N_2} = 10.62 \text{ atmospheres}.$$

Hence,

$$P_{H_2} = 3 \times 10.62 = 31.86 \text{ atmospheres}$$

and

$$P_{NH_3} = 50 - (10.62 + 31.86)$$

$$= 7.52 \text{ atmospheres}.$$

Since the partial pressure of ammonia = its mole fraction $\times$ 50,

$$\text{the mole fraction of ammonia} = \frac{7.52}{50} = 0.1504$$

$$= 0.15,$$

or the percentage of ammonia by moles is approximately 15.

It is thus seen that the increase of pressure from 10 to 50 atmosphere increases the proportion of ammonia formed from 3.85 to 15 per cent of the equilibrium mixture. This value is in agreement with the value shown in Table 1.

Also, the results listed in this Table are in harmony with the prediction that the increase of temperature favours the endothermic reaction, that is, the backward reaction, whereas the increase of pressure favours the forward reaction, since it is accompanied by a decrease in volume or decrease of pressure, and thus, the proportion of ammonia in the equilibrium mixture is increased.

It is important to point out that the law of chemical equilibrium does not deal with the time required for attaining equilibrium. The equilibrium constant can be calculated only from the concentrations in equilibrium mixture, however long the equilibrium takes to be established. Rise of temperature, apart from its effect on equilibrium concentrations in this exothermic reaction, accelerates chemical reactions in the sense that it increases the rates of both forward and backward reactions, although not to the same extent, and accordingly accelerates the attainment of equilibrium, but *displaces the point of equilibrium* in the direction of the endothermic compounds.

In a technical exothermic reactions, like that of the synthesis of ammonia which is not favoured by raising the temperature as is required by Le Chatelier's principle, it is more profitable to obtain a *low* percentage of ammonia formed rapidly at a *higher* temperature than to obtain a *higher* percentage formed slowly at a *lower* temperature. A moderate temperature is, therefore, used in industry, namely about 500°C, and the reaction is further accelerated by the use of a suitable catalyst. High pressure of about 200 atmospheres is also applied, since increase of pressure, as explained above, favours the formation of ammonia; this is *Haber's process.*

Corollary

The relation between the equilibrium constants K_P and K_C in gas reactions

For a reversible gas reaction represented by the equation

$$aA + bB + \dots = xX + yY + \dots,$$

the equilibrium constant expressed in terms of molecular concentrations, K_C, is given by

$$\frac{C_X^x \cdot C_Y^y \dots}{C_A^a \cdot C_B^b \dots} = K_C \qquad (1)$$

Since in homogeneous gas reactions, assuming that the gases behave ideally, the partial pressure *P* of each gas is proportional to its concentration *C* in the gas mixture, the law of chemical equilibrium can also be expressed in terms of the *partial* pressures in this case. Therefore, for the reaction represented by the above equation, we have

$$\frac{P_X^x \cdot P_Y^y \dots}{P_A^a \cdot P_B^b \dots} = K_P, \qquad (2)$$

where K_P is the equilibrium constant expressed in terms of partial pressures, usually in atmospheres unless otherwise specified. In gaseous equilibria, K_P is *used more often than* K_C

The relation between the two equilibrium constants K_P and K_C can easily be determined. Thus if 1 gram molecule of a gas is existing in a volume *V* litres, its concentration *C* is equal to $1/V$. Then according to the general gas reaction, $PV = RT$, we have

$$C = \frac{1}{V} = \frac{P}{RT}.$$

Substituting for this value of *P* in Eq. (2), we obtain

$$K_P = \frac{(C_X\ RT)^x \cdot (C_Y\ RT)^y \dots}{(C_A\ RT)^a \cdot (C_B\ RT)^b \dots}$$

$$= \frac{C_X^x \cdot C_Y^y \dots}{C_A^a \cdot C_B^b \dots} = (RT)^{(x+y)-(a-b)}$$

or

$$= K_C \, (RT)^{\Delta n},$$

where Δn is the change in the number of gram molecules of gases during the reaction; that is,

Δn = number of gram molecules of resultants – number of gram molecules of reactants.

It is evident that in reactions where the same number of moles occur on both sides of the chemical equation, $\Delta n = 0$, and thus, $K_P = K_C$.

Example:

For the reaction represented as

$$N_2 + 3H_2 = 2NH_3 , \qquad (1)$$

at 400°C and $K_C = 0.507$. Calculate the equilibrium constant in terms of partial pressures.

Let us denote this constant by K'_P. Therefore,

$$K'_P = K_C (RT)^{\Delta n} = 0.507 \times (0.08205 \times 673.15)^{-2}$$

$$= 1.66 \times 10^{-4},$$

where *R* is substituted by 0.08205 litre atm/deg mole.

It is to be noted that this value represents the equilibrium constant in terms of partial pressures of the given equation. That is,

$$K'_P = \frac{P^2_{NH_3}}{P_{N_2} \cdot P^3_{H_2}} = 1.66 \times 10^{-4}.$$

If, however, we represent the reaction in the usual experimental way as

$$1/2\ N_2 + 3/2\ H_2 = NH_3 , \qquad (2)$$

the equilibrium constant K_P is given by

$$K_P = \frac{P^2_{NH_3}}{P^{1/2}_{N_2} \cdot P^{3/2}_{H_2}} .$$

It is thus evident that in order to predict the value of K_P for reaction (2), we just take the square root of the given value of K'_P for reaction (1). Thus $K_P = 0.0129$. This value is experimentally verified, as has been shown above, provided the external pressure applied to the system of ammonia synthesis is not too high to assume that the gases at equilibrium at a given temperature behave as perfect gases for the calculation of partial pressures.

v. The vapour-pressure equation. The equilibrium between a solid or liquid and its vapour can be similarly treated as in the previous section. Let G_S be the molar free energy of a solid at the temperature *T* and G_g that of the vapour. For equilibrium, it is necessary that $G_S = G_g$. If G^o_g is the free energy of the vapour at this temperature and at *unit* pressure, and *P* is the pressure of vapour at this temperature, which is the equilibrium vapour pressure, we assume that the vapour is a perfect gas

$$G_g = G_g^o + RT\ln P. \qquad (1)$$

This is the equation we obtained above; it represents the free energy per gram molecule of a perfect gas at a pressure *P* and temperature *T*. Therefore, by substituting in this equation for G_g by G_S owing to the equilibrium between the two phases, we have

$$G_S = G_g^o + RT\ln P.$$

Differentiating, we obtain

$$R\left\{\frac{\partial \ln P}{\partial T}\right\}_P = \left\{\frac{\partial (G_S/T)}{\partial T}\right\}_P - \left\{\frac{\partial \left(G_g^o/T\right)}{\partial T}\right\}_P. \qquad (2)$$

But it has also been shown above that

$$\left\{\frac{\partial (G/T)}{\partial T}\right\}_P = -\frac{H}{T^2}, \qquad (3)$$

Therefore, by substituting Eq. (3) in Eq. (2) we have

$$R\left\{\frac{\partial \ln P}{\partial T}\right\}_P = -\frac{H_S}{T^2} + \frac{H_g^o}{T^2}$$

or

$$\left\{\frac{\partial \ln P}{\partial T}\right\}_P = -\left(\frac{H_S - H_g^o}{RT^2}\right) \qquad (4)$$

For the conversion of a solid into a gas, that is,

solid = gas

the latent heat of the reaction ΔH is given by

$$\Delta H = \left(H_g - H_S\right) \text{ and}$$

$$-\Delta H = \left(H_S - H_g\right) = \left(H_S - H_g^o\right), \qquad (5)$$

assuming that the vapour is a perfect gas. Therefore, the quotient $\left(\frac{\partial \ln P}{\partial T}\right)_P$ in Eq. (4) can be expressed as

$$\frac{d\ln P}{dT} = \frac{\Delta H}{RT^2}. \qquad (6)$$

The dissociation equilibrium between solids and a gas can be similarly treated. Thus we consider the reaction

$$CaCO_3 = CaO + CO_2(g).$$

The condition of equilibrium in a reaction is

$$\Sigma G_{\text{reactants}} = \Sigma G_{\text{resultants}},$$

so in this reaction, the equilibrium condition is

$$G_{CaCO_3} = G_{CaO} + G_{CO_2},$$

where G_{CaCO_3} , G_{CaO} , G_{CO_2} are the free energies at a given temperature, of the *quantities of these substances, which are represented in the chemical equation.* Writing

$$G_{CO_2} = G^{o}_{CO_2} + RT\ln P_{CO_2},$$

we thus have, for equilibrium,

$$\begin{aligned} G_{CaCO_3} - G_{CaO} - G_{CO_2} &= G_{CaCO_3} - G_{CaO} \\ &\quad - \left(G^{o}_{CO_2} + RT\ln P_{CO_2}\right) \\ &= G_{CaCO_3} - G_{CaO} \\ &\quad - G^{o}_{CO_2} - RT\ln P_{CO_2} \\ &= 0. \end{aligned}$$

Therefore,

$$G_{CaCO_3} - G_{CaO} - G^{o}_{CO_2} = RT\ln P_{CO_2}$$

or

$$\ln P_{CO_2} = \frac{1}{R}\left(\frac{G_{CaCO_3}}{T} - \frac{G_{CaO}}{T} - \frac{G^{o}_{CO_2}}{T}\right)$$

so by differentiating this equation with respect to temperature at constant pressure, we obtain

$$\left\{\frac{\partial \ln P_{CO_2}}{\partial T}\right\}_P = \frac{1}{RT^2}\left(- H_{CaCO_3} + H_{CaO} + H^{o}_{CO_2}\right),$$

that is,

$$\frac{d\ln P_{CO_2}}{dT} = \frac{\Delta H}{RT^2}, \qquad (7)$$

where ΔH is the heat content change in the reaction, and CO_2 is assumed as a perfect gas, as noted above, so that the heat content at unit pressure H^{o} is the same as that determined at any convenient pressure and denoted by *H*.

Thus if ΔH for the dissociation reaction

$$CaCO_3 = CaO + CO_2,$$

is known as a function of temperature, the integration of Eq. (7). gives the variation of the dissociation pressure P_{CO_2} with the temperature over the same range of temperature to which the heat content change ΔH applies.

At the same time, we must know that when the law of chemical equilibrium is applied to reactions taking place in *heterogeneous systems* such as the dissociation of calcium carbonate, it can be proved that this system is *only in equilibrium at a given temperature when a definite pressure of carbon dioxide is present*, so this pressure can be considered to represent the equilibrium constant of the dissociation reaction, and thus, it is called *dissociation pressure of the system.*

In the application of law of chemical equilibrium to reactions involving solids, it is, at the first sight, difficult to give a definite meaning to the active mass of a solid phase. We may, however, suppose that equilibrium is attained in the *gaseous phase* on the basis that each solid has a definite value of vapour pressure, which however small, is constant at constant temperature and is *independent of the amount of the solid.* The law of chemical equilibrium may thus be applied to this heterogeneous reaction as follows:

$$\frac{P_{CaO} \cdot P_{CO_2}}{P_{CaCO_3}} = K_P, \qquad (8)$$

where P_{CaO} and P_{CaCO_3} represent the very small vapour pressures of calcium oxide and calcium carbonate. Since these two pressures are independent of the respective amounts of the solids but are functions of temperature only, the above equation reduces to

$$P_{CO_2} = K_P \frac{P_{CaCO_3}}{P_{CaO}} = \text{constant},$$

or

$$P_{CO_2} = K'_P. \qquad (9)$$

It is thus clear that the system is only in equilibrium at a given temperature when a definite pressure of carbon dioxide is present. This pressure is, therefore, called *dissociation pressure of the system* as noted above. It increases with increase of temperature since the dissociation of calcium carbonate is endothermic. Thus, if the temperature of some calcium carbonate in a confined space is raised to a certain degree which is then kept constant, dissociation takes place until the dissociation pressure at that temperature is attained. To drive the reaction in the forward direction, such as in the manufacture of quicklime, CaO, the CO_2 must be removed from the sphere of the reaction, and this is done by passing a current of air in the ignition furnace.

The fact that the dissociation pressure of a solid substance is independent of its amount but only on the temperature is known as *Deville's law.*

Considering now the relation between the heat content change ΔH of this dissociation reaction and the dissociation pressure P_{CO_2}, which is obtained above, namely,

$$\frac{d \ln P_{CO_2}}{dT} = \frac{\Delta H}{R T^2}. \qquad (7)$$

We may modify this expression into

$$\frac{d \ln K'_P}{dT} = \frac{\Delta H}{R T^2}. \qquad (10)$$

It may be of interest to correlate between ΔH and ΔU of this endothermic reaction at a given temperature, in the sense that if one of these changes in heat content (enthalpy) or in internal energy is experimentally obtained, then the other can be calculated with the help the first law $(q = \Delta U + W)$. Thus if ΔH of this reaction at 20° C has a value equal to $+42{,}900$ calories, ΔU can be directly calculated from the relation $\Delta H = \Delta U + P\Delta V$ The pressure–volume work, , $P\Delta V$ in this case, assuming the carbon dioxide gas to be a perfect gas, is

$$P\,\Delta V = RT$$
$$= 1.987 \times 298 = 582 \text{ calories}.$$

Therefore,

$$\Delta U = \Delta H - P\Delta V$$
$$= +42.900 - 582 = 42.318 \text{ calories}.$$

4.3. Examples of engineering second-law analysis

1. Two idealised systems. Often, the system of interest is not isolated. Thus in order to investigate the reversibility, irreversibility, or impossibility of a process within the system, we can imagine that the system is in contact with *a hypothetical environment in which all processes are reversible. The system under study and this hypothetical environment then form an isolated system*, and any increase in the entropy within the combined isolated system must be due to *irreversibilities* within the system under study. The amount by which the entropy of the combined system increases during the process is then the *entropy production due to irreversibilities* within the system under study.

Since the interactions between the system and the *hypothetical reversible environment* include heat and work, we need to conceive two 'reservoirs' to which or from energy can be transferred *reversibly* to the system under study, as heat or work.

We shall conceive of a thermal energy reservoir **(TER)** as some *system of fixed mass that can undergo only heat interactions with its environment*. Any energy transferred into the TER as heat will appear as an *increase* in its internal energy. The TER is further idealised as having a *uniform internal temperature*; a TER is always *in an equilibrium state*. We usually conceive the TER as being very large, so that the temperature *remains constant* for the interactions we consider.

Internal energy represents '*disorganised' molecular energy*, and energy transfer as heat can be viewed as '*disorganised microscopic work'*. The TER can, therefore, be thought of as a source or sink for '*disorganised energy'*.

In contrast, the *mechanical energy reservoir* **(MER)** is some *system that possesses energy only in some fully organised mechanical form such as in a raised weight*. The only energy transfer made for a MER is the *reversible work*. All motions within a MER are assumed to be frictionless so that any energy put into the MER as work can be completely recovered as work. The MER can be thought of *as a source or sink for 'fully organised energy'* A MER can have but *one state* for a given energy; given its energy, we know the position of the weight exactly, without uncertainty.

These two conceptual systems would be difficult to build exactly, but they can be closely approximated and hence are reasonable concepts. A block of copper can be a good approximation to a TER; it has a large capacity to store internal energy and can be idealised as being incompressible (Fig. 1).

The MER and TER are useful concepts because they are two *non-isolated systems for which we can easily compute the entropy change.*

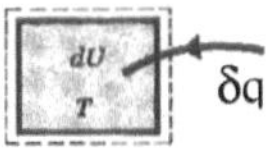

(a) The TER is a reservoir for 'disorganised energy'.

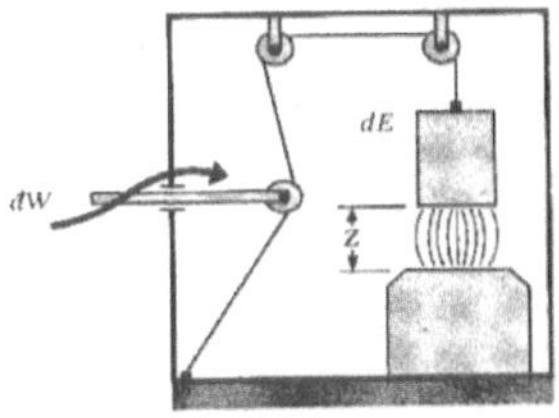

(b) The MER is a reservoir for 'organised energy'.

Fig. 1. Two conceptual systems.

The TER is a chunk of matter held at a fixed volume. The infinitesimal increase in its entropy associated with an infinitesimal increase in its internal energy is

$$dS = \frac{1}{T}\, dU. \tag{1}$$

This equation can be obtained as follows:
The analytical formulation of the combined first and second laws of thermodynamics for pressure–volume work only is

$$TdS = dU + PdV$$

$$TdS = dU \qquad \text{(for a TER } dV = 0\text{)}$$

$$dS = \frac{1}{T}\, dU. \tag{1}$$

An energy balance (Fig. 1a) reveals that the internal energy change is due solely to the energy transfer as heat to the TER

$$dU = d'q_{rev.} \tag{2}$$

Hence, the entropy change for a TER can be calculated from

$$dS = \frac{d'q_{rev.}}{T}. \tag{3}$$

The energy transfer to the TER results in an increase in the randomness inside, that is, an increase in our uncertainty about the microscopic state. The entropy of a TER can be decreased by removal of energy as heat ($dq < 0$). This will reduce the randomness inside, and hence, the entropy will decrease ($dS < 0$).

In contrast, the MER is conceptually a *perfectly organised* system. It has *one most probable macrostate* for each energy. The energy transfer to the MER as work appears as an increase in the mechanical energy stored inside. This organisation makes the energy fully recoverable as work. Given the energy, the macroscopic state of the MER is precisely known. The probability of its single macroscopic state is, therefore, exactly one, and hence, its entropy is always exactly zero. This means that

$$dS = 0 \text{ for a MER.}$$

2. Heat engines. The energy-conversion efficiency of a heat engine is defined, as shown before, as the ratio of the useful work output to the energy input as heat,

$$\eta = \frac{W}{q_A}$$

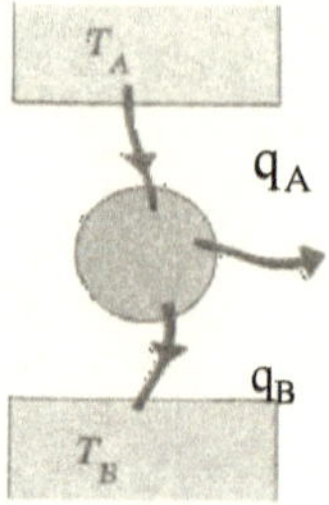

Fig. 2. A 2T heat engine.

This ratio will not remain constant when the engine is reversed, except for a reversible 2T engine (see Fig. 3). Note that there is no energy-storage

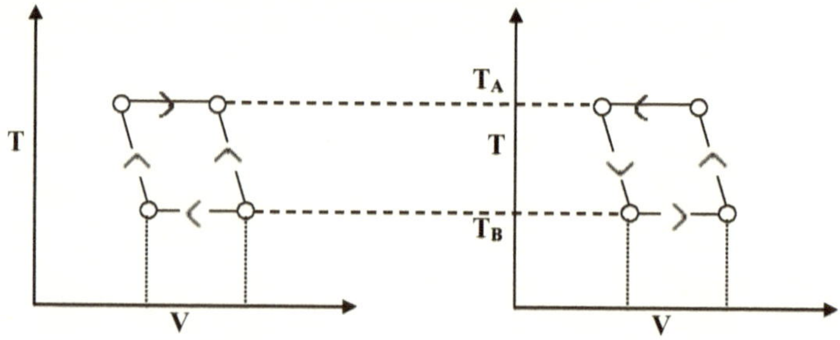

(a) As a heat engine. **(b) As a heat pump.**

Fig. 3. A reversible 2T engine.

terms have been shown in these two figures. The symbols q and W are to be interpreted as being 'for a cycle', and since the engines operate cyclically, there is no change in the energy within an engine over a cycle.

A simple expression for the limiting efficiency of a reversible 2T engine can be derived by considering the system of Fig. 4 and Fig. 5

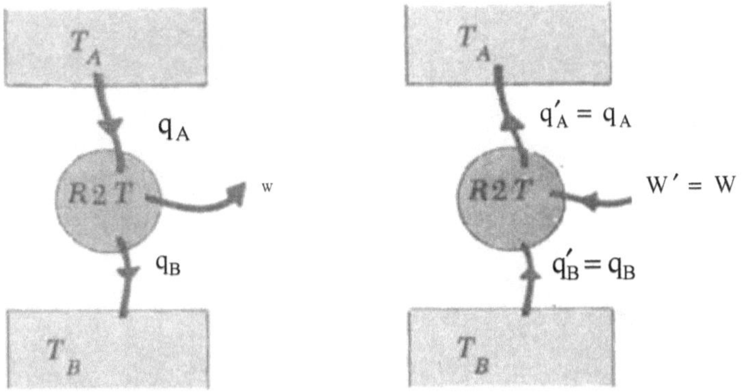

$$\eta = \frac{W}{q_A} = \frac{W'}{q'_A}$$

Fig.4. The efficiency of a reversible 2T heat engine is independent of the direction of operation.

being shown *a priori* to indicate η independence of the operation's direction.

The heat engine receives energy as heat from a thermal energy reservoir at temperature T_A and rejects energy as heat to a second thermal energy reservoir at temperature T_B. The energy output as work is stored in a mechanical energy reservoir. The combined system is isolated, and the second law says that the entropy of this

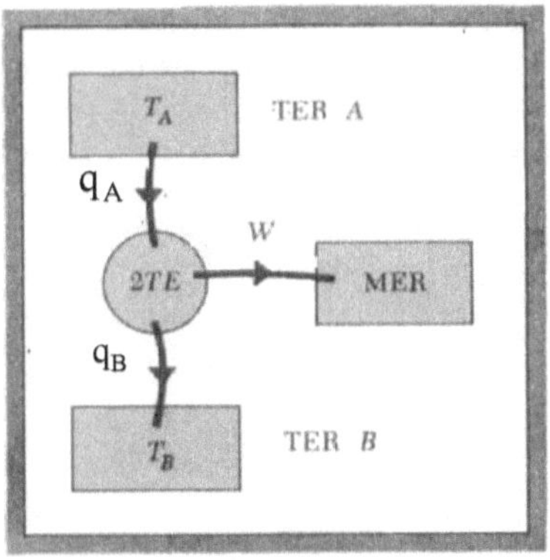

Fig. 5. Determination of the efficiency of a reversible 2T heat engine in an isolated system.

isolated system can never decrease. Now since the engine executes a cycle, all matter in the engine returns to the initial state, and hence, the *entropy is unchanged* over one cycle. The MER undergoes *no change in entropy*. The entropy changes of the two TERs are

$$\Delta S_A = \frac{q_A}{T_A} \quad \text{and} \qquad \Delta S_B = \frac{q_B}{T_B},$$

and hence, the entropy change produced, that is, entropy production, P_S, when the engine executes one cycle is

$$P_S = \left(\frac{q_B}{T_B} - \frac{q_A}{T_A}\right) \geq 0. \tag{5}$$

The equality sign holds only if the processes within the isolated system are reversible, that is, if the heat engine is a reversible 2T engine, that is, 2RT. In this case, according to Eq. (5), we have

$$\frac{q_B}{T_B} - \frac{q_A}{T_A} = 0$$

or

$$\frac{q_A}{q_B} = \frac{T_A}{T_B} \quad \text{for a reversible 2T engine.} \tag{6}$$

Applying the first law over the period of a cycle, we have

$$W = q_A - q_B, \tag{7}$$

so

$$\eta_{R2T} = \frac{q_A - q_B}{q_A} = \frac{T_A - T_B}{T_A} \quad \text{for any reversible 2T engine.} \tag{8}$$

It is common practice to call this efficiency as the *Carnot efficiency*. This is because Carnot engine is one type of a reversible 2T engine. However, we have said nothing about the nature of the engine, and thus, Eq. (8) holds good for any reversible 2T engine.

On the other hand, the inequality in Eq. (5)

$$P_S = \left(\frac{q_B}{T_B} - \frac{q_A}{T_A}\right) \geq 0$$

pertains to irreversible processes within the engine, that is, to an *irreversible* 2T engine, for which

$$\frac{q_B}{T_B} > \frac{q_A}{T_A}$$

or

$$\frac{q_B}{q_A} > \frac{T_B}{T_A}, \qquad \text{for } q_A > 0.$$

Therefore,

$$\left(1 - \frac{q_B}{q_A}\right) < \left(1 - \frac{T_B}{T_A}\right)$$

or

$$\frac{q_A - q_B}{q_A} < \frac{T_A - T_B}{T_A}.$$

Hence,

$$\eta_{irrev} = \left(1 - \frac{q_B}{q_A}\right) < \left(1 - \frac{T_B}{T_A}\right)$$

but

$$\left(1 - \frac{T_B}{T_A}\right) = \eta_{Carnot},$$

therefore,

$$\eta_{irrev} = \left(1 - \frac{q_B}{q_A}\right) < \left(1 - \frac{T_B}{T_A}\right) = \eta_{Carnot} \qquad (9)$$

This result is based on the fact that the entropy change in the isolated combined system when the 2T heat engine operates irreversibly is greater then zero.

Accordingly, the Carnot efficiency can be regarded as an upper limit for the performance of any real 2T heat engine. Highest efficiencies will be obtained when the ratio T_B/T_A is as small as possible; one would thus like to add the energy as heat at as high a temperature as possible and reject energy as heat at the lowest possible temperature. However, nature places physical limitations on man's capabilities, the energy which must be rejected as heat must flow to an environment which is cooler than T_B. This means that we are limited to T_B of the order of 60°F (520°R or 16°C). The energy transfer as heat to the engine must come into it from a region at a temperature greater than T_A. Temperatures of the order of 3500°R $\{\text{i.e.} = (3500 - 491) \times 5/9 \simeq 1670°\text{C}\}$ can be obtained by combustion reactions, but metallurgical considerations normally require that the device must be kept much cooler. Modern steam power plants operate at about 1100°F $\{\text{i.e.}=(1100-32)\times 5/9 \simeq 590°\text{C}\}$ not on the Carnot cycle. Advanced nuclear power systems are being designed and tested in the range of 1500–2500°F, that is, in the range of 815°≃1370°C. These high-temperature systems employ exotic metals and are not intended for long life or for production of low-cost electrical power. At 1500°F, that is 816°C, the Carnot cycle efficiency is

$$1 - \frac{T_B}{T_A} = 1 - \frac{(16°\text{C} + 273°\text{C})}{(816°\text{C} + 273°\text{C})} = 1 - \frac{289\text{K}}{1089\text{K}} = 1 - \frac{520°\text{R}}{1960°\text{R}}$$

$$= 1 - 0.265 = 0.735,$$

where T_B is put equal to 520°R, that is, =16°C = 289K, and T_A equal to 1960°R, that is, = 816°C = 1089K. It is evident that the same result of 0.735 for the Carnot efficiency could have directly obtained if we express T_B and T_A on the *Rankine scale of temperature*, that is, if we put T_B equal to 520°R and T_A = 1960°R. This scale is a *second absolute scale of temperature*, the first being the *thermodynamic temperature scale which was realised by Kelvin* and *called absolute temperature scale,* which will be considered in Chapter 5. The *Rankine scale* is defined by setting

$$1\text{K} = (9/5)^\circ\text{R}, \tag{10}$$

or

$$1^\circ\text{R} = 5/9\text{K}.$$

This makes the H_2O triple-point temperature correspond to 491.69°R.

Associated with the *Kelvin* and *Rankine* scales are the *Celsius (formerly), centigrade,* and *Fahrenheit relative* temperature scales.

The Celsius scale is such that

$$1^\circ\text{C} = 1\text{K}, \tag{11}$$

but it has its zero point (i.e. ice point) equal to 273.15K; this makes 0K correspond to –273.15°C, that is,

$$0\text{ K} = -273.15^\circ\text{C}. \tag{12}$$

This relation will be proved in Chapter 5 as noted above when we consider the realisation of the absolute thermodynamic scale of temperature by *Kelvin.*

Similarly,

$$1^\circ\text{F} = 1^\circ\text{R}, \tag{13}$$

but 0°F corresponds to – 459.67°F, that is,

$$0^\circ\text{F} = +459.67^\circ\text{R} \tag{14}$$

These four temperature scales are compared in Fig. 6.

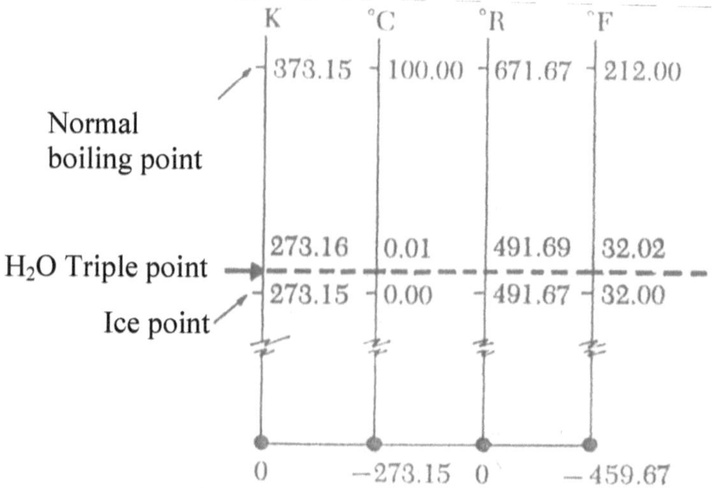

Fig. 6. The °R and K scales are 'absolute' temperature scales.

It should be pointed out that the empirical gas scale can be used only at temperatures *above* the boiling point of the *gas* and *below* the melting point of the *container*. Outside this range, other empirical temperatures must be used. The thermodynamic definition of temperature given before in terms of the specific intensive properties *u*, *s*, and *v* is

$$\left(\frac{\partial u}{\partial s}\right)_V = T. \qquad (15)$$

That is, the temperature is defined as *the rate of change of internal energy with entropy at constant volume*. This is a continuous definition of temperature valid over all ranges, and therefore, provides *an essential link between the several measures of temperature*. It is derived simply from the exact differential form of the internal energy. Using specific quantities this form is

$$du = Tds - Pdv. \qquad (16)$$

The temperature *T* as used here is the *absolute* temperature, which would be *measured on the Kelvin (K) scale or the Rankine (R°) scale*. The *relative* temperatures (°C and °F) can be used when temperature *differences* are involved. It might be added here that an important equation *stems* from Eq. (16) when rewritten in the form

$$ds = \frac{1}{T}du + \frac{P}{T}dv. \qquad (16)$$

This equation is known as the *Gibbs equation* for *simple compressible* substances; it provides a means for the evaluation of the *entropy* of substances from macroscopic laboratory data.

Returning now to Eq. (9) written in the form

$$\eta_{irrev.} < \left(1 - \frac{T_B}{T_A}\right) = \eta_{Carnot} \qquad (9)$$

and calculating η for a Carnot engine working reversibly between $T_A = 1960°R$ (= 1089K) and $T_B = 520°R$ (= 289 K), we obtain a value of 0.735. This indicates that even under such extreme conditions, this most ideal cycle could convert only 73.5 per cent of the energy transferred in the engine as heat into useful work. Realistic devices of present technology operate in the range of 15–40 per cent, and a heat engine having a thermal energy-conversion efficiency of even 50 per cent has yet to be built.

It is evident from the foregoing that it would be impossible to devise a 100 per cent efficient heat engine even if we really had reversible processes at our disposal. A 100 per cent efficiency of a heat engine or in other words a 1T heat engine means that all the energy absorbed as heat is converted into useful work as is necessitated by the first law, but however, this is denied by the second law. This statement that a continuously operating IT heat engine is impossible is known as the *Kelvin–Planck statement* and is taken as the starting point in many classical developments of the second law.

The second law places *restrictions* on systems that continuously transfer energy as heat from a region of low temperature to one of higher temperature such as *a refrigerator or a heat pump.* Also, the statement that a zero-work heat pump is impossible is called *the Clausius statement.*

3. Refrigerators and heat pumps. By the expenditure of mechanical work a Carnot engine was made to run backwards extracting heat from a cold reservoir and reflecting heat to a hot one (Fig. 7). Any device which, by the use of mechanical work transfers heat *continually* from a cold to a warmer body is called a refrigerator or heat pump.

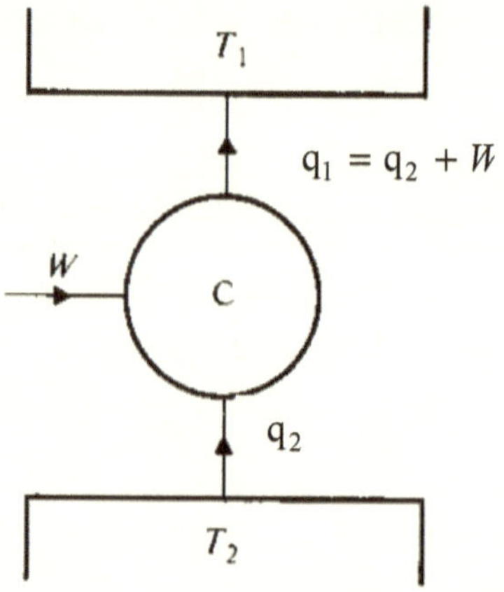

Fig. 7. A Carnot engine driven backwards absorbs heat at the cold reservoir and rejects heat at the hot.

a. Refrigerators. The function of a refrigerator is to extract heat q_2 from a body which is at a lower temperature T_2 than the surrounding bodies and rejects this heat and the heat equivalent to the *expended* mechanical work to a hotter body, which is the surroundings at T_1. The efficiency of a refrigerator should, therefore, be defined in terms of the *amount of heat extracted for a given expenditure of mechanical work.* For a perfect refrigerator, using a Carnot engine,

$$\eta_{\text{refrigerator}} = \frac{q_2}{W} = \frac{T_2}{T_1 - T_2}$$

$$= \frac{T_2/T_1}{1 - T_2/T_1}.$$

$$= \frac{T_2/T_1}{1 - T_2/T_1}. \qquad (1)$$

The efficiency of the ideal refrigerator is shown in Fig. 8.

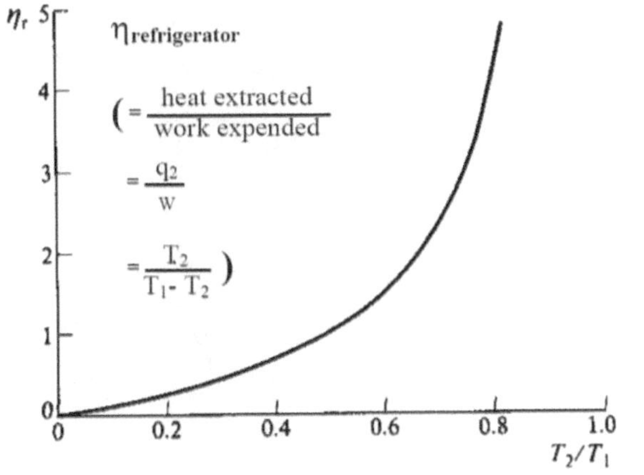

Fig. 8. The efficiency of an ideal refrigerator.

For moderate degrees of cooling, the efficiency is high. Down to $T_2/T_1 = 0.5$ more heat is absorbed than work required. But for a given extraction of heat q_2 the work required becomes very large as the temperature ratio T_2/T_1 is decreased. For a domestic refrigerator, the upper temperature T_1 is usually very close to room temperature by exchange of heat with the surroundings through cooling fins, while the lower temperature T_2 is kept somewhat below freezing point. With $T_1 = 300\,\text{K}$ and $T_2 = 270\text{ K}$ and $\eta = 9\left(= \frac{270}{300 - 270}\right)$. On the other hand, to absorb 4 watts of heat at 1K with a refrigerator working at room temperature T_1, which is the room temperature taken as 300K, would require an amount of power W given by

$$\frac{q_2}{W} = \frac{T_2}{T_1 - T_2},$$

$$\frac{4}{W} = \frac{1}{300 - 1} \simeq \frac{1}{300}$$

$$W = 300 \times 4 = 1200 \quad \text{watts},$$

that is,

$$W > 1 \text{ kilowatt} .$$

This example shows why it becomes increasingly difficult to obtain cooling at very low temperatures. In practice, efficiencies will fall well below these ideal figures.

b. Heat pumps. The function of a heat pump is to *deliver* heat to some body which is at a *higher* temperature than its surroundings. The efficiency of a heat pump should, therefore, be defined in terms of *the amount of heat delivered at the higher temperature for a given expenditure of mechanical work*. For a perfect heat pump using a Carnot engine,

$$\eta_{\text{heat pump}} = \frac{q_1}{W} = \frac{T_1}{T_1 - T_2} = 1 + \frac{T_2}{T_1 - T_2}$$

$$= 1 + \eta_{\text{refrigerator}}. \qquad (2)$$

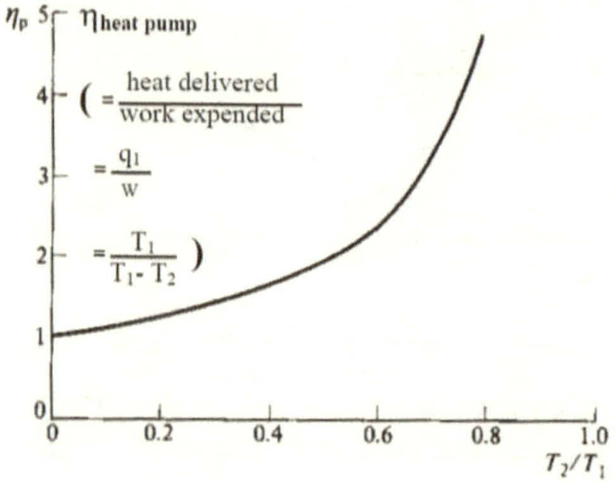

Fig. 9. The efficiency of an ideal heat pump.

The variation of $\eta_{\text{Heat pump}}$ as a function of the temperature ratio T_2/T_1 is represented graphically in Fig. 9. For small temperature differences between T_1 and T_2, considerably *more heat* is supplied than *power* consumed. This makes the heat pump very *attractive* as a device for heating buildings – heat being extracted from the surrounding atmosphere and delivered to building at a little above room temperature. Taking for example $T_1 = 320\text{K}$ and $T_2 = 280\text{K}$, we have

$$\eta_{\text{Heat pump}} = \frac{q_1}{W} = \frac{T_1}{T_1 - T_2} = \left(1 + \frac{T_2}{T_1 - T_2}\right)$$

$$= \frac{320}{320 - 280} = \frac{320}{40}$$

$$= 8,$$

which means that 8 kilowatts of heating would involve only a power consumption of only 1 kilowatt. Unfortunately, the high cost and low efficiency of any practical plant make this method of heating of doubtful economic advantage. When $T_2 \to 0$ the heat delivered becomes equal to the work required. In this case, the heat pump has no advantage over a simple device which turns the work directly into heat such as a simple electric heater.

4. Real heat engines. In completely general terms, little can be said about the efficiencies of real heat engines, although a crude comparison with a Carnot cycle is sometimes possible.

If we know the extremes of the temperature involved in the cycle of the real engine, then certainly its efficiency must be less than that of a Carnot engine operating between reservoirs at these extremes. Such a simple comparison is sufficient to show why the early steam engines were so inefficient. Steam was available at somewhat *above* atmospheric pressure, say at 390K, and was condensed by water at a temperature somewhat *below* the normal boiling point, say 350K. The efficiency of a Carnot engine operating between these temperatures would only be 10 per cent, and of course, for the steam engines, it was much smaller still. In the modern steam engines, the efficiency has been improved by using high pressure steam and forcing T_1 up; but the steam engine is still an *inefficient* means of generating mechanical power from heat because of the comparatively limited temperature range which is practicable. In contrast, one would expect the internal combustion engine to be capable of much higher efficiencies because of the extremely high temperature which is involved in the explosion.

To discuss a real heat engine in any detail, it is always necessary to invent an idealised cycle which may be used as a reasonable representation of the cycle of the real engine. Calculations based on such an idealised cycle will give an *upper limit* to the efficiency of the real heat engine. This idealisation involves *two basic approximations*. The first is that the working substance is a single pure substance. In the case of the *internal combustion engine*, this is clearly far from the truth. The working substance is, in fact, a mixture of *gases and vapours,* and its composition changes during the cycle. For the internal combustion engine, *air* is usually chosen to represent the working substance. Cycles based on air are known as '*air standard cycles*'. The second approximation consists of replacing the real cycle by a reversible cycle. Again, this is clearly far from the truth. Most real cycles proceed rapidly, and the conditions are for from quasistatic: heat flows through finite temperature gradients and there is friction and turbulence. In practice, it is usually this approximation which introduces the greatest errors.

We shall illustrate the use of an idealised cycle by considering one example, that of the petrol engine.

A. The petrol engine. In this engine, the cycle consists of six parts. Four of these involve motion of the piston and are called *strokes*. The cycle proceeds as follows:

1. *Intake stroke*. The mixture of petrol and air is drawn into the cylinder through the intake value by the movement of the piston.
2. *Compression stroke*. The intake valve now closes, and the piston moves up the cylinder compressing the mixture *rapidly*. The compression is nearly *adiabatic,* and there is a considerable temperature rise.
3. *Explosion.* When fully compressed, the mixture is caused to explode. There is negligible movement of the piston during the explosion so that the volume remains unchanged, but a very *high temperature and pressure* are reached.
4. *Power stroke*. The *hot* combustion products expand, doing mechanical work on the piston. There is a considerable *drop* in the pressure and temperature.

5. *Valve exhaust.* At the end of the power stroke, the exhaust valve opens. The combustion products which are still at a high pressure flow out rapidly into the atmosphere. There is a sudden drop in pressure.
6. *Exhaust stroke.* The piston moves up and making the cylinder forcing the remaining gases out, into the atmosphere. The exhaust value then closes, and the intake value opens in readiness for the *next intake stroke.*

The petrol engine cycle is clearly highly *irreversible.* The idealised cycle which replaces it is known as the *Air Standard Otto Cycle,* and it is illustrated in Fig. 10.

B. The air standard Otto cycle.

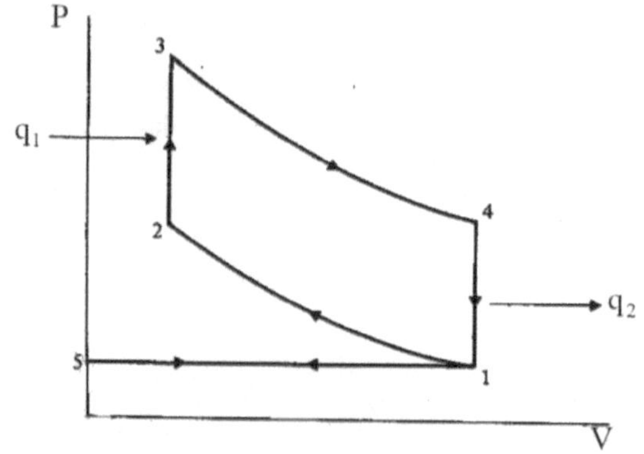

Fig. 10. The air standard Otto cycle.

Air is taken as the working substance and is assumed to obey the ideal gas laws with constant specific heats. All processes are assumed to be *reversible.* The various parts of the petrol engine cycle are then represented as follows:

5 ⟶ 1. *The intake stroke.* A quasistatic isobaric intake of air at P_0 to a volume V_1.
1 ⟶ 2. *The compression stroke.* A quasistatic adiabatic compression from V_1 to V_2 during which the temperature rises from T_1 to T_2 according to the perfect gas equation,

$$T_1 V_1^{\gamma-1} = T_2 V_2^{\gamma-1}, \tag{1}$$

where γ is the ratio of the specific heats.

2 ⟶ 3. *The explosion.* A quasistatic isovolumic rise of temperature and pressure brought about by the absorption of heat from a *series of reservoirs* between T_2 and T_3. It is to be pointed out that the series of reservoirs is necessary so that no temperature difference ever occurs between the system and the reservoir supplying heat. If a temperature difference was set up, the flow of heat would then become thermodynamically *irreversible.*
3 ⟶ 4. *The power stroke.* A quasistatic adiabatic expansion producing a drop in temperature according to

$$T_3 V_2^{\gamma-1} = T_4 V_1^{\gamma-1}. \tag{2}$$

$4 \longrightarrow 1$. *The valve exhaust.* A quasistatic isovolumic drop of temperature to T_1 (and of pressure to p_0) brought about by exchange of heat with a series of reservoirs between T_4 to T_1.
$1 \longrightarrow 5$. *The exhaust stroke.* A quasistatic isobaric expulsion of the air.

Clearly, the two isobaric processes $5 \longrightarrow 1$ and $1 \longrightarrow 5$ cancel one another out, and in calculating the efficiency, we only need to consider the rest of the cycle.

The heat absorbed along $2 \longrightarrow 3$ is

$$q_1 = \int_{T_2}^{T_3} C_v \, dT = C_V (T_3 - T_2), \tag{3}$$

and that rejected along $4 \longrightarrow 1$ is

$$q_2 = -\int_{T_4}^{T_1} C_v \, dT = C_V (T_4 - T_1). \tag{4}$$

Applying the first law, the efficiency is

$$\eta = \frac{W}{q_1} = \frac{q_1 - q_2}{q_1} = 1 - \frac{(T_4 - T_1)}{(T_3 - T_2)}. \tag{5}$$

From Eqs. (1) and (2)

$$(T_4 - T_1)\, V_1^{\gamma-1} = (T_3 - T_2) V_2^{\gamma-1}$$

or

$$\frac{(T_4 - T_1)}{(T_3 - T_2)} = \left(\frac{V_2}{V_1}\right)^{\gamma-1}. \tag{6}$$

Hence,

$$\eta = 1 - \left(\frac{V_2}{V_1}\right)^{\gamma-1}. \tag{7}$$

If we denote the *compression* ratio V_1/V_2 by *r*, then

$$\eta = 1 - \left(\frac{1}{V_1/V_2}\right)^{\gamma-1} = 1 - \frac{1}{r^{\gamma-1}}. \tag{8}$$

To obtain the highest efficiency, the compression ratio has to be as large as possible. However, it cannot be made too large because eventually regions in the fuel mixture detonate during the combustion rather than burning smoothly. The resulting *pinking* or *knocking* is mechanically bad for the engine and also reduces the efficiency. With modern fuels, a compression ratio of about ten can be used. Taking $\gamma = 1.4$ for air, this gives a theoretical maximum efficiency of about 61 per cent. In a real engine, the efficiency probably only reaches half this value.

5. A solar engine. It is proposed that solar energy be used to warm a large '*collector plate*'; this energy would, in turn, be transferred as heat to a fluid within a *heat engine*, and the engine would reject energy as heat to the atmosphere. Experiments indicate that about 200 Bt $u/h - ft^2$ of energy can be 'collected' when the plate is operating at 190°F.

As an example, we may estimate the *minimum* collector area that would be required for a plant producing 1 kilowatt of useful shaft power, that is, power output.

We must first estimate the maximum energy-conversion efficiency of this system, using the Carnot efficiency as an upper limit. The atmospheric temperature is assumed to be 10°F. Thus

$$\eta_{max} = 1 - \frac{70 + 460}{190 + 460} = 0.184 \quad . \qquad (1)$$

The efficiency of any real heat engine operating between the collector place $(T_1 = 190°F)$ and atmospheric temperature $(T_2 = 70°F)$ would be less than this, owing to *irreversiblilites* in real devices. The minimum rate at which energy must be collected is related to the required power output and maximum energy-conversion efficiency, Therefore, since the required power output is 1 kilowatt and the maximum energy-conversion efficiency is 0.184, we have

$$\dot{q}_{min} = \frac{\dot{W}}{\eta_{max}} = \frac{1}{0.184} = 5.44 \text{ kilowatt}$$

$$= 18{,}600 \text{ Btu/h}. \qquad (2)^*$$

Subsequently, the *minimum* collector area is

$$A_{min} = \frac{18{,}600}{200} = 93 \text{ ft}^2. \qquad (3)$$

A real system might be expected to need twice or three times this area, since the efficiency would probably be considerably less than 18.4 per cent.

* 1 Bt u = 1.055 × 1010 ergs = 252 cal ., kilowatt = 3413 Btu/h. See "Selected Dimensional Equivalents" Appendix B, Section 8, Table 5.

CHAPTER 5

Temperature and Some of Its Effects

5.1. The absolute thermodynamic scale of temperature

1. Development of a temperature scale. The temperature of a body is meant the number which expresses, on some definite scale, how hot or how cold the body is. The general notions which we have of temperature are gained through our thermal senses, but these yield neither quantitative results nor objective means of making measurements. In fact, temperature cannot be measured directly, for it is impossible to express 'hotness' in terms of another in the same sense that length or mass may be expressed in terms of another. *Temperature, like force, can be measured only in terms of its effects.*

As regards the early thermometers, it can be said that although the ancients were familiar with some of the effects that heating produces in a body, there was no application of this knowledge to the estimation of temperature until *Galileo* invented the *thermoscope*. This instrument was merely a glass bulb containing air and having a long stem which extended downward into a vessel of water. As the temperature changed, the air in the bulb expanded or contracted, and the water in the stem rose or fell. The changes in the atmospheric pressure also affected the height of the water column and were clearly recognised by *Pascal* and *Boyle*. But thermometers of this kind continued to be used throughout the first half of the seventeenth century.

It may be interesting to know that *Italy was the home of the first organised scientific academy, the Accademia del Cimento of Florence* (1657–1667). This academy was founded at Florence, Italy, in 1657, as a direct result of *Galileo's teaching* and to carry on his method of *investigating truth by experiment alone*. It may be pointed out that among the moving spirits of this academy was *Viliani, one of Galileo's most distinguished disciples*. In spite of its brief existence (only ten years), the academy achieved remarkable results and exerted an enormous influence in spreading the *experimental method of investigating truth* all over Europe.

The first use of the expansion of a liquid for the estimation of temperature was made in 1631 by a *French physician, Jean Rey*, who determined the body temperatures of his fevered patients by means of a glass bulb *similar* to one that was used by Galileo except that it was *inverted* and *filled* with water. Important improvements, which consisted of sealing the end of the tube to prevent evaporation and in substituting other liquids, such as alcohol for water, were made at about 1657 by the Accademia del Gimento at Florence, Italy. All that was needed to turn these *Florentine thermometers* into modern instruments was *a scale* that could be accurately duplicated, so that measurements of a given temperature made at various places and times with different instruments would be the same.

In setting up a scale of temperature, it eventually *became* clear that the first step is always to choose at least two temperatures which can be accurately and easily reproduced. The two temperatures that are now ordinarily chosen are the melting point of pure ice and the temperature of condensing steam, under a pressure of one standard atmosphere. These two particular fixed points are termed as the *ice point* and the *steam point,* respectively. *Hooke* was the first who suggested these points in 1664, and *Huygens*, the second, a few months later. The first thermometer that *graduated, by the use of two fixed points* with which useful observations were made, was probably that of *Newton*. The modern *mercury-in-glass thermometer* with a scale depending upon the ice point and the steam point was introduced in 1714 by *Gabriel Daniel Fahrenheit*.

After the selection of the fixed points, the next problems to be considered are of the *subdivision of the interval between them and the measurements of temperature outside this interval*. Here use is made of the principle that as a body grows hotter to the touch, many of its physical properties change. Next is selecting a particular substance and a suitable property, such as volume, pressure, or electrical resistance, whose variations can be accurately and conveniently measured; we use the changes in this property to indicate changes in the temperature of the body or of any other body with which it is in thermal equilibrium. For example, we could observe the volume of a given piece of *iron* first at the ice point and afterwards at the steam point and then divide the computed increase in volume into any convenient number of equal parts. Such a change in temperature, as will produce a volume change equal to one of these parts, is then defined as the *degree*. On the *centigrade system*, which was proposed in 1742 by the *Swedish astronomer Anders Celsius*, the number of equal parts is made 100, and the ice point is taken as the zero of this scale (0°C).

It is to be noted, however, that the choice of the zero of the scale and of the size of the degree should be entirely independent of the choice of a *thermometric substance* and of a *thermometric property*. It is found that the properties of different substances are not generally the same function of temperature, and therefore, thermometers constructed from different thermometric substances do not agree exactly with one another at temperatures other than the two fixed points. Hence, it becomes necessary to choose some particular thermometric substance and some particular property of this substance and to agree that the changes in the latter with change of temperature shall be taken as the measure of temperature. We shall see that certain gases possess peculiar advantages for this purpose.

3. **The constant-volume hydrogen thermometer.** The *constant-volume* gas thermometer is based on the principle that a given mass of gas contained in a closed vessel of constant volume assumes a pressure which is entirely determined by the temperature. The bulb, b (Fig. 1), which contains the gas, is connected by a bent capillary tube to the tube c through a flexible tube f containing mercury.

4.

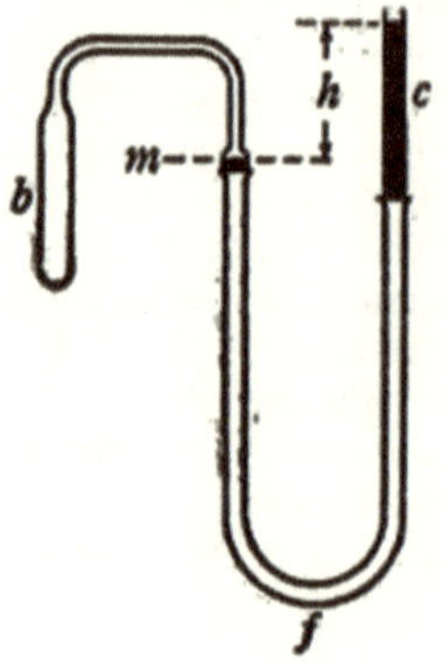

Fig. 1. A simple constant-volume gas thermometer.

By raising or lowering the tube, c, the surface of mercury in the other tube is always brought to a certain fiducial mark, m. This is done first when the bulb, b, is in melting ice, then when it is in the vapour arising from boiling water, and finally when it is at the unknown temperature *t,* which is to be determined. The pressure of the confined gas in each case can be measured by reading the difference in level, *h,* of the two mercury columns and adding to this the atmospheric pressure as determined by a barometer.

The *centigrade degree* on the *constant-volume hydrogen* thermometer is defined as any temperature change which, starting from any temperature whatever, will produce in the confined hydrogen a change in pressure amounting to 1/100 of that observed when the gas is heated from the ice point to the steam point. Accordingly we *define* any centigrade temperature, *t*, measured on the constant-volume hydrogen scale, as

$$t = \frac{P_t - P_0}{\frac{1}{100}(P_{100} - P_0)}, \qquad (1)$$

where P_t, P_0 and P_{100} are the pressures of the gas at $t°$, $0°$ and $100°C$, respectively. This equation can be simply obtained as follows:

Since a temperature equal to 100°C corresponds to a pressure difference equal to $P_{100} - P_0$, then any temperature *t*°C corresponds to a pressure difference $P_t - P_0$, so *t*°C is given by

$$t°C = \frac{100\,(P_t - P_0)}{P_{100} - P_0}$$

or

$$= \frac{(P_t - P_0)}{\frac{1}{100}(P_{100} - P_0)}.$$

The expression $(P_{100} - P_0)/100\,P_0$ applied to any constant-volume apparatus is called the *coefficient of increase of pressure* or simply the *pressure coefficient* of the gas. Evidently it represents the ratio of increase of pressure to the pressure at 0°C for 1° change in temperature. If this coefficient is denoted by β, that is,

$$\beta = \frac{(P_{100} - P_0)}{100\,P_0}. \qquad (2)$$

we may write Eq. (1) in the form

$$t = \frac{P_t - P_0}{\beta\,P_0}, \qquad (3)$$

where β is the *pressure coefficient for hydrogen.*

All gases increase in pressure if heated at constant volume. Experiments show that the pressure of a gas at any temperature, *t*, when *t* is measured by a constant-volume hydrogen thermometer, is given approximately by the equation

$$P_t = P_0\,(1 + \beta t). \qquad (4)$$

The pressure coefficient β is found to have somewhat different values for different gases. It will be noted that Eq. (4) can be put into the same form as Eq. (3).

Regnault selected the constant-volume hydrogen thermometer as his ultimate standard of reference in practical thermometry, and this instrument is universally used for this purpose. Eq. (3), as applied to hydrogen and also sometimes to helium, is therefore to be regarded as the *accepted practical definition of temperature*. Experiments show that β for hydrogen is approximately $1/273.04$. Thus 1°C is, by definition, *such a temperature variation as will result in a pressure change of* $1/273.04$ *of the pressure of 0°C in a given mass of hydrogen kept at a constant volume*. This definition is *based* on Eq. (3), when the *numerator becomes equal to the denominator*. Table 1 shows a comparison of hydrogen, air, and mercury-in-glass thermometer temperatures. The temperatures are reckoned from the ice point, 0°C.

Table 1

Hydrogen	Air	Mercury-in-Jena normal glass
0°	0°	0°
20	20.008	20.091
40	40.001	40.111
60	59.990	60.086
80	79.987	80.041
100	100	100

Among liquids, mercury is found to agree fairly closely with the gas scale, and mercury-in-glass thermometers doubtlessly will continue to be employed in cases where facility of observation is more important than the highest attainable degree of precision. As will be seen from Table 1, the variations between corresponding readings on the hydrogen, air, and mercury-in-glass thermometers due to *irregularities of expansion* are for many purposes negligible.

The equations concerning the constant-volume hydrogen thermometer may be summarised as follows:

i. The definition of any centigrade temperature *t* is

$$t = \frac{P_t - P_0}{\frac{1}{100}(P_{100} - P_0)} \tag{1}$$

ii. The expression

$$\frac{(P_{100} - P_0)}{100\ P_0}$$

that is applied to any constant-volume apparatus is called the coefficient of increase of pressure or simply the pressure coefficient of the gas and denoted by β.

$$\beta = \frac{(P_{100} - P_0)}{100\ P_0}. \tag{2}$$

iii. Eq. (1) may thus be written in the form

$$t = \frac{P_t - P_0}{\beta\ P_0}. \tag{3}$$

iv. The pressure of a gas at any temperature *t*, when *t* is measured with a constant-volume hydrogen thermometer, is given approximately by the equation

$$P_t = P_0\ (1 + \beta t), \tag{4}$$

where the pressure coefficient β is found to have different values for different gases.

3. The expansion of gases with temperature. The simplest way to approach the problem of expansion with temperature is to maintain the pressure constant and to measure the changes in volume of a given mass of the gas that occur with variations in the temperature alone.

Consider then the bulb, b, in Fig. 2, in which a given mass of gas is confined by means of an indicating globule of mercury, m,

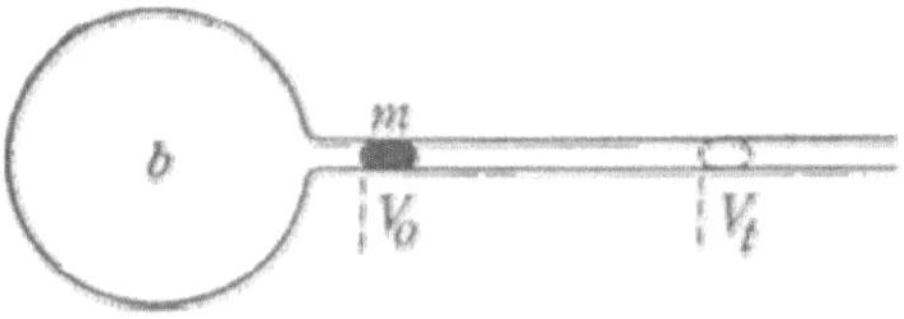

Fig. 2. A simple device for studying the thermal expansion of a gas under constant pressure. Evidently, it may also be used as a constant-pressure gas thermometer.

which moves with little friction forward or backward in the stem as the temperature rises or falls. The end of the stem is open so that the *constant pressure* used is that of the atmosphere. Experiments made with this apparatus show that the increase in volume is approximately proportional to the original volume and to the rise in temperature. If we always agree to take the volume at 0°C as the 'original volume', it then follows that

$$V_t - V_0 = \alpha V_0\, t$$

or

$$V_t = V_0\ (1 + \alpha t), \tag{5}$$

where V_t and V_0 are the volumes at t° and 0°C, respectively, and α is a proportionality constant called *expansivity* or *coefficient of expansion* of the gas. Eq. (5) can be rewritten as

$$t = \frac{(V_t - V_0)}{\alpha\ V_0}. \tag{5'}$$

In 1787, a Frenchman by the name of *Charles discovered that all gases have the same expansion coefficient* α *when heated through the same temperature range.* This was confirmed experimentally some fifteen years later by *Dalton* and by *Gay-Lussac,* who used the apparatus of Fig. 2. It is to be noted that *Charles* did not publish his results. This law, like that of *Boyle,* has been shown by the later experiments of *Renault* and others to be approximately correct. It is established experimentally that the gases which show the largest deviations from Boyle's law, such as carbon dioxide and nitrous oxide, show also the largest variations in the expansivity, α, with pressure and temperature.

The fact that α *is not, in general, a constant* makes it desirable to speak of the *mean* expansivity between the temperature t_1 and t_2, and this is defined by the equation

$$\alpha = \frac{(V_2 - V_1)}{V_0\,(t_2 - t_1)} \tag{6}$$

V_1 and V_2 being the values at t_1 and t_2, respectively. This equation is derived from Eq. (5) as follows:

$$V_1 = V_{t_1} = V_0\,(1 + \alpha t_1) \text{ and}$$

$$V_2 = V_{t_2} = V_0\,(1 + \alpha t_2),$$

Therefore,

$$V_1 - V_0 = V_0\,\alpha\, t_1 \text{ and}$$

$$V_2 - V_0 = V_0\,\alpha\, t_2,$$

so by subtraction, we have

$$\alpha = \frac{(V_2 - V_1)}{V_0\,(t_2 - t_1)}.$$

The limit which this quantity approaches as $t_2 - t_1$ approaches zero is defined as the *true expansivity* at the temp *t* in the interval $t_2 - t_1$; that is,

$$\alpha = \operatorname*{Lt}_{(t_2 - t_1)\to 0} \frac{(V_2 - V_1)}{V_0\,(t_2 - t_1)}$$

or

$$\alpha = \frac{1}{V_0}\frac{dV}{dt},$$

with the pressure remaining constant. Thus the true expansivity can be expressed as

$$\alpha = \frac{1}{V_0}\left(\frac{\partial V}{\partial t}\right)_P. \tag{7}$$

4. The equation of state of an ideal gas. Any equation that expresses the relation between the pressure, volume, and temperature of a substance is called the *equation of state of the substance.* Let us suppose that the ideal gas is initially in the state described by the three gas coordinates P_0, V_0, $0°C$, and that it is heated at *constant volume* V_0 until it arrives at the state P', V_0, t. Then by Eq. (4), we have

$$P' = P_0\,(1 + \beta t).$$

Next we assume that the volume V_0 is changed, *without changing the temperature*, until the gas arrives at a third and final state, which is characterised by the coordinates *P*, *V*, and *t*. For this isothermal change, by Boyle's law,

$$PV = P'V_0.$$

By eliminating the intermediate pressure, P', which occurs in both of these preceding equations, we finally obtain,

$$PV = P_0 V_0 (1 + \beta t). \tag{8}$$

We can prove that *for an ideal gas, the expansivity,* α, *and the pressure coefficient,* β, *are identical quantities* as can be seen from the following:

We assume that the gas is in the state P_0, V_0, $0°C$, and that it is heated at *constant pressure* P_0 until it arrives at the state P_0, V', t. Then by Eq. (5)

$$V' = V_0 (1 + \alpha t). \tag{9}$$

Again we assume that the pressure is changed without *changing the temperature* until the gas arrives at a third and final state which is of the coordinates *P*, *V*, *t*. For this isothermal change, by Boyle's law,

$$PV = P_0 V'.$$

Thus $V' = PV/P_0$, and subsequently, Eq. (9) can be rewritten as

$$PV = P_0 V_0 (1 + \alpha t). \qquad (9)'$$

A comparison of this equation with Eq. (8) yields $\alpha = \beta$.

5. The absolute thermodynamic scale. In 1824, *Carnot* published his remarkable memoir* memoir* on the theory of heat engines, in which he showed that *the maximum efficiency of an engine working through any fixed small range of temperature is independent of the nature of the working substance used in the engine and is a function of the temperature alone*. Twenty-four years later, Lord *Kelvin*† made the brilliant suggestion that a temperature scale based on Carnot's engine would *rid* the definition of temperature of the peculiarities, characteristic of any one real substance. This new scale is *absolute* in the sense of being independent of any particular property of any particular thermometric substance, and is therefore, called the *absolute thermodynamic scale*. Now it can be shown that an *ideal* gas would give this thermodynamic scale precisely *if it were used in a gas thermometer*. Thus either Eq. (3)

$$t = \frac{(P_t - P_0)}{\beta P_0} \tag{3}$$

* N. L. S. Carnot, a translation of this memoir, together with 'a biography of Carnot and an account of his theory', the latter by Lord Kelvin, have been published under the title *Reflections on the Motive Power of Heat*, ed. by Thurston (Wiley, 1897).

† William Thomson, who became Lord Kelvin in 1892, was one of the great English physicists of the nineteenth century. His first ideas regarding the thermodynamic scales were published in 1848 but were vastly improved upon in 1851.

or Eq. (5)′

$$t = \frac{(V_t - V_0)}{\alpha V_0} \tag{5'}$$

when applied to a thermometer containing an *ideal* gas, may, for our purposes, be regarded as the *definition of temperature on the thermodynamic scale*. Since, however, an ideal gas thermometer *cannot be realistic in practice*, it becomes of importance to see how much the temperatures determined by a thermometer containing a *real* gas differ from the absolute temperature. With this object in mind, *Kelvin* devised his famous porous plug experiment (which will be given below). The work was carried out in conjunction with *Joule* and, finally, resulted in showing that hydrogen behaves so nearly like an ideal gas that the *hydrogen scale* may be taken for all practical purposes as agreeing exactly with the *absolute scale* at ordinary temperatures.*

6. The absolute zero and the absolute temperature. Let us write Eq. (8) of Section 4, namely,

$$PV = P_0 V_0 (1 + \beta t)$$

in the form

$$PV = \frac{P_0 V_0 \left(\frac{1}{\beta} + t\right)}{\frac{1}{\beta}}. \tag{1}$$

By putting $(1/\beta) + t = T$, we introduce a temperature *T*, corresponding to *t* but referred to a zero point lying $\frac{1}{\beta}$ below the ice point. This zero point is called the *absolute zero*,† and temperatures reckoned from it are termed *absolute temperatures*. The *centigrade absolute scale* is commonly called the *Kelvin scale*, and is referred to by the abbreviation *TK*. On the *ordinary centigrade scale*, the temperature of the ice point is $t_0 = 0°C$, but on the *Kelvin scale*, it is found to be‡ $T_0 = 1/\beta = 273.18K$; in other words, it turns out that the pressure coefficient β for an ideal gas is $1/273.18$ and hence the *absolute zero* corresponds to $-273.18°C$.

The absolute zero and absolute temperature are concepts that greatly facilitate the mathematical expression and application of the gas laws and should be treated as such and

---

* Data on the departure of various gas scales from the thermodynamic scale are given in the *International Critical Tables* (1926), Vol. 1, p. 53.

† For an interesting history of the notion of the absolute zero, see the *collected papers* of Sir *Janes Dewar* (Cambridge University Press, 1927, vol. II, pp. 768–775). There due credit is given to the French physicist *Amontons* for having had the conception of an absolute zero as early as 1703 and for having calculated its value as – 240°C. Note, however, that the concept of an absolute zero arises merely from the way in which temperature has been defined and from the behaviour of an ideal gas, and that it can be entirely avoided simply by defining temperature in a different way.

‡ The latest values reported from Leiden and from Berlin are 273.144 K and 273.16 K, respectively.

nothing else. If we attempt to apply the gas laws at the *absolute zero*, we are led to the *meaningless equation* $PV = 0$, from which we deduce that at absolute zero, a gas either has no volume or exerts no pressure. We know, however, that all gases liquefy before this point is reached, so at the absolute zero, the gaseous state probably no longer exists.

In practice, many attempts have been made to attain temperatures close to the absolute zero. Up to 1877, the lowest temperature attained was - 110°C, produced by *Michael Faraday*, by the rapid evaporation in vacuum of ether and solid carbon dioxide, a mixture that ordinarily has a temperature of about $-80\,^{\circ}C$. In 1877, a Swiss, *Raoul Pictet*, and a Frenchman, *Louis Cailletet*, independently liquefied oxygen, which has a boiling point at $-182.97^{\circ}\ C$. But as neither of these experimenters obtained the liquid oxygen in a *static condition*, they could make no observation of its temperature. The lowest measured temperature was $-140\,^{\circ}C$, obtained by *Pictet*, by the rapid evaporation of nitrous oxide. Following these, the two poles, *Wróblewski* and *Karol Olszewski*, and the Britisher *James Dewar* (1842–1923) accomplished the liquefaction of oxygen, nitrogen, and air in quantity, and by their evaporation in vacuum produced and measured temperatures as low as $-210^{\circ}C$. In 1885, *Olszewski* liquefied hydrogen and located its boiling point as - 243.5°C. Hydrogen has since not only been liquefied in quantity (1898) but also solidified by *Dewar* (1900); the boiling and melting points of hydrogen are - 252.7°C and $-259\ .1^{\circ}C$, respectively, according to recent determinations. Of all the gases, helium was the last to resist liquefaction and solidification. In 1908, *Kamerlingth Onnes* (1853–1926), working in his celebrated cryogenic laboratory at Leiden, accomplished its liquefaction at a temperature of only 4.3°C above the absolute zero. *Onnes* made an unsuccessful attempt in 1921 to solidify helium but did succeed in reaching a temperature of 0.82K by evaporating the liquid in high vacuum. Four months after the death of *Onnes*, in 1926, helium was successfully solidified in the same laboratory, by placing it in a brass tube in a helium bath and subjecting it to pressure. In this work, temperatures as low as 0.71K were reached. Recently *Debye* and *Giauque* have suggested a method of reaching very low temperatures by means of the *adiabatic demagnetisation* of certain magnetic substances kept at the temperature of liquid helium. The essential idea is that if *heat is prevented from entering the substance while it is being demagnetised, the work of demagnetisation must be furnished by the substance itself, with a consequent lowering of its temperature.* This method has been applied with success at the University of California and at Leiden in 1933 and at Oxford in 1934. At Leiden, temperature of less than 0.005K is actually been reached.

7. The gas constants In view of the definition of absolute temperature, we may write Eq. (1) of Section 6, namely,

$$PV = \frac{P_0 V_0 (1/\beta + t)}{1/\beta} \tag{1}$$

in the form

$$\frac{PV}{T} = \frac{P_0 V_0}{T_0}. \tag{1'}$$

But V_0 is equal to m/ρ_0, where *m* is the mass of the gas, and ρ_0 is its density at the ice point T_0; hence Eq. (1′) can be rewritten in the form

$$\frac{PV}{T} = \frac{P_0 m}{\rho_0 T_0} = m\left(\frac{P_0}{\rho_0 T_0}\right)$$

or

$$\frac{PV}{T} = mR'. \qquad (4)$$

The constant R', which has made its appearance here in the equation of state of an ideal gas, can be obviously interpreted as the value of PV/T, or of $P/\rho T$, for *unit mass* of the particular gas considered, It is, therefore, called the *gas constant for unit mass*. The fact that its value is different for different gases should be noted, as it is expressed by

$$R' = \frac{P_0}{\rho_0 T_0}. \qquad (5)$$

Example:

Compute the value of the gas constant R' for air, expressing it in the cgs units. The data you are supplied with are: dry air at $T_0 = 273$ and 1 atm $= 1.0132 \times 10^6$ dyne/cm^2, has a density $\rho_0 = 0.0012930$ g/cm^2.

Solution:

Using Eq. (5), we obtain

$$R' = \frac{P_0}{\rho_0 T_0} = \frac{1.0132 \times 10^6}{1.2930 \times 10^{-3} \times 273}$$

$$\simeq 2.87 \times 10^6 \text{ erg/deg/g}.$$

In order to obtain a final form of the equation of state of an ideal gas, one must take into account a thoughtful guess made in 1811 by the Italian physicist *Amedeo Avogadro* to the effect that *equal volumes of different gases at the same temperature and pressure contain equal numbers of molecules.* This important law, now known to be exact for *ideal gases,* although approximately true for real ones, leads immediately to the conclusion that *moles* or *gram molecular weights* of the various gases will occupy the same volume at the same temperature and pressure. Consequently, if M_1, be the number of grams in a mole of one kind of a gas, M_2, that of another kind, etc, from Eq. (4), one has the following equation by replacing the mass *m* of a gas by its molecular weight in grams, that is

$$\frac{PV}{T} = M_1 R_1' = M_2 R_2' = \ldots = R, \qquad (6)$$

where *V is the volume of a mole* and *R* is a new constant called the *gas constant per mole.* Evidently *R* is a *universal constant*, that is, it is the same for all gases, its numerical value depending only upon the units in which *P*, *V*, and *T* are expressed. Its value in cgs units, as computed from experimental data, is $R = (8.313 \pm 0.001) \times 10^7$ erg/deg/mole. When the pressure is expressed in atmospheres, *R* has the easily remembered numerical value of 82 c.c.atm/deg/mole (approximately).

If we put the gas constant per unit mass, $R' = R/M$, in Eq. (4), namely,

$$\frac{PV}{T} = mR', \tag{4}$$

we produce

$$\frac{PV}{T} = \frac{m}{M}R.$$

But $\dfrac{m}{M} = \dfrac{\text{weight}}{\text{molecular weight}}$ represents the number of moles of the gas used, which is denoted by *n*, therefore, Eq. (8) can be rewritten in the form

$$PV = nRT. \tag{7}$$

Eq. (7) is the *equation of state of an ideal gas* in the final form that is commonly used in physical science and in engineering, *V* being the volume of *n* moles.

8. Model of an ideal gas. The fact that at very low pressures, the physical behaviour of all gases is represented by simple laws indicates that all have a common and simple structure, and thus, the construction of a *model* of an ideal gas that would yield these laws became the first objective in the kinetic theory. In the light of the atomic molecular theory of chemistry and the experiments on the nature of heat, which showed heat to be simply the energy involved in the random motions of molecules, the following postulates came to be formulated:

1. *A chemically homogeneous gas is composed of identical molecules which are moving in a random fashion that the number moving in one direction is on the average the same as that moving in any other.*
2. *The actual space occupied by the molecules is negligible in comparison with the space between them.*
3. *The molecules exert no forces on one another except when they actually collide; that is, their velocities are changed only by collision with other molecules or with the walls of the containing vessel.*
4. *The impacts between molecules and with the walls must be perfectly elastic; otherwise, there would be a continual loss of energy as time progresses.* It would be better to say that, *on the average, the impacts must be perfectly elastic*

5.2. The properties of vapours

1. Variations of the product *PV* with the pressure *P*. To just what extent the ideal gas laws, which we have been considering, failed to describe the behaviour of the gases found in nature was investigated first by the most skilled and patient experimenter *Regnault,* in 1847, and later by *Amagat,* in 1899. *Boyle* himself had noticed small departures from *constancy* of the product of pressure and volume, but he ignored the failure of his law to hold accurately because of its simplicity and general usefulness. Now if Boyle's equation, PV = constant, actually did apply accurately to a real gas, the curve connecting PV and P, for a given mass of gas kept at a constant temperature, would be a *straight* line parallel to the axis of pressures. Regnault and Amagat, however, found that when a gas is tested over a wide range of pressure and at different temperatures, curves like those of Figs. 3 and 4 are obtained. In general, the value of *PV* diminishes at first, but when

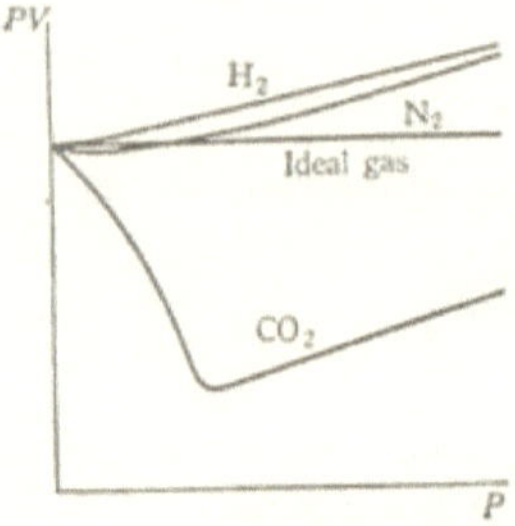

Fig. 3. Variations of the product *PV* with pressure for a given mass of various gases kept at constant temperature.

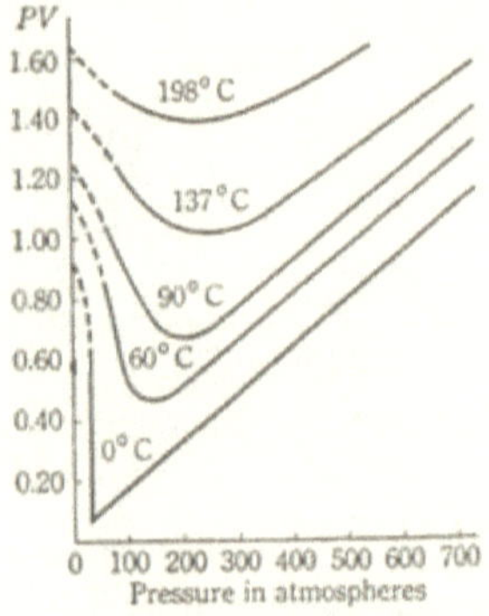

Fig. 4. Variations of *PV* with pressure for a given mass of carbon dioxide at various constant temperatures.

the pressure reaches a certain value, which depends on both the gas and the temperature, it steadily increases. It will also be observed that as the temperature increases, the marked drops in the curves disappear (Fig. 4). At high enough temperatures, depending on the gas, the curves are found to show only an upward slope. In the case of hydrogen (Fig. 3) and helium, this is true even at room temperature.

In considering experiments such as these, naturally there arises the question of what happens to the properties of a gas when it approaches close to the liquid state. As early as 1822, *Charles Cagniard De La Tour* had observed that when a liquid was heated in a hermetically sealed tube, it evaporated silently as the temperature rose up to a certain point, and then the boundary between liquid and gas grew indistinct, and then it completely disappeared. The densities of the liquid and gas had become the same, and the distinction between the two states of aggregation had vanished. A more thorough study of this effect was made later by *Thomas Andrews* (1813–1885) during the course of his classical experiments on the behaviour of carbon dioxide under pressure at different temperatures. *Andrews* had begun these experiments in an attempt to solve one of the great problems of his period, the liquefaction of what were then called the 'permanent gases', an extensive study of which had been made by *Faraday* from 1823 to 1845.

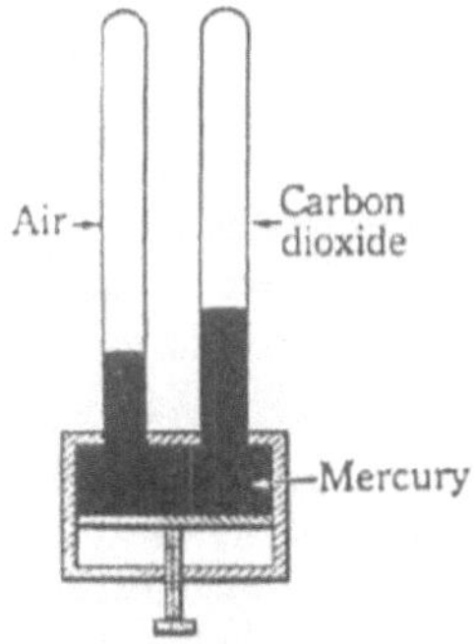

Fig.5. Schematic diagram of ANDREWS'S apparatus. The gas is compressed by forcing the mercury up into the tubes. One of the tubes contains air, which is used to measure the pressure exerted on the carbon dioxide.

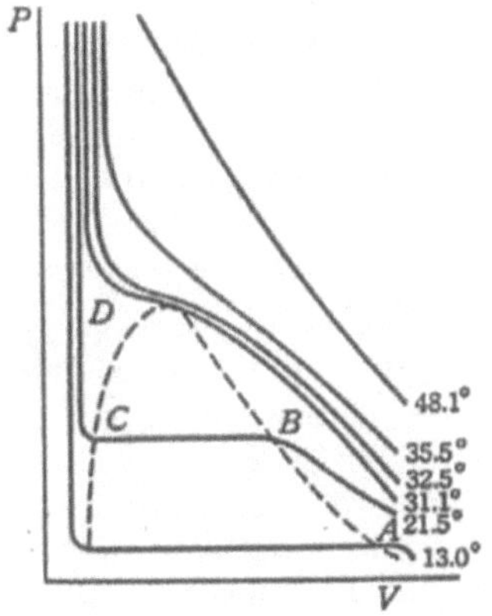

Fig 6. Andrews' experimental isotherms for a given mass of carbon dioxide.

2. Andrews's experiments. The apparatus used by Andrews is shown in Fig. 5, and results, typical of the kind which he obtained, are illustrated by the isothermal curves in Fig. 6. It is seen that the curve for 48.1°C is nearly a perfect hyperbola. So at this temperature, carbon dioxide behaves almost like an ideal gas. But at lower temperatures, the deviations from Boyle's law become greater, and the curves change in character, taking a form similar to that for the isotherm marked 32.5°C. In this latter curve, a jog is apparent, and for the higher pressures, the curve is very steep. Indeed, in this respect, the curve resembles the isotherm for a liquid, since a relatively large increase in pressure causes only a small decrease in volume. Yet no sign of liquefaction of the carbon dioxide could be noted at this temperature or above it, however much the pressure was increased.

But for temperatures below 31.1°C, a radical change occurs in the nature of the isotherms. Taking the curve for 21.5°C and starting from the point A, we see that as the pressure is increased, the volume diminishes rapidly until the point B is reached; between A and B we are dealing with a gas. After reaching the point B, if an attempt be made to compress the gas further, it is found that some of it liquefies and a considerable change of volume takes place with no change of pressure. In other words, in the portion BC, gas and liquid exist together with a visible meniscus between them, more and more liquid being formed as C is approached. At C, the carbon dioxide is again homogeneous – all of it now being in the liquid state – beyond this point, the gradient is very steep, since a liquid is not easily compressed.

3. The critical constants of a substance. It will be seen from Fig. 6 that a smaller contraction in volume upon liquefaction takes place at 21.5°C than at 13.0°C, and thus, the difference in density between the liquid and its vapour will be smaller, the higher the temperature. When a temperature of 31.1°C is reached, the horizontal part of the isotherm disappear altogether, and no separation into liquid and vapour can be effected, how much ever the pressure is increased. The temperature at which this occurs is called the *critical temperature* T_c, and the pressure that is just sufficient to liquefy the gas at its critical temperature is called the *critical pressure* P_c. Above the critical temperature, a gas cannot be liquefied by pressure alone; that is to say, a separation of the gaseous and liquid states cannot be effected. Indeed, it is possible to pass from any point A, where the substance would undoubtedly be regarded as a gas, to a point D, where it is in the dense state, almost an incompressible condition that one naturally call as liquid, without having the liquid distinct from the vapour at any time. To do this, it is only necessary to vary the pressure, temperature, and volume in such a way as not to pass through the region bounded by the dotted curve, inside which alone heterogeneity is possible. In other words, above the critical temperature, the two states become identical, a property which is referred to as the *continuity of the liquid and gaseous states*.

The properties represented by the isotherms for carbon dioxide (Fig. 6) are characteristic of all substances that have been studied, but the critical constants differ widely. Thus the critical temperature of water is 374°C; of air it is $-141°C$; and of helium it is $-268°C$. It is customary to give the name *vapour* to a substance in the gaseous condition when it is below its critical temperature and to confine the term *gas* to a substance, it should be above the critical temperature. This distinction is not important, however.

4. Finite volume of the molecules. Since the liquid and gaseous states are continuous, it should be possible to obtain a general equation of state concerning the pressure, volume, and temperature that will apply to a substance whether it is in the liquid or the gaseous state. The simple equation $PV = nRT$ obviously will not hold for this more general case, but it might be possible to modify this equation by removing some of the simplifying assumptions used in deriving it in the kinetic theory. A hint as to one of the ways in which this can be done is furnished by the nature of the curves in Fig. 4. It is found that for very high pressures, these curves became, for any given substance, a system of *sensibly parallel lines*. The equation for any one of these lines evidently will be $PV = bP + c$, where the quantity b, which is the slope of the curves, is found to depend upon the nature of the substance, and the quantity c, which is the PV-intercept, depends upon the temperature. By writing this equation in the form

$$P(V-b) = c,$$

one sees that P becomes *infinite* for $V = b$, and hence, b can be interpreted as the *least volume into which the substance can be compressed.* Thus $V - b$ is the whole space in which the gas is enclosed, diminished by the least volume of the substance. If $V - b$, rather than V, be considered as the volume of the gas, then Boyle's law may be said to apply to all gases even at high pressures.

From the point of view of kinetic theory, as *Clausius* himself recognised, $V - b$ is the volume which is available for the free motion of the molecules. In other words, the assumption that the volume occupied by the molecules themselves is negligible compared with the space between them is not exactly true for actual gases.

5. Influence of intermolecular actions; the Joule-Kelvin experiment. In attempting to account theoretically for the departure of actual gases from the laws of ideal gases exhibited in Figs. 4 and 6, one must also take into account the fact that the forces which the molecules exert on one another will not be entirely negligible. In the case of liquids and solids, the existence of such forces is sufficiently evident to be familiar to every one. The first attempt to detect their existence between gas molecules were made by *Gay-Lussac,* in 1807, and later, by *Joule,* in 1845. Based on his observation that no change in temperature had taken place in the free expansion of air from a pressure of 22 atmospheres into a vacuum, Joule concluded that if intermolecular forces exist, his experiments were too insensitive to detect them. Lord *Kelvin* thought that this matter should be tested by a better method, and this led him to devise his classical porous plug experiment, which he carried out in conjunction with Joule, in 1852. The improvement consisted in

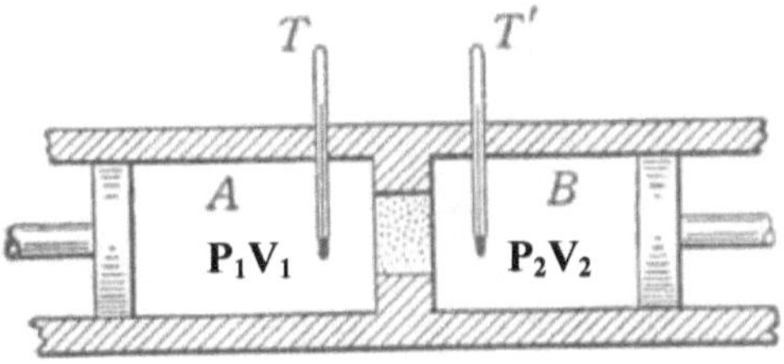

Fig. 7. The porous plug experiment.

making the expansion of the gas continuous instead of intermittent. The gas, which was kept at constant pressure by a pump, was allowed to flow continuously through *a porous plug of tightly packed cotton* as shown diagrammatically in Fig. 7. In this way, the resistance offered by the plug to the flow of the gas was so large that the kinetic energy of the flow was negligible, and consequently, the gas merely expanded through the plug. The lowering of temperature which was actually found for all gases, except hydrogen and helium at ordinary temperatures, can be explained by assuming that there is a very slight *attraction* between the gas molecules. So there is an increase of *potential energy* upon expansion, which takes place at the expense of its *kinetic energy,* since the system is thermally insulated. The fall of temperature was found to be nearly proportional to the difference between the pressures on the two sides of the plug and to increase as the initial temperature of the gas decreases. These results are of great importance, because from them can be calculated the corrections necessary to reduce the readings of a gas thermometer to absolute thermodynamic temperatures. They also form the basis of the method developed by *Karl Ritter von Linde,* in 1895, for liquefying air and the other so-called permanent gases on a large scale for commercial purposes. It has been found that even hydrogen and helium are cooled by their own expansion if their initial temperatures are below a certain value called the *temperature of inversion,* which is $-80.5°C$ for hydrogen and about $-238°C$ for helium. This fact is of great importance in the liquefaction of these gases.

In order to explain the *cooling* that was observed during the expansion of all gases except hydrogen and helium at ordinary temperatures, it is necessary to analyse this experiment in detail.

Let $W_t + W_i$ be, for the moment, the total *internal* energy per mole of the gas (kinetic plus potential) as it enters the plug, shown diagrammatically in Fig. 7, and let

$W_t' + W_i'$ be its total *internal* energy as it leaves the plug. Similarly, let P_1 and V_1 be the pressure and the volume per mole of the gas on side A, and P_2 and V_2 be the pressure and the volume per mole on side B. If no heat enters or leaves the gas from or to the surroundings, and if the speeds of inflow and outflow of the gas are equal, then the decrease in total internal energy per mole in passing through the plug must be equal to the external work per mole done *by* the gas, or

$$(W_t + W_i) - (W_t' + W_i') = P_2 V_2 - P_1 V_1 .$$

i. If Boyle's law holds so that

$$P_1 V_1 = P_2 V_2 ,$$

then

$$W_t + W_i = W_t' + W_i'$$

Consequently, if there are no forces between the molecules so that

$$W_i = W_i' ,$$

then also

$$W_t = W_t' ,$$

and there will be no temperature change. But if there are intermolecular forces of any sort whatever, the potential energy, W_i, must change on expansion, and W_i will not equal W_i'. Consequently, W_t cannot equal W_t', and the temperature, which is a function of W_t, must change. When the intermolecular forces are attractive, the gas must cool on passing through the plug.

ii. If, however, the product *PV* increases as the pressure is reduced, as Amagat found – it did up to a certain point for all gases except hydrogen and helium (Fig. 3) – then $P_2 V_2$ is greater than $P_1 V_1$, and $W_t + W_i$ must be greater than $W_t' + W_i'$. In this case, we should expect a cooling even if there are no intermolecular forces.

The deviations from Boyle's law, caused by the finite size of the molecules thus produced in this experiment temperature, changes and which must be added (algebraically, of course) to the cooling produced by the inner work done by the gas against intermolecular attractions. In the case of hydrogen and helium, at ordinary temperatures, the heating due to this cause, where the external work done on the gas is greater than that done by the gas, is sufficient to produce a *rise* in temperature since the intermolecular forces are small. At sufficiently low temperatures, however, the cooling effect predominates.

The Joule-Kelvin effect is of great importance for the following reasons.

1. All the methods of liquefaction of air and the other so-called permanent gases are based on it.
2. This effect, together with the entropy function which stemmed from Carnot's theorem, had led Lord Kelvin to establish his suggested absolute thermodynamic scale of temperature, by proving in an analytical way that the temperature T_o of ice point which is arbitrarily taken, on the centigrade scale, to be equal to 0°C, is on the

Kelvin scale, equal to reciprocal of the coefficient of expansion, or expansitivity, α of an ideal gas. That is

$$T_0 = \frac{1}{\alpha},$$
$$= 273.16\ \mathrm{K},$$

α being equal to $1/273.16$.

This important result which had been achieved by Lord Kelvin can be arrived at as will be shown below in Section 5.4.

It having been established that intermolecular forces exist, the next question is how they affect the pressure and volume of a gas. In the interior of the gas, the resultant effect is indeed negligible, for the molecules are attracted equally in all directions by the other molecules. But not so in the layers next to the walls of the containing vessel, where the resultant attraction is directed inward towards gas, thus tending to *reduce* the volume occupied by the gas, just as an increase of pressure would do. The effect of molecular attraction may therefore be represented by adding to the pressure *P*, which is applied to the gas externally, a quantity P', representing the pressure due to internal attraction. Our gas equation obtained above, namely,

$$P(V - b) = c$$

then becomes

$$(P + P')(V - b) = nRT. \quad (1)$$

6. Equations of state of a fluid. Van der Waals' equation. Of the numerous attempts that have been made to deduce a general equation holding for any substance throughout the liquid and gaseous states, the most celebrated is that developed by *Johannes Diderik Van Der Waals* (1837–1923). His equation may be derived from Eq. (1), in the previous Section, by assuming that the quantity P' is proportional both to the *number of molecules striking a unit area of the wall in unit time* and *to the number of molecules attracting any molecule.* Since both of these factors are proportional to the density of the fluid, P' will vary directly as the *square of the density* or *inversely as the square of the volume*. Hence, denoting P' by a/V^2, Eq. (1) in Section 5 takes the form

$$\left(P + \frac{a}{V^2}\right)(V - b) = nRT. \quad (1)$$

This is known as the *van der Waals equation.* The coefficients *a* and *b* depend upon the amount of gas as well as upon its nature, *a* being proportional to the square, and *b* to the first power, of the *mass of the gas.** When one considers the simplicity of this equation, its general agreement with experiment is remarkable. *Clausius*, *C. Dieterici*, and many others have proposed equations that are improvements in one respect or another, on the van der Waals equation. But the use of the equations in practical applications introduces complications that tend to offset any advantages gained.

---

* It simplifies the use of the van der Waals's equation if the pressure *P* is expressed in atmospheres and the volume *V* as the ratio between the actual volume and the volume which the substance would occupy if it were an ideal gas at the normal conditions, which are 0°C and 1 atmosphere.

7. Solutions of the van der Waals equation. Since this is cubic equation in *V*, it can have three roots, that is, a given value of *P* may correspond to three values of *V*, this will be more evident if we refer to Fig. 8, as compared with Andrews' experimental isotherms shown previously in Fig. 6, each of which representing the isotherms of carbon dioxide.

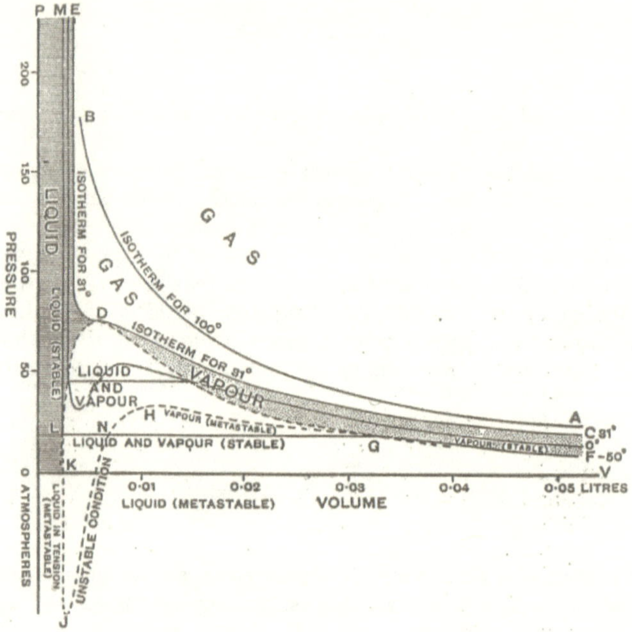

Fig. 8. Features of the theoretical isotherms of van der Waals equation for carbon dioxide.

It will be seen from this figure that the three values of *V* for a given value of *P* at $-50\,°C$ are *L*, *N*, and *G*. If *a* and *b* are fixed, then for low values of *T* the loops LJN and NHG (which should be equal in area) are large, and the three values of *V* differ widely from one another. As *T* becomes larger, the loops become smaller; and finally there is one particular value of *T* at which the loops just vanish, and the three values of *V* become identical. For still higher values of *T*, there is only one real root of the equation; the other two values of *V* are described mathematically as 'imaginary' quantities and have no physical significance.

As pointed out before (Fig. 6), that when a temperature of 31.1°C is reached, the horizontal part of the isotherm has disappeared altogether, and that the temperature at which this occurs is called the critical temperature. Obviously, the critical isotherm can be regarded as the one which for a particular value of *P*, namely, the critical pressure P_c, gives three equal values of *V*, all equal to the critical volume V_c. Expressed algebraically, this means that at the critical point, van der Waals equation

$$\left(P + \frac{a}{V^2}\right)(V - b) = RT \text{ per mole} \qquad (1)$$

must become identical with the particular solution represented by

$$(V - V_c)^3 = 0. \quad (2)$$

For the purpose of comparing the two equations, we must expand Eq. (2) as in Eq. (3).

$$V^3 - V_c V^2 + 3V_c^2 V - V_c^3 = 0 \quad (3)$$

We must also insert the special value $T = T_c$, and $P = P_c$ in the expanded form (1) of van der Waals equation, which then assumes the form

$$V^3 - \left(\frac{RT_c}{P_c} + b\right)V^2 + \frac{aV}{P_c} - \frac{ab}{P_c} = 0. \quad (4)$$

We can now equate coefficients of equal powers of *V* in Eqs. (3) and (4), and deduce the following relations:

Equating the coefficients of V^2

$$3V_c = \frac{RT_c}{P_c} + b. \quad (5)$$

Equating the coefficients of *V*

$$3V_c^2 = \frac{a}{P_c}. \quad (6)$$

Equating the coefficients of the terms not containing *V*

$$V_c^3 = \frac{ab}{P_c}. \quad (7)$$

From these three equations, the values of the *critical constants* in terms of *a* and *b* can be deduced as follows

i. Dividing Eq. (7) by Eq. (6), we obtain

$$V_c = 3b, \quad (8)$$

that is, the *critical volume* is equal to 3*b*; or conversely, van der Waals constant b is 1/3 of the critical volume.

ii. Substituting this value of V_c in Eq. (6), we find that we can express the *critical pressure* in terms of the two constants of van der Waals equation as follows:

$$P_c = \frac{a}{27b^2}. \quad (9)$$

iii. Finally, by substituting P_c and V_c in Eq. (5), it is found that the *critical temperature* is given by the expression

$$T_c = \frac{8a}{27Rb}. \quad (10)$$

Hence, if *a* and *b* are determined from measurements on compressed gases, they can be used to predict the critical constants. The converse process can also be used in order to deduce the values of *a* and *b* from the measured critical constants. Thus from Eqs. (8), (9), and (10), it is readily shown that

$$b = \frac{1}{3}V_c = \frac{RT_c}{8P_c}, \tag{11}$$

and

$$a = 3\,P_c V_c^2 = \frac{9RV_cT_c}{8} = \frac{27R^2T_c^2}{64P_c} \tag{12}$$

It may be added here that van der Waals equation is not numerically exact. This can be seen from the following:

An important general relation is revealed when the values of the critical constants given by Eqs. (8), (9), and (10) are used to calculate the ratio RT_c/P_cV_c, thus

$$\frac{RT_c}{P_cV_c} = \frac{R \times 8a \times 27b^2}{27Rb \times 3b \times a} = \frac{8}{3} = 2.67.$$

If, therefore, van der Waals equation give a true representation of the transition from the liquid to the gaseous state, this ratio should be constant for all substances and should have the value 2.67. It was found experimentally that this ratio is nearly constant for many substances, but it usually has the value 3.7, which is much larger than the figure predicted by van der Waals equation. There are two groups of substances which give abnormal values. The ratio of hydrogen and helium is lower, but it is difficult to determine V_c for these substances. Water, alcohols, and acetic acid – all give values higher than 3.7. They all belong to the group of 'associated' liquids, and high values of the ratio RT_c/P_cV_c are probably the characteristic of this group.

A further consideration of the van der Waals constants *a* and *b* for a gas, concerning the role they interplay in producing a temperature change or not in the Joule-Kelvin experiment, shall be taken into account below, when an application of the Joule--Kelvin coefficient $_\mu$ is applied to a van der Waals gas.

5.3. The Joule-Kelvin coefficient

A. Expressions and B. Applications.

A. Expressions.

1. The first expression for the Joule-Kelvin coefficient. A description and an analysis in detail of the porous plug experiment carried out by Kelvin and Joule together, to detect whether or not there is a variation of the energy of a gas with its pressure (or volume) have been given above (Section 5). On the other hand, by applying the analytical formulation of the first law for a finite change,

$$q = \Delta U + W,$$

in this experiment, we have $q = 0$ and $W = P_2 V_2 - P_1 V_1$ by assuming that the *external* work done by the gas against the surroundings is only mechanical work (i.e. pressure–volume work) so that this fundamental equation can be rewritten as

$$0 = (U_2 - U_1) + (P_2 V_2 - P_1 V_1).$$

Therefore,

$$(U_1 + P_1 V_1) = (U_2 + P_2 V_2),$$

or

$$H_1 = H_2.$$

This indicates that the heat content or enthalpy of the gas does not alter on passing through the plug; that is, the effect of change of pressure and temperature on the enthalpy *H* compensate each other. Therefore, this porous plug experiment is *isoenthalpic*.

We may put

$$H = H(P, T),$$

so

$$dH = \left(\frac{\partial H}{\partial T}\right)_P dT + \left(\frac{\partial H}{\partial P}\right)_T dP,$$

and subsequently in the case considered,

$$\left(\frac{\partial H}{\partial T}\right)_P dT = \left(\frac{\partial H}{\partial P}\right)_T dP = 0,$$

or

$$\left(\frac{\partial T}{\partial P}\right)_H = -\frac{(\partial H/\partial P)_T}{(\partial H/\partial T)_P}$$

$$= -\frac{\{\partial (U + PV)/\partial P\}_T}{C_P}.$$

Therefore,

$$\left(\frac{\partial T}{\partial P}\right)_H = -\frac{1}{C_p}\cdot\left(\frac{\partial U}{\partial P}\right)_T - \frac{1}{C_P}\cdot\left\{\frac{\partial (PV)}{\partial P}\right\}_T. \qquad (1)$$

The first term on the right-hand side of this equation is of positive sign. This is because $(\partial U/\partial P)_T$ is always negative, owing to the fact that the work done by the gas against the attractive forces between the molecules when the pressure is decreased (or the volume is increased) is associated with an increase in the potential energy of the gas, and since this negativity is preceded by a negative sign, the first term is always positive. The energy required for this *inner work* is supplied by its kinetic energy because the gas is thermally insulated, and thus the factor $\partial U/\partial P$ gives rise to a cooling effect when the gas is expanded. The second term represents the work done in the expansion, owing to the change in *PV*. For most gases at ordinary temperatures, except for hydrogen and helium, *PV* first diminishes and then increases as the pressure is increased at a given temperature, as noted before. Consequently, in such cases at moderate pressures when *PV* increases as *P* decreases, that is, the factor $\partial(PV)/\partial P$ is negative, or P_2V_2 greater than P_1V_1, work is done by the gas, which again due to the thermal insulation, will be at the expense of kinetic energy. Thus the second term on the right-hand side of the above equation is of positive sign, and accordingly, in such cases, when the gas expands through the plug a cooling effect is produced, this cooling being due to the external work done by the gas when the product PV increases with decrease of pressure. Finally, In the case of hydrogen and helium (except at very low temperatures), *PV* increases with increase of pressure, and the result of expansion is a heating effect which may be greater than the cooling effect of the first term.

The term $\left(\frac{\partial T}{\partial P}\right)_H$ on the left-hand side of the above equation is known as the *Joule-Kelvin coefficient and denoted by* μ. It is to be noted that expressions for μ by equations other than this equation, can be obtained as will be shown below. Such expressions are useful in the sense that they can be employed either to predict the value of μ for a given gas provided all terms in the equation applied are measurable, or conversely, to deduce the value of any of the factors or partial derivatives included in the applied equation, which is not measurable, by using the experimental value of the coefficient μ of the gas considered.

2. The second expression for the Joule-Kelvin coefficient. Starting with the first expression we have obtained above,

$$\left(\frac{\partial T}{\partial P}\right)_H = -\frac{1}{C_P}\left(\frac{\partial U}{\partial P}\right)_T - \frac{1}{C_P}\left\{\frac{\partial(PV)}{\partial P}\right\}_T , \qquad (1)$$

we can express the partial derivative $\left(\frac{\partial U}{\partial P}\right)_T$ as

$$\left(\frac{\partial U}{\partial P}\right)_T = \left(\frac{\partial U}{\partial V}\right)_T \cdot \left(\frac{\partial V}{\partial P}\right)_T ,$$

so that we have

$$\left(\frac{\partial T}{\partial P}\right)_H = -\frac{1}{C_P}\left(\frac{\partial U}{\partial V}\right)_T \cdot \left(\frac{\partial V}{\partial P}\right)_T - \frac{1}{C_P}\left\{\frac{\partial(PV)}{\partial P}\right\}_T$$

or

$$-\mu C_P = \left(\frac{\partial U}{\partial V}\right)_T \cdot \left(\frac{\partial V}{\partial P}\right)_T + \left\{\frac{\partial(PV)}{\partial P}\right\}_T .$$

This is an alternative form of Eq. (1), which can be made use of, for calculating the partial derivative $(\partial U/\partial V)_T$ for an actual gas, the remaining terms in this form being measurable.

The second expression for μ can be obtained from Eq. (1) if we introduce the enthalpy change with pressure at constant temperature. In other words, if we can express the partial derivative of enthalpy with respect to pressure at constant temperature, $\left(\frac{\partial H}{\partial P}\right)_T$, in an equation similar to that of the derivative $\left(\frac{\partial T}{\partial P}\right)_H$. This can be achieved as follows:

Let *H* be a function of *P* and *T*,

$$H = H(P,T),$$

therefore,

$$dH = \left(\frac{\partial H}{\partial P}\right)_T dP = \left(\frac{\partial H}{\partial T}\right)_P dP.$$

Since $H = U + PV$, we have

$$dH = dU + PdV + VdP$$

or

$$dH - VdP = dU + PdV.$$

According to the first law,

$$d'q = dU + d'W$$

for an infinitesimal change. If $d'W$ represents only pressure-volume work done reversibly by the system at constant pressure, the analytical formulation of first law takes the form

$$d'q_{rev.} = dU + PdV.$$

But according to the second law, we have

$$d'q_{rev.} = TdS.$$

Therefore,

$$dH - VdP = d'q_{rev}$$

or

$$TdS = dH - VdP.$$

Hence,

$$dS = \frac{1}{T}dH - \frac{V}{T}dP.$$

Substituting for dH in this equation by the partial derivatives of *H* with respect to *P* and *T*, which are given above, we have

$$dS = \frac{1}{T}\left\{\left(\frac{\partial H}{\partial P}\right)_T dP + \left(\frac{\partial H}{\partial T}\right)_P dT\right\} - \frac{V}{T}dP,$$

$$= \frac{1}{T}\left\{\left(\frac{\partial H}{\partial P}\right)_T - V\right\}_T dP + \frac{1}{T}\left(\frac{\partial H}{\partial T}\right)_P dT.$$

It is shown in calculus (as noted before) that if a variable *x* is a function of two independent variables *y* and *z*, then

$$\frac{\partial}{\partial z}\left(\frac{\partial x}{\partial y}\right)_z = \frac{\partial}{\partial y}\left(\frac{\partial x}{\partial z}\right)_y$$

or

$$\frac{\partial^2 x}{\partial z\, dy} = \frac{\partial^2 x}{\partial y\, \partial z},$$

provided that the function x for an infinitesimal change can be represented by an exact differential, and that it is continuous and also its partial derivatives are continuous. Since these conditions are satisfied by the entropy function S and its partial derivatives, we have

$$\left[\frac{\partial}{\partial T}\left(\frac{\partial S}{\partial T}\right)_T\right]_P = \left[\frac{\partial}{\partial P}\left(\frac{\partial S}{\partial T}\right)_P\right]_T.$$

That is, the so-called mixed second-order partial derivatives are independent of the order of differentiation. Therefore, applying this principle to the dS equation obtained above, we have

$$\frac{\partial}{\partial T}\left[\left\{\frac{1}{T}\left(\frac{\partial H}{\partial P}\right)_T - \frac{V}{T}\right\}\right]_P = \frac{\partial}{\partial P}\left[\left\{\frac{1}{T}\left(\frac{\partial H}{\partial T}\right)_P\right\}\right]_T.$$

Then differentiating this equation, we obtain

$$\frac{1}{T}\left(\frac{\partial^2 H}{\partial T\,\partial P}\right)_P - \frac{1}{T^2}\left(\frac{\partial H}{\partial P}\right)_T + \frac{V}{T^2} - \frac{1}{T}\left(\frac{\partial V}{dT}\right)_P = \frac{1}{T}\left(\frac{\partial^2 H}{\partial P\,\partial T}\right)_T.$$

At the same time, what has been said for S can be said for H, that is,

$$\left(\frac{\partial^2 H}{\partial T\,\partial P}\right)_P = \left(\frac{\partial^2 H}{\partial P\,\partial T}\right)_T.$$

Accordingly, the last equation reduces to

$$-\frac{1}{T^2}\left(\frac{\partial H}{\partial P}\right)_T + \frac{V}{T^2} - \frac{1}{T}\left(\frac{\partial V}{\partial T}\right)_P = 0$$

or

$$\left(\frac{\partial H}{\partial P}\right)_T = V - T\left(\frac{\partial V}{\partial T}\right)_P.$$

This is the equation we are looking for, which is similar to the derivative $\left(\frac{\partial T}{\partial P}\right)_H$ as noted above.

Now use can be made of this equation to obtain the second expression for μ as follows: Again, let H be a function of P and T, so that we have

$$dH = \left(\frac{\partial H}{\partial P}\right)_T dP + \left(\frac{\partial H}{\partial T}\right)_P dP,$$

Then substituting for $\left(\frac{\partial H}{\partial P}\right)_T$ by the right-hand side of the last equation derived above, we have

$$dH = \left\{V - T\left(\frac{\partial V}{\partial T}\right)_P\right\}dP + C_P\,dT.$$

But in the Joule-Kelvin porous plug experiment, dH = 0, therefore,

$$C_P dT = -\left\{V - T\left(\frac{\partial V}{\partial T}\right)_P\right\}dP.$$

Hence,

$$\mu = \left(\frac{\partial T}{\partial P}\right)_H = \frac{\left\{T\left(\frac{\partial V}{\partial T}\right)_P - V\right\}}{C_P}. \qquad (2)$$

This is the second expression for μ .

3. The third expression for the Joule-Kelvin coefficient. From the preceding equation, a third expression for μ can be obtained as follows:

Put P as a function of P and T, that is,

$$P = P(V, T),$$

therefore,

$$dP = \left(\frac{\partial P}{\partial V}\right)_T dV + \left(\frac{\partial P}{\partial T}\right)_V dT$$

And at constant pressure, we obtain the relation

$$\left(\frac{\partial V}{\partial T}\right)_P = -\frac{\left(\frac{\partial P}{\partial T}\right)_V}{\left(\frac{\partial P}{\partial V}\right)_T}.$$

Then by substituting for $\left(\frac{\partial V}{\partial T}\right)_P$ in the numerator of expression (2) by the right-hand side of this equation, we produce

$$\mu = -\frac{\left[T\left\{\left(\frac{\partial P}{\partial T}\right)_V \Big/ \left(\frac{\partial P}{\partial V}\right)_T\right\} + V\right]}{C_P}.$$

Hence,

$$\mu = -\frac{\left\{V\left(\frac{\partial P}{\partial V}\right)_T + T\left(\frac{\partial P}{\partial T}\right)_V\right\}}{C_P\left(\frac{\partial P}{\partial V}\right)_T}. \qquad (3)$$

This is the third expression for μ .

Since the sign of the derivative $\left(\frac{\partial P}{\partial V}\right)_T$ is negative, then it follows that the sign of μ is that of the numerator.

From the above, it will be evident that the Joule-Kelvin coefficient, $\mu = \left(\frac{\partial T}{\partial P}\right)_H$, can be represented by the following three equations

$$\left(\frac{\partial T}{\partial P}\right)_H = -\frac{1}{C_P}\left(\frac{\partial U}{\partial V}\right)_T\left(\frac{\partial V}{\partial P}\right)_T - \frac{1}{C_P}\left\{\frac{\partial (PV)}{\partial P}\right\}_T. \qquad (1)$$

$$\left(\frac{\partial T}{\partial P}\right)_H = \frac{\left\{T\left(\frac{\partial V}{\partial T}\right)_P - V\right\}}{C_P}. \qquad (2)$$

$$\left(\frac{\partial T}{\partial P}\right)_H = -\frac{\left\{V\left(\frac{\partial P}{\partial V}\right)_T + T\left(\frac{\partial P}{\partial T}\right)_V\right\}}{C_P\left(\frac{\partial P}{\partial V}\right)_T}. \qquad (3)$$

It is to be noted that Eqs. (2) and (3) are usually used, for their convenience, to calculate the Joule-Kelvin coefficient μ for different gases, provided the terms on the right-hand side of these equations are measurable, or conversely, if one of these terms cannot be obtained experimentally it will be estimated from a measurement of the corresponding Joule-Kelvin coefficient. Before applying these equations to different gases, it seems worthwhile to express the partial derivative $\left(\frac{\partial S}{\partial P}\right)_T$ in standard form, for its important in our subsequent analysis.

We recall that in the derivation of the equation

$$\left(\frac{\partial H}{\partial P}\right)_T = V - T\left(\frac{\partial V}{\partial T}\right)_P,$$

which was shown above. We had first derived an equation expressing the exact differential d*S* in terms of the two independent variables *P* and *T*, namely

$$dS = \frac{1}{T}\left\{\left(\frac{\partial H}{\partial P}\right)_T - V\right\}dP + \frac{1}{T}\left(\frac{\partial H}{\partial T}\right)_P dT.$$

Now if we substitute for $\left(\frac{\partial H}{\partial P}\right)_T$ in this equation by the right-hand side of the preceding one, we have

$$dS = \frac{1}{T}\left\{V - T\left(\frac{\partial V}{\partial T}\right)_P - V\right\}dP + \frac{1}{T}\left(\frac{\partial H}{\partial T}\right)_P dT.$$

Hence,

$$\left(\frac{\partial S}{\partial P}\right)_T = -\left(\frac{\partial V}{\partial T}\right)_P.$$

From this equation it will be evident that the isothermal increase of pressure results in a decrease of entropy.

There is another method for deriving this relation which is much easier than the above method. It is based on the rule applied before; that is, since *G* and its partial derivatives are continuous, then when it is expressed in its exact differential form as

$$dG = -SdT + VdP,$$

where *T* and *P* are its natural or proper independent variables, it follows mathematically, as noted above, that the 'mixed second-order partial derivatives' are independent of the order of differentiation; that is

$$\frac{\partial^2 G}{\partial T\,\partial P} = \frac{\partial^2 G}{\partial P\,\partial T}.$$

Applying this to the above differential form of *G*, we first obtain the first-order partial derivatives

$$\left(\frac{\partial G}{\partial P}\right)_T = V \text{ and}$$

$$\left(\frac{\partial G}{\partial T}\right) = -S$$

and then the second-order partial derivatives,

$$\frac{\partial^2 G}{\partial T \partial P} = \left(\frac{\partial V}{\partial T}\right)_P \text{ and}$$

$$\frac{\partial^2 G}{\partial P \partial T} = -\left(\frac{\partial S}{\partial P}\right)_T,$$

so

$$\left(\frac{\partial S}{\partial P}\right)_T = -\left(\frac{\partial V}{\partial T}\right)_P.$$

Furthermore, the same result can also be obtained *immediately* by applying the condition for d*G* to be an exact differential to the coefficients on the right-hand side the equation $dG = -SdT + VdP$. This method was applied before for deriving *Maxwell's* relations*

* See Section 2.4.

B. Applications.

The application of the Joule-Kelvin coefficient equation to a van der Waals gas. We may choose Eq. (3) for this purpose, that is,

$$\mu = -\frac{\left\{V\left(\frac{\partial P}{\partial V}\right)_T + T\left(\frac{\partial P}{\partial T}\right)_V\right\}}{C_P\left(\frac{\partial P}{\partial V}\right)_T}. \qquad (1)$$

According to Van der Waals equation, the pressure P can be expressed as

$$P = \frac{RT}{V-b} - \frac{a}{V^2}, \qquad (2)$$

so by differentiation, we have

$$\left(\frac{\partial P}{\partial V}\right)_T = -\frac{RT}{(V-b)^2} + \frac{2a}{V^3} \qquad (3)$$

and

$$\left(\frac{\partial P}{\partial T}\right)_V = \frac{R}{V-b}. \qquad (4)$$

Substituting these two partial derivatives, the corresponding right-hand sides of Eqs. (3) and (4), in Eq. (1) and considering only the numerator, we get

$$\mu = -\frac{\left\{\frac{2a}{V^2} - \frac{RTb}{(V-b)^2}\right\}}{C_P\left(\frac{\partial P}{\partial V}\right)_T}. \qquad (5)$$

Clearly, the sign of μ in Eq. (5) is determined by the sign of the numerator term $\{2a/V^2 - RTb/(V-b)^2\}$, since the denominator is of negative sign preceded by a negative. It is to be pointed out that the *van der Waals constants a and b for a gas work in two opposite directions,* in the sense that while the constant *a* is a measure of the attraction between the molecules so that *higher* the value of *a, easier* will be the liquefaction, the constant *b* being a function of the volume of the gas molecules so that *smaller* the value of b, *easier* will be the liquefaction. This means that the inversion temperature T_i is directly proportional to the constant *a* and inversely proportional to the constant *b* of a given gas. The value of the inversion temperature of a real gas, which is defined as the temperature T_i at which the Joule-Kelvin coefficient μ is equal to zero, which is approximately given by

$$T_i = \frac{2a}{Rb}. \qquad (6)$$

Eq. (6) can simply be obtained from Eq. (5) by substituting, in the numerator, for T by T_i and neglecting the constant *b* with respect to the volume *V*.

Also Eq. (2) can be put in an approximated form by using the expansion

$$(1-\delta)^{-1} = 1 + \delta + \delta^2 \ldots,$$

where $\delta << 1$. Thus considering the denominator of the first term on the right hand-side of Eq. (2), we have the approximation

$$\frac{1}{V-b} = \frac{1}{V\,(1-b/V)} = \frac{1}{V}\left(1 + \frac{b}{V} + \frac{b^2}{V^2} + \ldots\right).$$

Substituting this result in Eq. (2) gives

$$P = \frac{RT}{V} + \frac{bRT}{V^2} - \frac{a}{V^2},$$

which is the approximated form of Eq. (2) we are looking for.

The substitution in Eq. (5) for *T* by a temperature less than T_i, that is, less than $2a/Rb$, and then by a temperature higher than $2a/Rb$, gives a positive sign in the farmer case and a negative sign in the latter. That is, a cooling effect is produced in the Joule-Kelvin experiment if the initial temperature of the gas is *less* than its inversion temperature, and a *heating* effect if it is *higher* than this temperature.

On the other hand, the application of Eq. (2) to an ideal gas will result in a zero value of μ; that is, neither cooling nor heating takes place in the Joule-Kelvin experiment if the gas was ideal, as can be seen from the following:

Since, for an ideal gas, the equation $PV = RT$ (per mole) holds, then by differentiation, we obtain $PdV + VdP = RdT$, so that

$\left(\frac{\partial P}{\partial V}\right)_T = -\frac{P}{V}$, and $\left(\frac{\partial P}{\partial T}\right)_V = \frac{R}{V}$, and by substituting this in Eq. (1), we have

$$\mu = -\frac{\left\{-\frac{P}{V} \times V + \frac{RT}{V}\right\}}{-C_P \times \frac{P}{V}} = -\frac{(-P+P)}{-C_P \times \frac{P}{V}} = 0. \qquad (9)$$

5.4. The realisation of the absolute thermodynamic scale of temperature by Lord Kelvin

The idea of absolute zero was discovered as a result of the law of thermal expansion of a gas at constant pressure

$$V = V_0\left(1 + \frac{t}{273}\right), \qquad (1)$$

where V_0 is the volume at $0\ ^\circ C$, and $\frac{1}{273}$ is the coefficient of thermal expansion (or expansivity) of the gas at constant pressure, and is usually denoted by α. In this relation if $t = -273\ ^\circ C$, *V* will be equal to zero. This temperature is, therefore, called *absolute zero*.

The law of isobaric thermal expansion as expressed by Eq. (1) was discovered by *Charles*. Evidently, it describes the manner in which the volume of a gas varies when it is heated at constant pressure. It expresses the fact that all gases expand and contract uniformly and to a similar extent when the temperature is changed. Thus, the volume of all gases increases by about 37 per cent of its value at 0°C between the freezing point and the boiling point of water.

Equation (1) can also be rewritten as

$$V = V_0 \ (1 + \alpha t),$$

or

$$V = V_0 \ \alpha \left(\frac{1}{\alpha} + t \right),$$

so by putting

$$\left(\frac{1}{\alpha} + t \right) = T,$$

we introduce a temperature T corresponding to t but referred to a new zero point lying $\frac{1}{\alpha}$ below the ice point. Since α is equal to 1/273, as pointed out above, then the zero point corresponds to $-273\ ^\circ C$. which is the *absolute zero*, and the temperature to $T = (t + 273)$, which is reckoned from the zero point, that is, the absolute zero is the *absolute temperature*.

Accordingly, Charles's law can be algebraically expressed in the following manner

$$V = k\ (\ t + 273) = kT, \qquad (2)$$

where the numerical value of the constant k depends upon (i) the quantity of the gas considered, (ii) the pressure, and (iii) the units in which V and T are expressed. Hence, Charles's law may be stated as follows:

> *At constant pressure, the volume of a given quantity of any gas is proportional to its absolute temperate.*

The centigrade absolute scale, whose zero point corresponds to $-273\ ^\circ C$ on the ordinary centigrade scale, is called the *Kelvin scale* or *the absolute thermodynamic scale*, and is referred to by the abbreviation TK.

Again, it may be pointed out as noted before that William Thomson (who became Lord Kelvin in 1892) made the brilliant suggestion in 1848 as a consequence of the remarkable theorem of *N. L. Carnot* published in 1824, which says

> *The efficiency of a heat engine working through any fixed small range of temperature is independent of the nature of the working substance used in the engine and is a function of temperature alone.*

That is, a temperature scale based on Carnot's engine would be realised. This new temperature scale is called *absolute* in the sense of being independent of any particular property of any particular thermometric substance, and is therefore, called the absolute thermodynamic scale.

The absolute zero and absolute temperature are concepts that greatly facilitate the mathematical expression and application of gas laws, and should be treated as such and nothing else. If we attempt to apply the gas laws at the absolute zero, we are led to the meaningless equation $PV = 0$, from which, as pointed out before, we deduce that at the absolute zero, a gas either has no volume or exerts no pressure; we know, however, that all gases liquefy before this point is reached, so at absolute zero, the gaseous state probably no longer exists.

It may be noted that until 1845, the absolute temperature scale, whose zero point is $-273\,^{\circ}C$, was defined in terms of a perfect gas thermometer. But practically, there is no real gas that approaches an ideal or perfect gas closely enough at practically useful pressures which should not be very low pressures to permit the construction of an accurate gas thermometer for the measurement of the absolute temperature.

On the other hand, Lord Kelvin was able to realise this absolute scale of temperature in the sense that its zero point was proved, in a thermodynamical way, to be equal to $-273.1\,^{\circ}C$, instead of being deduced from Charles's law on the basis that at this temperature, the volume of the gas is zero. The zero point of this scale was called the *absolute zero*, and temperatures reckoned from it were termed as *absolute temperatures* by him.

The efficiency η of Carnot's engine, as expressed in terms of the absolute temperatures is (as shown before in Section 1.5) given by

$$\eta = \frac{W_r}{q_2} = \frac{q_2 + q_1}{q_2} = \frac{T_2 - T_1}{T_2} \tag{3}$$

where q_2 is the amount of heat absorbed *reversibly* at the absolute temperature T_2, q_1 is that absorbed *reversibly* at T_1, where T_1 is less than T_2, and W_r is the reversible work obtained from the cyclic process, and q_1, in this equation, being of negative sign.

There is another expression for the efficiency η of the engine, namely,

$$\eta = \frac{W_r}{q_2} = \frac{q_2 - q_1}{q_2} = \frac{T_2 - T_1}{T_2}\,, \tag{4}$$

where q in this expression is the heat rejected by the engine to the reservoir which is of lower temperature T_1 than the heat reservoir temperature T_2. The absolute temperatures T_2 and T_1 are the temperatures obtained by a perfect gas thermometer. A real gas thermometer approaches a perfect gas closely enough at practically useful pressures which should not be very low pressures in order to permit the construction of an accurate gas thermometer for the measurement of the absolute temperature.

The remarkable memoir of the French scientist *Carnot*, in 1824, on the theory of heat engines had led not only Lord Kelvin to the suggestion of a new scale of temperature, which he called absolute scale of temperature but also had led *Clausius* to the definition of a new concept called *entropy S*, as a state function of the system by its exact differential d*S* as follows:

$$dS = \frac{d'q_{rev.}}{T}, \tag{5}$$

where $d'q_{rev}$ is the infinitesimal amount of heat absorbed reversibly by the system from the surroundings (considering the surroundings as the source of heat). T is the absolute temperature of the source of heat at which absorption takes place, and $1/T$ is known as the *integrating factor*, that is, the factor that converts an inexact differential $d'q_{rev.}$ into exact. Eq. (5) is the analytical formulation of the second law of thermodynamics. It states that for a reversible process, the differential element of heat is given by

$$d'q_{rev.} = TdS, \tag{5'}$$

which is Eq. (5) written in a different way. It should be noted here that any quantity that can be derived from that second law is expressible in a relation between the absolute temperature and other variables. The role which the absolute temperature plays in the analytical formulation of the second law makes it a quantity of general thermodynamic importance and gives a practical way of determining it indirectly, since the perfect gas thermometer is only an *abstraction* in the sense that it is theoretical rather than practical. Therefore, one should expect that the absolute temperature can be expressed in terms of other measurable variables.

On this basis, Lord Kelvin was able to realise the absolute thermodynamic scale of temperature, considering the following points.

1. Use was made of a phenomenon that can be expressed as a function of the absolute temperature T, and at the same time, it can be measured in terms of an arbitrary scale as the centigrade scale. This phenomenon is the Joule-Kelvin effect expressed by the coefficient μ as

$$\mu = \left(\frac{\partial T}{\partial P}\right)_H . \tag{6}$$

2. A combination of this relation with that of the entropy is defined by Clausius by

$$dS = \frac{d'q_{rev.}}{T}. \tag{5}$$

3. The Joule-Kelvin coefficient, $\left(\frac{\partial T}{\partial P}\right)_H$, was introduced in the expression for $\left(\frac{\partial H}{\partial P}\right)$, which was obtained above in Section 3, namely,

$$\left(\frac{\partial H}{\partial P}\right)_T = V - T\left(\frac{\partial V}{\partial P}\right)_P \tag{7}$$

Accordingly, the following analysis can be tentatively made

i. Clearly the Joule-Kelvin coefficient $\left(\frac{\partial T}{\partial P}\right)_H$ does not appear in Eq. (7), therefore, in order to make its appearance, we make use of the relation

$$dH = \left(\frac{\partial H}{\partial T}\right)_P dT + \left(\frac{\partial H}{\partial P}\right)_T dP \tag{8}$$

and substitute for $\left(\frac{\partial H}{\partial P}\right)_T$ on the right-hand side of Eq. (7). Thus we have

$$dH = \left(\frac{\partial H}{\partial T}\right)_P dT + \left\{V - T\left(\frac{\partial V}{\partial T}\right)_P\right\} dP.$$

ii. Putting $dH = 0$, which is the case of the Joule-Kelvin porous plug experiment, we obtain

$$\left(\frac{\partial H}{\partial T}\right)_P dT = \left\{T\left(\frac{\partial V}{\partial T}\right)_P - V\right\} dP$$

or

$$\left(\frac{\partial H}{\partial T}\right)_P \left(\frac{\partial T}{\partial P}\right)_H = T\left(\frac{\partial V}{\partial T}\right)_P - V.$$

Thus,

$$C_P\left(\frac{\partial T}{\partial P}\right)_H + V = T\left(\frac{\partial V}{\partial T}\right)_P$$

or

$$\frac{T\left(\frac{\partial V}{\partial T}\right)_P}{V + C_P\left(\frac{\partial T}{\partial P}\right)_H} = 1. \tag{10}$$

Hence,

$$\frac{1}{T} = \frac{\left(\frac{\partial V}{\partial T}\right)_P}{V + C_P\left(\frac{\partial T}{\partial P}\right)_H}. \tag{11}$$

iii. Multiplying both the sides of this equation by the exact differential dT of the thermodynamic coordinate T to represent an infinitesimal amount of temperature change, we obtain

$$\frac{1}{T} dT = \left\{\frac{\left(\frac{\partial V}{\partial T}\right)_P}{V + C_P\left(\frac{\partial T}{\partial P}\right)_H}\right\} dT. \tag{12}$$

iv. This equation can be rewritten in another way by leaving the temperature as the absolute temperature on the absolute scale, and changing it on the right-hand side into temperatures on the centigrade scale, that is, by replacing T by t, bearing in mind that each temperature t on the centigrade scale corresponds to a temperature T on the absolute scale, and that the difference between any two temperatures on one scale is the same as that on the other scale. Thus Eq. (12) takes the form

$$\frac{1}{T}\,dT = \left\{ \frac{\left(\frac{\partial V}{\partial t}\right)_P}{V + C_P\left(\frac{\partial t}{\partial P}\right)_H} \right\} dt\,. \tag{13}$$

v. The right-hand side of this equation contains readily measurable quantities, and consequently, the absolute temperature *T* on the left-hand side can be determined by integrating *t* on the right-hand side.
Thus we have

$$\int \frac{1}{T}\,dT = \int \left\{ \frac{\left(\frac{\partial V}{\partial t}\right)_P}{V + C_P\left(\frac{\partial t}{\partial P}\right)_H} \right\} dt\,. \tag{14}$$

vi. The initial temperature for the limit of integration on the right-hand side of this equation can be taken as the ice point on the centigrade scale which is $t_0 = 0°C$. Let the corresponding temperature on the absolute scale be T_0. Subsequently, if the final temperature on the centigrade scale be taken as 100°C, then the corresponding point on the absolute scale is $T_0 + 100$. That is, we take the lower and upper limits of integration on the right-hand side of Eq. (14) as the ice point and the steam point on the centigrade scale, which are by definition 0°C and 100°C, respectively, and subsequently, the corresponding limits of integration on the left-hand side on the absolute scale are T_0 and $T_0 + 100$, respectively. Hence, Eq. (14) may be rewritten as

$$\int_{T_0}^{T_0+100} \frac{1}{T}\,dT = \int_0^{100} \left\{ \frac{\left(\frac{\partial V}{\partial t}\right)_P}{V + C_P\left(\frac{\partial t}{\partial P}\right)_H} \right\} dt\,, \tag{15}$$

from which it is clear that we increase the two temperatures 0 and *T* by the same quantity of degrees, namely, 100.

vii. Applying now Charlie's law

$$V = V_0\,(1 + \alpha t),$$

where the isobaric expansivity, α, is given by $\alpha = \frac{1}{V_0}\left(\frac{\partial V}{\partial t}\right)_P$, we obtain $\left(\frac{\partial V}{\partial t}\right)_P = \alpha V_0$. Accordingly, substituting αV_0 for $\left(\frac{\partial V}{\partial t}\right)_P$ on the right-hand side of Eq, (15), then the integration of the left-hand side of this equation gives

$$[\ln T]_{T_0}^{T_0+100} = \int_0^{100} \left\{ \frac{\alpha V_0}{V + C_P\left(\frac{\partial t}{\partial P}\right)_H} \right\} dt\,. \tag{16}$$

viii. Putting $V = V_0(1 + \alpha t)$, and $\left(\frac{\partial t}{\partial P}\right)_H = \frac{\Delta t}{\Delta P}$ and then dividing both the numerator and denominator on the right-hand side of Eq. (16) by V_0, we produce

$$[\ln T]_{T_0}^{T_0 + 100} = \int_0^{100} \left\{ \frac{\alpha}{(1 + \alpha t) + \frac{C_P}{V_0} \cdot \frac{\Delta t}{\Delta P}} \right\} dt \qquad (17)$$

ix. The measurements of $\frac{\Delta T}{\Delta P}$ need not to be carried out at very low temperatures as it is small. In the range of $t = 0°C$ to $t = 100°C$, the value of $C_P \frac{\partial t}{\partial P}$ or $C_P \frac{\Delta t}{\Delta P}$ for a real gas is pretty small, and in the case of helium, it is less than 0.001 as found by Lord Kelvin in any point of interval, and so it can be neglected without introducing an error greater than 0.1 per cent. Accordingly, Eq. (17) reduces to

$$\ln \frac{T_0 + 100}{T_0} = \int_0^{100} \left\{ \frac{\alpha}{1 + \alpha t} \right\} dt = [\ln(1 + \alpha t)]_0^{100}.$$

Therefore,

$$\ln \frac{T_0 + 100}{T_0} = \ln(1 + 100\alpha),$$

that is,

$$1 + \frac{100}{T_0} = 1 + 100\alpha$$

or

$$\frac{1}{T_0} = \alpha .$$

Hence,

$$T_0 = \frac{1}{\alpha}. \qquad (18)$$

Eq. (18) is the equation we are looking for. It means that the zero degree centigrade on the centigrade scale corresponds to $1/\alpha$ on the absolute scale of temperature. Also, this equation means that the absolute temperature is expressed in a measurable quantity. With the experimental value of $\alpha = 0.003659$ the value of $T_0 = 273.32$. The symbol or abbreviation K is used to indicate degrees on the absolute or Kelvin scale. Thus

$$T_0 = 273.32\ K . \qquad (19)$$

The most accurate determination of α was carried out by *Henning* and *Heuse*; the mean results gave a value of $T_0 = 273.18 \pm 03K$. The value $T_0 = 273.1K$ is *usually adopted*, and accordingly, the zero of the absolute scale is $-273.1°C$, that is,

$$0\ K = -273.1°C \qquad (20)$$

Temperatures in the vicinity of 0K are established by means of the magnetocaloric effect. Helium has the lowest boiling point; it boils at 4.2K (= – 268.9°C). If helium is boiled under low pressure, a temperature of about 0.82K is reached. As pointed out before, a lower temperature has been reached by the adiabatic demagnetisation of a paramagnetic substance, previously cooled with liquid helium, it is about 0.001 K.

From the foregoing, it is evident that Kelvin made great use of Carnot's theorem by his suggestion of a new scale of temperature which he called as the absolute scale, and that he was able, after making his famous porous plug experiment with Joule, to realise this absolute thermodynamic scale of temperature.

An ideal gas would give this thermodynamic scale precisely if it were used in a gas thermometer. But since an ideal gas thermometer cannot be realised in practice, it was necessary to see how much the temperatures determined by a thermometer containing a real gas differ from absolute temperatures. The work of Kelvin resulted in showing that hydrogen behaves so nearly like an ideal gas that the hydrogen gas thermometer or the hydrogen scale may be taken for all practical purposes, as agreeing with the absolute scale at ordinary temperatures.

CHAPTER 6

Change of Phase

6.1. Systems of more than one phase

When a system consists of more than one phase, each phase may be considered as a separate system within the whole. The thermodynamic parameters of the whole system may then be constructed out of these component phases. If we allow new degrees of freedom within the system, such as mass transport between phases or chemical reaction between constituents, the conditions for thermodynamic equilibrium do lead to new results which are related to the restrictions which equilibrium places on the new degrees of freedom. In this chapter, we shall consider systems whose chemical composition is uniform (for example, systems of one component) but in which more than one phase is present. For simplicity, we shall develop the general results for a system subjected to work by hydrostatic pressure only, that is, pressure–volume work.

6.2. Availability and the general conditions for thermodynamic equilibrium

Suppose that a system interacts with its surroundings. Then if heat enters, the entropy change of the system for an infinitesimal reversible process as given by the second law of thermodynamics is

$$dS = d'q_{rev.}/T_0, \qquad (1)$$

where T_0 is the temperature of the surroundings, and the equality sign necessarily holds when the change is reversible. If, however, the change is irreversible, that is, the heat enters the system from the surroundings in an irreversible way, the entropy change dS of the system will be exactly the same as that in the case of reversible heat absorption since the entropy is a state function or a thermodynamic coordinate of the system, and thus it will be given by

$$dS > d'q_{irr.}/T_0. \qquad (2)$$

Accordingly, the entropy change of the system is, in general, related to the heat flow by

$$dS \geq d'q/T_0, \qquad (3)$$

bearing in mind that the equality sign holds when the change is reversible.

If the surroundings exert a pressure P_0, and they are only the source of work that is done *on* the system, then for an infinitesimal reversible work, which is a pressure–volume work given by $-P_0 dV$, the first law of thermodynamics is expressed in the usual formulation as

$$d'q = dU + d'W, \qquad (4)$$

where $d'W$ represents the work done *by* the system against the surroundings, which takes the form

$$d'q = dU - d'W \quad , \qquad (5)$$

where d'W here should represent the work done *on* the system. Hence, substituting for d'W in Eq. (5) by $-P_0dV$, we have

$$d'q = dU - (-P_0dV)$$

$$d'q = dU + P_0dV. \tag{6}$$

Combining Eqs. (3) and (6), we obtain

$$T_0\,dS \geq dU + P_0dV \tag{7}$$

or

$$dF' = dU + P_0dV - T_0dS \leq 0, \tag{8}$$

where F' is a new function defined by

$$F' = U - T_0S + P_0V\ . \tag{9}$$

The quantity F' is known as the *availability* of the system. It should be noted that it contains T_0 and P_0 of the surroundings and may be quite different from the temperature and pressure of the system.

Eq. (8) expresses the fact that in any *natural* change, the availability of a system *cannot increase*. It follows that the general condition for equilibrium of a system *in given surroundings* is that the availability F' be a *minimum*. Then we must have

$$dF' = dU + P_0dV - T_0dS = 0 \tag{10}$$

for all possible infinitesimal displacements from equilibrium. We have obtained this result *directly* from the second law of thermodynamics, which is given above by Eq. (3), namely,

$$dS \geq d'q/T_0, \tag{3}$$

by considering the interaction of a system with its surroundings.

The quantity F is known as the availability of the system, since it gives a measure of the *maximum* amount of work which may be extracted from a system *in given surroundings*. We may see this to be so by the following argument. Suppose that we place the system in a cylinder fitted with a piston so that we may subject it to a pressure P different from P_0 and suppose that *we isolate it thermally from the surroundings* so that its temperature T_0 is different from the temperature T_0 of the surroundings. Then, in a given change of state, we shall extract from it the greatest possible amount of work, provided the change is performed reversibly. According to the first law, for an infinitesimal reversible change, we have

$$dU = TdS - PdV\ , \tag{5}$$

where PdV is the work done by the system against the surroundings, so by substituting for dU from Eq. (5) in Eq. (10), we find

$$dF' = (T - T_0)dS - (P - P_0)\,dV. \tag{11}$$

Now suppose that we change the entropy of the system by operating a *reversible heat engine* between it and the surroundings. Then, since the process is *reversible* and the entropy remains constant, the work done by the engine $d'W_e$ is

$$\begin{aligned} d'W_e &= \text{net amount of heat absorbed} \\ &= d'q - d'q_0 \\ &= TdS - T_0dS = (T - T_0)dS. \end{aligned}$$

Therefore, the first term in Eq. (11) represents the maximum work which can be *obtained from* the system in the entropy change. Similarly, $(P - P_0)dV$ represents the *net* mechanical work done on the piston. As long as $T \neq T_0$ and $P \neq P_0$, we may continue to extract work and to reduce the value of F′. Thus $(F' - F'_{min})$ is equal to the maximum amount of work which may be extracted from the system in the given surroundings.

Now, the general condition for equilibrium, that the availability be a minimum, reduces to simpler forms in several important cases. We again suppose the system to be thermally isolated from the surroundings so that we may explore how F' varies near equilibrium by displacing the system *reversibly* and *infinitesimally* from equilibrium, and using Eq. (11) to examine the consequences of the displacement. According to this equation for F′ to be a minimum, we have

$$dF' = (T - T_0)dS - (P - P_0)\,dV = 0, \tag{11}$$

where d*S* and d*V* are *independent* variables, so the coefficient of each of them must vanish separately in an infinitesimal displacement when equilibrium is attained.

Now we consider *four special cases.*

1. *Thermally isolated isovolumic system.* Since the system is thermally isolated, *T* will, in general, be different from T_0. So for the first term, $(T - T_0)\,dS$, in Eq. (11) to be zero, we must have $dS = 0$. (*S* will, of course, be a maximum.) Since d*V* = 0, the second term $(P - P_0)\,dV$ is necessarily zero, and *P* is not directly defined. Then Eq. (10),

$$dF' = dU + P_0 dV - T_0 dS = 0, \tag{10}$$

reduces to

$$dF' = dU = 0.$$

Therefore, the appropriate set of conditions for equilibrium of the system is.

$$dS = 0, \quad dV = 0, \quad dU = 0.$$

2. *Thermally isolated isobaric system.* Again, for the first term in Eq. (11) to be zero, we must have d*S* = 0. Since the volume may now change, for the second term in the equation to be zero, we require $P = P_0$, or $dP = 0$ in any infinitesimal reversible change.

Then Eq. (10) reduces to

$$dF' = dU + PdV = dH = 0$$ since *P* is constant.

Thus appropriate set of conditions for equilibrium is

$$dS = 0, \quad dP = 0, \quad dH = 0$$

3. *Not thermally isolated isovolumic system.* The entropy may now change; so for the first term in Eq. (11) to be zero, we must have $T = T_0$, or $dT = 0$ in any infinitesimal reversible change.

Since $dV = 0$, the second term in this equation is necessarily zero, and *P* is not directly defined.

Then Eq. (10) reduces to

$$dF' = dU - TdS = dF = 0,$$ since *T* is constant.

Here *F* is the *Helmholtz free energy function.*

The appropriate set of conditions is

$$dT = 0, \quad dV = 0, \quad dF = 0.$$

4. *Not thermally isolated isobaric system.* The entropy may now change, and thus, for the first term in Eq. (11) to be zero, we must have $T = T_0$, or $dT = 0$. Since the volume may change, equilibrium, according to this equation, requires $P = P_0$, or $dP = 0$.

Eq. (10) now reduces to

$dF' = dU + PdV - TdS = dG = 0$, since *T* and *P* are constants.

Here *G* is the *Gibbs free energy function.*

The appropriate set of conditions is

$$dT = 0, \quad dP = 0, \quad dG = 0$$

The set of four conditions for equilibrium is

$$\text{Proper variables} \rightarrow \left\{ \begin{matrix} dS = 0, & dV = 0, & dU = 0 \\ dS = 0, & dP = 0, & dH = 0 \\ dT = 0, & dV = 0 & dF = 0 \\ dT = 0, & dP = 0 & dG = 0 \end{matrix} \right\} \leftarrow \text{Potentials.} \tag{12}$$

It should be noted that each thermodynamic function, or potential, appears with its *proper variables*. It must be emphasised that these sets of conditions are entirely equivalent in the sense that they lead to identical physical results; which of these to use is entirely a matter of convenience. If a system is kept at constant temperature and pressure, the obvious choice is to minimise the Gibbs function, since its accompanying conditions are *automatically fulfilled.*

It is important to be clear about the significance of the results we have derived. In arriving at the general condition for equilibrium that the *availability be a minimum*, we placed no restrictions on the internal complexity of the system. We may generally expect dU to contain, besides *T*, *S*, *P*, and *V*, other variables related to degrees of freedom, which are *internal* to the system. Corresponding terms do not appear in Eq. (11) because the system, as a whole, changes its internal energy only by exchange of heat and work with its surroundings.

On the other hand, it is helpful to the memory to note that the potential functions *U*, *H*, *F*, and *G* are always coupled with their proper variables and that when the proper variables are the *quantities held constant*, the corresponding potential function is always a minimum at equilibrium. It is also worth pointing out that if the appropriate potential is a maximum, then we have a situation of *unstable equilibrium.* These conditions provide the basis for the treatment of phase change and underlie much of chemical thermodynamics.

6.3. The condition for equilibrium between phases

Let us first consider a one-component system of two phases, maintained at constant pressure and temperature. This might be a liquid in contact with its vapour. If we ignore any possible surface effects at the interface, both temperature and pressure will be uniform throughout. Suppose that the masses present in each phase are m_1 and m_2 and that the *specific* Gibbs functions are g_1 and g_2. Then,

$$G = m_1 g_1 + m_2 g_2, \tag{13}$$

where *G* is a function of *P*, *T*, m_1, and m_2. On the other hand, g_1 and g_2 are functions of *P* and *T* only. Thus, since the differential of Eq. (13) at constant *T* and *P* gives

$$dG = g_1\,dm_1 + g_2\,dm_2, \tag{14}$$

then the condition of equilibrium at constant *T* and *P* reduces to

$$dG = g_1\,dm_1 + g_2\,dm_2 = 0. \tag{15}$$

As we are considering a *closed system,* in which mass is conserved,

$$m_1 + m_2 = \text{constant.}$$

When the system is subject to the constraint, then

$$dm_1 + dm_2 = 0. \tag{16}$$

Substituting for dm_1 in Eq. (15) from Eq. (16), we have

$$\left(g_2 - g_1\right) dm_2 = 0.$$

Hence,

$$g_1 = g_2. \tag{17}$$

This argument may be generalised to cases where more than two phases are present, with the result that *for equilibrium, the specific Gibbs functions are all equal.*

Although this condition was derived for a system subject to constant temperature and pressure, it holds, as we might in fact expect, whatever be the external constraints, as seen from the following:

For example, if we have a system subject to constant volume and temperature, the appropriate condition for equilibrium is $dF = 0$. The new constraint gives:

$$m_1v_1 + m_2v_2 = \text{constant}$$

or

$$m_1dv_1 + v_1\,dm_1 + m_2dv_2 + v_2dm_2 = 0\ . \tag{18}$$

The equilibrium condition at constant *T* and *V* gives

$$dF = f_1\,dm_1 + f_2\,dm_2 + m_1df_1 + m_2df_2 = 0. \tag{19}$$

Since *F* by definition is

$$F \equiv U - TS,$$

$$dF = dU - TdS - SdT$$

or

$$dF = d'q - PdV - TdS - SdT$$

$$= PdV - SdT.$$

So for an isothermal reversible change, we have

$$dF = -PdV$$

or

$$df + Pdv = 0 \quad \text{(specific values).} \tag{20}$$

On the other hand, since *G*, by definition, is

$$G \equiv H - TS$$

or

$$G = U + PV - TS,$$
$$G = F + PV$$

(21)

or

$$g = f + Pv \qquad \text{(specific values).} \qquad (22)$$

Multiplying Eq. (18) by *P* and adding the resulting equation to Eq. (19), we obtain

$$(f_1 + Pv_1)dm_1 + (f_2 + Pv_2)dm_2 +$$
$$(df_1 + Pdv_1)m_1 + (df_2 + Pdv_2)m_2 = 0.$$

This equation, according to Eq. (20), reduces to

$$(f_1 + Pv_1)\,dm_1 + (f_2 + Pv_2)\,dm_2 = 0, \qquad (23)$$

which, in turn, because of Eq. (22) takes the form of Eq. (15), namely

$$g_1\,dm_1 + g_2\,dm_2 = 0 \qquad (15)$$

Since conservation of the total mass still holds, that is,

$$m_1 + m_2 = \text{constant}$$

or

$$dm_1 + dm_2 = 0, \qquad (16)$$

we obtain

$$g_1 = g_2 \qquad (17)$$

A similar calculation yields the same result for any conditions of constraint for the whole system.

6.4. The Clausius-Clapeyron Equation*

A simple substance normally has two degrees of freedom. However, if we require that two phases of the substance coexist in equilibrium, then only one degree of freedom remains. The pressure and temperature of a given mass of water may be chosen at will; but if water is to be in equilibrium with its vapour, then the pressure, which is now by definition the *vapour pressure*, becomes a unique function of the temperature. If the pressure is increased above the vapour pressure, then the vapour will condense. If it is reduced below, then the liquid will evaporate. The equation

$$g_1 = g_2, \qquad (17)$$

obtained before, leads immediately to an important result connecting the pressure and temperature when two phases are in equilibrium.

Consider the *boundary between two phases of a substance* (Fig. 1). At any point on the boundary, the specific Gibbs functions of the two phases must be equal. In particular, this must be true at the neighbouring points a and b.

---

* This equation has been derived before (Section 1.5) using Carnot's theorem; nevertheless it has seemed worthwhile to rederive it on the basis of the condition for thermodynamic equilibrium between the phases.

and
$$\left.\begin{aligned}(g_1)_a &= (g_2)_a\\(g_1)_b &= (g_2)_b\end{aligned}\right\}. \qquad (24)$$

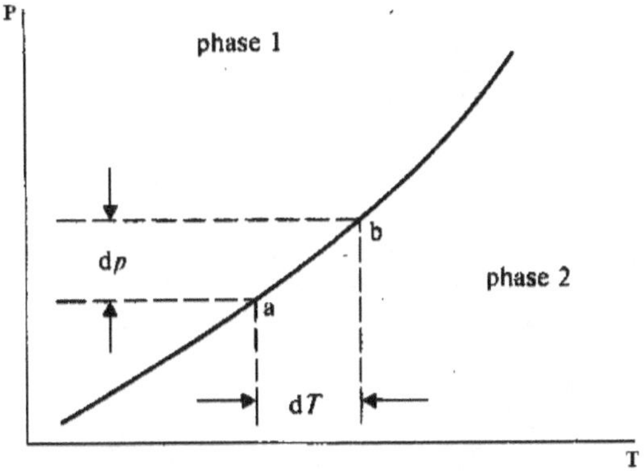

Fig. 1. Derivation of the Clausius-Clapeyron equation.

In passing from a to b, for phase 1 and for phase 2, we may write, respectively,

$$\left.\begin{aligned}dg_1 &= \left(\frac{\partial g_1}{\partial T}\right)_P dT + \left(\frac{\partial g_1}{\partial P}\right)_T dP\\ &\text{and,}\\ dg_2 &= \left(\frac{\partial g_2}{\partial T}\right)_P dT + \left(\frac{\partial g_2}{\partial P}\right)_T dP\end{aligned}\right\}. \qquad (25)$$

Since $dg = -sdT + vdP$, Eq. (25) is reduced to

$$\left.\begin{aligned}dg_1 &= -s_1\,dT + v_1\,dP,\\ &\text{and}\\ dg_2 &= -s_2\,dT + v_2\,dP,\end{aligned}\right\}. \qquad (26)$$

Also, since according to Eq. (24), g_1 for phase 1 is equal to g_2 for phase 2, at either a or b, we have from Eq. (26)

$$-s_1\,dT + v_1\,dP = -s_2\,dT + v_2\,dP.$$

Therefore,

$$(v_2 - v_1)dP = (s_2 - s_1)dT$$

or

$$\frac{dP}{dT} = \frac{\Delta s}{\Delta v}. \qquad (24)$$

For quantities other than specific quantities, the quotient Δs/`Δv is replaced by ΔS/`ΔV, so the rate of change of pressure with temperature is given by

$$\frac{dP}{dT} = \frac{\Delta S}{\Delta V}$$

or

$$\frac{dP}{dT} = \frac{L}{T\,\Delta V}, \qquad (28)$$

where L, ΔS, and ΔV are the latent heat (absorbed), the change in entropy, and the change in volume on passing from phase 1 to phase 2. Eq. (28) is the *Clausius-Clapeyron equation.* It gives the rate at which the pressure must change with temperature for two phases to *remain* in equilibrium. It gives the gradient of the phase boundary in the P–T plane. It applies to all phase changes in which there is a *discontinuity in entropy and volume at the transition.* These are known as *first-order* phase changes for reasons which will be explained below. This class includes all solid–liquid, liquid–vapour, and solid–vapour transitions.

The phase diagram for a simple substance is shown in Fig. 2(a). The lines represent the unique relationships which must exist between the pressure and temperature if two phases are to coexist. For all three phases to coexist, there can remain *no freedom* in the system, and the condition $g_1 = g_2 = g_3$ leads to a *unique temperature and pressure which define the triple point* T_r.

For most substances, the gradient of the solid–liquid line is positive. The Clausius-Clapeyron equation shows that this is associated with the fact that most substances expand on melting

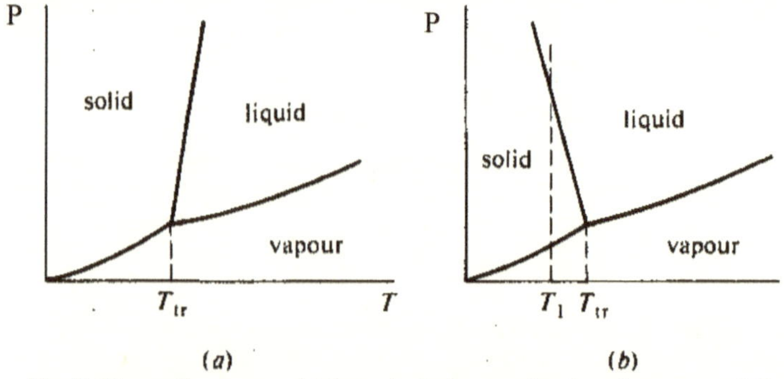

Fig. 2. Phase diagrams – (a) is typical of most simple substances, while (b) shows the behaviour of water, which expands on freezing

and, therefore, have ΔV positive (ΔS must, of course, always be positive because of the increase in disorder associated with melting* . Water is an exception with respect to ΔV in that it *expands* on freezing so that the solid–liquid boundary has a *negative* slope. Thus, in the case of water, it is possible by increasing the pressure *isothermally* to pass from *vapour*

* The transition in the ^3He is an exception. This shall be discussed in Section 7.6.

to solid to liquid, as can be seen, for example, at T_1 in Fig. 2 (b), whereas for most substances, the solid is the *high-pressure phase*.

The Clausius-Clapeyron equation has been checked experimentally over a wide range of conditions in experiments on the vapour pressure of solids and liquids and in measurements of melting curves. All the measurements have shown it to be obeyed to a high order of accuracy. Its validity provides one of the *most direct experimental tests of the truth of the second law*.

6.5. Integration of the Clausius-Clapeyron equation

It is sometimes useful to have an explicit functional form which may be used as an approximation for the relationship between vapour pressure and temperature. The Clausius-Clapeyron equation itself is exact. By making approximations, which are reasonable under certain conditions, it is possible to integrate it to obtain an explicit expression for the vapour pressure.

If the pressure is not too high, and we are not near to the critical point, it is reasonable to assume that the vapour obeys the *perfect gas law* and that the specific volume of the condensed phase V_c is negligible in comparison with that of the vapour V_V. With these assumptions, the Clausius–Clapeyron equation,

$$\frac{dP}{dT} = \frac{1}{T\,\Delta V} = \frac{\Delta S}{\Delta V},$$

becomes

$$\frac{dP}{dT} = \frac{L}{TV_v} = \frac{LP}{RT^2}, \tag{29}$$

when one mole of the vapour is considered.

To proceed further, we need to assume some fundamental form for the latent heat. There are two degrees of approximation which heat. There are two degrees of approximation which is useful to make:

a. The *crudest* approximation is to take *L* as constant. Over a sufficiently small temperature interval, this is not unreasonable. Eq. (29) then integrates to give

$$R \ln P = -\frac{L}{T} + A, \tag{30}$$

where *A* is a constant, or

$$\ln P = -\frac{L}{RT} + \frac{A}{R},$$

so that

$$P = P_0\, e^{-L/RT}. \tag{31}$$

It should be noted that this expression has the form of the *Maxwell-Boltzmann factor,*[*] $e^{-\epsilon/kt}$, representing the probability that a molecule is thermally excited out of the liquid into the vapour over a potential barrier $\epsilon = L/N_a$, where N_a is Avogadro's number.

* The Maxwell-Boltzmann statistics shall be seen in Section 9.4.

b. Instead of assuming that the latent heat is constant, we may make the better approximation that the specific heats of the two phases are constant. For this purpose, we express the latent heat *L* in terms of the entropies of the two phases at the temperature of transition from the condensed phase to the vapour phase. We already know that

$$dS = d'q_{rev.}/T,$$

so using the definitions

$$d'q_p = C_p dT$$

and

$$C_P = \left(\frac{\partial H}{\partial T}\right)_P$$

and integrating

$$\int_c^v dS = \frac{1}{T} \int_c^v \left(\frac{\partial H}{\partial T}\right)_P dT ,$$

we have

$$(S_v - S_c) = \frac{1}{T}(H_v - H_c) = \frac{\Delta H}{T} = \frac{L}{T}$$

or

$$T(S_v - S_c) = L. \tag{32}$$

Then substituting for *L* in Eq. (29) from Eq. (32), we obtain

$$\frac{dP}{dT} = \frac{(S_v - S_c)P}{RT}. \tag{33}$$

Now we find the derivative of L/T at constant *P*,

$$\left\{\frac{d}{dT}\left(\frac{L}{T}\right)\right\}_P = \frac{T\left(\frac{\partial L}{\partial T}\right)_P - L}{T^2}$$

$$= \frac{1}{T}\left(\frac{\partial L}{\partial T}\right)_P - \frac{L}{T^2}. \tag{34}$$

Since

$$L = \Delta H = H_v - H_c,$$

we produce

$$\left(\frac{dL}{dT}\right)_P = \left(\frac{dH_v}{dT}\right)_P - \left(\frac{dH_c}{dT}\right)_P = (C_P)_v - (C_P)_c \tag{35}$$

so that Eq. (34) may be rewritten in the form

$$\left\{\frac{d}{dT}\left(\frac{L}{T}\right)\right\} = \frac{(C_P)_v - (C_P)_c}{T} - \frac{L}{T^2}. \tag{36}$$

Eq. (36) can be derived in another way as follows:
Let *L* be a function of pressure *P* and temperature T, that is,

$$L = L(P, T).$$

Therefore,

$$dL = \left(\frac{\partial L}{\partial P}\right)_T dP + \left(\frac{\partial L}{\partial T}\right)_P dT$$

or

$$\frac{dL}{dT} = \left(\frac{\partial L}{\partial P}\right)_T \frac{dP}{dT} + \left(\frac{\partial L}{\partial T}\right)_P.$$

Thus we can, in general, write the expression

$$\frac{d}{dT} = \left(\frac{\partial}{\partial T}\right)_P + \left(\frac{\partial}{\partial P}\right)_T \times \frac{dP}{dT}.$$

Returning now to Eq. (32), namely,

$$L/T = (S_v - S_c), \qquad (32)$$

we may apply this expression to the quotient L/T. Thus, we have

$$\frac{d(L/T)}{dT} = \left\{\frac{\partial(L/T)}{\partial T}\right\}_P + \left\{\frac{\partial(L/T)}{\partial P}\right\}_T \frac{dP}{dT}$$

$$= \left\{\frac{\partial(S_v - S_c)}{\partial T}\right\}_P + \left\{\frac{\partial(S_v - S_c)}{\partial P}\right\}_T \frac{dP}{dT}. \qquad (37)$$

But*

$$\left(\frac{\partial S}{\partial T}\right)_P = \frac{C_P}{T}, \text{ and } \qquad \left(\frac{\partial S}{\partial P}\right)_T = -\left(\frac{\partial V}{\partial T}\right)_P,$$

therefore, using these two partial derivatives, Eq. (37) can be rewritten as

$$\frac{d(L/T)}{dT} = \frac{(C_P)_v - (C_P)_c}{T} - \left\{\frac{\partial(V_v - V_c)}{\partial T}\right\}_P \frac{dP}{dT}. \qquad (38)$$

Again, applying the condition that $V_v >> V_c$, and assuming that the perfect gas law applies, then substituting for $\frac{dP}{dT}$ in Eq. (38) from Eq. (29), we obtain

$$\frac{d(L/T)}{dT} = \frac{(C_P)_v - (C_P)_c}{T} - \frac{L}{T^2}, \qquad (36)$$

which is Eq. (36), derived above in another way.
Now comparing Eqs. (34) and (36), we produce

$$\frac{1}{T}\left(\frac{dL}{dT}\right)_P = \frac{(C_P)_v - (C_P)_c}{T}$$

---

*See the Maxwell relations, Section 2.4.

or

$$dL = \{(C_P)_v - (C_P)_c\} dT,$$

which by integration, taking the difference between heat capacities as a constant, gives

$$L = L_0 + L_1 T. \tag{39}$$

Substituting back in Eq. (29), namely,

$$\frac{dP}{dT} = \frac{LP}{RT^2}, \tag{29}$$

or

$$\frac{dP}{P} = \frac{L}{R}\frac{dT}{T^2},$$

we obtain

$$= \frac{1}{R}\frac{(L_0 + L_1 T)dT}{T^2}.$$

Then integrating this equation, that is,.

$$\int \frac{dP}{P} = \frac{1}{R}\int \frac{L_0}{T^2} dT + \int \frac{L_1}{T} dT ,$$

we produce

$$R \ln P = -\frac{L_0}{T} + L_1 \ln T + A', \tag{40}$$

where A' is a constant. This equation when compared with Eq. (30) shows that the two right-hand sides should be equivalent.

It should be noted that the *assumption of constant heat capacities is equivalent to introducing a linear form in the temperature dependence of the latent heat, as evidenced by Eq. (39).*

6.6. The Gibbs functions in first-order transitions

It is instructive to examine how the Gibbs functions of two phases behave in the neighbourhood of a transition. The *specific* function of a single phase must be a continuous function of pressure and temperature. This may be represented as a *surface* in three-dimensional, $g - P - T$, space. The surfaces of two different phases will, in general, intersect in a *line* along which the *specific Gibbs functions are equal.* Along this line, the two phases will be in *equilibrium*, while away from it, the phase with the *lower* g value will be the *stable* one, as is required by the condition for equilibrium.

If we consider a simple substance which may exist in the solid, liquid, and vapour states, there will be three g surfaces which intersect in pairs to give three lines representing equilibrium between the corresponding pairs of phases. In general, there will be one point lying in all three surfaces, known as the *triple point*, at which all three phases are in equilibrium. The phase diagram of Fig. 3 is, therefore, the projection on the P–T plane of the lines of intersection of the g surfaces for the solid, liquid, and vapour phases. The dotted extensions through the triple point *represent the continuation of the boundary between two of the phases into the region where the third phase becomes more stable than either of the original two.* For example, the point X lies on the *intersection* of the *solid* and *vapour,* g surfaces, but does not represent a *stable state* because, for the value of *P* and *T*, the liquid g surface lies below the other two.

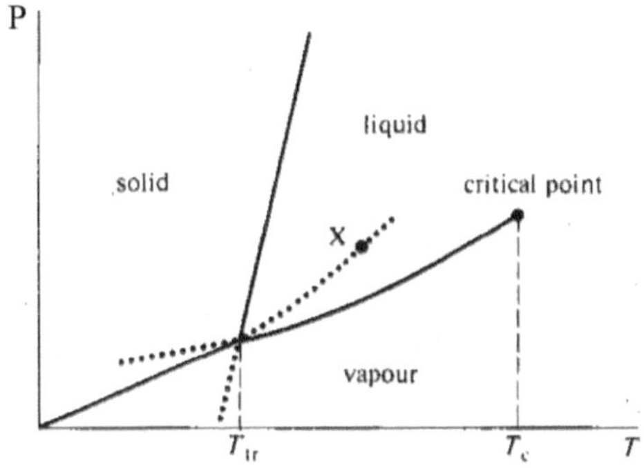

Fig. 3. A phase diagram as the projection of the intersection of the g surfaces of the three component phases.

Let us now examine sections through the g surfaces. Fig. 4 shows a section in a plane of constant temperature *T*. Since for a given *P* and *T*, the *stable state* is that of the *lowest* g,

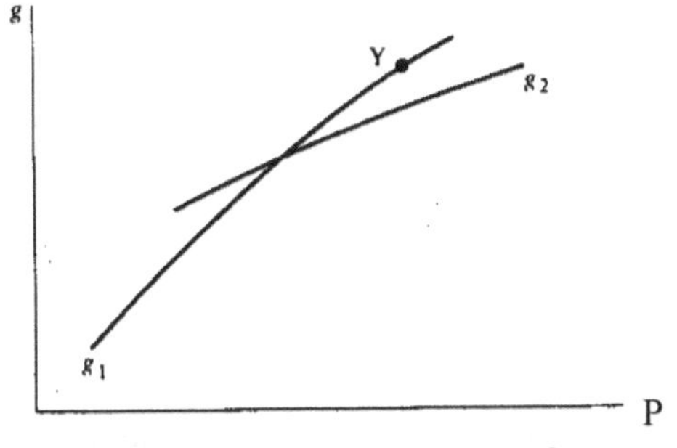

Fig. 4. A section through the g surfaces in a plane of constant temperature.

states such as Y are not stable, but may often be realised as *metastable states*. For example, if no nuclei are present to initiate condensation, a vapour may be compressed to a pressure well above the vapour pressure of the liquid without condensation taking place. It is then said to be a *supersaturated vapour*. Similarly, if a liquid is very pure, it may be heated well above its boiling point without boiling taking place to produce a *superheated liquid.* The relative stability of these states is a result of *surface effects*. Such states do exist, so we are justified in extending the g functions into regions which do not correspond to a stable configuration.

Since $\left(\partial g/\partial P\right)_T = v$, the gradient of g as a function of *P* in a constant *T* section must always be *positive*. In the case of *solid–liquid* transition, either the *solid* or the *liquid* may be the *high-pressure phase*; it is that phase which has above intersection the *smaller specific volume*. This follows directly by the simple *topological* argument that since the gradients are *positive*, the phase with *lower* g value at pressures above the intersection must have the *smaller* gradient, and thus, in this transition, the *liquid is the high-pressure phase*. In contrast, for transitions from the *vapour to the solid or liquid*, the *vapour must always be the low-pressure phase* since, at a given temperature, its *specific volume* is always greater than that of the solid or the liquid.

A section in a plane of constant *P* is shown in Fig. 5. Because $\left(\partial g/\partial T\right)_P = -s$, the gradients are always *negative*; and since specific heats at constant pressure are always *positive*, the curvatures must always be negative. The topological argument shows that the *high-temperature phase* is always that of *greater entropy*.

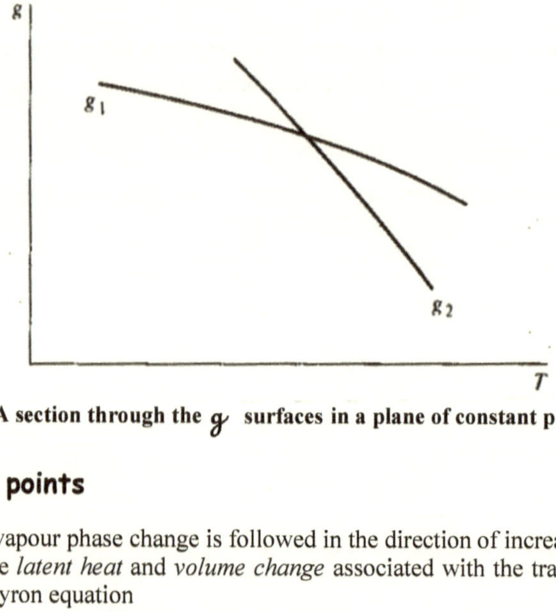

Fig. 5. A section through the g surfaces in a plane of constant pressure.

6.7. Critical points

If the liquid–vapour phase change is followed in the direction of increasing temperature, it is found that the *latent heat* and *volume change* associated with the transition, according to Clausius-Clapeyron equation

$$\frac{dP}{dT} = \frac{\Delta S}{\Delta V} = \frac{L}{T\Delta V}, \qquad (1)$$

become smaller until they eventually vanish and it is no longer possible to identify a transition from one phase to another. The point at which this occurs is known as the *critical point* (Figs. 3 and 6).

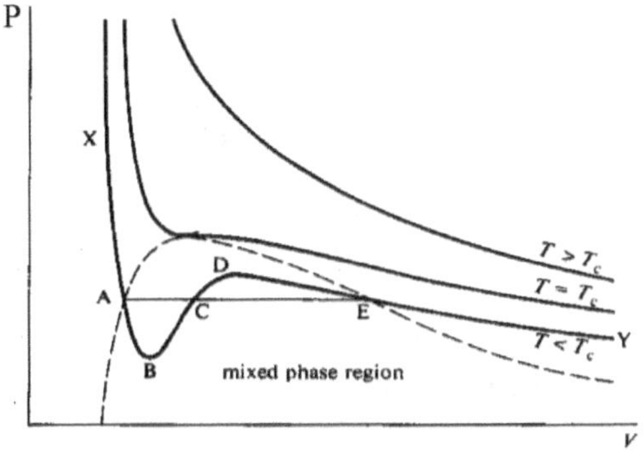

Fig. 6. Van der Waals isotherms, near the critical temperature.

Above the critical temperature, it is possible to pass continuously from liquid to vapour. Remembering that at a phase change, the latent heat per unit mass is given by

$$\left(\frac{\partial g}{\partial T}\right)_P = -s \tag{2}$$

or

$$T\Delta s = -T\Delta\left(\frac{\partial g}{\partial T}\right)_P, \tag{3}$$

and the volume change by

$$v = \left(\frac{\partial g}{\partial P}\right)_T \tag{4}$$

or

$$\Delta v = \Delta\left(\frac{\partial g}{\partial P}\right)_T \tag{5}$$

then the absence of latent heat or volume change above the critical temperature shows that we no longer have *intersecting* g surfaces, but that the system passes *continuously* along a smooth g surface . How, then, do we change from the situation of having separate g surfacesabove it? We may gain insight into this by analysing the behaviour of a van der Waals fluid.

The form of the isotherms of a van der Waals' fluid near the critical temperature is shown in Fig. 6. Consider the isotherm for $T < T_c$. We know that the whole of the curve YEDCBAX is not traced by any real fluid and that as the volume is decreased and the system passes along the curve from Y, we reach some point E, where the liquid begins to condense out. As the volume is further decreased, more of the substance passes into the liquid, and the pressure remains constant at the liquid vapour pressure until *no more vapour* remains at A. The section ECA spans the mixed phase region. The system then follows the van der Waals isotherm along AX. In the mixed phase region, we know that the specific Gibbs functions of the liquid and vapour are equal and also that they are constant since the pressure and temperature do not change along ECA. Then, in particular, the Gibbs function for the system at E, where it is all vapour, and at A, when it is all liquid, must also be equal, Hence,

$$\mathcal{g}_A = \mathcal{g}_E \, . \tag{6}$$

Now, let us suppose that the whole of the van der Waals' isotherm has physical meaning. Then we may calculate how $\mathcal{g}$ varies along the isotherm, beginning from the point E, where $P = P_o$ is the equilibrium vapour pressure of the condensed liquid, using:

$$\mathcal{g}(P,T) = \mathcal{g}(P_0,T) + \int_{P_0}^{P} \left(\frac{\partial \mathcal{g}}{\partial P}\right)_T dP$$

$$= \mathcal{g}(P_0,T) + \int_{P_0}^{P} v\,dP \, , \tag{7}$$

where we may substitute for v from van der Waals equation. The behaviour of $\mathcal{g}$ calculated in this way is illustrated* in Fig. 7. The

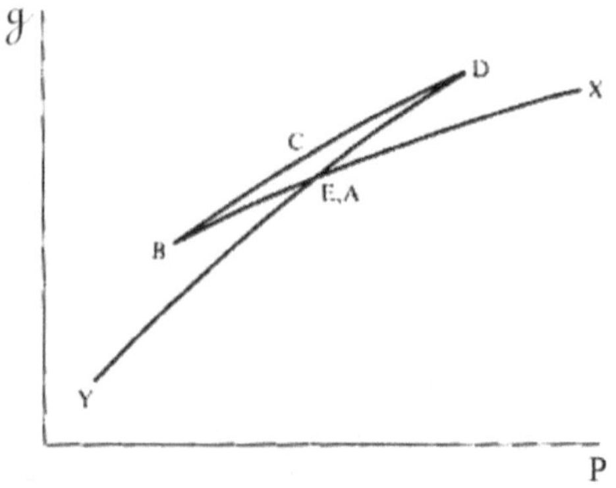

Fig. 7. The Gibbs function of a van der Waals fluid below the critical temperature.

* See *Equilibrium Thermodynamics* by C. J. Adkins (Cambridge University Press, Cambridge 1983, Section 10.7.

states represented by BCD in Figs. 6 and 7 are *mechanically unstable*, for this part of the van der Waals isotherm has a *negative* coefficient of isothermal compressibility k $\left(= -1/v\left(\frac{\partial v}{\partial P}\right)_T\right)$; but the regions BA and ED may, in principle, be traced out as *metastable states*. Points B and D represent the limits of metastability for the van der Waals fluid.

Now, as the temperature is raised, the mixed phase region of the fluid EA becomes smaller and eventually disappears at the critical temperature. As a consequence, the size of the closed loop in the Gibbs function (Fig. 7) becomes smaller, and the difference in the gradients at the intersections (E, A) (Fig. 6) also becomes smaller and disappear at the critical point. Thus, below the critical temperature, the Gibbs surface of the *stable state* has a *crease* which becomes shallower as the critical point is approached, eventually vanishing there to give a smooth surface above it. The form of the van der Waals g surfaces is illustrated in Fig. 8.

It is worth noting that the *vapour pressure* of a van der Waals fluid any be found from Eqs. (6), (7), and van der Waals equation itself. Since

$$g_A = g_E$$

and

$$\left(\frac{\partial g}{\partial P}\right)_T = v,$$

then

$$\int_E^A dg = \int_E^A v\,dP = 0.$$

This means that the area ABC is equal to the area CDE in Fig. 6.

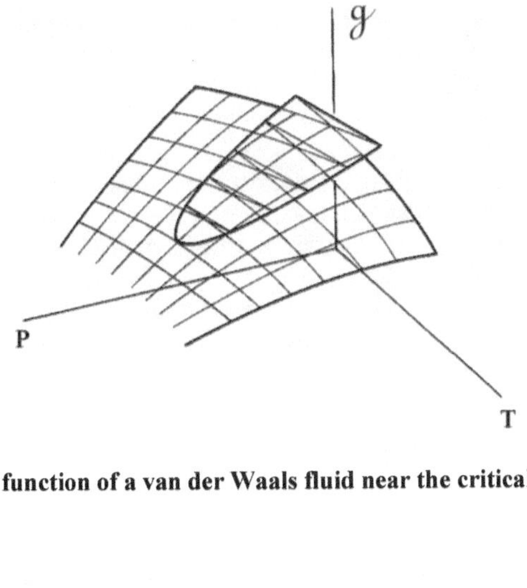

Fig. 8. Gibbs function of a van der Waals fluid near the critical temperature.

Once the vapour pressure is found, the volume change associated with the phase change can be calculated. Also, with the help of the Clausius-Clapeyron equation, the latent heat of the transition may be calculated.

6.8. Higher order change of phase

The kind of phase change we have analysed in the earlier part of this chapter is characterised by discontinuous changes in the entropy and volume at the transition. These are related to discontinuous changes in the *first derivatives* of the Gibbs function for the stable state of the system with respect to its proper variables P and T. Not all phase changes are of this type. It is convenient to adopt a classification scheme *first introduced by Ehrenfest, whereby* the *order* of a transition is defined as:

The order of the lowest derivative of the Gibbs function which shows a discontinuity at the transition.

Fig. 9 illustrates schematically the behaviour of the Gibbs function and its first and second derivatives in first- and second-order transitions.

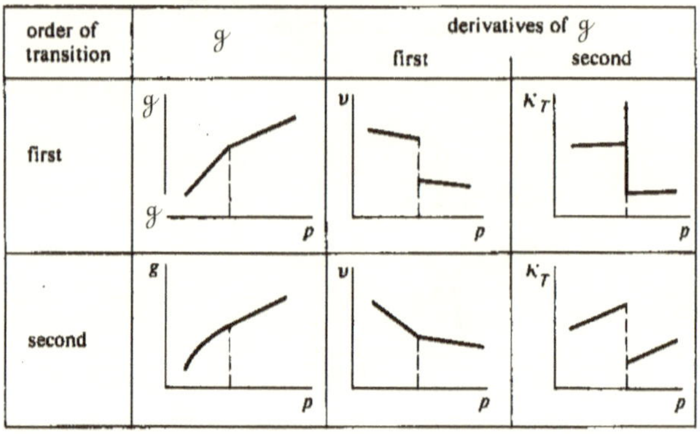

Fig. 9. The behaviour of the Gibbs function and its first and second derivatives in first- and second-order transitions.

The Clausius-Clapeyron equation was obtained above for the gradient dP/dT of the phase boundary by using the equality of the specific Gibbs function for the phases in equilibrium. If we try to apply this equation to transitions of order higher than first, we obtain *indeterminate* result for both numerator and denominator as zero. We may, however, obtain analogous equations for second-order transitions by using the equality of entropies or volumes at the transition. We proceed as follows:

Consider the volume v as a function of the two variables P and T,

$$v = v(T, P),$$

therefore,

$$dv = \left(\frac{\partial v}{\partial T}\right)_P dT + \left(\frac{\partial v}{\partial P}\right)_T dP,$$

so for the two phases, we have

$$dv_1 = \left(\frac{\partial v_1}{\partial T}\right)_P dT + \left(\frac{\partial v_1}{\partial P}\right)_T dP$$

and

$$dv_2 = \left(\frac{\partial v_2}{\partial T}\right)_P dT + \left(\frac{\partial v_2}{\partial P}\right)_T dP.$$

Taking $dv_1 = dv_2$ for an infinitesimal change along the boundary, we obtain

$$\frac{dP}{dT} = -\frac{\left(\frac{\partial v_2}{\partial T}\right)_P - \left(\frac{\partial v_1}{\partial T}\right)_P}{\left(\frac{\partial v_2}{\partial P}\right)_T - \left(\frac{\partial v_1}{\partial P}\right)_T} = \frac{\beta_2 - \beta_1}{k_2 - k_1}, \tag{8}$$

where β *and* k are the isobaric cubic expansivity and isothermal compressibility

In like manner, let s be a function of P and T,
therefore,

$$ds_1 = \left(\frac{\partial s_1}{\partial T}\right)_P dT + \left(\frac{\partial s_1}{\partial P}\right)_T dP$$

and

$$ds_2 = \left(\frac{\partial s_2}{\partial T}\right)_P dT + \left(\frac{\partial s_2}{\partial P}\right)_T dP$$

Taking $ds_1 = ds_2$ for an infinitesimal change along the boundary, we have

$$\left(\frac{\partial s_1}{\partial T}\right)_P dT + \left(\frac{\partial s_1}{\partial P}\right)_T dP = \left(\frac{\partial s_2}{\partial T}\right)_P dT + \left(\frac{\partial s_2}{\partial P}\right)_T dP,$$

therefore,

$$\frac{dP}{dT} = -\frac{\left(\frac{\partial s_2}{\partial T}\right)_P - \left(\frac{\partial s_1}{\partial T}\right)_P}{\left(\frac{\partial s_2}{\partial P}\right)_T - \left(\frac{\partial s_1}{\partial P}\right)_T}. \tag{9}$$

We already know* that

$$\left(\frac{\partial s}{\partial T}\right)_P = \frac{c_P}{T}, \quad \text{and} \quad \left(\frac{\partial s}{\partial P}\right)_T = -\left(\frac{\partial v}{\partial T}\right)_P = -\beta v.$$

Therefore, by substitution, Eq. (9) takes the form

$$\frac{dP}{dT} = \frac{1}{vT}\,\frac{c_{P_2} - c_{P_1}}{\beta_2 - \beta_1}. \tag{10}$$

Now combining Eqs. (8) and (10), we find

$$\frac{dP}{dT} = \frac{\beta_2 - \beta_1}{k_2 - k_1} = \frac{1}{vT}\,\frac{c_{P_2} - c_{P_1}}{\beta_2 - \beta_1}$$

* See the Maxwell relations, Section 2.4.

or

$$\frac{dP}{dT} = \frac{\Delta\beta}{\Delta k} = \frac{1}{vT}\,\frac{\Delta c_P}{\Delta\beta}. \tag{11}$$

Eq. (11) is known as *Ehrenfest's equation.* It is worth noting that, in effect, we have used the usual procedure for expressions which are *indeterminate* because numerator and denominator tend to zero; namely, we have replaced the numerator and denominator by their first derivatives.

Table 1. First- second- and third-order phase transitions.

The table lists, for each order of transition, the derivatives of g and the most closely related experimental quantities in which discontinuity appears.

	Discontinuity appears in:				
Order	**Derivatives of** g		**Corresponding experimental quantities**		
First	$\left(\frac{\partial g}{\partial T}\right)_P$	$\left(\frac{\partial g}{\partial P}\right)_T$	s		v
Second	$\left(\frac{\partial s}{\partial T}\right)_P$ $\left(\frac{\partial s}{\partial P}\right)_T$	$\left(\frac{\partial v}{\partial T}\right)_P$ $\left(\frac{\partial v}{\partial P}\right)_T$	c_P	β	k
Third	$\left(\frac{\partial^2 s}{\partial T^2}\right)_P$ $\frac{\partial^2 s}{\partial P\,\partial T}$ $\left(\frac{\partial^2 s}{\partial P^2}\right)_T$	$\left(\frac{\partial^2 v}{\partial T^2}\right)_P$ $\frac{\partial^2 v}{\partial P\,\partial T}$ $\left(\frac{\partial^2 v}{\partial P^2}\right)_T$	$\left(\frac{\partial c_P}{\partial T}\right)_P$ $\left(\frac{\partial c_P}{\partial P}\right)_T$	$\left(\frac{\partial \beta}{\partial T}\right)_P$ $\left(\frac{\partial \beta}{\partial P}\right)_T$	$\left(\frac{\partial k}{\partial T}\right)_P$ $\left(\frac{\partial k}{\partial P}\right)_T$

6.9. Some examples of phase changes of various orders

Unfortunately, very few systems showing higher-order transitions approach the idealised behaviour illustrated in Fig. 9. Usually, the gradient of the specific heat becomes infinite on one or both sides of the transition, and it is always difficult to decide to which of the idealised classes a particular system best belongs. Some examples of transitions of various orders are given below.

a. First order

i. Solid–liquid, solid–vapour, and liquid–vapour phase changes.
ii. The superconducting transition in a magnetic field.
iii. Some allotropic transitions in solids.

b. Second order

i. The superconducting transition in zero magnetic field.
ii. The superfluid transition in liquid helium.
iii. The order–disorder transition in β – brass .

c. Third order

The Curie point of many ferromagnets as in the case of iron.

We shall now discuss some of these in detail.

6.10. Phase changes in solid iron

Iron is an interesting example in that it shows both first- and third-order transitions as can be seen form Fig. 10.

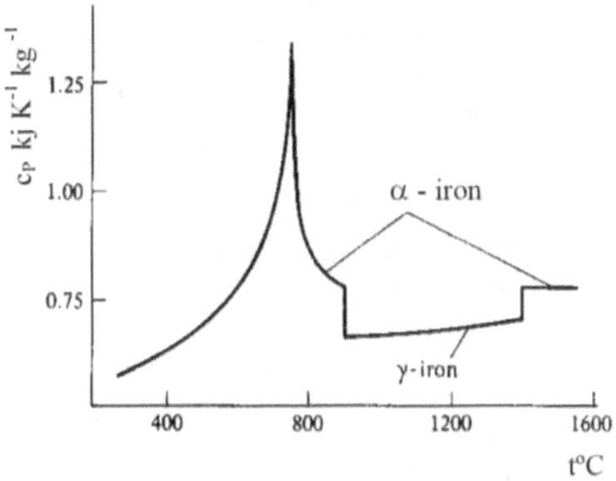

Fig. 10. The specific heat of iron.

The third-order transition is the change from the *magnetically ordered ferromagnetic state to the disordered paramagnetic state*. The total area under the specific heat anomaly is related to the entropy change associated with this transition. At higher temperatures, the two first-order transitions are associated with changes in *crystalline structure*. Below 906°C and above 1400°C, the α – phase is stable, while between these temperatures, the γ – phase is stable. The $\mathcal{g}$ surfaces for the α – and γ – phases, therefore, intersect at these temperatures, while the third-order transition, corresponding to a discontinuity in $(\partial c_P/\partial T)_P$, also corresponds to a *discontinuous* change in the curvature of the $\mathcal{g}$ surface at the *Curie point.*

6.11. The superfluid transition in liquid helium

The phase diagram for the common isotope of ^4He is shown in Fig. 11. Unlike all other elements, ^4He and the lighter

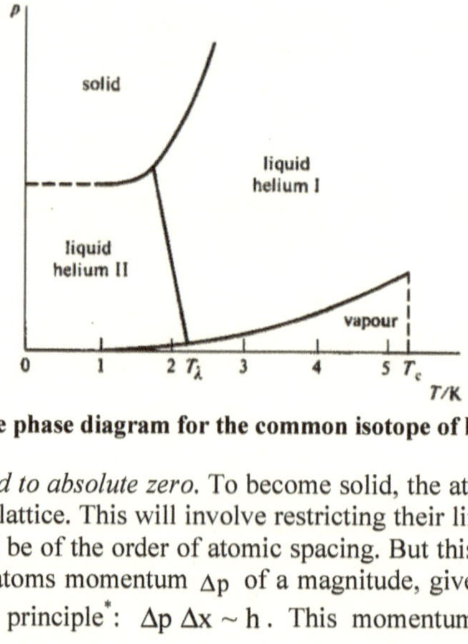

Fig. 11. The phase diagram for the common isotope of helium, ^4He

isotope ^3He *remain liquid to absolute zero.* To become solid, the atoms must be confined to proper sites on a crystal lattice. This will involve restricting their linear motion within some length, Δx, which will be of the order of atomic spacing. But this restriction may only be achieved by giving the atoms momentum Δp of a magnitude, given experimentally by the *Heisenberg uncertainty* principle*: $\Delta p\, \Delta x \sim h$. This momentum is associated with an energy $E = (\Delta p)^2/2m$, This in turn will be larger for helium, which is a small, light atom, than for elements higher in the Periodic Table. Helium is also an inert gas with a closed outer shell of electrons, so the interatomic forces are very weak, and the energy available for restricting the atoms to their proper positions for the solid is correspondingly small. In the case of helium, the zero-point energy is greater than the energy available for bringing about solidification, and unless the effect of interatomic forces is enhanced by applying a large pressure, the helium *remains liquid to absolute zero.* The uncertainty principle also explains why the vapour pressure of the *lighter* isotope is *higher* than that of the *heavier*, the respective normal boiling points for ^3He and ^4He being 3.19K and 4.21K; that is, the boiling point of ^3He is *lower* than that of ^4He.

---

* See Section 8.1.

Unlike the lighter isotope ^3He, ^4He has two liquid phases known as helium I and helium II (Fig. 11). The former, the high-temperature phase, is in all respects a normal liquid. Helium II, however, has an *extremely high thermal conductivity* and, in some respects, behaves as if it has an *extremely small viscosity*, for which reasons it has been called '*superfluid*'. The transition to the superfluid state is well defined, but there is no change in *density* nor *can any latent heat be detected.* The specific heat, however, shows strong anomaly (Fig. 12). It *rises rapidly* below the transition and, apparently, falls *discontinuously* at it.

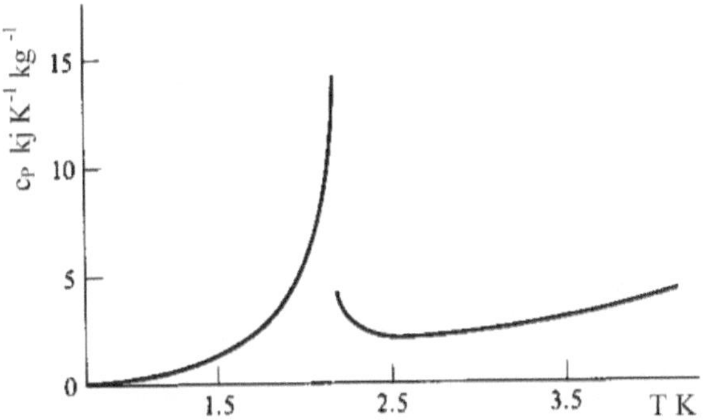

Fig. 12. The specific heat of liquid helium, ^4He, under its vapour pressure.

When a specific heat anomaly has a shape like that for the superfluid phase transition in helium, *which is of the second order*, or *that of the third-order transition in iron*, the phase change is known, on account of the shape of the specific heat curve, as a λ *transition*, and the temperature at which it occurs as the λ *point*.

6.12. The superconducting phase change

We shall give a fairly detailed analysis of the supercomputing phase change both because of its intrinsic interest, and also because it provides the only example of an ideal *second-order transition.*

Many metals, when cooled to a sufficiently low temperature, become superconducting. In the superconducting state they are characterised by two properties:

a. Zero electrical resistance.
b. Perfect diamagnetism; that is, complete exclusion of magnetic flux from the interior.

The superconducting state may be *destroyed* and the normal state may be restored by the following:

a. Raising the temperature;
b. Applying a magnetic field greater than some *critical value* or a combination of both (a) and (b).

Perfect diamagnetism is not implied by infinite electrical conductivity. It is true that a changing magnetic filed induces *eddy currents* in the surface of a *conductor* and that these currents act to *screen* the field changes from the interior; but even in a perfect conduction, the surface currents decay, and the magnetic field changes eventually penetrate. In contrast with a superconductor, surface currents *induced* by a magnetic field *persist indefinitely* so long as the material remains *superconducting*, and what is more remarkable is that if the superconductivity is destroyed by raising the *field* above the critical value and then restored by reducing the *field*, then the surface currents *reappear*, and the flux is *expelled* from the interior. This is called *Meissner effect.*

The perfect diamagnetism of a superconductor is illustrated in

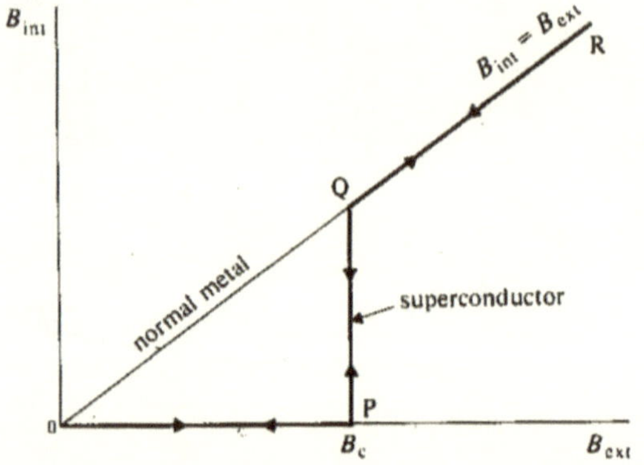

Fig. 13. The perfect diamagnetism of a superconductor.

Fig. 13. As the external field, B_{ext}, is raised, induced surface currents *screen* the field changes from the interior, and the magnetic induction in the material *remains strictly zero,* until the critical field B_c is reached at P. Superconductivity is then *destroyed* and flux *enters* until $B_{int} = B_c$ at Q. For $B_{ext} > B_c$, $B_{int} = B_{ext}$, the system moves along QR, as it would for a *normal metal.* If the field is reduced, the surface currents reappear with the restoration of superconductivity at $B_{ext} = B_c$, the flux is expelled and the system returns along QPO. Of the two basic characteristics of superconductivity, *perfect conduction* and *perfect diamagnetism,* the latter is, in fact, the more fundamental.

The sharpness of the transition to the superconducting state is strongly affected by the presence of strains and impurities in the superconductor. In the absence of such extraneous effects, the transition is sharp and reversible. For example, it has been shown by direct experiment that the transition of tin in zero magnetic field takes place reversibly at $T_c = 3.73\text{K}$, over a temperature interval smaller than 10^{-4} K.

We must consider the superconductor to be a system of three degrees of freedom.

, $P, V; T, S; \boldsymbol{B}, \boldsymbol{m}.$

In the thermodynamic analysis below, we shall retain all of these; but superconductivity is not strongly affected by hydrostatic pressure, and it is often possible to disregard the first two variables. *We shall simplify the mathematics by dropping the vector notation for the magnetic quantities.* Also, here and in what follows, we use B for B_{ext}.

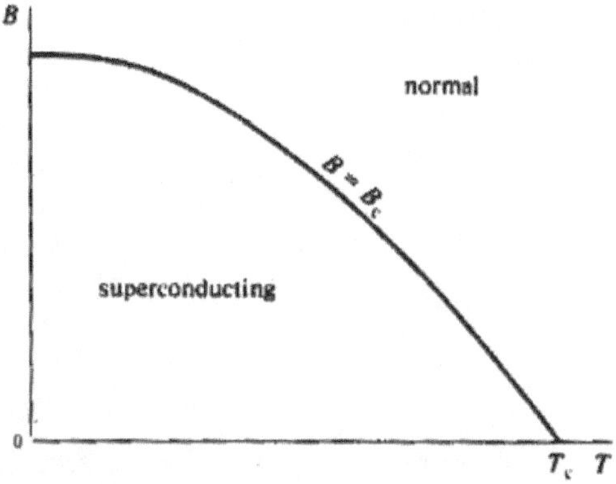

Fig. 14. The phase diagram of a superconductor or the temperature dependence of the critical field B_c.

Fig. 14 shows the phase diagram for a superconductor in the $B-T$ plane. At all points on the boundary between the superconducting and normal phases, the transition is *first order,* except at $B = 0$ and $T = 0$. This is because at this temperature, the entropy is equal to zero as is required by the third law of thermodynamics.* As noted above, the value of the magnetic field, which, at a given temperature, destroys superconductivity is the *critical field,* B_c. Now, in contrast, the *critical temperature* T_c is normally taken as the *transition temperature in zero field,* at which transition from superconducting phase to normal phase takes place.

We now apply the thermodynamics which we have developed for change of phase to prove that on the phase boundary, $g_s = g_n$, where the subscripts s and n refer to the superconducting and normal phases, respectively.

It is convenient to choose *T*, *B*, and *P* as independent variables. The appropriate Gibbs potential function is

$$g = u - Ts + Pv - Bm, \tag{12}$$

where *m* is the *magnetic moment per unit mass,* and *u* is defined as

$$du = Tds - Pdv + B\,dm. \tag{13}$$

Taking the differential of Eq. (12), we have

$$dg = du - Tds - sdT + Pdv + vdp$$
$$- B\,dm - m\,dB\,.$$

Substituting for d*u* in this equation by Eq. (13), we obtain

$$dg = Tds - Pdv + Bdm - Tds - sdT$$
$$+ Pdv + vdP - B\,dm - m\,dB$$
$$dg = -sdT + vdP - mdB \tag{14}$$

* The third law of thermodynamics shall be seen in Chapters 7 and 20.

At equilibrium, $d\mathcal{g} = 0$, since $dT = dP = dB = 0$, which gives the condition that on the *phase boundary*

$$\mathcal{g}_n = \mathcal{g}_s. \tag{15}$$

Before discussing properties on the phase boundary, it is worth digressing for a moment to derive a simple expression for the difference of the Gibbs function of a conductor in the normal and superconducting states. From Eq. (14),

$$\mathcal{g}(B) = \mathcal{g}(0) - \int_0^B m\,dB \quad \text{(at constant } T \text{ and } P\text{)}$$

For the superconductor, as long as it remains superconducting, we may put

$$m = -Bv/\mu_o,$$

since this gives $B_{int} = 0$ and, therefore, *corresponds to perfect diamagnetism*, μ_o being the *permeability of a vacuum*. Therefore, $\mathcal{g}_s(B)$ is given by

$$\mathcal{g}_s(B) = \mathcal{g}_s(0) + \int_0^B \frac{Bv}{\mu_o},$$

which by integration for $B \leq B_c$ gives

$$\mathcal{g}_s(B) = \mathcal{g}_s(0) + \frac{vB^2}{2\mu_o}. \tag{16}$$

At the transition, using Eq. (15),

$$\mathcal{g}_n(B_c) = \mathcal{g}_s(B_c),$$

Eq. (16) can be rewritten in the form

$$\mathcal{g}_n(B_c) = \mathcal{g}_s(0) + \frac{vB_c^2}{2\mu_o}. \tag{17}$$

But the normal metal has a negligible susceptibility, so to a very good approximation,

$$\mathcal{g}_n(B) = \mathcal{g}_n(0),$$

and hence,

$$\mathcal{g}_n(0) = \mathcal{g}_s(0) + \frac{vB_c^2}{2\mu_o} \tag{18}$$

Returning now to consider the phase boundary, we may derive from Eq. (14) three analogues of the Clausius-Clapeyron equation, taking the three independent variables *T*, *P*, and *B* in pairs and applying the condition of equilibrium on the phase boundary, namely, Eq. (15) as we did for a two-parameter system in Section 4. This gives the following:

1. *We take the independent variables P and T at a constant B.*

Eq. (14) thus reduces to

$$d\mathcal{g} = -s\,dT + v\,dP$$

so that at equilibrium *on the phase boundary*, we have

$$-s_n dT + v_n dP = -s_s dT + v_s dP$$

$$(v_s - v_n) dP = (s_s - s_m) dT$$

or

$$\left(\frac{\partial P}{\partial T}\right)_B = \frac{s_n - s_s}{v_n - v_s} = \frac{\Delta s}{\Delta v} \tag{19}$$

2. *We take the independent variables B and T at a constant P*
Eq. (14) thus reduces to

$$d\mathcal{g} = -sdT - mdB.$$

Taking *B* as the critical field B_c, we have

$$d\mathcal{g} = -sdT - mdB_c.$$

At equilibrium *on the phase boundary,* we have

$$-s_n dT - m_n dB_c = -s_s dT - m_s dB_c$$

$$(m_n - m_s) dB_c = -(s_n - s_s) dT$$

or

$$\left(\frac{\partial B_c}{\partial T}\right)_P = -\frac{s_n - s_s}{m_n - m_s} = -\frac{\Delta s}{\Delta m}. \tag{20}$$

But we already know, as noted above, that the magnetic moment *m* per unit mass is

$$m = -Bv/\mu_\circ.$$

Therefore,

$$m_n = -B_c v_n/\mu_\circ,$$

$$m_s = -B_c v_s/\mu_\circ,$$

and

$$m_n - m_s = -B_c(v_n - v_s)/\mu_\circ.$$

Again, since the normal metal has a negligible susceptibility, to a good approximation, by neglecting v_n, we have

$$m_n - m_s = v_s B_c/\mu_\circ.$$

Accordingly, Eq. (20) can be rewritten as

$$\left(\frac{\partial B_c}{\partial T}\right)_P = -\frac{\Delta S}{\Delta m} = -\mu_\circ\left(\frac{s_n - s_s}{v_s B_c}\right) \tag{21}$$

3. *We finally take the independent variables B and P at constant T.*

Eq. (14) thus reduces to

$$d\mathcal{g} = vdP - mdB,$$

and at equilibrium on the boundary surface, we have

$$v_n dP - m_n dB = v_s dP - m_s dB.$$

Putting *B* as equal to B_c, we obtain

$$(v_n - v_s)dP = (m_n - m_s)dB_c.$$

Therefore,

$$\left(\frac{\partial B_c}{\partial P}\right)_T = \frac{v_n - v_s}{m_n - m_s} = \frac{\Delta v}{\Delta m},$$

Considering again the difference $m_n - m_s$ as approximately equal to $-m_s = v_s B_c/\mu_\circ$, we have

$$\left(\frac{\partial B_c}{\partial P}\right)_T = \frac{v_n - v_s}{m_n - m_s} = \frac{\Delta v}{\Delta m}$$

$$= \mu_\circ \frac{v_n - v_s}{v_s B_c} \qquad (22)$$

Of the above Eqs. (19), (21), and (22), the first is of identical form to the Clausius-Clapeyron equation for a system with *two* degrees of freedom and subjected to work by hydrostatic pressure. The values of s_n, s_s, v_n and v_s, which are appropriate to these equations should, of course, be those at the transition; that is, with $B = B_c$. However, these quantities are *virtually independent of field*, as may be seen by examining two of the *Maxwell* relations generated from Eq. (14).

$$d\mathcal{g} = -\,sdT + vdP - mdB, \qquad (14)$$

$$\left(\frac{\partial s}{\partial B}\right)_{T,P} = \left(\frac{\partial m}{\partial T}\right)_{B,P} \quad \text{and} \qquad (23)$$

$$\left(\frac{\partial v}{\partial B}\right)_{T,P} = -\left(\frac{\partial m}{\partial P}\right)_{B,T}. \qquad (24)$$

In the absence of ferromagnetism (and ferromagnets do not become superconducting), a normal metal is only very weakly magnetic, so m_n is essentially zero, and consequently, s_n and v_n are essentially field independent. On the other hand, in the superconducting state, we have $m_s = -\,v_s B_{ext}/\mu_\circ$, which is almost independent of temperature and pressure if the field is constant. Hence s_s and v_s are essentially field independent. It is, therefore, sufficient to take *for these quantities their values in zero field.*

Rearranging now Eq. (21) obtained above, namely,

$$\left(\frac{\partial B_c}{\partial T}\right)_P = -\,\mu_\circ \frac{s_s - s_s}{v_s B_c} \qquad (21)$$

we see that the change in entropy at the transition is

$$s_n - s_s = -\frac{v_s B_c}{\mu_\circ}\left(\frac{\partial (B_c)}{\partial T}\right)_P$$

$$= -\frac{v_s}{2\mu_\circ}\left(\frac{\partial B_c^2}{\partial T}\right)_P . \tag{25}$$

This change *vanishes* at T_c where $B_c \to 0$ with a finite slope, and also, it vanishes at $T = 0$, where $(\partial B_c/\partial T)_P = 0$ (Fig. 14), the latter behaviour being required by the third law.* *Except in these limits*, the transition is, therefore, *first order*.

Differentiating Eq. (25) with respect to *T* at constant *P*, we obtain the difference of the specific heats

$$\left(\frac{\partial s_n}{\partial T}\right)_P - \left(\frac{\partial s_s}{\partial T}\right)_P = -\frac{v_s}{2\mu_\circ}\frac{\partial^2}{\partial T^2}\left(B_c^2\right)_P ,$$

that is,

$$(c_P)_s - (c_P)_n = \frac{Tv_s}{2\mu_\circ}\frac{\partial^2}{\partial T^2}\left(B_c^2\right)_P . \tag{26}$$

From *the typical temperature dependence of the critical field* B_c, *shown* in Fig. 14, we see that Eqs. (25) and (26) yield entropy difference, $s_n - s_s$, and specific heat difference, $(c_P)_s - (c_P)_n$, of the form shown in Fig. 15. Experiment shows that these equations are obeyed well by real superconductors.

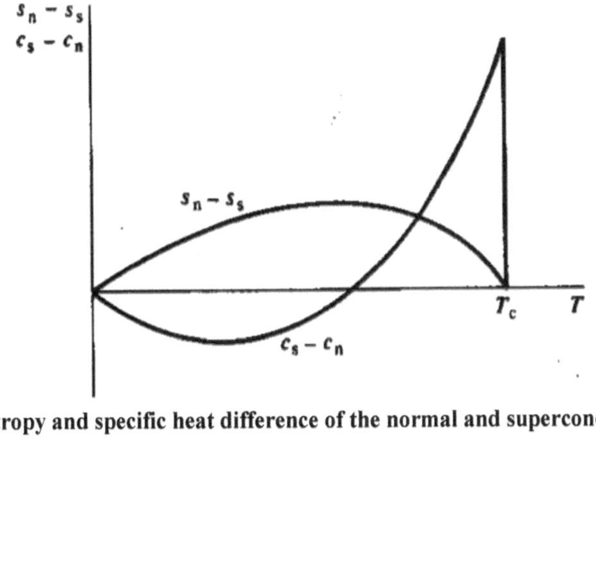

Fig. 15. The entropy and specific heat difference of the normal and superconducting phases.

* The third law shall be seen in Chapters 7 and 20.

At $T = T_c$, where the latent heat becomes *zero*, the transition in zero field becomes *second order*, the discontinuity being in c_p.

Now considering Eq. (19) obtained above, namely,

$$\left(\frac{\partial P}{\partial T}\right)_B = \frac{\Delta s}{\Delta v} = \frac{s_n - s_s}{v_n - v_s}, \tag{19}$$

which is of identical form to the Clausius-Clapeyron equation, and evaluating its limiting forms by taking the derivatives of the numerator with respect to *T* and the denominator with respect to *P*, we obtain

$$\left(\frac{\partial T_c}{\partial P}\right)_{B=0} = vT_c \frac{\beta_n - \beta_s}{(c_p)_n - (c_p)_s} = \frac{k_n - k_s}{\beta_n - \beta_s}. \tag{27}$$

This equation can be derived as follows:

Let *s* be a function of *P* and T_c, therefore, for the superconductor we have

$$ds_s = \left(\frac{\partial s_s}{\partial T_c}\right)_P dT_c + \left(\frac{\partial s_s}{\partial P}\right)_{T_c} dP$$

and for the normal conductor,

$$ds_n = \left(\frac{\partial s_n}{\partial T_c}\right)_P dT_c + \left(\frac{\partial s_n}{\partial P}\right)_{T_c} dP$$

Taking $ds_s = ds_n$ for an infinitesimal change along the boundary, we obtain

$$\left(\frac{\partial P}{\partial T_c}\right)_{B=0} = -\frac{\left(\frac{\partial s_s}{\partial T}\right)_P - \left(\frac{\partial s_n}{\partial T}\right)_P}{\left(\frac{\partial s_s}{\partial P}\right)_T - \left(\frac{\partial s_n}{\partial P}\right)_T} = \frac{(c_P)_s - (c_p)_n}{\beta_s - \beta_n}\frac{1}{vT_c}$$

In like manner, let *v* be a function of *P* and T_c, therefore,

$$dv_s = \left(\frac{\partial v_s}{\partial T_c}\right)_P dT_c + \left(\frac{\partial v_s}{\partial P}\right)_{T_c} dP,$$

and

$$dv_n = \left(\frac{\partial v_n}{\partial T_c}\right)_P dT_c + \left(\frac{\partial v_n}{\partial P}\right)_{T_c} dP.$$

Taking again $dv_s = dv_n$ for an infinitesimal change along the boundary, we have

$$\left(\frac{\partial P}{\partial T_c}\right)_{B=0} = -\frac{\left(\frac{\partial v_s}{\partial T_c}\right)_P - \left(\frac{\partial v_n}{\partial T_c}\right)_P}{\left(\frac{\partial v_s}{\partial P}\right)_{T_c} - \left(\frac{\partial v_n}{\partial P}\right)_{T_c}} = \frac{\beta_s - \beta_n}{k_s - k_n}$$

Combining the previous two equations after taking their reciprocals, we have

$$\left(\frac{\partial T_c}{\partial P}\right)_{B=0} = vT_c \frac{\beta_s - \beta_n}{(c_p)_s - (c_p)_n} = \frac{k_s - k_n}{\beta_s - \beta_n},$$

or

$$\left(\frac{\partial T_c}{\partial P}\right)_{B=0} = vT_c \frac{\beta_n - \beta_s}{(c_p)_n - (c_p)_s} = \frac{k_n - k_s}{\beta_n - \beta_s}. \tag{27}$$

Eq. (27), as might be expected, simply the *Ehrenfest equation*, examples of which are those derived before in Section 8, namely,

$$\frac{dP}{dT} = \frac{\beta_2 - \beta_1}{k_2 - k_1} = \frac{1}{vT}\frac{(c_p)_2 - (c_p)_1}{\beta_2 - \beta_1}$$

Since the superconducting transition in zero field is of the *ideal second order form*, as pointed out above, the specific heat having a simple discontinuity (Fig. 15), it would be of interest to verify these equations in this case. Unfortunately, this test does not seem to have been made, partly because recent interest has been more concerned with a microscopic interpretation of changes at the transition and also because the changes in the expansivities and compressibilities are so small as to make the experiments difficult to do with sufficient precision. For this, for example, it is found that

$$\left(\frac{\partial T_c}{\partial P}\right)_{B=0} = -5 \times 10^{-5}\ \mathrm{K\ atm}^{-1},$$

and

$$\Delta c_p = 9 \times 10^{-2}\ \mathrm{jK^{-1}\ kg^{-1}}$$

which imply a change in the expansivity of about $5 \times 10^{-8}\ \mathrm{K}^{-1}$ and a fractional change in the isothermal compressibility about 10^{-5}.

It is to be noted that in a first-order transition, the gradients of the Gibbs function for the stable state of the system *change discontinuously* at the transition, as the system passes from the g surface of one phase to that of another phase at their intersection.

Now in a second-order transition, the system cannot pass from one g surface to another as we may show by the following argument:

Suppose that the system did pass from one g surface to another.

1. The absence of latent heat or volume change would require the gradients of the Gibbs function to be identical at the transition.
2. The discontinuous change in the second-order differential of g function would require one curvature to be greater than the other.

But then, the surfaces would touch and not cross at the transition. The same phase would always be more stable, and there would be no transition.

We are, therefore, forced to the conclusion that a single g surface is involved in the second-order transition and that some property of the system caused the second derivatives in g to change discontinuously.

Explanations for why there are different kinds of behaviour in high-order transitions are found in microscopic theory.*

6.13. Cooling by adiabatic demagnetisation

1. Statistical approach and the unattainability of absolute zero. Although the production of low temperatures may require great experimental skill and resources, the theoretical principle involved is extremely simple. Let us suppose that the state of an assembly† is defined by the temperature and one or more other parameters $\zeta_1, \zeta_2 \ldots$, so that, in particular, the entropy S is of the form

$$S = S(T, \zeta_1, \zeta_2 \ldots). \quad (1)$$

Provided one of the quantities $\zeta_1, \zeta_2 \ldots$, say ζ can be *controlled and varied reversibly*, then evidently, by varying ζ_1 under *reversible adiabatic conditions*, one can obtain either a *rise* or *fall* of T according to the direction of the change in ζ_1 From a practical point of view, the important question is whether there is a suitable ζ_1 which will produce a sufficiently great change of T to make the process worth while. We shall now explain why the choice of a magnetic field H as the *variable* ζ_1 is a particularly useful one at low temperatures and provides an efficient means of covering the temperature range with an upper end between 4K and 0.1K and a lower end between 0.01K and 0.001K. This method of attaining such low temperatures was proposed almost simultaneously by *Giauque* and by *Debye.*

In an assembly, whose state can be completely defined by the *temperature* T and *the magnetic field* H (all other degrees of freedom such as the pressure and composition being either irrelevant or held constant), the equation for a reversible adiabatic process, that is, process constant entropy, has the form

$$S(H, T) = \text{constant}. \quad (2)$$

In a crystal containing *paramagnetic ions*, at not too high temperature, there are only two significant contributions to the entropy, a contribution S^{ac} of the *acoustical modes* and a contribution S^{sp} of the *electronic spins* of the paramagnetic ions, which is due to the orientation of the resultant electronic spins of the paramagnetic ions. We, therefore, rewrite Eq. (2) in the more explicit form

$$S^{ac}(T) + S^{sp}(H, T) = \text{constant}. \quad (3)$$

---

* See *Equilibrium Thermodynamics* by C. J. Adkins (Cambridge University Press, 1983, Section 10.9.

† The term assembly is used to represent a group of systems, either of the same type or of different types N_1, $N_2 \ldots$ in number. This term will be considered in detail when we deal with statistical mechanics in Section 17.1.

For an understanding of the efficiency of the use of a magnetic field for producing changes of temperature, the point of fundamental importance is the following. Where $S^{ac}(T)$ is *positive* and decreases as the temperature decreases, $S^{sp}(H, T) - S^{sp}(0, T)$ is *negative*, and for a given value of *H*, its *magnitude increases as the temperature decreases.* It is, therefore, obvious that lower the temperature more important become the variations of the term S^{sp} relative to the variations of term S^{ac}. We may, therefore, expect *changes of temperature associated with adiabatic variation of the magnetic field to be more important at low temperatures than at high temperatures.* We now verify this expectation and obtain quantitative relations.

Based on Debye's limiting law for the acoustical modes, the equation

$$S^{ac} + S^{sp} = \text{constant} \qquad (3)'$$

is valid in the temperature range below 10K for a reversible adiabatic process, since all other contributions to the entropy are either zero (e.g. *rotations* and *internal vibrations* of molecules or ions) or *independent of temperature* (e.g. *nuclear spin* and other *nuclear internal degrees of freedom*). Usually S^{ac} become *inappreciable* somewhere between 5K and 1K. Eq. (3) then reduces to

$$S^{sp}(H, T) = \text{constant}. \qquad (4)$$

Also in virtue of the *partition function** of the electronic orientation for an ion in the *S* state (or in an effective *S* state) in a magnetic field *H*, it can be proved, without going into mathematical details, that Eq. (4) is equivalent to

$$T \ \alpha \ H \qquad (S = \text{constant}). \qquad (5)$$

Now suppose that a magnetic field is applied to a suitable paramagnetic crystal, say ammonium ferric alum, and is cooled to 4K by means of liquid helium. If the crystal is then insulated, and the magnetic field is reduced to a *quarter* of its original value, the temperature will, according to Eq. (5), fall to 1K. This is clearly illustrated in Fig. 1 (1) which is due to Simon.† If we compare the field strengths at 4K, and 1K for *equal entropy*, we notice that the former are almost exactly four times the latter. At higher temperatures, the relation (5) between *T* and *H* is no longer accurate, because the contribution of the lattice vibrations to the entropy.

The method just described has been used in recent years, especially by *De Haas, Wiersma*, and *Kramers* at Leiden; by *Giauque‡ and Mac-Dougall* at Berkeley; and by *Simon* and *kurti* at Oxford. It has been used to reach temperatures below 0.01K.

* Partition functions or sum of states shall be seen in detail in Chapter 9.

† Simon, Conference on Magnetism (Strasburgy, May 1939).

‡ For detailed references, see for example, Giauque, Proc. 7th. Int. Cong. Refrigerations (1936); Ind. Eng. Chem. **28**, 743 (1936).

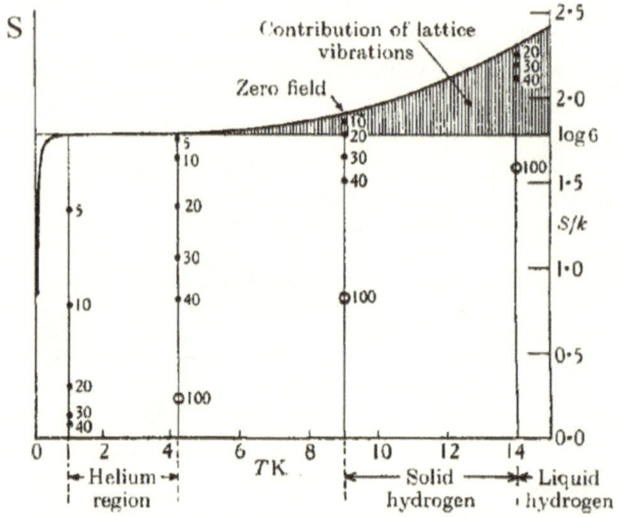

Fig. 1. Molecular entropy of $NH_4\,Fe(SO_4)12H_2O$ due to Simon. Numbers scattered over diagram denote magnetic field strength in *k* gauss.

The question that might arise here is whether the absolute zero is attainable or not? The reply is that it is unattainable as can be seen from the following:

If Eq. (5) is accurate, then by adiabatic reduction of the magnetic field to zero, one would reach $T = 0$, in *contradiction* to the principle called the *third law of thermodynamics*, the statistical discussion of which will be found in Chapter 20. But how we can suffice ourselves by the precise enunciation of this principle which is

> *It is impossible by any procedure, no matter how idealised, to reduce any assembly to the absolute zero in a finite number of operations.*

Eq. (4), however, from which Eq. (5) was derived, becomes *inaccurate* before $T = 0$ is *reached.* Actually there are two different effects which invalidate Eqs. (4) and (5).

Eq. (4) depends on the assumption that in the absence of a magnetic field each paramagnetic ion is in a $(2S+1)$-fold degenerate state. This is not strictly true; it is true only as far as one ignores the *Stark effect* of the crystalline field on the spins.

The other effect which invalidates the Eq. (4) at sufficiently low temperatures is the *interaction* between the paramagnetic ions. We have hitherto regarded these ions as completely independent of one another, so far as their orientations are concerned. Actually, there is an interaction energy between every pair of magnetic ions depending on their *mutual orientations*. Consequently, it is only an approximation to regard the assembly as composed of N *independent magnets*, but the approximation is *allowable* as long as the interaction energies between pairs of ions is *small* compared with kT. The temperature at which this condition *fails* will be lower as *farther apart* the paramagnetic ions are. By choosing a crystal in which the paramagnetic ions are *greatly diluted*, one can make this temperature as *low* as one pleases, but the practical utility of this is limited; for the increase in the volume of crystal containing a given number of ions would soon become a disadvantage, outweighing any advantage of the diminution of the magnetic interaction between the ions.

2. The magnetocaloric effect. Before discussing this effect, which is the basis of the process of cooling by adiabatic demagnetisation, it is helpful to give, as an introduction, a brief account of some basic relations in magnetism as follows.

In magnetism, the *magnetic induction B* is related to the *magnetic field strength H* and *magnetisation M* (using SI units) by

$$B = \mu_\circ \mu_r H = \mu_\circ (H + M), \tag{1}$$

where $\mu_\circ$ is the permeability of a vacuum and μ_r is the relative permeability. The *magnetic susceptibility* X_m is given by

$$X_m = \mu_r - 1 = M/H_{ext}\,. \tag{2}$$

It is important to remember that *B* is the induction *in the absence of the specimen.* If the material is weakly magnetic $(X << 1)$, then the internal and external fields will be approximately equal, but if the material is strongly magnetic, then the internal field may be much smaller than the applied. This difference is treated by introducing the idea of a *demagnetising factor* n, defined by

$$H_{int} = H_{ext} - nM, \tag{3}$$

in which *n* varies from 0 to 1 for differently shaped bodies. The total magnetic moment *m* of the specimen is related to the magnetisation *M* by

$$M = \frac{m}{V} \text{ (in magnitude)}, \tag{4}$$

where *V* is the volume of the specimen. Owing to the significant role which the partial derivative $(\partial\chi_m/\partial T)_B$ plays in the process of adiabatic demagnetisation, a convenient expression for it seems likely. Thus, in the absence of the material (i.e. $\mu_r = 1$), the applied induction *B* according to Eq. (1) is given by

$$B = \mu_\circ H_{ext},$$

or

$$H_{ext} = \frac{B}{\mu_\circ} \text{ (in magnitude)}. \tag{5}$$

The susceptibility χ_m according to Eqs. (2), (4), and (5) is, therefore, given by

$$\chi_m = \frac{M}{H_{ext}} = \frac{m/V}{H_{ext}} = \frac{m/V}{B/\mu_\circ} = \frac{\mu_\circ m}{BV}. \tag{6}$$

It is evident from this equation that the denominator $B/\mu_\circ$ stands for the external field H_{ext}. Thus if we neglect the demagnetising* effects, then it follows from Eq. (3) that H_{int} is equal to H_{ext}. So Eq. (6) can be considered as an applicable equation in this case, where

* The demagnetising effects can be neglected if χ_m is not too large.

χ_m is $\dfrac{m/V}{H_{int}}$, that is, $\chi_m = \dfrac{\mu_\circ m}{BV}$. Clearly, from this equation, the variation of χ_m with temperature yields the following important equation at constant induction.

$$\left(\frac{\partial \chi_m}{\partial T}\right)_B = \frac{\mu_\circ}{BV}\left(\frac{\partial m}{\partial T}\right)_B . \tag{7}$$

Now by making use of the above relations (1 to 7), a discussion of the interdependence of the thermal and magnetic properties of a material, which is known as the *magnetocaloric* effect, might be achieved.

When the magnetisation of a material is *changed isothermally*, heat is usually exchanged with the *surroundings*. If the change in magnetisation is performed under *adiabatic conditions*, then the temperature will change. At low temperatures, the effect may become *very large* and is of considerable importance in providing the basis of the method of obtaining temperatures below 1K. In our analysis, we shall ignore the hydrostatic pressure work, which is normally negligible if the magnetic material is a solid. Thus neglecting works other than the magnetic work and assuming the material to be isotropic, the first law becomes

$$TdS = dU - B\,dm \tag{8}$$

Applying the condition for dU to be an exact differential to the coefficients on the right-hand side of

$$dU = TdS + B\,dm, \tag{3}$$

we have

$$\left(\frac{\partial T}{\partial m}\right)_B = \left(\frac{\partial B}{\partial S}\right), \tag{9}$$

where B and T are our independent variables. This Maxwell relation may be put in the form

$$\left(\frac{\partial S}{\partial B}\right)_T = \left(\frac{\partial m}{\partial T}\right)_B . \tag{10}$$

In a reversible isothermal change of magnetisation, the heat absorbed by the magnetic material, which is denoted by $C_T^{(B)}$, is given by

$$c_T^{(B)} = \frac{d'q}{dB} = T\left(\frac{\partial S}{\partial B}\right)_T , \tag{11}$$

or by

$$c_T^{(B)} = \frac{d'q}{dB} = T\left(\frac{\partial S}{\partial B}\right)_T = T\left(\frac{\partial m}{\partial T}\right)_B , \tag{12}$$

when Eq. (10) is used.

On the other hand; in a reversible *adiabatic* change of magnetisation, the change in temperature can be deduced as follows:

Put

$$T = T(B,S),$$

therefore,

$$dT = \left(\frac{\partial T}{\partial B}\right)_S dB + \left(\frac{\partial T}{\partial S}\right)_B dS. \tag{13}$$

For dT = 0, we have

$$\left(\frac{\partial T}{\partial B}\right)_S = -\left(\frac{\partial T}{\partial S}\right)_B \left(\frac{\partial S}{\partial B}\right)_T. \tag{14}$$

But according to Eq. (10), we have

$$\left(\frac{\partial S}{\partial B}\right)_T = \left(\frac{\partial m}{\partial T}\right)_B. \tag{10}$$

and since the heat capacity (absorbed) at constant induction, C_B, is expressed as

$$C_B = \frac{d'q_B}{dT} = T\left(\frac{\partial S}{\partial T}\right)_B, \tag{15}$$

then the substitution of Eqs. (10) and (15) in Eq. (14) yields

$$\left(\frac{\partial T}{\partial B}\right)_S = -\left(\frac{\partial T}{\partial S}\right)_B \left(\frac{\partial S}{\partial B}\right)_T = -\frac{T}{C_B}\left(\frac{\partial m}{\partial T}\right)_B. \tag{16}$$

If the susceptibility $\chi_m = m/V\,H_{int}$ is not too large, we may neglect demagnetising effects and put $H_{ext} \simeq H_{int} = B/\mu_\circ$. Hence, in terms of $_\chi$, Eqs. (10) and (16) become

$$T\left(\frac{\partial S}{\partial B}\right)_T = \frac{TVB}{\mu_\circ}\left(\frac{\partial \chi_m}{\partial T}\right)_B \tag{17}$$

and

$$\left(\frac{\partial T}{\partial B}\right)_S = -\frac{TVB}{\mu_\circ C_B}\left(\frac{\partial \chi_m}{\partial T}\right)_B, \tag{18}$$

respectively. Eq. (18) is known as the *magnetic cooling equation.*

It should be noted that both Eqs. (17) and (18) contain the temperature derivative of the magnetic susceptibility, so materials in which the susceptibility does not vary with temperature show so magnetocaloric (or magnetothermal) effects; that is, there is *no interdependence* of the thermal and magnetic properties in these materials. This is the case with *simple diamagnetism*, where the magnetic response results from the *perturbation by the applied field of the electronic eigenstates in the atoms*. These are essentially unaffected by temperature. Paramagnetism, on the other hand, results from the *presence of microscopic magnetic dipoles* in the material, which may be aligned by the application of a magnetic field. Thermal motions tend to *misalign* the diploes in a paramagnetic material, so that the extent of the alignment *decreases* with *increasing* temperature. Thus in such a condition, $(\partial\chi_m/\partial T)_B$ is always *negative*, and, according to Eq. (17), *heat is evolved in an isothermal magnetisation*. This is in agreement with what would be expected from the connection between *entropy* and *order*, for the magnetisation of a paramagnet increases the magnetic *ordering* and thus *decreases* the magnetic contribution to the *entropy*. In an *isothermal* change, heat will, therefore, be *evolved*. The dependence of the magnetic susceptibility of a paramagnetic material on temperature increases *rapidly* as the temperature is *reduced* towards the point at which *spontaneous* magnetic ordering sets in. It is, therefore, in this region, that strong magnetothermal effects are to be expected. They are only important at low temperatures where specific heats are generally *small*, and relatively large changes of temperature may be produced.

The magnetocaloric effect has been and still is of great importance as an experimental tool for obtaining temperatures below 0.3K.

3. The entropy of a paramagnetic salt as a function of the two independent variables *T* and *B*. Fig. 2 shows the variation of entropy with temperature and magnetic field for a typical paramagnetic salt. *In zero field,* the fall in entropy at the

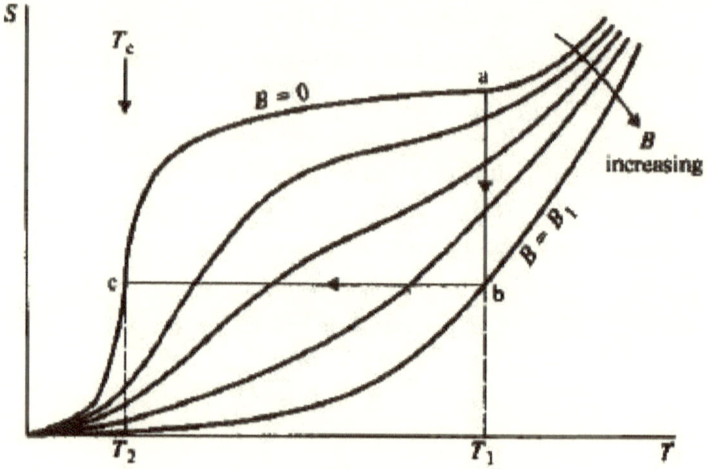

Fig. 2. The entropy of a paramagnetic salt as a function of temperature and magnetic induction.

Curie temperature, T_c, *corresponds to the onset of spontaneous ordering.* At higher temperatures, the entropy may be reduced by applying a magnetic field and hence increasing the magnetic order. The process of cooling the salt is illustrated in the figure. The salt is first magnetised by applying a field of induction B_1 at an initial temperature T_1, which is usually obtained by evaporation of liquid ^4He or liquid ^3He under reduced pressure. The *heat evolved during magnetisation* is conducted away from the helium bath, and the entropy *falls* as the salt goes from state a to state b. The salt is then *thermally insulated and demagnetised.* If the demagnetisation is performed *sufficiently slowly*, then the process is *reversible*, and thus the *entropy remains constant,* whereas the temperature *falls*. If the field is reduced to *zero,* then the final state of the salt will be at *c* with temperature T_2. Clearly, the lowest temperature to which the salt can be cooled by adiabatic demagnetisation is *effectively the Curie temperature.*

4. The arrangements for adiabatic demagnetisation experiment. One experimental arrangement for adiabatic demagnetisation is illustrated in Fig. 3. The salt is suspended in a

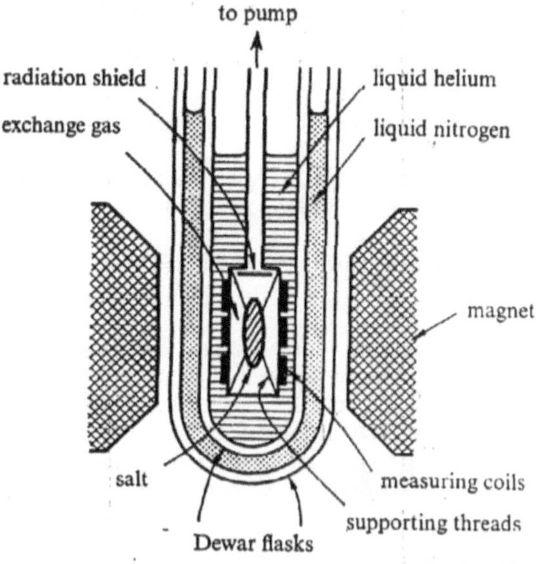

Fig. 3. A typical arrangement for adiabatic demagnetisation experiment.

chamber which is immersed in the helium bath which produces the initial cooling. During the isothermal magnetisation, thermal contact with the bath is provided by helium 'exchange' gas in the chamber. After magnetisation, the gas is removed thus insulating the sample thermally, and the field is then reduced to zero.

Another arrangement for adiabatic demagnetisation experiment is shown in Fig. 4.

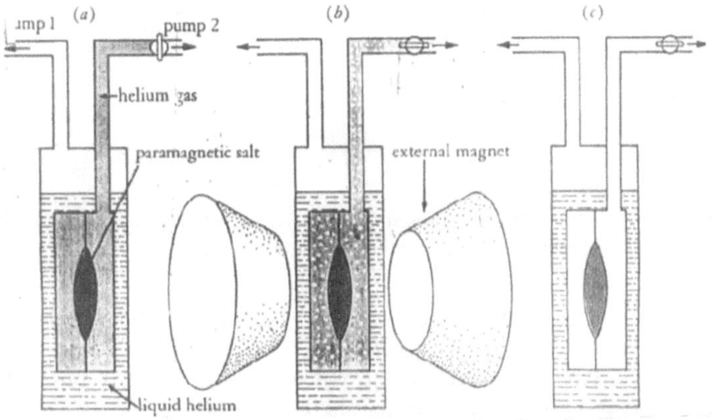

Fig. 4. Another arrangement for adiabatic demagnetisation experiment.

The experimental aspects of the situations diagrammed in Fig. 4. (a), (b), and (c).

a. Pump 1 keeps pumping on liquid helium, thus maintaining the temperature at its boiling point. The helium gas surrounding the salt serves to maintain the thermal contact between the salt and liquid helium that acts as a constant-temperature bath.
b. The magnetic field is applied. After the field has reached full strength, the helium gas is pumped off by pump 2 so that the paramagnetic salt becomes thermally insulated.
c. The magnetic field is slowly removed, and finally, the salt assumes a lower temperature.

5. The microscopic significance of adiabatic demagnetisation. The microscopic significance of the adiabatic demagnetisation is illustrated in Fig. 5. In zero field (a), the energy levels of the microscopic dipoles are close together, their *separation* being determined by the strength of the interaction between neighbouring dipoles and between the dipoles and the lattice. If the *separation is small in comparison with kT* at the initial temperature, the levels will be *nearly equally populated.* The relative populations of the levels

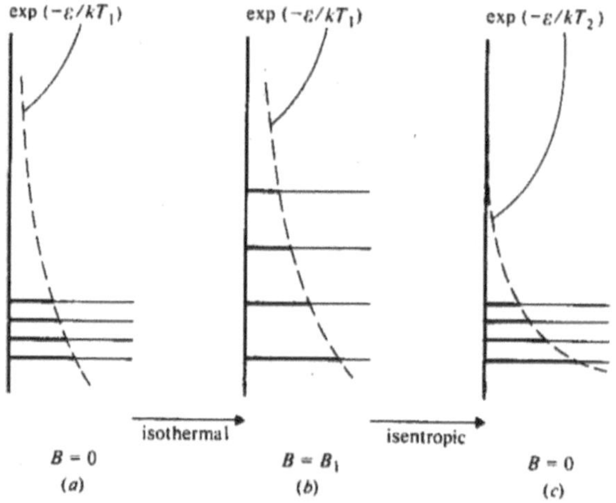

Fig. 5. The microscopic significance of adiabatic demagnetisation. Energy is plotted vertically, and occupation of the levels is indicated by the length of heavy lines.

will vary as $e^{-\epsilon/kT}$, where ϵ is the energy of the level. This will be discussed later when we deal with the *Bose-Einstein quantum statistics.** It may be noted here that the different levels correspond to different orientations. The application of a field (b) *causes the levels to separate*. Transitions then take place between the levels in which the *magnetic subsystem* loses energy to the surroundings (to the helium bath, via the crystal lattice and the exchange gas, as shown in Fig. 3), and a *new distribution* of the microscopic dipoles among the levels

* See Section 9.4.

is established – characteristic of the same temperature – but with the levels *differently* populated because of their changed energies. Under the adiabatic change, which is carried out reversibly, there is no entropy change, and thus *transitions do not occur* (this is the microscopic significance of 'adiabatic'). So when the levels return to their initial or original separations on removal of the field (c), the populations are characteristic of *a much lower temperature*.

Spontaneous magnetic ordering occurs when the thermal energies become *smaller* than the energy differences between the various possible orientations of the dipoles in the absence of an applied field. These energy differences result from the interactions of the dipoles with one another and with the lattice. The *stronger* they are, *higher the Curie temperature,* T_c, *will be, and higher the temperature reached by demagnetisation.* This is one of the reasons why the paramagnetic salts used for '*adiabatic*' demagnetisation are usually chemically complex. By 'diluting' the active magnetic dipoles, so that their mean separation is greater, their interaction energy is reduced. This helps to *lower* the Curie temperature, but it also means that the *magnetic entropy* of the salt is *smaller*, so if heat has to be absorbed from elsewhere, then the salt will become a less *powerful cooling agent*. This is of importance in experiments where the salt is used as a means of *cooling* some other *experimental system*. In such a case, the entropy of the system must be added to that of the salt in calculating the effect of the isentropic process.

When the salt is used as a cooling agent, the condition of reversibility in the demagnetisation process becomes much more *stringent*. *Reversibility requires thermal equilibrium at all times throughout the salt and all that is in thermal contact with it.* When the salt *alone is being cooled*, the only *relaxation time* involved is that for the spin system (on which the applied magnetic field acts directly) to reach *equilibrium with the crystal lattice* in which it is situated. This time varies rapidly with temperature, but for magnetically diluted salts, it is of the order of 1 sec at 1.5K. When another system has to be cooled, however, the limiting factor is usually the thermal contact between the salt and the rest of the system. At very low temperatures, the boundaries between materials offer a high *thermal resistance*, and it may take hours to achieve thermal equilibrium.

In order to calculate the magnetic cooling for a particular substance using the equation

$$\left(\frac{\partial T}{\partial B}\right)_S = -\left(\frac{\partial T}{\partial S}\right)_B \left(\frac{\partial S}{\partial B}\right)_T,$$

$$= -\frac{T}{C_B}\left(\frac{\partial m}{\partial T}\right)_B, \qquad (16)$$

which has been obtained above, it is enough to know its equation of state, $M = M(B, T)$, and its specific heat in zero field over the relevant temperature range. For illustration, we take a highly simplified model.

Sufficiently far above the Curie point, the susceptibility of a paramagnetic material is essentially *independent of magnetic field and obeys Curie's law*, $\chi = a/T$, where a is a constant. This takes no account of the interactions which bring about *spontaneous ordering* at the Curie temperature close to which the susceptibility will rise sharply. A better approximation is the *Curie-Weiss law,*

$$\chi = \frac{a}{T - T_c}, \qquad (19)$$

where T_c is the Curie temperature. This is, in reality, a poor approximation close to the Curie point, where χ becomes large but not infinite and is no longer independent of *B*. However, for our model, we will take the susceptibility to obey the Curie-Weiss law as follows:

We first calculate $C_B\ (B, T)$.

The heat capacity at constant *B* is given, as shown above, by

$$C_B = \frac{d' q_B}{dT} = T\left(\frac{\partial S}{\partial T}\right)_B, \tag{15}$$

therefore,

$$\begin{aligned}\left(\frac{\partial C_B}{\partial B}\right)_T &= \left.\frac{\partial}{\partial B}\right)_T \left[T\left(\frac{\partial S}{\partial T}\right)_B\right] \\ &= T\frac{\partial^2 S}{\partial B\,\partial T} = T\frac{\partial^2 S}{\partial T\,\partial B} \\ &= T\left.\frac{\partial}{\partial T}\right)_B \left(\frac{\partial S}{\partial B}\right)_T .\end{aligned} \tag{20}$$

Using the Maxwell relation

$$\left(\frac{\partial S}{\partial B}\right)_T = \left(\frac{\partial m}{\partial T}\right)_B, \tag{10}$$

which has been obtained above from the differential d*U*

$$dU = TdS + B\,dm, \tag{8}$$

then the substitution for $\left(\frac{\partial S}{\partial B}\right)_T$ in Eq. (20) from relation (10) yields

$$\begin{aligned}\left(\frac{\partial C_B}{\partial B}\right)_T &= T\left.\frac{\partial}{\partial T}\right)_B \left(\frac{\partial m}{\partial T}\right)_B \\ &= T\left(\frac{\partial^2 m}{\partial T^2}\right)_B .\end{aligned} \tag{21}$$

This is one form of the partial derivative $\left(\frac{\partial C_B}{\partial B}\right)_T$. Another form may be obtained by substituting for $\left(\frac{\partial S}{\partial B}\right)_T$ in Eq. (20) by Eq. (17) derived above, which is

$$\left(\frac{\partial S}{\partial B}\right)_T = \frac{VB}{\mu_\circ}\left(\frac{\partial \chi_m}{\partial T}\right)_B, \tag{17}$$

assuming that the demagnetising effects are negligible. Thus we obtain

$$\begin{aligned}\left(\frac{\partial C_B}{\partial B}\right)_T &= T\left.\frac{\partial}{\partial T}\right)_B \frac{VB}{\mu_\circ}\left(\frac{\partial \chi_m}{\partial T}\right)_B \\ &= \frac{TVB}{\mu_\circ}\left(\frac{\partial^2 \chi_m}{\partial T^2}\right)_B .\end{aligned} \tag{22}$$

But from Eq. (19)

$$\left(\frac{\partial \chi_m}{\partial T}\right)_B = \frac{a}{(T - T_c)^2}. \quad (19)$$

Therefore, we have

$$\left(\frac{\partial^2 \chi_m}{\partial T^2}\right)_B = \frac{2a}{(T - T_c)^3}. \quad (23)$$

So by substituting Eq. (23) in Eq. (22), we obtain

$$\left(\frac{\partial C_B}{\partial B}\right)_T = \frac{2a\,TVB}{\mu_\circ (T - T_c)^3}. \quad (24)$$

The integration of Eq. (24) gives

$$C_B(B, T) = C_B(0,T) + \int_0^B \frac{2a\,TVB}{\mu_\circ (T - T_c)^3}\, dB$$

$$= C_B(0,T) + \frac{a\,TVB^2}{\mu_\circ (T - T_c)^3}. \quad (25)$$

The second term on the right-hand side of this equation is associated with the change of magnetic order brought about by the applied field. The first term contains all other contributions to the heat capacity. These are:

a. The contributions of the lattice containing the magnetic ions and of any material cooled by the salt. These are often very small at low temperatures, and we shall neglect them.

b. The contribution is derived from the spontaneous change in magnetic order which takes place in zero field near the *Curie point*. It is not small, but the contribution is peaked close to the Curie temperature, and at higher temperatures, it is relatively small.

Provided then that we do not come too close to the Curie point, we may neglect $C_B(0,T)$ and take $C_B(B,T)$ as

$$C_B(B,T) = \frac{a\,TVB^2}{\mu_\circ (T - T_c)^3}. \quad (26)$$

Substituting Eq. (26) in the expression for the magnetic cooling, Eq. (18), we have

$$\left(\frac{\partial T}{\partial B}\right)_S = \frac{TVB\,(T - T_c)^3}{\mu_\circ \,.\, a\,TVB^2}\left(\frac{\partial \chi_m}{\partial T}\right)_B$$

$$= \frac{(T - T_c)^3}{a\,\mu_\circ\, B}\left(\frac{\partial \chi_m}{\partial T}\right)_B. \quad (27)$$

But

$$\left(\frac{\partial \chi_m}{\partial T}\right)_B = \frac{a}{(T - T_c)^2}, \tag{21}$$

therefore,

$$\left(\frac{\partial T}{\partial B}\right)_S = \frac{T - T_c}{\mu_\circ B}. \tag{28}$$

Integrating this equation, we obtain

$$\frac{dT}{T - T_c} = \frac{dB}{\mu_\circ B}$$

$$[\ln T - T_c]_{T_1}^{T_2} = \frac{1}{\mu_\circ}[\ln B]_1^2$$

$$\frac{T_1 - T_c}{T_2 - T_c} = \frac{B_1}{\mu_\circ B_2}. \tag{24}$$

In this approximation, demagnetising to zero field cools the salt to Curie temperature. In practice, the temperature does not drop as much as this because the approximations we have made cease to be valid near this temperature.

It is clear from the foregoing discussion that the temperature reached in adiabatic demagnetisation is the result of a compromise. The salt must be sufficiently diluted magnetically for the interactions between the magnetic atoms to be small and the Curie temperature to be low. On the other hand, if the salt is too dilute, the entropy change associated with magnetisation becomes small in comparison with the entropy of the rest of the system, and this again, restricts the cooling which can be obtained.

To achieve the lowest temperatures, sophisticated experimental techniques have to be used. With cerium magnesium nitrate, it is possible to reach a few millikelvins. For tower temperatures, weaker magnetic systems have to be used. Adiabatic demagnetisation of *nuclear* magnetic moments, previously cooled by an ordinary paramagnetic salt, have produced nuclear spin temperature of a few tens of microkelvins.

CHAPTER 7

Classical Approach to the Third Law of Thermodynamics and Nernst Heat Theorem

7.1. Introduction

During the nineteenth century, the first and second laws of thermodynamics were applied to many scientific problems. It was in the first decade of the nineteenth century that investigations at very low temperatures had progressed sufficiently to lay the basis for the generalisation which constitutes the third law of thermodynamics.

It has been found experimentally that the heat capacities of solids decrease very rapidly to zero as the absolute zero of temperature is approached. On the other hand, according to Einstein's explanation of heat capacities, the internal energy of a solid is *quantised* so that only finite quanta of energy can be absorbed. Accordingly, as the solid is cooled down to absolute zero, all its constituent particles fall into their lowest quantum states; in other words, we can say that the entire crystal is in its *lowest quantum state*. This is a simple statement which we may take as the *absolute unit of probability*. Conversely, we may define the absolute unit of probability of a crystal as *the lowest quantum state of the crystal which is acquired at absolute zero*.

The second law of thermodynamics is a *consequence of probability*. The equation for the relationship of entropy difference to the probability ratio is

$$S_B - S_A = \frac{R}{N_a} \ln \frac{W_B}{W_A},$$

where N_a is Avogadro's number, W is the probability of the state, and R is the general gas constant.

If we adopt a new probability state with $W = 1$, that is, with unit probability and entropy equal to zero for each element in a crystalline form at absolute zero, we find the entropy of all substances to be positive at all temperatures above the absolute zero, and the probability in these *other states* is, therefore, *greater than unity*.

Now let us consider an element such as sulphur which has more than one crystalline form. If we define the entropy $S = 0$, at absolute zero for rhombic sulphur, we may determine experimentally the entropy of monoclinic sulphur at absolute zero. This experiment requires that we *supercool* the monoclinic sulphur below the temperature of its transition to rhombic, which is 368.5K.

In 1937, *Eastman and McGanock* measured with great care the heat capacity of rhombic sulphur and supercooled monoclinic sulphur in the same calorimeter over a temperature range from 365K down to 13K. In 1959, *West* measured the heat capacity of rhombic sulphur from 298K to the transition temperature of 368.5K. Their results are

1. By graphical integration of $C_P \, d \ln T$ for rhombic sulphur from 0K to 368.5K, the entropy change ΔS is equal to 8.810 ± 0.05 cal K^{-1}; that is,

$$\Delta S_1 = S_{368.5}(\text{rhombic}) - S_0(\text{rhombic})$$

$$= 8.810 \pm 0.05 \text{ cal K}^{-1}. \quad (1)$$

2. The entropy change ΔS_2 of transition from rhombic to monoclinic at $T = 368.5$K is given by

$$\Delta S_2 = S\ (\text{monoclinic}) - S\ (\text{rhombic})$$

$$= \frac{\Delta H}{T} = \frac{96.5}{368.5} = 0.261 \pm 0.002 \text{ cal K}^{-1}. \quad (2)$$

3. By graphical integration of $C_P \, d\ln T$ for monoclinic sulphur from 0 to 368.5K, the entropy change ΔS_3 is given by

$$\Delta S_3 = S_{368.5}\ (\text{monoclinic}) - S_0\ (\text{monoclinic}) \quad (3)$$

$$= 9.041 \pm 0.01 \text{ cal K}^{-1}. \quad (4)$$

Now adding Eqs. (1) and (2), we have

$$S_{368.5}\ (\text{monoclinic}) - S_0\ (\text{rhombic}).$$

$$= 9.07 \pm 0.05 \text{ cal K}^{-1}.$$

This value, as given by Eq. (4), represents the total entropy difference between *S* (monoclinic) at 368.5K and *S* (rhombic) at 0K. Evidently, it is the same within the experimental error, as the entropy difference between *S* (monoclinic) at 368.5K and S (monolclinic) at0K, which is 9.041 ± 0.01 cal K^{-1}, as given by Eq. (3). This means that since we have taken the entropy of rhombic sulphur to be zero at 0K, then the entropy of monoclinic sulphur should also be zero at this temperature.

The same can be said to *any two crystalline modifications* of a compound, that is, the entropy is the same for the two modifications at the absolute zero of temperature.

7.2. The Temperature dependence of ΔS, ΔG, and ΔH

1. ΔS variation with temperature. For an infinitesimal reversible process, the second law states

$$dS = \frac{d'q_{rev}}{T},$$

where $d'q_{rev}$ is the differential amount of heat absorbed by the system reversibly from the source of heat, at the temperature *T*.

Since $C_V = \frac{d'q}{dT}$ at constant *V*, and $C_P = \frac{d'q}{dT}$ at constant *P*, then the infinitesimal entropy change under these conditions is

$$dS = C_V \frac{dT}{T},$$

and

$$dS = C_P \frac{dT}{T},$$

respectively.

For a finite change of the system from T_1 to T_2, we have

$$\Delta S = \int_{T_1}^{T_2} C_V \frac{dT}{T} = \int_{T_1}^{T_2} C_V \frac{dT}{T} \quad (1)$$

or

$$= \int_{T_1}^{T_2} C_V \, d\ln T,$$

and

$$\Delta S = \int_{T_1}^{T_2} C_P \frac{dT}{T}, \quad (3)$$

or

$$\Delta S = \int_{T_1}^{T_2} C_P \, d\ln T \cdot \quad (4)$$

Let us consider a *constant-pressure process*. If C_P is constant in the temperature interval T_1 to T_2, we have

$$\Delta S = C_P \int_{T_1}^{T_2} d\ln T$$

$$= C_P \ln \frac{T_2}{T_1}$$

If C_P is not constant, its dependence on temperature may be handled graphically. For example, Fig. 1 shows the molar specific heat of crystalline PbI_2 between 20 and 100K.

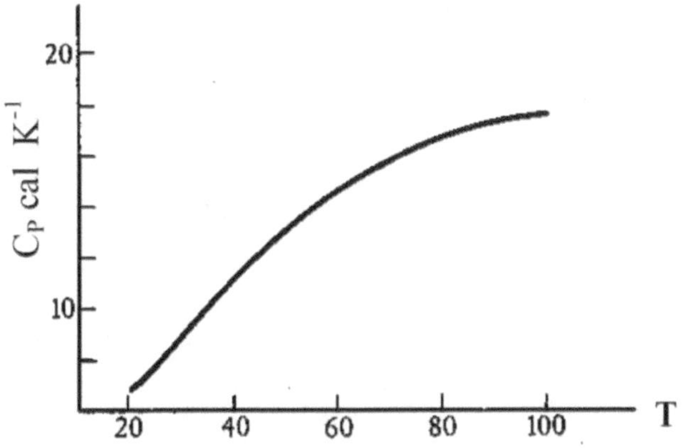

Fig. 1. Variation of C_P with *T* for PbI_2 between 20 and 100 K.

There are two ways for calculating ΔS caused by a temperature increase, based on Eqs. (3) and (4).

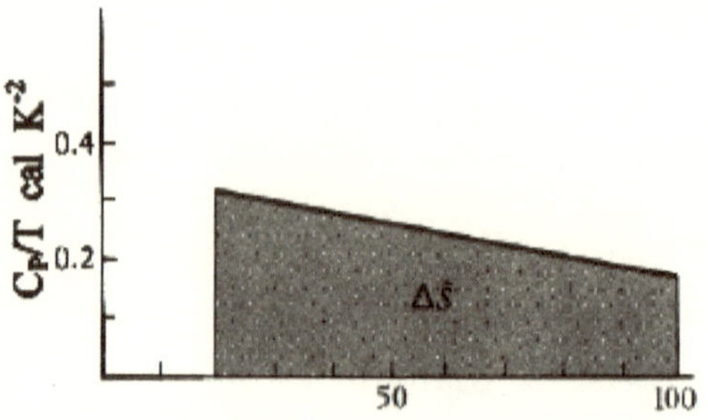

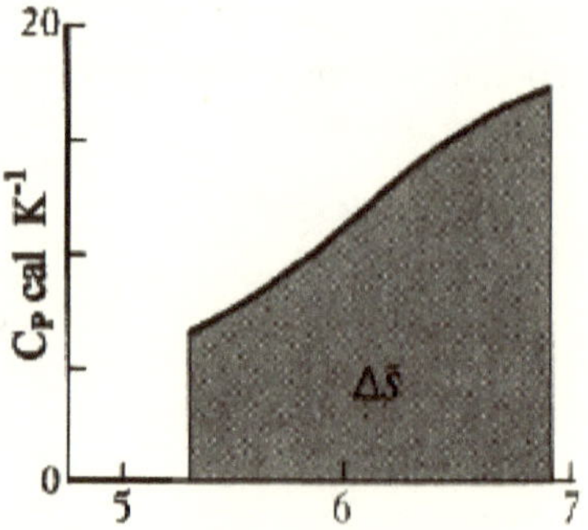

According to Eq. (3), C_P / T is plotted against *T* as shown in Fig. 2, so that the area shaded is equal to ΔS. The other way for calculating ΔS is based on Eq. (4) where C_P is plotted against $\ln T$ as shown in Fig. 3. The area shaded is again equal to ΔS.

Similarly, for heating at constant volume, we have

$$\Delta S = \int_{T_1}^{T_2} \frac{C_v}{T}\, dT \tag{5}$$

or

$$\Delta S = \int_{T_1}^{T_2} C_v \; d \ln T \quad . \tag{6}$$

It is important to know that the *increase of temperature is not always accompanied by an increase in the entropy*. For example, when a gas is *compressed adiabatically* and *reversibly*, there is no entropy change since

$$dS = \frac{d'q_{rev.}}{T} = \frac{0}{T} = 0,$$

despite the fact that *there is a rise in temperature arising from the compression process*. This can be accounted for statistically as follows:

According to statistical thermodynamics,* the entropy *S* of an isolated system is related to the thermodynamic probability *W* of this system by the equation

$$S = k \ln W, \tag{7}$$

which is known as the *Planck–Boltzmann equation*. This relationship may be made use of in a qualitative way to understand the nature of entropy. Thus, the entropy is a measure of molecular disorder, that is, the increase in entropy parallels an increase in disorder. The second law in which the entropy plays the important role is, therefore, a matter of *probability*. Under the *randomising influence* of thermal motion, an *isolated* system tends to assume the *macrostate of maximum* thermodynamic probability. This means that the arrow of time is in the direction of *spreading the system* over as many energy levels as possible. At the same time, if the increase of entropy is a matter of probability, there would always be

---

* The statistical approach to the third law shall be seen in detail in Chapter 20.

the chance that the second law may be broken in the sense that if an isolated system is in a state of maximum probability or maximum entropy, then this state is not a static one, since the constituents of this system are in continuous motion. So a state will result for which the probability and hence the entropy is less than its maximum value. Small changes are more likely than large ones, but large ones are not impossible, they are only highly improbable. We shall consider this question in more detail when we deal with 'Density Fluctuations in a Gas' in Chapter 15.

Now we can attribute exclusively the *increase* of entropy with *increase* of temperature, at *constant volume,* to an increase in *randomness of molecular motion.* That is, as the temperature increases, the spread of atomic or molecular velocities in *velocity space* increases. *At constant pressure,* there is an *additional* increase in randomness connected with the increase in volume. Here, there is, of course, larger heat input which is responsible for this additional increase in randomness.

Suppose that the *increase* of temperature at constant pressures is accompanied by a *decrease* in volume as in the case of water at temperatures 'between 0° and 4°C'. In this case, C_P is still greater then C_V as is required by the equation

$$C_P - C_V = \frac{\beta^2 T V}{k},$$

which has been derived before, where β is the coefficient of volume expansion at constant pressure, and k which is positive is the isothermal compressibility. It is evident from this equation that even when β is negative, β^2 is positive, so C_P is *never smaller* than C_V; it is equal to C_V at 4°C, when β is zero. This means that the *increase* in the randomness of molecular motion with increase of temperature at constant pressure is still *larger* than the *decrease,* that would be if the volume decreases, and also *larger* than that which would be at a constant volume.

Now we return to the *isentropic* process of the *reversible adiabatic compression* of a gas noted above. This process, in the first place, means that the randomness of molecular motion or the randomness on the molecular level *remains constant.* But owing to the *decrease* in volume resulting from compression, there is a *decrease* in randomness of the *spatial distribution* of the molecules. To compensate for this, there must be an *equal increase* in randomness of molecular motion, arising from the *increase of temperature,* which is caused by the adiabatic compression process. This process is thus called *adiabatic heating process.*

Conversely, in the *adiabatic reversible expansion* of a gas, the temperature is *lowered,* and the *entropy remains unchanged.* This isentropic process is thus called *adiabatic cooling process.* The expansion, by itself, *increases the randomness of the spatial distribution of molecules*, and because $\Delta S = 0$, there must be an *equal decrease in the randomness of the molecular motion.* This in turn implies a decrease of temperature.

The fact that the adiabatic reversible *compression* of water below its maximum density is accompanied by slight *cooling, that* may be accounted for as follows:

Since C_p, in the temperature range below 4°C, where the density is maximum, is always higher than C_V, it is concluded that the *randomness* of molecular arrangement (or spatial distribution) at the final volume, which is less than the initial volume due to *compression*, is still greater than it would be at constant volume. This property, in turn, is associated with an *entropy increase*. But since $\Delta S = 0$, in this process of reversible adiabatic compression, there must be an equal *decrease* in the *randomness* of molecular motion. This implies a *decrease* in temperature which is observed.

A question might arise here, which is, 'Does the entropy of a system subjected to an *increase* of temperature as well as a *decrease* of volume caused by compression, undergo a decrease?'

We know that the increase of temperature results in an increase in the randomness of molecular motion, and hence to an increase of entropy. If the increase of temperature is accompanied by a *sufficient decrease* in volume, so that the increase in entropy resulting from increase in molecular motion *does not outweigh* the decrease in entropy resulting from the decrease in molecular arrangement or molecular distribution accompanying the volume decrease, then there would be an *entropy decrease*. This necessitates that there should be a sufficient decrease in volume.

As an illustration of the above question, we might calculate ΔS for the following change of state:

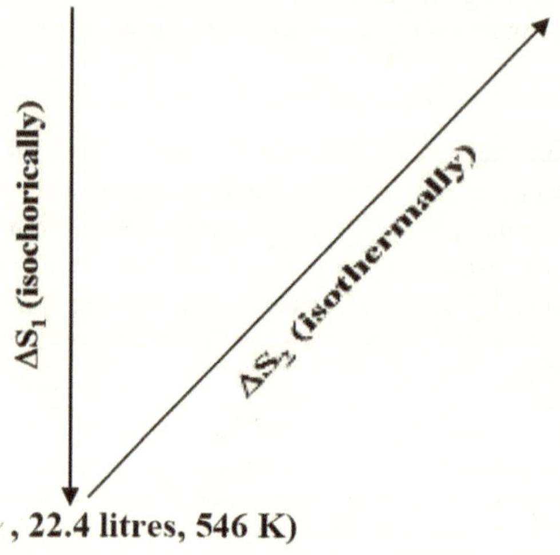

He (g, 22.4 litres, 273K) = He (g, 2.24 litres, 546K) $C_v = 2.97\,\text{cal/mole}$

From this sketch, we have

$$\Delta S_1 = \int_{273}^{546} C_V \, d\ln T = 2.97 \ln \frac{546}{273} = 2.97 \times 0.693$$

$$= 2.06 \text{ cal K}^{-1},$$

and

$$\Delta S_2 = \int_{22.4}^{2.24} R \frac{dV}{V} = R \ln \frac{2.24}{22.4}$$

$$= -4.58 \text{ cal K}^{-1},$$

so

$$\Delta S = \Delta S_1 + \Delta S_2 = -2.52 \text{ cal K}^{-1}$$

Hence, it is quite evident that in such a case, in spite of the increase of temperature, there is a net decrease of entropy.

2. ΔG and ΔH variation with temperature. The relation between ΔG and ΔH for an isothermal process represented in the form

$$\Delta G = \Delta H - T\Delta S, \tag{1}$$

can be represented in another form as can be seen from the following:

Starting with the second law

$$dS = \frac{d'q_{rev.}}{T},$$

we have

$$dS = C_p \frac{dT}{T},$$

$$= \left(\frac{\partial H}{\partial T}\right)_P \frac{dT}{T}.$$

Therefore,

$$T\left(\frac{\partial S}{\partial T}\right)_P = \left(\frac{\partial H}{\partial T}\right)_P,$$

and also

$$T\left(\frac{\partial \Delta S}{\partial T}\right)_P = \left(\frac{\partial \Delta H}{\partial T}\right)_P.$$

According to Eq. (1) and by differentiation, we have

$$\left(\frac{\partial \Delta G}{\partial T}\right)_P = \left(\frac{\partial \Delta H}{\partial T}\right)_P - T\left(\frac{\partial \Delta S}{\partial T}\right)_P - \Delta S,$$

therefore,

$$\left(\frac{\partial \Delta G}{\partial T}\right)_P = -\Delta S.$$

Subsequently, Eq. (1) can be written in the form

$$\Delta G = \Delta H + T\left(\frac{\partial \Delta G}{\partial T}\right)_P. \tag{2}$$

Hence, *the free energy change and heat content change in a chemical reaction are related by the two equivalent equations (1) and (2).*

It was long suspected that some general relation between ΔG and ΔH might be found which would enable free energy changes to be determined from thermal data. *T. W. Richards,* in 1902, studied free energy and heat content changes in a number of galvanic cells and found that they rapidly approached each other as the temperature was lowered. *Van't Hoff* further studied the matter in 1904, and *Nernst,* in 1906, formulated the relations known as the *Nernst heat theorem.*

According to Eqs. (1) and (2), it is evident that if $(\partial \Delta G/\partial T)_P$ (or ΔS) is finite, $\Delta G = \Delta H$ when $T = 0$. Nernst postulated that not only are ΔG and ΔH equal at absolute zero, but they approach equality at this temperature *asymptotically*. Some of the possibilities are shown in the following figures (a), (b), and (c).

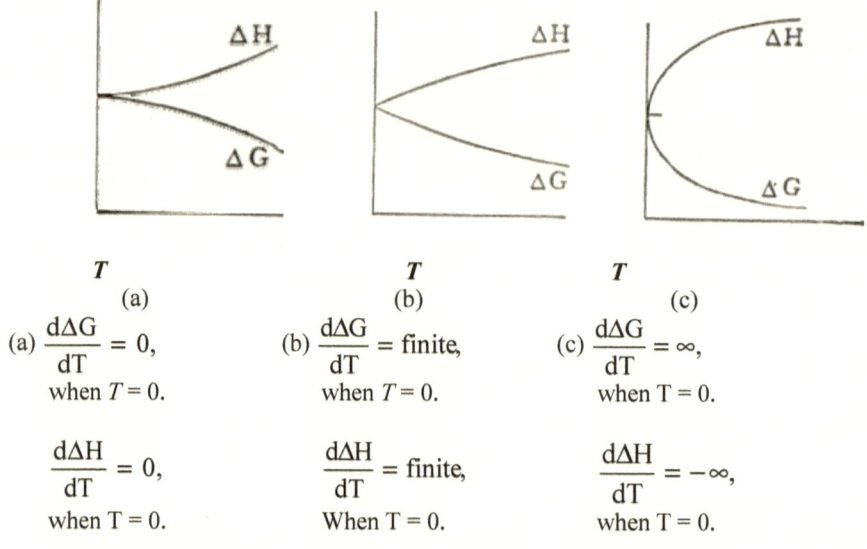

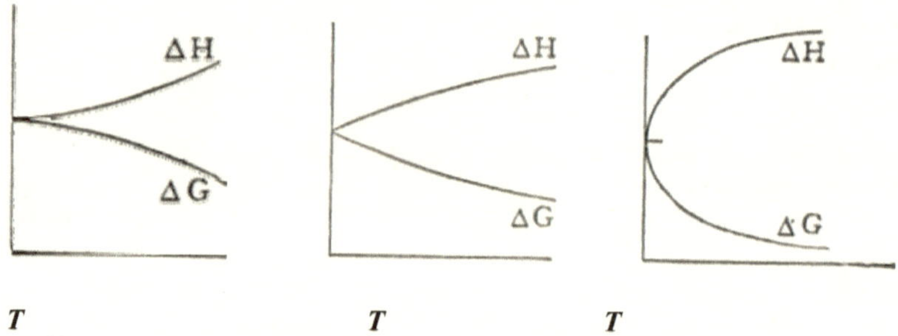

(a) (b) (c)

(a) $\frac{d\Delta G}{dT} = 0,$ when $T = 0$.

$\frac{d\Delta H}{dT} = 0,$ when $T = 0$.

(b) $\frac{d\Delta G}{dT} = \text{finite},$ when $T = 0$.

$\frac{d\Delta H}{dT} = \text{finite},$ When $T = 0$.

(c) $\frac{d\Delta G}{dT} = \infty,$ when $T = 0$.

$\frac{d\Delta H}{dT} = -\infty,$ when $T = 0$.

Nernst postulated that the actual behaviour of condensed systems (solids or liquids) was that of case (a), that is,

$$\text{(a)} \quad \frac{d\Delta H}{dT} = 0, \text{ in the limit when } T = 0, \tag{3}$$

and

$$\text{(b)} \quad \frac{d\Delta G}{dT} = -\Delta S = 0, \text{ in the limit when } T = 0, \tag{4}$$

Since $\left(\frac{d\Delta H}{dT}\right)_P = \Delta C_P$, Eq. (3) implies that the heat capacity change for all reactions in *condensed* systems is zero when $T = 0$. If this is the case, then the *heat capacity of compounds must be equal to that of the elements from which they are formed.* Nernst thought that the heat capacities of all liquids and solids approached the value $C_P = 1.5$ calories per gram molecule in the vicinity of absolute zero. In 1907, *Einstein* predicted that the heat capacities would approach *not a finite value but zero, at the absolute zero,* and this prediction has since been verified for numerous substances.

The second relation (Eq. 4) means that the entropy change for all reactions of condensed systems is zero at absolute zero. Nernst supposed that his theorem applied to all reactions of condensed systems, but *Planck* pointed out that it could not apply to reactions involving solutions. Just as there is an increase of entropy in a mixture of two different gases,* there is also an entropy increase in the formation of a solution from its components, and there is no reason to suppose that this does not persist to the absolute zero. *Lewis and Gibson*, in 1920, pointed out that in all probability, the entropy of a *supercooled liquid* is greater than that of a *crystalline solid* at absolute zero. There is no sharp dividing line between pure liquids and solutions since liquids may contain more than one molecular species. A test of this point can be made by measuring the heat capacities of the solid and the supercooled liquid forms of the substance from the melting point, say at temperature *T* to the vicinity of absolute zero. If the entropy change of fusion at *T* is ΔS_T, that is,

$$\Delta S_T = S_T\,(\text{liquid}) - S_T\,(\text{solid}),$$

and ΔS_0 the entropy difference between the solid and supercooled liquid at absolute zero, that is,

$$\Delta S_0 = S_0\,(\text{supercooled}) - S_0\,(\text{solid}),$$

then it follows that

$$\Delta S_T = \Delta S_0 + \Sigma\, S_0^T, \tag{5}$$

where S_0^T is the entropy difference of the substance between the temperature *T* of fusion and the absolute zero.

This equation is modified into another form concerning a chemical reaction; that is,

$$\Delta S = \Delta S_0 + \int_0^T \frac{1}{T}\Delta\, C_P\, dT, \tag{6}$$

where ΔS_0 is the change in entropy of the reaction at the zero of absolute temperature, ΔC_P, the difference between the heat capacities of the resultants and reactants and, instead of the summation ΣS_0^T, the integral $\int_0^T \frac{1}{T}\Delta C_P\, dT$ is used on the basis that

$$\left(\frac{\partial \Delta S}{\partial T}\right)_P = \frac{1}{T}\left(\frac{\partial \Delta H}{\partial T}\right)_P = \frac{1}{T}\Delta C_P. \tag{7}$$

Hence, substituting Eq. (6) in Eq. (1), we obtain

$$\Delta G = \Delta H - T\int_0^T \frac{1}{T}\Delta C_P dT - T\Delta S_0. \tag{8}$$

The third law is primarily concerned with the quantity ΔS_0.

Since the term ΔS_0 appears in the equations pertaining to chemical reactions, and since the first and second laws permit only the computation of *differences in properties*, the need of some additional principle is apparent.

On the basis of experiments, Nernst proposed, as has been pointed out above, that for all chemical changes, $\Delta S_0 = 0$. *Gibson and Giauque* determined the heat capacities of glycerine as a solid and as a supercooled liquid and found that their results required that ΔS_0 should be an appreciable positive quantity. *Simon and Lange* have also shown that amorphous silica has an appreciably greater entropy at absolute zero than the crystalline

---

* This property shall be proved in Section 13.2, using quantum statistics.

variety. It is, therefore, necessary *to limit the scope of the third law to crystalline substances.* For this reason, Nernst heat theorem can be regarded as a *partial statement* of this law, a statement of which is that

> *the entropy change of a reaction between crystalline solids is zero at the absolute zero of temperature.*

This statement is due to *Gibson* (1917) and to *Gibson and Latimer* (1922).

The third law was also expressed in another way as follows:

> *If the entropy changes of all reactions between crystalline solids are zero at absolute zero, then the entropy of a crystalline compound must be the same as that of its crystalline elements at this temperature Consequently, if we take the entropies of the elements in the crystalline form at absolute zero to be zero, the entropies of crystalline compounds must also be zero at the absolute zero of temperature.*

The fully satisfactory statement of the third law is that of *Lewis and Randall* (1923); it says

> *If the entropy of each element in some crystalline state be taken as zero at the absolute zero, the entropy of any pure substance may become zero at absolute zero, provided the substance is perfectly crystalline.*

Historically, before this statement, Einstein, in 1907, predicted that $C_P \rightarrow 0$, at the absolute zero of temperature for all substances. This prediction has been verified for solids since the *Debye* rule $C_P = \alpha T^3$ is applicable at low temperatures and shows that $S_T - S_0$ remains finite. In 1912, Planck went further and proposed that the entropy of *pure* solids and liquids is zero at the absolute zero of temperature. This last statement has long accepted as an expression of the third law and includes the Nernst statement.

We now ask what evidence statistical thermodynamics offers in support or denial of the Nernst's and Planck's statements of the third law.

Before considering the statistical analysis, we must clarify certain points.

Classical thermodynamics and *calorimetric* (i.e. *calorific*) experiments permit the calculation or measurement of *entropy difference* only. Thus adopting Planck's statement of the third law, we find that the *calorific value of entropy* in given by

$$S_{cal.} = \int_0^T \frac{C_P}{T} dT. \tag{9}$$

The entropy calculated from *spectroscopic* data by means of statistical* formulae is known as the *spectroscopic entropy*. Denoting the *thermodynamic probability of the lowest energy state by* W_0, the difference in entropy between any state and the lowest energy state is given by

$$S - S_0 = k \ln W - k \ln W_0 . \tag{10}$$

Therefore, the *calorimetric* value of entropy can be expressed as

$$S_{cal.} = k \ln W - k \ln W_0, \tag{11}$$

* The statistical approach to the third law and Nernst's theorem shall be seen in Chapter 20.

whereas the *spectroscopic* value S_{spec} is given by

$$S_{spec} = k \ln W \quad , \tag{12}$$

on the basis that the thermodynamic probability at the absolute zero of temperature is presumably equal to unity. Hence, S_{cal} and S_{spec} are related by

$$S_{cal} = S_{spec} - k \ln W_0 \tag{13}$$

or

$$S_{spec} = S_{cal} + k \ln W_0 \quad . \tag{14}$$

Now in order that spectroscopic and calorimetric values of entropy is identical, the thermodynamic probability, W_0, at the absolute zero of temperature must be equal to unity. However, there are some practical difficulties which make the comparison meaningless. In the first place, the absolute zero of temperature is *unattainable in a practical sense,* as has been shown before.* This is one of the important weaknesses of Nernst heat theorem, since he had to extrapolate empirical equations for the specific heat at low temperatures. The Debye T^3 rule, which predicts that C_P is proportional to T^3 as T approaches the absolute zero, is a decided improvement in the validation of Nernst's theorem but is by no means a complete validation. If we denote the lowest achievable absolute temperature by T_1, then extrapolation (based on Debye's rule) between this temperature and the absolute zero of temperature assumes that the only energy to be frozen out is that due to low-frequency vibrations. Also, at this temperature, T_1, there is a possibility that there exists a state of metastable equilibrium in which intermolecular forces prevent the molecules of the system from moving into a position of lower energy states. Also, there is a possibility that the lowest energy level is degenerate. Furthermore, energy levels lower than the thermal increment kT may possibly exist, and if they do, the colorimetric value of entropy will be considerably higher than expected, a fact which could account for differences between $S_{cal.}$ and $S_{sepc.}$ that are not connected with the value of W_0, as required by Eq. (14).

According to Nernst, the statistical formulation of the third law is

$$\Delta(\Sigma \ln W_0) = 0 \tag{15}$$

for all chemical changes, as the absolute zero of temperature is approached

According to Planck, it is

$$W_0 = 1, \tag{18}$$

at the absolute zero of temperature, for all systems. These two principles cannot, of course, be proved.

Now returning to Eq. (10)

$$S - S_0 = k \ln W - k \ln W_0, \tag{10}$$

we note that the value of the calorimetric entropy S_{cal}, given by Eq. (11), namely,

$$S_{cal} = k \ln W - k \ln W_0 \tag{11}$$

---

* See Section 6.13.

is unchanged if W and W_0 are multiplied by the same factor. Yet we always omit factors in evaluating W, and despite this fact, experiment indicates that for many substances there is an agreement between the practical values of spectroscopic entropy and the calorific values so that one is led to the conclusion that $W_0 = 1$ at the absolute zero of temperature. Thus, a reasonable statement of the third law is

> *As the temperature T approaches absolute zero, the entropy change of a chemical reaction at the temperature T also approaches zero.*

This statement is unobjectionable, but it, certainly, is *not universal* in the sense of the first and second laws, and one may still question the desirability of regarding it as a law at all.

In some special cases, in which more than one arrangement of the atoms in the crystal can occur at absolute zero, a further qualification is necessary as will be seen below.

7.3. Some properties at the absolute zero of temperature

In order to discuss some of the thermodynamic properties of substances at 0K, use could be made of Maxwell's relations, the derivation of which has been given before in Chapter 2.

Consider a process at 0K in which an entropy change results from an infinitesimal pressure change at constant temperature. Therefore,

$$\left(\frac{\partial S}{\partial P}\right)_T dP = S_2 - S_1 = \Delta S_0.$$

From the Nernst heat theorem, $\Delta S_0 = 0$, so that $\underset{T\to 0}{\mathrm{Lt}} \left(\frac{\partial S}{\partial P}\right)_T = 0,$

but

$$\left(\frac{\partial S}{\partial P}\right)_T = -\left(\frac{\partial V}{\partial T}\right)_P,$$

which is one of Maxwell's relations, therefore

$$\left(\frac{\partial V}{\partial T}\right)_P = 0,$$

in the limit $T \to 0$. This equation means that the coefficient of thermal expansion, $\beta = \frac{1}{V}\left(\frac{\partial V}{\partial T}\right)_P$, vanishes at $T = 0$; that is, the volume becomes temperature independent as long as the pressure remains constant.

In like manner, consider a process at 0K in which an entropy change results from an infinitesimal volume change at constant temperature. Therefore,

$$\left(\frac{\partial S}{\partial V}\right)_T dV = S_2 - S_1 = \Delta S_0.$$

Again, according to Nernst heat theorem, $\Delta S_0 = 0$ so that

$$\mathop{Lt}_{T\to 0}\left(\frac{\partial S}{\partial V}\right)_T dV = 0.$$

But

$$\left(\frac{\partial S}{\partial V}\right)_T = \left(\frac{\partial P}{\partial T}\right)_V,$$

which is also another Maxwell's relation, therefore,

$$\mathop{Lt}_{T\to 0}\left(\frac{\partial P}{\partial T}\right)_V = 0.$$

This equation means that the pressure remains constant, independent of the temperature near $T = 0$, when the volume is kept constant.

Similarly, the surface tension, σ, is independent of the temperature in the vicinity of $T = 0K$. This can be proved as follows:

The increase of the Gibbs free energy ΔG with the increase of surface area ΔA is expressed by the equation

$$\Delta G = \sigma.\Delta A$$

Differentiating this equation with respect to T, we get

$$\left(\frac{\partial \Delta G}{\partial T}\right)_P = \left(\frac{\partial \sigma}{\partial T}\right)_P . \Delta A,$$

since the surface area, A, is not a function of temperature. According to Nernst heat theorem,

$$\frac{\partial \Delta G}{\partial T} = 0,$$

in the limit when $T = 0$, therefore,

$$\mathop{Lt}_{T\to 0}\left(\frac{\partial \sigma}{\partial T}\right)_P = 0$$

That is, the surface tension is independent of temperature near $T = 0K$.

This equation may be tested with liquid helium which remains liquid under its own vapour pressure at the lowest temperature 0.82 K. Thus the surface tension of liquid helium is independent of temperature near 0K, as predicted by Nernst heat theorem. Furthermore, liquid helium may be frozen at pressures higher than it own vapour pressure so that when the system containing solid and liquid helium is subjected to a change of temperature, the change of equilibrium pressure tends to zero at 0K; that is

$$\mathop{Lt}_{T\to 0}\left(\frac{\partial P}{\partial T}\right)_V = 0.$$

Now considering again the equation

$$\mathop{Lt}_{T\to 0}\left(\frac{\partial V}{\partial T}\right)_P = 0,$$

which requires that the coefficient of thermal expansion, $\beta = \frac{1}{V}\left(\frac{\partial V}{\partial T}\right)_P$, becomes zero at $T = 0$, *Latimer* found in confirmation with this conclusion that this coefficient, for a number of crystalline solids, approach zero in the vicinity of absolute zero and change more rapidly than in the case of heat capacities with temperature.

The agreements with predictions made from the Nernst's heat theorem or strictly from the third law of thermodynamics can be taken as a validity of this law.

7.4. Absolute entropy, multiplicity, and residual entropy

In order to have absolute values of entropy, we make use of the fact that we have absolute basis for probability. Instead of the difference in entropy and the ratio of probabilities shown above as

$$S_B - S_A = \frac{R}{N_a}(\ln W_B - \ln W_A),$$

we write the equation

$$S = \frac{R}{N_a}\ln W = k \ln W,$$

which is commonly called the *Planck–Boltzmann equation*, where W, as noted above, is known as the thermodynamic probability of a given state. It is not necessary here to define W with full precision since this shall be done later,* when we deal with the *Bose-Einstein quantum statistics*. Thus we may indicate W as *the number of quantum states of the system which are accessible under the conditions of energy, volume, etc, which are applicable.* With the simultaneous development of quantum mechanics and statistical mechanics, it became possible to calculate the entropy of ideal gases in terms of partition functions using quantum-statistical methods.

For some range of temperature, above the lowest temperature of measurement, the molecules of a given substance are distributed among the available quantum states in accordance with the normal distribution law, and that as the temperature is lowered, the distribution of molecules among the available energy levels changes progressively in such a way that in the limit, at 0K, the molecules *will have available one state of existence.* If, however, the arrangement of molecules in the crystal is fully *random* with ρ states for each molecule, then the total number of available configurations, W, for one mole of molecules is given by

$$W = \rho^{N_a},$$

where W is also known as the number of states available for each molecule, and N_a is Avogadro's number. Accordingly, the residual entropy arising from such randomness is given by

$$S_{resid.} = S_{rand.} = k \ln W$$

$$= k \ln \rho^{N_a}$$

$$= R \ln \rho. \quad \text{(per mole)} \qquad (19)$$

---

* See Sections 9.3 and 9.6.

For *linear molecules* with only two positions or states available, and these two states are of the same energy, many molecules will exist in these two states; that is, $p = 2$, and thus

$$S_{rand} = 1.38 \text{ cal K}^{-1} \text{ mole}^{-1}.$$

It is found that $S = 0.66$ in the case of NO, and $S = 1.1$ in the case of CO. It, therefore, appears that some movement away from complete randomness has occurred in each case, and that the molecule NO moved *closer* to the *orderly* arrangement than the CO molecule.

If a given ensemble of molecules is in a *single quantum state* at the absolute zero of temperature, then the entropy of the ensemble is zero at this temperature. Accordingly, the third law may be put in a formal statement as follows:

> *The entropy of any substance of which the component parts are in complete internal equilibrium becomes zero at the absolute zero of temperature.*

The meaning *of complete internal equilibrium* is that each of the species (atoms or molecules), comprising the given substance *has free access* to all of the permitted quantum states in accordance with the distribution law. In the case of a solid substance, such as graphite for example, which is in internal equilibrium among all the states permitted, will have a zero entropy at the absolute zero of temperature.

Let us now consider the case of imperfect crystals, glasses, and solid solutions. In the Lewis and Randall statement (1923) of the third law, as given above, the term 'perfect crystalline substances' was used. Let us now consider the limitations which this term introduces.

Any state in which the arrangement of atoms is disordered will have greater probability than a perfectly ordered arrangement. Accordingly, we might expect that if any type of *disorder remains at the absolute zero of temperature, the entropy will remain larger than zero.*

The atomic arrangement in a glassy solid is essentially similar to that in a liquid which is less regular than the solid structure. Thus we find it very reasonable that the entropy of a glassy solid *fails* to approach zero at the absolute zero of temperature. The classic investigation of this question was the study of glycerine by *Gibbson and Giauque* (1923) and by *Simon and Lange* (1926). They found that the entropy of glassy glycerine (supercooled glycerine) at zero absolute was $5.6 \pm 0.1 \text{ cal K}^{-1} \text{ mole}^{-1}$, greater than that of crystalline glycerine, as was expected.

In solid solutions, different atoms and molecules are distributed *randomly* over certain sites in solid or crystal. If this randomness is still present at zero absolute, we expect the solid solution to have an entropy greater than zero at 0 K. *Zahman and Milner* (1933) found this to be true for the solid solution of AgCl and AgBr. The entropy of formation of solid solution of AgCl and AgBr was $1.21 \pm 0.1 \text{ cal K}^{-1} \text{ mole}^{-1}$ at 298K. Since the heat capacity of the solution did not differ significantly from that of its pure components (or constituents) any where in the range 15 to 298 K, it would be presumably present at 0K.

As an example of a solid solution, we consider a mixture of isotopes of chlorine. Chlorine at 0K is a solid solution of

$$Cl^{35} - Cl^{35}$$
$$Cl^{35} - Cl^{37}$$
$$Cl^{37} - Cl^{37}$$

molecules. Because of the close similarity of chemical properties of these isotopes, we expect that the two types of chlorine atoms are distributed with *complete randomness.* Consequently, in a strict sense, the entropy of this material is *not zero* at the absolute zero of temperature.

For most purposes, it is convenient to ignore the entropy of mixing isotopes because the same entropy is present in any substance in which the element may occur. That is, the contribution of the entropy of mixing isotopes will be the same for the products as the reactants in any process and hence will cancel.

In addition to the rather straight forward examples of glasses and solid solutions there are few cases where the crystals of a pure substance fail to attain the perfect order at the zero absolute of temperature. Examples of such imperfect crystals at the absolute zero are CO and NO, which have been considered before, to which we may now add $N_2SO_4 . 10H_2O$, 1-olefins with more than 10 carbon atoms. In each of these substances, there are different geometrical orientations of the molecule which may be expected to have almost exactly the same energy. For example, the carbon monoxide molecule has practically zero dipole moment, and therefore, may easily form a crystal with random, end-for-end orientation. The spectroscopic entropy of carbon monoxide is $47.2 \text{ cal K}^{-1} \text{ mole}^{-1}$ at 298K. This value is confirmed by chemical equilibrium studies. However, *Clayton* and *Giauque* (1932) found from heat capacity measurements that the entropy of solid carbon monoxide at 0K is only $46.2 \text{ cal K}^{-1} \text{ mole}^{-1}$, that is, less than that at 298K. This means that an entropy of $1.0 \text{ cal K}^{-1} \text{ mole}^{-1}$ remains at 0K, indicating, when compared with the value $1.38 \text{ cal K}^{-1} \text{mole}^{-1}$ calculated before, that the molecular orientation of CO at 0K is *not quite freely random.*

Added to the above examples of the presence of residual entropy associated with randomness in the structure of crystals of some compounds at the absolute zero of temperature is the example of water or heavy water, the observed S_0 values being 0.83 and $0.69 \text{ cal K}^{-1} \text{mole}^{-1}$, respectively. The explanation of this randomness of arrangement at absolute zero was first given by *Pauling*[*] in terms of the *randomness of arrangement of hydrogen bonds in ice.*

The problem of the structure of liquid water is an interesting one that it is still for from a complete solution. There is no doubt that liquid water, like other liquids, has a structure that involves a great deal of *randomness*, and yet it is likely that there are certain configurations of groups of water molecules that occur with high frequency in the liquid. *Bernal* and *Fowler*[†] suggested that water retains *in part* as a hydrogen-bonded structure similar to that of ice, and that as more and more hydrogen bonds are broken, with the increase of temperature, the oxygen molecules may arrange themselves in a manner approximating more and more closely to closest packing of spheres, and that there would be a significant *increase* in density for this sort of packing, as compared with the open packing of the completely hydrogen-bonded structure of ice. On the other hand, the increase in density of water that occurs, as noted before, on warming from 0°C to 4°C may be attributed to a *decrease* in the concentration of aggregates with the ice-like structure and an *increase* in the number of complexes with some denser structure.

---

[*] Pauling. *J. Am. Chem.*, **57**, 2680 (1935).

[†] J.D. Bernal and R. H. Fowler, *J. Chem. Phys.*, **1**, 515 (1933).

The crystal structure of water ice has been shown by X-ray

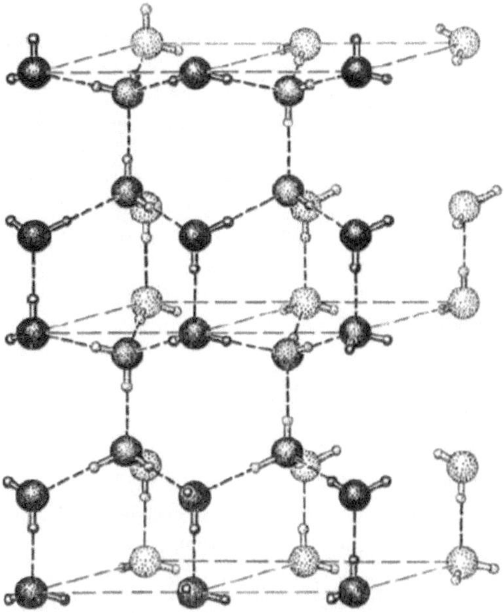

Fig. 1. The arrangement of molecules in the ice crystal. The orientation of the water molecules, as represented in the drawing, is arbitrary; there is one proton along each oxygen–oxygen axis, closer to one or the other of the two oxygen atoms.

investigation that each oxygen atom is surrounded tetrahedrally by four other oxygen atoms at the distance 2.76 Å, as shown in Fig. 1. This is a very open structure, which causes ice to have a low density, that is, ice is less dense than water. Hydrogen sulphide, for example crystallises in a closely-packed arrangement, each sulphur atom (hydrogen sulphide molecule) having twelve equivalent neighbours. The ice structure is, however, just that expected in case O – H ... O hydrogen bonds are formed, with each bond making greater or less use of *one* of the four valence electron *pairs* of each of the two bonded oxygen atoms.

The question now arises as to whether a given hydrogen atom is midway between the two oxygen atoms it connects, or closer to one than to the other. The answer to this is that it is closer to one than to the other, and that (with few exceptions) each oxygen atom has two hydrogen atoms bonded to it by strong bonds. In the *gas* molecule, the $O-H$ distance is 0.96 Å. This distance is increased to 1.38 Å in ice. An accurate value, 1.01, Å for the $O-D$ bond distance has been obtained by neutron diffraction of deuterium oxide ice.

An interesting verification of the existence of discrete water molecules in ice is provided by the discussion of the *residual entropy,* which also, according to Pauling, gives *definite* information about the orientation of the water molecules in the crystal as has been pointed out above. Pauling emphasised that if each water molecule in the ice crystal were oriented in a definite way, permitting the assignment of a *unique configuration* to the crystal, such as that suggested by Bernal and Fowler, the residual entropy would vanish. Accordingly, Pauling assumed that each water molecule is so oriented that the two hydrogen atoms are dissected approximately towards two of the four surrounding oxygen atoms and that only one hydrogen atom lies along each oxygen–oxygen line and also that under ordinary conditions, the interaction of non-adjacent molecules is such as not to stabilise appreciably any one of the many configurations satisfying these conditions with reference to the others. Thus an assumption can be made that an ice crystal can exist in any one of a large number of configurations, each configuration corresponding to certain orientations of the water molecules. It can change from one configuration to another by the rotation of some of the molecules or by motion of the same of the hydrogen nuclei, each moving 0.76 Å from a position 1.00 Å, from one oxygen atom to the similar position near the other bonded atom. It is assumed that the protons tend to jump in this way in groups, so as to leave each oxygen atom with two protons attached.

When a crystal of ice is cooled to very low temperatures, it is *caught* in some one of the many possible configurations, but it does not assume (in a reasonable period of time) a *uniquely determined configuration with no randomness of molecular orientation.* It accordingly retains the *residual entropy* $k \ln W$, in which *W* is the number of configurations accessible to the crystal.

Let us now calculate *W*. In a mole of ice there are $2N_a$ hydrogen nuclei. If each had the choice of *two* positions along the $O - O$ axis, one closer to one oxygen atom and the other closer to the second oxygen atom, there would be 2^{2N_a} *configurations*. However, many of these are ruled out by the condition that each *oxygen atom has two attached hydrogen atoms*. Let us consider a particular *oxygen atom and the four surrounding hydrogen nuclei.* There are 16 arrangements of this OH_4 group; *one* for all four hydrogen nuclei close to the oxygen atom, corresponding to the ion $(H_4O)^{++}$, *four* corresponding to the ion $(H_3O)^{+}$, *six* to H_2O, *four* to $(OH)^{-}$, and *one* to O^{--}. The acceptable arrangements, assigning *two strongly bonded hydrogen nuclei to this oxygen atom* accordingly, comprise six-sixteenth or three-eighths of the total. Of these, only three-eighth are suitable with respect to the *second oxygen atom* and so on. Hence, the number of configurations, *W*, accessible to the crystal is

$$W = 2^{2N_a}\left(\frac{3}{8}\right)^{N_a}$$

or

$$= \left(\frac{3}{2}\right)^{N_a}.$$

This leads to the theoretical value for the *residual entropy* of ice; that is

$$S_{residual} = k \ln W$$

$$= k \ln\left(\frac{3}{2}\right)^{N_a} = R \ln\frac{3}{2}$$

$$= 0.806 \text{ cal K}^{-1} \text{ mole}^{-1}.$$

The experimental values are $0.82\ \text{cal K}^{-1}\ \text{mole}^{-1}$ for ordinary ice and $0.77\ \text{cal K}^{-1}\ \text{mole}^{-1}$ for heavy ice. The agreement with the theoretical value provides strong support for the postulated structure involving *hydrogen bonds with the hydrogen nucleus asymmetrically placed between the two bonded oxygen atoms.*

It may be added here that an investigation of single crystals of deuterium oxide at $-50\,^\circ\text{C}$ and $-150\,^\circ\text{C}$ by neutron diffraction has led to the determination of the $\text{O}-\text{D}$ distance, which is 1.01 Å, and the $\text{D}-\text{O}-\text{D}$ angles as close to tetrahedral $(100.5^\circ \pm 0.5^\circ)$.

Of the enthalpy of sublimation of ice, $12.2\ \text{K cal mole}^{-1}$, about one-fifth can be attributed to the van der Waals forces (as estimated from values for other substances); the remainder, $10\ \text{k cal mole}^{-1}$, represents the rupture of hydrogen bonds and leads to the value $5\ \text{k cal mole}^{-1}$ for the energy of the $\text{O}-\text{H}\ldots\text{O}$ hydrogen bond in ice. The small value, $1.44\ \text{k cal mole}^{-1}$, of the enthalpy of fusion of ice shows that on melting, only about 15 per cent of the hydrogen bonds are broken.

7.5. Practical basis for absolute entropies

Absolute entropies measured and tabulated for practical chemical applications are those obtained by experimental measurement down to temperatures of 15K and extrapolation to absolute zero is then carried out in accordance with Deby's theory of heat capacity. Depending on the temperature at which the *absolute calorimetric value* is required, the entropy changes of fusion and vaporisation, if necessary, are added to the value of *S* obtained by graphical integration for the area enclosed.

The heat capacity has been measured down to very low temperatures for most chemical elements and for a substantial number of compounds. From the data obtained, the entropy values have been calculated by the method indicated above. If it is convenient to tabulate the entropy of a substance in its *standard state,* as done for heat capacities, enthalpy, and free energy change for solids and liquids, the standard state being taken at one atmosphere, as usual, and indicated by S°. To obtain the *standard entropy* of a perfect gas at a temperature *T,* we first note that the entropy difference for a *perfect gas* between two pressures is generally given by the following partial derivative of *S* with respect to *P* at constant temperature is given, as shown before, by

$$\left(\frac{\partial S}{\partial P}\right)_T = -\beta V = \left(\frac{\partial V}{\partial T}\right)_P$$

$$= -\frac{R}{P}$$

$$dS = -\frac{R}{P}dP$$

$$S_{II} - S_I = R\ln\frac{P_I}{P_{II}}.$$

Then taking P_{II} as the standard pressure of one atmosphere and denoting P_I by P, the entropy $_s$ of a gas at a pressure *P* for a given temperature *T* is

$$S^\circ - S_P = R \ln P$$

or

$$S_P = S^\circ - R \ln P.$$

7.6. Elementary physical consequences of the third law related to phase change

1. An important consequence of the third law is that all heat capacities tend to zero as the temperature approaches zero. This may be seen by writing the heat capacity C_P in the form

$$C_P = T\left(\frac{\partial S}{\partial T}\right)_P = \left(\frac{\partial S}{\partial \ln T}\right)_P. \tag{1}$$

This equation has been obtained above in the form

$$T\left(\frac{\partial S}{\partial T}\right)_P = \left(\frac{\partial H}{\partial T}\right)_P.$$

As $T \to 0$, $\ln T = -\infty$, and $S \to 0$, so the derivative also tends to zero. We may reach the same conclusion by considering the total entropy change on cooling from *T* to absolute zero

$$\Delta S = \int_T^0 \frac{C_P}{T} dT. \tag{2}$$

The integral is only finite if $C \to 0$ as $T \to 0$. Thus, all heat capacities tend to zero as the temperature approaches absolute zero. This result emphasises the connection between the third law and quantum statistics, for classical heat capacities do not vary with temperature. It is, therefore, impossible to construct a classical interpretation of the third law.

2. We consider the Clausius-Clapeyron equation

$$\frac{dP}{dT} = \frac{\Delta S}{\Delta V} = \frac{L}{T\Delta V}, \tag{1}$$

where L, ΔS and ΔV are the latent heat (absorbed), the change in entropy, and the change in volume in passing from phase 1 to phase 2. According to the third law, $\Delta S \to 0$ as $T \to 0$, so that in this limit the gradient of the phase-boundary must become parallel to the *T*-axis. The only substance on which careful experiments have been made is ^{4}He, for which *solid and liquid may coexist at absolute zero*. These have shown that the gradient of the melting curve, dP/dT, *does approach zero at the lowest temperatures.** The explanation that a liquid exists at absolute zero is that the entropy of the system does not only depend on the positions of the atoms but also on their momenta, and for precisely the

* See Fig. 11, Section 6.11.

same reason, helium may remain liquid to absolute zero. The state of zero entropy of the liquid corresponds to a configuration *in which the ordering is dominant in momentum rather than position.* Again this is a consequence of quantum effects. A similar agreement must apply to ^3He which can also exist as a liquid at absolute zero. The zero entropy configuration must again be one in which there is *ordering* in the momenta of the particles, and at about 3mK ^3He also become a superfluid, but by a mechanism similar to that which operates with electrons in metals to produce superconductivity. However, ^3He is also interesting for other reasons. Below about 0.3K, the slope of the melting curve becomes *negative* as can be seen from Fig. 2, the solid *becoming the high temperature phase.* According to Clausius-Clapeyron equation,

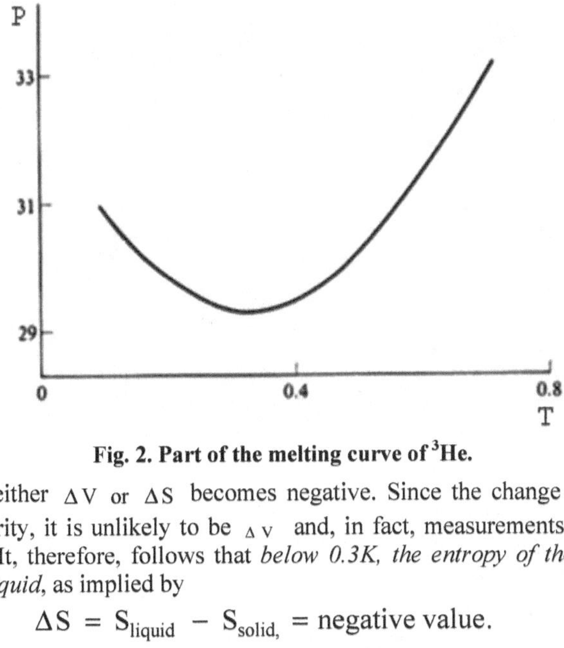

Fig. 2. Part of the melting curve of ^{3}He.

this means that either ΔV or ΔS becomes negative. Since the change is not associated with any singularity, it is unlikely to be ΔV and, in fact, measurements show that this is nearly constant. It, therefore, follows that *below 0.3K, the entropy of the solid is greater than that of the liquid,* as implied by

$$\Delta S = S_{liquid} - S_{solid,} = \text{negative value.}$$

The reason for this is that the ^3He nucleus, consisting of two protons and one neutron, has a *net spin.* and thus associated with a *nuclear magnetic moment.* Now the entropy must, of course, include a contribution related to the degree of order in the spin system. Magnetic measurements show that as the liquid is compressed towards solidification, it obeys *Curie's law* to progressively lower temperatures, implying that magnetic ordering sets in *more readily* at liquid densities than it does at solid. This is presumably because the *uncertainty* in atomic position in the liquid allows a *stronger interaction* between nuclear moments. Thus for a certain range of temperature, the liquid is *more ordered* than the solid, and the gradient of the melting curve becomes *negative.* Eventually, of course, it must again become zero in accordance with the third law.

3. A relation analogous to the Clausius-Clapeyron equation describing the variation of the critical field B_c of a *superconductor* with temperature has been derived before,* namely,

$$\left(\frac{\partial B_c}{\partial T}\right)_P = -\frac{\Delta S}{\Delta m} = -\mu_o \left(\frac{S_n - S_s}{V_S\, B_c}\right),$$

so the third law requires the critical field to become *constant* as the temperature approaches absolute zero.

All these elementary consequences of third law are well supported by experiments.

Finally, the third law of thermodynamics and Nernst's heat theorem shall be seen in Chapter 20, using statistical mechanics.

---

* See Eq. (21), Section 6.12..

CHAPTER 8

Elementary Wave Mechanics – The Rotation of Molecules and Nuclei

8.1. Elementary wave mechanics

1. Introduction. Several chapters of various books in physical chemistry are concerned with the derivation of molecular shape and size from observations on spectra. The interpretation of the experimental results would have been almost impossible without the guiding principles of quantum mechanics. The new quantum mechanics or wave mechanics was developed to explain the details of atomic spectra. The advancement that followed from the solution of the problems of spectra left no one in doubt that the correct formula for quantisation had been discovered.

To keep the account of the principles of wave mechanics within bounds, the attention shall be confined here to one-dimensional systems. This has the advantage that we are cut off from the consideration of physically real systems; but when the mathematical techniques are the same, the manipulation of the algebra is enormously simplified when we deal with only one coordinate instead of three.

2. Mathematical preliminaries. As the concept of an operator plays an important part in the mathematics of wave mechanics, it is instructive to introduce an account of wave-mechanical principles with a summary of some of the properties of operators. We shall consider functions of the variable x and will designate them by $u(x)$ or simply by u; for the present it will be sufficient to visualise $u(x)$ as an elementary function, such as $u(x) = \sin x$, $u(x) = e^{-\frac{1}{2}x^2}$, though this is not, of course, a necessary restriction. Expression

$$\frac{d}{dx}u(x) \tag{1}$$

can be thought of having two parts: the operator d/dx and the operand $u(x)$. Likewise the expression $x\,u(x)$ and $a\,u(x)$, where x and a may be regarded as operators governing the operand $u(x)$. The process of operating with the operators d/dx, x, and a is, respectively, that of differentiation with respect to x, multiplication by x, and multiplication by a constant a. Thus if $u(x) = e^{-\frac{1}{2}ax^2}$, the product of operating with d/dx is $-axe^{-\frac{1}{2}ax^2}$.

Operators can be combined into sums and products. The term $\left(-d^2/dx^2 + x^2\right)$ in the expression $\left(-d^2/dx^2 + x^2\right)u(x)$ is an operator representing the sum of the operators $-d^2/dx^2$ and x^2. As we shall see, $\left(-d^2/dx^2 + x^2\right)$ is an operator with properties of its own, distinct from those of $-d^2/dx^2$ and x^2. Similarly, the product of the operator x and d/dx is a new operator x(d/dx), and the product of the operators x and a is a new operator xa. It is important to realise that when a differential operator, say $\frac{d}{dx}$, is compounded into a product, the order of the factors is significant; for instance $x\left(\frac{d}{dx}\right)$ is an operator with properties different from $\left(\frac{d}{dx}\right)x$. If, for example, $u(x) = e^{-\frac{1}{2}ax^2}$, the result of operating with $x\frac{d}{dx}$ is $xe^{-\frac{1}{2}ax^2}\left(-\frac{1}{2}\times 2ax\right) = -ax^2e^{-\frac{1}{2}ax^2}$, whereas the product of operating with $\left(\frac{d}{dx}\right)x$ is

$$\frac{d}{dx}\left(xe^{-\frac{1}{2}ax^2}\right) = e^{-\frac{1}{2}ax^2} + x \times -ax\,e^{-\frac{1}{2}ax^2}$$

$$= e^{-\frac{1}{2}ax^2} - ax^2e^{-\frac{1}{2}ax^2}$$

$$= \left(1 - ax^2\right)e^{-\frac{1}{2}ax^2}.$$

Note that the operation by a product, such as $x\left(\frac{d}{dx}\right)$, entails consecutive operations; first one operates on $u(x)$ with the operator $\frac{d}{dx}$, and then one operates on the result of this operation with the operator x.

Finally, two operators α and β are defined to be equal when, for every operand $u(x)$, on which one can operate,

$$\alpha u(x) = \beta u(x). \tag{2}$$

It will be clear from the example above that the operators $x\left(\frac{d}{dx}\right)$ and $\left(\frac{d}{dx}\right)x$ are not equal,

$$x\left(\frac{d}{dx}\right) \neq \left(\frac{d}{dx}\right)x \tag{3}$$

This means that differential operators do not obey the commutative law of multiplication.

The operator $(\alpha\beta - \beta\alpha)$ is known as the commutator of the operators α and β. The commutator of the operators $\left(\frac{d}{dx}\right)x$ and $x\left(\frac{d}{dx}\right)$ is the operator

$$\frac{d}{dx}x - x\frac{d}{dx} \tag{4}$$

The usual rule for differentiation of a product tells us that

$$\frac{d}{dx}x = 1 + x\frac{d}{dx} \tag{5}$$

and hence that

$$\frac{d}{dx}x - x\frac{d}{dx} = 1 \tag{6}$$

This equation states that the commutator of the operators $\frac{d}{dx}$ and x is the operator +1. To regard a numeral, or a constant, as an operator is largely a mathematical formality – the convention adopted so that expressions like (5) and (6) can be described as operator equations.

Certain classes of operators and operands are of special interest. When an operator α and an operand *u(x)* (which must obey certain supplementary conditions that are specified below) are such that operating by α regenerates the function *u(x)* multiplied by a constant *a*,

$$\alpha\, u(x) = a\, u(x), \tag{7}$$

we say that *u(x)* is an *eigenfunction* of the operator α belonging to the *eigenvalue a*. The supplementary conditions imposed on *u(x)* are that it must be *finite, continuous, and single-valued throughout the whole range of possible values of x, from* $-\infty$ *to* $+\infty$. Only these solutions *u(x)* of Eq. (7), which confirm to these boundary conditions, are accepted as *eigenfunctions* of the operator α.

To develop the concept of an eigenfunction, let us suppose that α is the operator $-d^2/dx^2$. Although the operand $u(x)=e^x$, that is,

$$-\frac{d^2e^x}{dx^2} = -1.e^x \tag{8}$$

the function e^x is not acceptable as an eigenfunction because it violates the boundary conditions by becoming infinite as $x\rightarrow\infty$. On the other hand, *u(x)* = sin 3*x* is a possible eigenfunction of $-d^2/dx^2$, since

$$-\frac{d^2}{dx^2}\sin 3x = 9\sin 3x, \tag{9}$$

and sin3x is a function that satisfies the boundary conditions. The eigenvalue belonging to the eigenfunction sin3x is the number 9. Likewise sin4x, sin5x, . . . We see that the operator $-d^2/dx^2$ possesses a whole spectrum of eigenfunctions and eigenvalues.

It frequently happens that a symbol must be used in two senses – to denote an operator and to denote a variable. For instance, the letter x may represent the operator x or the coordinate *x*. To comprehend the equations, it will be necessary to read into them the correct significance of the symbol, as operator or variable.

Compound operators, like $\left(-d^2/dx^2+x^2\right)$, will generally be written in parentheses to distinguish them clearly from the operand.

3. The operator $\left(-d^2/dx^2+x^2\right)$. To illustrate the methods of operator algebra, let us determine the eigenfunctions and eigenvalues of the operator.

$$-\frac{d^2}{dx^2}+x^2, \tag{1}$$

which is of central importance in the wave-mechanical treatment of the linear oscillator. The problem is to discover the functions *u* = *u(x)* which must satisfy the boundary conditions and the possible values of the constant *a*, for which the differential equation

$$\left(-\frac{d^2}{dx^2} + x^2\right)u = a\,u \qquad (2)$$

has solutions. We enquire first what form a adopts for large values of x. In Eq. (2), x^2 and a are both multiples of u, consequently, a can be neglected in comparison with x^2 in any region where x is large, and then Eq. (2) simplifies to

$$\frac{d^2u}{dx^2} + x^2u = 0. \qquad (3)$$

An approximate solution of Eq. (3) is

$$u = e^{\pm\frac{1}{2}x\,2},$$

for

$$\left(\frac{d^2}{dx^2}\right)e^{\pm\frac{1}{2}x^2} = \left(x^2 \pm 1\right)e^{\pm\frac{1}{2}x^2}$$

and the term ± 1 can be neglected when x is large. The solution $e^{\frac{1}{2}x^2}$ violates the boundary conditions by becoming infinite as $x \to \infty$ and is of no further interest. The solution $u = e^{-\frac{1}{2}x^2}$ does satisfy the boundary conditions (see Fig. 1), but as yet is only known to be a solution of Eq. (2) when x is large. However, by substituting $u = e^{-\frac{1}{2}x^2}$ in Eq. (2), we obtain

$$\left(-\frac{d^2}{dx^2} + x^2\right)e^{-\frac{1}{2}x^2} = 1.e^{-\frac{1}{2}x^2} \qquad (4)$$

so the function $e^{-\frac{1}{2}x^2}$ is one possible eigenfunction of the operator (1) belonging to eigenvalue +1. Likewise, if we try the substitution

$$u = x\,e^{-\frac{1}{2}x^2},$$

we have

$$\left(-\frac{d^2}{dx^2} + x^2\right)x\,e^{-\frac{1}{2}x^2} = 3.x\,e^{-\frac{1}{2}x^2} \qquad (5)$$

and thus $x\,e^{-\frac{1}{2}x^2}$ is also an eigenfunction of Eq. (1), belonging to the eigenvalue +3. These trial substitutions suggest that the *general solution of Eq. (2) may have the form*

$$u = V\,e^{-\frac{1}{2}x^2}, \qquad (6)$$

where $V = V(x)$ is normally a function of x, though it may reduce to a constant as in Eq. (4) in a special case.

To discover when Eq. (6) is a solution of Eq. (2), we substitute Eq. (6) into the differential Eq. (2).

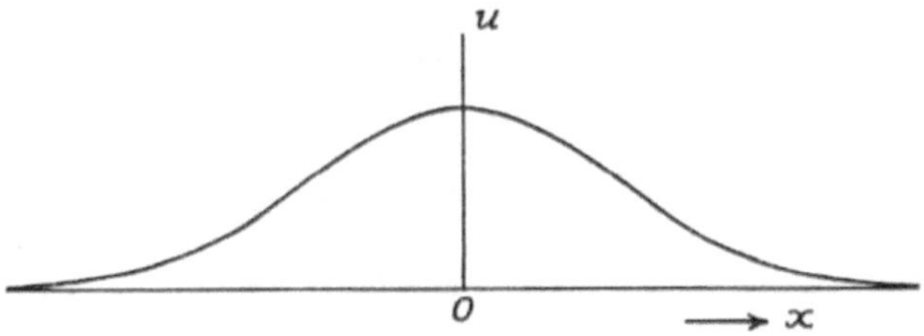

Fig. 1. Graph of the function $u = e^{-\frac{1}{2}x^2}$. The function is continuous, finite, and single-valued throughout the range $\infty > x > -\infty$

The condition for compatibility that emerges is

$$\frac{d^2V}{dx^2} - 2x\frac{dV}{dx} + (a-1)V = 0. \qquad (7)$$

This equation, which is a general solution of Eq. (2), can be obtained by substituting $Ve^{-\frac{1}{2}x^2}$ for u in Eq. (2), as can be seen from the following:

$$-\frac{d^2(u)}{dx^2} = -\frac{d^2}{dx^2}\left(Ve^{-\frac{1}{2}x^2}\right) = -\frac{d}{dx}\left[Ve^{-\frac{1}{2}x^2} \times -x + e^{-\frac{1}{2}x^2}\frac{dV}{dx}\right]$$

$$= -\frac{d}{dx}\left[-xVe^{-\frac{1}{2}x^2} + \frac{dV}{dx}e^{-\frac{1}{2}x^2}\right]$$

$$= \frac{d}{dx}\left[xVe^{-\frac{1}{2}x^2} - \frac{dV}{dx}e^{-\frac{1}{2}x^2}\right]$$

$$= Ve^{-\frac{1}{2}x^2} + \frac{dV}{dx}xe^{-\frac{1}{2}x^2} - x^2Ve^{-\frac{1}{2}x^2} - \frac{d^2V}{dx^2}e^{-\frac{1}{2}x^2} + \frac{dv}{dx}\dot{x}e^{-\frac{1}{2}x^2}$$

$$= -\frac{d^2V}{dx^2}e^{-\frac{1}{2}x^2} + 2xe^{-\frac{1}{2}x^2}\frac{dV}{dx} - x^2Ve^{-\frac{1}{2}x^2} + Ve^{-\frac{1}{2}x^2}.$$

Therefore,

$$-\frac{d^2\left(Ve^{-\frac{1}{2}x^2}\right)}{dx^2} + x^2Ve^{-\frac{1}{2}x^2} = -\frac{d^2V}{dx^2}.e^{-\frac{1}{2}x^2} + 2x\frac{dV}{dx}.e^{-\frac{1}{2}x^2} + Ve^{-\frac{1}{2}x^2}$$

$$+ x^2Ve^{-\frac{1}{2}x^2} - x^2Ve^{-\frac{1}{2}x^2}$$

or

$$-\frac{d^2}{dx^2}.\left(Ve^{-\frac{1}{2}x^2}\right) + x^2\left(Ve^{-\frac{1}{2}x^2}\right) = -\frac{d^2V}{dx^2}.\left(e^{-\frac{1}{2}x^2}\right) + 2x\frac{dV}{dx}.e^{-\frac{1}{2}x^2} + Ve^{-\frac{1}{2}x^2}.$$

Thus, the substitution of *u* by $V e^{-\frac{1}{2}x^2}$ in Eq. (2) gives

$$-\frac{d^2V}{dx}.e^{-\frac{1}{2}x^2} + 2x\frac{dV}{dx}.e^{-\frac{1}{2}x^2} + Ve^{-\frac{1}{2}x^2} + aVe^{-\frac{1}{2}x^2} = 0.$$

Hence,

$$\frac{d^2V}{dx^2} - 2x\frac{dV}{dx} + (a-1)V = 0. \tag{7}$$

Since this is a general solution, it must embrace the two solutions discussed earlier. Thus when $V = x$, Eq. (7) tells us that $a = 3$.

Therefore, $u = x e^{-\frac{1}{2}x^2}$ is an eigenfunction of the operator $\left(-\frac{d^2}{dx^2} + x^2\right)$ belonging to the eigenvalue +3 as we have previously established by Eq. (5):

$$\left(-\frac{d^2}{dx^2} + x^2\right) x e^{-\frac{1}{2}x^2} = 3 x e^{-\frac{1}{2}x^2}. \tag{5}$$

Also, Eq. (7) embraces the solution $u = e^{-\frac{1}{2}x^2}$ discussed earlier. This is because when we put $V = 1$, Eq. (7) says that

$$(a-1) = 0$$

or

$$a = 1.$$

Accordingly, $u = e^{-\frac{1}{2}x^2}$ is an eigenfunction of the operator $\left(-\frac{d^2}{dx^2} + x^2\right)$ belonging to the eigenvalue +1.

Eq. (7) is similar in form to a classical differential equation known as *Hermite's equation*,

$$\frac{d^2V}{dx^2} - 2x\frac{dV}{dx} + 2vV = 0, \tag{8}$$

the solution of which can be proved to be

$$V = (-1)^v e^{x^2} \frac{d^v}{dx^v} e^{-x^2} \tag{9}$$

for all positive integers *v*, including $v = 0$. The expression on the right-hand side of Eq. (9) is usually denoted by the symbol $H_v(x)$ and is described as a *Hermite polynomial of degree v*, that is,

$$V = H_v(x) = (-1)^v e^{x^2} \frac{d^v}{dx^v} e^{-x^2}. \tag{9'}$$

The first six Hermite polynomials for $v = 0$ to $v = 5$ are

$$\left.\begin{array}{ll} H_0(x)=1 & H_1(x)=2x \\ H_2(x)=4x^2-2 & H_3(x)=8x^3-12x \\ H_4(x)=16x^4-48x^2+12 & H_5(x)=32x^5-160x^3+120x \end{array}\right\} \tag{10}$$

By substituting $V = H_0(x)$ and $V = H_1(x)$ in Eq. (6), which was considered as the general solution of Eq. (2), we see that, apart from a numerical constant, the first two Hermite polynomials are the factors of $e^{-\frac{1}{2}x^2}$ in the trial solutions Eqs. (4) and (5) of the differential Eq. (2).

Hence, the most general solution of Eq. (2), that is, the manifold of possible eigenfunctions of the operator $\left(-d^2/dx^2 + x^2\right)$ is

$$u_v = N_v H_v(x) e^{-\frac{1}{2}x^2}, \tag{11}$$

where N_v is a numerical constant. It is to be noted here that if *u* is an eigenfunction of an operator belonging to an eigenvalue *a*, *N u* is also an eigenfunction belonging to this same eigenvalue, provided *N* is simply a numerical constant. This result can be simply verified. Graphs of the function u_v with

$$N_v = \left(2^v v! \sqrt{\pi}\right)^{-\frac{1}{2}}$$

for $v = 0$ to 3 are given in Fig. 2. Evidently u_v satisfies the standard boundary conditions by being everywhere finite, continuous, and single-valued.

The eigenvalues of the operator $\left(-d^2/dx^2 + x^2\right)$ are obtained from the condition that Eq. (11) is a solution of the differential Eq. (2) for all positive integral values of *v* including the value zero. Comparing Eqs. (8) and (7), we see that *v* is related to the eigenvalue *a* by the expression.

$$2v = a - 1,$$

or

$$a = 2v + 1, \qquad \text{where } v = 0, 1, 2, \ldots \tag{12}$$

The eigenvalues of the operator are, therefore, the odd positive integers,

$$a = 1, 3, 5 \ldots$$

The graphs in Fig. 2 for $v = 0$ to $v = 3$, correspond, respectively, to the following eigenfunctions:

1. $\boldsymbol{v = 0}$, therefore, $u_0 = N_0 H_0(x) e^{-\frac{1}{2}x^2}$,

where $H_0(x) = 1$, and $N_0 = \left(\sqrt{\pi}\right)^{-\frac{1}{2}}$,

so that $$u_0 = (\pi)^{-\frac{1}{4}} e^{-\frac{1}{2}x^2}.$$

2. $\boldsymbol{v = 1}$, therefore,

$$u_1 = N_1 H_1(x) e^{-\frac{1}{2}x^2},$$

where $H_1(x) = 2x$, and $N_1 = \left(2\sqrt{\pi}\right)^{-\frac{1}{2}}$,
so that

$$u_1 = \left(2\sqrt{\pi}\right)^{-\frac{1}{2}} .2x\, e^{-\frac{1}{2}x^2} = \sqrt{2}\,(\pi)^{-\frac{1}{4}} e^{-\frac{1}{2}x^2}.$$

3. $\nu = 2$, therefore,

$$u_2 = N_2 H_2(x) e^{-\frac{1}{2}x^2},$$

where $H_2(x) = 4x^2 - 2$, and $N_2 = \left(2^2 .2!\sqrt{\pi}\right)^{-\frac{1}{2}}$,
so that

$$u_2 = \left(8\sqrt{\pi}\right)^{-\frac{1}{2}} . \left(4x^2 - 2\right) e^{-\frac{1}{2}x^2}$$
$$= \frac{1}{2\sqrt{2}} \pi^{-\frac{1}{4}} \left(4x^2 - 2\right) e^{-\frac{1}{2}x^2} .$$

4. $\nu = 3$, therefore,

$$u_3 = N_3 H_3(x) e^{-\frac{1}{2}x^2},$$

where $H_3(x) = 8x^3 - 12x$, and $N_2 = \left(2^3 .3!\sqrt{\pi}\right)^{-\frac{1}{2}}$,
so that

$$u_3 = \left(48\sqrt{\pi}\right)^{-\frac{1}{2}} . \left(8x^3 - 12x\right) e^{-\frac{1}{2}x^2}$$
$$= \frac{1}{4\sqrt{3}} \pi^{-\frac{1}{4}} \left(8x^3 - 12x\right) e^{-\frac{1}{2}} .$$

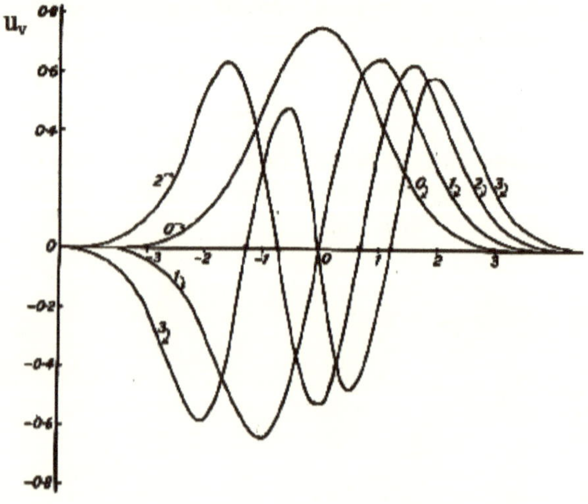

Fig. 2. Eigenfunctions u_v of the operator $\left(-\frac{d^2}{dx^2} + x^2\right)$ ($\nu = 0$ to 3)

4. Some remarks on classical mechanics.* In Newtonian mechanics, a particle of mass *m* in motion along a coordinate *x* and subjected to a force *f* obeys the equation

$$f = m\ddot{x}, \tag{1}$$

where $\ddot{x}$ stands for $\frac{d^2x}{dt^2}$. For systems which are of interest to us, the force *f* to which the particle is subject is the negative gradient of the potential energy, which we denote by *V(x)*,

$$f = -\frac{dV(x)}{dx}. \tag{2}$$

Elimination of *f* between (1) and (2) yields

$$\frac{dV(x)}{dx} + m\ddot{x} = 0. \tag{3}$$

Eq. (3) can be directly integrated by conversion to the variable $\dot{x} = dx/dt$,

$$dV(x) + m\dot{x}\, d\dot{x} = 0. \tag{4}$$

This equation integrates to

$$V(x) + \frac{1}{2}m\dot{x}^2 = E. \tag{5}$$

E in Eq. (5) appears as the constant of integration; it is the sum of the kinetic energy $T = \frac{1}{2}m\dot{x}^2$ and the potential energy *V(x)*, and is known as the total or *Hamiltonian energy of the system.*

For a linear harmonic oscillator (Fig. 3), in classical mechanics, *V(x)* is given by

$$V(x) = \frac{1}{2}Fx^2, \tag{6}$$

in which *F* is known as the force constant of the oscillator, and

$$T = \frac{1}{2}m\dot{x}^2, \tag{7}$$

and so, the Hamiltonian energy *E* is given by

$$E = T + V = \frac{1}{2}m\dot{x}^2 + \frac{1}{2}Fx^2. \tag{8}$$

---

* In this chapter, the symbol *E* will be used instead of the common symbol *U* for the internal energy.

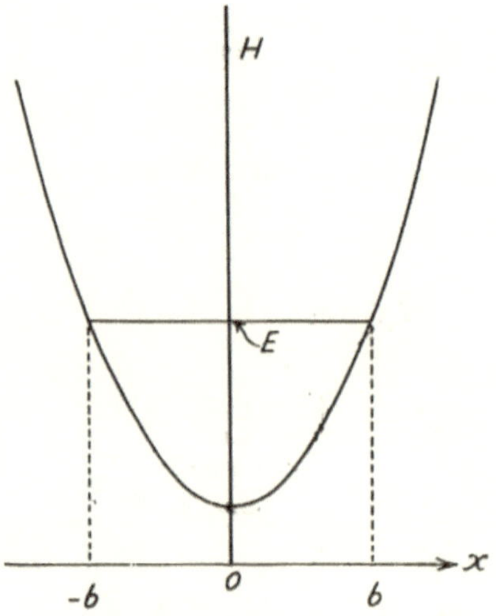

Fig. 3. Energy diagram of the linear harmonic oscillator, according to classical mechanics. The division of *E* between kinetic and potential energies varies with the phase of motion. At the midpoint $x = 0$, the energy is wholly kinetic; at the turning points $x = \pm b$, it is wholly potential energy.

Once the oscillator is set in motion, it continues to vibrate with the total energy *E*, provided there is no exchange of energy with the surroundings. The condition that there shall be no loss or gain of energy is implicit in Eq. (5) which shows that *E* is a constant of motion. The classical frequency of the oscillator can be obtained directly from the equation of motion in its differential form, which is Eq. (3) obtained above, namely,

$$\frac{dV(x)}{dx} + m\ddot{x} = 0. \qquad (3)$$

For a simple oscillator, $dV(x)/dx = Fx$,
so Eq. (3) becomes

$$m\ddot{x} + Fx = 0, \qquad (9)$$

a possible solution of which is

$$x = A\sin\left(\sqrt{\frac{F}{m}}\,t\right), \qquad (10)$$

where *A* is the amplitude of oscillation. This is because

$$\dot{x} = A\cos\left(\sqrt{\frac{F}{m}}\,t\right)\cdot\sqrt{\frac{F}{m}},$$

and

$$\ddot{x} = -A\sin\left(\sqrt{\frac{F}{m}}\,t\right)\cdot\frac{F}{m}$$

or
$$\ddot{x} = -x.\frac{F}{m}, \qquad (11)$$
so the substitution in Eq. (9) yields
$$-x\frac{F}{m}\times m + Fx = 0.$$
Putting the quotient *F/m* in Eq. (10), as equal to the square of the angular frequency ω, that is,
$$\omega^2 = \frac{F}{m} \qquad (12)$$
or
$$(2\pi\nu)^2 = \frac{F}{m}, \qquad (13)$$
where ν is the classical frequency of oscillator, we obtain Eq. (10) in the form
$$x = A\sin(2\pi\nu t). \qquad (14)$$
This is the standard equation of a wave motion in a simple form containing no arbitrary phase angles. Eq. (13) gives the classical frequency of oscillation.
$$\nu = \frac{1}{2\pi}\sqrt{\frac{F}{m}}. \qquad (15)$$
The coordinate, *x*, the velocity, $\dot{x}$, the momentum, $P(= mv)$, the kinetic energy, $T\left(=\frac{1}{2}m\dot{x}^2\right)$, etc, of any dynamical system are described as *dynamical variables*. It is important to realise that the Hamiltonian energy, besides being a constant of motion, is also a dynamical variable. This is because although a dynamical system possessing the Hamiltonian energy *E* maintains this energy indefinitely, there is an infinite range of values to which *E* may be adjusted initially. *When we refer to the Hamiltonian energy as a dynamical variable, we use the symbol H, to distinguish the meaning dearly from the numerical magnitude of the energy which we shall continue to designate by E. The classical expression for the Hamiltonian energy as a dynamical variable* is that given by Eq. (5),
$$E = \frac{1}{2}m\dot{x}^2 + V(x), \qquad (5)$$
but with H substituted for E, it becomes
$$H = \frac{1}{2}m\dot{x}^2 + V(x). \qquad (16)$$
Further, we shall write $\frac{1}{2}m\dot{x}^2 = P^2/2m$, where the momentum $P = m\dot{x}$, whence
$$H = \frac{P^2}{2m} + V(x). \qquad (17)$$
Differentiation of Eq. (17) with respect to *P* yields
$$\frac{\partial H}{\partial P} = \frac{P}{m} = \dot{x}, \qquad (18)$$

and with respect to *x* yields,

$$\frac{\partial H}{\partial x} = \frac{dV(x)}{dx} = -f. \tag{19}$$

But according to Eq. (1),

$$f = m\ddot{x} = \dot{P}, \tag{1}$$

therefore Eq. (19) can be rewritten as

$$\frac{\partial H}{\partial x} = -\dot{P}. \tag{20}$$

Eqs. (18) and (20) are known as *Hamilton's canonical equations*. In general, two dynamical variables *q* and *s* are said to be *canonically conjugate* if

$$\frac{\partial H}{\partial q} = \dot{s}, \text{ and } \frac{\partial H}{\partial s} = -\dot{q}. \tag{21}$$

5. Schrödinger operators. We shall introduce the system of mechanics known as Schrödinger mechanics or wave mechanics through a series of postulates. These postulates like *Newton's laws* of motion are not proved; thus the reader is asked to take them for granted. These postulates as stated here are adequate for an elementary treatment.

Postulate I. *To every dynamical variable q, there must be assigned an operator q. The physical properties of a dynamical variable are deducible from the mathematical properties of the operator assigned to the variable. In particular, the possible results of an exact experimental measurement of a dynamical variable q are the eigenvalues of the operator q, and conversely.*

The choice of operator to be associated with a dynamical variable requires a *second postulate* which introduces Planck's constant and is known as the *quantum condition.*

Postulate II. *The operators q and r associated with two canonically conjugate dynamical variables q and r must satisfy the equation.*

$$qr - rq = \frac{ih}{2\pi} \tag{1}$$

In Eq. (1), *h* is Planck's constant and $i^2 = -1$. The left-hand side of the equation represents the *commutator* of the operators q and r.

Special interest attaches to the canonically conjugate variables *x* and *P* which characterise a one-dimensional dynamical system. If we assign the operator x to the variable *x*, and the operator $-(ih/2\pi)d/dx$ to the variable *P*, it is evident that we have a solution of Eq. (1), for then,

$$xP - Px = x\left[-\left(\frac{ih}{2\pi}\right)\frac{d}{dx}\right] - \left[-\left(\frac{ih}{2\pi}\right)\frac{d}{dx}\right]x$$

$$= \left(\frac{ih}{2\pi}\right)\left(\frac{d}{dx}x - x\frac{d}{dx}\right), \tag{2}$$

and it has already been shown that the operator $\frac{d}{dx}x - x\frac{d}{dx}$ equals the operator +1. In the Schrödinger method, therefore, the operator x is associated with the variable *x*, and the operator $-(ih/2\pi)\,d/dx$ with the momentum *P* (along *x*).

Schrödinger operators assigned to other dynamical variables of a system are derived from the basic x and P operators by ordinary processes of algebra. Thus the operator assigned to the variable x being x, that assigned to x^2 is the operator x^2; likewise the operator associated with the variable P^2 is $\left[\left(-\frac{ih}{2\pi}\right)\frac{d}{dx}\right]^2 = -\frac{h^2}{4\pi^2}\frac{d^2}{dx^2}$, and so forth. Some of our conclusions up to this point are summarised in Table 1.

Table 1

Schrödinger Operators for a One-Dimensional System

Dynamical variable	Operator
x	X
x^2	x^2
$P = m\dot{x}$	$-\left(\frac{ih}{2\pi}\right)\frac{d}{dx}$
P^2	$-\left(\frac{h^2}{4\pi^2}\right)\frac{d^2}{dx^2}$
xP	$x\left[-\left(\frac{ih}{2\pi}\right)\frac{d}{dx}\right]$
H	$-\left(\frac{h^2}{8\pi^2 m}\right)\frac{d^2}{dx^2} + V(x)$
φ	φ
P_z	$-\left(\frac{ih}{2\pi}\right)\frac{d}{d\varphi}$
E	$\left(\frac{ih}{2\pi}\right)\frac{\partial}{\partial t}$
T	T

The operator H assigned to the Hamiltonian energy as a dynamical variable denoted by H is of central importance in the Schrödinger scheme; it is known as the *Hamiltonian variable H.*

The classical expression for the Hamiltonian energy H is that given above by Eq. (17), namely,

$$H = \frac{P^2}{2m} + V(x). \qquad (3)$$

This equation is an operator equation, which may be expanded by substituting in the right-hand side the operators P and V(x) expressed in terms of x. The operator V(x) changes from one dynamical system to another, and for the moment, will be left unspecified. Introducing $P^2 = -\frac{h^2}{4\pi^2}\frac{d^2}{dx^2}$, Eq. (3) becomes

$$H = -\frac{h^2}{8\pi^2 m}\frac{d^2}{dx^2} + V(x). \qquad (4)$$

Eq. (4) is the *general form of the expression for the Hamiltonian operator of a system possessing one degree of freedom.*

According to postulate I, the results of an exact experimental measurement of the energy of a system, whose classical Hamiltonian energy is a dynamical variable, is given by the equation,

$$H = \frac{P^2}{2m} + V(x),$$

are the eigenvalues E of the operator H in Eq. (4); hence we may write *symbolically,*

$$H\psi = E\psi. \qquad (5)$$

This is *Schrödinger's first equation,* sometimes called *Schrödinger's amplitude equation.* The operands ψ are *eigenfunctions of the operator H* and are subject to the boundary conditions outlined above; that is, they must be *finite, continuous, and single-valued through the whole range of possible values of x, from* $X = -\infty$ *to* $x = +\infty$. The eigenfunctions of the Hamiltonian operator are referred to as *amplitude wave functions.*

6. The linear harmonic oscillator. For the linear oscillator, possessing one degree of freedom, the classical potential energy, *V (x)*, is equal to $\frac{1}{2}Fx^2$, where F is the force constant, and the corresponding Schrödinger operator *V(x)* is, therefore, the operator $\frac{1}{2}Fx^2$. The Hamiltonian operator for the oscillator is obtained by substituting $\frac{1}{2}Fx^2$ for *V(x)* in Eq. (4) of Section 5, whence

$$H = -\left(\frac{h^2}{8\pi^2 m}\right)\frac{d^2}{dx^2} + \frac{1}{2}Fx^2. \qquad (1)$$

The amplitude wave functions, ψ, and permitted energy levels, E, of the oscillator are those which satisfy the Schrödinger amplitude equation with *H* given by Eq. (1); namely the equation,

$$\left[-\left(\frac{h^2}{8\pi^2 m}\right)\frac{d^2}{dx^2} + \frac{F}{2}x^2\right]\psi = E\psi. \qquad (2)$$

The algebra of this equation is, to a large extent, already solved above in Section 3, when we consider the differential equation

$$\left(-\frac{d^2}{dx^2} + x^2\right)u = au$$

to discover the functions $u = u(x)$ – which must satisfy the boundary conditions – and the possible values of the constant *a*. To take advantage of the results obtained there, we make the substitutions

$$\eta = \frac{x}{r}, \tag{3}$$

and

$$E = \frac{ah}{4\pi}\sqrt{\frac{F}{m}}, \tag{4}$$

where $a = \sqrt[4]{h^2/4\pi^2 Fm}$. (5)

Here, as in Section 3, *m* represents the mass, and *F* the force constant of the oscillator. Introducing the substitutions (3)–(5) into the Schrödinger Eq. (2), after some simplification yields,

$$\left(-\frac{d^2}{d\eta} + \eta^2\right)\psi = a\psi, \tag{6}$$

which in form is identical with the differential equation shown above in Section 3, namely,

$$\left(-\frac{d^2}{dx^2} + x^2\right)u = au. \tag{6$'$}$$

Evidently, the eigenvalues of Eq. (6) are the same as those of Eq. (6)′, that is, they are the odd positive integers 1, 3, 5, . . . Hence,

$$a = 2v + 1; \qquad v = 0, 1, 2 \ldots. \tag{7}$$

If we now substitute this result in Eq. (4), we obtain the energy eigenvalues, E_v, of the Hamiltonian operator

$$E_v = \frac{(2v+1)h}{4\pi}\sqrt{\frac{F}{m}}$$

$$= \left(v + \frac{1}{2}\right)\frac{h}{2\pi}\sqrt{\frac{F}{m}}, \tag{8}$$

the subscript *v* being attached to E_v as a reminder that Eq. (8) specifies a manifold of energy levels defined by the running number $v = 0, 1, 2 \ldots$. The Eq. (8) for E_v can be reduced, for it was shown before that the angular frequency ω of the oscillator in classical mechanics is related to *F* and *m* by

$$2\pi\nu = \omega = \sqrt{\frac{F}{m}}.$$

Therefore, we have

$$\nu = \frac{1}{2\pi}\sqrt{\frac{F}{m}}, \tag{9}$$

and

$$E_v = \left(v + \frac{1}{2}\right)h\nu, \qquad (10)$$

therefore, according to wave mechanics the minimum energy E_0 of the oscillator is not zero, for setting $v = 0$, we find

$$E_0 = \frac{1}{2}h\nu. \qquad (11)$$

E_0, the *energy of the lowest state of the oscillator, is termed as the zero-point energy.*

The amplitude wave functions ψ_v of Eq. (2), namely,

$$\left(-\frac{h^2}{8\pi^2 m}\frac{d^2}{dx^2} + \frac{F}{2}x^2\right)\psi = E\psi,$$

are the eigenfunctions given by Eq. (11) in Section 3, namely,

$$u_v = N_v H_v(x) e^{-\frac{1}{2}x^2};$$

these functions must be modified in order to take account of the change of variable, that is

$$\psi_v = N_v H_v(\eta) e^{-\frac{1}{2}\eta^2}, \qquad (12)$$

where $\eta = x/r$ is substituted for x. The functions ψ_v for $v = 0$ to 3 are represented in Fig. 4.

We shall now proceed to evaluate the constant N_v. This requires a *third postulate,* which we shall present in two parts.

Postulate III A. *One-dimensional Schrödinger wave functions,* ψ, *have the property that* $\psi^*\psi\, dx$ *is the probability that the variable x lies within the limits x and x + dx.* (ψ^* denotes the complex conjugate of ψ, i.e. the function obtained by replacing i by $-i$ whenever it occurs. This is not a detail of immediate concern, for the wave functions (Eq. 12) of the oscillator are wholly real, and thus $\psi^* \equiv \psi$.)

In a one-dimensional system, the variable x must lie somewhere within the totality of points comprising the x axis, and hence

$$\int_{-\infty}^{\infty} \psi^* \psi\, dx = 1, \qquad (13)$$

since the probability that x lies between the limits of $x = \pm\infty$ is unity. Eq. (13) is known as the *normalising condition.* It will be shown below that the amplitude wave functions given by Eq. (12) obey Eq. (13) if

$$N_v = \left(r\, 2^v\, v!\sqrt{\pi}\right)^{-\frac{1}{2}}; \qquad v = 0, 1, 2, \ldots \qquad (14)$$

The wave functions belonging to the energy eigenvalue E_0 of the oscillator in its lowest, zero-point energy state is, therefore,

$$\psi_0 = \left\{ \left(r\sqrt{\pi}\right)^{-\frac{1}{2}} \right\} e^{-\frac{1}{2}(x/r)^2} . \qquad (15)$$

Now we set ourselves to derive Eq. (14):

The normalising condition for the wave functions of the harmonic oscillator is

$$\int_{-\infty}^{\infty} \psi_v^2 \, dx = 1, \qquad (1)$$

with

$$\psi_v = N_v H_v(\eta) e^{-\frac{1}{2}\eta^2}; \quad \eta = \frac{x}{r}. \qquad (2)$$

As $r d\eta = dx$, Eq. (1) becomes

$$N_v^2 r \int_{-\infty}^{\infty} [H_v(\eta)]^2 e^{-\eta^2} d\eta = 1. \qquad (3)$$

It can be proved that

$$\int_{-\infty}^{\infty} [H_v(x)]^2 e^{-x^2} dx = (-1)^{2v} 2^v v! \int_{-\infty}^{\infty} e^{-x^2} dx , \qquad (4)$$

and since $(-1)^{2v} = 1$, and $\int_{-\infty}^{\infty} e^{-x^2} dx = \sqrt{\pi}$,

Eq. (4) becomes

$$\int_{-\infty}^{\infty} [H_v(x)]^2 e^{-x^2} dx = 2^v v! \sqrt{\pi}, \qquad (5)$$

and

$$\int_{-\infty}^{\infty} [H_v(\eta)]^2 e^{-\eta^2} d\eta = 2^v v! \sqrt{\pi}. \qquad (6)$$

Thus, substituting for the integral in Eq. (3) by its value given in Eq. (6), we obtain

$$N_v^2 r 2^v v! \sqrt{\pi} = 1,$$

so

$$N_v = \left(r 2^v v! \sqrt{\pi}\right)^{-\frac{1}{2}}.$$

Now we return to *postulate III A*. A *corollary* of this postulate is that *if* ψ_v *is a normalised amplitude* wave function, $\psi_v^* \psi_v$ *specifies the probability distribution of a one-dimensional dynamical system in the vth state with respect to the coordinate x.* Functions $\psi_v^* \psi_v \equiv \psi_v^2$ for the harmonic oscillator are represented by dotted lines in Fig. 4. In each state of the oscillator, there is a finite (though small) probability that the particle will be found by experiment *outside* the classical limits, that is, outside the area bounded by the parabola $V = \frac{1}{2} F x^2$. The probability of penetrating into the non-classical region is *greater* for *smaller* value of *v*.

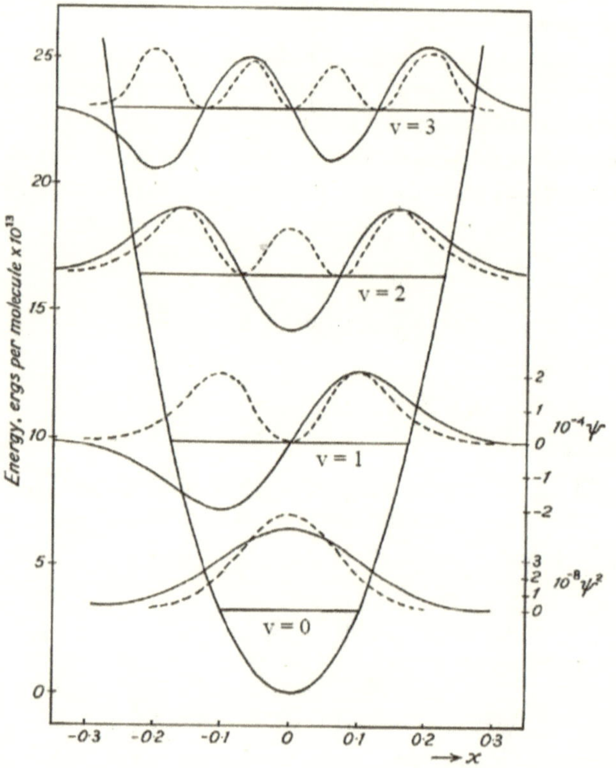

Fig. 4. Energy levels, wave functions (solid curves), and probability distribution (broken curves) for the harmonic oscillator. The wave functions, (Eq. 12), are calculated for *m* =1.008 atomic weight units, and $\sigma = 3000\ \text{cm}^{-1}$ wave number, that is, they are appropriate to the vibration of a hydrogen nucleus against a much larger mass.

It seems instructive to evaluate the amplitude wave functions of the linear harmonic oscillator for the states $v = 0$ to $v = 4$ together with the corresponding energy eigenvalues and the probability distributions in these states. The equations used are respectively

$$\psi_v = N_v H_v(\eta) e^{-\frac{1}{2}\eta^2},$$

with

$$\eta = x/r,$$

$$N_v = \left(r\, 2^v\, v!\sqrt{\pi}\right)^{-\frac{1}{2}}$$

$$H_v(\eta) = (-1)^v e^{\eta^2} \frac{d^v}{dx^v} e^{-\eta^2}$$

$$r = \sqrt[4]{h^2/4\pi^2 F m}$$

$$E_v = \left(v + \frac{1}{2}\right) h\nu$$

$$\nu = \frac{1}{2\pi}\sqrt{\frac{F}{m}}.$$

The calculations are carried out, as follows:

1. For $v = 0$:

$$\psi_0 = N_0 H_0 (\eta) e^{-\frac{1}{2}\eta^2}$$
$$H(\eta) = 1$$
$$N_0 = \left(r\sqrt{\pi}\right)^{-\frac{1}{2}}.$$

Therefore,

$$\psi_0 = \left(r\sqrt{\pi}\right)^{-\frac{1}{2}} e^{-\frac{1}{2}\left(\frac{x}{r}\right)^2}$$

$$\psi_0^* \psi_0 = \psi_0^2 = \left(r\sqrt{\pi}\right)^{-1} e^{-\left(\frac{x}{r}\right)^2}$$

$$E_0 = \frac{1}{2} h\nu$$

$$= \frac{1}{2} \times 6.625 \times 10^{-27} \times 2.998 \times 10^{10} \times 3 \times 10^3$$

$$= 2.969 \times 10^{-13} \text{ erg / molecule}.$$

2. For $v = 1$:

$$\psi_1 = N_1 H_1 (\eta) e^{-\frac{1}{2}\eta^2}$$
$$H_1(\eta) = 2\eta$$
$$N_1 = \left(r\, 2.1!\, \sqrt{\pi}\right)^{-\frac{1}{2}}$$
$$= \left(2\, r\sqrt{\pi}\right)^{-\frac{1}{2}}.$$

Therefore,

$$\psi_1 = \left(2\, r\sqrt{\pi}\right)^{-\frac{1}{2}} . 2\left(\frac{x}{r}\right) e^{-\frac{1}{2}\left(\frac{x}{r}\right)^2}$$

$$\psi_1^* \psi_1 = \psi_1^2 = \left(2r\sqrt{\pi}\right)^{-1} . 4\left(\frac{x}{r}\right)^2 e^{-\left(\frac{x}{r}\right)^2} .$$

Thus the amplitude wave function curve as well as the probability distribution curve passes through the origin at $x = 0$ as shown in Fig. 4.

$$E_1 = \frac{3}{2} h\nu = 3 \times 2.969 \times 10^{-13}$$

$$= 8.91 \times 10^{-13} \text{ erg / molecule}.$$

3 For $v = 2$:

$$\psi_2 = N_2 H_2 (\eta) e^{-\frac{1}{2}\eta^2}$$

$$H_2(\eta) = 4\eta^2 - 2$$

$$= 4\left(\frac{x}{r}\right)^2 - 2$$

$$N_2 = \left(r\, 2^2\, 2!\, \sqrt{\pi}\right)^{-\frac{1}{2}}$$

$$= \left(8\, r \sqrt{\pi}\right)^{-\frac{1}{2}}.$$

Therefore,

$$\psi_2 = \left(8 r \sqrt{\pi}\right)^{-\frac{1}{2}} \left\{4\left(\frac{x}{r}\right)^2 - 2\right\} e^{-\frac{1}{2}\left(\frac{x}{r}\right)^2}.$$

Accordingly, the curve representing ψ_2 does not pass through the point corresponding to $x = 0$, but when x attains the values $x = \pm \frac{1}{\sqrt{2}} r$, it acquires two zero amplitude values as shown in Fig. 4.

$$\psi_2^2 = \left(8 r \sqrt{\pi}\right)^{-1} \left\{4\left(\frac{x}{r}\right)^2 - 2\right\}^2 e^{-\left(\frac{x}{r}\right)^2}.$$

Thus the probability distribution curve resembles the amplitude wave function curve in acquiring a zero value at the same x points, namely at $x = \pm \frac{1}{\sqrt{2}} r$, but differs, as is expected, in possessing a *crest* in the region between these two limits $+\frac{1}{\sqrt{2}} r$ and $-\frac{1}{\sqrt{2}} r$.

$$E_2 = \frac{5}{2} h\nu = 5 \times 2.969 \times 10^{-13}$$

$$= 14.85 \times 10^{-13} \text{ erg/molecule}.$$

4. For $v = 3$:

$$\psi_3 = N_3 H_3 (\eta) e^{-\frac{1}{2}\eta^2}$$

$$H_3(\eta) = 8\eta^3 - 12\eta,$$

$$= 8\left(\frac{x}{r}\right)^3 - 12\left(\frac{x}{r}\right)$$

$$N_3 = \left(r\, 2^3\, 3!\, \sqrt{\pi}\right)^{-\frac{1}{2}}$$

$$= \left(48\, r \sqrt{\pi}\right)^{-\frac{1}{2}}.$$

Therefore,

$$\psi_3 = \left(48\,r\sqrt{\pi}\right)^{-\frac{1}{2}} \left\{8\left(\frac{x}{r}\right)^3 - 12\left(\frac{x}{r}\right)\right\} e^{-\frac{1}{2}\left(\frac{x}{r}\right)^2} .$$

Accordingly, the curve representing the amplitude wave function intersects with the x axis at three points: $x = 0$, $x = +\sqrt{\frac{3}{2}}\,r$, and $x = -\sqrt{\frac{3}{2}}\,r$. In the x range 0 to $+\sqrt{\frac{3}{2}}\,r$, it, thus, represents a *trough*, and in the range 0 to $-\sqrt{\frac{3}{2}}\,r$, it represents a *crest*, as shown in Fig. 4.

$$\psi_3^2 = \left(48\,r\sqrt{\pi}\right)^{-1} \left\{8\left(\frac{x}{r}\right)^3 - 12\left(\frac{x}{r}\right)^2\right\}^2 e^{-\left(\frac{x}{r}\right)^2} ,$$

therefore, the probability distribution curve resembles, as is expected, the amplitude wave function curve in its dependence on the x coordinate for acquiring zero amplitude values at $x = 0$, $+\sqrt{\frac{3}{2}}\,r$, and $-\sqrt{\frac{3}{2}}\,r$ but differs in having no *troughs* with respect to this coordinate as shown in Fig. 4.

$$E_3 = \frac{7}{2} h\nu = 7 \times 2.969 \times 10^{-13}$$

$$= 20.78 \times 10^{-13} \text{ erg/molecule.}$$

7. The free particle: degenerate eigenvalues. The classical Hamiltonian function of a particle of mass m in motion along the coordinate x and subject to no force is

$$H = \frac{P^2}{2m} + V. \tag{1}$$

Since $f = 0$, the potential energy V is independent of x

$$f = -\frac{dV}{dx}. \tag{2}$$

Thus we are at liberty to choose the scale of potential energy so that $V = 0$. The Schrödinger amplitude equation,

$$H\psi = E\psi , \tag{3}$$

is then

$$\left(-\frac{h^2}{8\pi^2 m}\right)\frac{d^2}{dx^2}\psi = E\psi \cdot \tag{4}$$

Since the operator $h^2/8\pi^2 m$ *commutes* with d^2/dx^2, Eq. (4) rearranges to

$$-\frac{d^2}{dx^2}\psi = \frac{8\pi^2 mE}{h^2}\psi . \tag{5}$$

We can see from this equation that the wave functions and energy eigenvalues of Eq. (4), apart from a multiplicative constant, which is $8\pi^2 m / h^2$, are those of the simple operator $-d^2/dx^2$.

The operation $-d^2/dx^2$ was discussed briefly before, where it was shown that $\sin\sqrt{a}\,x$, with a constant *a*, is an acceptable eigenfunction. Equally, $\cos\sqrt{a}\,x$ is a possible eigenfunction, for sine and cosine functions differ only in phase. A property of eigenfunctions, easily verified by trial, is that if two functions $u_1(x)$ and $u_2(x)$ are eigenfunctions of the same operator, their sum $u_1(x)+u_2(x)$ is also an eigenfunction. Further, recalling that an eigenfunction can be multiplied by any arbitrary constant, *the solution of* Eq. (5) is seen to have the general form

$$N_1 \sin\sqrt{a}\,x \;+\; N_2 \cos\sqrt{a}\,x, \tag{6}$$

with

$$a = \frac{8\pi^2 mE}{h^2}, \tag{7}$$

so we may write

$$\psi = N_1 \sin\left(\frac{2\pi\sqrt{2mE}}{h}\right)x + N_2 \cos\left(\frac{2\pi\sqrt{2mE}}{h}\right)x\,\cdot \tag{8}$$

The only restriction on *E* stems from the requirement that ψ must conform to the boundary conditions; in particular, that ψ must remain finite as $x \to \pm\infty$. This condition is satisfied if

$$\frac{2\pi\sqrt{2mE}}{h}$$

is wholly real or, in other words, if *E* is any positive number since, both, *m* and *h* are positive and real. Our first conclusion, therefore, is that the energy eigenvalues *E* of Eq. (4) include all positive numbers.

Secondly, if *E* = 0, Eq. (8) becomes

$$\psi = N_2 \tag{9}$$

Since N_2 is simply a constant, the wave function (Eq. 9) satisfies the boundary conditions, and hence, this equation is an *eigenfunction* of Eq. (4), and we can add *E* = 0 to the list of permitted energy values. If *E* is negative, however, $\sqrt{E}$ is a pure imaginary quantity. To develop the arguments in this situation let

$$\frac{2\pi\sqrt{2mE}}{h}$$

be equal to $i\gamma$

$$\frac{2\pi\sqrt{2mE}}{h} = i\gamma \tag{10}$$

so that γ is *wholly real*. Now

$$\cos(i\gamma x) = \frac{1}{2}\left(e^{\gamma x} + e^{-\gamma x}\right), \tag{11}$$

and

$$\sin(i\gamma x) = \left(\frac{i}{2}\right)\left(e^{\gamma x} - e^{-\gamma x}\right). \tag{12}$$

Substitution of Eq. (10) in Eq. (8) gives

$$\psi = N_1 \sin(i\gamma x) + N_2 \cos(i\gamma x),$$

and then substitution of the expanded sine and cosine functions given by Eqs. (11) and (12) yields

$$\psi = N_1\left[\frac{i}{2}\left(e^{\gamma x} - e^{-\gamma x}\right)\right] + N_2\left[\frac{1}{2}\left(e^{\gamma x} + e^{-\gamma x}\right)\right]$$

$$= \frac{1}{2}(iN_1 + N_2)e^{\gamma x} - \frac{1}{2}(iN_1 - N_2)e^{-\gamma x}. \quad (13)$$

When $E^{1/2}$ is imaginary, neither of the terms on the right-hand side of Eq. (13) is acceptable as a wave function, for the first becomes infinite as $x \rightarrow \infty$, and the second as $x \rightarrow -\infty$. Hence, the free particle energy cannot be negative; that is,

$$E \nless V,$$

since, by convention, $V$was set equal to zero. The permitted energy eigenvalues are, therefore,

$$E \geq 0. \quad (14)$$

Wave mechanics places no other restriction on the energy levels, which may be thought of as *infinitely fine grained.* More detailed considerations yield the same result for motion in three dimensions; the free translation of molecules is *unquantised.*

Two further aspects of Eq. (8) are noteworthy. First, the energy eigenvalue $E = 0$ is a solution of the *cosine part only* of the wave function (Eq. 8); all other energies *occur twice*, once as a solution of the sine term and another as a solution of the cosine term. Because of this property the non-zero energy eigenvalues are described as *doubly degenerate.* Physically, every non-zero eigenvalue occurs twice because the particle may move with the same energy in either the positive or the negative direction of x; but the value $E = 0$ occurs only once, because in this state, the particle is *motionless.* Second, both the terms on the right-hand side of Eq. (8) are periodic with a 'frequency',

$$\nu = \frac{\sqrt{2mE}}{h}. \quad (15)$$

We recall that the energy of the free particle is wholly kinetic, whence

$$E = \frac{1}{2}m\dot{x}^2,$$

and therefore,

$$\sqrt{2mE} = \sqrt{2m\frac{1}{2}m\dot{x}^2}$$

$$= \sqrt{m^2\dot{x}^2}$$

$$= m\dot{x}$$

$$= P. \quad (16)$$

Accordingly, Eq. (15) becomes

$$\nu = \frac{m\dot{x}}{h} = \frac{P}{h} \quad (17)$$

or

$$\lambda = \frac{h}{m\dot{x}} = \frac{h}{P}, \quad (18)$$

since $\lambda = \frac{1}{\nu}$. Hence, we may associate a *characteristic wave length* λ with the particle given by Eq. (18). This equation is the *de Broglie relation* between the wavelength λ of a particle of mass *m* and its velocity $\dot{x}$. *The idea that particles obey the laws of wave motion underlies the whole of wave mechanics and is strikingly verified by observation of the diffraction of electrons.*

8. The fixed-axis rotator. A rigid body able to rotate about an axis fixed in space is a system of *one degree of freedom*. The classical kinetic energy $T = \frac{1}{2} I\omega^2$ involves the moment of inertia *I* and the angular velocity, ω, in place of the mass, *m*, and the linear velocity, $\dot{x}$, of the free particle. As the total energy of the rotator can be taken as wholly kinetic, without loss of generality, the *classical expression for the Hamiltonian energy as a dynamical variable, or in other words, the classical Hamiltonian function is*

$$H = \frac{1}{2} I\omega^2 = \frac{(I\omega)^2}{2I}. \tag{1}$$

Let the fixed axis of rotation be denoted by *z;* then the angular momentum of the system is

$$I\omega = P_z, \tag{2}$$

and thus, Eq. (1) takes the form

$$H = (P_z)^2 / 2I \tag{3}$$

The *angular momentum operator* associated with the classical *dynamical variable* P_z can be shown to be

$$P_z = \left(-\frac{ih}{2\pi}\right)\frac{d}{d\phi}, \tag{4}$$

where ϕ is the angular coordinate defining the position of the rotator.

Hence, the Schrödinger amplitude equation,

$$H\psi = E\psi, \tag{5}$$

for the fixed-axis rotator is

$$\left(-\frac{h^2}{8\pi^2 I}\frac{d^2}{d\phi^2}\right)\psi = E\psi, \tag{6}$$

with the operator *H* for the rotator being

$$H = -\frac{h^2}{8\pi^2 I}\frac{d^2}{d\phi^2}. \tag{7}$$

Eq. (6) is an equation precisely like the Schrödinger amplitude equation for the free particle

$$\left(-\frac{h^2}{8\pi^2 m}\frac{d^2}{dx^2}\right)\psi = E\psi,$$

which has been obtained above in Section 7.

The solution of Eq. (6) can be written as a sum of sine and cosine functions as in Eq. (8) in Section 7, but it is here more convenient to set it in the alternative form

$$\psi = N e^{iM\phi}, \tag{8}$$

in which

$$M = \pm \frac{2\pi\sqrt{2IE}}{h}. \qquad (9)$$

However, a significant difference between the free particle and the rotator is that, whereas the coordinate x of the particle may lie anywhere between $x = \pm\infty$, ϕ is confined to the limits 0 and 2π. Therefore, in order that ψ shall be single-valued, each time ϕ becomes a multiple of 2π, ψ must begin to repeat itself – a restriction that has no analogue in the treatment of the free particle. This means that for ψ to be a single-valued function, we must have

$$\psi(\phi) = \psi(\phi + 2\pi), \qquad (10)$$

that is,

$$e^{iM\phi} = e^{iM(\phi + 2\pi)}, \qquad (11)$$

or

$$e^{iM\phi} = e^{iM\phi}\ e^{i2\pi M}, \qquad (11')$$

Therefore, in order that ψ shall be a single-valued function between the limits 0 and 2π for the angular coordinate ϕ, the exponential $e^{i2\pi M}$ must be equal to unity. This is only true when M is a positive or negative integer, that is

$$\ldots +2, +1, 0, -1, -2, \ldots$$

as required by the condition equation

$$e^{i2\pi M} = \cos 2\pi M + i \sin 2\pi M = 1. \qquad 12)$$

The energy eigenvalues obtained from Eq. (9) are then given by

$$E = M^2 \frac{h^2}{8\pi^2 I}; \quad M = 0, \pm 1, \pm 2, \ldots \qquad (13)$$

That is, M should be any positive or negative integer, zero included.

The lower part of the energy manifold is shown below in Fig. 5. Owing to the cyclic character of the angular coordinate ϕ, the spectrum of energy values is *discrete (i.e. quantised), quite unlike that of the free particle.* Yet in other respects, there is a close resemblance between the rotator and the free particle – thus all energy eigenvalues except the lowest are *doubly degenerate*. This is because rotation may be clockwise or counterclockwise with the same energy and negative energy eigenvalues are excluded.

For the evaluation of the numerical constant, N, in Eq. (8), we apply the normalisation condition equation

$$\int_0^{2\pi} \psi^* \psi \, d\phi = 1, \qquad (14)$$

so that the substitution for ψ^* and ψ yields

$$\int_0^{2\pi} N\, e^{-iM\phi} \,.\, N\, e^{iM\phi}\, d\phi = 1$$

$$N^2 \int_0^{2\pi} d\phi = 1$$

$$N^2 \,[\phi]_0^{2\pi} = 1$$

$$N^2 \,.\, 2\pi = 1.$$

Therefore,

$$N = \frac{1}{\sqrt{2\pi}}. \tag{15}$$

Hence, Eq. (8) is expressed as

$$\psi = \frac{1}{\sqrt{2\pi}} e^{iM\phi}, \tag{16}$$

in which *M* is given above by Eq. (9), namely,

$$M = \pm \frac{2\pi\sqrt{2IE}}{h}. \tag{9}$$

The wave function of the lowest state ψ_0 is thus equal to $\frac{1}{\sqrt{2\pi}}$, that is independent of ϕ. This means that *all orientations are equally probable*.

The wave functions of the fixed-axis rotator, Eq. (8), besides being eigenfunctions of the operator *H* of the rotator as given by Eq. (16), also *happen to be eigenfunctions of the angular momentum operator given by Eq. (4)*. This can be seen from the following:

$$-\frac{ih}{2\pi}\frac{d}{d\phi}\left(\frac{1}{\sqrt{2\pi}} e^{iM\phi}\right) = -\frac{ih}{2\pi}\left(\frac{1}{\sqrt{2\pi}} e^{iM\phi}\right) iM$$

$$= \frac{Mh}{2\pi}\left(\frac{1}{\sqrt{2\pi}} e^{iM\phi}\right).$$

It is evident from this equation that the eigenvalues $\mathbf{P_z}$ of the operator P_z are

$$\mathbf{P_z} = \frac{Mh}{2\pi}, \qquad M = 0, \pm 1, \pm 2, \ldots. \tag{17}$$

That is, the angular momentum of the rotator is *restricted* in wave mechanics to values which are *integral multiples* of h/2π. The reason that the energy *E* and the angular momentum $\boldsymbol{P_z}$ are simultaneous eigenvalues is that the classical expression of the Hamiltonian function given above, namely,

$$H = \frac{1}{2} I\omega^2 = \frac{(I\omega)^2}{2I} = \frac{P_z^2}{2I}, \tag{1}$$

which relates the total energy to the square of the angular momentum is *also valid for the eigenvalues* given by Eqs. (13) and (17), respectively. This means that if we substitute for *H* in the left-hand side of Eq. (1) by the energy eigenvalues

$$E = M^2 \frac{h^2}{8\pi^2 I},$$

and for $\boldsymbol{P_z}$ in the right-hand side of the same equation by the eigenvalues

$$\mathbf{P_z} = \frac{Mh}{2\pi},$$

we get respectively,

$$M^2 \frac{h^2}{8\pi^2 I},$$

and

$$\left(\frac{Mh}{2\pi}\right)^2 \frac{1}{2I} = M^2 \frac{h^2}{8\pi^2 I},$$

that is, we get the same result.

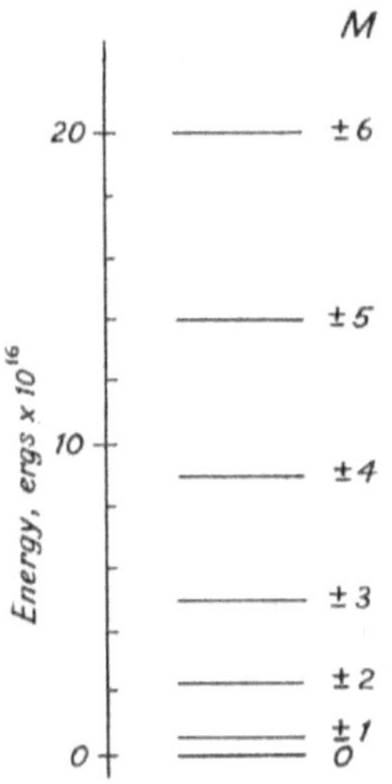

Fig. 5. Energy levels of a fixed-axis rotator. The energy scale corresponds to $I = 10^{-40}$ g.cm^2.

9. The Heisenberg uncertainty principle. To introduce the *Heisenberg uncertainty principle*, it is necessary to complete the third postulate, the first part of which was given in Section 6.

Postulate III B*. For a system confined to a linear coordinate x in a state described by a normalised amplitude wave function* ψ, *the average value* $\overline{q}_v$ *of a dynamical variable* q_v *is*

$$\overline{q}_v = \int_{-\infty}^{\infty} \psi_v^* q \psi_v \, dx, \tag{1}$$

where q is the operator associated with the dynamical variable q. The integration of Eq. (1) is taken over all the whole range of possible values of x between $x = \pm\infty$.

Let us consider the linear harmonic oscillator for the calculation of the average value of x in the vth quantum state of the oscillator, that is, $\overline{x}_v$, which is given by

$$\overline{x}_v = \int_{-\infty}^{\infty} \psi_x^* \, x \, \psi_v \, dx. \tag{2}$$

In this instance, it happens that $\overline{x}_v$ can be obtained by a simple yet powerful argument. Referring to Eq. (12) in Section 6, Eq. (10) in Section (3), or Fig. 4, we see that the functions ψ_v fall into two classes. For v even, substitution of $-x$ for x whenever it occurs in the function ψ_v leaves ψ_v unchanged in magnitude and sign, whereas for v odd, the substitution leaves ψ_v unaltered in magnitude but changes its sign. The first class of functions is said to be *symmetric in x*, and the second *asymmetric in x*, the substitution of $-x$

for x being equivalent to the simple operation of reflection at the origin of coordinates. Now the integrand in Eq. (2) contains three factors: ψ_v^*, x, and ψ_v; two of them, ψ_v^* and ψ_v, are either both symmetric (v even) or both asymmetric (v odd), while the third, x, is asymmetric. Therefore, the integrand $\psi_v^* x \psi_v$ in Eq. (2) must change sign when the operation of reflection is carried out. But the limits of integration are symmetrical about $x = 0$, and so the magnitude of the integral cannot depend on whether the operation of reflection has been applied. This is only possible if the value of the integral is identically zero. Thus, for the linear oscillator

$$\overline{x}_v = 0 \tag{3}$$

A similar argument applied to the average value $\overline{P}$ of the momentum P yields.

$$\overline{P}_v = \int_{-\infty}^{\infty} \psi_v^* \left(-\frac{ih}{2\pi} \frac{d}{dx} \right) \psi_v \, dx = 0, \tag{4}$$

since the operator $-\left(\frac{ih}{2\pi}\right)\frac{d}{dx}$, like the operator x, changes sign under the operation of reflection at the coordinate's origin.

Symmetry arguments do not require the average of the dynamical variables x^2 and P^2 of the linear oscillator to be equal to zero. Therefore, to determine $\overline{x^2}$ and $\overline{P^2}$, it is necessary to evaluate the integrals.

$$\overline{x_v^2} = \int_{-\infty}^{\infty} \psi_v^* \, x^2 \, \psi_v \, dx \tag{5}$$

and

$$\overline{P_v^2} = \int_{-\infty}^{\infty} \psi_v^* \left\{ -\left(\frac{h^2}{4\pi^2} \right) \frac{d^2}{dx^2} \right\} \psi_v \, dx \cdot \tag{6}$$

The integrals are solved easily for the zeroth ($v = 0$) state of the oscillator. In this case, we use Eq. (15) in Section 6, that is,

$$\psi_0^* = \psi_0 = \frac{1}{\left(r\sqrt{\pi}\right)^{1/2}} e^{-\frac{1}{2}\left(\frac{x}{r}\right)^2}$$

or

$$= \left(r\sqrt{\pi}\right)^{-\frac{1}{2}} e^{-\frac{1}{2}\left(\frac{x}{r}\right)^2}.$$

Therefore,

$$\overline{x_0^2} = \int_{-\infty}^{\infty} \frac{1}{\left(r\sqrt{\pi}\right)^{\frac{1}{2}}} e^{-\frac{1}{2}\left(\frac{x}{r}\right)^2} \, x^2 \, \frac{1}{\left(r\sqrt{\pi}\right)^{\frac{1}{2}}} e^{-\frac{1}{2}\left(\frac{x}{r}\right)^2} dx \cdot$$

As the operator x^2 commutes with ψ_0, the order of the factors under the integral sign in Eq. (5) is immaterial. Thus

$$\overline{x_0^2} = \left(\frac{1}{r\sqrt{\pi}} \right) \int_{-\infty}^{\infty} x^2 \, e^{-\left(\frac{x}{r}\right)^2} dx \cdot$$

But we know from the tables of standard integrals* that

$$\int_{-\infty}^{\infty} x^2 e^{-ax^2} dx = 2\int_0^{\infty} x^2 e^{-ax^2} dx = 2\left(\frac{1}{4}\sqrt{\frac{\pi}{a^3}}\right),$$

Therefore,

$$\int_0^{\infty} x^2 \ e^{-\frac{x^2}{r^2}} = \frac{1}{4}\sqrt{\pi r^6}$$
$$= \frac{r^3}{4}\sqrt{\pi},$$

and

$$\overline{x_0^2} = 2\times\left(\frac{1}{r\sqrt{\pi}}\times\frac{r^3}{4}\sqrt{\pi}\right),$$
$$= \frac{r^2}{2}. \qquad (7)$$

As to Eq. (6), the presence of the differential operator

$$-\left(\frac{h^2}{4\pi^2}\right)\frac{d^2}{dx^2}$$

renders the order of its factors unimportant. But as

$$-\frac{d^2}{dx^2}\ e^{-\frac{1}{2}(x/r)^2} = \frac{1}{r^2}\left\{1-\left(\frac{x}{r}\right)^2\right\} e^{-\frac{1}{2}(x/r)^2},$$

Eq. (6) for the zeroth state is

$$\overline{P_0^2} = \frac{h^2}{4\pi^2}\int_{-\infty}^{\infty}\left(\frac{1}{r\sqrt{\pi}}\right)^{\frac{1}{2}} e^{-\frac{1}{2}\left(\frac{x}{r}\right)^2}\left(\frac{1}{r^2}\right)\left\{1-\left(\frac{x}{r}\right)^2\right\} e^{-\frac{1}{2}\left(\frac{x}{r}\right)^2}\left(\frac{1}{r\sqrt{\pi}}\right)^{\frac{1}{2}} dx$$
$$= \frac{h^2}{4r^3\pi^{5/2}}\int_{-\infty}^{\infty}\left\{1-\left(\frac{x}{r}\right)^2\right\} e^{-\left(\frac{x}{r}\right)^2} dx. \qquad (8)$$

Since

$$\int_{-\infty}^{\infty} e^{-\left(\frac{x}{r}\right)^2} dx = 2\times\frac{1}{2}\sqrt{\pi r^2} = r\sqrt{\pi}$$

and

$$\int_{-\infty}^{\infty}\left(\frac{x}{r}\right)^2 e^{-\left(\frac{x}{r}\right)^2} dx = \frac{1}{r^2}\int_{-\infty}^{\infty} x^2 e^{-\left(\frac{x}{r}\right)^2} dx = \frac{1}{2}r\sqrt{\pi},$$

Eq. (8) becomes

$$\overline{P}_0^2 = \frac{h^2}{4r^3\pi^{5/2}}\left(r\sqrt{\pi} - \frac{1}{2}r\sqrt{\pi}\right) = \frac{h^2}{8\pi^2 r^2}. \qquad (9)$$

The results obtained in Eqs. (3), (4) (7), and (9) are summarised in Table 1. The quantities $\overline{x}_0$, $\overline{P}_0$, $\overline{V}_0^2$, and $\overline{P}_0^2$ are to be thought of as the average values of a large number

* See Appendix E, p. 1141.

of experimental measurements of the variable, made on an oscillator which was in the state ψ_0 before each measurement.

Table 1

Time Average and Uncertainty in x and P for the zeroth($v = 0$) State of the Oscillator.

Variable, q	Time average, $\bar{q}$	Uncertainty, Δq
x_0	0	$r/\sqrt{2}$
x_0^2	$r^2/2$	
P_0	0	$h/2\sqrt{2}\,r\pi$
P_0^2	$h^2/8\pi^2 r^2$	

According to statistical theory, when the repeated measurement of some variable, q, gives rise to a spread of values, the *uncertainty*, Δq, *associated with an individual measurement is taken to be*

$$\Delta q = \sqrt{\overline{q^2} - (\bar{q})^2}\,, \tag{10}$$

in which $\overline{q^2}$ is the average value of the variable q^2, and $(\bar{q})^2$ is the square of the average value of the variable q. Clearly both $\overline{q^2}$ and $\bar{q}$ must be available in order to obtain Δq.

Let us now use the results in Table 1 to determine uncertainties Δx_0 and ΔP_0 in a possible measurement of the coordinate x and the momentum P of an oscillator in the ground state. For the coordinate x, we find

$$\Delta x_0 = \sqrt{r^2/2 - 0} = r/\sqrt{2}, \tag{1}$$

and for the momentum P

$$\Delta P_0 = \sqrt{h^2/8\pi^2 r^2} = h/2\sqrt{2}\;r\pi; \tag{12}$$

Δx_0 and ΔP_0 measure the spread of repeated experimental determinations of x and P made on an oscillator in its ground state. The product of the uncertainties,

$$\Delta x_0 \,.\, \Delta P_0 = \frac{r}{\sqrt{2}} \cdot \frac{h}{2\sqrt{2}\,r\pi} = \frac{h}{4\pi}, \tag{13}$$

is then a constant, independent of the parameters of the system. Eq. (13) represents the *Heisenberg Uncertainty Principle* applied to the ground state of a linear harmonic oscillator.

Whatever experiment may be devised to measure simultaneously the values of x and P pertaining to a given oscillator (in its ground state), the limit of accuracy is always determined by Eq. (13). Thus if measurements of P are relatively precise so that ΔP is *small*, the uncertainty in x must be *proportionately* great in order that Eq. (13) is satisfied.

The scope of the uncertainty principle is much broader than its application to the harmonic oscillator, by which we have chosen to introduce it. The principle establishes a relation between any pair of *canonically conjugate variables* belonging to a given dynamical system. These include the variables x and P of the oscillator and the free particle, the coordinate ϕ, and the momentum $\mathbf{P_z}$ of the fixed-axis rotator, and the energy E and the time t in systems when t is considered explicitly. The form of the general relation between the uncertainties in any two canonically conjugate variables q and s then is

$$\Delta q \,.\, \Delta s \approx h\,. \tag{14}$$

The Heisenberg principle helps to explain why certain dynamical systems are endowed with zero-point energy. Of the three systems we have considered, the harmonic oscillator, the fixed-axis rotator and the free particle, only the first has energy greater than zero in the lowest state. This is in accord with the principle: for if the oscillator were allowed zero energy, it would be located exactly at the position of minimum potential energy, and then the uncertainty in x and P *would simultaneously be zero*, which is contrary to Eq. (13). The existence of zero-point energy can, therefore, be regarded as necessary if the oscillator has to satisfy the Heisenberg principle. As to the rotator, the wave function $\psi_0 = 1/(2\pi)^{\frac{1}{2}}$ of the lowest energy state is independent of ϕ and corresponds to a situation in which all orientations ϕ are equally probable. Therefore, the uncertainty in position is infinite, and the momentum and hence the energy can have the *precise value zero*. The free particle, likewise, has no zero-point energy.

Now we are at a stage to consider the generalised rotation of a molecule, making use of the elementary wave mechanics we introduced.

8.2. The rotation of molecules

1. Introduction. A molecule has the capacity to execute several types of dynamical motion simultaneously. *Translation* is the movement of the molecule as a whole, that is, of the centre of mass but not of the position of atoms relative to the centre of mass. In *rotation*, the atoms move about the centre of mass as a fixed point, maintaining their positions with respect to one another; while *vibration* involves the oscillation of atoms about equilibrium positions without displacement of the latter or of centre of mass. This preliminary description emphasises the degree to which the basic types of molecular motion are independent of one another. Complete independence would mean that we could discuss, for instance, the rotation of a molecule quite separately from its translation and vibration; and obviously, this approach is must simpler than if the three forms of motion had to be considered indivisible.

2. Moments of inertia of a rigid molecule. The moment of inertia of a *rigid* molecule about an axis passing through the centre of mass is defined by

$$I = \sum_i m_i r_i^2, \tag{1}$$

in which r_i is the perpendicular distance from the axis and m_i the mass of the ith nucleus. It is a theorem of mechanics that the locus of points found by plotting $I^{-\frac{1}{2}}$ radially from the centre of mass in the direction of the axis of rotation is the surface of a triaxial ellipsoid known as the *momental ellipsoid.*

The three mutually perpendicular axes of the momental ellipsoid coincide with the three *principal axes of inertia* of the molecule, about which rotation is dynamically balanced. It is customary to label the axes of the ellipsoid *a*, *b*, and *c*, in order of decreasing length, so *a* is the major axis, *b* the intermediate, and *c* the minor axis of the figure. Since the axial length is proportional to $I^{-\frac{1}{2}}$, it follows that $I_a < I_b < I_c$ always. Sections through the momental ellipsoid of a simple molecule, CH_2O, are shown in Fig. 1.

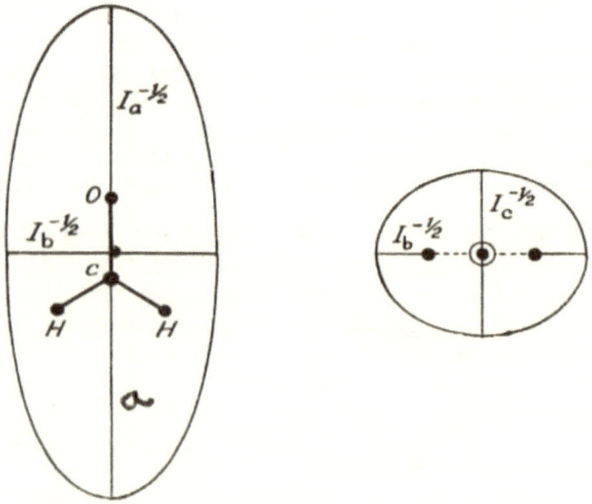

Fig. 1. Principal sections of the momental ellipsoid of CH_2O.

In symmetrical molecules the direction of the principal axes of inertia can often be determined by inspection. Thus the fact that *a*, *b*, and *c* are symmetry axes of the momentum ellipsoid means that a symmetry axis present in the molecule must coincide with one of these axes. Likewise a plane of symmetry in the molecule must necessarily be a principal section of the ellipsoid; that is, it must contain two of the principal axes of inertia of the molecule and be perpendicular to the third. Let us use these ideas to find the orientation of the principal axes of inertia in the formaldehyde, CH_2O, molecule, that is, the axes *a*, *b*, and *c*.

The elements of symmetry in CH_2O are the two-fold axis C_2 and two mutually perpendicular planes of symmetry, σ_v, containing the C_2 axis. One of the axes of inertia must then coincide with the C_2 axis; while of the second and third axes, one must lie in each of the planes σ_v. Which of these moments of inertia is the least (I_a) and which the greatest (I_c) is a decision to be made by calculation rather than from symmetry considerations alone. It turns out that I_a is coincident with C_2 axis of symmetry, that I_b lies in the molecular plane of the molecule and normal to the C_2 axis, and that I_c is normal to the molecular plane. Indeed the direction of I_a is self-evident, for only the light hydrogen atoms contribute to the moment for rotation about C_2 axis.

When the orientation of the three principal axes of inertia *a*, *b*, and *c* can be established in this way, it is relatively easy to calculate the corresponding moments of inertia using Eq. (1). However, if the molecule lacks the necessary symmetry so that the positions of the axes are not obvious, the moments are obtained from the determinantal equation,

$$\begin{vmatrix} I_{xx}-I & -I_{xy} & -I_{xz} \\ -I_{xy} & I_{yy}-I & -I_{yz} \\ -I_{xz} & -I_{yz} & I_{zz}-I \end{vmatrix} = 0. \qquad (2)$$

Here I_{xx}, I_{xy}, . . . are moments and products of inertia with respect to any convenient frame of Cartesian coordinates, having the centre of mass as origin. Thus,

$$I_{xx} = \sum_i m_i \left(y_i^2 + z_i^2\right) \text{ and} \tag{3}$$

$$I_{xy} = \sum_i m_i \, x_i \, y_i , \tag{4}$$

where x_i, y_i, and z_i are the coordination of the *i*th atom of mass m_i, and the sum is taken over all atoms present in the molecule. The three roots of Eq. (2) are the principal moments of inertia. Usually, some simplification is possible. For instance, if a molecule possesses a symmetry plane, one of the principal axes must be perpendicular to the plane. Let this be the x direction, then the products of inertia I_{xy} and I_{xz} are zero, and the determinant factorises into one linear and one quadratic expression. In the extent that x, y, and z coincide with the principal axes of inertia, all the products of inertia are zero, and the determinant factorises into three linear equations, which are precisely like Eq. (1).

3. Rotational energy levels. In classical mechanics, the total Hamiltonian energy of a rotator with one degree of freedom is

$$H = \frac{(I\,\omega)^2}{2I}, \tag{1}$$

as shown above. The generalised rotation of a molecule has components of the total angular momentum vector along three mutually perpendicular directions, that is, it has three degrees of freedom, and hence total energy of rotation is

$$E_R = \frac{\mathbf{P}_a^2}{2I_a} + \frac{\mathbf{P}_b^2}{2I_b} + \frac{\mathbf{P}_c^2}{2I_c}, \tag{2}$$

in which $\mathbf{P}_a\left(=I_a\,\omega_a\right)$, $\mathbf{P}_b\left(=I_b\,\omega_b\right)$, and $\mathbf{P}_c\left(=I_c\,\omega_c\right)$ are the components of the total angular momentum, $\boldsymbol{P}$, along the three principal axes of inertia of the molecule, which are a, b, and c, respectively. The total angular momentum is given by

$$\mathbf{P}^2 = \mathbf{P}_a^2 + \mathbf{P}_b^2 + \mathbf{P}_c^2 . \tag{3}$$

Before we consider the quantisation of the energy, let us describe the possible simplification of the classical Eq. (2) for molecules that possess some degree of symmetry. The basis of simplification is the symmetry of the momental ellipsoid. For real molecules, there are four separate classes.

1. The highest symmetry of the momental ellipsoid is attained when the principal axes of inertia of the molecule, about which motion is dynamically balanced, are of equal lengths, $a = b = c$. In this case, the momental ellipsoid is then a sphere. The moment of inertia is the same about any axis through the centre of mass, so that we may drop the subscripts formerly attached to the moments. Hence Eq. (2) becomes

$$E_R = \left(\mathbf{P}_a^2 + \mathbf{P}_b^2 + \mathbf{P}_c^2\right)/2I . \tag{4}$$

Introducing Eq. (3), we find

$$E_R = \frac{\mathbf{P}^2}{2I} \tag{5}$$

Molecules in this class are known as *spherical tops*. Examples are methane and carbon tetrachloride.

2. For a *linear molecule*, the moment of inertia about the longest axis *a*, which is the internuclear axis is zero, that is, $I_a = 0$, and $I_b = I_c$. Accordingly, the momental ellipsoid is a circular cylinder of infinite length. The component of the total angular momentum along the axis *a*, that is, $\mathbf{P}_a\ (=I_a\omega) =$ zero also. Thus Eq. (3) reduces to

$$\mathbf{P}^2 = \mathbf{P}_b^2 + \mathbf{P}_c^2, \tag{6}$$

and Eq. (2) gives

$$\begin{aligned} E_R &= \left(\mathbf{P}_b^2 + \mathbf{P}_c^2\right)/2I_b \\ &= \frac{\mathbf{P}^2}{2I_b}. \end{aligned} \tag{7}$$

3. For a *symmetric top* molecule, there are two moments of inertia equal to one another but different from the third. The symmetric top molecule is described as *prolate* if $I_a < I_b = I_c$, or *oblate* if $I_a = I_b < I_c$.

A prolate top molecule is shaped like a shuttlecock; examples are CH_3Cl and CH_3CN. The molecule can be thought of as elongated along the top axis.

An oblate top molecule resembles a discus; examples are BF_3 and C_6H_6. The molecule can be thought of as flattened about the top axis.

Any molecule with a single three-fold or higher principal axis of symmetry is necessarily a symmetric top; the principal axis of symmetry of the molecule being coincident with the unique axis of the top. In the examples mentioned above, the principal axis is C_3 except for the benzene molecule, where it is C_6. The momental ellipsoid of a symmetric top is an ellipsoid of revolution.

For a prolate symmetric top $(I_b = I_c)$, the Eq. (2)

$$E_R = \frac{\mathbf{P}_a^2}{2I_a} + \frac{\mathbf{P}_b^2}{2I_b} + \frac{\mathbf{P}_c^2}{2I_c} \tag{2}$$

can be arranged on the basis $I_b = I_c$ that

$$\begin{aligned} E_R &= \frac{\mathbf{P}_a^2}{2I_a} + \frac{\mathbf{P}_b^2 + \mathbf{P}_c^2}{2I_b} \\ &= \frac{\mathbf{P}_a^2}{2I_a} + \frac{\mathbf{P}^2 - \mathbf{P}_a^2}{2I_b} \\ &= \frac{\mathbf{P}^2}{2I_b} + \frac{\mathbf{P}_a^2}{2}\left\{\frac{1}{I_a} - \frac{1}{I_b}\right\}. \end{aligned} \tag{8}$$

This equation can be verified easily from the vector diagram in Fig. 2.

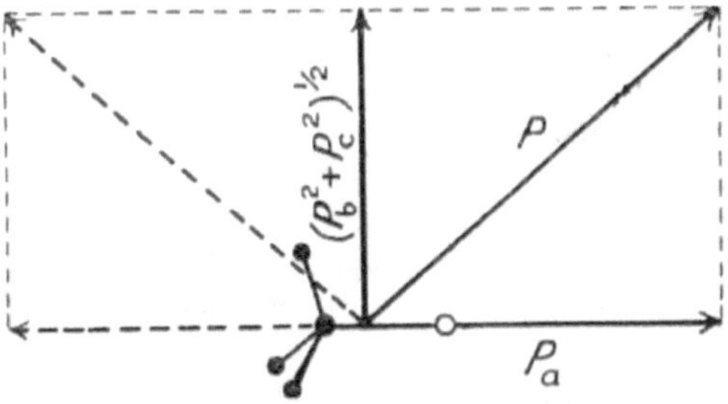

Fig. 2. Vector diagram for a prolate symmetric top. The dotted lines show the effect of reversing the sign of rotation about the top axis. In wave mechanics, *P* = *J* and *P* = *K*.

If the symmetric top is oblate the only effect is that the subscript *c* replaces *a* in Eq. (8), as can be seen from the following:

$$E_R = \frac{P_a^2}{2I_a} + \frac{P_b^2}{2I_b} + \frac{P_c^2}{2I_c}$$

$$I_a = I_b$$

$$E_R = \frac{P_a^2 + P_b^2}{2I_a} + \frac{P_c^2}{2I_c}$$

$$P^2 = P_a^2 + P_b^2 + P_c^2$$

$$P_a^2 + P_b^2 = P^2 - P_c^2$$

$$E_R = \frac{P^2 - P_c^2}{2I_a} + \frac{P_c^2}{2I_c}$$

$$= \frac{P^2}{2I_a} + \frac{P_c^2}{2}\left\{\frac{1}{I_c} - \frac{1}{I_a}\right\}$$

$$= \frac{P^2}{2I_b} + \frac{P_c^2}{2}\left\{\frac{1}{I_c} - \frac{1}{I_b}\right\}. \tag{9}$$

4. The *asymmetric top* class of molecules has all three principal moments of inertia unequal. Therefore, the momental ellipsoid has three unequal axes, and the classical expression for the total Hamiltonian energy of a rotation, given by Eq. (2), cannot be reduced to a simpler form.

A molecule is an asymmetric top unless it possesses at least one three-fold or higher symmetry axis. A majority of molecules found in nature are asymmetric tops.

Quantisation of rotation. The quantisation of rotation restricts the rotational energy of molecules to the sharply defined values allowed by wave mechanics. In a rigorous approach, one must determine the energy eigenvalues of the operator corresponding to the classical Hamiltonian energy E for rotation, given by Eq. (2). This treatment can be found in standard texts on wave mechanics. It leads to a number of rules, which may be used retrospectively to adapt the *classical equations to quantised conditions.*

The first rule relates to the total angular momentum of a molecule. For quantised rotation, we use the letter $\boldsymbol{J}$ to denote the total angular momentum vector, in place of the classical symbol $\boldsymbol{P}$. The quantum rule is that *the permitted values of* $\boldsymbol{J}$ *are multiples of* $h/2\pi$, *given by the expression*

$$\mathbf{J} = \sqrt{J(J+1)}\, h/2\pi\,, \qquad J = 0, 1, 2, \ldots. \tag{10}$$

where J is a quantum number that may be any positive integer, zero included.

The second rule refers to the component $\boldsymbol{K}$ of the total angular momentum $\boldsymbol{J}$ along the axis of a symmetric top. We recall the discussion in the *case of the fixed-axis rotator*, where it was found that the angular momentum $\boldsymbol{P}_z$ about an axis z fixed in space is quantised being given by

$$\mathbf{P_z} = Mh/2\pi\,, \qquad M = 0, \pm 1, \pm 2, \ldots \tag{11}$$

The substance of the rule is that, for symmetric top molecules, the component of the total angular momentum $\boldsymbol{J}$ parallel to the top axis is quantised, obeying an expression of the same form as that in the case of the fixed-axis rotator about an axis z, which is given by Eq. (11). Therefore, the axial component of $\boldsymbol{J}$ which is denoted by $\boldsymbol{K}$, in place of $\mathbf{P_z}$, may be written as

$$\mathbf{K} = \pm\, Kh\ /\ 2\pi\,, \tag{12}$$

wherein K is a *second quantum number*, which may assume integral positive values including the value zero. However, since $\dot{\mathbf{K}}$ is the axial component of $\boldsymbol{J}$, the quantum number K cannot be *greater* than the quantum number J; therefore, the spectrum of possible values of K is

$$K = 0, 1, 2, \ldots J. \tag{13}$$

The *simultaneous quantisation* of the total angular momentum and of the component parallel to the *top axis* means that the orientation θ of the total angular momentum with respect to the axis is also *quantised.* That is, the angle θ can assume only the *discrete values* given by

$$\cos\theta = \frac{\mathrm{K}}{\mathrm{J}} = \pm\frac{K}{\sqrt{J(J+1)}}. \tag{14}$$

The spatial quantisation for $J = 3$ is illustrated in Fig.3. Since the magnitude $\sqrt{J(J+1)}\, h/2\pi$ of the angular momentum vector $\boldsymbol{J}$ exceeds the maximum value of the component vector $\boldsymbol{K}$, which is $kh/2\pi$, it is clear that $\boldsymbol{J}$ can never point exactly in the direction of the top axis.

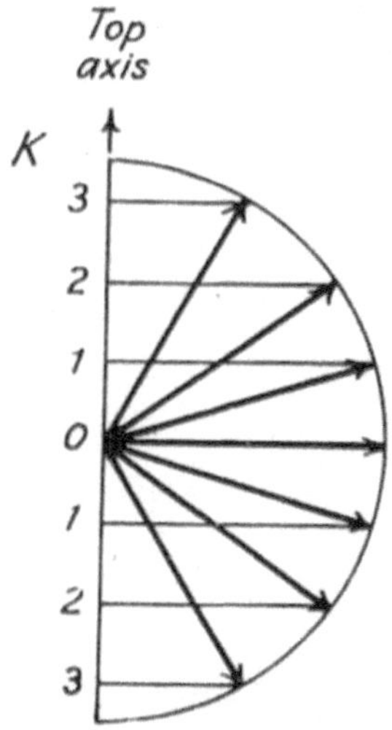

Fig. 3. Components of J parallel to the axis of a symmetric top. The diagram is drawn for J = 3, when J = $\sqrt{3\times4}$ = 3.464 in units of h/2π.

Let us determine the quantised rotational energy levels for a *prolate symmetric top.* Eq. (8) obtained above gives the classical energy. Thus substituting ***J*** and ***K*** for the classical symbols ***P*** and ***P****a*, respectively, we have

$$E_R = \frac{\mathbf{J}^2}{2I_b} + \frac{\mathbf{K}^2}{2}\left\{\frac{1}{I_a} - \frac{1}{I_b}\right\}. \tag{15}$$

Introducing the quantisation of *J* and *K* by Eqs.(10) and (12), respectively, they yields.

$$E_R = J(J+1)\frac{h^2}{8\pi^2 I_b} + K^2\left\{\frac{1}{I_a} - \frac{1}{I_b}\right\}\frac{h^2}{8\pi^2} \tag{16}$$

This is the equation we require. If both sides are divided by *h*, and then *A* is substituted for $\frac{h}{8\pi^2 I_a}$ and B for $\frac{h}{8\pi^2 I_b}$, the equation simplifies to

$$\frac{E_R}{h} = J(J+1)B + K^2(A-B), \tag{17}$$

where A and B are known as *rotational constants*; they are proportional to reciprocal moments of inertia, the factor of proportionality being $h/8\pi^2$. The rotational energy levels of a prolate symmetric top are shown diagrammatically in Fig. 4. When the top is oblate, the subscript c replaces a in Eq. (8), giving Eq. (9) as shown above, and subsequently, for an oblate symmetric top, E_R is given by

$$E_R = J(J+1)\frac{h^2}{8\pi^2 I_b} + K^2\left\{\frac{1}{I_c} - \frac{1}{I_b}\right\}\frac{h^2}{8\pi^2}. \quad (18)$$

Therefore,

$$E_R/h = J(J+1)\frac{h}{8\pi^2 I_b} + K^2\left\{\frac{1}{I_c} - \frac{1}{I_b}\right\}\frac{h}{8\pi^2}$$

$$= J(J+1)B + K^2(C - B), \quad (19)$$

where $C = \dfrac{h}{8\pi^2 I_c}$ is also a *rotational constant.*

This equation can be deduced directly from Eq. (17) when the rotational constant A is substituted for by C.

From the convention $I_a < I_b < I_c$, it follows that $A > B > C$, and hence the quantity $(C - B)$ in Eq. (19), which expresses the rotational energy levels of an oblate top, is always negative.

The rotational energy levels of a linear molecule, whose classical energy, E_R, as shown above as

$$E_R = \frac{\mathbf{P}_b^2 + \mathbf{P}_c^2}{2I_b} = \frac{\mathbf{P}^2}{2I_b} \quad (7)$$

can be obtained in like manner by substituting $\boldsymbol{J}$ for the classical symbol $\boldsymbol{P}$. Thus, for a linear molecule, we have

$$E_R = \frac{J(J+1)h^2}{8\pi^2 I_b}, \quad (20)$$

so

$$E_R/h = \frac{J(J+1)h}{8\pi^2 I_b} = J(J+1)B. \quad (21)$$

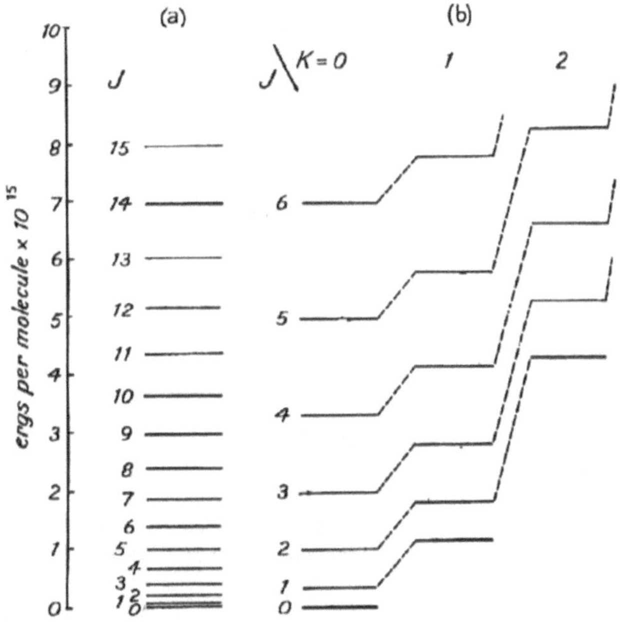

Fig. 4. Rotational energy levels for a linear molecule (a) and a prolate symmetric top (b). The energies are calculated for B = 5000 Mc sec^{-1} (a) and A = 150. 000 and B = 25. 000 Mc sec^{-1} (b); that is, they are roughly to scale for carbon oxysulphide and methyl bromide, respectively.

Let us calculate E_R for the linear molecule carbon oxysulphide COS, given the rotational constant $B\left(=\dfrac{h}{8\pi^2 I_b}\right)$ equal to 5000 Mc/sec. and the quantum number $J = 5$

$$E_R / h = J(J+1)\frac{h}{8\pi^2 I_b}$$

$$\begin{aligned} E_R &= 5 \times 6 \times 5000 \times 10^6 \times 6.625 \times 10^{-27} \\ &= 15 \times 10^{10} \times 6.625 \times 10^{-27} \\ &= 10^{12} \times 10^{-27} \\ &= 1 \times 10^{-15} \text{ ergs/molecule} \end{aligned}$$

This value is approximately equal to that given in Fig. 4.

Again, we can check the value calculated from Eq. (17) for the prolate symmetric top molecule methyl chloride CH_3Cl, given the rotational constants A and B as 150,000 and 25,000 Mc/sec, respectively, and $J = 5$.

$$E_R / h = J(J+1)B + K^2(A-B). \quad (17)$$

When $J = 5$, $\boldsymbol{K}$ has the possible values 0, 1, 2, 3, 4, 5 corresponding to eleven orientations of the total angular momentum vector $\boldsymbol{J}$ with respect to the top axis of molecule. This means that there are permitted eleven axial components of $\boldsymbol{J}$ belonging to the vector $\boldsymbol{K}$.
Therefore, for $K = 0$, we have

$$E_R/h = 5 \times 6 \times 25000 \; / \; 10^6$$

$$E_R = 30 \times 25 \times 10^9 \times 6.625 \times 10^{-27}$$

$$= 75 \times 10^{10} \times 6.623 \times 10^{-27}$$

$$= 500 \times 10^{10} = 5 \times 10^{-15}$$

$$\simeq 5 \times 10^{-15} \text{ ergs / molecule.}$$

This value is almost the same as that given in Fig. 4 for the quantum numbers $J = 5$, and $K = 0$,

In like manner, the energy values for the other values of the quantum number K, namely, 1, 2, 3, 4, 5 corresponding to the quantum number $J = 5$, can be obtained using the same Eq. (17).

That the rotational constants A, B, and C have, in the dimensional cgs system, the units cycles/sec. is because any of these constants is the quotient

$$\frac{h}{8\pi^2 I},$$

whose dimensions are $[\text{erg. sec}] \,/\, \text{g - cm}^2$

$$= \left[\frac{\text{dyne .cm .sec .}}{\text{g .cm}^2}\right]$$

$$= \left[\frac{\text{g} . \dfrac{\text{cm}}{\text{sec}^2} . \text{cm . sec}}{\text{g . cm}^2}\right]$$

$$= \text{sec}^{-1}$$

If we now refer to Eq. (12),

$$\mathbf{K} = \pm \, Kh / 2\pi, \qquad (12)$$

as well as to Fig. 3, we see that the axial component $\boldsymbol{K}$ of the angular momentum $\boldsymbol{J}$ of a symmetric top may be positive or negative. However, it follows from Eq. (15),

$$E_R = \frac{\mathbf{J}^2}{2I_b} + \frac{\mathbf{K}^2}{2}\left\{\frac{1}{I_a} - \frac{1}{I_b}\right\}, \qquad (15)$$

that E_R depends upon the magnitude of K but not upon its sign, so every energy level for which $K \neq 0$ is *doubly degenerated*. The degeneracy corresponds physically to opposite directions of rotation about the top axis. The total number of sub-levels belonging to any value of J is $2J + 1$ of which, therefore, $2J$ sub-levels occur as degenerate pairs and one corresponding to $K = 0$ is non-degenerated.

For a linear molecule, the quantised energy levels are given as shown above, by Eq. (21)

$$E_R / h = J(J+1)B. \tag{21}$$

When this equation is compared with Eq. (17),

$$E_R / h = J(J+1)B + K^2(A-B), \tag{17}$$

it can be seen that the energy levels of a linear molecule and those of a symmetric top molecule (whether prolate or oblate) are given by the same expression when K is equal to zero.

Eqs. (20) and (21) apply also to spherical top molecules. Considerations of wave mechanics for rotation of a linear molecule and of a spherical top shows that the energy levels are $(2J + 1)$-fold degenerate; that is, there are $(2J + 1)$ coincident sub-levels for each value of J.

4. The thermal population of rotational energy levels for linear molecules. In a sample of a gas or vapour, we have to deal with a very large number of molecules, some of which will be in the lowest ($J = 0$) rotational energy level, and others in excited levels. According to the *Maxwell-Boltzmann statistics,* the number of molecules having the classical energy E_R is proportional to $e^{-E_R/kT}$, where k is Boltzmann's constant (the gas constant per molecule) and T is the absolute temperature. In quantum mechanics, we have to multiply $e^{-E_R/kT}$ of the level by the number of times the level occurs so that the population of the level E_R is proportional to

$$g\, e^{-E_R/kT}. \tag{1}$$

The factor g in Eq. (1) is known as the *statistical weight* of the energy level E_R.

Any other development is somewhat difficult except for linear unsymmetrical molecules and the discussion that follows is restricted to molecules of this type (all of which belong to the point group $(C_{\infty v})$. For them, *the statistical weight of a level of given J is simply the degeneracy*, $2J + 1$, of the level. Substituting the statistical weight $g = 2J + 1$ and the energy E_R as given by Eq. (21) in Section 3, namely,

$$E_R = \frac{J^2}{2I_b} = J(J+1)\frac{h^2}{8\pi^2 I_b} = J(J+1)hB, \tag{2}$$

where $B = h/8\pi^2 I_b$ (known as the rotational constant) and I_b is the moment of inertia about the axis b, perpendicular to the internuclear axis a in Eq. (1), we obtain, for the number of molecules N_J present in the level J, N_g, the equation

$$N_J \alpha\, (2J+1)\, e^{-J(J+1)hB/kT}. \tag{3}$$

The total number of molecules, N, is the sum of those present in all the levels, and thus

$$N = \sum_J N_J. \tag{4}$$

Since the proportionality constant in Eq. (3) is the same for every level, we have

$$N \propto \sum_J (2J+1)\, e^{-J(J+1)hb/kT}. \qquad (5)$$

The expression on the right-hand side of Eq. (5) is termed as the *rotational partition function*, f_R, or Z_{rot}. Written out explicitly, Z_{rot} for a linear *unsymmetric molecule* is given by

$$Z_{rot} = 1 + 3\,e^{-2hB/kT} + 5\,e^{-6hB/kT} + \ldots \qquad (6)$$

If the levels are fairly close and the temperatures are not too low, the sum may be expressed approximately by the integral

$$\int_0^{\infty} (2J+1)\, e^{-J(J+1)hB/kT}\, dJ. \qquad (7)$$

Let $J(J+1) = y$, then the integral becomes

$$Z_{rot} = \int_0^{\infty} e^{-(hB/kT)y}\, dy = \frac{kT}{hB}. \qquad (8)$$

Hence, for this simple case, $Z_{rot} = kT/hB$. As an example, Z_{rot} for the molecule HCN ($B = 44.316$ Mc sec^{-1}) at 300K is, according to Eq. (8), 141.06. The exact value calculated from the summation given by Eq. (6) is 141.39, that is, $Z_{rot} = 141.39$.

The next step in evaluating the relative population of the levels N_J/N is to divide Eq. (3) by Eq. (5), when the proportionality constant cancels, and we obtain

$$\frac{N_J}{N} = \frac{(2J+1)\, e^{-J(J+1)hB/kT}}{Z_{rot}}. \qquad (9)$$

Introducing the value of Z_{rot}, given by Eq. (8), we have

$$\frac{N_J}{N} = \frac{hB}{kT}(2J+1)\, e^{-J(J+1)hB/kT}; \qquad (10)$$

N_J/N is the fractional number of molecules in the *J*th rotational level.

Fig. 5 illustrates the variation of the relative population of the energy levels with the quantum number J (which may be any positive integer, zero included) for the molecule HCN at 300K. The relative population of the levels passes through a maximum owing to the linearly increasing degeneracy given by $2J + 1$ with ascending J. The location of the maximum can be obtained by differentiating Eq. (10); the result is

$$J_{max} + \frac{1}{2} = (kT/2hB)^{1/2}. \qquad (11)$$

Hence, the energy per molecule of the level, which has the greatest population of molecules, is

$$\begin{aligned} E_R(max) &= J_{max}(J_{max}+1)hB \\ &= \frac{1}{2}kT - \frac{1}{4}hB. \end{aligned} \qquad (12)$$

Except at very low temperatures, $\frac{1}{2}kT \gg \frac{1}{4}hB$. So we can consider that the energy per molecule in the most populous level is approximately $\frac{1}{2}kT$, *which is the classical equipartition energy for a single degree of freedom.*

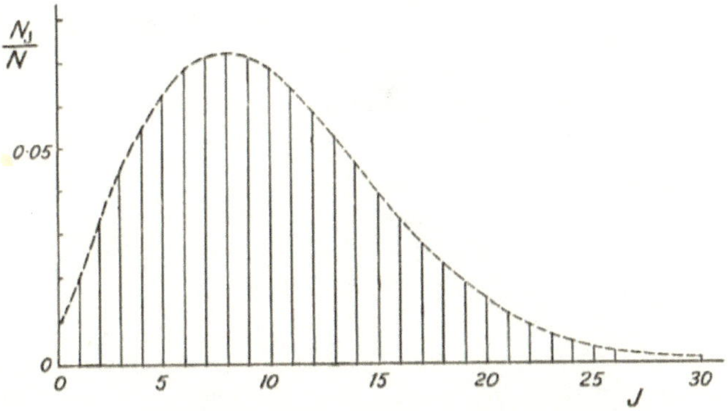

Fig. 5. Thermal population of rotational levels.

It is important to point out here that Eq. (10) can be immediately obtained by using the *Maxwell-Boltzmann distribution law*

$$N_i = \frac{N}{Z} g_i \, e^{-\epsilon_i / kT}, \tag{13}$$

where ε_I is the energy of the molecule in level i, g_i is the degeneracy of this level, N_i is the number of molecules in level *i*, and *N* is the total number of the molecules. It will be shown shortly that in the derivation of this law, which has a fundamental significance in statistical mechanics since it gives the most probable distribution of the molecules among the possible energy levels in the final steady state of the system which is known as the state of thermodynamic equilibrium; no restrictions were made on the nature of energy provided the total energy of the ensemble of molecules is constant. That is the law is equally applicable to any distribution of the total energy of the system and to any particular form of energy which is constant for the entire system. Originally, the Maxwell-Boltzmann distribution law stemmed from the Bose-Einstein distribution *function,* which is based on *quantum statistics*, namely,

$$\frac{N_i}{g_i} = \frac{1}{B\, e^{\beta \epsilon_i} - 1}. \tag{14}$$

This function leads to the Maxwell-Boltzmann statistics as a limiting case when we have a system in which the number of particles in a cell, N_i, is very much smaller than the number of compartments, g_i. For such a system, the left-hand side of Eq. (14) is very much less than unity, and hence the denominator on the right side is much greater than unity. The term 1 in the denominator can, therefore, be neglected, and Eq. (14) for such a system reduces to

$$\frac{N_i}{g_i} = \frac{1}{B\ e^{\beta \epsilon_i}}, \tag{15}$$

with $B >> 1$. This is the *Maxwell-Boltzmann distribution law* for particles among cells.

An expression for *B* can be obtained from the requirement that the sum of all the N_is must be equal to the total number, N_i, of the particles,

$$\sum N_i = N = \frac{1}{B} \sum g_i \, e^{-\beta \epsilon_i}. \tag{16}$$

The sum $\sum g_i\ e^{-\beta\epsilon_i}$ is the *partition function* or *sum of state* and is usually represented by the letter *Z*,

$$Z = \sum g_i\, e^{-\beta\epsilon_i}. \tag{17}$$

From Eqs. (16) and (17), the constant *B* is given by

$$B = \frac{Z}{N}.$$

It can also be proved, as will be shown in Chapter 9, that

$$\beta = \frac{1}{kT} \qquad (k = \text{Boltzmann's constant})$$

Hence, the number of particles in the cell, *i*, in the state of maximum thermodynamic probability, according to Eq. (15) is

$$N_i = \frac{N}{Z} g_i\, e^{-\epsilon_i/kT}. \tag{13}$$

This is the equation we have shown above; it is one form of the Maxwell-Boltzmann distribution law.

Use can be made of this equation to evaluate the relative population of the levels in the case of linear molecules

$$Z = \sum g_i\, e^{-\epsilon_i/kT}$$
$$= \sum (2J + 1)\, e^{-J(J+1)hB/kT},$$

where *B* here represents the rotational constant:

$$B = \frac{h}{8\pi^2 I_b}.$$

Expressing the sum in *Z* given by Eq. (17) approximately by the integral, we have

$$Z = \int_0^\infty (2J+1)\, e^{-J(J+1)hB/kT}\, dJ,$$

its value being, as shown above is, kT/hB.

Hence,

$$\frac{N_i}{N} = \frac{hB}{kT}(2J+1)\, e^{-J(J+1)hB/kT},$$

which is Eq. (10) obtained above.

8.3. Nuclear magnetic resonance

1. Introduction. To understand the pure rotation spectra of gases, it is sufficient to regard the nucleus of an atom as having mass (*m*) and position (*r*), add to these the fact that it carries a charge (Ze), and we have a list of three properties that adequately describe the nucleus for most chemical purposes. However, in order to interpret the detailed structure of rotational spectra, it is necessary to consider three more properties of the nucleus, namely, its spin (*I*), magnetic moment (μ), and electric quadrupole moment (*Q*).

These three properties have an essentially simple explanation. Historically, the first to be appreciated was the nuclear spin. The substance of this idea is that a nucleus is capable of rotation, or spin, about an axis. The spin generates a quantised angular momentum ***I*** whose magnitude is given by

$$\mathbf{I} = \sqrt{I(I+1)}\,h/2\pi\,. \qquad (1)$$

Here, *I*, the quantum number for nuclear spin, can assume integral, half-integral, or zero values. Eq. (1) is a typical wave-mechanical equation for the quantisation of spin angular momentum. It may be compared with the equation

$$\mathbf{J} = \sqrt{J(J+1)}\;h/2\pi\,,$$

which represents the quantisation of the angular momentum ***J*** associated with molecular rotation discussed before. Values of the spin quantum number, *I*, for some common nuclei are given in Table 1.

It should be emphasised that in chemical applications we are interested in a single value of *I*, which is given in the table, for each nucleus. To understand why this is so requires some explanation of nuclear structure. The nuclear spin angular momentum ***I*** can be thought of as the *resultant of the angular momenta of the protons and*

Table 1
Spin Quantum Numbers of Some Common Nuclei

Element	Atomic number	Isotope, (Mass number)	Natural abundance %	Spin quantum number I
Hydrogen	1	^{1}H $^{2}H(D)$	99.985 0.015	1/2 1
Carbon	6	^{12}C ^{13}C	98.89 1.11	0 1/2
Nitrogen	7	^{14}N ^{15}N	99.63 0.37	1 1/2
Oxygen	8	^{16}O ^{17}O ^{18}O	99.76 0.04 0.20	0 5/2 0
Fluorine	9	^{19}F	100	1/2
Phosphorous	15	^{31}P	100	1/2
Chlorine	17	^{35}Cl ^{37}Cl	75.4 24.6	3/2 3/2

neutrons of which the nucleus is composed. Each of these fundamental particles is considered to have a spin quantum number $I = \frac{1}{2}$, the value of the proton being experimentally observable. As the deuterium nucleus, or deuteron, possesses one proton and one neutron, the resultant spin must be 1 or 0, depending on whether the individual spins are aligned parallel or antiparallel. Experiment shows that the deuteron in the ground state has $I = 1$, corresponding to the parallel alignment. $I = 0$ and represents an excited state of the deuteron, but the excitation energy is so high that it actually exceeds the dissociation energy of the nucleus. Many heavier nuclei do possess excited spin states which are stable to dissociation; yet it is true that the energy of such states is so great that they are of no interest in chemistry.

It may be noted here that it was *Purcell at Harvard University* and *Block at Stanford University* who announced independently this far-reaching discovery of rotation or spin of the nucleus about an axis. This axis of spin (or the spin axis) is the axis about which the nucleus has axial symmetry.

The assignment of a spin quantum number $I = \frac{1}{2}$ to protons or neutrons has one obvious consequence: nuclei with even mass number – hence with an even total number of protons and neutrons – must have zero or integral values of I, whereas nuclei of odd mass number have half-integral values. A glance at the table will show that this is true. A useful rule, more difficult to understand theoretically, is that nuclei with even mass number (even total number of protons and neutrons) and even charge (even number of protons, that is,. even atomic number) have zero spin. A theory of nuclear structure should predict the numerical value of I for nuclei of any mass number, but at present, the situation is mysterious and nuclear spin quantum numbers are accessible only by experiment Finally, let us digress for a moment on the use of the phrase 'fundamental particles' as applied to protons and neutrons. Suppose that we were to investigate a small molecule, say N_2, with energy source at our disposal up to 10 eV. We could then observe the dissociation of the molecule into atoms; but we could not detect the formation of ions because the energy available would be insufficient to remove an electron from either the molecule or the separated atoms. Under such conditions, we should justifiably consider the nitrogen atom as a fundamental particle and speculate on the nature of the excited states of the atom and molecule observable below 10 eV. Lacking higher energies, we could not record the existence of electrons. At present, the knowledge of nuclear structure is restricted: probably the energy necessary to excite or dissociate a proton has not yet been attained. Because the energy released in chemical and even in radiochemical processes is insignificant next to nuclear excitation energies, the chemist's idea of protons and neutrons as ultimate particles has some logic though the limitations of the description should be borne in mind.

2. Nuclear magnetic moments and nuclear quadrupole moments.

Nuclei with spin greater than zero, invariably, possess a magnetic moment, μ; that is to say, the nucleus behaves like a small magnet, whose axis coincides with the axis of spin. According to quantum electrodynamics, the absolute magnetic moment μ_P of the *proton* is proportional to the product of the angular momentum, which is $\mathbf{I} = \sqrt{\frac{1}{2}\left(\frac{1}{2}+1\right)}\frac{h}{2\pi} = \frac{1}{2}\sqrt{3}\frac{h}{2\pi}$ times $\frac{e}{2\,m_P\,c}$, the factor of proportionality being denoted by g_P.

Thus,

$$\mu_P = g_P \frac{\sqrt{3}}{2} \frac{e}{2m_P c} \frac{h}{2\pi}. \qquad (1)$$

Here, m_P is the mass of the proton and e is its charge; and g_P is an irrational* number, which from experiment is known to have the value 5.585. The group of universal constants,

$$\frac{e}{2m_P c} \frac{h}{2\pi},$$

is known as the *nuclear magneton* and is denoted by μ_n, its magnitude being 5.050×10^{-24} erg gauss^{-1}. Thus

$$\mu_P = 5.585 \frac{\sqrt{3}}{2} \times 5.050 \times 10^{-24}$$
$$= 24.398 \times 10^{-24} \text{ erg gauss}^{-1}.$$

For *nuclei other than the proton*, the absolute magnetic moment μ is given by the general expression

$$\mu = g\sqrt{I(I+1)} \frac{e}{2m_P c} \frac{h}{2\pi}, \qquad (2)$$

or in terms of the nuclear magneton,

$$\mu_n = \frac{e}{2m_P c} \frac{h}{2\pi},$$

by

$$\mu = g \mu_n \sqrt{I(I+1)}. \qquad (3)$$

Nuclei with spin $I = 0$, therefore, have no magnetic moment. The dimensionless quantity $g \times I$ is often referred to as the 'magnetic moment' of the nucleus. Clearly, $g \times I$ is not equal to the absolute magnetic moment μ, though μ can be calculated from it by use of Eq. (3). We shall show later that $g \times I$ is actually the maximum component of the absolute magnetic moment in the direction of the lines of magnetic forces of an external magnetic field. At the present time, nuclear g factors can be found by experiment only; they are not calculable by any theory.

Some nuclei have negative g (and hence negative moments), which signifies that the magnetic moment and angular momentum vectors, *while both coincide with the spin axis, actually point in opposite directions.*

---

* An irrational number cannot be expressed as the quotient of two integers.

We have just discussed above two of the properties of the nucleus, namely, its spin, *I*, and magnetic moment, μ. The third property is the electric quadrupole moment, *Q*. To make plain the nature of a quadrupole, consider the arrangements in Fig. 1(*a*).

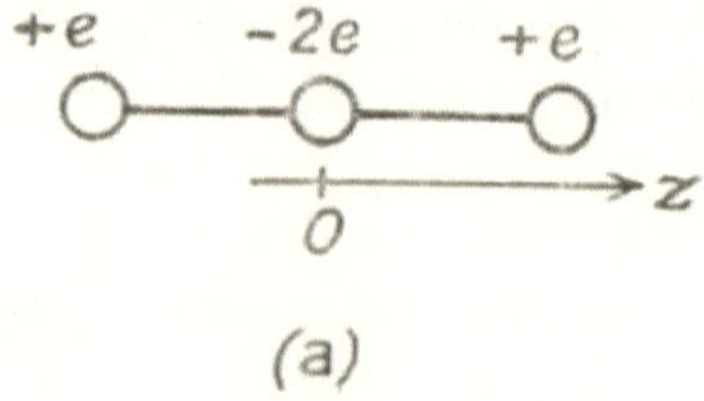

Fig. 1 (*a*). A simple quadrupole.

This quadrupole system has no dipole moment since $\sum e z_i = 0$, yet it gives an external field, which, however, falls off *more rapidly than that of a dipole*. Its action is characterised by a *quadrupole moment* which has the dimensions of charge *x* (length)^2, and in this example, it is simply

$$\sum_i e_i \, (z_i)^2 ,$$

where *z* is measured along the *z* axis on which the charges are located.

Nuclear quadrupoles do not have the linear form shown in Fig. 1 (*a*). They are more complex shapes which arise when the distribution of nuclear charge is not spherical. In such cases, the nucleus can be thought of as elongated (Fig. 1 (*b*)) or flattened (Fig. 1 (*c*)) with respect to the axis of spin about which the nucleus has axial symmetry. The charge distribution can then be regarded as equivalent to a quadrupole, indicated in the figure by the + and – charges superposed on a spherical charge *Ze*.

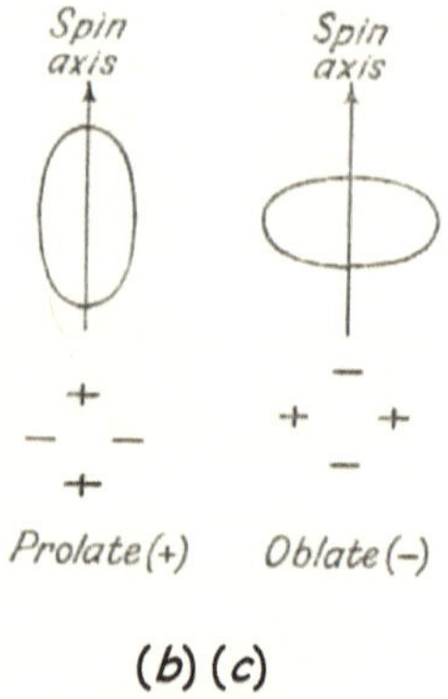

Fig. 1. Nuclear quadrupoles. The nucleus has a net positive charge, and thus the negative signs in (*b*) and (*c*) merely indicates regions of low density of positive charge.

The analytical expression for the quadrupole moment need not concern us, for the nuclear electric quadrupole moments, like the nuclear magnetic moments, must be found by experiment. It emerges that nuclei with $I = 0$ or $\frac{1}{2}$ have zero quadrupole moments; such nuclei must have a perfectly spherical charge distribution. When $I \geq 1$ the nuclear electric quadrupole moment is always different from zero, though its variation from one nucleus to another is seemingly irregular.

The nuclear quadrupole moment is designated as eQ, where e is the proton charge and Q has the dimensions of cm^2. Values of Q in units of 10^{-24} cm^2 are usually given in tables for some common nuclei. To illustrate the use of data, consider the nucleus of iodine of mass number 127, that is, ^{127}I. Its natural abundance = 100 per cent, spin quantum number, $I = \frac{5}{2}$, magnetic moment, gI (in units of the nuclear magneton) = + 2.794, and nuclear quadrupole moment expressed as Q is equal to -0.75×10^{-24} cm^2. Hence, the quadrupole moment $eQ = -4.80 \times 10^{-10} \times 0.75 \times 10^{-24} = 3.60 \times 10^{-34}$ esu. The negative sign signifies that the nuclear charge distribution is oblate.

The concepts of nuclear spin, magnetic moment, and electric quadrupole moment are now sufficiently developed to discuss the *phenomena of nuclear magnetic resonance spectra.*

3. Spatial quantisation of the nuclear spin angular momentum. The orientation of the angular momentum vector

$$\mathbf{I} = \sqrt{I(I+1)}\ \frac{h}{2\pi}$$

is quantised with respect to an axis fixed in space. Let us denote this fixed axis by *z*. Theory allows $\mathbf{I}_z$, the component of $\boldsymbol{I}$ along this space-fixed axis to assume any one of the values given by the expression.

$$\mathbf{I}_z = M_I \frac{h}{2\pi}. \tag{1}$$

Here M_I is a quantum number that may adopt certain positive or negative integral or half-integral values, zero included. Since $\boldsymbol{I}_z$ is the axial component of $\boldsymbol{I}$, the quantum number M_I cannot be greater than I, the limits of M_I are $\pm I$, and the spectrum of values of M_I is

$$M_I = I,\ (I-1), \ldots - (I-1), - I. \tag{2}$$

(Note that the values of M_I are either all integral or all half-integral, depending on whether I is integral or half-integral). The total number of possible values of M_I is $2I + 1$, and hence the vector $\boldsymbol{I}$ is allowed $2I + 1$ discrete orientations with respect to a space-fixed axis. The simultaneous quantisation of the spin angular momentum and of its component along a space-fixed axis z means that the orientation θ_i of the vector $\boldsymbol{I}$ with respect to this axis is also quantised. Thus the angle θ_I between $\boldsymbol{I}$ and the fixed axis can assume only the discrete values given by

$$\cos \theta_i = \frac{\mathbf{I}_z}{\mathbf{I}} = \frac{M_I}{\sqrt{I(I+1)}}. \tag{3}$$

Since the magnitude $\sqrt{I(I+1)}\,h/2\pi$ of the vector ***I*** exceeds the maximum value $Ih/2\pi$ of its component I_z, it is clear that ***I*** can never point exactly in the direction of I_z.

It is evident that the situation here is analogous to the spatial quantisation of the total angular momentum ***J*** of a molecule due to its overall rotation which was discussed before.

Fig. 1 shows the allowed orientations in space of the nuclear angular momentum ***I*** with a spin quantum number $I = 1$. The components, I_z, are the projections of ***I*** on the fixed axis set vertically in the figure. As the nuclear magnetic moment vectors, μ, coincide with the axis of spin, their axes are also inclined at θ_i to the fixed axis z. Using Eq. (3) and the equation expressing the absolute magnetic moment μ by

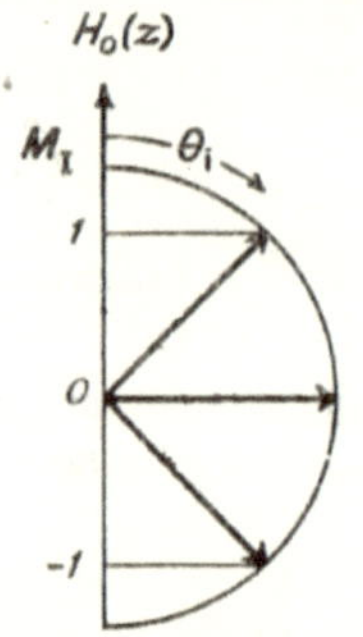

Fig. 1. Spatial quantisation of *I* for a nucleus with *I* = 1. The length of the vector *I* is proportional to $[I(I+1)]^{\frac{1}{2}} = 2^{\frac{1}{2}}$.

$$\mu = g\sqrt{I(I+1)}\,\frac{e}{2m_p c}\,\frac{h}{2\pi}$$
$$= g\mu_n\sqrt{I(I+1)}, \qquad (4)$$

which was obtained above in Section 2; the possible values of the z component μ_z of the absolute magnetic moment are given by

$$\mu_z = \mu\cos\theta_i,$$
$$= g\mu_n\sqrt{I(I+1)}\,\frac{M_I}{\sqrt{I(I+1)}}$$
$$= gM_I\mu_n.$$

The limits of M_I are $\pm$ I. Hence the *maximum value* of μ_z is $gI\mu_n$, that is, the product of the 'magnetic moment' is $g\times I$ times the nuclear magneton, μ_n. Put in another way, the 'magnetic moment' $g\times I$ is the *maximum value*, in units of the nuclear magneton, of the component of the absolute moment parallel to a space-fixed direction.

The $2I + 1$ orientations of a nucleus may differ in energy for either of the two different reasons. First, if $I \geq \frac{1}{2}$, a nucleus possesses a magnetic moment μ, whose axis coincides with the vector $\boldsymbol{I}$ (though it may be opposite in sign). Hence, if the nucleus is placed in a uniform magnetic field, there is a contribution E_H to the potential energy of the nucleus that is given by

$$E_H = -\mu H_o \cos\theta_i. \qquad (6)$$

Here E_H is the potential energy of a magnetic dipole μ set at an angle θ_i to a field of strength H_o. The lines of magnetic force define a space-fixed direction, and consequently, we may substitute for $\cos\theta_i$ from Eq. (3), namely,

$$\cos\theta_i = \frac{M_I}{\sqrt{I(I+1)}}. \qquad (3)$$

Thus,

$$E_H = -\mu H_o \frac{M_I}{\sqrt{I(I+1)}}. \qquad (7)$$

Also, introducing the expression

$$\mu = g\mu_n \sqrt{I(I+1)}, \qquad (4)$$

for μ in terms of the nuclear magneton, μ_n, Eq. (7) becomes

$$E_H = -g\mu_n M_I H_o. \qquad (8)$$

This equation tells us that each allowed orientation of a nucleus in a magnetic field corresponds to a different value of E_H; or in other words, that the $(2I + 1)$-fold degeneracy of the nuclear energy level is lifted in favour of a set of $2I + 1$ sub-levels of different energy. The process is represented for a nucleus with $I = \frac{1}{2}$ in Fig. 2.

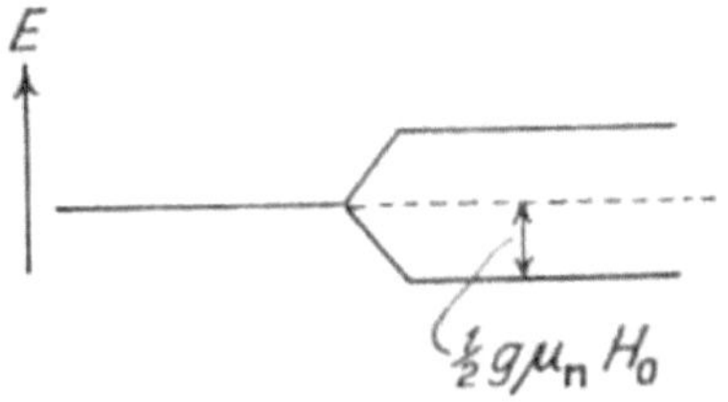

Fig. 2. Energy sub-levels of a nucleus with $I = \frac{1}{2}$ in a pure magnetic field. For nuclei with negative magnetic moment, the sub-levels occur in the reverse order to that shown, E_H being equal to $-g\mu_n M_I H_o$.

Transition between these sub-levels involve a reorientation of the nucleus with respect to the lines of magnetic force of the applied field. It turns out that such transitions can be observed; they constitute the *nuclear magnetic resonance* spectra (NMR) of atoms and molecules.

And the nuclear energy may depend on orientation if the nucleus possesses a quadrupole moment. The potential energy of an electric quadrupole in a non-uniform electric field varies according to the angle between the axis of the quadrupole and the lines of electric force that passes through it. (In a *uniform* electric filed, however, all orientations are of equal energy.) Now at every nucleus, there is an electric field due to all other charged particles in the molecule. Hence, if this field happens to be non-uniform and the nucleus is quadrupolar, we have a situation in which the degeneracy of the nuclear energy level is partly lifted without the application of any external field whatever. Transitions between the sub-levels involve the reorientation of the nucleus with respect to the internal electric field of the molecule. Such transitions can be observed in the crystalline state; they constitute the *pure nuclear quadrupole resonance* spectra of solids.

CHAPTER 9

Bose-Einstein Quantum Statistics

9.1. Phase space

The simplest system to consider, from the statistical viewpoint; is the *monatomic gas*. A complete specification of the state of the gas is given when the position and momentum of each atom are prescribed. The position of the atom can be given in terms of its rectangular coordinates x, y, and z. If m is the mass of an atom moving with a velocity v, its momentum p is

$$p = mv.$$

The momentum of each atom can be given is terms of the rectangular coordination of momentum

$$p_x = m v_x, \qquad p_y = m v_y, \qquad p_z = m v_z$$

That is, if the six coordinates

$$x, y, z, p_x, p_y, p_z$$

are specified for each atom, the *state of the gas is determined.*

In dealing with the kinetic theory of an ideal gas, it is helpful for the derivation of the *Maxwell* velocity-distribution function to introduce the concept of *velocity space* and to speak of the number of representative points per unit volume in velocity space. This imposes no great strain on the imagination, since we have but three velocity components to deal with. We now wish to consider a six-dimensional space based on the three position coordinates and the three momentum coordinates given above. Although the entire argument can be carried out on a purely mathematical basis, most people find it helpful to use the *language of geometry* and to speak of the three position coordinates and the three momentum coordinates of an atom as determining the *position of a point in a six-dimensional hyperspace or phase space.*

A more general approach is to consider *a large number of identical systems* of N atoms each, and a phase space of $6N$ dimensions. The coordinates of a single point in this space then give the position and momentum of *all N atoms of any one system*, and the methods of statistics are applied to the large number of points representing the large number of systems.

Let us subdivide the phase space into small six-dimensional elements of volume, which for brevity we call *cells*, with sides of length dx, dy, dx, dp_x, dp_y, and dp_z. The differentials $d\underline{x}$, . . . , dp_x are small compared with the dimensions of the system and the range of linear momenta of the atoms but are large enough that each cell contains a large number of atoms. The volume of each all (i.e. the product of the six quantities above) will be represented by

$$H = dx\, dy\, dz\, dp_x\, dp_y\, dp_z,$$

so its units are those of $(\text{length})^3 \times (\text{momentum})^3$.

Every atom of the gas must be in a cell. Imagine the cells to be numbered 1, 2, 3, . . . *i*, . . ., and let N_1, N_2, . . . N_i, . . . be the number of atoms in each cell. All the N_i's >> 1. The fundamental problem of statistical mechanics is to find how the N_i depend upon the coordinates of the *i*th cell.

9.2. Microstates and macrostates

In studying the quantum mechanics of a free particle, we find that the position and momentum of the particle can not be prescribed to any degree of precision. That is, the *Heisenberg uncertainty principle* places a quantitative limit on the product of the uncertainties in the position and momentum of the particle,

$$\Delta x \Delta p_x \approx h, \quad \Delta y \Delta p_y \approx h, \quad \Delta z \Delta p_z \approx h,$$

where *h* is *Planck's constant,* equal to 6.6237×10^{-34} joule. sec. In phase space, the uncertainty principle requires that the six coordinates of a particle can be specified only to the extent that the point representing the position and momentum of the particle lies *somewhere within an element of phase space of volume* h^3. That h^3 has the dimensions of a volume in phase space is easy to see, since

$$(\text{joule. second})^3 = (\text{newton. metre. second})^3$$
$$= (\text{metre})^3 \times (\text{newton. second})^3,$$

where the newton-second is the mks unit of impulse, which is equal to that of momentum. Thus the units of h^3 are those of $(\text{length})^3 \times (\text{momentum})^3$.

It is to be noted that *by definition,* the product of the average value of a force *f* and the time interval during which it acts is the *impulse* of the force is equal to the change in momentum produced by it. That is, for infinitesimal quantities,

$$d\bar{f}\, dt$$

represents the impulse of a force.

Now we speak of an clement of volume h^3 as a compartment, to distinguish it from a cell of volume *H.* The volume *H* is arbitrary, subject only to the restriction that

$$dx, dy, dz$$

be small compared with the linear dimensions of the system, and that

$$dp_x\,,\ dp_y,\ dp_z$$

be small compared with the range of momenta of the particles. Even with this restriction, *H* can be made very much larger than h^3, so that we can superpose on the subdivision of phase space into cells of volume *H,* a still linear subdivision of each cell into compartments of volume h^3. The number of compartments per cell, g, is therefore. equal to H/h^3. It is instructive to consider the magnitude of the element h^3 in phase space.

Take $\Delta v_x = 10^{-3}$ m / sec as a small increment of velocity v_x and take *m* as equal to the mass of an oxygen atom which is equal to 2.66×10^{-26} kg. (This value is calculated by dividing the mass of one mole of O in the *mks* system, which is equal to 16kg, by *Avogadro's number* 6.025×10^{26}.) Then applying the uncertainty principle

$$\Delta x \, . \, \Delta p_x \approx h$$

or

$$\Delta x . \, m \Delta v_x \approx h,$$

we have

$$\Delta x = \frac{h}{m\Delta V_x},$$

$$= \frac{6.624 \times 10^{-34} \text{ joule.sec}}{2.66 \times 10^{-26} \text{ kg } 10^{-3} \text{ m / sec}}$$

$$= 2.5 \times 10^{-5} \left[\frac{\text{newton m sec}}{\text{kg m / sec.}} \right]$$

$$= 2.5 \times 10^{-5} \left[\frac{\text{kg } \frac{\text{m}}{\text{sec}^2} \text{ m sec.}}{\text{kg m / sec.}} \right]$$

$$= 2.5 \times 10^{-5} \text{ m},$$

which is a very small distance. Since the compartments are so small, we may easily take the cells to be sufficiently large that $H / h^3 >> 1$.

A complete specification of the *six coordinates for the compartment* (having an element of volume h^3) in which each molecule of the system lies, is said to define a *microstate* of the system. This detailed description is quite unnecessary to determine the *observable properties* of the gas. For example, the density, which in the ordinary sense is equal to the mass per unit volume, is the same if the number of atoms which are present in each element of volume, *dx dy dz*, of ordinary space is the same. Similarly, the pressure exerted by the gas depends only on how many atoms have specified momenta, that is how many atoms lie within each element, $dp_x \, dp_y \, dp_z$, of momentum space. In other words, the observable properties of a gas depend only on how many particles lie in each *cell of phase space. A specification of the number of particles in each cell* N_i *of phase space, that is, of the numbers* N_i*'s, is said to define a macrostate of the system.*

The distinction between microstates and macrostates is illustrated in Fig. 1. The cells in phase space are numbered 1, 2, 3, etc. For simplicity, we assume that each cell has four compartments which are denoted by i, ii, iii, and iv. The particles are denoted by *P*. The particular microstate given in this figure is specified by stating that there is one particle in cell 1, compartment i, two particles in cell 1,

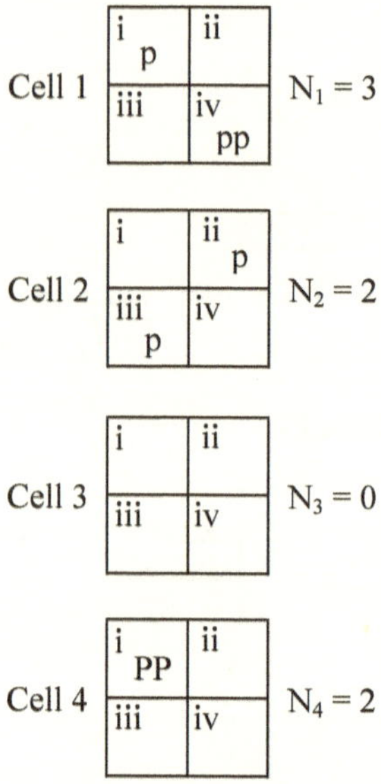

Fig. 1. A particular microstate of a gas corresponding to the macrostate $N_1 = 3, N_2 = 2, N_3 = 0, N_4 = 2$.

compartment iv; one particle in cell 2, compartment ii, one particle in cell 2, compartment iii, and so on.* The corresponding macrostate is specified, as pointed out above, by stating the number of total particles in each cell of phase space; that is, the number $N_1 = 3$ in cell 1, the number $N_2 = 2$ in cell 2, and, in general, the number N_i in the *i*th cell.

Whether we can specify a *microstate* of the system or we cannot (and of course we cannot), it exists for a given gas at every instant. But no microstate can *persist without change* owing to the fact that all atoms are continually in motion. In the intervals between collisions, every particle moves in phase space due to the continuous change in the coordinates of its position *x*, *y*, and *z*, although the component of its momentum p_x, p_y, and p_z remain the same. When a collision between two atoms takes place, the components of the momentum of each atom change very suddenly, and as a consequence, the two atoms jump to two other volume elements in phase space having the same position coordinates but different momentum coordinates. The continuous shifting around of atoms in phase space is similar to the motion of the atoms of a gas in ordinary space, except that it is more

* This statement concerning a microstate is considered as equivalent to the specification of the six coordinates for each of the compartments of a given cell, in which particles lie, so that it is said to define a microstate of the system.

complicated. The important point is that the gas is continually and spontaneously changing from one *microstate to another.*

One of the fundamental hypothesis of statistical mechanics is that *all microstates are equally probable*. That is, over a long period of time, any one microstate occurs as often as any other microstate. At first, this does not seem reasonable. Let us consider, for example, one particular microstate in which every atom in a box is *located in a small element of volume in one corner with all atoms travelling in the same direction, and in another microstate, the atoms are uniformly distributed in small elements of volume throughout the box, with a random distribution of velocities.* Although these two microstates are very different, and although, the second appears more probable than the first, in both cases, the compartment in phase space in which each atom lies is *specified*. The specifications are different but they are complete, and it is the fact that *they are complete is what makes these two microstates equally probable*.

Picture, if you can, the microstate of a simple gas at the present moment and imagine how long you would have to wait, as a result of collisions, until the particles are distributed in phase space exactly as they were originally. You would wait just as long to find all the molecules gathered in one corner of the box moving in the same direction. This is according to the fundamental hypothesis of statistical mechanics saying that all microstates are equally probable, that is they have equal probability of occurrence, the probabilities for various macrostates are far from equal.

It is important to understand that we cannot speak of any one case of a sample gas as a microstate of the gas, unless this microstate is defined by a complete specification of the six coordinates of the compartment in which each molecule lies, as has been stated above. If, however, such a complete specification is not fulfilled, we can no longer speak of the case considered for the system as a microstate of this system.

It can be easily seen that many different microstates of a system correspond to the same macrostate. Any shift of the phase points in phase space that does not change the number of particles in each cell leaves unaltered the macrostate of the gas and its observable properties. As time goes on, and the microstates of a gas continually change, because the atoms are all in motion, the macrostate that occurs *most frequently will be that for which there are many more microstates than for any other macrostates*, and that macrostate *will be practically the only one ever observed.* Other macrostates will be observed occasionally, however, and these rare occurrences are responsible for, among other things, the *scattering of the blue light from the earth's atmosphere.* This phenomenon will be discussed later when we consider the density fluctuations in a gas in Chapter 15.

9.3. Bose-Einstein distribution function

We now set ourselves the problem of determining how many microstates correspond to any given macrostate, and whether there is any particular macrostate for which this number is a maximum. The number of microstates corresponding to any given macrostate is called the *thermodynamic probability* of the macrostate and is represented by W. In general, W is a very large number.

Let us take a simple example. Suppose that there are just two calls in phase space, *i* and *j*, that each cell is divided up into four compartments; and that there are four identical particles. Let N_i and N_j represent the number of particles, that is, the number of phase points in the respective cells. Accordingly, there are five possible macrostates which are illustrated in the following table:

	Macrostates				
	1	2	3	4	5
N_i	4	3	2	1	0
N_g	0	1	2	3	4

Before considering how many microstates correspond to each of these macrostates we stop to consider two important related points. The first is a *restriction on the number of particles in a compartment.* This restriction is the *Pauli exclusion principle.* It applies to electrons but not to the *molecules of a gas* or to *photons.* This principle governs the arrangement of electrons in an atom; *it states that no two electrons in the same atom can have the same set of quantum numbers.* The same principle as it applies to an electron gas, asserts *that there can be no more than two phase points in each of the compartment of volume* h^3. This is the restriction on the magnitude of any N_i in the *Fermi-Dirac statistics*, as we shall consider later in Chapter 10, N_i being the number of phase points in the *i*th cell.

In this section concerning thermodynamic probability of a macrostate, we shall consider only particles to which the exclusion principle does not apply, and for which there may be any number of phase points in a compartment. The theory was developed independently by *Einstein and by Bose* and is called the *Bose-Einstein statistics.*

Our analysis is also restricted to the case in which all the particles are identical. Therefore, if any two particles are interchanged, the same microstate is obtained. The statistics of a gas mixture (for example, helium and argon atoms) are different and the thermodynamic properties differ, as we shall see in Chapter 13. These differences account for many of the *paradoxes* encountered is classical thermodynamics, an example of which is the *Gibbs paradox,* concerning the entropy of mixing of gases.

It may be noted here that the sciences of Classical thermodynamics and Statistical mechanics are based on the existence of equilibrium systems whose coordinates are time *invariant.*

We now return to the determination of the number of microstates in a macrostate for the problem previously posed in the above table. In particular, we look at the case $N_i = 3$ and $N_j = 1$. This case is illustrated in Fig. 2. We see that there are 20 different ways of arranging the three particles among the four compartments in cell *i*, and four different ways of arranging the single particle among the four compartments in cell *j*. We can therefore assign a *thermodynamic probability* to each cell, which is taken as equal to the number of possible ways of arranging the phase points within that cell. If W_i and W_j represent the probabilities of the respective cells, then, in this example

$$W_i = 20, \quad W_j = 4$$

For any one of the arrangements in cell i, we can have any one of those in cell j, so that the total possible number of arrangements, which we now call the *thermodynamic probability* of the macrostate, $N_i = 3$, and $N_j = 1$, as

$$W = W_i W_j$$
$$= 20 \times 4 = 80.$$

In general, when there are any number of *occupied* cells, we have

$$W = \Pi \ W_i, \qquad (1)$$

where the product extends over all occupied cells of phase space.

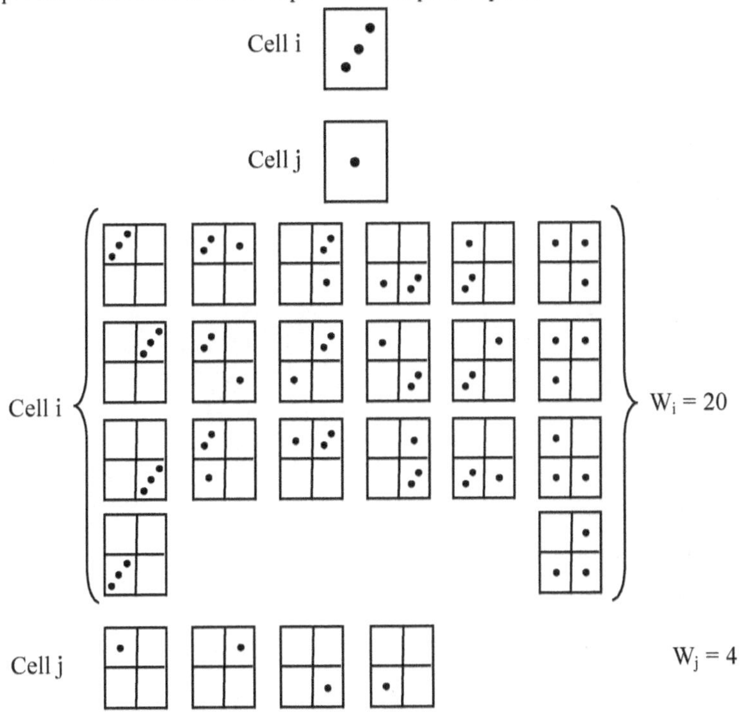

Fig. 2. Different ways of arranging phase points within a cell of phase space in the Bose-Einstein statistics.

We now derive the *general expression* for the thermodynamic probability of a cell *i*, W_i, in terms of N_i and g_i, that is, the general expression for the number of microstates of cell *i* having a number of particles N_i and of compartments g_i.

Suppose that the compartments in the *i*th cell are numbered 1, 2, 3, . . . up to g_i, and the particles are lettered a, b, c . . . up to N_i. Although the particles are taken indistinguishable, we assign letters to them temporarily as an aid in explaining how the thermodynamic probability is computed. In some arrangement of the particles in cell i, we might have particles a and b in compartment 1 and particle c in compartment 2; compartment 3 might be empty, while compartment 4 contained particles d, e, f, etc. This state is represented by the mixed sequence.

1 a b	2 c	3	4 d c f

where the letters following a number designate the particles in the compartment of that number. If the numbers and letters are arranged in all possible sequences, each sequence will represent a *microstate*, provided the sequence begins with a number. Since we have g_i compartments in each cell, there are g_i *ways* in which the sequences can begin with a number, that is, one for each of the g_i compartments. In each of these ways the remaining $\left(g_i + N_i - 1\right)$ numbers and letters can be arranged in any order. We know that the number of different ways in which x things can be arranged in sequences is x!; so the number of different sequences that begin with a number is

$$g_i\left(g_i + N_i - 1\right)! \cdot \qquad (3)$$

Although each sequence represents a microstate, many of them represent the same microstate. For example, let us arrange the blocks in Eq. (2) in a different sequence, such as

3	1 a b	4 d e f	2 c

This sequence does not change the microstate, since *the same compartments contain the same particles*. There are g_i such blocks in the sequence one for each compartment, the number of different sequences of blocks is g_i!. We must divide Eq. (3) by g_i! and in order to avoid counting the same microstate more than once, we must divide Eq. (3) by g_i!. From this, we see that g_i is the statistical weight.

Also, since the particles are *indistinguishable*, a different sequence of the letters such as

1 c a	2 e	3	4 d b f

represents the same microstate as Eq. (2) because any given compartment contains the same number of particles. Since the N_i letters can be arranged in sequence in $N_i!$ different ways, therefore, Eq. (3) must also be divided by $N_i!$.

Hence, the number of microstates for the cell *i*, that is, the thermodynamic probability of the cell *i*, is given by

$$W_i = \frac{g_i\left[\left(g_i + N_i - 1\right)!\right]}{g_i!\, N_i!},$$

which is more conveniently written as

$$W_i = \frac{\left(g_i + N_i - 1\right)!}{\left(g_i - 1\right)!\ N_i!} \tag{4}$$

since

$$g_i! = g_i\left(g_i - 1\right)!$$

Applying Eq. (4) to the case illustrated in Fig. (2), where $g_i = 4$, $N_i = 3$ and $N_j = 1$, we get

$$W_i = \frac{(4+3-1)!}{(4-1)!\,3!} = \frac{6!}{3!3!} = 20 \text{ and}$$

$$W_j = \frac{(4+1-1)}{(4-1)!\,1!} = \frac{4!}{3!} = 4,$$

which agrees with the result obtained by counting.

For any one of the microstates of cell *i*, we may have any one of the microstates of cell *j*, so that the total number of microstates including all cells or the thermodynamic probability *W* of the corresponding macrostate is

$$W = \Pi W_i$$

$$= \Pi \frac{\left(g_i + N_i - 1\right)!}{\left(g_i - 1\right)!\ N_i!}. \tag{5}$$

We now turn to the problem of evaluating *W* for a gas, where the number *N* and all the N_i's are very large. The factorial of a large number can be found with sufficient precision from *Sterling's approximation* which we now derive.

The natural logarithm of factorial *x* is

$$\ln(x!) = \ln 2 + \ln 3 + \ldots + \ln x.$$

This is exactly equal to the *area* between $x = 1$ and $x = x$ under the step curve shown by dashed lines in Fig. 3. This is because each rectangle is of unit width and the height of the first is $\ln 2$, that of the

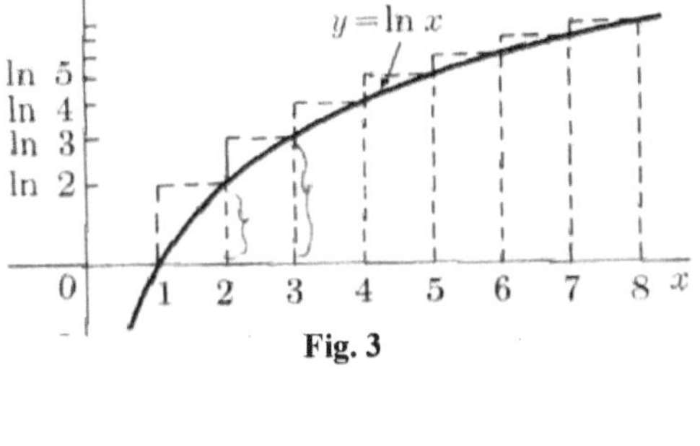

Fig. 3

second is $\ln 3$, etc. This area is approximately equal to the area under the smooth curve $y = \ln x$ between the same limits, provided x is large. For small values of x, the step curve differs appreciably from the smooth curve, but the latter becomes more and more horizontal as x increases. Hence, approximately, for large values of x, we have

$$\ln x! = \int_1^x \ln x \, dx .$$

Integration of the right-hand side by parts by parts yields

$$\ln x! = x \ln x - \int_1^x x \, d \ln x$$

$$= x \ln x - \int_1^x x . \frac{1}{x} dx$$

$$= x \ln x - x + 1,$$

and if x is large, we may neglect 1, so that we finally have

$$\ln x! = x \ln x - x . \tag{6}$$

This is *Sterling's approximation.*

An exact analysis of factorial x leads to the following infinite series.

$$x! = \sqrt{2\pi x}\left(\frac{x}{e}\right)^x \left[1 + \frac{1}{12x} + \frac{1}{288 x^2} - \frac{139}{51{,}840 x^2} + \ldots\right].$$

If all terms in the series except the first are neglected, we obtain by taking the natural logarithm of both sides of this equation

$$\ln x! = \frac{1}{2} \ln 2\pi + \frac{1}{2} \ln x + x \ln x - x .$$

If x is very large compared with unity, the first two terms of this expression are negligible also, and we obtain Eq. (6).

Taking now the natural logarithm of both sides of Eq. (5),

$$W = \Pi \frac{\left(g_i + n_i - 1\right)!}{\left(g_i - 1\right)! \, N_i!}, \tag{5}$$

we obtain

$$\ln W = \sum\left[\ln\left(g_i + N_i - 1\right)! - \ln\left(g_i - 1\right)! - \ln N_i!\right],$$

and using the Sterling approximation, we find

$$\ln W = \sum\left[\left(N_i + g_i\right)\ln\left(N_i + g_i\right) - g_i \ln g_i - N_i \ln N_i\right.$$

$$\left. - \left(N_i + g_i\right) + g_i + N_i\right]$$

$$= \sum\left[\left(N_i + g_i\right)\ln\left(N_i + g_i\right) - g_i \ln g_i - N_i \ln N_i\right], \tag{7}$$

where 1 has been neglected in comparison with g_i and N_i.

As time goes on and the particles in the cells of phase space shift around, the numbers N_i's will change. If the system is in a state of *maximum thermodynamic probability* which we shall denote by W°, the first variation of W° arising from variations in the N_i's is zero. We shall use the symbol δ to represent a small change arising from the continual motion of the particles in phase space. If the thermodynamic probability for a given macrostate is a maximum, that is, $W = W^\circ$, its logarithm is also a maximum. Accordingly, the *condition for maximum thermodynamic probability is*

$$\delta \ln W^\circ = \sum \Big[(N_i + g_i) \frac{1}{(N_i + g_i)} \delta N_i + \ln (N_i + g_i) \delta N_i - N_i \times \frac{1}{N_i} \delta N_i - \ln N_i \, \delta N_i \Big] = 0$$

$$= \sum \Big[\delta N_i + \ln (N_i + g_i) \delta N_i - \delta N_i - \ln N_i \, \delta N_i \Big] = 0$$

$$= \sum \left[\ln \left(\frac{N_i + g_i}{N_i} \right) \delta N_i \right] = 0.$$

Denoting N_i, which represents the number of particles in the *i*th cell, by $\overset{\circ}{N_i}$ in the state of maximum thermodynamic probability, we have the following expression,

$$\sum \left[\ln \left(\frac{\overset{\circ}{N_i} + g_i}{\overset{\circ}{N_i}} \right) \right] \delta N_i = 0, \tag{8}$$

representing the condition for this state of maximum probability, which is the state of *thermodynamic equilibrium* for the system.

Writing out the first few terms of Eq. (8), we find that this is equivalent to

$$\ln \left(\frac{\overset{\circ}{N_1} + g_1}{\overset{\circ}{N_1}} \right) \delta N_1 + \ln \left(\frac{\overset{\circ}{N_2} + g_2}{\overset{\circ}{N_2}} \right) \delta N_2 + \ldots = 0$$

The quantities $\delta N_1, \delta N_2$, etc, are the small increases (or decreases) in the numbers N_1, N_2, etc which occur as a result of particle motions or collisions. If these were all *independent, the coefficient* of *each would have to vanish separately*. But the δN_i's are not independent because the total number of particles is constant, so any increases in the population of some cells must just balanced by decrease in the population of others; that is,

$$\delta N = \sum \delta N_i = \delta N_1 + \delta N_2 + \ldots = 0. \qquad (9)$$

This is *one condition equation imposed on the* δN_i's. But there is also another one. The system under consideration is presumed to be *isolated* so that its internal energy *U* remains constant. Therefore, any shifts in the populations of some cells that take some particles into cells of greater energy must be balanced by shifts that take other particles into cells of lower energy. Let ϵ_i represent the energy of a particle when it lies in the *i*th cell. The quantity ϵ_i depends, in general, on all the coordinates of the cell. The total energy of all the N_i particles which lie in the *i*th cell is $\epsilon_i N_i$, and the total internal energy *U* of the system is, therefore,

$$U = \sum \epsilon_i N_i . \qquad (10)$$

When the number of particles in the *i*th cell changes by δN_i, the change in internal energy is $\epsilon_i \delta N_i$, and since the total internal energy remains constant, the sum of all the these changes must be zero. Hence

$$dU = \sum \epsilon_i \delta N_i = \epsilon_1 \delta N_1 + \epsilon_2 \delta N_2 + \ldots = 0. \qquad (11)$$

This a *second condition equation imposed on* $\delta N_i{}'s$.

Using a method invented by *Lagrange*, known as *Lagrange's method of undetermined multipliers,* we can *in effect,* combine Eq. (8) and the condition equations to obtain an equation in which $\delta N_i{}'s$ become *independent*. Thus, let us multiply the first condition equation, Eq. (9), by a *constant* which, for convenience, we write as $-\ln B$, whose value we shall determine later ($-\ln B$ is then our first undetermined multiplier); and multiply the second condition equation, Eq. (11), by a constant $-\beta$, whose value we shall also determine later, and add the resulting equations to Eq. (8). This gives

$$\sum \left[\ln\left(\frac{N_i^o + g_i}{N_i^o} \right) - \ln B - \beta\epsilon_i \right] \delta N_i = 0,$$

and, since *in effect,* the δN_i's are now *independent*, we have

$$\ln\left(\frac{N_i^o + g_i}{N_i^o} \right) = \ln B - \beta\epsilon_i$$

or

$$\frac{N_i^o + g_i}{N_i^o} = B\, e^{\beta\epsilon_i},$$

$$1 + \frac{g_i}{N_i^o} = B e^{\beta\epsilon_i}.$$

Hence,

$$\frac{N_i^o}{g_i} = \frac{1}{B \exp(\beta\epsilon_i) - 1} \qquad (12)$$

This is the *Bose-Einstein distribution function.*

9.4. The Maxwell-Boltzmann statistics

Let us consider a system in which the number of particles in a cell, N_i, is very much *smaller* than the number of compartments, g_i. For such a system, the left-hand side of Eq. (12) is very much less than unity, and hence the denominator on the right-hand side is very much greater unity. The term_1 in the denominator can then be neglected, and for such a system,

$$\frac{N_i^o}{g_i} = \frac{1}{B\exp(\beta\epsilon_i)}, \tag{13}$$

with B >> 1. This is the *Maxwell-Boltzmann distribution* of particles among cells.

For simplicity we shall drop the superscript 'o' in our subsequent analysis, as our analysis will be restricted to *equilibrium distribution.*

An expression for B can be obtained from the requirement that the sum of all the N_i's must be equal to the total number of particles N. Thus

$$\sum N_i = N = \frac{1}{B}\sum g_i \exp(-\beta\epsilon_i). \tag{14}$$

The sum $\sum g_i \exp(-\beta\epsilon_i)$ over energy levels plays an important role in the statistical theory. It is called the *partition function or sum of state* and is represented by the letter *Z* (German; *Zustandssumme*).

$$Z = \sum g_i \exp(-\beta\epsilon_i) \ . \tag{15}$$

The partition function depends on β, and on the way in which the energy ϵ varies from cell to cell. Since the latter varies from problem to problem, no general expression can be written for *Z* other than that given above; it must be evaluated for each special case.

It is evident from Eqs. (14) and (15) that

$$BN = \sum g_i \exp(-\beta\epsilon_i)$$
$$= Z,$$

so we can express the constant *B* in terms of *Z as*

$$B = \frac{Z}{N}.$$

Returning now to Eq. (13),

$$\frac{N_i}{g_i} = \frac{1}{B\exp(\beta\epsilon_i)}.$$

We may rewrite it in the form

$$N_i = \frac{g_i}{B}\exp(-\beta\epsilon_i), \tag{13}$$

therefore,

$$N_i = \frac{N g_i}{Z}\exp(-\beta\epsilon_i). \tag{16}$$

Eq. (16) thus gives the number of particles in the *i*th cell in the state of maximum thermodynamic probability, that is, the state of thermodynamic equilibrium.

Eqs. (13) and (16) are forms of the Maxwell-Boltzmann distribution law, and they have a fundamental significance in statistical mechanics. If the total energy of the ensemble of particles is constant, these equations give the probable distribution of the particles among the possible energy levels. In deriving these equations, no restrictions were placed on the nature of the energy, and therefore, they are equally applicable to the distribution of the total energy of the system, and to any particular form of energy which is constant for the entire system.

9.5. Entropy and probability

Eq. (16) obtained above, namely,

$$N_i = \frac{Ng_i}{Z} e^{-\beta\epsilon_i}$$

provides an expression for the number of particles N_i in the *i*th cell of phase space, for the macrostate of maximum thermodynamic probability. *Presumably* this is the state towards which the *isolated system* will tend. From the thermodynamic point of view, *the state of equilibrium for an isolated system is the state of maximum entropy*. If the system is not in equilibrium, then changes take place within it until the *state of maximum entropy has been attained*; this is the *state of thermodynamic equilibrium or the final steady state.* Therefore, in the equilibrium state of an isolated system, both, *entropy* and the *thermodynamic probability,* have their *maximum* values. This leads us to expect some correlation between them. We might assume the entropy *S* to be proportional to the thermodynamic probability *W*, but it turns out that in order to obtain agreement between the *thermodynamic* and *statistical definitions*, we must take the entropy as proportional to the *logarithm* of the thermodynamic probability. Of course, if the probability is a maximum, its logarithm is a maximum, also. We, therefore, set the following important relation which is known as the Planck–Boltzmann equation, as noted before,

$$S = k \ln W . \tag{17}$$

Here *k* is a proportionality constant that we shall identify later. (It turns out to be none other than the *Boltzmann constant*).

Therefore, statistical mechanics interprets the increase of entropy in an isolated system as a consequence of the *natural trend of a system from a less probable to a more probable state.*

It is often helpful to phrase the concept of probability in other terms, such as the '*mixed upness*' or the '*disorder*' of a system. Greater the *disorder*, greater the *thermodynamic probability* and *entropy*. The greatest degree of order of the atoms of a gas in phase space results if all are in a single cell; that is, if all are in a *very small volume* in ordinary space and all are travelling with the *same* velocity in velocity space. The thermodynamic probability *W* of such a state has its *minimum* value of unity, and the entropy, $S = k \ln W$, is zero. The more the particles are spread out in ordinary space and more their velocities are spread out in velocity space, the greater the disorder and entropy.

Consider a vessel divided into two equal parts by a partition, with an equal number of atoms of two different monatomic gases on opposite sides of the partition as shown in Fig. 4.

N atoms gas (1)	N atoms gas (2)

Fig. 4.

Such a system has a *certain degree of order*, in that all atoms of one gas are on one side of the partition and all those of the other gas are on the other side. If the partition is now removed, then the gases diffuse into each other. Eventually, both kinds of atoms are uniformly distributed throughout the total volume. Thus, the original order has disappeared and the *disorder* of the system or *its mixed upness* has *increased.* But the *entropy* of the system has also *increased* owing to the fact that the volume occupied by each go has been doubled* at *constant temperature* if the gases are ideal.

In a reversible adiabatic expansion of a gas, the volume *increases,* but the temperature *decreases*. The entropy change in this process, which is given by

$$\int dS = \int \frac{d\,q_{rev}}{T},$$

is equal to *zero,* since $d\,q_{rev} = 0$. That is, the entropy of the gas *remains constant.* This constancy is explained by statistical mechanics in the sense that the increase in disorder, resulting from the *increase* in volume, is compensated for by a *decrease* in disorder resulting from a *smaller* velocity spread out in velocity apace at the lower temperature.

According to the laws of classical thermodynamics, only those processes can take place in an isolated siptem, for which the *entropy* of this system *increases* or, in the limit, *remains constant.* Any process in which the *entropy* would *decrease* is prohibited. Put another way, according to classical thermodynamics there are *two rules* governing the changes that can occur in an isolated system, namely, (1) when an *irreversible* (i.e. *spontaneous*) change takes place in any part of an isolated system, the *total entropy increases*, (2) when a *reversible* (i.e. *equilibrium*) change takes place in any part of an isolated system, the *total entropy remains unchanged.*

The statement outlined by classical thermodynamics that any process in which the entropy of the isolated system would *decrease* does not take place is a *dogmatic* statement in the light of the statistical interpretation of entropy, and thus, it must be *modified.* Consider an isolated system in a state of maximum probability or maximum entropy. This state is not a *static* state since the particles are continuously shifting around in phase space as a result of their continuous motions and collisions. Thus, *occasionally*, a state of the system will result for which the thermodynamic probability, and hence the entropy, is less than its *maximum* value. Small changes are more likely than large ones, but large ones are not impossible; they are only *highly improbable*. The rare occurrences of these macrostates

---

$$\left(\frac{\partial S}{\partial V}\right)_T = \left(\frac{\partial P}{\partial T}\right)_X = \frac{\beta}{k} = \text{positive quantity, since}$$

* $\beta = \frac{1}{V}\left(\frac{\partial V}{\partial T}\right)_D$ and $k = -\frac{1}{V}\left(\frac{\partial V}{\partial P}\right)_T$.

whose thermodynamic probability, and hence the entropy, is less than the maximum value are responsible for, among other things, the scattering of the blue light from the earth's atmosphere, as pointed out before. We shall consider this question in more detail later when we discuss the density fluctuations in a gas in Chapter 13.

Let us now return to Eq. (17)

$$S = k \ln W. \tag{17}$$

For large values of N_i and g_i, the thermodynamic probability is given by Eq. (7) above, namely,

$$\ln W = \left[\left(N_i + g_i\right)\ln\left(N_i + g_i\right) - g_i \ln g_i - N_i \ln N_i\right]. \tag{7}$$

We again consider the Maxwell-Boltzmann limit, $g_i >> N_i$. In this limit, the following approximation may be made as

$$\ln\left(g_i + N_i\right) = \ln\left[g_i\left(1 + \frac{N_i}{g_i}\right)\right]$$

$$= \ln g_i + \ln\left(1 + \frac{N_i}{g_i}\right)$$

$$= \ln g_i + \frac{N_i}{g_i} + \ldots$$

Substitution of this result in Eq. (7) gives

$$\ln W = \sum\left[\left(N_i + g_i\right)\left(\ln g_1 + \frac{N_i}{g_i}\right) - g_1 \ln g_1 - N_i \ln N_i\right]$$

$$= \sum\left(N_i \ln g_i + g_i \ln g_i + N_i g_i \ln g_i - N_i \ln N_i\right)$$

$$= \sum\left(N_i \ln g_i + N_i - N_i \ln N_i\right), \tag{18}$$

where the term N_i^2/g_i is of higher order and thus has been neglected.

But

$$\sum N_i = N$$

and

$$\sum \epsilon_i N_i = U,$$

so substituting Eq. (16),

$$N_i = \frac{N g_i}{Z} e^{-\beta \epsilon_i}, \tag{16}$$

in Eq. (18), we have

$$S = \ln W = k \sum N_i \left(\ln g_i + 1 - \ln N_i \right)$$

$$= k \sum N_i \left(\ell n \frac{g_i}{N_i} + 1 \right).$$

But from Eq. (16), we obtain

$$\ln \frac{g_i}{N_i} = \ln \frac{Z}{N} + \beta \epsilon_i ,$$

therefore,

$$S = k \sum N_i \left(\ln \frac{Z}{N} + \beta \epsilon_i + 1 \right)$$

$$= k N \ln \frac{Z}{N} + k \beta U + k N$$

$$= k N \left(\ln \frac{Z}{N} + 1 \right) + k \beta U \cdot \qquad (19)$$

Up to this point, the concept of *temperature* has not appeared in our development of statistical theory. It can be introduced as follows:
Classical thermodynamics gives the relation

$$\left(\frac{\partial U}{\partial S} \right)_V = T, \quad \text{or} \quad \left(\frac{\partial S}{\partial U} \right)_V = \frac{1}{T},$$

which is obtained from the relation

$$d U = T d S - P d V$$

which is the combined first and second laws, as shown before. This relation indicates that the *natural or proper independent variables* for the internal energy *U* is the entropy *S* and the volume *V*; that is

$$S = S (S, V).$$

If we now differentiate Eq. (19) with respect to *U* at constant *V*, taking into consideration the fact that the partition function *Z*, which is defined by

$$Z = \sum g_i \, e^{-\beta \epsilon_i} , \qquad (15)$$

depends on β and on the way in which the ϵ varies from cell to cell, as pointed out above, we obtain

$$\left(\frac{\partial S}{\partial U} \right)_V = \frac{k N}{Z} \left(\frac{\partial Z}{\partial U} \right)_V + k \beta + k U \left(\frac{\partial \beta}{\partial U} \right)_V .$$

But from the definitions of ϵ and *Z*, we obtain

$$\frac{d Z}{d \beta} = \sum g_i \, e^{-\beta \epsilon_i} \, X - \epsilon_i ,$$

$$= - \sum \epsilon_i \, g_i \, e^{-\beta \epsilon_i} = - \frac{U Z}{N} . \qquad (20)$$

The equation representing the variation of entropy with internal energy at constant volume can be rewritten in the following form, expressing the variation of the partition function with the variation of β

$$\left(\frac{\partial S}{\partial U}\right)_V = \frac{kN}{Z}\cdot\frac{dZ}{d\beta}\left(\frac{\partial \beta}{\partial U}\right)_V + k\beta + kU\left(\frac{\partial \beta}{\partial U}\right)_V .$$

Subsequently, substituting for $\frac{dZ}{d\beta}$ by the right-hand side of Eq. (20),

we have

$$\left(\frac{\partial S}{\partial U}\right)_V = \frac{kN}{Z} \times -\frac{UZ}{N}\left(\frac{\partial \beta}{\partial U}\right)_V + k\beta + kU\left(\frac{\partial \beta}{\partial U}\right)_V .$$

$$= k\beta. \tag{21}$$

But $\left(\frac{\partial S}{\partial U}\right)_V$ from classical thermodynamics, as pointed out above is given by

$$\left(\frac{\partial S}{\partial U}\right)_V = \frac{1}{T},$$

therefore,

$$\frac{1}{T} = k\beta,$$

or

$$\beta = \frac{1}{kT}. \tag{22}$$

The constant β is now determined, except that we have not yet shown that *k* is the *Boltzmann constant*

Now we can express the number of particles in the *i*th cell in terms of *T*. Thus considering Eq. (16)

$$N_i = \frac{N g_i}{Z} e^{-\beta \epsilon_i} \tag{16}$$

which is one form of the *Maxwel–Boltzmann distribution law*, the other form being

$$\frac{N_i}{g_i} = \frac{1}{B\, e^{\beta \epsilon_i}}, \tag{13}$$

as given above, we have

$$N_i = \frac{N g_i}{Z} e^{-\epsilon_i / kT}, \tag{23}$$

where the partition function *Z* from Eq. (15),

$$Z = \sum g_i\, e^{-\beta \epsilon_i}, \tag{15}$$

is

$$Z = \sum g_i\, e^{-\epsilon_i / kT}. \tag{24}$$

Since the internal energy of the system is given by

$$U = \sum \epsilon_i N_i, \tag{15}$$

the substitution for N_i from Eq. (23) in this equation gives

$$U = \sum \epsilon_i \frac{N g_i}{Z} e^{-\epsilon_i / kT}$$

$$= \frac{N}{Z} \sum g_i \epsilon_i \, e^{-\epsilon_i / kT}. \tag{25}$$

The derivative of *Z* with respect to *T* is

$$\frac{dZ}{dT} = \frac{d}{dT}\left[\sum g_i \, e^{-\epsilon_i / kT}\right]$$

$$= \sum g_i \, e^{-\epsilon_i / kT} . \epsilon_i / kT^2$$

$$= \frac{1}{kT^2} \sum g_i \epsilon_i \, e^{-\epsilon_i / kT}.$$

Introducing Eq. (25) in this equation, we obtain

$$\frac{dZ}{dT} = \frac{1}{kT^2} . \frac{UZ}{N},$$

therefore,

$$U = \frac{NkT^2}{Z} \frac{dZ}{dT},$$

or

$$= NkT^2 \frac{d(\ln Z)}{dT}. \tag{26}$$

Also, from Eq. (19),

$$S = Nk\left[\ln\left(\frac{Z}{N}\right) + 1\right] + k\beta U, \tag{19}$$

we have

$$S = Nk\left[\ln\left(\frac{Z}{N}\right) + 1\right] + \frac{U}{T}, \tag{27}$$

where β is substituted for by $1/kT$.

The *Helmholtz function* which is defined as

$$F \equiv U - TS$$

is given by the substitutions for *U* and *S* by the respective Eqs.(26) and (27), thus we obtain

$$F = NkT^2 \frac{d(\ln Z)}{dT} - TNk\left[\ln\left(\frac{Z}{N}\right) + 1\right] - U,$$

that is,

$$F = -NkT\left[\ln\left(\frac{Z}{N}\right) + 1\right]. \tag{28}$$

It is evident from the foregoing that once the partition function *Z* has been evaluated, all the thermodynamic properties of the system can be evaluated. Note that while in classical thermodynamics only differences in internal energy and entropy can be defined, statistical methods provide for both of these quantities' expressions containing *no undetermined constants*.

As an example, let us consider a system of *N* particles and a phase space of just two cells, 1 and 2, and take $g_1 = g_2 = 1$. Suppose that the energy ϵ of a particle is the same in both cells; that is,

$$\epsilon_i = \epsilon_2 = \epsilon$$

Then, from the partition function, *Z*, given by Eq. (24),

$$Z = \sum g_1 \, e^{-\epsilon_i/kT}, \qquad (24)$$

we have

$$Z = e^{-\epsilon/kT} + e^{-\epsilon/kT}$$
$$= 2\, e^{-\epsilon/kT}.$$

From Eq. (23),

$$N_i = \frac{N g_i}{Z} . e^{-\epsilon_i/kT}, \qquad (23)$$

we have

$$N_1 = \frac{N\, e^{-\epsilon/kT}}{2\; e^{-\epsilon/kT}} = \frac{N}{2},$$

and

$$N_2 = \frac{N\, e^{-\epsilon/kT}}{2\; e^{-\epsilon/kT}} = \frac{N}{2}.$$

The particles are, therefore, distributed uniformly between the two cells, as would be expected. This is true for any number of cells of equal energy.

The internal energy *U* as given by Eq. (26)

$$U = NkT^2 \frac{d(\ln Z)}{dT}, \qquad (26)$$

is

$$U = NkT^2 \frac{d\left(\ln 2\, e^{-\epsilon/kT}\right)}{dT}$$
$$= NkT^2 \frac{d}{dT}\left(\ln 2 - \frac{\epsilon}{kT}\right)$$
$$= NkT^2 . \frac{\epsilon}{k} \times \frac{1}{T^2} = N\epsilon .$$

This is an obvious conclusion, since the energy of each of the *N* particles is ϵ.

Next suppose that

$$\epsilon_1 = 0 , \quad \text{and} \quad \epsilon_2 = \epsilon .$$

This means that all particles which lie in cell 1 have no energy, and all particles which lie in cell 2 have energy ϵ. The system in this state has a partition function *Z* given by

$$Z = \sum g_i \, e^{-\epsilon_i/kT} \qquad (24)$$

$$= e^{-\epsilon_i/kT} + e^{-\epsilon_i/kT}$$

$$= 1 + e^{-\epsilon_i/kT}.$$

Putting θ as an *abbreviation* for ϵ / k, which has the dimensions of temperature, we have

$$Z = 1 + e^{-\theta/T},$$

where θ is called the *characteristic temperature.*

Let us evaluate the number of particles in each cell in this state, using the relation

$$N_i = \frac{N g_i}{Z} e^{-\epsilon_i/kT}. \qquad (23)$$

Therefore,

$$N_1 = \frac{N}{1 + e^{-\theta/T}}, \quad \text{and} \quad N_2 = \frac{N e^{-\theta/T}}{1 + e^{-\theta/T}}.$$

N_2 can be rewritten as

$$N_2 = \frac{N}{e^{\theta/T} + 1}$$

$$= \frac{N}{1 + e^{\theta/T}}.$$

At temperatures which are very small compared with the characteristic temperature θ, θ / T is very large, that is,

$$\theta / T >> 1,$$

and subsequently $e^{-\theta/T} \left(= 1/e^{\theta/T}\right)$ is also very small, whereas $e^{\theta/T}$ is very large.

Accordingly, N_1 is very nearly equal to N, and N_2 is very small. That is, nearly all the particles are in cell 1.

At temperatures that are large compared with the characteristic temperature θ, θ / T is very small, that is,

$$\theta / T << 1,$$

the two exponential terms $e^{\theta/T}$ and $e^{-\theta/T}$ are nearly equal to unity. Accordingly, N_1 and N_2 are nearly equal to N/2.

When the temperature is equal to the characteristic temperature, that is,

$$T = \theta,$$

then

$$N_1 = \frac{N}{1 + e^{-1}},$$

and

$$N_2 = \frac{N}{1 + e}.$$

Thus, substituting for the exponential* e in the last two equations by 2.7169, we obtain

$$N_1 = \frac{N}{3.7169} \times 2.7169$$

$$= 0.73 \ N,$$

* See Appendix A.

and

$$N_2 = \frac{N}{3.7169}$$
$$= 0.27\ N.$$

It is, therefore, evident that as the temperature T is raised from a value where $\frac{\theta}{T} >> 1$, that is, $T << \theta$, or where $kT << \epsilon$, some particles move from cell 1 of zero energy to cell 2 of energy ϵ. But even at very high temperature, where $T >> \theta$, half of the particles remain in a state of zero energy. Consequently, it can be concluded that 'high' and 'low' temperatures have a meaning so far T is large or small compared with the characteristic temperature θ, or as the product kT is large or small compared with the energy ϵ. This will be more evident in the following table, which illustrates the population of the energy levels as a function of θ/T or ϵ/kT.

***T* is very small compared with θ**	***T* is large compared with θ**
$\frac{\theta}{T} >> 1$, or $T << \theta$, or $\frac{\epsilon}{kT} >> 1$, or $kT << \epsilon$ $N_1 \approx N$ N_2 is very small	$\frac{\theta}{T} << 1$, or $T >> \theta$, or $\frac{\epsilon}{kT} << 1$, or $kT >> \epsilon$ $N_1 \approx N/2$, $N_2 \approx N/2$,

$T = \theta$

$N_1 = 0.71N$

$N_2 = 0.27N$

9.6. The monatomic ideal gas

We now return to our consideration of a monatomic gas consisting of N atoms, each is characterised by its position coordinates x, y, z, and its velocity coordinates v_x, v_y, v_z. The energy ϵ_i of an atom in the ith cell is the sum of its potential and kinetic energies. If the atoms exert no forces on one another, there is no mutual potential energy between them. For the present, we neglect any effect of a gravitation force field. The potential energy is, then, a constant is all cells of phase space whose coordinates x, y, z lie within the space occupied by the gas, and we shall consider this constant to be zero. The fact that the walls of the container are *impenetrable* to atoms can be taken into account by considering the potential energy to be infinite in all cells for which x, y, z lie outside the container. Then, in all such cells

$$\epsilon_i = \infty \quad \text{and} \quad e^{-\epsilon_i/kT} = 0$$

The number of atoms in such cells, given by Eq. (23),

$$N_i = \frac{N g_i}{Z} e^{-\epsilon_i/kT}, \tag{23}$$

is then zero, and these cells contribute *nothing* to the partition function.

If the atoms can be considered point masses, the kinetic energy is *only translational.* For cell *i*, whose velocity coordinates are v_x, v_y, v_z, the kinetic energy is

$$\frac{1}{2} m \left(v_x^2 + v_y^2 + v_z^2 \right) = \frac{1}{2} m v_i^2,$$

for an atom of mass *m* in this *i*th cell. Then for any cell in the space occupied by the gas, the kinetic energy ϵ_i is

$$\epsilon_i = m v_i^2,$$

and for any cell outside this space

$$\epsilon_i = \infty.$$

The partition function *Z* is, therefore,

$$Z = \sum g_i e^{-m v_i^2/2kT},$$

where the sum extends only over those cells in the space occupied by the gas. The statistical weight g_i, which represents the number of compartments in the *i*th cell, is given, as shown before, by

$$g_i = \frac{H}{h^3} = \frac{1}{h^3} dx\, dy\, dz\, dp_x\, dp_y\, dp_z$$

$$= \frac{m^3}{h^3} dx\, dy\, dz\, dv_x\, dv_y\, dv_z. \tag{29}$$

This result is substituted into the preceding equation, and the sum is replaced by integrals to give

$$Z = \sum \frac{m^3}{h^3} e^{-m v_i^2/2kT} dx\, dy\, dz\, dv_x\, dv_y\, dv_z$$

$$= \frac{m^3}{h^3} \int\int\int\int\int\int e^{-m v_i^2/2kT} dx\, dy\, dz\, dv_x\, dv_y\, dv_z$$

$$= \frac{m^3}{h^3} \int\int\int dx\, dy\, dz \int_{-\infty}^{\infty} e^{-m v_i^2/2kT} dv_x \int_{-\infty}^{\infty} e^{-m v_y^2/2kT} dv_y$$

$$\times \int_{-\infty}^{\infty} e^{-m v_i^2/2kT} dv_z.$$

The triple integral over *x*, *y*, *z* is simply the total volume *V* occupied by the gas. Each of the single integrals is $\left(\frac{2\pi kT}{m}\right)^{\frac{1}{2}}$; the integral* $\int_{-\infty}^{\infty} e^{-ax^2}$ is $\sqrt{\frac{\pi}{a}}$, where *a*, in our case, is

* See Appendix E, Section 3, some standard integrals.

equal to $\frac{m}{2kT}$. Hence, the partition function Z, which is the *translational partition function*, is given by

$$Z = \frac{Vm^3}{h^3}\left(\frac{2\pi kT}{m}\right)^{3/2}$$

$$= \frac{V}{h^3}\left(2\pi m kT\right)^{3/2}. \tag{30}$$

In writing down the number of atoms in a cell, we replace N_i by d^6N_i. Therefore, the equation,

$$N_i = \frac{N g_i}{Z} e^{-\epsilon_i/kT}, \tag{23}$$

which expresses the number of particles in the *i*th cell in terms of the temperature, *T*, as shown above, can be put in the following form

$$d^6N_i = \frac{N.h^3}{V(2\pi m kT)^{3/2}} \cdot \frac{m^3}{h^3} e^{-mv^2/2kT}$$
$$dx\, dy\, dz\, dv_x\, dv_y\, dv_z$$

$$= \frac{N}{V}\left(\frac{m}{2\pi kT}\right)^{3/2} e^{-mv^2/2kT}\, dx\, dy\, dz\, dv_x\, dv_y\, dv_z. \tag{31}$$

The distribution of particles in ordinary space is now obtained by integrating Eq. (31) over all values of v_x, v_y, v_z. This gives

$$d^3N_{x,y,z} = \frac{N}{V}\left(\frac{m}{2\pi kT}\right)^{3/2}\left(\frac{2\pi kT}{m}\right)^{3/2} dx\, dy\, dz, \tag{32}$$

or

$$\frac{d^3N}{dx\, dy\, dz} = \frac{N}{V}. \tag{33}$$

That is, the number of atoms per unit volume of ordinary space is a constant, independent of position and equal to the total number of atoms, *N*, divided by the total volume, *V*. In other words, the atoms are uniformly distributed throughout the space occupied by the gas.

To find the distribution in velocity space, we integrate Eq. (31) over *x*, *y*, *z*, the limits being chosen to include the space occupied by the gas. The integral is just the total volume *V*; so

$$d^3N_{v_x,v_y,v_z} = N\left(\frac{m}{2\pi kT}\right)^{3/2} e^{-mv^2/2kT}\, dv_x\, dv_y\, dv_z. \tag{34}$$

A comparison can be made now between this equation and the equation

$$d^3N_{v_x,v_y,v_z} = \frac{N}{\pi^{3/2}}\left(\frac{m}{2kT}\right)^{3/2} e^{-mv^2/2kT}\, dv_x\, dv_y\, dv_z, \tag{35}$$

which also represents the velocity distribution function from kinetic theory of an ideal gas using classical statistics.

It is evident that Eq. (34), which is obtained using *quantum statistics*, is precisely Eq. (35).

It is, therefore, evident that *quantum statistics* leads to the same velocity distribution as *kinetic theory* when the assumption $N_i/g_i << 1$ is made. To see how well this inequality is satisfied, we find from Eqs. (23) and (30) that

$$\frac{N_i}{g_i} = \frac{N h^3}{V(2\pi m k T)^{3/2}} e^{-\epsilon_i/kT}.$$

Substituting ϵ_i by $1/2\ m v_i^2$ in this equation, we have

$$\frac{N_i}{g_i} = \frac{N h^3}{V(2\pi m k T)^{3/2}} e^{-m v_i^2/2kT} \quad (36)$$

Since we have a Maxwell velocity distribution, the energies of the atoms are grouped around the mean energy so that the quantity ϵ_i/kT can be considered to be of the order of unity and so is $e^{-m v_i^2/2kT}$. Taking as a specific example, helium at standard conditions, we have

$$m = \frac{4\,\text{kg}}{6.025 \times 10^{26}} \approx 6.7 \times 10^{-27}\ \text{kg}$$

$$\frac{N}{V} = \frac{6.025 \times 10^{26}\ \text{atoms kg}^{-1}.\ \text{mole}^{-1}}{22.414\ \text{m}^3\ \text{kg}^{-1}.\ \text{mole}^{-1}}$$

$$= 0.3 \times 10^{26}\ \text{atoms m}^{-3}$$

$$= 3 \times 10^{25}\ \text{atoms m}^{-3}.$$

Thus the substitution of these values in Eq. (36) gives

$$\frac{N_i}{g_i} \approx 4 \times 10^{-6},$$

which is evidently much less than unity. However, as the temperature is lowered, the number of molecules per unit volume increases. Both of these effects make N_i/g_i larger, as is evidenced from the above equation (Eq. 36), providing the gas can be cooled to very low temperatures without condensing, the *Maxwell-Boltzmann statistics* may cease to be applicable.

The internal energy *U* of the gas is

$$U = \frac{NkT^2}{Z}\frac{dZ}{dT},$$

or

$$= NkT^2 \frac{d(\ln Z)}{dT}$$

In the case of the monatomic ideal gas, the partition function is given by

$$Z = \frac{V}{h^3}(2\pi m k T)^{3/2},$$

so

$$\ln Z = \ln V + \frac{3}{2}\ln T - \frac{3}{2}\ln(2\pi k) - 3\ln h,$$

then

$$\frac{d\ln Z}{dT} = \frac{3}{2}\cdot\frac{1}{T}.$$

Hence,

$$U = \frac{3}{2}NkT$$

or

$$= \frac{3}{2}nRT$$

and

$$u = \frac{U}{n} = \frac{3}{2}RT,$$

where *u* is the molar specific internal energy, *small letters* being used for specific extensive variables.

The *molar specific heat at constant volume*, c_v is

$$c_v = \left(\frac{\partial u}{\partial T}\right)_v = \frac{3}{2}R.$$

This result is in agreement with that derived from kinetic theory and the equipartition principle.

Now we want to obtain the equation of state for an ideal gas. For this purpose, we can use the *Helmholtz function F*, which is

$$F = -NkT\left[\ln\frac{Z}{N} + 1\right],$$

together with the thermodynamic relation,

$$P = -\left(\frac{\partial F}{\partial V}\right)_T.$$

This relation can be obtained as shown before as follows:

According to the definition of the Helmholtz function, which is

$$F \equiv U - TS,$$

we have

$$dF = dU - TdS - SdT,$$

and from the *combined* first and second laws, we have the relation

$$dU = TdS - pdV,$$

so the exact differential d*F* is expressed as

$$dF = -PdV - SdT,$$

and thus,

$$\left(\frac{\partial F}{\partial V}\right)_T = -P.$$

Now expanding *F* expression, we obtain

$$F = -NkT(\ln z - \ln N + 1)$$

$$= -NkT\left(\ln V - \ln N + \frac{3}{2}\ln(2\pi kT) - 3\ln h + 1\right).$$

Therefore,

$$\left(\frac{dF}{dV}\right)_T = -NkT \cdot \frac{1}{V},$$

and hence,

$$P = \frac{NkT}{V}$$

$$= \frac{nRT}{V}. \qquad (37)$$

This is the familiar equation of state of an ideal gas.

The entropy *S* of a monatomic ideal gas can be obtained directly from the general expression of the entropy derived above, namely,

$$S = Nk\left[\ln\left(\frac{Z}{N}\right) + 1\right] + \frac{U}{T}.$$

by the substitution for *Z* by its expression for translational motion

$$Z = \frac{V}{h^3}(2\pi mkT)^{3/2},$$

and for *U* by

$$3/2\, nRT.$$

Thus we obtain

$$S = Nk\left[\ln\left(\frac{V}{N}\right) + 3/2\ln T + 3/2\ln(2\pi mk) - 3\ln h + 1\right] + 3/2\, nR,$$

or

$$S = Nk\left[\ln\left(\frac{V}{N}\right) + 3/2\ln T + 3/2\ln(2\pi mk) - 3\ln h + 5/2\right], \qquad (38)$$

and with $Nk = nR$,

$$S = nR\left[\ln V + 3/2\ln T + 3/2\ln\left(2\pi mk\right) - 3\ln h - \ln N + 5/2\right]. \qquad (39)$$

Denoting the sum of all terms that do not depend on *T* or *V* by A; that is

$$A = 3/2\ln\left(2\pi mk\right) - 3\ln h - \ln N + 5/2,$$

we have

$$S = nR\left[\ln V + 3/2\ln T + A\right] \tag{40}$$

Hence, the *molar specific entropy*, $s\ (=S/n)$, of a monatomic ideal gas is given by

$$s = R\ln v + 3/2R\ln T + RA. \tag{41}$$

It seems instructive for the present to find out, using *classical thermodynamics*, whether the dependence of entropy of a system on temperature and volume leads to an equation analogous to Eq. (5). For this purpose, let us consider *s* as a function of *T* and *v* for a system, that is,

$$s = s(T, v),$$

therefore,

$$ds = \left(\frac{\partial s}{\partial T}\right)_v dT + \left(\frac{\partial s}{\partial v}\right)_T dv\ \cdot \tag{i}$$

From the combined first and second laws equation,

$$du = Tds - Pdv,$$

and the equation

$$du = \left(\frac{\partial u}{\partial T}\right)_v dT + \left(\frac{\partial u}{\partial v}\right)_T dv,$$

in which *u* is considered as a function of *T* and *v*, we obtain

$$Tds = \left(\frac{\partial u}{\partial T}\right)_v dT + \left\{P + \left(\frac{\partial u}{\partial v}\right)_P\right\}dv,$$

or

$$ds = \frac{1}{T}\left(\frac{\partial u}{dT}\right)_v dT + \frac{1}{T}\left\{P + \left(\frac{\partial u}{\partial v}\right)_T\right\}dv. \tag{ii}$$

Since d*T* and d*v* are independent variables in Eq. (i) as well as in Eq. (ii), their coefficients in these two equations must be equal; that is,

$$\left(\frac{\partial s}{\partial T}\right)_v = \frac{1}{T}\left(\frac{\partial u}{\partial T}\right)_v, \tag{iii}$$

and

$$\left(\frac{\partial s}{\partial v}\right)_T = \frac{1}{T}\left\{P + \left(\frac{\partial u}{\partial v}\right)_T\right\}. \tag{iv}$$

Mathematically, as pointed out before, if a variable *z* is a function of other two independent variables *x* and *y*, then the so-called 'mixed' second-order partial derivatives are independent of the order of differentiation, provided the function and the partial derivations are continuous. Since these conditions are satisfied by *s* and its partial derivatives, we have from Eqs. (iii) and (iv)

$$\left\{\frac{\partial}{\partial v}\left(\frac{\partial s}{\partial T}\right)\right\}_T = \left\{\frac{\partial}{\partial T}\left(\frac{\partial s}{\partial v}\right)\right\}_v,$$

that is,

$$\frac{\partial^2 s}{d v \partial T} = \frac{\partial^2 S}{\partial t \partial v}.$$

Hence, differentiating Eq. (iii) partially with respect to *v* at constant *T*, and Eq. (iv) partially with respect to *T* at constant *v*, we obtain the following important equation

$$\left\{\left(\frac{\partial u}{\partial v}\right)_T + p\right\} = T\left(\frac{\partial p}{\partial T}\right)_V, \qquad \text{(v)}$$

which represents the dependence of internal energy on volume, at constant temperature for any substance. Thus, this equation can be rewritten as

$$\left(\frac{\partial u}{\partial v}\right)_T = T\left(\frac{\partial p}{\partial T}\right)_V - p \cdot \qquad \text{(vi)}$$

Also, Eq. (iv) takes the form

$$\left(\frac{\partial s}{\partial v}\right)_T = \frac{1}{T} \times T\left(\frac{\partial p}{\partial T}\right)_V,$$

that is,

$$\left(\frac{\partial s}{\partial v}\right)_T = \left(\frac{\partial p}{\partial T}\right)_V. \qquad \text{(vii)}$$

It can be proved mathematically, as also shown before, that if we have an equation in which the variables are already independent, or made *in effect* independent, using the method of *Lagrange* of undetermined multiplies, their coefficients must vanish separately. Based on this principle, we can put $v = v(T, P)$, and $P = P(T, v)$, then eliminate d*P* between the resulting two equations and collect the coefficients of d*T* and d*v*, each coefficient thus vanishes separately, since the changes d*T* and d*v* are independent; ultimately we obtain

$$(\partial v/\partial P)_T = 1\Big/\left(\frac{\partial P}{\partial v}\right)_T, \qquad \text{(viii)}$$

and

$$\left(\frac{\partial p}{\partial T}\right)_v = \frac{\beta}{k}, \qquad \text{(ix)}$$

where β is the coefficient of volume expansion,

$$\beta = \frac{1}{v}\left(\frac{\partial v}{\partial T}\right)_P,$$

and *k* is the coefficient of isothermal compressibility,

$$k = -\frac{1}{v}\left(\frac{\partial v}{\partial P}\right)_T.$$

Hence, substituting Eq. (ix) in Eq. (vii), we obtain

$$\left(\frac{\partial s}{\partial v}\right)_T = \frac{\beta}{k}. \tag{x}$$

Also, replacing the partial derivative $\left(\frac{\partial u}{\partial T}\right)_v$ by the molar specific heat at constant volume ç, we write Eq. (iii) as

$$\left(\frac{\partial s}{\partial T}\right)_v = \frac{c_v}{T}. \tag{xi}$$

Hence, the substitution of Eqs. (x) and (xi) in Eq. (i) gives

$$ds = \frac{c_v}{T}\,dT + \frac{\beta}{k}\,dv. \tag{42}$$

For an ideal gas, β/k is obtained as follows

$$Pv = RT, \qquad \text{per mole}$$

$$P\,dv + v dP = RdT\ ,$$

therefore,

$$\left(\frac{\partial v}{\partial T}\right)_P = \frac{R}{P}, \text{ and } \beta = \frac{R}{Pv} = \frac{1}{T}$$

$$\left(\frac{\partial v}{\partial P}\right)_T = -\frac{v}{P}, \text{ and } k = \frac{1}{P},$$

so

$$\frac{\beta}{k} = \frac{P}{T} = \frac{R}{v}.$$

Now substituting for $\frac{\beta}{\kappa}$ in Eq. (42) by $\frac{R}{v}$, we have

$$ds = c_V\frac{dT}{T} + R\frac{dv}{v}.$$

The integration of this equation gives

$$s - s_o = c_V \ln\frac{T}{T_o} + R\ln\frac{v}{v_o},$$

or

$$s = s_o + 3/2\ R\ln\frac{T}{T_o} + R\ln\frac{v}{v_o}, \tag{43}$$

where c_v is replaced for the monatomic ideal gas by $3/2\ R$. This is the equation we are looking for.

Comparison of Eq. (43), which is obtained from *classical thermodynamic,* with Eq. (41), namely,

$$s = R \ln v + 3/2\, R \ln T + R A, \qquad (41)$$

which is obtained above using *quantum statistics,* shows that the two expressions for *s* are identical as far as the dependence on temperature and volume is concerned. In Eq. (39), there are no arbitrary or undetermined constants. This brings up the question whether this equation can ever be verified experimentally since, apart from this equation, we have developed no other explicit expression for the entropy of a gas. Thermodynamic reasoning defines only entropy differences. The absolute value of the entropy of a system is a matter of the greatest importance, and we shall be able to go into detail in Chapters 7 and 20. Suffice it to say that there are methods of determining absolute values of entropy, and Eq. (39) seems to hold within the limits of experimental accuracy.

Utilising the equation of state of an ideal gas, we obtain from Eq. (39)

$$S = n R \left[\ln V + 3/2 \ln T + 3/2 \ln \left(2 \pi m k \right) - 3 \ln h - \ln N + 5/2 \right], \qquad (39)$$

the following alternative equation for the *molar specific entropy*, $S/n = s$,

$$s = R \left[5/2 \ln T - \ln P + 5/2 \ln k + 3/2 \ln \left(2\pi m \right) - 3 \ln h + 5/2 \right], \qquad (40)$$

in which we substituted for $\ln V$ by

$$\ln V = \ln N k T - \ln P$$
$$= \ln N + \ln k + \ln T - \ln P,$$

utilising the ideal gas equation

$$PV = N k T.$$

Equation (40) is known as the *Sackur – Tetrode equation* for the absolute molar specific entropy of a monatomic ideal gas.

9.7. The barometric equation

In the preceding section, the energy of an atom was considered to be wholly kinetic. Now let as consider the effect of *a gravitational fore field*, as illustrated by the vertical distribution of molecules in the earth's atmosphere. We take the origin of space coordinates at the earth's surface, with the *z* axis vertically upward, and consider a column of air of horizontal cross section A and uniform temperature *T*. Near the earth's surface the temperature decreases with increasing elevation, but it is fairly constant in the *stratosphere*, which is a layer of the earth's atmosphere between about 10 and 60 km above the earth's surface. A particle in a cell in this column whose vertical position coordinate is *z* then has a gravitational potential energy $m g z$ in addition to its kinetic energy $m v^2/2$; so

$$\epsilon = m g z + m v^2/2,$$

the acceleration g due to gravity being essentially constant in the atmosphere.

The partition function *Z* which is generally given by

$$Z = \sum g_i \, e^{-\epsilon_i/kT}$$

can be determined in this case as follows:

$$\begin{aligned} g_i &= \frac{H}{h^3} \\ &= \frac{1}{h^3} \, dx \, dy \, dz \, dv_x \, dv_y \, dv_z \qquad (1) \\ &= \frac{m^3}{h^3} \, dx \, dy \, dz \, dp_x \, dp_y \, dp_z \end{aligned}$$

This result is substituted into the preceding equation, and ϵ_i is replaced by $\left(m g z + \frac{1}{2} m v^2\right)$, and then the summation is replaced for, by integrals to give

$$\begin{aligned} Z &= \frac{m^3}{h^3} \iiiiiint e^{-mgz/kT} \, . \, e^{-mv^2/2kT} \, dx \, dy \, dz \, dv_x \, dv_y \, dv_z \\ &= \frac{m^3}{h^3} \iint dx\,dy \int e^{-mgz/kT} \, dz \\ &\quad \times \int_{-\infty}^{\infty} e^{-mv_x^2/2kT} \, dv_x \, . \int_{-\infty}^{\infty} e^{-mv_y^2/2kT} \, dv_y \\ &\quad \times \int_{-\infty}^{\infty} e^{-mv_z^2/2kT} \, dv_z \end{aligned}$$

The double integral over *x* and *y* gives the area of the horizontal cross section A. The integral over *z* between the limits $z = 0$ and $Z = \infty$, yields

$$\begin{aligned} \int_0^{\infty} e^{-mgz/kT} &= \left[-e^{-mgz/kT} \, . \, \frac{kT}{mg} \right]_0^{\infty} \\ &= -(0-1) \frac{kT}{mg} = \frac{kT}{mg}. \end{aligned}$$

Each of the single integral is $\left(\frac{2\pi kT}{m}\right)^{1/2}$

Hence,

$$\begin{aligned} Z &= \frac{A}{h^3} . \frac{m^3 \, kT}{mg} . \left(\frac{2\pi k T}{m^{3/2}}\right)^{3/2} \\ &= \frac{A}{h^3} . \frac{kT}{mg} . (2\pi m k T)^{3/2}. \qquad (2) \end{aligned}$$

The step that follows after the determination of the partition function *Z* is the determination of the number of particles N_i in the *i*th cell. For this purpose, we begin with the general expression for N_i using the *Maxwell-Boltzmann statistics* as a limiting case of the *Bose-Einstein distribution function* when $N_i/g_i \ll 1$, this expression being

$$N_i = \frac{N}{Z} g_i \, e^{-\epsilon_i/kT} \qquad (3)$$

as shown before. Thus, substituting for Z in this equation by Eq. (2), for g_i by Eq. (1), and then replacing N_i by d^6N, we find

$$d^6N = \frac{N}{A\,.\,kT} \cdot \frac{h^3 m g}{(2\pi m k T)^{3/2}} \cdot \frac{1}{h^3} \,.\, m^3$$

$$\times\ e^{\left(-\frac{m g z + m v^2/2}{k T}\right)} \quad dx\ dy\ dx\ dv_x\ dv_y\ dv_z$$

$$= \frac{N m g}{A.kT}\left(\frac{m}{2\pi kT}\right)^{3/2} e^{\left(-\frac{m g z + mv^2/2}{kT}\right)} \quad dx\ dy\ dx\ dv_x\ dv_y\ dv_z$$

$$= \frac{N m g}{A.kT}\left(\frac{m}{2\pi kT}\right)^{3/2} e^{-m g z/kT}\ dx\ dy\ dz \times e^{-m v^2/2kT}\ dv_x\ dv_y\ dv_z. \qquad (4)$$

Now we want to find out the *distribution of particles in ordinary space*, we integrate this equation over all values of v_x, v_y, and v_z; this gives

$$d^3N = \frac{N m g}{A.kT}\left(\frac{m}{2\pi kT}\right)^{3/2} e^{-mgz/kT}\ dx\ dy\ dz$$

$$\times \left(\frac{2\pi kT}{m}\right)^{3/2}$$

$$= \frac{N m g}{A k T} \,.\, e^{-m g z/kT}\ dx\ dy\ dz,$$

or

$$\frac{d^3N}{dx\,dy\,dz} = \frac{N m g}{A k T}\ e^{-m g z/k T}.$$

The left-hand side of this equation is the number of particles per unit volume, $n\left(=\frac{N}{V}\right)$. But $PV = N k T$, or $P = n k T$, then the pressure *P* at a height *z* is given according to this equation by

$$P = \frac{d^3N}{dx\,dy\,dz} k T,$$

$$= \frac{N m g}{A k T} \,.\, k T\ e^{-m g z/k T}$$

$$= \frac{N m g}{A}\ e^{-m g z/k T}.$$

When $z = 0$, the pressure at the earth's surface denoted by P_o is

$$P_o = \frac{N m g}{A},$$

which is evidently correct. This is because $N m g$ is the total weight of all molecules in the column. We can, therefore, write

$$P = P_0 \, e^{-m g z/kT} \qquad (5)$$

This equation is known as the *law of atmospheres.* It can also be derived from the principles of hydrostatics and the equation of state of an ideal gas.

It can be easily shown that at any height *z*, the distribution in velocity space has the same form as that in the *absence of gravitational force field.* Thus let us consider again Eq. (4), namely

$$d^6N = \frac{N m g}{A k T}\left(\frac{m}{2\pi k T}\right)^{3/2} e^{-m g z/kT} \, dx \, dy \, dz$$
$$\times \, e^{-m v^2/2kT} \, dv_x \, dv_y \, dv_z,$$

and integrate it over values of *x*, *y*, and *z*, the limits being chosen as shown above between 0 and ∞ for *z*.

$$d^3N = \frac{N m g}{A k T}\left(\frac{m}{2\pi k T}\right)^{3/2} \iint dx \, dy$$
$$\times \int_0^\infty e^{-m g z/kT} \, dz \, . \, e^{-mv^2/2kT} \, dv_x \, dv_y \, dv_z$$
$$= \frac{N m g}{A k T}\left(\frac{m}{2\pi k T}\right)^{3/2} . A . \frac{kT}{mg} .$$
$$e^{-m v^2/2kT} \, dv_x \, dv_y \, dv_z$$
$$= N\left(\frac{m}{2\pi k T}\right)^{3/2} e^{-m v^2/2kT} \, dv_x \, dv_y \, dv_z .$$

Evidently, this equation which represents the distribution in velocity space is identical with Eq. (34), obtained in Section 6, for a monatomic ideal gas for which we neglect any effect of a gravitational force field. This means that distribution in ordinary place and velocity space are *independent* so long as the expression for the energy ϵ can be written as a *sum of terms*, one of which contains only the *space or position coordinates* and the other only the *velocity coordinates.*

9.8. The principle of equipartition of energy

The principle of equipartition of energy was first obtained merely as *an inference from some of the results of the kinetic theory of an ideal gas.* Now we want to show how this principle follows from the *Maxwell-Boltzmann statistics*, and what its *limitations* are.

The energy ϵ of a particle is, in general, a function of all the coordinates of the cell in phase space in which it is located. Let *z* represent any arbitrary coordinate and ϵ_z the energy associated with that coordinate. If the distribution of particles in phase space can be represented by a *continuous* function of the coordinates, as in the preceding sections, the distribution in the coordinate *z* can be obtained by setting up the general expression for the distribution or partition function Z and integrating over all coordinates *except z*. The result will have the form

$$Z = \Sigma g_i \; e^{-\epsilon_i/kT}$$

$$g_i = \frac{H}{h^3}$$

$$= \frac{m^3}{h^3} \, dx \; dy \; dz \; dv_x \; dv_y \; dv_z$$

$$Z = \frac{m^3}{h^3} \int\int\int\int\int\int e^{-\epsilon_i/kT} \, dx\, dy\, dz\, dv_x \, dv_y \, dv_z \; .$$

Subsequently, integrating over all coordinates except z we obtain

$$Z = A \int e^{-\epsilon_z/kT} \; dz,$$

$$= A \; Z_z,$$

where Z_z is given by

$$Z_z = \int e^{-\epsilon_z/kT} \; dz \; ,$$

and A is *not* a function of z.

We want now to determine the total energy U_z associated with the coordinate z. According to Eq. (26), derived before in Section 6, the total internal energy of a system is

$$U = \frac{NkT^2}{Z} \cdot \frac{dZ}{dT}$$

$$= NkT^2 \; \frac{d(\ln z)}{dT},$$

therefore, in our case U_z is given by

$$U_z = NkT^2 \frac{d}{dT} \ln \left[\int e^{-\epsilon_z/kT} \, dz \right].$$

The mean energy $\bar{\epsilon}_z$ of a single particle associated with the coordinate z is

$$\bar{\epsilon}_z = U_2/N$$

If the energy ϵ_z is a *quadratic function* of the coordinate z, that is, it has the form $\epsilon_z = a z^2$, where a is a constant, and if the limits on the integration of z are from 0 to ∞ or from $-\infty$to ∞, then we have

$$\bar{\epsilon}_z = \frac{U_z}{N} = kT^2 \frac{d}{dT} \ln \left[\int_{-\infty}^{\infty} e^{a z^2/kT} \, dz \right] .$$

But

$$\int_{-\infty}^{\infty} z^n \, e^{-a z^2} \, dz = \sqrt{\frac{\pi}{a}} \qquad\qquad \text{if } n = 0$$

therefore,

$$\int_{-\infty}^{\infty} e^{-az^2/kT} \, dz = \sqrt{\frac{\pi k T}{a}},$$

and

$$\bar{\epsilon}_z = kT^2 \frac{d}{dT} \ln \sqrt{\frac{\pi k T}{a}}$$

$$= kT^2 \frac{d}{dT}\left\{\frac{1}{2}(\ln\pi + \ln k + \ln T - \ln a)\right\}.$$

Hence,

$$\bar{\epsilon}_z = kT^2 \cdot \frac{1}{2} \cdot \frac{1}{T} = \frac{1}{2} kT.$$

This means that for every coordinate, for which the conditions pointed out above are fulfilled, *the mean internal energy per particle, in an assembly of particles in thermodynamic equilibrium at a temperature T, is* $1/2$ *kT*. This is the general statement of the *equipartition principle*.

The conditions above are fulfilled for the *translational velocity coordinates* v_x, v_y, and v_z, since the kinetic energy associated with, say v_x, is $1/2\, m v_x^2$ and the range of each is from $-\infty$ to ∞. They are also fulfilled for the *displacement x of a simple harmonic oscillator*, since the potential energy associated with *x* is $1/2\, F x^2$, where *F* is the *force constant*. However, these conditions are *not* fulfilled for the vertical coordinate *z* of a gas in a gravitational field, where the potential energy $m g z$ is a linear function of *z*. This explains that the *mean gravitational potential energy is not* $1/2\, kT$. Also they are not fulfilled for the energies associated with *molecular rotation, vibration* and *electronic excitation*. This is because the spacing of the energy levels is a *significant fraction of kT*, so the *summation over the energy levels cannot be replaced by an integral*.

9.9. Theory of paramagnetism

Consider a system of *n* particles per unit volume, each of which has a spin quantum number *s*, and an associated magnetic moment of magnitude μ. The particles may be *electrons*, *atoms* having *one unpaired electron*, or *nuclei* of *odd mass number*.

The rotation or spin of an electron generates a spin angular momentum *vector* **s** given by

$$\mathbf{s} = \sqrt{s(s+1)}\, \frac{h}{2\pi}, \tag{1}$$

where *s* is the spin quantum number whose value is equal to $1/2$, so that

$$\mathbf{s} = \frac{\sqrt{3}}{2} \frac{h}{2\pi} \ . \tag{2}$$

The axis of spin of an electron is the axis about which the electron has an axial symmetry. The spin angular momentum vector *coincides* with the spin axis. The orientation of **s** is *quantised with respect to an axis fixed in space*. The spin of an electron also generates a magnetic moment; that is, the electron behaves like a small bar magnet whose axis *coincides* with the *axis of the spin*. According to quantum electrodynamics the absolute magnetic moment vector μ of the electron is *proportional* to the product of the spin angular momentum $\mathbf{s} = \frac{\sqrt{3}}{2}\frac{h}{2\pi}$ times $e/2m_e c$, the factor of proportionality being dented by g_e. Thus the magnitude of the vector μ is given by

$$\mu = g_e \frac{\sqrt{3}}{2}\frac{e}{2m_e c}\frac{h}{2\pi}. \tag{3}$$

Here m_e is the mass of the electron, e its charge c velocity of light and g_e is an irrational* number. The group of universal constants $\frac{e}{2m_e c}\frac{h}{2\pi}$ is known as the *Bohr magneton* which is denoted by μ_B. Thus in terms of μ_B, the absolute magnetic moment μ of the electron is given by

$$\mu = g_e \mu_B \frac{\sqrt{3}}{2} . \tag{4}$$

The spin angular momentum vector ***s*** and the spin magnetic moment vector μ both *coincide* with the spin axis of rotation. The orientation of both vectors is quantised with respect to an axis fixed in space. In this situation, we take the direction of an applied magnetic field as the direction of the space-fixed axis denoted by z. Theory allows the component s_z of ***s*** along z to assume generally any of the values

$$s_z = M_s\, h/2\pi , \tag{5}$$

where M_s a *second quantum number adopting the values* $2s + 1 = 2$.

Since the quantum number s of the spin angular momentum vector ***s*** is equal here to $1/2$ the quantum number M_s has the two values $+1/2$ and $-1/2$. Hence, the vector ***s*** is allowed *two discrete orientations* with respect to the space-fixed axis z. These orientations are

$$s_z = \frac{1}{2}\cdot\frac{h}{2\pi} = +\frac{h}{4\pi} \tag{6}$$

and

$$s_z = -\frac{1}{2}\cdot\frac{h}{2\pi} = -\frac{h}{4\pi}. \tag{7}$$

Since the absolute magnetic moment vector μ coincides with the spin angular momentum vector ***s***, then it follows that the former vector μ is also allowed two discrete orientations with the z axis when the quantum number s is equal to $1/2$.

Let the angle between the spin angular momentum vector of the electron and the space-fixed axis z be θ_i, therefore,

$$\cos\theta_i = \frac{s_z}{s}$$

---

* An irrational number cannot be expressed by the quotient of two integers; 3/4, 5/6, 22/7, etc. are rational numbers, whereas $\sqrt{2}$, $\sqrt{3}$, 5.585, etc. are irrational numbers.

$$= \frac{M_s}{\sqrt{s(s+1)}} \tag{8}$$

The simultaneous quantisation of ***s*** and s_z along the axis *z* means that the orientation θ_i of the spin angular momentum vector with respect to this axis is also quantised. Thus the angle, θ_i, can assume only the following discrete values when $s = 1/2$.

$$\cos\theta_i = \frac{1/2}{\sqrt{3}} = \frac{1}{2\sqrt{3}} \tag{9}$$

$$\cos\theta_i = \frac{-1/2}{\sqrt{3}} = -\frac{1}{2\sqrt{3}}. \tag{10}$$

Since the spin magnetic moment μ of the electron coincides with the axis of spin as does the spin angular momentum ***s***, the z components of μ are given by

$$\mu_z = \mu\cos\theta_i$$

$$= g_e \mu_B \frac{\sqrt{3}}{2}\frac{1}{2\sqrt{3}}$$

$$= \frac{1}{4} g_e \mu_B, \tag{11}$$

and

$$\mu_z = -\frac{1}{4} g_e \mu_B. \tag{12}$$

These two equations may be compared with Eqs. (6) and (7) representing the components s_z of the spin angular momentum ***s*** along the *z* axis.

Taking the direction of the space-fixed axis *z* as the direction of the applied magnetic field *B*, the component of the spin angular momentum s_z is either $+h/4\pi$ or $-h/4\pi$ as given by Eqs. (6) and (7), and the corresponding values of the magnetic moment component of the particle is either $+\mu$ or $-\mu$. For $+\mu_z$, the magnetic moment component is said to be *parallel* to the applied magnetic field, and for $-\mu_z$ the magnetic moment component is said to be *antiparallel* to the applied filed.

We now compute the number of particles per *unit volume* aligned parallel and antiparallel to the field. The energy of a particle whose magnetic moment μ_z is *parallel* to the field whose magnetic flux density is *B*, is

$$\epsilon_1 = -\mu B\cos\theta_i = -\mu_z B, \tag{13}$$

and if the magnetic moment μ_z is *antiparallel* to the field, the energy is

$$\epsilon_2 = -\mu B\cos\theta_i = +\mu_z B. \tag{14}$$

Now, since the are only two possible energy states, the partition function *Z*, considering only the magnetic energy, is

$$Z = \Sigma g_i e^{-\epsilon_i/kT}$$

$$= g_1 e^{\mu_z B/kT} + g_2 e^{-\mu_z B/kT}.$$

Taking for simplicity $g_1 = g_2 = 1$, and *replacing* $\mu_z B/kT$ by *x*, the partition function *Z* reduces to

$$Z = e^x + e^{-x} = 2\cosh x. \qquad (15)$$

We recall that the number of particles in *i*th cell, in the state of maximum thermodynamic probability, that is, the state of thermodynamic equilibrium, is given by

$$N_i = \frac{N}{Z} g_i e^{-\epsilon_i/kT}. \qquad (16)$$

Therefore, in our situation, the number of particles per unit volume, n_i, in each state using Eqs. (13) and (14) is given by

$$n_1 = \frac{n}{2\cosh x} e^{\mu_z B/kT},$$

$$= \frac{n}{2\cosh x} e^x \quad \text{(parallel to } B\text{)}, \qquad (17)$$

and similarly,

$$n_2 = \frac{n}{2\cosh x} e^{-x} \quad \text{(antiparallel to } B\text{)}. \qquad (18)$$

The total *magnetisation* parallel to *B* is $n_1\mu_z$, and that antiparallel to *B* is $n_2\mu_z$ so that the net magnetisation *M* in the direction of *B* is

$$M = n_1\mu_z - n_2\mu_z$$

$$= \frac{n}{2\cosh} \mu_z \left(e^x - e^{-x}\right)$$

$$= n\mu_z \frac{e^x - e^{-x}}{e^x + e^{-x}}$$

$$= n\mu_z \tanh x$$

$$= n\mu_z \tanh \frac{\mu_z B}{kT}. \qquad (19)$$

Let us investigate the limiting value of Eq. (19) in the two *extreme cases*:

1. In *strong* fields and at low temperatures when $x = \mu_z B/kT$ is very large.

2. In *weak* fields and at *high temperatures* when x is very small.

When $x >> 1$,

$$\tan h\ x \approx 1,$$

so that Eq. (19) reduces to

$$M_{sat} \approx n\mu_z. \qquad (20)$$

This is simply the *saturation magnetisation* which occurs when all the magnetic moments μ_z of the particles are *aligned parallel to the magnetic field.*

In *weak fields and at high temperatures*, $x << 1$, so that we can make the approximation

$$\tanh x \approx x ,$$

or,

$$\tanh \frac{\mu_z B}{kT} \approx \frac{\mu_z B}{kT}.$$

Hence, Eq. (19) becomes

$$M = n\mu_z \left(\frac{\mu_z B}{kT}\right)$$

$$= \frac{n\mu_z^2}{kT} B \tag{21}$$

or

$$= \frac{n\mu_z^2 \mu_0 H}{kT}. \tag{22}$$

Here μ_0 is the permeability of free space, and H is the magnetic intensity; and since the magnetisation is so small in all *paramagnetic materials*, we have set $B = \mu_0 H$.

Experimental measurements of magnetisation show that for many substances, over a wide range of magnetic intensity and temperature, magnetisation is directly proportional to H and inversely proportional to T; that is,

$$M = C\frac{H}{T}. \tag{23}$$

This equation is known as the *Curie's law*, and the proportionality constant C is called the *Curie constant*. The experimental analysis shows that Curie's law is valid *for weak fields and high temperatures*, and that the constant C, on comparing the empirical Eq. (23) with Eq. (22), is given by

$$C = \frac{n(\mu_z)^2 \mu_0}{k}. \tag{24}$$

The curve in Fig.1. is a graphical representation of Eq. (19) rewritten in the form, which is favoured in weak fields and at high temperature,

$$M = n\mu_z\left(\frac{\mu_z B}{kT}\right).$$

In this figure, $\frac{M}{n\mu_z}$ is plotted against $\frac{\mu_z B}{kT}$. The straight dashed line is the tangent at $x\left(=\frac{\mu_z \beta}{kT}\right)=0$, where Eq. (19) reduces to Curie's law expressed by Eq. (23).

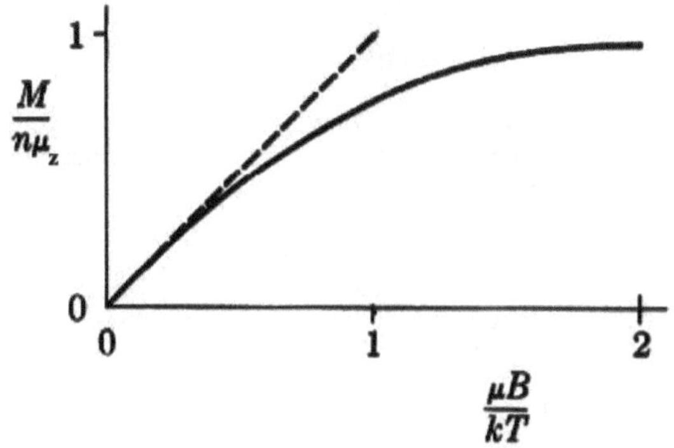

Fig. 1. Dependence of the magnetisation on the magnetic field; Curie's law is given by a dashed live.

In a region where Curie's law is obeyed, Eq. (24) can be used to compute the magnetic moment component along the direction of the applied magnetic field, which is μ_z. The values obtained are of the order of magnitude of a few *Bohr magnetons. A Bohr magneton,* μ_B, *is the magnetic moment of an electron revolving in the Bohr orbit of atomic hydrogen, and it is also equal to the magnetic moment of an electron arising from its spin or rotation, about its own axis.* According to quantum electrodynamics, the absolute magnetic moment μ of an electron is given by Eq. (3), as shown above, namely

$$\mu = g_e \frac{\sqrt{3}}{2} \frac{e}{2m_e c} \frac{h}{2\pi}. \tag{3}$$

and the group of the universal constants

$$\frac{e}{2m_e c} \frac{h}{2\pi} = \mu_B$$

is the *Bohr magneton,* μ_B; its magnitude in mks system is

$$\mu_B = 0.9274 \times 10^{-23} \text{ amp.m}^2,$$

and in cgs system is

$$\mu_B = 0.9274 \times 10^{-20}\ \text{erg. gauss}^{-1},$$

since 1 amp. m^2 is equal to 10^3 erg gauss^{-1} as will be shown below.

Thus, for oxygen gas, which is paramagnetic and obeys Curie's law down to very low temperatures, experiments have shown that

$$\mu_z \approx 3\mu_B$$

Now we can estimate the extent to which the molecular magnets in oxygen as an example are aligned by an external field. If Curie's law is obeyed, the ratio of actual magnetisation *M*, given above by Eq. (21),

$$M = \frac{N\mu_z^{\ 2}}{Kt} B, \tag{21}$$

to the magnetisation at saturation given by

$$M_{set} = n\mu_z, \tag{20}$$

which indicates

$$\frac{M}{M_{set}} = \frac{n\mu_z^{\ 2}}{kT} \cdot \frac{B}{n\mu_z}$$

$$= \frac{\mu_z B}{kT}$$

$$= x\ . \tag{25}$$

Let *B*, the magnetic flux density, = 2 weber*/m^2 (20.000 gauss), *T* = 300 K, and $\mu_z = 3\mu_B$. Then using cgs units, we have

$$\frac{M}{M_{sat}} = x = \frac{3 \times 0.927 \times 10^{-23} (\text{amp.m}^2) \times 2 \left(\frac{\text{weber}}{\text{m}^2}\right)}{1.38 \times 10^{-23} (\text{joule/K}) \times 300\ (\text{K})}$$

$$= 1.35 \times 10^{-2} \frac{[\text{amp.weber}]}{[\text{joule}]}$$

$$= 1.35 \times 10^{-2}.$$

Let us digress for a moment to make a remark on the units used here.

A remark on units used

We have, as we shall see in detail in Appendix B concerning dimensions and units that

$$1\,\text{ampere} \times 1\,(\text{metre})^2 = 1\left\{\frac{\text{joule}}{\text{weber}}\right\}(\text{metre}^2), \tag{i}$$

or

$$1\,\text{ampere} = \frac{1(\text{joule})}{1(\text{weber})}, \tag{ii}$$

that is,

$$1 \text{ ampere} \times 1 \text{ weber} = 1 \text{ joule}. \qquad \text{(iii)}$$

Also, that

$$\frac{1\,\text{weber}}{1\,(\text{metre})^2} = 1\,\text{gauss} \times 10^4, \qquad \text{(iv)}$$

therefore, rewriting Eq. (i) in the form

$$1 \text{ ampere} \times 1\,(\text{metre})^2 = \frac{1 \text{ joule}}{1\,\text{weber}/1\,(\text{metre})^2},$$

we have

$$1\,\text{ampere} \times 1\,(\text{metre})^2 = \frac{10^7 \text{ erg}}{10^4 \times \text{gauss}}$$

$$= 10^3 \text{ erg gauss}^{-1}.$$

The weber is the unit of magnetic flux, and it is related to the gauss by the relation

$$1 \text{ unit of magnetic flux density} = 1 \text{ tesla} = \frac{1\,\text{weber}}{1\,\text{m}^2} = 10^4 \text{ gauss}.$$

Now we return to the estimation of the extent to which the molecular magnets in oxygen gas are aligned by an external field. Under the conditions assumed and taking M_{sat} as equal to $n\mu_z$, we see that even a strong magnetic field is capable of producing only a relatively small alignment, equivalent to that which would result of one or two molecules per thousand were exactly parallel to the magnetic field while the other molecules remained oriented at random. In other words, *at a temperature of 300 K the disorienting effect of thermal motions far outweighs the aligning influence of the field.*

9.10. The statistics of a photon gas

One of the most important applications of the Bose-Einstein statistics is the study of radiation. It is a familiar fact that a hot body becomes luminescent and emits *radiation in the form of heat*. It is also known that the thermal radiation consists of electromagnetic waves. This heat radiation is able to travel through a vacuum as readily as through a material medium. The thermal radiation emitted at any temperature is distributed over a continuous spectrum of wavelengths, the distribution changing with temperature. *When this radiation is in thermodynamic equilibrium, it is known as blackbody radiation.*

Blackbody radiation can be considered to be *a 'gas' made up of photons.* The photons can be treated statistically as particles; however, in this case the *number of particles (photons) is not conserved.* The constraint applied in the derivation of the *Bose-Einstein* distribution function that the total number of particles be constant, *must be relaxed in order to treat a photon gas statistically.* The condition for maximum thermodynamic probability and the constraint on the total internal energy must be applied. Multiplying the equation

$$\partial U = \sum \epsilon_i \,\delta N_i = 0 \qquad (1)$$

by the constant $-\beta$, and adding this to the equation

$$\sum \ln\left(\frac{g_i + N_i^\circ}{N_i^\circ}\right)\delta N_i = 0, \qquad (2)$$

we obtain

$$\sum\left[\ln\left(\frac{g_i + N_i^\circ}{N_i^\circ}\right) - \beta\epsilon_i\right]\delta N_i = 0; \qquad (3)$$

and, since in effect the δN_i's are arbitrary, it is necessary that

$$\ln\left(\frac{g_i + N_i^\circ}{N_i^\circ}\right) - \beta\epsilon_i = 0$$

or

$$\frac{N_i^\circ}{g_i} = \frac{1}{e^{\beta\epsilon_i} - 1} \qquad (4)$$

This is the *distribution function for a photon gas*. Once again we identify

$$\beta = 1/kT.$$

The energy ϵ of a photon of frequency ν is

$$\epsilon = h\nu,$$

and its momentum, *p*, is

$$p = \frac{\epsilon}{c} = \frac{h\nu}{c},$$

where *c* is the velocity of light. The degeneracy or statistical weight, g_i, is the number of compartments in the *i*th cell, and for photons it is given by

$$g_i = \frac{2}{h^3}\, dx\, dy\, dz\, dp_x\, dp_y\, dp_z, \qquad (5)$$

where the factor 2 is equivalent to doubling the number of compartments per cell; it must be included in order to account for *both* right-handed and left-handed polarised photons since light can have either polarisation *independently*.

The number of photons N_i° in cell *i* is given by

$$N_i^\circ = g_i \frac{1}{e^{\beta\epsilon_i - 1}},$$

and the substitution for g_i in this equation by the right-hand side of Eq. (5) gives in the state of maximum thermodynamic probability.

$$N_i^\circ = \frac{2}{h^3}\frac{1}{e^{\beta\epsilon_i} - 1}\, dx\, dy\, dz\, dp_x\, dp_y\, dp_z.$$

Replacing in this equation the number of photons N_i in cell *i* by d^6N, and the constant β by $1/kT$, we get

$$d^6N = \frac{2}{h^3}\frac{1}{e^{\beta\epsilon_i/kT}-1}\,dx\,dy\,dz\,dp_x\,dp_y\,dp_z.$$

Since the energy of a photon in cell *i* is equal to the product of momentum and velocity of light, as shown above, the preceding equation takes the form

$$d^6N = \frac{2}{h^3}\frac{1}{e^{Pc/kT}-1}\,dx\,dy\,dz\,dp_x\,dp_y\,dp_z. \qquad (6)$$

The integration of this equation over *x*, *y*, and *z*, the limits being chosen so as to include the space occupied by the photon gas, gives the total volume *V*, and the resulting equation

$$d^3N = V\frac{2}{h^3}\frac{1}{e^{Pc/kT}-1}\,dp_x\,dp_y\,dp_z,$$

gives the distribution of photons in momentum space. The division of this equation by *V* yields the equation

$$\frac{d^3N_{P_xP_yP_z}}{V} = \frac{2}{h^3}\frac{1}{e^{Pc/kT}-1}\,dp_x\,dp_y\,dp_z, \qquad (7)$$

which represents the distribution of photons in momentum space per unit volume of ordinary space. This distribution function is spherically symmetric, that is, isotropic, owing to the fact that it only contains *p*, so the number of photons in a thin spherical shell in momentum space, of radius *p* and thickness d*p*, per unit volume of ordinary space is

$$dn_p = \frac{2}{h^3}\frac{4\pi p^2\,dp}{e^{Pc/kT}-1}. \qquad (8)$$

Finally, we want to express the distribution function in terms of frequency rather than momentum, through the relations

$$p = \frac{h\nu}{c}, \quad dp = \frac{h}{c}d\nu,$$

we obtain

$$dn_\nu = \frac{8\pi}{c^3}\frac{\nu^2\,d\nu}{e^{h\nu/kT}-1}. \qquad (9)$$

According to Eq. (9), dn_ν represents the number of photons *per unit volume* of ordinary apace, and since the energy of a photon of a frequency ν is $h\nu$, then the energy *per unit volume* (that is, the energy density) *within* a frequency range (or bandwidth) $d\nu$, taking into consideration that u_ν is the energy density per unit frequency, is given by

$$u_\nu\,d\nu = h\nu\,dn_\nu$$

$$= h\nu\frac{8\pi}{c^3}\frac{\nu^2\,d\nu}{e^{h\nu/kT}-1}.$$

Here, dn_ν is substituted for by the right-hand side of Eq. (9). Thus

$$u_\nu \, d\nu = \frac{8\pi h}{c^3} \frac{\nu^3 \, d\nu}{e^{h\nu/kT} - 1}. \tag{10}$$

This is known as the *Planck distribution.* The energy density per unit frequency and volume is a function of frequency, temperature, and fundamental constants. This distribution, in *dimensionless form*, is plotted in Fig. 2

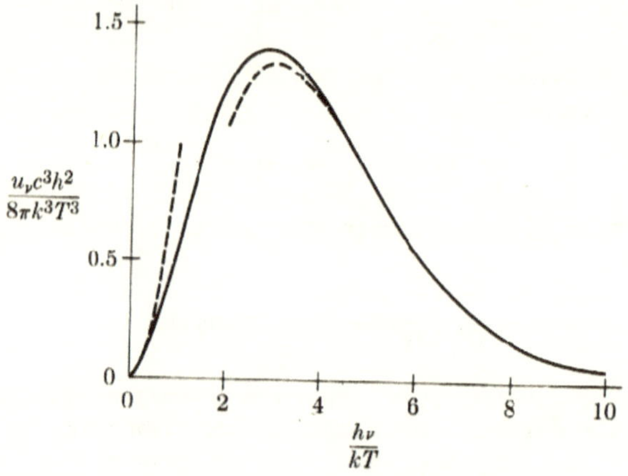

Fig. 2. The Planck distribution of energy density on frequency. The dashed lines are the Rayleigh-Jeans formula at low frequencies and Wien's displacement law at high frequencies.

As the temperature increases, the radiation shifts to higher frequencies (from red to blue), as noted when observing hot bodies.

At low frequencies, $h\nu/kT << 1$, the exponential, function in the denominator of Eq. (10) can be expanded, with the result that

$$\begin{aligned} u_\nu &= \frac{8\pi h}{c^3} \frac{\nu^3}{\{(1 + h\nu/kT + ...) - 1\}} \\ &= \frac{8\pi h}{c^3} \frac{\nu^3 \, kT}{h\nu} \\ &= \frac{8\pi\nu^2}{c^3} kT + ... \end{aligned} \tag{11}$$

This is known as the *Rayleigh–Jeanes formula.*

At high frequencies, $h\nu/kT >> 1$, so the exponential in the denominator in Eq. (10) is much greater than unity. Hence, Eq. (10) reduces to

$$u_\nu = \frac{8\pi h \nu^3}{c^3} e^{-\frac{h\nu}{kT}} \quad . \tag{12}$$

This is known as *Wein's displacement law*. Both the *Rayleigh-Jeans formula* and the *Wien's displacement law* are compared with the *Planck distribution* in Fig. 2.

The total energy per unit volume or energy density of a radiation field is obtained by integrating Eq. (10) over all frequencies. Thus

$$u = \int_0^\infty u_\nu \, d\nu = \frac{8\pi h}{c^3} \int_0^\infty \frac{\nu^3}{e^{h\nu/kT} - 1} \, d\nu .$$

This gives

$$u = \frac{4\sigma T^4}{c}, \tag{13}$$

where σ is a group of universal constants expressed as

$$\sigma = \frac{2\pi^5 k^4}{15 h^3 c^2}$$
$$= 5.670 \times 10^{-8} \text{ joule/m}^2 \, K^4 \, \text{sec}.$$

It is called the *Stefan–Boltzmann constant*. From Eq. (13), it is evident that the energy density is proportional to the fourth power of temperature; this is known as *Boltzmann's law*.

As regards the dimensionless form of the plot of the Planck distribution of energy shown in Fig. 2, it will be evident if the dimensions of the numerator and the denominator, for the ordinate, and for the abscissa, are determined as can be seen from the following:
For the ordinate we have

$$\frac{\dfrac{\text{joule}}{m^3} \text{ see.} \dfrac{m^3}{\text{sec}^3} . (\text{joule})^2 . (\text{sec})^2}{\left(\dfrac{\text{joule}}{K}\right)^3 . K^3} = \frac{(\text{joule})^3}{(\text{joule})^3},$$

and for the abscissa

$$\frac{\text{joule. sec.} \dfrac{1}{\text{sec.}}}{\dfrac{\text{joule}}{K} . K} = \frac{(\text{joule})}{(\text{joule})}.$$

PART II

CHAPTER 10

Fermi-Dirac Quantum Statistics

10.1. The Fermi-Dirac statistics

We can now consider the statistics of an electron gas, a system to which the *Pauli exclusion principle* applies. As before, let N_i be the number of phase points in the *i*th cell. In neither the *Maxwell-Boltzmann* nor the *Bose-Einstein* statistics was there an *a priori* restriction on the magnitude of any N_i, but there is a very definite restriction in the *Fermi-Dirac statistics*. The exclusion principle, as it applies to an electron gas, asserts that *there can be no more than two phase points in each of the compartments of volume* h^3. The same principle governs the arrangement of electrons in an atom; it states that *no two electrons in the same atom can have the same set of quantum numbers*. The *coordinates of a compartment* in phase apace correspond to *quantum numbers*. The reason that there can be two points *only* in one compartment is that the electrons which the points represent have oppositely directed spins. Therefore, the maximum number of representative points in a cell is *twice* the number of compartments; of course, the actual number may be less than this if not all the compartments are occupied. Let us imagine that each compartment is divided into two and that there can be *no more than one phase point in each half*. Therefore, the number of *half-compartments* in each cell, which can also be denoted as g_i is

$$g_i = 2\frac{H}{h^3}, \tag{1}$$

and the *maximum* number of phase points in each cell is equal to the *number of half-compartments*, that is, equal to g_i.

For brevity, we shall use the term 'compartment' for these half-compartments.

Let us take as a specific example a system with just two cells *i* and *j*, each divided into four compartments (bearing in mind that in fact each of these *four* compartments is a half-compartment that cannot accommodate more than one election) and consider the macrostate.

$$N_i = 3, \quad N_j = 1.$$

Fig. 1 shows the cells *i* and *j*, and we see that if there cannot

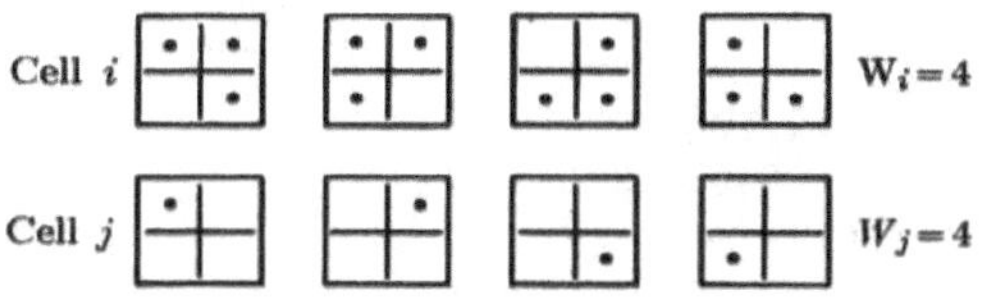

Fig.1. Different ways of arranging phase points within a cell of phase space in the Fermi-Dirac statistics.

be more than one phase point per compartment, there are *four* different ways of arranging the *three phase points* in cell *i*, and *four* ways of arranging the single point in cell *j*. Therefore, defining, as shown above, the thermodynamic probability of a cell as the number of different ways of arranging the particles within that cell, that is, the number of microstates of the cell, the thermodynamic probability W_i of cell *i* is four and that of cell *j*, W_J, is also four. For any one of the arrangements in cell *i*, we can have any one of those in cell *j* so that the number of possible arrangements, or the thermodynamic probability *W* of the given macrostate of having three electrons in cell *i* and one electron in cell *j*, and which is defined as the number of microstates corresponding to it, is

$$\begin{aligned} W &= W_i \; W_J \\ &= 4 \times 4 = 16 . \end{aligned}$$

This contrasts with $W = 80$, the result of Bose-Einstein statistics.

In general, when this is any number of cells which are occupied, we have

$$W = \Pi W_i, \tag{2}$$

where the product extends over all occupied cells of phase space.

Now we want to find out the general expression for any W_i. The derivation of this in the Fermi-Dirac statistics is much simpler than it is with the Bose-Einstein statistics as can be seen from the following.

Of the g_i compartments of a cell, N_i are occupied and $(g_i - N_i)$ are empty. The problem therefore consists of counting the number of ways *in which* g_i *compartments* can be divided into two groups, with the *occupied compartments* in one group and the *empty compartments* in the other.

We already know that the number of different ways in which *n indistinguishable objects* can be placed in two categories, m_1 in the first category and m_2 in the second category with $n = m_1 + m_2$ is

$$\frac{n!}{m_1! \, m_2!}. \tag{3}$$

The thermodynamic probability W_i for a given cell *i*, which is defined as the number of possible ways in which the phase points can be arranged in the compartments, can be defined in the *special case of Fermi-Dirac statistics* as the number of different ways the compartments in a given cell *i* can be divided into two groups, one of which is *the group of occupied compartments* and the other is the *group of unoccupied compartments*; the indistinguishable objects to be placed are the compartments of a cell so that in the above expression $m = g_i$. The number of objects in the first category is the number of *occupied* compartments, which is equal to the number of phase points in cell, that is,

$$m_1 = N_i .$$

The number of objects in the second category is the number of empty compartments, that is,

$$m_2 = g_i - N_i .$$

Hence, the number of ways of dividing the compartments in the *i*th cell into occupied and unoccupied compartments, or the thermodynamic probability W_i of the cell *i*, is

$$W_i = \frac{g_i!}{(g_i - N_i)!\, N_i!}. \tag{4}$$

In our simple example, $N_i = 3$, $N_j = 1$, $g_i = g_j = 4$. Hence,

$$W_i = \frac{4}{1!\,3!} = 4,$$

$$W_j = \frac{4}{3!\,1!} = 4,$$

which agrees with the result obtained by counting.

The general expression for the thermodynamic probability of a given macrostate in the Fermi-Dirac statistics is therefore

$$W = \Pi \frac{g_i!}{(g_i - N_i)!\, N_i!}. \tag{5}$$

Next, as in any type of statistics, we assume that the entropy is proportional to the logarithm of the thermodynamic probability, and that the equilibrium state of maximum entropy is also that for which $\ln W$ is a maximum or $\delta \ln W = 0$. From Eq. (5), we have

$$\ln W = \Sigma\left[\ln g_i! - \ln\left(g_i - N_i\right)! - \ln N_i!\right]. \tag{6}$$

Since the cells are large enough so that g_i and N_i are large numbers we can use Sterling's approximation so that Eq. (6) takes the form

$$\ln W = \Sigma\Big[g_i \ln g_i - g_i - \left(g_i - N_i\right)\ln\left(g_i - N_i\right) +$$
$$\left(g_i - N_i\right) - N_i \ln N_i + N_i\Big],$$

$$\ln W = \Sigma\Big[g_i \ln g_i - g_i \ln\left(g_i - N_i\right) + N_i \ln\left(g_i - N_i\right)$$
$$- N_i \ln N_i\Big]. \tag{7}$$

Now, applying the condition for maximum thermodynamic probability, we get

$$\delta \ln W^{\circ} = \Sigma\left[\left(g_i - N_i\right)\frac{1}{\left(g_i - N_i\right)} \times \delta N_i + \ln\left(g_i - N_i\right)\delta N_i\right.$$

$$\left. - N_i \times \frac{1}{N_i}\delta N_i - \ln N_i \, \delta N_i\right],$$

$$= \Sigma\left[\ln\left(g_i - N_i\right)\delta N_i - \ln N_i \, \delta N_i\right],$$

$$= \Sigma\left[\ln\left(\frac{g_i - N_i^{\circ}}{N_i^{\circ}}\right)\right]\delta N_i = 0 \,. \tag{8}$$

The quantities $d N_i$'s, which are due to the random motions of the electrons, are not independent, for they are subject to the conditions that the total number of electrons and the total energy remain constant. We, therefore, have the condition equations as follows:

$$\delta N = \Sigma \delta N_i = 0, \tag{9}$$

$$\delta U = \Sigma \epsilon_1 \delta N_i = 0 \,. \tag{10}$$

Using the method invented by *Lagrange*, as shown before, we can, in effect, obtain an equation in which the δN_1's are independent. Thus, we multiply the first condition equation, that is, Eq. (9), by a constant which, for convenience, we write as $\ln B$, and the second condition equation, that is, Eq. (10), by $-\beta$, and add the products to Eq. (8). This gives

$$\Sigma\left[\ln\left(\frac{g_i - N_i^{\circ}}{N_i^{\circ}}\right) - \ln B - \beta\epsilon_i\right]\delta N_i = 0, \tag{11}$$

and, since in effect the δN_i's are now independent, we must have

$$\ln\left(\frac{g_i - N_i^{\circ}}{N_i^{\circ}}\right) - \ln B - \beta\epsilon_i = 0,$$

or

$$\ln\left(\frac{g_i - N_i^\circ}{N_i^\circ}\right) = \ln B + \beta\epsilon_i ,$$

so that

$$\frac{g_i - N_i^\circ}{N_i^\circ} = B\, e^{\beta\epsilon_i} , \tag{12}$$

or

$$\frac{g_i}{N_i^\circ} - 1 = B e^{\beta\epsilon_i} .$$

Hence,

$$\frac{N_i^\circ}{g_i} = \frac{1}{B\, e^{\beta E_i} + 1} . \tag{13}$$

This is the *Fermi-Dirac distribution function* for the state of maximum thermodynamic probability. It should be compared with the corresponding equation obtained by the *Bose-Einstein statistics*

$$\frac{N_i^\circ}{g_i} = \frac{1}{B\, e^{\beta E_i} - 1} ,$$

and with that derived by the Maxwell-Boltzmann statistics

$$\frac{N_i^\circ}{g_i} = \frac{1}{B\, e^{\beta E_i}} .$$

The next step is to evaluate the quantities *B* and *β*. To determine *β*, we again make use of the thermodynamic relation that for a system in equilibrium, and for a process at constant volume, we have

$$dU = TdS ,$$

so that

$$\left(\frac{\partial S}{\partial U}\right)_V = \frac{1}{T} .$$

On the other hand, the substitution of Eq. (7) into the equation relating the entropy to the thermodynamic probability

$$S = k \ln W ,$$

gives

$$S = k \sum \Big[g_i \ln g_i - N_i \ln N_i - g_i \ln\left(g_i - N_i\right) + N_i \ln\left(g_i - N_i\right) \Big] . \tag{14}$$

If we assume that $g_i >> N_i$, then in this limit the following approximation can be made:

$$\ln\left(g_i - N_i\right) = \ln\left[g_i\left(1 - \frac{N_i}{g_i}\right)\right]$$

$$= \ln g_i + \ln\left(1 - \frac{N_i}{g_i}\right)$$

$$= \ln g_i - \frac{N_i}{g_i} + \ldots$$

Substitution of this result into Eq. (14) gives

$$S = \Sigma\left[k\left(g_i \ln g_i - N_i \ln N_i\right) - \left(g_i - N_i\right)\right.$$

$$\left.\times \ln\left(g_i - N_i\right)\right],$$

$$= \Sigma\left[k\left(g_i \ln g_i - N_i \ln N_i - (g_i - N_i)\right.\right.$$

$$\left.\left.\times\ (\ln g_i - \frac{N_i}{g})\right)\right]$$

$$= \Sigma\left[k\left(g_i \ln g_i - N_i \ln N_i - g_i \ln g_1\right.\right.$$

$$\left.\left. + N_i + N_i \ln g_1\right)\right]$$

$$= \Sigma\left[k\ N_i\left(\ln \frac{g_i}{N_i} + 1\right)\right]. \tag{15}$$

According to Fermi-Dirac distribution function for the state of equilibrium which is given by

$$\frac{N_i^\circ}{g_i} = \frac{1}{B\ e^{\beta\epsilon_i} + 1}, \tag{13}$$

the number of electrons N_i in the *i*th cell is

$$N_i^\circ = \frac{g_i}{B\ e^{\beta\epsilon_i} + 1}.$$

If this number is very much smaller than the number of compartments g_i, the denominator on the right-hand side of Eq. (13) is much greater than unity, and subsequently, the term '1' in this denominator can be neglected so that the number N_i is given by

$$N_i^{\circ} = \frac{g_i}{B\, e^{\beta \epsilon_i}},$$

or, dropping for simplicity the superscript 'o' bearing in mind that all our analysis is restricted to equilibrium distribution, we have

$$N_i = \frac{g_i}{B}\, e^{-\beta \epsilon_i} \,. \tag{16}$$

Therefore,

$$\left(\frac{g_i}{N_i}\right) = B\, e^{\beta \epsilon_i},$$

$$\ln\left(\frac{g_i}{N_i}\right) = \ln B + \beta \epsilon_i \tag{17}$$

Substituting for $\ln\left(\frac{g_i}{N_i}\right)$ in Eq. (15) by Eq. (17), we obtain

$$\begin{aligned} S &= \Sigma\left[k N_i \left(\ln B + \beta \epsilon_i + 1\right)\right] \\ &= k N + k N \ln B + k \beta U . \end{aligned} \tag{18}$$

Also, from Eq. (16), we have

$$\Sigma N_i = N = \frac{1}{B} \Sigma\, g_i\, e^{-\beta \epsilon_i}, \tag{19}$$

so that the constant *B* is given by

$$B = \frac{\Sigma g_i\, e^{-\beta \epsilon_i}}{N}. \tag{20}$$

Now substituting for *B* in Eq. (18) by Eq. (20), we have

$$\begin{aligned} S = k N + k N &\left\{ \left(\ln \Sigma g_i\, e^{-\beta \epsilon_i} \right) - \ln N \right\} \\ &+ k\, B U . \end{aligned} \tag{21}$$

Therefore,

$$\left(\frac{\partial S}{\partial U}\right)_V = kN \frac{1}{\Sigma g_i e^{-\beta \epsilon_i}} \times \frac{d\left(\Sigma g_i e^{-\beta \epsilon_i}\right)}{dU} + kB + kU\left(\frac{\partial \beta}{\partial U}\right)_V$$

$$= \frac{kN}{\left(\Sigma g_i e^{-\beta \epsilon_i}\right)} \frac{d\left(\Sigma g_i e^{-\beta \epsilon_i}\right)}{d\beta}\left(\frac{\partial \beta}{\partial U}\right)_V + k\beta + kU\left(\frac{\partial \beta}{\partial U}\right)_V .$$

But,

$$\frac{d\left(\Sigma g_i e^{-\beta \epsilon_i}\right)}{d\beta} = -\Sigma \epsilon_i g_i e^{-\beta \epsilon_i} = -\frac{U}{N} \Sigma g_i e^{-\beta \epsilon_i} ,$$

hence,

$$\left(\frac{\partial S}{\partial U}\right)_V = -\frac{kNU}{N} \frac{\Sigma g_i e^{-\beta \epsilon_i}}{\Sigma g_i e^{-\beta \epsilon_1}} \left(\frac{\partial \beta}{\partial U}\right)_V + k\beta + kU\left(\frac{\partial \beta}{\partial U}\right)_V ,$$

$$= k\beta . \tag{22}$$

The assumed relation between *S* and *W* will therefore lead to the same value of $\left(\frac{\partial S}{\partial U}\right)_V$ as that given by classical thermodynamics which has been shown above, namely

$$\left(\frac{\partial S}{\partial U}\right)_V = \frac{1}{T},$$

if

$$k\beta = \frac{1}{T} \quad \text{or} \quad \beta = \frac{1}{kT}. \tag{23}$$

The quantity *B* is determined, as in any statistics, from the fact that $\sum N_i = N$, the total number of particles. To evaluate $\sum N_i$, we approximate the discontinuous distribution of phase points by a continuous function and replace the sum by an integral. In Eq. (13), let us replace g_i by

$$g_i = \frac{2H}{h^3} = \frac{2}{h^3}\, dx\, dy\, dz\, dp_x\, dp_y\, dp_z,$$

and change, as done before, the notation from N_i° to d^6N and from ϵ_i to ϵ. Then

$$d^6N = \frac{2}{h^3}\,\frac{1}{B\,e^{\epsilon/kT}+1}\,dx\,...dp_z.$$

Next we integrate over *x*, *y*, and *z*. In the absence of a force field, these quantities appear only in the differentials, and the integral over these variables is the total volume of the system, *V*. Therefore, we have

$$d^3N_{P_x,P_y,P_z} = \frac{2V}{h^3}\,\frac{1}{B\,e^{\epsilon/kT}+1}\,dp_x\,dp_y\,dp_z. \qquad (24)$$

We now have the distribution function in three-dimensional momentum space. The next step is to express ϵ in terms of *p*, or vice versa, and set the integral of d^3N over all values of *p* (or ϵ) equal to *N*.

If $Be^{\epsilon/kT} >> 1$, the term 1 in the denominator can be neglected and, as with the Bose-Einstein statistics, we obtain the Maxwell-Boltzmann statistics as a limiting case. *This in the Fermi-Dirac statistics presents nothing new of interest.* At the same time, this approximation cannot be made for an electron gas, and thus the constant *B* must be evaluated from Eq. (24).

The expression for *B*, when *B* is small, was first derived by *Sommerfeld.* The calculation is long and complicated, and we shall only state the result. *For reasons that will be evident later*, let us write

$$B = e^{-\epsilon_m/kT}, \qquad (25)$$

where ϵ_m is a *reference energy* that can be a function of *temperature*. Substituting for *B* in Eq. (24), by Eq. (25), we obtain

$$d^3N_{P_x,P_y,P_z} = \frac{2V}{h^3}\,\frac{1}{e^{(\epsilon-\epsilon_m)/kT}+1}\,dp_x, dp_y, dp_z. \qquad (26)$$

Denoting the expression

$$\frac{2V}{h^3}\,\frac{1}{e^{(\epsilon-\epsilon_m)/kT}+1}$$

by ρ, we have

$$d^3N_{P_x,P_y,P_z} = \rho\, dp_x, dp_y, dp_z. \qquad (27)$$

It is thus evident that the symbols ρ in this equation represents the number of phase points per unit volume or the density in momentum space.

When $T = 0$ K in Eq. (26), this distribution function reduces to a very simple one as can be seen from the following:

Let the reference energy ϵ_m at 0 K be denoted by ϵ_{m_0}. Then for a cell in momentum space for which ϵ is less than ϵ_{m_0}, the term $(\epsilon - \epsilon_{m_0})$ in Eq. (26) is negative, and, subsequently $(\epsilon - \epsilon_m)/kT = -\infty$, and since $e^{-\infty} = 0$, it follows that:

$$\rho_0 = \frac{2V}{h^3} \qquad \left(T = 0\,K, \epsilon < \epsilon_{m_0}\right). \tag{28}$$

This means that at absolute zero, the density of representative points in momentum space is constant and equal to $2V/h^3$ *in all cells for which* $\epsilon < \epsilon_{m_0}$.

On the other hand, if $\epsilon > \epsilon_{m_0}$ and $T = 0$ K, the term $(\epsilon - \epsilon_{m_0})$ in Eq. (26) is positive and thus $(\epsilon - \epsilon_{m_0})/kT = +\infty$, and since $e^{\infty} = \infty$, *the density in momentum space is zero in all cells for which* $\epsilon > \epsilon_{m_0}$, *that is*

$$\rho_0 = 0, \qquad \left(T = 0\,K, \epsilon > \epsilon_{m_0}\right). \tag{29}$$

The physical significance of ϵ_{m_0}, then, is *the maximum energy of the electron at absolute zero*, which is the reason for the choice of this symbol.

The energy ϵ and the momentum p are related by the equation

$$\frac{1}{2} m v^2 = \epsilon = \frac{p^2}{2m}, \quad p^2 = 2m\epsilon.$$

The maximum energy at absolute zero ϵ_{m_0} corresponds to a maximum momentum p_{m_0} given by

$$p_{m_0} = \left(2m\,\epsilon_{m_0}\right)^{1/2}. \tag{30}$$

Therefore, in geometrical language, we can say that at absolute zero, momentum space is *uniformly populated within a sphere of radius* p_{m_0} and that *there are no phase points outside this sphere*. The process of integrating the density over all of momentum space, therefore, reduces to the multiplication of the constant density $\rho_0\left(=\frac{2V}{h^3}\right)$ by the volume of a sphere of radius p_{m_0}, and this product equals the total number of electrons, N,

$$\frac{2V}{h^3} \times \frac{4}{3}\pi p_{m_0}^3 = N. \tag{31}$$

From this equation, we find that

$$p_{m_0} = \left(\frac{3Nh^3}{8\pi V}\right)^{1/3}, \tag{32}$$

and

$$\epsilon_{m_0} = \frac{(p_{m_0})^2}{2m} = \left(\frac{3Nh^3}{8\pi V}\right)^{2/3} \Big/ 2m$$

$$= \frac{h^2}{8m}\left(\frac{3N}{\pi V}\right)^{2/3},$$

or

$$\epsilon_{m_0} = \frac{h^2}{2m}\left(\frac{3N}{8\pi V}\right)^{2/3}. \tag{33}$$

We next estimate the magnitude of ϵ_{m_0} from Eq. (33). The constant *h* is Planck's constant. 6.62×10^{-34} J s, and *m* is the mass of the electron 9×10^{-31} kg m. The number of electrons per unit volume cannot be measured directly. Since the atoms are all alike, the most reasonable assumption is that each atom *contributes the same (integral) number of electrons to the electron gas.* We would also expect this number to be small, probably 1 for monovalent atoms, 2 for bivalent, etc. There is good indirect evidence that this assumption is correct. Let us make our calculation for *silver* and let us assume *one electron per atom.* Then taking* $(N/V) = 5.86 \times 10^{28}$ free electrons per m^3, the substitution in Eq. (33) gives

$$\epsilon_{m_0} = \frac{\left(6.62 \times 10^{-34}\right)^2}{2 \times 9 \times 10^{-31}}\left(\frac{3 \times 5.86 \times 10^{28}}{8\pi}\right)^{2/3}$$

$$= 9.0 \times 10^{-19} \text{ joule}.$$

It many be convenient to express this energy in electron volts, (abbreviated eV), where by definition, one electron volt is equal to 1.602×10^{-19} J, thus

$$\epsilon_{m_0} = 9.0 \times 10^{-19} \times \frac{1}{1.602 \times 10^{-19}}$$

$$= 5.6 \text{ eV}.$$

This is the maximum kinetic energy of free electron at absolute zero. It will be proved below that the average energy $\bar{\epsilon}$ of the electron at absolute zero is $3/5$ of ϵ_{m_0}, so that its value is

$$\bar{\epsilon} = \frac{3}{5} \times 5.6 = 3.36 \text{ ev} = 5.39 \times 10^{-19} \text{J}.$$

According to the Maxwell-Boltzmann statistics, the average kinetic energy of the molecules of a gas is $3kT/2$ is zero at absolute zero, and the temperature at which $3/2\,kT = 5.37 \times 10^{-19}$ J is 26,800 K. Hence, if the *new statistics of Fermi-Dirac is correct*, and there is an ample evidence that it is, *the concept of absolute zero, as a state in which all molecular (or electronic) motion has ceased is far from the truth.* The mean kinetic energy of the electrons in a metal, even at absolute zero, is much *greater* than that of the molecules of an ordinary gas, even at a temperature of thousands of degrees.

Let us now return to the evaluation of ϵ_m at a temperature other than 0 K. The result obtained by Sommerfeld is

$$\epsilon_m = \epsilon_{m_0}\left\{1 - \frac{\pi^2}{12}\left(\frac{kT}{\epsilon_{m_0}}\right)^2 + \ldots\right\}. \tag{34}$$

---

* See J. F. Lee, P. W. Sears, D. L. Turcotte, *Statistical Thermodynamics*, Addison-Wesley Publishing Company, Inc. 1972, Chapter 7.

When $T = 0$ K, $\in_m = \in_{m_0}$. Even at elevated temperatures the difference between $\in_m$ and $\in_{m_0}$ is small, since the term kT is only a few tenth of an electron volt, while $\in_{m_0}$ is of the order of 2–10 eV. Hence, in evaluating the distribution function ρ in Eq. (27), even at temperatures as high as several thousand degrees kelvin, we make but a slight error in substituting $\in_{m_0}$ for $\in_m$.

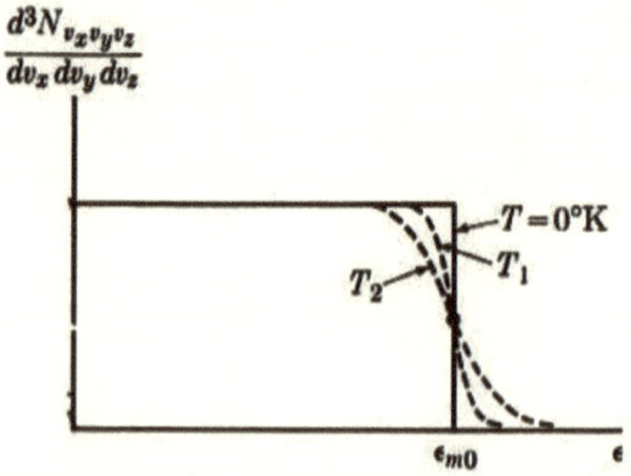

Fig. 2. Distribution in momentum space in the Fermi-Dirac statistics, at $T = 0$ K (full line) and at two higher temperatures T_1 and T_2.

Fig. 2 is a graph of the distribution function, ρ, plotted as a function of $\in$. The ordinate of the curve is the number of representative points per unit 'volume' of momentum space, which is given by

$$\rho = \frac{d^3 N_{v_x v_g v_z}}{d_{v_x} d_{v_y} d_{v_z}}.$$

The solid line is the distribution at $T = 0$ K. The density is constant at all points for which $\in < \in_{m_0}$ (or $p < p_{m_0}$) and is zero beyond this value. The dashed lines are the distributions at higher temperatures T_1 and T_2. If $T \neq 0$ K, the function falls off asymptotically to zero as the energy increases, and there is no sharp upper limit to the energy or momentum. This means that $\in_m$ does not represent the maximum energy at a temperature T in the way that $\in_{m_0}$ represents the maximum energy at $T = 0$ K. Note also that only the more energetic electrons are affected by a rise of temperature.

10.2. Velocity, speed, and energy distribution functions

To obtain the *velocity distribution function* from Eq. (26) obtained above, namely

$$d^3 N_{p_x p_y p_z} = \frac{2V}{h^3} \frac{1}{e^{(\in - \in_m)/kT} + 1} dp_x\, dp_y dp_z. \tag{26}$$

we need only replace dp_x by $d(m v_x) = mdv_x$, etc. This gives

$$d^3 N_{v_x v_y v_z} = \frac{2m^3V}{h^3} \frac{1}{e^{(\epsilon - \epsilon_m)/kT} + 1} dv_x \, dv_y dv_z \cdot \tag{35}$$

Of course ϵ may be written out explicitly as

$$\epsilon = \frac{1}{2} m \left(v_x^2 + v_y^2 + v_z^2\right),$$

in order to obtain an expression involving velocity components only. The coefficient of the volume element $dv_x \, dv_y dv_z$ in Eq. (35), which is

$$\frac{2m^3V}{h^3} \frac{1}{e^{(\epsilon - \epsilon_m)/kT} + 1},$$

represents *the number of phase points per unit volume or the density in velocity space*; that is, Eq. (35) represents the *velocity distribution function.*

The speed distribution function is not readily derived from the fact that this distribution is spherically symmetrical as it is the case for the velocity or the momentum distribution. Accordingly, the number of representative points in a thin spherical shell in velocity space of *radius v*, that is, dN_v, is equal to the product of the density in velocity space at this radius and the volume of the shell is $4\pi v^2 \, dv$. The density in velocity space at this radius, dv being the thickness of this shell. The density is a function of *v* only, since $\epsilon = \frac{1}{2} m v^2$.

Therefore, using the velocity distribution function, Eq. 35, and replacing ϵ by $\frac{1}{2} mv^2$, we have

$$\begin{aligned} dN_v &= \frac{2m^3V}{h^3} \frac{1}{e^{\left(\frac{1}{2}mv^2 - \epsilon_m\right)/kT} + 1} \times 4\pi v^2 \, dv \\ &= \frac{8\pi m^3V}{h^3} \frac{v^2}{e^{\left(\frac{1}{2}mv^2 - \epsilon_m\right)/kT} + 1} dv \cdot \end{aligned} \tag{36}$$

This is the *speed distribution function.* Substituting for v^2 in this equation by $2\epsilon/m$, we obtain

$$dN_v = \frac{16\pi m^2V}{h^3} \frac{\epsilon}{e^{(\epsilon - \epsilon_m)/kT} + 1} dv. \tag{37}$$

At absolute zero, we have

$$\frac{dN_v}{dv} = \frac{8\pi m^3 V}{h^3} v^2 \left(v < v_{m_\circ}\right),$$

from Eq. (36),

$$\frac{dN_v}{dv} = \frac{16\pi m^2V}{h^3} \epsilon, \left(\epsilon < \epsilon_{m_\circ}\right),$$

from Eq. (37);
and

$$\frac{dN_v}{dv} = 0. \qquad \left(v > v_{m_\circ}, \in > \in_{m_\circ}\right).$$

from either of Eq. (36) or Eq. (37).

The *speed distribution function* is plotted in Fig. 3(a) as a function of *v*, and in Fig. 3(b) as a function of $\in$, at $T = 0\,K$, and at two higher temperatures, the equations used being Eqs. (36) and (37), respectively. It is to be noted that the ratio dN_v/dv in either Eq. (36) or (37), or what is the same thing, namely, the coefficient of d*v* in Eq. (36) is also spoken of as the *distribution function of electronic speeds.* Unlike the velocity distribution function given by Eq. (35), namely,

$$d^3 N_{v_x v_y v_z} = \frac{2m^3 V}{h^3} \frac{1}{e^{(\in - \in_m)/kT} + 1} dv_x\, dv_y dv_z, \tag{35}$$

the speed distribution function given by Eq. (36) or (37) does not represent the number of phase points per unit volume in velocity space, but represents the number of points per unit range of speed d*v*. Also that because of the factor v^2, which does not occur in the velocity distribution function, Eq. (35), the graphs in Figs. 3(a) and 3(b) are zero when $v = 0$.

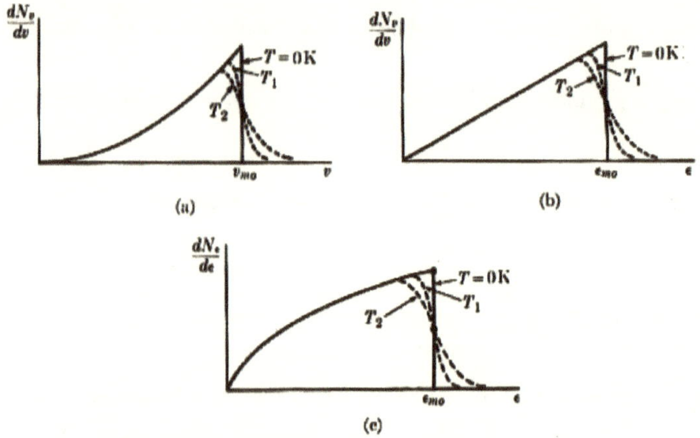

Fig. 3. (a) and (b) Speed distribution function as a function of *v* and $\in$; (c) energy distribution function as a function of $\in$.

The *energy distribution function* can be obtained from Eq. (37)

$$dN_v = \frac{16\,\pi m^2 V}{h^3} \frac{\in}{e^{(\in - \in_m)/kT} + 1} dv \tag{37}$$

by inserting the expressions

$$\in = \frac{1}{2} m v^2, \quad v^2 = \frac{2\in}{m}, \; dv = (2\in m)^{-1/2}\, d\in.$$

Hence,

$$dN_{\epsilon} = \frac{4\pi V}{h^3}\left(2^2\ 2^{-1/2}\ m^2\ m^{-1/2}\right)\frac{\epsilon^{1/2}}{e^{(\epsilon-\epsilon_m)/kT}+1}\,d\epsilon$$

$$= \frac{4\pi V}{h^3}(2m)^{3/2}\frac{\epsilon^{1/2}}{e^{(\epsilon-\epsilon_m)/kT}+1}\,d\epsilon. \qquad (38)$$

This is the *energy distribution function.* The ratio $dN_{\epsilon}/d\epsilon$, or, what is the same thing, namely, the coefficient of $d\epsilon$ in this equation, is also spoken of as the energy distribution function; it is potted as a function of ϵ in Fig. 3(c) to represent graphically this distribution.

Let us find the distribution in any *one velocity component*, say v_x. For this purpose, we return to Eq. (35)

$$d^3 N_{v_x v_y v_z} = \frac{2m^3V}{h^3}\frac{1}{e^{(\epsilon-\epsilon_m)/kT}+1}\,dv_x\,dv_y dv_z, \qquad (35)$$

which represents the velocity distribution function, and integrate over all values of v_y and v_z. Then,

$$dN_{v_x} = \frac{2m^3V}{h^3}\left[\int_{-\infty}^{\infty}\int_{-\infty}^{\infty}\frac{1}{e^{(\epsilon-\epsilon_m)/kT}+1}dv_y\,dv_z\right]dv_x, \qquad (39)$$

where,

$$\epsilon = \frac{1}{2}m\left(v_x^2 + v_y^2 + v_z^2\right).$$

To evaluate the integral, let us consider v_y and v_z as variables in a rectangular coordinate system. Then $dv_y\ dv_z$ is an element of area, and the double integral is the surface integral of a function of v_y and v_z over the entire v_yv_z plane. We now transform to polar coordinates, *r* and *θ*, as shown in Fig. 4. Thus,

$$v_y^2 + v_z^2 = r^2,$$

$$\epsilon = \frac{1}{2}m\,(r^2 + v_x^2)$$

and an element of area $dv_y\ dv_z$ is

$$dv_y\,dv_z = r\,dr\,d\theta.$$

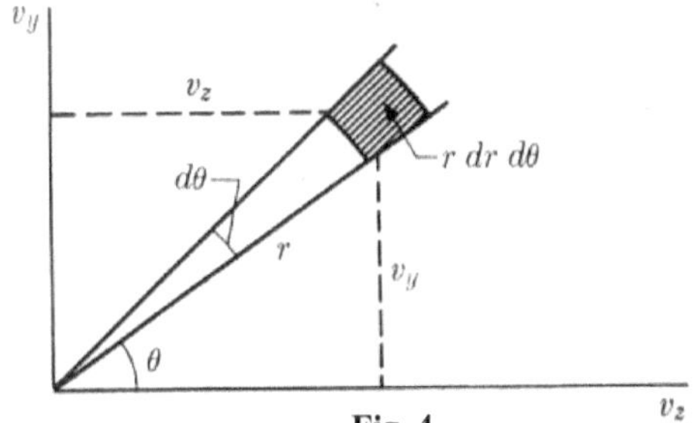

Fig. 4.

Therefore, Eq. (39) can be put in the form

$$dN_{v_x} = \frac{2m^3V}{h^3}\left[\int_0^{\infty}\int_0^{2\pi}\frac{r\ dr\ d\theta}{e^{\left\{\frac{1}{2}m\left(r^2+v_x^2\right)-\epsilon_m\right\}/kT}+1}\right]dv_x .$$

Now,

$$\frac{\left\{\frac{1}{2}m\left(r^2+v_x^2\right)-\epsilon_m\right\}}{kT} = \frac{m r^2}{2kT} + \frac{\frac{1}{2}m v_x^2 - \epsilon_m}{kT},$$

let

$$\frac{m r^2}{2kT} = x, \qquad r\,dr = \frac{kT}{m}dx, \qquad e^{mr^2/2kt} = e^x, \quad \text{and}$$

$$e^{\left(\frac{1}{2}m v_x^2 - \epsilon_m\right)/kT} = a.$$

Accordingly, dN_{v_x} is given by

$$dN_{v_x} = \frac{2m^3V}{h^3}\frac{kT}{m}\left[\int_0^{\infty}\int_0^{2\pi}\frac{dx\,d\theta}{ae^x+1}\right]dv_x .$$

The integral

$$\int_0^{\infty}\frac{dx}{a\,e^x+1} = \left[-\ln\left(a+e^{-x}\right)\right]_0^{\infty}$$

$$= \ln\left(1+\frac{1}{a}\right),$$

and the integral over θ is simply 2π.
Hence,

$$dN_{v_x} = \frac{4\pi V m^2 kT}{h^3}\ln\left[\left(1+\frac{1}{a}\right)d_{v_x},\right.$$

$$= \frac{4\pi V m^2 kT}{h^3}\ln\left[\frac{1}{e^{\left(\frac{1}{2}m v_x^2 - \epsilon_x\right)/kT}}+1\right]dv_x,$$

$$= \frac{4\pi V m^2 kT}{h^3}\ln\left[e^{\frac{\left(\epsilon_m - \frac{1}{2}m V_x^2\right)}{kT}}+1\right]dv_x, \tag{40}$$

or, since $\epsilon_x = \frac{1}{2}m v_x^2$, we have

$$dN_{v_x} = \frac{4\pi V m^2 kT}{h^3}\ln\left[e^{\frac{(\epsilon_m-\epsilon_x)}{kT}}+1\right]dv_x . \tag{41}$$

To see what form this expression takes at absolute zero, we note that *T* appears in the coefficient of the logarithmic term, and also in the denominator of the exponential term, so that

$$dN_{v_x} = 0 \times \infty,$$

which is an undefined product. Therefore, let *T* be some *finite but very small number*. Then if $\in_x$ is less than $\in_m$, the exponential term is large and we may neglect the 1. But

$$\ln e^x = x,$$

so that when *T* is very small, the *T* in Eq. (41) *drops out*, and this equation reduces to

$$dN_{v_x} = \frac{4\pi^2 V m^2}{h^3}\left(\in_{m_0} - \in_x\right)dv_x. \qquad (42)$$

Hence, this equation, which is obtained at sufficiently low temperature in the neighbourhood of the absolute zero, may be taken as the *distribution function in any one velocity component, say* v_x, at $T = 0$ K.

A graphical representation of this distribution function can thus be made by plotting the *coefficient* of dv_x in Eq. (42), which stands for the ratio dN_{v_x}/dv_x, against $\in_x$ as shown in Fig. 5.

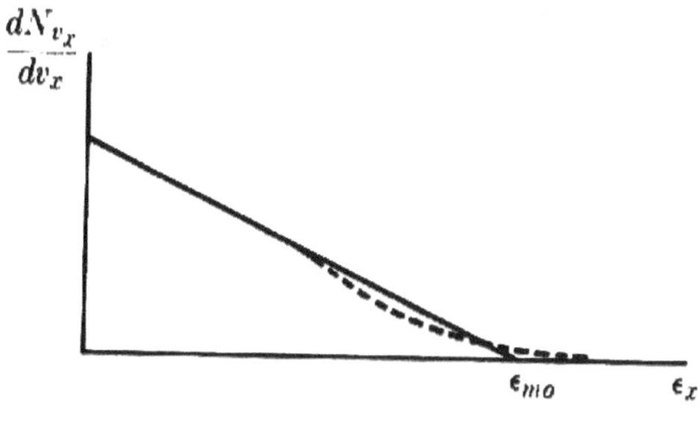

Fig. 5.

It is seen that the distribution decreases linearly with $\in_x$ with a very gradual tailing off at higher temperatures.

10.3. Specific heat of the electron gas

An example must be given to justify the hypotheses introduced in setting up the Fermi-Dirac statistics. One of the outstanding failures of the Maxwell-Boltzmann statistics was the prediction that the specific heat of a metal should be much greater than the *Dulong-Petit value* of 3R. That is, if there is equipartition of energy between the atoms and the free electrons, the electrons should *contribute an amount of* 3/2 *R per mole* to the specific heat, like any monatomic gas. We shall now see that although the Fermi-Dirac statistics *attributes* a much greater energy to the electrons than does the older molecular kinetic theory of gas molecules, *the change in the energy of electrons with temperature is very small*, and it is only the *change in energy that influences the heat capacity*.

The average energy of an electron is defined in the usual way as

$$\overline{\epsilon} = \frac{\int_0^\infty \epsilon \, dN_\epsilon}{\int_0^\infty dN_\epsilon}.$$

At absolute zero, the upper limit of the integral can be taken as ϵ_{m_0}, since at this temperature there are no electrons with energies greater than ϵ_{m_0}. For $\epsilon < \epsilon_{m_0}$ and at $T =$ 0 K, we have from the equation expressing the energy distribution function which is

$$dN_\epsilon = \frac{4\pi V}{h^3}(2m)^{3/2} \frac{\epsilon^{1/2}}{e^{(\epsilon - \epsilon_m)/kT} + 1} d\epsilon \tag{38}$$

the following reduced equation:

$$dN_\epsilon = \frac{4\pi V}{h^3}(2m)^{3/2} \epsilon^{1/2} \, d\epsilon. \tag{43}$$

Hence, at absolute zero, the average energy $\overline{\epsilon}_0$ of an electron is

$$\overline{\epsilon}_0 = \frac{\int_0^{\epsilon_{m0}} \epsilon^{3/2} \, d\epsilon}{\int_0^{\epsilon_{m0}} \epsilon^{1/2} \, d\epsilon}$$

$$= \frac{3}{5}\epsilon_m; \tag{44}$$

that is, the average energy at absolute zero equals 3/5 at the maximum energy.

The average energy $\overline{\epsilon}$ at any temperature T is obtained in the same way using the series expansion for ϵ_m as a function of T, which is

$$\epsilon_m = \epsilon_{m_0}\left\{1 - \frac{\pi^2}{12}\left(\frac{kT}{\epsilon_{m_0}}\right)^2 + \ldots\right\}, \tag{34}$$

as shown above, and then integrating between 0 and ∞. Details, for convenience, are omitted, but the result is

$$\overline{\epsilon} = \frac{3}{5}\epsilon_{m_0}\left[1 + \frac{5\pi^2}{12}\left(\frac{kT}{\epsilon_{m_0}}\right)^2 + \ldots\right]. \tag{45}$$

The total internal energy U of N electrons is $U = N\overline{\epsilon}$, and the heat capacity at constant volume is

$$C_v = \frac{dU}{dT}$$

$$= N\frac{d\overline{\epsilon}}{dT}$$

$$= N\frac{\pi^2 k^2 T}{2\epsilon_{m_0}}.$$

If *N* is equal to Avogadro's number, then $Nk = R$ and C_v becomes the molar specific heat which, we denote by c_v, is given by

$$c_v = \frac{\pi^2 kT}{2\epsilon_{m_0}} R. \qquad (46)$$

The Maxwell-Boltzmann statistics predicts a value

$$c_v = \frac{3}{2} R,$$

independent of temperature, whereas the predicted Fermi-Dirac value is temperature dependent being proportional to *T*. The numerical value of $\pi^2 k/2\epsilon_{m_0}$ for silver as an example is

$$\frac{\pi^2 k}{2\epsilon_{m_0}} = \frac{\pi^2 \times 1.38 \times 10^{-23}}{2 \times 9 \times 10^{-19}}$$

$$= 7.6 \times 10^{-5} \text{ deg}^{-1}.$$

Hence, at a temperature of 300 K, the coefficient of *R* in Eq. (46) is only $7.6 \times 10^{-5} \times 300 = 0.023$, compared with $3/2$, or 1.5, predicted by the older theory. The contribution of the electrons to the specific heat is *therefore extremely small in agreement with experiment.*

It may be pointed out here that since metals as well as non-metals obey the Dulong-Petit law, the electrons apparently do not share in the thermal energy. This was a very puzzling phenomenon for many years, but now it has, as shown above, a very satisfactory explanation when quantum statistical methods are used.

CHAPTER 11

Partition Functions

11.1. Introduction

The partition function is the essential *link* between the coordinates of microscopic systems and the thermodynamic properties. Once it has been evaluated, all these physical properties can be readily computed. The partition function also permits the computation of many other parameters, such as equilibrium constants and activity coefficients.

Partition functions can be evaluated for a large number of systems. However, for many complex systems it is not possible to derive the expression for the partition function. Furthermore, in same instances where the form of partition function is known, it is extremely difficult or impossible to calculate. Nevertheless, as it stands, the partition function is a *powerful, thermodynamic tool,* and widespread investigation in this comparatively new field and the availability of high-speed computers promise to extend its application to many areas in engineering.

11.2. The Partition Function

In Chapter 9, we found from the Maxwell-Boltzmann statistics that the partition function is expressed as

$$Z = \sum e^{-\beta \epsilon_i},$$

and we noted that it depends upon β and the way in which the energy ϵ_i varies from cell to cell in phase space. Since the latter is different in different problems, no general expression can be written for Z other than that given above, and therefore, it must be evaluated for each special case. Our first task is to identify the energy ϵ_i as the *total energy* of a microscopic system expressed in terms of all the contributing energies. In addition to the *kinetic energy* of a particle associated with the *translational motion* of its mass, there are *internal contributions* to the total energy. There are the *electronic*, *rotational*, *vibrational*, *nuclear*, and *chemical* contributions. The internal energy of a particle in a particular state can then be expressed as

$$\epsilon_i = \epsilon_{trans} + \epsilon_{int}\,.$$

It is to be noted that for some systems, such as liquids, it will not be possible to separate the translational and internal contributions to the total energy. Thus for the present, we restrict ourselves to gases where such a separation is appropriate.

As an example, let us consider a diatomic molecule such as HCl. Such a molecule can be thought of as two atomic centres H and Cl bound together *ionically* and separated by a distance r_0. However, this distance between the atomic centres is not rigidly fixed at the value r_0 since the atoms move in a mutual potential field $\epsilon_P(r)$ as shown in Fig. 1. The distance r_0 corresponds to the separation of the atomic centres when the potential energy is *minimum.* Other values of r correspond to a *positive* potential energy of distortion. As the atoms, separated by some distance *r*, come closer to each other, they are attracted, until at r_0 a molecule is formed with *liberation of energy in the amount* ϵ_0. The energy ϵ_0 is the *chemical energy* and must be taken into account when chemical reactions take place. We already know that the zero-point

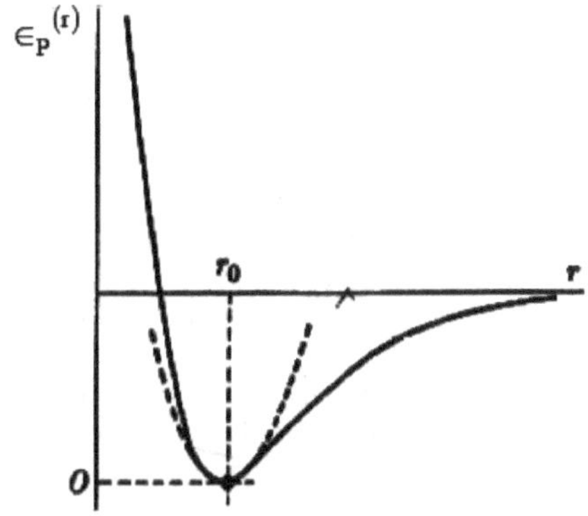

Fig. 1.

vibrational energy is equal to 1/2 $h\nu$, and if the potential energy is designated by ϵ_P, the chemical energy ϵ_0 is given by

$$\epsilon_0 = -(\epsilon_P - 1/2\ h\nu).$$

Since ϵ_P is always positive and greater than $\frac{1}{2}h\nu$, ϵ_0 is always *negative in the above equation in keeping with thermodynamic convention.* We shall postpone further consideration of the

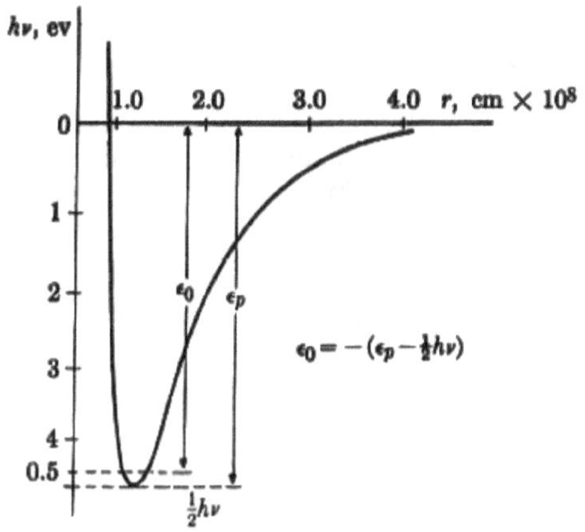

Fig. 2. Potential energy diagram for HCl.

chemical energy until we deal with chemical systems in Chapter 14. Fig. 2 shows the potential energy diagram for the diatomic HCl molecule.

Let us now return to the several types of dynamical motion which a molecule has the capacity to execute simultaneously concerning their contributions to the total energy. In *translation*, the molecule moves as a whole; that is, there is a movement of the centre of mass but not the position of the atoms relative to the centre of mass. In *rotation*, the atoms move about the centre of mass as a fixed point, maintaining their positions with respect to one another. *Vibration* involves the oscillation of the atoms about equilibrium positions without displacement of the latter or of the centre of mass. This preliminary description emphasise the degree to which the basic types of molecular motion are independent of one another. Complete independence would mean that we could discuss, for instance, the rotation of a molecule quite separately from its translation and vibration; and obviously this approach is much simpler than if the three forms of motion had to be considered as *indivisible*. From the outset, then, we shall assume the *separability* (or *independence*) of these three types of motion and take account of their mutual interaction only when it is necessary to do so. It may be added that the vibrational and rotational energy can be regarded independent provided (*a*) the centrifugal force due to rotation has little effect or no effect on the vibrations, and (*b*) the moment of inertia of the rotating molecule is insensitive to the vibrational fluctuations in r. These two conditions are usually met very nearly in the case of diatomic molecules.

In Chapter 10, we have seen that there is an electronic contribution to the energy of an atom. Normally, the excitation energies are much too great for thermal excitation to occur at terrestrial temperature. In some molecular species, however, notably O_2 and NO, there are *low-lying excited states* and the electronic contribution to the internal energy, ϵ_{int}, must be taken into account. Finally, there is a contribution from the existence of *nuclear spin*. This contribution cancels out in equilibrium reactions, but when the absolute value of entropy is to be computed, this contribution must be included for exactness, as will be shown later in Chapter 20.

It is now possible to write the equation for internal contributions to energy as

$$\epsilon_{int} = \epsilon_{rot} + \epsilon_{vib} + \epsilon_{elec} + \epsilon_{nuc} + \epsilon_{chem},$$

where the terms on the right-hand side of the equation stand for the rotational, vibrational, electronic, nuclear, and chemical contributions. The total energy of a particle in a particular state is written as

$$\epsilon_i = \epsilon_{trans} + \epsilon_{rot} + \epsilon_{vib} + \epsilon_{elec} + \epsilon_{nuc} + \epsilon_{chem}, \qquad (1)$$

where it is understood that any one contribution may not be present, or may be negligible for the particular state, or may cancel out in an equilibrium reaction.

If we leave unanswered for the present the question of *statistical weights* (g_i's), the partition function can be written in general form

$$Z = \Sigma \exp\left[-\beta(\epsilon_{trans} + \epsilon_{rot} + \epsilon_{vib} + \epsilon_{elec} + \epsilon_{nuc} + \epsilon_{chem})\right],$$

or

$$Z = \Sigma\left[\left(e^{-\beta\epsilon_{trans}}\, e^{-\beta\epsilon_{rot}}\, e^{-\beta\epsilon_{vib}}\, e^{-\beta\epsilon_{elec}}\, e^{-\beta\epsilon_{nuc}}\, e^{-\beta\epsilon_{chem}}\right)\right],$$

which in a concise form is

$$Z = Z_{trans}\; Z_{rot}\; Z_{vib}\; Z_{elec}\; Z_{nuc}\; Z_{chem}, \qquad (2)$$

where the separation is of particular value when the contributions are completely independent. In such a case, to determine the total partition function of a molecule, it is necessary to find each of contributing partition functions. These shall be discussed in the sections which follow.

11.3. Translational partition function

We want now to derive the partition function for translational motion, Z_{trans}, not as before using *quantum statistics* and considering the case of an ideal monatomic gas, where the kinetic energy associated with translation was the only contribution to the energy of each particle, but directly from *Schrödinger's equation*, using the solution for a free particle confined to a volume *V*, and whether the gas is a monatomic or polyatomic ideal gas. The energy levels for a free particle in a rectangular box with edges of length *a*, *b*, and *c* are derived with the result that the translational energy levels of the particle are quantised, the permitted values being

$$\in_{trans} = \frac{h^2}{8m}\left(\frac{n_x^2}{a^2} + \frac{n_y^2}{b^2} + \frac{n_z^2}{c^2}\right),$$

with $n_x = 1, 2, 3, \ldots, n_y = 1, 2, 3, \ldots, n_z = 1, 2, 3, \ldots$.

Now we return to quantum statistics, where the translational partition function per particle is defined by

$$Z_{trans} = \Sigma\, e^{-\in_{tarns}/kT},$$

and substitute for $\in_{trans}$ by the above expression. Thus, we obtain

$$Z_{trans} = \Sigma\left\{ e^{-\frac{h^2}{8mkT}\left(\frac{n_x^2}{a^2} + \frac{n_y^2}{b^2} + \frac{n_z^2}{c^2}\right)} \right\}$$

$$= \sum_{n_x=1}^{\infty} e^{-\frac{h^2 n_x^2}{8mkTa^2}} \times \sum_{n_y=1}^{\infty} e^{-\frac{h^2 n_y^2}{8mkTb^2}} \times \sum_{n_z=1}^{\infty} e^{-\frac{h^2 n_z^2}{8mkTc^2}}. \qquad (3)$$

The quantities

$$\frac{h^2}{8mkTa^2},\ \frac{h^2}{8mkTb^2},\ \frac{h^2}{8mkTc^2},$$

are very small for reasonable values of *a*, *b*, *c*, and *T* so that it is appropriate to replace the summation in Eq. (3) by integrals, with the result

$$\sum_{n_x=1}^{\infty} e^{-\frac{h^2 n_x^2}{8mkTa^2}} = \int_0^{\infty} e^{-\frac{h^2 n_x^2}{8mkTa^2}}\, dn_x = \frac{a(2\pi mkT)^{1/2}}{h},$$

$$\sum_{n_y=1}^{\infty} e^{-\frac{h^2 n_y^2}{8mkTb^2}} = \int_0^{\infty} e^{-\frac{h^2 n_y^2}{8mkTb^2}}\, dn_y = \frac{b(2\pi mkT)^{1/2}}{h},$$

$$\sum_{n_z=1}^{\infty} e^{-\frac{h^2 n_z^2}{8mkTc^2}} = \int_0^{\infty} e^{-\frac{h^2 n_z^2}{8mkTc^2}} dn_z = \frac{c(2\pi mkT)^{1/2}}{h}.$$

The translational partition function in this limit is given by

$$Z_{trans} = \frac{abc}{h^3}(2\pi mkT)^{3/2}.$$

The volume of the rectangular box is $V = abc$ so that we have

$$Z_{trans} = \frac{V}{h^3}(2\pi mkT)^{3/2}. \tag{4}$$

This result is identical to the partition function for an atom in a monatomic ideal gas as derived from *quantum statistics*.

This is automatically the case of a monatomic gas where the motion of the particle in only translational, whereas in the case of diatomic or polyatomic gas, Z_{trans} is separated from the product

$$Z = Z_{trans}\ Z_{vib}\ Z_{rot}\ Z_{elec}\ Z_{nuc}\ Z_{chem},$$

when the contributions are completely independent, as noted above.

11.4. Rotational partition function

In a *diatomic* molecule, the atomic nuclei are on the axis of zero moment of inertia $I_a = 0$ and $I_b = I_c$. It has been pointed out before,* that for the generalised rotation of a rigid molecule, whether linear or non-linear, we make use of a certain theorem in mechanics for the determination of the corresponding rotational energy levels. This theorem says that the locus of points formed by plotting $I^{-\frac{1}{2}}$ radially from the centre of mass of a molecule in the direction of the axis of rotation is the surface of a triaxial ellipsoid known as the *momental ellipsoid.* The three mutually perpendicular axes of the momental ellipsoid coincide with the principal axes of inertia of the molecule, about which rotation is dynamically balanced. It is customary to label the axes of the ellipsoid *a*, *b*, and *c* in order of decreasing length so that *a* is the major axis, *b* the intermediate, and *c* the minor axis of the ellipsoid. Thus, we have $I_a < I_b < I_c$ always.

Accordingly, for a linear molecule, I_b and I_c are mutually perpendicular, and perpendicular to the internuclear axis *a*, and the momental ellipsoid is thus a circular cylinder of infinite length.

Now based on Schrödinger mechanics, or wave mechanics, the energy levels associated with the rotational motion of a linear molecule are

$$\epsilon_{rot} = J(J+1)\frac{h^2}{8\pi^2 I_b}, \tag{5}$$

where J is a quantum number that may be any positive integer, zero included; the rotation of the molecule generating a quantised angular momentum **J** vector given by the expression

* See Section 8.3.

$$\mathbf{J} = \sqrt{J(J+1)}\,\frac{h}{2\pi}. \tag{6}$$

This is the first rule implied by wave mechanics. The second rule, which can also be used retrospectively to adapt the classical equations to quantised conditions, is that the component **K** of the total angular momentum vector **J** along an axis *z* fixed in space is quantised; that is, it assumes any one of the values given by the expression

$$\mathbf{K} = K\,\frac{h}{2\pi}, \tag{7}$$

where *K* is a second quantum number which may adopt certain positive or negative integral values, including the value zero. However, since **K** is the axial component of **J**, the quantum number *K* cannot be greater than *J*; therefore, the permitted values of *K* or the spectrum of possible values of *K* is

$$K = J,\ J-1,\ \dots,\quad -J+1,\ -J.$$

This means that for each value of the quantum number *J*, there are $2J + 1$ values of the quantum number *K*. Consideration of the wave functions for rotation of a linear molecule shows that the energy levels are $(2J + 1)$-fold degenerate, that is, there are $(2J + 1)$ coincident sub-levels for each value of *J*.

The rotational partition function Z_{rot} is given by

$$Z_{rot} = \Sigma\, g_{rot}\, e^{-\beta\epsilon_i}. \tag{8}$$

Substituting for the statistical weight g_{rot} of the rotational energy level in the case of a diatomic molecule by $2J + 1$, and for the energy ϵ_{rot} by Eq. (5), we obtain

$$Z_{rot} = \sum_{J=0,1,2,\dots} (2J+1)\, e^{-J(J+1)h^2/8\pi^2 I_b kT}. \tag{9}$$

The value of *kT* at room temperature is $\frac{1}{40}$ eV, and the spacing of the rotational energy states is comparatively small so that at this temperature and considerably below the sum in Eq. (9) can be replaced by an integral, and we have

$$Z_{rot} = \int_{J=0}^{J=\infty} (2J+1)\left(e^{-J(J+1)h^2/8\pi^2 I_b kT}\right) dJ. \tag{10}$$

On the basis of a *rotational characteristic temperature* defined as

$$\Theta_{rot} = \frac{h^2}{8\pi^2 I_b k},$$

which is seen to have the dimensions of temperature. Eq. (10) can be rewritten as

$$Z_{rot} \int_{J=0}^{J=\infty} (2J+1)\, e^{-J(J+1)\Theta_{rot}/T}\, dJ. \tag{11}$$

Letting

$$x = \frac{\Theta_{rot}\, J(J+1)}{T},$$

we have

$$\frac{T}{\Theta_{rot}}dx = (2J+1)\,dJ,$$

and the partition function given in Eq. (11) assumes the form

$$Z_{rot} = \frac{T}{\Theta_{rot}}\int_0^{\infty} e^{-x}\,dx$$

$$= \frac{T}{\Theta_{rot}} = \frac{8\pi^2 I_b kT}{h^2}. \tag{12}$$

Eq. (12) is the *classical rotational partition function*, and is shown here to be valid for $T/\Theta_{rot} >> 1$. An asymptotic expansion of the series form of Z_{rot}, Eq. (9), which is valid for $T/\Theta_{rot} > 1$, yields

$$Z_{rot} = \frac{T}{\Theta_{rot}}\left[1 + \frac{1}{3}\left(\frac{\Theta_{rot}}{T}\right) + \frac{1}{15}\left(\frac{\Theta_{rot}}{T}\right)^2 + \ldots\right], \tag{13}$$

which permits calculation of the *first deviation* from the classical value given by Eq. (12). At high temperatures, the classical and quantum equations for the rotational partition function are very nearly the same. On the other hand, Eq. (9) shows that at very low temperatures the rotational contribution disappears.

The results obtained above are valid for *diatomic molecules with different atoms*. However, the results must be modified when the two atoms are *identical*. At high temperatures, that is, in the *classical limit*, the required modification is *simple*. When a diatomic molecule of two identical atoms is rotated through 180 degrees, its orientation after rotation cannot be distinguished from its original orientation. Thus, such a molecule of two identical atoms has only *one-half* as many distinguishable orientations as a diatomic molecule with different atoms, and consequently, the classical rotational partition function for diatomic molecules with two identical atoms is *one-half* the partition function for diatomic molecules with different atoms.

To distinguish the different cases, it is convenient to introduce *a symmetry number** k ($\text{or } \sigma$) which is the number of different but *indistinguishable spatial orientations that a molecule can have*. For a diatomic molecule of two *identical* atoms, $k = 2$; for a diatomic molecule of two *different* atoms, $k = 1$. Hence, the classical rotational partition function for all diatomic or linear molecules can be expressed as

$$Z_{rot} = \frac{8\pi^2 I_b kT}{k h^2}. \tag{14}$$

The rotational partition function for polyatomic molecules with three principal moments of inertia I_a, I_b, and I_c are given by

$$Z_{rot} = \frac{8\pi^2\left(8\pi^3 I_a I_b I_c\right)^{1/2}}{k h^3}(kT)^{3/2}. \tag{15}$$

---

* The symmetry number of a rotating molecule in classical statistics was first introduced by *Ehrenfest* and *Trkal* in 1920, and it was denoted by σ. See Section 19.11, pp. 907–908.

Symmetry numbers for various molecules are given in Table 1.

Table 1
Symmetry numbers

Molecules	Symmetry number	Geometry
HCl, HI, etc.	1	Dumb-bell
H_2O, H_2, Cl_2, C_6H_{10}, etc.	2	Isosceles triangle
NH_3, etc.	3	Triangular pyramid
C_2H_4, etc.	4	Rectangle
C_6H_{12}, etc.	6	Chair form
CH_4, CCl_4, etc.	12	Regular tetrahedral
C_6H_6, etc.	12	Hexagon

11.5. Vibrational partition function

If the vibrations of a molecule are sufficiently small, the curve which is shown in Fig. 1 (in Section 11.2) may be replaced by a parabola, as indicated by the dashed curve. This is also equivalent to taking the leading term in a *Taylor's expansion* about the equilibrium position. Clearly, such an approximation is not valid for the higher vibrational levels. We shall assume here that the vibrations occur only at the bottom of the curve, and may, therefore, be treated as a simple harmonic vibration.

The vibration of a diatomic molecule is equivalent to the one-dimensional oscillator, since only the single radial coordinate determines the separation of the nuclei. We have shown before,* using wave mechanics, that the energy of a linear harmonic oscillator is given by

$$\epsilon_{vib} = \left(v + \frac{1}{2}\right)h\nu, \tag{16}$$

where ν is the classical absolute frequency of the oscillator. It is evident from Eq. (16) that the minimum energy of the oscillator is not zero; for letting the quantum number $v = 0$, we find

$$(\epsilon_0)_{vib} = \frac{1}{2}h\nu. \tag{17}$$

This energy of the lowest state of the oscillator is termed by some authors the *zero-point energy* and by others the *residual vibration energy*. Since the statistical weight is unity, the vibrational partition function is

$$Z_{vib} = \sum_{v=0}^{v=\infty} e^{-\left(v+\frac{1}{2}\right)h\nu/kT}.$$

(18)

It should be pointed out that the upper limit on the summation is in fact finite, since the molecule will dissociate. However, at low temperatures, at which the simple harmonic approximation is valid, the contribution for large v values is insignificant and Eq. (18) is

* See Section 8.5.

appropriate. Since the energy ϵ_v given by Eq. (17) is not a continuous function of a coordinate, the summation in Eq. (18) must be evaluated directly. If we write

$$Z_{vib} = e^{-\frac{1}{2}h\nu/kT} + e^{-\frac{3}{2}h\nu/kT} + e^{-\frac{5}{2}h\nu/kT} + \ldots$$

$$= e^{-\frac{1}{2}h\nu/kT}\left(1 + e^{-h\nu/kT} + e^{-(h\nu/kT)^2} + \ldots\right)$$

and note that

$$1 + x + x^2 + \ldots = \frac{1}{1-x},$$

then the partition function may be expressed as

$$Z_{vib} = \frac{e^{-h\nu/2kT}}{1 - e^{-h\nu/kT}}. \tag{19}$$

The vibrational partition function given in Eq. (19) is clearly restricted to diatomic molecules because of the original assumption of a simple oscillator. The partition function for *rigid* polyatomic molecules can be easily derived. The limitation to rigid molecules *does not imply that vibrations are absent*, and fortunately, the majority of polyatomic molecules (including nearly all molecules composed of three or four atoms) are essentially rigid. A molecule containing N atoms requires *3N spatial coordinates* to determine the positions of the atoms. This number is the number of *degrees of freedom* possessed by the molecule. Of them, three are accounted for by the *translational motion of the* centre of mass, and, in general, *a further three by the rotation of the molecule*. The remainder, $3N–6$, must be associated with motions of the nuclei relative to one another, or, in other words, with the *internal modes of vibration of the molecule*. Linear molecules form a special case, for rotation about the internuclear axis *does not change the nuclear coordinates*; a linear molecule therefore has two rotational and $3N–5$ vibrational degrees of freedom.

The fundamental vibration frequencies corresponding to the normal modes can be obtained from spectroscopic analysis. For each normal frequency ν, we have the partition function Z_{vib} expressed by Eq. (19). Hence, the vibrational partition function for a *rigid polyatomic molecule* is

$$Z_{vib} = \Pi\left[\frac{e^{-h\nu_i/2kT}}{1 - e^{-h\nu_i/kT}}\right],$$

where the product is taken over all $3N–6$ (or $3N–5$ for linear molecules) modes of vibration.

Considering the rotational and vibrational contributions together, we can write a general expression for the partition function for a rigid polyatomic molecule with three large principal moments of inertia, at elevated temperatures, and with the modes of vibration taken into account. We obtain

$$Z_{vib,rot} = \frac{8\pi^2\left(8\pi^2 I_a I_b I_c\right)^{1/2}}{k h^3}(kT)^{3/2}\,\Pi\left[\frac{e^{-h\nu_i/2kT}}{1 - e^{-h\nu_i/kT}}\right]. \tag{21}$$

The data needed for the solution of the equations for the rotational and vibrational partition functions are usually tabulated in the literature in terms of the wave number $1/\lambda = \nu/c$, where c is the velocity of light.

11.6. Electronic partition function

In a system composed of an assembly of single atoms (monatomic), the electron is capable to rotate in an orbit around the nucleus and thus generates an orbital angular momentum vector which is specified or quantised by the orbital quantum number ℓ, and is capable also to spin and thus generates a spin angular momentum **s** which is specified by the spin quantum number *s*. Coupling between these momenta takes place, giving rise to a resultant or total angular momentum **J** for the electron, specified by the quantum number *J* which is given by $J = \ell \pm s$; it is always positive. Clearly, it is this motion of the electron which makes contribution to the internal energy $\in_{int}$. For diatomic and polyatomic molecules, vibrational and rotational contributions are added to the electronic contribution. We shall consider now the latter type.

Under normal conditions, nearly all molecules and atoms are in their lowest electronic energy state or simply *ground state*; that is, the electrons are in their *lowest orbitals. The energy* $\in_0$ *of the electronic ground state is, by convention, taken to be equal to zero.* This ground-state energy is the chemical energy discussed above in Section 11.2, and because of the convention described above, it has to be taken into account by including an appropriate factor in the partition functions for systems involving certain chemical reactions. Even if the ground-state energy is taken to be zero, there is a statistical weight which must be considered.

Another factor is the possibility of the excited states. For atoms, with the exception of *thallium*, there is *very large gap* between the *lowest excited state* and the *ground state* so that thermal excitation cannot be significant except at extremely high temperatures. In the case of some molecular species, notably NO, O_2, and S_2, there are sometimes low-lying excited states which can be reached at normal temperatures.

Suppose that a molecule has the electronic energy levels $\in_0, \in_1, \in_2, \ldots$, and that the appropriate statistical electronic weights are $g_0, g_1, g_2, \ldots$, respectively. Then the *electronic partition function* Z_{elec} is given by

$$Z_{elec} = g_0 \, e^{-\in_0/kT} + g_1 \, e^{-\in_1/kT} + \ldots. \qquad (22)$$

If the electrons are not excited or, to give a rule of thumb, if the exponent $hc/\lambda k$ or $h\nu/k$ is greater then $5T$ for any term, then that term can be neglected. Under these conditions, and taking $\in_0 = 0$, Eq. (22) reduces to

$$Z_{elec} = g_0 \quad . \qquad (23)$$

It will be shown in Sections 19.4 and 19.6 that when the *lowest electronic level is chosen as energy zero*, the *electronic (or internal) partition function reduces to* υ_0. The symbol υ_0 thus *represents the electronic spin weight of the normal molecule* as does the symbol g_0.

The statistical weight in each of the terms of Eq. (22) is found from spectroscopic data. As noted above, the total angular momentum of an electron is quantised or specified by a rotational quantum number *J*, which is always positive, and is obtained by the coupling between the orbital quantum number ℓ and the electron spin quantum number *s*, that is,

$$J = \ell \pm s.$$

There are $2J + 1$ orientations of the total angular momentum vector **J** for each value of *J* giving an energy state. The statistical weight is therefore

$$g = 2J + 1.$$

If there are low-lying electronic states close to the ground state, the second term is included, and the partition function is evaluated by direct summation to give

$$Z_{elec} = g_0 + g_1 e^{-\epsilon_1/kT}. \quad (24)$$

The computation of the electronic partition function involves the knowledge of the electronic energy values in the excited states and of the statistical weights; which are given in *Thermochemical Tables.*[*]

11.7. Nuclear partition function

We have now to consider the energy states of the nucleus. The nuclei of atoms are not excited to energy states above the ground states at any temperatures achievable on earth. Hence, the nuclear energy levels are not of direct concern here, with notable exception. A nucleus in the ground state is capable of *rotation, or spin, about an axis*. The spin generates a quantised[†] angular momentum vector **I** whose magnitude in given by

$$\mathbf{I} = \sqrt{I(I+1)}\ h/2\pi. \quad (25)$$

Here, *I* the quantum number for the nuclear spin can assume integral, half-integral, or zero value. The angular momentum vector **I** can be thought of as the *resultant* of the angular momenta of the protons and neutrons of which the nucleus is composed. Each of these particles is *considered to have a spin quantum number* $I = 1/2$, the value of the proton being experimentally observable. Nuclei with resultant spin greater than zero possess a *magnetic moment*; that is to say, the nucleus *behaves like a small bar magnet* whose axis coincides with the axis of spin.

The *orientation* of the angular momentum vector **I** *is quantised with respect to an axis fixed in space*. The general principles of spatial quantisation of this kind allow $\mathbf{I}_z$, the component of **I** along a space-fixed axis *z*, to assume any one of the values given by the expression.

$$\mathbf{I}_z = M_I h/2\pi, \quad (26)$$

where M_I is a *second quantum number* that may adopt certain positive or negative integral or half-integral values, zero included. Since $\mathbf{I}_z$ cannot be greater than **I** *itself*, the limits of M_I are $\pm I$, and the spectrum of the allowed values of M_I is

$$M_I = I,\ I-1,\ \ldots\ -\ (I-1),\ -I. \quad (27)$$

(Note that the values of M_I are either all integral or all half-integral, depending on whether *I* is integral or half-integral). The total number of possible values of the quantum number M_I is $2I+1$ and hence the vector **I** is allowed $2I + 1$ discrete orientations with respect to a space-fixed axis.

---

[*] *JANAF Interim Thermochemical Tables*. The Dow Chemical Co., Midland, Michigan, 1961.

[†] See Section 8.2.

The electrically charged, spinning nucleus with a resultant spin quantum number I *greater than zero* produces a magnetic field as noted above so that such a field will interact with an externally applied magnetic field. However, in the absence of an applied external field, the oriented states cannot be detected, but these states which represent the $(2I + 1)$-fold degeneracy of the nuclear energy level exist and give the nucleus a statistical weight of $(2I + 1)$. This degeneracy will be lifted in favour of a set of $2I + 1$ sub-levels of different energy when an external magnetic field is applied.

Conventionally, the ground-state nuclear energy level is taken to be zero; therefore, the effect of the nuclear spin is to increase the number of quantum states in every energy level by the factor $2I + 1$ for each atom in the molecule. In terms of the nuclear partition function, this is given by

$$Z_{nuc} = g_0 = 2I + 1. \tag{28}$$

Since the number of atoms present in a gas is constant, the nuclear spin partition function *does not influence* the thermodynamic behaviour of a system. The nuclear spin contribution to the internal energy ϵ_{int} cancels out in *equilibrium reactions*. However, the nuclear partition function can affect the *values of the entropy and free energies* so that when *exact values* of these thermodynamic functions are required, the nuclear spin contribution to ϵ_{int} must be taken into account.

It will be shown in Section 19.12, that an extra factor denoted by ρ representing the number of possible nuclear spin eigenfunctions, that is, nuclear spin weight, is introduced into the partition function of *a monatomic gas* as a matter of amendment if we wish to keep its partition of function formulae into line with those of diatomic molecules. We have then

$$Z_{(monatomic)} = \frac{(2\pi m k T)^{3/2}}{h^3} \upsilon_0 \rho,$$

where υ_0 is the electronic weight of the normal state.

CHAPTER 12

Thermodynamic Properties

12.1. Introduction

In this chapter, we shall establish the link between the thermodynamic coordinates of a macroscopic system and the coordinates of the constituent microscopic systems. This link is provided by the partition function described in Chapter 11.

There we have been primarily concerned thus far in writing the partition function for a *single* particle, molecule, or microscopic system. The *total partition function*, in the sense that it is written for *N identical* particles, is

$$\mathbf{Z} = Z^N = \left(\sum e^{-\beta\epsilon_i}\right)^N.$$

However, this relation implies that the *N* particles are *distinguishable*. If we are considering a system composed of *N* weakly interacting particles with *identical energy levels*, then the total partition function **Z** must be modified to account for the *degeneracy of the total energy* of the system. It is already known that the number of different ways in which the same total energy can be obtained by interchanging the *N identical* particles is

$$\frac{N!}{\Pi\, N_i!},$$

where N_i is the number of particles in the *i*th energy state. However, for a weakly coupled gas at a reasonably high temperature, the probability that more than one particle would have the same energy should be negligible. Consequently, it is appropriate to assume that N_i's are either zero or one. While the product, $\Pi\, N_i!$ thus reduces to unity, the $N!$ must be retained. As a result, the total partition function for a system composed of *N identical, indistinguishable* particles at a reasonably high temperature must be written as

$$\mathbf{Z} = \frac{1}{N!} Z^N = \frac{1}{N!}\left(\sum e^{-\beta\epsilon_i}\right)^N. \tag{1}$$

12.2. Internal energy

The internal energy of a system comprising *N* atoms or molecules is equal to the sum of the energies of the individual particles. That is,

$$U = \sum \epsilon_i N_i = N\overline{\epsilon},$$

where the average energy $\overline{\epsilon}$ of a particle is defined by

$$\overline{\epsilon} = \frac{\sum \epsilon_i N_i}{\sum N_i}. \tag{2}$$

The number of particles in the *i*th cell in the state of thermodynamic equilibrium (or the state of maximum thermodynamic probability) is

$$N_i = \frac{N}{Z} e^{-\beta \epsilon_i}, \tag{3}$$

where the partition function *Z* for a single particle is

$$Z = \Sigma e^{-\beta \epsilon_i}. \tag{4}$$

Substituting Eqs. (3) and (4) into Eq. (2), we have

$$\overline{\epsilon} = \frac{\frac{N}{Z} \Sigma \epsilon_i e^{-\beta \epsilon_i}}{\frac{N}{Z} \Sigma e^{-\beta \epsilon_i}},$$

$$= \frac{\Sigma \epsilon_i e^{-\beta \epsilon_i}}{Z}. \tag{5}$$

Taking the derivative of *Z* with respect to β, at constant volume since the energy ϵ_i may depend upon the volume, using Eq. (4), we obtain

$$\left(\frac{\partial Z}{\partial \beta}\right)_V = \left\{\frac{\partial\left(\Sigma e^{-\beta \epsilon_i}\right)}{\partial \beta}\right\}$$

$$= - \Sigma \epsilon_i e^{-\beta \epsilon_i}. \tag{6}$$

Substituting Eq. (6) into Eq. (5), we have

$$\overline{\epsilon} = -\frac{1}{Z}\left(\frac{\partial Z}{\partial \beta}\right)_V = -\left(\frac{\partial \ln Z}{\partial \beta}\right)_V. \tag{7}$$

Therefore, the internal energy for a system of *N* particles is

$$U = N\overline{\epsilon} = -N\frac{1}{Z}\left(\frac{\partial Z}{\partial \beta}\right)_V = -N\left(\frac{\partial \ln Z}{\partial \beta}\right)_V. \tag{8}$$

Eq. (8) can also be put in the form

$$U = -\left(\frac{\partial \ln \mathbf{Z}}{\partial \beta}\right)_V. \tag{9}$$

This is because by taking the logarithm of Eq. (1), we obtain

$$\ln \mathbf{Z} = -\ln N_i! + N \ln Z,$$

so that

$$\frac{\partial \ln \mathbf{Z}}{\partial \beta} = N \frac{\partial \ln Z}{\partial \beta}.$$

Since $\beta = 1/kT$, and $Nk = nR$, where *n* is the number of moles, we obtain, according to Eq. (8),

$$U = NkT^2\left(\frac{\partial \ln Z}{\partial T}\right)_V = \frac{NkT^2}{Z}\left(\frac{\partial Z}{\partial T}\right)_V, \tag{10}$$

or

$$U = n\,R\,T^2\left(\frac{\partial \ln Z}{\partial T}\right)_V = \frac{n\,R\,T^2}{Z}\left(\frac{\partial Z}{\partial T}\right)_V,$$

where $\partial\beta$ is substituted for by $-\,dT/kT^2$. Also Eq. (9) takes the form

$$U = kT^2\left(\frac{\partial \ln \mathbf{Z}}{\partial T}\right). \tag{11}$$

It may be noted that if the derivative of the partition function *Z* or its logarithm ln *Z* is taken with respect to $1/T$ instead of *T*, a modification of the corresponding expressions of *U* in terms of *T* should be made. That is, since

$$\frac{d(1/T)}{dT} = -\frac{1}{T^2},$$

we have

$$\left(\frac{\partial Z}{\partial T}\right)_V = \left\{\frac{\partial Z}{\partial\, 1/T}\cdot\frac{\partial\, 1/T}{\partial T}\right\}_V = -\frac{1}{T^2}\left(\frac{\partial Z}{\partial\, 1/T}\right)_V,$$

and

$$\left(\frac{\partial \ln Z}{\partial T}\right)_V = \left\{\frac{\partial \ln Z}{\partial\, 1/T}\cdot\frac{\partial\, 1/T}{\partial T}\right\}_V = -\frac{1}{T^2}\left(\frac{\partial \ln Z}{\partial\, 1/T}\right)_V.$$

Accordingly, Eq. (10) can be rewritten in the form

$$U = -Nk\left(\frac{\partial \ln Z}{\partial\, 1/T}\right)_V = -\frac{Nk}{Z}\left(\frac{\partial Z}{\partial\, 1/T}\right)_V, \tag{12}$$

and Eq. (11) in the form

$$U = -k\left(\frac{\partial \ln \mathbf{Z}}{\partial\, 1/T}\right)_V. \tag{13}$$

12.3. Specific heat

By definition, the *molar specific heat* of a substance at constant volume is

$$c_V = \frac{1}{n}\left(\frac{\partial U}{\partial T}\right)_V, \tag{14}$$

where *n* is the number of moles. Therefore, using Eq. (12) we have

$$\left(\frac{\partial U}{\partial T}\right)_V = -Nk\left\{\frac{\partial(\partial \ln Z/\partial 1/T)}{\partial T}\right\}_V.$$

But

$$\left\{\frac{\partial(\partial \ln Z/\partial 1/T)}{\partial T}\right\}_V = \left\{\frac{\partial(\partial \ln Z/\partial 1/T)}{\partial 1/T}\right\}_V \frac{\partial 1/T}{\partial T},$$

and

$$\frac{\partial 1/T}{\partial T} = -\frac{1}{T^2},$$

so that

$$\left(\frac{\partial U}{\partial T}\right)_V = \frac{Nk}{T^2}\left\{\frac{\partial^2 \ln Z}{\partial (1/T)^2}\right\}_V. \tag{15}$$

Hence,

$$c_V = \frac{1}{n}\left(\frac{\partial U}{\partial T}\right)_V$$

$$= \frac{R}{T^2}\left\{\frac{\partial^2 \ln Z}{\partial (1/T)^2}\right\}_V. \tag{16}$$

12.4. Entropy

The entropy was evaluated before* in terms of the single particle partition function. The equation obtained was

$$S = Nk\left[\ln\left(\frac{Z}{N}\right) + 1\right] + \frac{U}{T}.$$

Substituting for U from

$$U = NkT^2\left(\frac{\partial \ln Z}{\partial T}\right)_V, \tag{10}$$

we obtain

$$S = Nk\left[\ln\left(\frac{Z}{N}\right) + T\left(\frac{\partial \ln Z}{\partial T}\right)_V + 1\right]. \tag{17}$$

Since N is a large number, *Stirling's approximation* may be used when we take the logarithm of Eq. (1), with the result

$$\mathbf{Z} = \frac{1}{N!}Z^N \tag{1}$$

$$\ln \mathbf{Z} = -\ln N! + N\ln Z$$

$$= -N\ln N + N + N\ln Z,$$

$$= N\left[\ln\left(\frac{Z}{N}\right) + 1\right], \tag{18}$$

so that

$$\left(\frac{\partial \ln \mathbf{Z}}{\partial T}\right)_V = N\left(\frac{\partial \ln Z}{\partial T}\right)_V. \tag{19}$$

---

* See Section 9.5.

Now, in order to express the entropy as a function of the *total* partition function **Z**, we substitute Eqs. (18) and (19) into Eq. (17). Thus, we obtain

$$S = k\left[\ln \mathbf{Z} + T\left(\frac{\partial \ln \mathbf{Z}}{\partial T}\right)_V\right]. \qquad (20)$$

12.5. Helmholtz function

The Helmholtz function can be readily obtained from the partition function equations for the internal energy and entropy. Substituting for U and S in the Helmholtz function defined as

$$F \equiv U - TS$$

by their corresponding expressions, namely Eqs. (10) and (17), we obtain

$$F = NkT^2\left(\frac{\partial \ln Z}{\partial T}\right)_V - NkT\left[\ln\left(\frac{Z}{N}\right) + T\left(\frac{\partial \ln Z}{\partial T}\right)_V + 1\right].$$

$$= -NkT\left[\ln\left(\frac{Z}{N}\right) + 1\right]. \qquad (21)$$

The same result can be readily obtained from the S expression without any substitution. That is, since, as shown above for a single particle partition function, S is given by

$$S = Nk\left[\ln\left(\frac{Z}{N}\right) + 1\right] + \frac{U}{T}.$$

Therefore,

$$TS - U = NkT\left[\ln\left(\frac{Z}{N}\right) + 1\right],$$

or

$$F = U - TS,$$

$$= -NkT\left[\ln\left(\frac{Z}{N}\right) + 1\right],$$

or

$$= -nRT\left[\ln\left(\frac{Z}{N}\right) + 1\right]. \qquad (21)$$

On the other hand, in order to express F in terms of the total partition function **Z**, we make an expansion of Eq. (21). Thus,

$$F = -NkT(\ln Z - \ln N) - NkT,$$

$$= -kT(N\ln Z - N\ln N + N),$$

$$= -kT\left[\ln Z^N - \left(N\ln N - N\right)\right]. \qquad (22)$$

Using Sterling's approximation,

$$\ln N! = N\ln N - N,$$

Eq. (22) can be rewritten as

$$F = -kT\left[\ln Z^N - \ln N!\right],$$

or

$$F = -kT\ \ln\frac{Z^N}{N!},$$

therefore,

$$F = -kT\ \ln \mathbf{Z}\,, \tag{23}$$

where the total partition function **Z** is given above by Eq. (1).

12.6. Pressure and Gibbs function

The Gibbs function by definition is

$$G \equiv H - TS$$

or

$$G \equiv U - TS + PV\,,$$

and in terms of the Helmholtz function is

$$G \equiv F + PV\,. \tag{24}$$

Differentiating Eq. (21) with respect to volume at constant temperature, we get

$$\left(\frac{\partial F}{\partial V}\right)_T = -NkT\left(\frac{\partial \ln Z}{\partial V}\right)_T. \tag{25}$$

But from classical thermodynamics,

$$\left(\frac{\partial F}{\partial V}\right)_T = -P, \tag{26}$$

as shown before; therefore, from the two previous equations – Eqs. (25) and (26) – we obtain for the pressure the expression

$$P = NkT\left(\frac{\partial \ln Z}{\partial V}\right)_T,$$

or

$$= nRT\left(\frac{\partial \ln Z}{\partial V}\right)_T. \tag{27}$$

Substituting this result and Eq. (21) into Eq. (24), we find that the Gibbs function is given by

$$G = -nRT\left[\ln\left(\frac{Z}{N}\right) + 1 - V\left(\frac{\partial \ln Z}{\partial V}\right)_T\right]. \tag{28}$$

In the case of an *ideal gas*, the free energy expression can be readily obtained as follows. The product *PV* in the definition of *G* given above is replaced by *NkT* so that

$$G = U - TS + NkT\,. \tag{29}$$

Then replacing *S* by its expression

$$S = Nk\left[\ln\left(\frac{Z}{N}\right) + 1\right] + \frac{U}{T},$$

Eq. (29) takes the form

$$G = U - NkT\left[\ln\left(\frac{Z}{N}\right) + 1\right] + NkT - U$$

$$= -NkT\left[\ln\left(\frac{Z}{N}\right)\right],$$

or,

$$G = -nRT\left[\ln\left(\frac{Z}{N}\right)\right] \qquad \text{(ideal gas).} \qquad (30)$$

12.7. Enthalpy

By definition the enthalpy *H* is

$$H \equiv U + PV .$$

The substitution for *U* and *P* by their expressions in terms of the partition function *Z* yields

$$H = NkT^2\left(\frac{\partial \ln Z}{\partial T}\right)_V + NkTV\left(\frac{\partial \ln Z}{\partial V}\right)_T . \qquad (31)$$

If the derivatives are considered with respect to $\ln T$ instead of *T* at constant volume, and with respect to $\ln V$ instead of *V* at constant temperature, Eq. (31) can be rewritten as

$$H = NkT\left[\left(\frac{\partial \ln Z}{\partial \ln T}\right)_V + \left(\frac{\partial \ln Z}{\partial \ln V}\right)_T\right],$$

or,

$$= nRT\left[\left(\frac{\partial \ln Z}{\partial \ln T}\right)_V + \left(\frac{\partial \ln Z}{\partial \ln V}\right)_T\right]. \qquad (32)$$

12.8. Work and heat

In deriving the equation for the internal energy of a system, we considered the system to be at constant volume. Hence, there could be no '*P* d*V*' work. We want now to consider the case when the volume is allowed to change.

As a particular example, we may consider a single particle in a rectangular box with dimensions *a*, *b*, and *c* as shown in Fig. 1. Let us

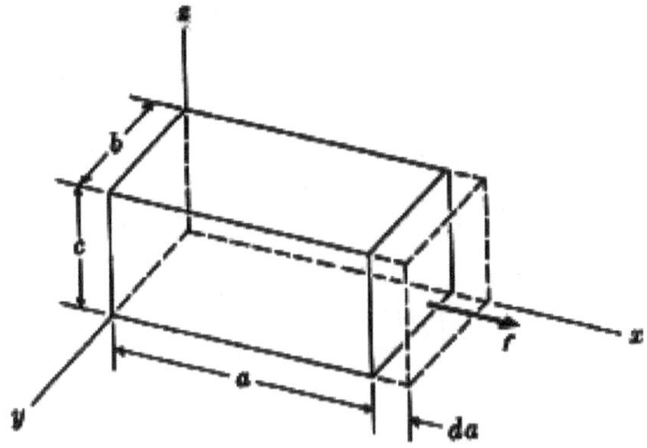

Fig.1.

denote by f the force in the *x*-direction exerted by the particle on the right wall; that is, the wall at $x = a$. Since the particle can be in any one quantum level, let us denote by f_i the force exerted by the particle when it is in quantum state *i* with an energy ϵ_i. Now we consider what happens when the right wall is displaced slowly by an amount d*a*, to the position

$$x = a + da.$$

In this process, the particle does an amount of work $f_i\,da$ which must be equal to *decrease* in the energy of the particle by an amount $d\epsilon_i$, that is,

$$-d\epsilon_i = f_i\,da,$$

or

$$f_i = -\frac{\partial \epsilon_i}{\partial a},$$

with the other dimensions of the box remaining constant.

The mean force $\bar{f}$ exerted against the wall by a particle which is a part of a system of *N* particles is

$$\bar{f} = \frac{\Sigma N_i f_i}{\Sigma N_i} = \frac{\Sigma f_i\, e^{-\epsilon_i/kT}}{\Sigma\, e^{-\epsilon_i/kT}},$$

$$= -\frac{\Sigma(\partial \epsilon_i/\partial a)\, e^{-\epsilon_i/kT}}{\Sigma\, e^{-\epsilon_i/kT}}.$$

The change of $Z = \left(\Sigma\, e^{-\epsilon_i/kT}\right)$ with respect to the dimension *a* at constant temperature is

$$\left(\frac{\partial Z}{\partial a}\right)_T = -\Sigma\, e^{-\epsilon_i/kT}\left(\frac{\partial \epsilon_i}{\partial a}\right)\frac{1}{kT},$$

$$= -\frac{1}{kT}\sum\left(\frac{\partial \epsilon_i}{da}\right)_T e^{-\epsilon_i/kT}.$$

This equation together with the preceding equation, gives

$$\bar{f} = \frac{kT}{Z}\left(\frac{\partial Z}{\partial a}\right)_T = kT\left(\frac{\partial \ln Z}{\partial a}\right)_T.$$

For *N* particles the work done when *a* is increased by d*a* is

$$d'W = N\bar{f}\ da = NkT\left(\frac{\partial \ln Z}{\partial a}\right)_T da.$$

Noting that $V = abc$, we may generalise the above result to

$$d'W = NkT\left(\frac{\partial \ln Z}{\partial V}\right)_T dV. \tag{33}$$

The substitution of *P* given above as

$$P = NkT\left(\frac{\partial \ln Z}{\partial V}\right)_T,$$

into Eq. (33) yields

$$d'W = P\,dV, \tag{34}$$

which is valid for a *reversible process*.

From Eq. (34) for the work done by the system against the surroundings, and the first law of thermodynamics written in its analytical formulation

$$d'q = dU + d'W,$$

we are able to write the equation for heat addition in terms of the partition function. Thus, substituting Eqs. (10) and (27), we obtain

$$d'q = Nk\,d\left[T^2\left(\frac{\partial \ln Z}{\partial T}\right)_V\right] + NkT\left(\frac{\partial \ln Z}{\partial V}\right)_T dV. \tag{35}$$

This equation in turn suggests that we may make use of the total differential of $(\ln Z)$, by expressing it as a function of temperature and volume. That is, we put

$$\ln Z = \ln Z(T, V),$$

therefore,

$$d(\ln Z) = \left(\frac{\partial \ln Z}{\partial T}\right)_V dT + \left(\frac{\partial \ln Z}{\partial V}\right)_T dV. \tag{36}$$

Thus, the substitution for $\left(\frac{\partial \ln Z}{\partial V}\right)_T dV$ in Eq. (35) from Eq. (36) gives

$$\begin{aligned} d'q &= Nk\,d\left[T^2\left(\frac{\partial \ln Z}{\partial T}\right)_V\right] + NkT\left[d(\ln Z) - \left(\frac{\partial \ln Z}{\partial T}\right)_V dT\right] \\ &= Nk\,d\left[T^2\left(\frac{\partial \ln Z}{\partial T}\right)_V\right] - NkT\left(\frac{\partial \ln Z}{\partial T}\right)_V dT \\ &\qquad + NkT\,d(\ln Z). \end{aligned} \tag{37}$$

We can also write*

$$d\left[T^2\left(\frac{\partial \ln Z}{\partial T}\right)_V\right] = \left[T\left(\frac{\partial \ln Z}{\partial T}\right)_V dT + \right.$$

$$\left. Td\left\{T\left(\frac{\partial \ln Z}{\partial T}\right)_V\right\}\right], \tag{38}$$

so that Eq. (37) can be re-written, after the substitution of Eq. (38), as

$$d'q = Nk\left[T\left(\frac{\partial \ln Z}{\partial T}\right)_V dT + Td\left\{T\left(\frac{\partial \ln Z}{\partial T}\right)_V\right\}\right]$$

$$-NkT\left(\frac{\partial \ln Z}{\partial T}\right)_V dT + NkT\, d(\ln Z)$$

$$= NkTd\left[(\ln Z) + T\left(\frac{\partial \ln Z}{\partial T}\right)_V\right]. \tag{39}$$

The differential of the entropy *S*, whose equation has been obtained above, namely

$$S = Nk\left[\ln\left(\frac{Z}{N}\right) + T\left(\frac{\partial \ln Z}{\partial T}\right)_V + 1\right],$$

is given by

$$dS = Nkd\left[\ln(Z) + T\left(\frac{\partial \ln Z}{\partial T}\right)_V\right]. \tag{40}$$

Hence, the substitution of this equation into Eq. (39) produces

$$d'q = TdS, \tag{41}$$

which is valid for *a reversible process.*

12.9. Properties of ideal gases

Let us return to the problem of a particle in a rectangular box which we considered in Section 12.8. Clearly, a large number of such particles constitute an ideal gas, since interactions between particles have not been considered. The force f exerted by each particle f_i does an amount of work against the right wall of the box, as shown in Fig. 1, for a displacement (da), is equal to the decrease in the energy of the particle so that the mean force $\overline{f}$ exerted by a single particle, which is a part of a system of *N* particles, as shown above, is

$$\overline{f} = -\frac{\Sigma\,(\partial \epsilon_i/\partial a)\, e^{-\epsilon_i/kT}}{\Sigma\, e^{-\epsilon_i/kT}}.$$

---

* Put. $d\left[T^2\left(\frac{\partial \ln Z}{\partial T}\right)_V\right]$ as $d\left[T.T\left(\frac{\partial \ln Z}{\partial T}\right)_V\right]$

Taking the partial derivative $\left(\dfrac{\partial Z}{\partial a}\right)_T$ with respect to *a*, we obtained the expression

$$\bar{f} = kT\left(\frac{\partial \ln Z}{\partial a}\right)_T,$$

for the mean force exerted by a particle.

Since the area of the end wall is *bc*, then the pressure exerted on this wall by an ideal gas of *N* molecules is

$$P = \frac{N\bar{f}}{bc}.$$

Substituting for $\bar{f}$ in this equation by the preceding equation, we obtain

$$P = \frac{NkT}{bc}\left(\frac{\partial \ln Z}{\partial a}\right). \tag{42}$$

For an ideal gas, only the translational partition function is a function of volume, it is given as shown before by

$$Z = \frac{V}{h^3}(2\pi mkT)^{3/2}.$$

Since *V* in our case is *abc*, we have

$$Z = \frac{abc}{h^3}(2\pi mkT)^{3/2}.$$

Accordingly, $\left(\dfrac{\partial \ln Z}{\partial a}\right)_T$ is

$$\left(\frac{\partial \ln Z}{\partial a}\right)_T = \frac{\partial}{\partial a}\left[\ln\left\{\frac{abc}{h^3}(2\pi mkT)^{3/2}\right\}\right]$$

$$= \frac{1}{a}.$$

Substituting this result into Eq. (42) gives

$$P = \frac{NkT}{abc},$$

or, with $V = abc$,

$$PV = NkT, \tag{43}$$

which is the equation of state for an ideal gas.

Again, with the use of the partition function, which is the link between the thermodynamic properties, or coordinates, of a macroscopic system and the coordinates of the constituent microscopic systems, the ideal gas equation of state can be deduced, as is expressed by Eq. (43).

For an ideal gas, we have obtained the expression for the Gibbs function *G* as shown by Eq. (30) above, namely

$$G \equiv -\, nRT\left[\ln\left(\frac{Z}{N}\right)\right].$$

Similarly, for the enthalpy *H* defined by

$$H \equiv U + PV = U + nRT,$$

and then using the definition of *G* as

$$G \equiv H - TS,$$

H can be expressed as

$$H \equiv G + TS,$$

the use of which will be shortly considered.

The two general expressions for *S* obtained above are

$$S = Nk\left\{\ln\left(\frac{Z}{N}\right) + 1\right\} + \frac{U}{T},$$

and

$$S = Nk\left\{\ln\left(\frac{Z}{N}\right) + 1\right\} + NkT\left(\frac{\partial \ln Z}{\partial T}\right)_V .$$

The latter is commonly written as

$$S = Nk\left\{\ln\left(\frac{Z}{N}\right) + T\left(\frac{\partial \ln Z}{\partial T}\right)_V + 1\right\}.$$

For an ideal gas, the translational partition function, as we know, is a function of volume so that from the expression of Z_{trans}

$$Z_{trans} = \frac{V}{h^3}(2\pi mkT)^{3/2},$$

the derivative of $\ln Z$ with respect to *T* at *constant volume* is given by

$$\ln Z_{trans} = \ln V - \ln h^3 + \ln(2\pi mk)^{3/2} + \frac{3}{2}\ln T,$$

$$\left(\frac{\partial \ln Z}{\partial T}\right)_V = \frac{3}{2}\cdot\frac{1}{T}.$$

Also the derivative of $\ln Z$ with respect to *T* at constant pressure is

$$\ln Z_{trans} = \ln NkT - \ln P - \ln h^3 + \ln(2\pi mk)^{3/2} + \frac{3}{2}\ln T$$

$$\left(\frac{\partial \ln Z}{\partial T}\right)_P = \frac{1}{T} + \frac{3}{2}\cdot\frac{1}{T}.$$

Therefore, while

$$T\left(\frac{\partial \ln Z}{\partial T}\right)_P = \frac{3}{2} + 1 = \frac{5}{2},$$

we have

$$T\left(\frac{\partial \ln Z}{\partial T}\right)_V = \frac{3}{2}.$$

That is,

$$T\left(\frac{\partial \ln Z}{\partial T}\right)_V + 1 = T\left(\frac{\partial \ln Z}{\partial T}\right)_P.$$

Thus, the substitution of this result into Eq. (17) gives

$$S = N k \left\{ \ln\left(\frac{Z}{N}\right) + T\left(\frac{\partial \ln Z}{\partial T}\right)_P \right\},$$

or

$$= nR \left\{ \ln\frac{Z}{N} + T\left(\frac{d \ln Z}{\partial T}\right)\right\}_P, \qquad (44)$$

in the case of an ideal gas.

Now when Eq. (44) is substituted together with Eq. (30) in the above enthalpy definition, namely

$$H \equiv U + PV = U + n R T = G + TS,$$

for S and G, respectively, we obtain

$$H = -\, n R T \ln\left(\frac{Z}{N}\right) + n R T \ln\left(\frac{Z}{N}\right) T \left\{ n R T \left(\frac{\partial \ln Z}{\partial T}\right)_P \right\},$$

$$= n R T^2 \left(\frac{\partial \ln Z}{\partial T}\right)_P \text{ (ideal gas).} \qquad (45)$$

From this equation, the specific heat at constant pressure of an ideal gas can be readily written as

$$c_p = \frac{\partial}{\partial T}\left[RT^2 \left(\frac{\partial \ln Z}{\partial T}\right)_p \right] \text{ (ideal gas). (46)}$$

12.10. Vibrational, rotational, electronic, and nuclear contributions to thermodynamic properties

Instead of computing the total partition function for *a molecule* ($Z = Z_{trans}\ Z_{rot}\ Z_{vib}\ \ldots$), it is often more convenient to determine the individual contributions of the partition functions. Such a breakdown is useful in comparing the relative magnitudes of the individual contributions. As an example, we shall consider the individual contributions in the *Helmholtz function.*

From Eq. (21), we have

$$F = -n R T\left[\ln\left(\frac{Z}{N}\right) + 1 \right], \qquad (21)$$

where $Z = Z_{trans}\ Z_{rot}\ Z_{vib}\ Z_{elec}\ Z_{nuc}\ Z_{chem}$. Therefore,

$$F = -nRT\left[\ln\left(\frac{Z_{trans}}{N}\right) + \ln Z_{rot} + \ln Z_{vib} + \ln Z_{elec} + \ln Z_{nuc} + \ln Z_{chem} + 1\right],$$

from which it is readily seen that

$$F_{rot} = -nRT\ln Z_{rot},$$
$$F_{vib} = -nRT\ln Z_{vib},$$
$$F_{elec} = -nRT\ln Z_{elec},$$
$$F_{nuc} = -nRT\ln Z_{nuc},$$
$$F_{chem} = -nRT\ln Z_{chem}.$$

The *Gibbs function* can be written as defined above by Eq. (24), that is,

$$G \equiv F + PV, \tag{24}$$

and since the *rotational and vibrational partition functions are independent of pressure and volume* [as can be seen respectively from Eqs. (15) and (19) in Chapter 11], they will make the same contribution to both the Gibbs and Helmholtz functions. For an ideal gas, all the *internal contributions* are independent of pressure and volume.

The individual contributions to the *entropy* can be obtained in a similar way. The equation of the entropy obtained above, namely

$$S = nR\left[\ln\left(\frac{Z}{N}\right) + T\left(\frac{\partial \ln Z}{\partial T}\right)_V + 1\right], \tag{17}$$

can be written as

$$S = nR\ln Z + nRT\left(\frac{\partial \ln Z}{\partial T}\right)_V + nR - nR\ln N. \tag{47}$$

Since we have

$$\left(\frac{\partial T\ln Z}{\partial T}\right)_V = T\left(\frac{\partial \ln Z}{\partial T}\right)_V + \ln Z,$$

Eq. (47) can be rewritten as

$$S = nR\left(\frac{\partial T\ln Z}{\partial T}\right)_V + nR - nR\ln N.$$

But

$$n(R - R\ln N) = -k(N\ln N - N) = -k\ln N!,$$

and this term is not appropriate to the rotational, vibrational, electronic, and nuclear partition functions. Therefore,

$$S = nR\left\{\frac{\partial(T\ln z_{trans})}{\partial T}\right\}_V + nR - nR\ln N + nR\frac{\partial(T\ln Z_{rot})}{\partial T} + nR\frac{\partial(T\ln Z_{vib})}{\partial T} + nR\frac{\partial(T\ln Z_{elec})}{\partial T} + nR\frac{\partial(T\ln Z_{nuc})}{\partial T} + nR\frac{\partial(T\ln Z_{chem})}{\partial T}. \tag{48}$$

1. The *translational contribution* to the entropy is

$$S_{trans} = nR\,\frac{\partial (T\ln Z_{trans})}{\partial T} + nR - nR\ln N.$$

But

$$\left\{\frac{\partial (T\ln Z_{trans})}{\partial T}\right\}_V = T\left(\frac{\partial \ln Z_{trans}}{\partial T}\right)_V + \ln Z_{trans},$$

therefore,

$$\begin{aligned} S_{trans} &= nR - nR\ln N + nR\ln Z_{trans} \\ &\quad + nRT\,\frac{\partial \ln Z_{trans}}{\partial T} \\ &= nR + nRT\,\frac{\partial \ln Z_{trans}}{\partial T} + nR\ln\frac{Z_{trans}}{N}. \end{aligned} \tag{49}$$

Since

$$Z_{trans} = (2\pi mkT)^{3/2}\,\frac{V}{h^3},$$

therefore,

$$\ln Z_{trans} = \frac{3}{2}(\ln 2\pi mk) + \frac{3}{2}\ln T + \ln V - 3\ln h,$$

and

$$\left(\frac{\partial \ln Z_{trans}}{\partial T}\right)_V = \frac{3}{2}\cdot\frac{1}{T}.$$

Hence, the substitution into Eq. (49) for $\left(\frac{\partial \ln Z}{\partial T}\right)$ by $\frac{3}{2}\cdot\frac{1}{T}$, and for Z_{trans} by $(2\pi mkT)^{3/2}\frac{V}{h^3}$ gives

$$\begin{aligned} S_{trans} &= \frac{5}{2}nR + nR\ln\frac{(2\pi mkT)^{3/2}V}{Nh^3} \\ &= nR\left[\ln\left(\frac{V}{N}\right) + \frac{3}{2}\ln T + \frac{3}{2}\ln(2\pi mk) - 3\ln h + \frac{5}{2}\right]. \end{aligned} \tag{50}$$

Utilising the ideal gas equation $PV = NkT$, Eq. (50) takes the form of the *Sackur-Tetrode* equation expressing the molar specific entropy which was obtained before.*

2. The *rotational contribution* to the entropy according to Eq. (48) for $T \gg \Theta_{rot}$ is

$$S_{rot} = nR\,\frac{\partial (T\ln Z_{rot})}{\partial T}. \tag{51}$$

---

* See Section 9.6, Eq. (40).

(a) For *diatomic or linear polyatomic molecules* the rotational partition function Z_{rot}, as shown before,* is

$$Z_{rot} = \frac{8\pi^2 I_b kT}{kh^3}.$$

But

$$\frac{\partial(T \ln Z_{rot})}{\partial T} = T\left(\frac{\partial \ln Z_{rot}}{\partial T}\right) + \ln Z_{rot}.$$

and $\ln Z_{rot}$ is

$$\ln Z_{rot} = \ln\left(\frac{8\pi^2 I_b}{kh^3}\right) + \ln T,$$

so that

$$\frac{\partial \ln Z_{rot}}{\partial T} = \frac{1}{T},$$

or,

$$T\frac{\partial \ln Z_{rot}}{\partial T} = 1.$$

Therefore,

$$\frac{\partial(T \ln Z_{rot})}{\partial T} = 1 + \ln Z_{rot}.$$

Hence, S_{rot} according to Eq. (51) is

$$S_{rot} = nR\left[1 + \ln\left(\frac{8\pi^2 I_b kT}{kh^3}\right)\right],$$

or

$$S_{rot} = nR\ln\left(\frac{8\pi^2 I_b kT}{kh^3}\right) + nR \tag{52}$$

(b) For *polyatomic molecules*, the rotational partition function, as again shown before, is

$$Z_{rot} = \frac{8\pi^2\left(8\pi^3 I_a I_b I_c\right)^{1/2}}{kh^3}(kT)^{3/2},$$

and the *rotational contribution* to the entropy according to Eq. (48) is

$$S_{rot} = nR\frac{\partial(T\partial \ln Z_{rot})}{\partial T}.$$

---

* See Section 11.4: Eqs. (14) and (15) for Z_{rot}; Eqs. (19) and (20) for Z_{vib}; Eq. (24) for Z_{elec}; and Eq. (28) for Z_{nuc}.

Therefore,

$$S_{rot} = nR\left\{T\left(\frac{\partial \ln Z}{\partial T}\right) + \ln Z_{rot}\right\}, \tag{53}$$

and

$$\ln Z_{rot} = \ln\left\{\frac{8\pi^2\left(8\pi^3 I_a I_b I_c\right)^{1/2}}{kh^3}\right\} + \frac{3}{2}\ln kT.$$

Also

$$\frac{\partial \ln Z_{rot}}{\partial T} = \frac{3}{2}\cdot\frac{1}{T},$$

or

$$T\frac{\partial \ln Z_{rot}}{\partial T} = \frac{3}{2}.$$

Hence, S_{rot} according to Eq. (53) is

$$S_{rot} = nR\left[\frac{3}{2} + \ln\left\{\frac{8\pi^2\left(8\pi^3 I_a I_b I_c\right)^{1/2}}{kh^3}(kT)^{3/2}\right\}\right],$$

$$= nR\ln\left\{\frac{8\pi^2\left(8\pi^3 I_a I_b I_c\right)^{1/2}(kT)^{3/2}}{kh^3}\right\} + \frac{3}{2}nR. \tag{54}$$

3. The *vibrational contribution* to entropy according toEq. (48) is

$$S_{vib} = nR\frac{\partial(T\ln Z_{vib})}{\partial T}. \tag{55}$$

(a) For a *diatomic molecule* the vibrational partition function, as shown before, is

$$Z_{vib} = \frac{e^{-h\nu/2kT}}{1 - e^{-h\nu/kT}}.$$

Therefore,

$$\ln Z_{veb} = \ln\left(\frac{e^{-h\nu/2kT}}{1 - e^{-h\nu/kT}}\right),$$

and

$$S_{veb} = nR\frac{\left\{\partial T\ln\left(e^{-h\nu/2kT}/1 - e^{-h\nu/kT}\right)\right\}}{\partial T}. \tag{56}$$

(b) For a *polyatomic molecule*, Z_{vib} is

$$Z_{vib} = \Pi\left(\frac{e^{-h\nu_i/2kT}}{1 - e^{-h\nu_i/kT}}\right).$$

Therefore,

$$\ln Z_{veb} = \ln \Pi \left(e^{-h\nu_i/2kT} / 1 - e^{-h\nu_i/kT} \right),$$

and

$$S_{vib} = nR \frac{\partial \left\{ T \ln \Pi \left(e^{-h\nu_i/2kT} / 1 - e^{-h\nu_i/kT} \right) \right\}}{\partial T}. \tag{57}$$

4. The *electronic contribution* to entropy according to Eq. (48) is

$$S_{elec} = nR \frac{\partial (T \ln Z_{elec})}{\partial T}. \tag{58}$$

Therefore,

$$S_{elec} = nR \left(\ln Z_{elec} + T \frac{\partial \ln Z_{elec}}{\partial T} \right).$$

Using Eq. (24) obtained before in Section 11.6, namely

$$Z_{elec} = g_e = g_0 + g_1 \, e^{-\epsilon_1/kT},$$

we find

$$S_{elec} = nR \ln g_e + nRT \frac{\partial \ln g_e}{\partial T}. \tag{59}$$

5. The *nuclear contribution* to entropy is

$$S_{nuc} = nR \frac{\partial (T \ln Z_{nuc})}{\partial T}, \tag{60}$$

where Z_{nuc} is given by Eq. (28) also obtained before in Chapter 11, namely

$$Z_{nuc} = g_0 = 2I + 1.$$

Therefore,

$$S_{nuc} = nR \frac{\partial \{ T \ln (2I+1) \}}{\partial T}$$

$$= nR \ln (2I + 1). \tag{61}$$

6. Finally, the *chemical contribution* to entropy is

$$S_{chem} = nR \frac{\partial (T \ln Z_{chem})}{\partial T}. \tag{62}$$

The chemical partition function Z_{chem} is

$$Z_{chem} = e^{-\epsilon_0/kT},$$

where ϵ_0 is the chemical energy as discussed before in Section 11.2. Therefore,

$$S_{chem} = nR \frac{\partial \left(T \ln e^{-\epsilon_0/kT} \right)}{\partial T}. \tag{63}$$

It is to be noted that the nuclear partition function Z_{nuc} can affect the values of entropy and free energy so that when absolute values of entropy and free energy are to be computed, the nuclear spin contribution must be included for exactness. But usually in the standard practice, the nuclear partition function contribution is omitted in evaluating entropies and free energies so that Eq. (63) can be equated to zero. Also the contribution from nuclear spin cancels out in equilibrium reactions.

12.11. Thermodynamic properties of polyatomic ideal gases

As examples of the general equations for the properties of ideal gases, we shall write the equations for the entropy of diatomic and polyatomic ideal gases.

The equation for the entropy can be written as

$$S = S_{trans} + S_{rot} + S_{vib} + S_{elec} + S_{nuc} + S_{chem}.$$

For a diatomic or linear polyatomic molecule of an ideal gas, we can use the *Sackur-Tetrode* equation for the *translational* entropy contribution. This equation has been obtained* before, it gives the absolute value of the *molar specific entropy for an ideal monatomic gas*; it is of the form

$$s_{trans} = R\left[\frac{5}{2}\ln T - \ln P + \frac{5}{2}\ln k + \frac{3}{2}\ln(2\pi m) - 3\ln h + \frac{5}{2}\right].$$

This equation which stands for 1 mole of an ideal monatomic molecule takes the following form for n moles of this gas:

$$S_{trans} = nR\left[\frac{5}{2}\ln k - \ln P + \ln\frac{(2\pi m)^{3/2}}{h^3} + \frac{5}{2}\ln T + \frac{5}{2}\right]$$

$$= \frac{5}{2}nR\ln k + \frac{5}{2}nR\ln T - nR\ln P + \frac{5}{2}nR + nR\ln\frac{(2\pi m)^{3/2}}{h^3}.$$

In the case of a diatomic or linear molecule, there is a *rotational* contribution to the entropy, this contribution being expressed by Eq. (52) which been derived above in section 12.10, namely

$$S_{rot} = nR\ln\left(\frac{8\pi I_b kT}{kh^3}\right) + nR.$$

---

* See Section 9.6, Eq. (40). Also see Eq. (50) in Chapter 12.

Accordingly, S for a diatomic or linear molecule considering only *translational and rotational contribution* is

$$S = S_{trans} + S_{rot} = \frac{5}{2} nR\ln k + \frac{7}{2} nR\ln T$$

$$- nR\ln P + \frac{7}{2} nR + nR\ln\left[\frac{(2\pi m)^{3/2}}{h^3} \cdot \frac{8\pi^2 I_b k}{k h^3}\right].$$

The total entropy for a diatomic or linear molecule considering all types of contributions is given by

$$S = \frac{5}{2} nR\ln k + \frac{7}{2} nR\ln T - nR\ln P + \frac{7}{2} nR$$

$$+ nR\ln\left[\frac{(2\pi m)^{3/2}}{h^3} \cdot \frac{8\pi^2 I_b k}{k h^3}\right] + S_{vib} + S_{elec} + S_{nuc} + S_{chem} .$$

For polyatomic molecules, the entropy is

$$S = \frac{5}{2} nR\ln k + \frac{7}{2} nR\ln T - nR\ln P$$

$$+ \frac{7}{2} nR + nR\ln\left[\frac{(2\pi m)^{3/2}}{h^3} \cdot \frac{8\pi^2 \left(8\pi^3 I_a I_b I_c\right)^{1/2} k^{3/2}}{k h^3}\right]$$

$$+ S_{vib} + S_{elec} + S_{nuc} + S_{chem} .$$

It is to be noted the generalisations we have made above cannot be extended to the vibrational, electronic, nuclear, and chemical contributions, which must be computed separately for each gas.

CHAPTER 13

Gas Mixtures

13.1. Dalton's law of partial pressures

Let us assume that we have a mixture of gases which do not react chemically with one another and that the temperature and density are such that their behaviour approximates that of *an ideal gas*. *Dalton* found experimentally that the total pressure of the mixture is the *sum* of the pressures which each component gas would have if it were to occupy the volume of the mixture alone. This pressure is known as the *partial pressure of the gas* and thus the experimental law is *Dalton's law of partial pressures.*

It is evident that according to this law the pressure which each component gas would exert occupying the whole volume of the mixture is the same whether it occupies this volume alone or not. Put another way, any component gas in the mixture behaves as if it were alone.

The question which might arise here is that whether the entropy of individual gases in a mixture of *non-reacting ideal gases* behaves in a similar way as the pretrial pressure does. For the reply, we consider the following experiment based on classical thermodynamics.

Let us suppose that we have two *ideal gases*, gas (1) and gas (2), filling initially the compartments V_1, and V_2 respectively, which are separated by a partition as shown in Fig. 1, and that they are at the same temperature and pressure initially

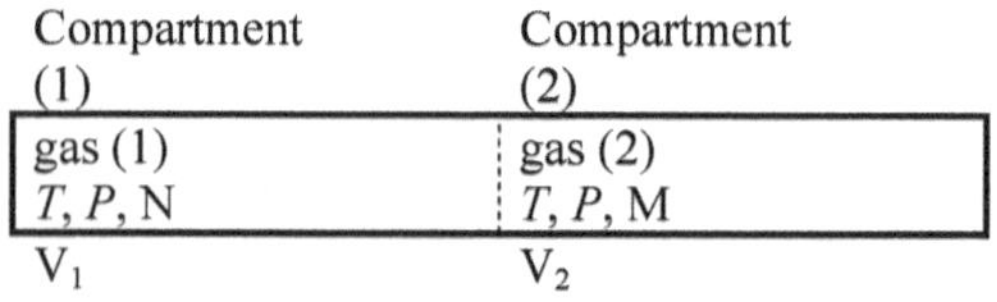

Fig. 1. Mixing of two gases.

We want to find out the entropy change $\Delta S \left(= S_{\text{mixture}} - S_1 - S_2\right)$ for the irreversible process of mixing when the partition is removed, that is, ΔS_{mixing}.

In the first place, the two gases *cannot* be initially in the two compartments V_1 and V_2 with number of molecules, say *N* and *M* of the gases (1) and (2), respectively, at the same temperature and pressure *unless* the relation

$$\frac{V_1}{V_2} = \frac{N}{M} \tag{1}$$

is satisfied. This is obvious on applying the general gas equation to each of the two ideal gases. That is, since

$$V_1 = \frac{NkT}{P}, \tag{2}$$

$$V_2 = \frac{MkT}{P}, \tag{3}$$

where *k* is *Boltzmann's* constant; then by dividing Eq. (2) by Eq. (3), we obtain Eq. (1).

Secondly, in order to find out the entropy change in this irreversible process of mixing the two gases, a reversible path must be *invented, even in thought*, for mixing these gases. This can be achieved by following a three-step procedure as can be seen from the following:

1. Gas (1) is expanded *isothermally* and *reversibly* from volume V_1 and pressure *P* to a volume $(V_1 + V_2)$. The entropy change is given by

$$dS_1 = \frac{d'q_{rev}}{T}, \tag{4}$$

where $d'q_{rev.}$ is

$$d'q_{rev} = dU + PdV = PdV. \tag{5}$$

Therefore,

$$d'q_{rev} = NkT\frac{dV}{V},$$

and

$$dS_1 = Nk\frac{dV}{V},$$

or

$$\Delta S_1 \int_{V_1}^{V_1 + V_2} \frac{dV}{V}$$

$$= Nk \ln \frac{V_1 + V_2}{V_1}. \tag{6}$$

Instead of expressing the result in terms of volumes, we may express it in terms of mole fractions by making use of Eq. (1). Thus the quotient $\frac{V_1 + V_2}{V_1}$ is given by

$$\frac{V_1 + V_2}{V_1} = \frac{N + M}{N},$$

and then Eq. (6) can be written as

$$\Delta S_1 = Nk \ln \frac{N + M}{N},$$

or,

$$= -\mathrm{N k} \ln \frac{\mathrm{N}}{\mathrm{N+M}},$$

$$= -\mathrm{N k} \ln f_1, \tag{7}$$

where f_1 is the *mole fraction* of gas (1).

2. Gas (2) is expanded in the same way, that is, *isothermally* and *reversibly*, from volume V_2 to volume $V_1 + V_2$. Thus,

$$\Delta S_2 = -\mathrm{M k} \ln f_2, \tag{8}$$

where f_2 is the *mole fraction* of gas (2).

The summation of the entropy changes in steps (1) and (2) and therefore is given by

$$\Delta S_1 + \Delta S_2 = -\mathrm{N k} \ln f_1 - \mathrm{M k} \ln f_2$$

$$= -n_1 \mathrm{R} \ln f_1 - n_2 \mathrm{R} \ln f_2, \tag{9}$$

where n_1 and n_2 are the *number of moles* of gas (1) and gas (2), respectively.

3. The two gases are brought together in the volume $V_1 + V_2$ *isothermally* and *reversibly*. This can be carried out by the use of a device suggested by *Planck* (the details of which are not required here). This device is used for a *thought experiment* only. At the same time, it can be constructed in principle.

Based on this device, when the two gases are allowed to combine *isothermally and reversibly*, the *total work done* is equal to *zero* so that the corresponding entropy change, which we denote by ΔS_3, is

$$\Delta S_3 = \int d S_3 = \int \frac{d' q_{rev}}{T} = \frac{1}{T} \int d' W = 0 .$$

Therefore, the entropy change of mixing, ΔS_{mixing}, is only given by

$$\Delta S_{mixing} = \Delta S_1 + \Delta S_2 . \tag{10}$$

This result signifies that the entropy of each component gas occupying a volume *V* remains the same whether it occupies this volume alone or not. Consequently, the entropy of a mixture of gases is the sum of the entropies each component gas would have if it were to occupy the total volume alone. Therefore, Dalton's law of partial pressures, according to which each component gas in a *perfect gas mixture* behaves as if it were present alone, applies to entropy, as it is the case to pressure.

Expectedly, the rule holds good for internal energy, enthalpy, and hence for free energy; all these quantities are additive.

In our case, this result given by Eq. (10) can be made more obvious if we sketch the following equations which represent the three-stage procedure suggested for the invented path in thought in order to calculate the entropy change of mixing the two gases, which takes place practically in an irreversible process.

Thus the main irreversible process together with the three-step reversible process can be represented as shown below.

$$\left(V_1, T, P, S_1 \right) \text{gas (1)} + \left(V_2, T, P, S_2 \right) \text{gas (2)} = \left(V_1 + V_2, T, P, S \right) \text{ mixture, } \Delta S_{\text{mixing}}.$$

$$\Delta S_1 = -N k \ln f_1 \qquad \text{Step 1} \qquad \Delta S_2 = -M k \ln f_2 \qquad \text{Step 2}$$

$$\left(V_1 + V_2, T, P_1', S_1' \right) \text{gas (1)} \qquad \left(V_1 + V_2, T, P_2', S_2' \right) \text{gas (2)}$$

According to Planck $\Delta S_3 = 0$

Step 3

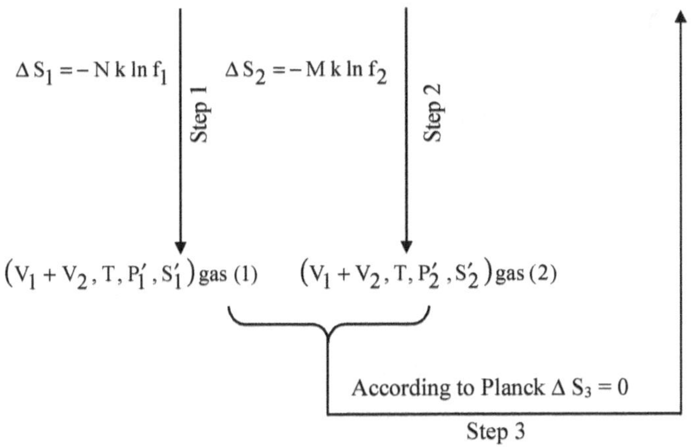

Therefore, the entropy change of mixing the two gases, ΔS_{mixing}, according to the main reaction is given by

$$\Delta S_{\text{mixing}} = S_{\text{mixture}} - S_1 - S_2, \tag{11}$$

and according to the three-step reversible process,

$$\Delta S_{\text{mixing}} = \Delta S_1 + \Delta S_2 + \Delta S_3$$
$$= \Delta S_1 + \Delta S_2, \tag{10}$$

since $\Delta S_3 = 0$.

On the other hand, Eq. (10) can be rewritten as

$$\Delta S_{\text{mixing}} = S_1' - S_1 + S_2' - S_2, \tag{12}$$

where S_1' is the entropy of gas (1) in the total volume $\left(V_1 + V_2 \right)$ before mixing, and S_2' is that of gas (2) in the volume $\left(V_1 + V_2 \right)$ before mixing.

Thus from Eqs. (11) and (12), we have

$$S_{\text{mixture}} - S_1 - S_2 = S_1' - S_1 + S_2' - S_2,$$

or

$$S_{\text{mixture}} = S_1' + S_2'. \tag{13}$$

Hence, a conclusion might now be made again that using classical thermodynamics which cannot distinguish between the two types of molecules whether they are identical or not identical, the last equation arrived at, namely Eq. (13), which is based on the experimental Dalton's law of partial pressures, makes it more apparent that, in general, whenever we have a mixture of gases which do not react chemically, the total entropy of that mixture occupying a volume V is the sum of the entropies which each component gas would have if it were to occupy the total volume of the mixture V.

In continuation, a combination of Eqs. (9) and (10) obtained above might be made in order to give a more general expression obtained by classical thermodynamics for the entropy of mixing of gases. Thus beginning with Eqs. (7), (8), and (10) for two gases, we have

$$\Delta S_{mixing} = \Delta S_1 + \Delta S_2 = -Nk\ln f_1 - Mk\ln f_2,$$

or

$$\Delta S_{mixing} = -n_1 R \ln f_1 - n_2 R \ln f_2. \qquad (9)$$

In terms of the total number n of moles,

$$n = n_1 + n_2, \qquad (14)$$

Eq. (9) may be put in the form

$$\Delta S_{mixing} = -nf_1 R\ln f_1 - nf_2 R\ln f_2,$$

$$= -nR\left(f_1 \ln f_1 + f_2 \ln f_2\right). \qquad (15)$$

Hence, *for more than two gases the entropy of mixing is*

$$\Delta S_{mixing} = -nR\sum_i f_i \ln f_i, \qquad (16)$$

where f_i is the mole fraction of component gas *i*, and *n* is the total number of moles.

It is to be noted that since ΔS_3 in Eq. (10) is equal to zero, the value of ΔS_{mixing} might be regarded as the entropy change accompanying the change in the volume of each of the component gases to the total final volume of the mixture.

Also note that ΔS_{mixing} for two gases (Eq. 15) and ΔS_{mixing} for more than two gases (Eq. 16) is *always positive*, because all mole fractions f_i's are fractions by definition so that $\ln f_i$ is always negative.

Eqs. (15) and (16) can be shown to apply to the mixing of liquids also, provided the mixed liquids form an ideal solution; that is, provided *Raoult's* law applies.

Eq. (15) is valid for the mixing of any two gases, even if the gases differ only because their molecules contain different isotopes of the same element. For example, it applies when the two gases are ^2He and ^4He or $^{235}UF_6$ and $^{238}UF_6$.

What happens then, upon removing the partition separating the two gases (Fig. 1) when they are identical?

Seeing that all molecules are *identical, that is, of the same kind*, we may *conclude that there is no macroscopic change* so that

$$\Delta S_{mixing} = 0,$$

in this case of identical molecules, even though Eq. (15), applied blindly, suggests a positive value of ΔS_{mixing} *which is wrong*, and hence this fact can be regarded as a *discrepancy of classical thermodynamics.*

It is to be pointed out that for gases that are as close to their properties as are $^{235}UF_6$ and $^{238}UF_6$, ΔS_{mixing} is *positive*, whereas for identical gases ΔS_{mixing} is *suddenly zero.* This *discontinuity* is known as *Gibbs's paradox*. Gibbs felt that it should be possible to choose gases that would be more and more alike so that the *identity of the two gases would be considered to be a limiting case*; in other words, that there should be a *continuous transition between different and identical gas.*

It was therefore disturbing that Eq. (15),

$$\Delta S_{mixing} = -nR\left(f_1 \ln f_1 + f_2 \ln f_2\right) \tag{15}$$

did not depend on any property of the gas molecules, such as their *masses*, but rather depended only on the *mole fractions* of the gases involved.

This *discontinuity* was one of the *harbingers* of quantum statistics, in which there is a fundamental difference between the treatment of particles of the same kind and that of particles of a different kind, because there is no way to tell particles of the same kind apart. Different isotopes of the same element may behave very similarly, but they are fundamentally different; they can be told apart.

If we could, for example, fill the two compartments of Fig. 1 with the *same gas* but mark all molecules in compartment 1 with green paint, the entropy change upon removing the partition would be given by Eq. (15), whereas for unmarked molecules in both compartments ΔS_{mixing} would be zero.

In brief, this matter is closely related to the fact that by statistical thermodynamics, *two microstates* in which particles of the *same kind* have been interchanged are indistinguishable; they are *one and the same microstate*, and thus they cannot be counted separately when counting up the microstates belonging to a given macrostate. But in the case of particles of *different kinds*, two microstates in which two different particles have been interchanged are *distinguishable* and thus must be so counted.

This subject will be discussed on the basis of quantum statistics in Section 13.2.

13.2. Calculation of entropy change of mixing gases using Bose-Einstein quantum statistics

Introduction

A summary of the previously* given hypotheses of quantum statistics, and of the corresponding derived equations will be first given, as can be seen from the following:

1. All microstates are squally probable, that is, over a long period of time; any one microstate occurs as often as any other.
2. Any shift of the phase points in phase space that does not alter the number of particles in each cell leaves unaltered the macrostate of the gas and its observable properties.
3. As time goes on, and the microstates of a gas continually change because the atoms are all in motion, the macrostate that occurs most frequently will be that for which there are many more microstates than for any other, and that macrostate will be practically the only one ever observed; that is, that macrostate will be the one responsible of the observable properties of the system. Other macrostates will be observed occasionally, however, and these rare occurrences are responsible for other things such as the density fluctuations in a gas.

---

* See Sections 9.2.–9.6.

4. The number of microstates corresponding to any given macrostate is called the thermodynamic probability of that macrostate, and is represented by *W*, which, in general, is a very large number. Similarly, the number of different ways of arranging phase points within a cell of phase space is the thermodynamic probability of that cell. Accordingly, for any number of occupied cells in phase space, the thermodynamic probability of any given macrostate of the gas is

$$W = \Pi W_i,$$

where the product extends over all occupied cells of phase space.

Based on these assumptions, together with the use of *Stirling's* approximation of $\ln x!$, and of *Lagrange's* method of undetermined multipliers, the following equations were derived consecutively considering the state of maximum thermodynamic probability that is the state of thermodynamic equilibrium,

$$W_i = \frac{\left(N_i + g_i - 1\right)!}{\left(g_i - 1\right)!\, N_i!}. \tag{1}$$

$$W = \Pi \frac{\left(N_i + g_i - 1\right)!}{\left(g_i - 1\right)!\, N_i!}, \tag{2}$$

where '1' has been neglected in comparison with g_i and N_i. Therefore,

$$\ln W = \Sigma\left[\left(N_i + g_i\right)\ln\left(N_i + g_i\right) - g_i \ln g_i - N_i \ln N_i\right], \tag{2)$'$}$$

Using the Maxwell-Boltzmann limit $g_i << N_i$, Eq. (2)′ reduces to

$$\ln W = \Sigma\left(N_i \ln g_i + N_i - N_i \ln N_i\right). \tag{3}$$

The condition for maximum* probability using Eq. (2)′ is

$$= \delta \ln W = \Sigma \ln\left(\frac{N_i + g_i}{N_i}\right)\delta N_i = 0. \tag{4}$$

$$\frac{N_i}{g_i} = \frac{1}{B e^{\beta \epsilon_i} - 1}. \tag{5}$$

Eq. (5) is the *Bose-Einstein distribution function.*

$$\frac{N_i}{g_i} = \frac{1}{B\, e^{\beta \epsilon_i}}. \tag{6}$$

---

* The superscript 'o' is dropped for simplicity.

Eq. (6) is the *Maxwell-Boltzmann distribution law* of particles among cells using again the approximation

$$N_i << g_i.$$

$$Z = \Sigma g_i e^{-\beta \epsilon_i}. \quad (7)$$

Eq. (7) is the *partition function per particle* of a system; it varies from problem to problem.

$$B = \frac{Z}{N}. \quad (8)$$

Eq. (8) determines the constant *B*.

$$N_i = \frac{N}{Z} g_i e^{-\beta \epsilon_i} \quad (9)$$

$$S = k \ln W, \quad (10)$$

the proportionality constant *k* is proved to be none other the *Boltzmann constant.* Based on statistical mechanics, Eq. (10) interprets the increase of entropy in a closed system as a consequence of the natural trend of a system to undergo a change from a less probable to a more probable state.

$$S = kN\left\{ \ln\left(\frac{Z}{N}\right) + 1\right\} + k\beta U \quad (11)$$

$$\beta = \frac{1}{kT}. \quad (12)$$

Eq. (12) determines the constant β.

$$U = \frac{NkT^2}{Z}\frac{dZ}{dT},$$

or

$$= NkT^2 \frac{d \ln Z}{dT}. \quad (13)$$

Eq. (13) is the expression for the internal energy *U.*

$$S = Nk\left\{ \ln\left(\frac{Z}{N}\right) + 1\right\} + \frac{U}{T}, \quad (14)$$

$$F = -NkT\left\{ \ln\left(\frac{Z}{N}\right) + 1\right\}. \quad (15)$$

Eqs. (14) and (15) are the expressions for the *entropy S* and the *Helmholtz free energy F* in terms of the partition function *Z.*
The equation

$$Z_{trans} = \frac{V}{h^3}\left(2\pi m k T\right)^{3/2} \quad (16)$$

represents the translational partition function for a monatomic ideal gas.

Hence, based on the above quantum statistical equations, we can now consider our problem of calculating the entropy change of mixing two *non-identical*, chemically inert gases. The mixing may be represented symbolically by the equation.

$$\begin{array}{ccccc} \text{gas(1)} & + & \text{gas(2)} & = & \text{mixtre of gas(1) and gas(2)}, \\ \text{N molecules} & & \text{M molecules} & & (\text{N}+\text{M}) \text{ molecules} \\ S_1 & & S_2 & & S_{\text{mixture}} \end{array}$$

so that ΔS_{mixing} is given by

$$\Delta S_{\text{mixting}} = S_{\text{mixture}} - S_1 - S_2,$$

or

$$\Delta S_{\text{mixing}} = S_1' - S_1 + S_2' - S_2 = \Delta S_1 + \Delta S_2,$$

as shown in Section 13.1, where S_1' and S_2' are the entropies of gas (1) and gas (2) after expansion from the initial volumes V_1 and V_2, respectively, to final volume V_1+V_2, before the process of mixing.

The discussion will be based on the assumption that the two gases are monatomic, and that the molecules of each gas are only weakly interacting or are free of intermolecular forces, so that we can treat the mixture as an *ideal gas mixture*, where the molecules of one gas can be distinguished from those of the other.

In the mixture, for each microstate of any of the cells occupied by the molecules of gas (1), we may have any one of the microstates of the cells occupied by the molecules of gas (2) so that by such an association, the mixture exhibits a macrostate that is different from that of each of its constituents before the mixing. Accordingly, the thermodynamic probability for *any given macrostate* of the mixture is given by the product of the thermodynamic probabilities, W_1' and W_2' in the mixture of the constituent gases. That is,

$$W_{\text{mixture}} = W_1' \cdot W_2'. \tag{17}$$

Since the molecules of the first gas are indistinguishable, the thermodynamic probability of a given macrostate of this gas, W_1', in the mixture is given by Eq. (3), which has been shown above, namely,

$$\ln W_1' = \Sigma\left(N_i \ln g_i + N_i - N_i \ln N_i\right), \tag{3}$$

where N_i is the number of particles of the first gas in the *i*th cell. Similarly, for the second gas in the mixture,

$$\ln W_2' = \Sigma\left(M_i \ln g_i + M_i - M_i \ln M_i\right),$$

where M_i is the number of particles of the second gas in the *i*th cell.

The thermodynamic probability for the mixture is therefore

$$\begin{aligned} \ln W_{\text{mixture}} &= \ln W_1' . W_2' = \ln W_1' + \ln W_2' \\ &= \Sigma\left(N_i \ln g_i + N_i - N_i \ln N_i\right) + \\ &= \Sigma\left(M_i \ln g_i + M_i - M_i \ln M_i\right). \end{aligned} \tag{18}$$

For the state of maximum thermodynamic probability, we have

$$0 = \delta \ln W = \Sigma \ln g_i \, \delta N_i + \delta N_i - N_i \frac{1}{N_i} \delta N_i - \ln N_i \, \delta N_i$$

$$+ \Sigma \ln g_i \,\delta M_i + \delta M_i - M_i \frac{1}{M_i} \delta M_i - \ln M_i \,\delta M_i$$

$$= \Sigma \left(\ln g_i - \ln N_i \right) \delta N_i + \Sigma \left(\ln g_i - \ln M_i \right) \delta N_i . \quad (19)$$

Therefore,

$$-\delta \ln W = \Sigma \left[\ln \left(\frac{N_i}{g_i} \right) \delta N_i + \ln \left(\frac{M_i}{g_i} \right) \delta M_i \right] = 0,$$

or,

$$\Sigma \ln \left(\frac{N_i}{g_i} \right) \delta N_i + \Sigma \ln \left(\frac{M_i}{g_i} \right) \delta M_i = 0 . \quad (20)$$

The variables δN_i and δM_i in Eq. (19) are *not independent* since the number of molecules of each gas is constant, and also the total energy of the two gases is *constant*, for the system is presumed to be isolated. That is,

$$N = \Sigma N_i = \text{const.}$$

$$M = \Sigma M_i = \text{const.}$$

$$U = \Sigma \epsilon_i N_i + \Sigma \epsilon_i' M_i = \text{const.},$$

where ϵ_i and ϵ_i' are the energies per one molecule of gas (1) and gas (2), respectively, in the *i*th cell of phase space.

Accordingly, the system has the following three condition equations imposed on δN_i's , δM_i's, and δU , respectively.

$$\Sigma \delta N_i = 0, \quad \text{(i)}$$

$$\Sigma \delta M_i = 0, \quad \text{(ii)}$$

$$\Sigma \delta U = \Sigma (\epsilon_i \delta N_i + \epsilon_i' \delta M_i)$$
$$= \epsilon_1 \delta N_1 + \epsilon_2 \delta N_2 + \ldots + \epsilon_i' \delta M_1$$
$$+ \epsilon_2' \delta M_2 + \ldots = 0 \quad \text{(iii)}$$

Now we can use *Lagrange's method* of undetermined multipliers described before.* We multiply Eq. (i) by a constant which, for convenience, we write as $-\ln \alpha$; multiply Eq. (ii) by another constant $-\ln \gamma$; and multiply Eq. (iii) by a third constant $-\beta$; and add to Eq. (20). This gives

$$\Sigma \left(\ln \frac{N_i}{g_i} - \ln \alpha + \beta \epsilon_i \right) \delta N_i + \Sigma \left(\ln \frac{M_i}{g_i} - \ln \gamma \right.$$
$$\left. + \beta \epsilon_i' \right) \delta M_i = 0,$$

* See Section 9.3.

and, since *in effect,* the $\delta N_i'$'s and δM_i's are *now independent*, we must have

$$\ln \frac{N_i}{g_i} = \ln \alpha - \beta \epsilon_i,$$

$$\ln \frac{M_i}{g_i} = \ln \gamma - \beta \epsilon_i'.$$

Also since it can again be shown that

$$\beta = \frac{1}{kT},$$

we have

$$N_i = \alpha g_i e^{-\epsilon_i/kT}, \tag{21}$$

$$M_i = \gamma g_i e^{-\epsilon_i'/kT}. \tag{22}$$

Using Eq. (18), we may write the total entropy of the mixture as follows:

$$S_{mixture} = k \ln W_{mixture} = k\left[\sum N_i \left(\ln g_i - \ln N_i + 1 \right) + \sum M_i \left(\ln g_i - \ln M_i + 1 \right) \right]. \tag{23}$$

But according to Eqs. (21) and (22), we have

$$\ln g_i - \ln N_i + 1 = -\ln \alpha + \frac{\epsilon_i}{kT} + 1,$$

and

$$\ln g_i - \ln M_i + 1 = -\ln \gamma + \frac{\epsilon_i'}{kT} + 1,$$

respectively so that substituting these two equations into the preceding equation of $S_{mixture}$ [Eq. (23)], we obtain

$$S_{mixture} = k\left[\sum N_i \left(-\ln \alpha + \frac{\epsilon_i}{kT} + 1 \right) + \sum M_i \left(-\ln \gamma + \frac{\epsilon_i'}{kT} + 1 \right) \right]. \tag{24}$$

Also, since

$$\sum \epsilon_i N_i + \sum \epsilon_i' M_i = U,$$

and

$$k\left(\sum N_i + \sum M_i \right) = k\left(N + M \right) = n_1 R + n_2 R = nR,$$

where *n* is the total number of moles of the two gases in the mixture, then it follows that

$$S_{mixture} = \frac{U}{T} + nR - k \ln \alpha \sum N_i - k \ln \gamma \sum M_i . \tag{25}$$

However, from Eqs. (21) and (22), we have

$$N = \Sigma N_i = \alpha \Sigma g_i e^{-\epsilon_i/kT} = \alpha Z_1', \quad (26)$$

and

$$M = \Sigma M_i = \gamma \Sigma g_i e^{-\epsilon_i'/kT} = \gamma Z_2', \quad (27)$$

where Z_1' is the partition function of the first gas, and Z_2', that of the second gas in the *mixture*. Thus substituting Eqs. (26) and (27) into Eq. (25), to drop $\ln\alpha$ and $\ln\gamma$, respectively, we obtain

$$S_{mixture} = \frac{U}{T} + nR + k\ln\left(\frac{Z_1'}{N}\right)^N + k\ln\left(\frac{Z_2'}{M}\right)^M. \quad (28)$$

Eq. (28) is the equation we are looking for; it represents the entropy of a mixture of *N* molecules of one gas and *M* molecules of another gas in terms of the partition function Z_1' and Z_2' for the two different molecules present.

The corresponding value of the Helmholtz function *F*, which is defined by

$$F \equiv U - TS,$$

can therefore be obtained by substituting for *S* in this equation by Eq. (28). Thus

$$F = -T\left[\left\{k\ln\left(\frac{Z_1'}{N}\right)^N + k\ln\left(\frac{Z_2'}{M}\right)^M\right\} + nR\right],$$

or

$$F = -T\left[k\ln\left\{\left(\frac{Z_1'}{N}\right)^N\left(\frac{Z_2'}{M}\right)^M\right\} + nR\right]. \quad (29)$$

Now it is essential to evaluate the entropy of a mixture of two *non-identical* gases when they are monatomic ideal gases. Thus if m_1 is the mass of one atom in the first gas, and m_2 that of one atom in the second, the partition functions of the two gases Z_1' and Z_2' *in the mixture of volume V* at temperature T can be calculated from the general expression given above for the partition function of an ideal monatomic gas. Thus we have

$$Z_1' = \frac{V}{h^3}(2\pi m_1 kT)^{3/2}, \quad (30)$$

and

$$Z_2' = \frac{V}{h^3}(2\pi m_2 kT)^{3/2}, \quad (31)$$

for the two gases (1) and (2), respectively. If we first substitute in Eq. (28) for the internal energy of the mixture *U* by the sum of the two internal energies of the constituents, which are $3/2\,NkT$ and $3/2\,MkT$, respectively, that is, by $3/2\,nRT$, and secondly for the partition functions from Eqs. (30) and (31), we obtain

$$S_{mixture} = \frac{5}{2}nR + (N+M)k\ln V - 3(N+M)k\ln h$$

$$+\frac{3}{2}(N+M)k\ln(2\pi k) + \frac{3}{2}(N+M)k\ln T$$

$$+ Nk\ln\frac{m_1^{3/2}}{N} + Mk\ln\frac{m_2^{3/2}}{M}. \tag{32}$$

Replacing $(N+M)k$ by nR and taking the latter as a common factor in Eq. (32), we have

$$S_{mixture} = nR\left[\frac{5}{2} - 3\ln h + \frac{3}{2}\ln 2\pi k + \frac{3}{2}\ln T + \ln V + \frac{Nk}{(N+M)k}\ln\frac{m_1^{3/2}}{N} + \frac{Mk}{(N+M)k}\ln\frac{m_2^{3/2}}{M}\right],$$

$$= nR\left[\frac{5}{2} - 3\ln h + \frac{3}{2}\ln 2\pi k + \frac{3}{2}\ln T + \ln V + \frac{N}{N+M}\ln\frac{m_1^{3/2}}{N} + \frac{M}{N+M}\ln\frac{m_2^{3/2}}{M}\right]. \tag{33}$$

This is the equation we require.

Since the entropy due to mixing ΔS_{mixing} can be determined by subtracting the sum of the separate entropies from the entropy of the mixture given by Eq. (33), we must determine the entropy of each of the two monatomic gases before mixing; that is,

$$\Delta S_{mixing} = S_{mixture} - S_1 - S_2.$$

The general expression for the entropy of an *ideal monatomic gas* in the separate form as given above by Eq. (14) is

$$S = Nk\left[\ln\left(\frac{Z}{N}\right) + 1\right] + \frac{U}{T}. \tag{35}$$

The partition function *Z* for this gas as also given above by Eq. (16) is

$$Z = \frac{V}{h^3}(2\pi mkT)^{3/2}.$$

Thus applying these two equations to each gas separately, we obtain

$$S_1 = Nk\left[\ln\left\{\frac{(2\pi m_1 kT)^{3/2}\,V_1/h^3}{N}\right\} + 1\right] + \frac{U_1}{T}, \tag{36}$$

where

$$U_1 = NkT^2\frac{d(\ln Z_1)}{dT}. \tag{37}$$

Therefore,

$$S_1 = Nk\left[\ln(2\pi m_1 k)^{3/2} + \frac{3}{2}\ln T + \ln V_1 - \ln Nh^3 + 1\right] + NkT^2 \times \frac{3}{2}\cdot\frac{1}{T},$$

$$= Nk\left[\frac{5}{2} + \frac{3}{2}\ln(2\pi m_1 k) + \frac{3}{2}\ln T + \ln V_1 - \ln Nh^3\right]. \tag{38}$$

In like manner S_2 is given by

$$S_2 = Mk\left[\frac{5}{2} + \frac{3}{2}\ln(2\pi m_2 k) + \frac{3}{2}\ln T + \ln V_2 - \ln N h^3\right]. \tag{39}$$

Having determined the entropies of the two gases in their separate forms S_1 and S_2, as given above by Eqs. (38) and (39) and the entropy of the mixture $S_{mixture}$, as given by Eq. (33), we have

$$\begin{aligned}
\Delta S_{mixing} = (N+M)k & \left[\frac{5}{2} - 3\ln h + \frac{3}{2}\ln 2\pi k + \frac{3}{2}\ln T \right.\\
& \left. + \ln V + \frac{N}{N+M}\ln\frac{m_1^{3/2}}{N} + \frac{M}{N+M}\ln\frac{m_2^{3/2}}{M}\right]\\
& - Nk\left[\frac{5}{2} + \frac{3}{2}\ln(2\pi m_1 k) + \frac{3}{2}\ln T + \ln V_1 - \ln N h^3\right]\\
& - Mk\left[\frac{5}{2} + \frac{3}{2}\ln(2\pi m_2 k) + \frac{3}{2}\ln T + \ln V_2 - \ln M h^3\right],\\
& = (N+M)k\ln(V_1+V_2) - Nk\ln V_1 - Mk\ln V_2\\
& \qquad + Nk\ln m_1^{3/2} - Nk\ln N\\
& \qquad - Nk\ln m_1^{3/2} + Nk\ln N\\
& \qquad + Mk\ln m_2^{3/2} - Mk\ln M\\
& \qquad - Mk\ln m_2^{3/2} + Mk\ln M,\\
& \qquad = (N+M)k\ln(V_1+V_2) - Nk\ln V_1\\
& \qquad\quad - Mk\ln V_2.
\end{aligned} \tag{40}$$

Hence,

$$\Delta S_{mixing} = k\ln\left[\frac{(V_1+V_2)^{N+M}}{V_1^N\, V_2^M}\right]. \tag{41}$$

This equation is of *central importance*.

It is of interest to find out whether the entropy change due to mixing of *any two gases* obtained by classical thermodynamics can be deduced using statistical mechanics for *non-identical gases*.

For one gas, we have according to Eq. (14) above

$$S_1 = \frac{U_1}{T} + Nk + Nk\ln\left(\frac{Z_1}{N}\right),$$

but on mixing, Z_1 changes to Z_1' so that

$$S_1' = \frac{U_1}{T} + Nk + Nk\ln\left(\frac{Z_1'}{N}\right).$$

Similarly, for the other gas

$$S_2 = \frac{U_2}{T} + Mk + Mk\ln\left(\frac{Z_2}{M}\right),$$

and

$$S_2' = \frac{U_2}{T} + Mk + Mk \ln\left(\frac{Z_2'}{M}\right).$$

Since no work is done on the system as pointed out above in the three-step procedure of reversible and isothermal mixing, ΔS_{mixing} can be obtained by summing the separate entropy changes of the two gases, that is

$$\Delta S_{mixing} = \Delta S_1 + \Delta S_2,$$

But

$$\Delta S_1 = S_1' - S_1 = Nk \ln\left(\frac{Z_1'}{N}\right) - Nk \ln\left(\frac{Z_1}{N}\right),$$

and

$$\Delta S_2 = S_2' - S_2 = Mk \ln\left(\frac{Z_2'}{M}\right) - Mk \ln\left(\frac{Z_2}{M}\right),$$

so that

$$\Delta S_{mixing} = Nk \ln\left(\frac{Z_1'}{Z_1}\right) + Mk \ln\left(\frac{Z_2'}{Z_2}\right). \tag{42}$$

Now substituting into the equation for the partition functions given by

$$Z_1 = \frac{V_1}{h^3}(2\pi m_1 kT)^{3/2}, \qquad \text{(separate)}$$

$$Z_1' = \frac{V}{h^3}(2\pi m_1 kT)^{3/2}, \qquad \text{(in the mixture)}$$

$$Z_2 = \frac{V_2}{h^3}(2\pi m_2 kT)^{3/2}, \qquad \text{(separate)}$$

$$Z_2' = \frac{V}{h^3}(2\pi m_2 kT)^{3/2}, \qquad \text{(in the mixture)}$$

we obtain,

$$\Delta S_{mixing} = Nk \ln\left(\frac{V}{V_1}\right) + Mk \ln\left(\frac{V}{V_2}\right). \tag{43}$$

Since the separate gases are initially at the same temperature and pressure, we have as shown before

$$PV_1 = NkT,$$

and

$$PV_2 = MkT,$$

so that

$$\frac{V_1}{V_2} = \frac{N}{M},$$

or

$$\frac{V_1}{V} = \frac{N}{N+M}.$$

and

$$\frac{V_2}{V} = \frac{M}{N+M}.$$

Hence, according to Eq. (43), we find

$$\begin{aligned} \Delta S_{mixing} &= -N k \ln\left(\frac{V_1}{V}\right) - M k \ln\left(\frac{V_2}{V}\right) \\ &= -N k \ln\left(\frac{N}{N+M}\right) - M k \ln\frac{M}{N+M} \\ &= -N k \ln f_1 - M k \ln f_2 \\ &= -n_1 R \ln f_1 - n_2 R \ln f_2. \end{aligned} \tag{44}$$

But

$$n_1 = f_1 n \ , \quad \text{and} \quad n_2 = f_2 n .$$

Therefore, the entropy due to mixing, ΔS_{mixing}, of two non-identical gases is

$$\begin{aligned} \Delta S_{mixing} &= -n R f_1 \ln f_1 - n R f_2 \ln f_2 \\ &= -n R \left(f_1 \ln f_1 + f_2 \ln f_2\right). \end{aligned} \tag{45}$$

This is the same equation obtained above using classical thermodynamics for any two gases whether non-identical or identical. In other words, statistical mechanics is in consistency with classical thermodynamics in the evaluation of the entropy change due to mixing of gases provided these gases are non-identical.

Now we want to utilise quantum statistics to determine the entropy change on mixing *two identical gases*. For this purpose, let us begin with the general expression for the entropy of a monatomic ideal gas, namely Eq. (14) shown above

$$S = N k \left[\ln\left(\frac{Z}{N}\right) + 1 \right] + \frac{U}{T}.$$

Therefore,

$$S_1 = N k \left[\ln\left(\frac{Z_1}{N}\right) + 1 \right] + \frac{U_1}{T}, \qquad \text{(before mixing)}$$

and

$$S_2 = M k \left[\ln\left(\frac{Z_2}{M}\right) + 1 \right] + \frac{U_2}{T}, \qquad \text{(before mixing)}$$

On mixing and assuming $\Delta S_3 = 0$ in the three-step procedure described before, we have

$$S_1' = N k \left[\ln\left(\frac{Z_1'}{N}\right) + 1 \right] + \frac{U_1}{T}, \qquad \text{(after mixing)}$$

and

$$S_2' = M k \left[\ln\left(\frac{Z_2'}{M}\right) + 1 \right] + \frac{U_2}{T}. \qquad \text{(after mixing)}$$

Since the mixture consists of *one type* of a gas whose molecules are identical, its entropy according to quantum statistics is

$$S_{\text{mixture}} = (N + M)k \left[\ln\left(\frac{Z}{N + M}\right) + 1 \right] + \frac{U}{T}. \tag{46}$$

But the entropy change due to mixing ΔS_{mixing} is given by

$$\Delta S_{\text{mixing}} = S_{\text{mixture}} - S_1 - S_2. \tag{47}$$

Therefore, using the expressions for the individual terms in this equation, we have

$$S_1 + S_2 = N k \ln Z_1 + Nk - N k \ln N + \frac{U_1}{T}$$

$$+ M k \ln Z_2 + Mk - M k \ln M + \frac{U_2}{T}$$

$$= (N + M)k + \frac{U}{T} + k \ln\left(\frac{Z_1^N . Z_2^M}{N^N M^M}\right),$$

and

$$S_{\text{mixture}} = (N+M)k + \frac{U}{T} + (N + M)k \ln\left(\frac{Z}{N + M}\right)$$

$$= (N+M)k + \frac{U}{T} + k \ln\left(\frac{Z}{N + M}\right)^{N + M}.$$

Subsequently, the substitution of the preceding two equations into Eq. (47) gives

$$\Delta S_{\text{mixing}} = k \ln\left(\frac{Z}{N + M}\right)^{N+M} - k \ln\left(\frac{Z_1^N . Z_2^M}{N^N M^M}\right). \tag{48}$$

The question which arises now concerns the relation between the two quotients.

$$\left(\frac{Z}{N+ M}\right)^{N+M} \quad \text{and} \quad \frac{Z_1^N . Z_2^M}{N^N M^M}.$$

Since the two gases were initially at the same temperature and pressure, we have as has been shown above.

$$\frac{V_1}{V_2} = \frac{N}{M}.$$

But

$$P V_1 = N k T, \quad \text{and} \quad PV_2 = M k T.$$

Therefore,

$$V_1 + V_2 = (N + M)\frac{kT}{P},$$

and

$$(V_1 + V_2)^{N+M} = \{(N+M)\}^{N+M} . \left(\frac{kT}{P}\right)^{N+M}. \tag{i}$$

Also

$$V_1^N \quad = \quad N^N\left(\frac{kT}{P}\right)^N, \tag{ii}$$

and

$$V_2^M \quad = \quad M^M\left(\frac{kT}{P}\right)^M. \tag{iii}$$

From Eqs. (i)–(iii), we obtain the following relation

$$\left\{\frac{(V_1 + V_2)^{N+M}}{V_1^{\,N} \,.\, V_2^{\,M}}\right\} = \left\{\frac{(N+M)^{N+M}}{N^N \,.\, M^M}\right\}. \tag{49}$$

We may now substitute into Eq. (48) for the partition function *Z* of the mixture, Z_1 of the first gas, and Z_2 for the second gas from the following equations, respectively

$$Z = \frac{V}{h^3}\,(2\pi m k T)^{3/2},$$

$$Z_1 = \frac{V_1}{h^3}\,(2\pi m k T)^{3/2},$$

$$Z_2 = \frac{V_2}{h^3}\,(2\pi m k T)^{3/2}.$$

Eq. (48) may be firstly rewritten as

$$\begin{aligned}\Delta S_{\text{mixing}} &= k\Big\{\ln\ Z^{N+M} - \ln\,(N+M)^{N+M} \\ &\qquad - \ln Z_1^{\,N} - \ln Z_2^{\,M} + \ln N^N.M^M\Big\} \\ &= k\left\{\ln \frac{Z^{N+M}}{Z_1^{\,N}.Z_2^{\,M}} + \ln \frac{N^N\, M^M}{(N+M)^{N+M}}\right\}.\end{aligned} \tag{50}$$

Then the substitution for the partition functions in the term

$$\ln\left(\frac{Z^{\,N+M}}{Z_1^{\,N} \,.\, Z_2^{\,M}}\right) \tag{51}$$

gives

$$\begin{aligned}\ln \frac{Z^{N+M}}{Z_1^N \,.\, Z_2^M} &= (N+M)\left\{\ln V - 3\ln h + \frac{3}{2}\ln\,(2\pi m k T)\right\}. \\ &\quad -N\left\{\ln V_1 - 3\ln h + \frac{3}{2}\ln\,(2\pi m k T)\right\} \\ &\quad -M\left\{\ln V_2 - 3\ln h + \frac{3}{2}\ln\,(2\pi m k T)\right\} \\ &= \{(N+M)\ln V - N\ln V_1 - M\ln V_2\} \\ &= \ln V^{(N+M)} - \ln V_1^{\,N} - \ln V_2^{\,M} \\ &= \ln\left\{\frac{(V_1 + V_2)^{N+M}}{V_1^{\,N}.V_2^{\,M}}\right\}.\end{aligned} \tag{52}$$

Now replacing the term of Eq. (51) in Eq. (50) by its equivalent given by Eq. (52), we obtain

$$\Delta S_{mixing} = k\left\{ \ln \frac{(V_1 + V_2)^{N+M}}{V_1^N . V_2^M} - \ln \frac{(N+M)^{N+M}}{N^N . M^M} \right\}, \tag{53}$$

and since according to Eq. (49), namely

$$\frac{(V_1 + V_2)^{N+M}}{V_1^N . V_2^M} = \frac{(N+M)^{N+M}}{N^N . M^M}, \tag{49}$$

we therefore have

$$\Delta S_{mixing} = \text{zero}, \tag{54}$$

in the case of mixing two identical gases; a result which cannot be arrived at using classical thermodynamics.

The change in entropy due to mixing of two non-identical monatomic ideal gases that are chemically inert as given above by

$$\Delta S_{mixing} = k \ln \left[\frac{(V_1 + V_2)^{N+M}}{V_1^N . V_2^M} \right] \tag{41}$$

is always positive. At first, it might seem strange that there is a change in entropy although the temperature, pressure, and internal energy remain constant. However, if the mixing is viewed from statistical grounds, the increase is *self-explanatory*. Clearly, when two gases are mixed, the *disorder* of the system is *increased* provided the two gases are different since the entropy is a measure of disorder. Also since the process is not *reversible*, it is not possible to separate the two gases into separate volumes with a decrease in the entropy of the system.

Another important equation is the Gibbs function of the mixture* $G_{mixture}$; it is given by

$$G_{mixture} = -kT \ln \left[\left(\frac{Z'_1}{N} \right)^N \left(\frac{Z'_2}{M} \right)^M \right],$$

$$= -nR \left[-3 \ln h + \frac{3}{2} \ln 2\pi k + \frac{3}{2} \ln T + \ln V + \frac{N}{N+M} \ln \frac{m_1^{3/2}}{N} + \frac{M}{N+M} \ln \frac{m_2^{3/2}}{M} \right]. \tag{55}$$

This equation shall be used when we consider chemical systems in Chapter 14.

* Eq. (55) can be derived by considering two chemically inert ideal gases, for which we have

CHAPTER 14

Chemical Systems

14.1. Introduction

Although the microscopic assemblies, or thermodynamic systems, discussed before excluded the possibility of chemical reactions, the statistical methods are readily extended to multiphase and multicomponent systems in which such variations occur. It is quite possible to develop a complete statistical theory of chemical equilibria with minor reference to classical thermodynamics. However, such a procedure is unnecessarily complex and the practical application of the theory to other than simple chemical systems is quite formidable. Therefore, we shall take advantage of relationships already established from classical thermodynamics and utilise statistical concepts whenever they give a clearer insight into reaction phenomena.

As shown before, the partition function is the essential link between the coordinates of microscopic systems and the thermodynamic properties. Once it has been evaluated, all the physical properties, such as internal energy, entropy, modulus of elasticity, and specific heats, can be readily computed. The partition function also permits the computation of many other parameters, such as equilibrium constants and activity coefficients.

14.2. Law of mass action

We have considered in Chapter 13 the case of a mixture of two chemically inert ideal gases and obtained the following two important equations – Eqs. (41) and (55),

$$\begin{aligned}\Delta S_{mixing} &= (N + M)k \ln (V_1 + V_2) - N \\ &\quad \times k \ln V_1 - M k \ln V_2, \\ &= k \ln \left[\frac{(V_1 + V_2)^{N+M}}{V_1^N \, V_2^M} \right],\end{aligned}$$

and

$$G = -kT \ln \left\{ \left(\frac{Z_1'}{N} \right)^N \left(\frac{Z_2'}{M} \right)^M \right\}.$$

These equations can be extended to a mixture composed of any number of constituents, provided they are ideal gases and chemically inert. However, the constituents must be indeed *different gases*.

We shall now take up mixtures of chemically *reactive* gases, considering first, for simplicity, the reaction

$$A + B = AB .$$

Let us consider a thermodynamic system in which N_a molecules of reactant A and N_b molecules of reactant B are mixed with N_{ab} molecules of the product AB. If the species A, B, and AB were inert, then any composition N_a, N_b, and N_{ab} would be allowed. However, since a chemical reaction, as given above, may take place in the system, a state of equilibrium between the *reactants* and the *products* will be achieved eventually, and it is this *equilibrium state* which we will determine.

Several features of the problem must be made clear as follows:

1. The A, B, and AB molecules are *different* species, and the molecules of a given species are *indistinguishable.*
2. If the energy of each constituent is measured from its *ground state*, the energy of formation is given by the energy difference
$$U_a + U_b - U_{ab}.$$
If, however, we measure *all* the energies from the *same zero point*, we must subtract the energy of formation from U_{ab}.
Thus the energy of formation will be taken into account implicitly.
3. For simplicity, the energy levels of all constituents will be tentatively assumed to be non-degenerate.

The energy levels of each molecule of constituent A are designated as

$$\epsilon_{a1}, \epsilon_{a2}, \dots, \epsilon_{ai}, \tag{1}$$

and similarly for constituents B and AB, we have

$$\epsilon_{b1}, \epsilon_{b2}, \dots, \epsilon_{bi}, \tag{2}$$

and

$$\epsilon_{ab1}, \epsilon_{ab2}, \dots, \epsilon_{abi}, \tag{3}$$

respectively. The corresponding distributions are

$$N_{a1}, N_{a2}, \dots, N_{ai}, \tag{4}$$

$$N_{b1}, N_{b2}, \dots, N_{bi}, \tag{5}$$

$$N_{ab1}, N_{ab2}, \dots, N_{abi}. \tag{6}$$

We will first determine the state of maximum thermodynamic probability for given values of the constituents A, B, and AB.

The number of different microstates corresponding to any given macrostate, that is, the thermodynamic probability W, is given by

$$W = \Pi \frac{g_i^{N_{ai}}}{N_{ai}!} \Pi \frac{g_i^{N_{bi}}}{N_{bi}!} \Pi \frac{g_i^{N_{abi}}}{N_{abi}!}. \tag{7}$$

This equation can be derived at in the following way:

It has been shown before[+] that for a system the number of microstates for the *i*th cell is

$$W_i = \frac{(g_i + N_i - 1)!}{(g_i - 1)!\ N_i!},$$

[+] See Section 9.3, Eqs. (4) and (5).

also that for each microstate of cell *i* we may have any one of the microstates of cell *j*. Therefore, the total number of microstates including all occupied cells, or the thermodynamic probability *W* for any given macrostate is

$$W = \Pi W_i = \Pi \frac{\left(g + N_i - 1\right)!}{\left(g_i - 1\right)! N_i!}. \tag{i}$$

Using the Maxwell-Boltzmann limit, $g_i >> N_i$, and taking the logarithm of this equation, it reduces to

$$\ln W = \Sigma\left[N_i \ln g_i + N_i - N_i \ln N_i\right], \tag{ii}$$

This equation in turn takes the form

$$\ln W = \Sigma\left[\ln g_i^{N_i} - \ln N_i!\right],$$

$$= \Sigma \ln \frac{g_i^{N_i}}{N_i!}, \tag{iii}$$

by using Stirling's approximation. In the form of a product, Eq. (iii) can be rewritten as

$$W = \Pi\left(\frac{g_i^{N_i}}{N_i!}\right).$$

Therefore, in general, the thermodynamic probability of a given macrostate is given by

$$W = \Pi\left(\frac{g_i^{N_i}}{N_i!}\right). \tag{iv}$$

We already know that if we have a mixture of two chemically inert ideal gases, the thermodynamic probability of the mixture comprising these two gases is the product of the thermodynamic probabilities acquired by the constituent gases, that is,

$$W = W_1 W_2.$$

The same can be said in our case where N_a molecules of A and N_b molecules of B are mixed with N_{ab} molecules of the product AB so that the thermodynamic probability for given values of the three constituents, or the number of different microstates corresponding to a specified macrostate, is given by

$$W = W_a W_b W_{ab}. \tag{v}$$

Hence, the substitution of the thermodynamic probability *W* for each constituent in the mixture by $\Pi\left(g_i^{N_i}/N_i!\right)$ yields Eq. (7) shown above.

The number of molecules of species A, N_a, of species B, N_b, and of species AB, N_{ab}, are given by the following *condition equations*:

$$N_a = \sum_i N_{ai} \tag{8}$$

$$N_b = \sum_i N_{bi} \tag{9}$$

$$N_{ab} = \sum_i N_{abi} \,. \tag{10}$$

We also have the *condition equations* which give the total number of molecules N_A of *type* A, whether they are present as *uncombined* molecules or *combined* with B to form AB molecules, and similarly for the B molecules:

$$N_A = N_a + N_{ab} = \text{const.} \tag{11}$$

$$N_B = N_b + N_{ab} = \text{const.} \tag{12}$$

The total energy is given by

$$U = \sum_i N_{ai} \epsilon_{ai} + \sum_i N_{bi} \epsilon_{bi} + \sum N_{abi} \epsilon_{ab} \,. \tag{13}$$

Eqs. (11) and (12) account for the number of molecules of A and the number of molecules of B, whether combined or uncombined. Eq. (13) defines the energy of the entire system, *noting that the energy of each constituent is measured from a common zero point.* Thus the energy of formation is allowed for *implicitly* in Eq. (13).

Taking the logarithm of Eq. (7), we obtain

$$\begin{aligned} \ln W = {} & \Sigma \left(N_{ai} \ln g_i - \ln N_{ai}! \right) \\ & + \Sigma \left(N_{bi} \ln g_i - \ln N_{bi}! \right) \\ & + \Sigma \left(N_{abi} \ln g_i - \ln N_{abi}! \right). \end{aligned}$$

Applying Sterling's approximation, we have

$$\begin{aligned} \ln W = {} & \Sigma\left(N_{ai} \ln g_i - N_{ai} \ln N_{ai} + N_{ai} \right) \\ & + \Sigma\left(N_{bi} \ln g_i - N_{bi} \ln N_{bi} + N_{bi} \right) \\ & + \Sigma\left(N_{abi} \ln g_i - N_{abi} \ln N_{abi} + N_{abi} \right), \\ = {} & \Sigma \left(N_{ai} \ln \frac{g_i}{N_{ai}} \right) + \Sigma \left(N_{bi} \ln \frac{g_i}{N_{bi}} \right) \\ & + \Sigma \left(N_{abi} \ln \frac{g_i}{N_{abi}} \right) + N_a + N_b + N_{ab} \,. \end{aligned} \tag{14}$$

For the state of maximum thermodynamic probability, we have

$$\delta \ln W = \Sigma\left(\ln g_i \,\delta N_{ai} - N_{ai}\times\frac{1}{N_{ai}}\delta N_{ai} - \ln N_{ai}\,\delta N_{ai} + \delta N_{ai}\right)$$

$$+\Sigma\left(\ln g_i \,\delta N_{bi} - N_{bi}\times\frac{1}{N_{bi}}\delta N_{bi} - \ln N_{bi}\,\delta N_{bi} + \delta N_{bi}\right)$$

$$+\Sigma\left(\ln g_i \,\delta N_{abi} - N_{abi}\times\frac{1}{N_{abi}}\delta N_{abi} - \ln N_{abi}\,\delta N_{abi} + \delta N_{abi}\right) = 0.$$

That is,

$$\delta \ln W = \Sigma\left(\ln\frac{g_i}{N_{ai}}\right)\delta N_{ai} + \Sigma\left(\ln\frac{g_i}{N_{bi}}\right)\delta N_{bi} + \Sigma\left(\ln\frac{g_i}{N_{abi}}\right)\delta N_{abi} = 0,$$

or

$$-\delta \ln W = \Sigma\left(\ln\frac{N_{ai}}{g_i}\right)\delta N_{ai} + \Sigma\left(\ln\frac{N_{bi}}{g_i}\right)\delta N_{bi} + \Sigma\left(\ln\frac{N_{abi}}{g_i}\right)\delta N_{abi} = 0, \tag{15}$$

where the quantities δN_{ai}'s, δN_{bi}'s, and δN_{abi}'s are not independent. In order to make them *in effect* independent, we proceed as follows:

We first obtain a maximum for given constant values of N_a, N_b, N_{ab}, hence we require that.

$$\delta N_a = \Sigma\,\delta N_{ai} = 0 \tag{16}$$

$$\delta N_b = \Sigma\,\delta N_{bi} = 0 \tag{17}$$

$$\delta N_{ab} = \Sigma\,\delta N_{abi} = 0. \tag{18}$$

Also, the total energy is constant as given by Eq. (13) so that

$$\delta U = \sum_i \epsilon_{ai}\,\delta N_{ai} + \sum_i \epsilon_{bi}\,\delta N_{bi} + \sum_i \epsilon_{abi}\,\delta N_{abi} = 0. \tag{19}$$

Eqs. (16)–(19) are the condition equations. Now applying Lagrange's method of undetermined multipliers, we multiply Eq. (16) by $-\ln\alpha_a$, Eq. (17) by $-\ln\alpha_b$, Eq. (18) by $-\ln\alpha_{ab}$, Eq. (19) by β, and add the products to Eq. (15). This gives

$$\Sigma\left(\ln\frac{N_{ai}}{g_i} - \ln\alpha_a + \beta\epsilon_{ai}\right)\delta N_{ai}$$

$$+\Sigma\left(\ln\frac{N_{bi}}{g_i} - \ln\alpha_b + \beta\epsilon_{bi}\right)\delta N_{bi}$$

$$+\Sigma\left(\ln\frac{N_{abi}}{g_i} - \ln\alpha_{ab} + \beta\epsilon_{abi}\right)\delta N_{abi} = 0\ .$$

Since in effect the δN_{ai}'s, δN_{bi}'s, and δN_{abi}'s are now independent, their *coefficients must vanish separately*; that is, we must have

$$\ln\left(\frac{N_{ai}}{g_i}\right) = \ln\alpha_a - \beta\epsilon_{ai}\,, \qquad (20)$$

$$\ln\left(\frac{N_{bi}}{g_i}\right) = \ln\alpha_b - \beta\epsilon_{bi}\,, \qquad (21)$$

$$\ln\left(\frac{N_{abi}}{g_i}\right) = \ln\alpha_{ab} - \beta\epsilon_{abi}\ . \qquad (22)$$

The values of N_{ai} , N_{bi} and N_{abi} that can be obtained from these equations are

$$N_{ai} = \alpha_a g_i e^{-\beta\epsilon_{ai}}\,, \quad \text{or} \quad \Sigma g_i e^{-\beta\epsilon_{ai}} = N_a/\alpha_a\,,$$

$$N_{bi} = \alpha_b g_i e^{-\beta\epsilon_{bi}}\,, \quad \text{or} \quad \Sigma g_i e^{-\beta\epsilon_{bi}} = N_b/\alpha_b\,,$$

and

$$N_{abi} = \alpha_{abi} g_i e^{-\beta\epsilon_{abi}}\,, \quad \text{or} \quad \Sigma g_i e^{-\beta\epsilon_{abi}} = N_{abi}/\alpha_{abi}\,,$$

respectively. When these values of N_{ai}, N_{bi}, and N_{abi} are substituted in the expression for $\ln W$, which is given above by Eq. (14), we get an equation representing *the maximum thermodynamic probability for the given values of; this expression is*

$$\ln W = \beta U + N_a \ln\left(\Sigma g_i e^{-\beta\epsilon_{ai}}\right) + N_b \ln\left(\Sigma g_i e^{-\beta\epsilon_{bi}}\right)$$

$$+ N_{ab} \ln\left(\Sigma g_i\, e^{-\beta\epsilon_{abi}}\right)$$

$$+ N_a - N_a \ln N_a + N_b - N_b \ln N_b$$

$$+ N_{ab} - N_{ab} \ln N_{ab}\,. \qquad (23)$$

We have now determined the maximum thermodynamic probability for any given values of N_a, N_b, and N_{ab}, but have not yet attempted to determine which values of N_a, N_b, and N_{ab} correspond to *chemical equilibrium.*

It is evident from Eq. (23) that the constants α_a, α_b and α_{ab} or their logarithms $\ln \alpha_a$, $\ln \alpha_b$ and $\ln \alpha_{ab}$, which have been used as the multipliers of three condition equations, Eqs. (16)–(18), respectively, do not appear in this expression for $\ln W$. That these constants had cancelled one another when they were added to Eq. (15) thus leading to Eq. (23), might lead one to suggest that Eq. (23) in this form is intended to be used for the derivation of another function, which is essential. This is actually the case as will be shown below. At the same time, Eq. (23) can be utilised for other purposes such as *S* and *G* calculations.

On the other hand, it may be interesting to verify that the substitution into Eq. (23) of the values of N_a, N_b and N_{ab} in terms of the constants α_a, α_b and α_{ab}, respectively, should definitely result in Eq. (14). This may be proved by first replacing the *partition functions* in Eq. (23) by the corresponding number of the molecules shown above so that this equation takes the following form:

$$\begin{aligned}\ln W = \beta U &+ N_a \ln (N_a/\alpha_a) + N_b \ln (N_b/\alpha_b) \\ &+ N_{ab} \ln (N_{ab}/\alpha_{ab}) - N_a \ln N_a \\ &- N_b \ln N_b - N_{ab} \ln N_{ab} \\ &+ N_a + N_b + N_{ab},\end{aligned}$$

thus,

$$\begin{aligned}\ln W = \beta U &- N_a \ln \alpha_a - N_b \ln \alpha_b - N_{ab} \ln \alpha_{ab} \\ &+ N_a + N_b + N_{ab}.\end{aligned} \qquad (23)'$$

But

$$\alpha_a = \frac{N_{ai}}{g_i} e^{\beta \epsilon_{ai}},$$

$$\alpha_b = \frac{N_{bi}}{g_i} e^{\beta \epsilon_{bi}},$$

$$\alpha_{ab} = \frac{N_{abi}}{g_i} e^{\beta \epsilon_{abi}},$$

Therefore,

$$-\ln \alpha_a = \ln \frac{g_i}{N_{ai}} - \beta \epsilon_{ai},$$

$$-\ln \alpha_b = \ln \frac{g_i}{N_{bi}} - \beta \epsilon_{bi},$$

$$-\ln \alpha_{ab} = \ln \frac{g_i}{N_{abi}} - \beta \epsilon_{abi}.$$

Subsequently, multiplying these three preceding equations by N_{ai}, N_{bi} and N_{abi}, respectively, and then taking the sum, we get the following three equations:

$$-N_a \ln \alpha_a = -\Sigma N_{ai} \ln \alpha_a = \Sigma N_{ai} \ln \frac{g_i}{N_{ai}} - \Sigma N_{ai} \epsilon_{ai} \beta, \qquad \text{(i)}$$

$$-N_b \ln \alpha_b = -\Sigma N_{bi} \ln \alpha_b = \Sigma N_{bi} \ln \frac{g_i}{N_{bi}} - \Sigma N_{bi} \epsilon_{bi} \beta, \qquad \text{(ii)}$$

$$-N_{ab} \ln \alpha_{ab} = -\Sigma N_{abi} \ln \alpha_{ab} = \Sigma N_{abi} \ln \frac{g_i}{N_{abi}} - \Sigma N_{abi} \epsilon_{abi} \beta, \qquad \text{(iii)}$$

where

$$\Sigma N_{a_i} \epsilon_{ai} \beta + \Sigma N_{bi} \epsilon_{bi} \beta + \Sigma N_{abi} \epsilon_{abi} \beta$$
$$= \beta U.$$

Substituting now the right-hand sides of the above three equations, Eqs. (i)–(iii) into Eq. (23)′ for $(-N_a \ln \alpha_a, -N_b \ln \alpha_b, -N_{ab} \ln \alpha_{ab})$
respectively, we eventually find

$$\ln W = \Sigma \left(N_{ai} \ln \frac{g_i}{N_{ai}} \right) + \Sigma \left(N_{bi} \ln \frac{g_i}{N_{bi}} \right)$$
$$+ \Sigma \left(N_{abi} \ln \frac{g_i}{N_{abi}} \right)$$
$$+ N_a + N_b + N_{ab},$$

which is Eq. (14) as is expected.

Now again by means of Eq. (23), we have determined the maximum thermodynamic probability for given values of N_a, N_b and N_{ab} but have not attempted to determine which values of N_a, N_b and N_{ab} correspond to chemical equilibrium. The maximum thermodynamic probability, which we shall denote by W°, corresponding to chemical equilibrium can be determined by varying N_a, N_b and N_{ab} with temperature while keeping N_A and N_B fixed. Since it can be proved that $\beta = 1/kT$, then using Eq. (23), we have

$$\delta \ln W^{\circ} = \left[\ln \Sigma g_i \, e^{-\epsilon_{ai}/kT} - \ln N_a \right] \delta N_a$$
$$+ \left[\ln \Sigma g_i \, e^{-\epsilon_{bi}/kT} - \ln N_b \right] \delta N_b$$
$$+ \left[\ln \Sigma g_i \, e^{-\epsilon_{abi}/kT} - \ln N_{ab} \right] \delta N_{ab} = 0. \qquad (24)$$

Also, since the total number of molecules of type A, denoted by N_A, and those of type B, denoted by N_B, are fixed, as noted above according to Eqs. (11) and (12), then the following two equations

$$\delta N_A = \delta N_a + \delta N_{ab} = 0 \tag{25}$$

$$\delta N_B = \delta N_b + \delta N_{ab} = 0 \tag{26}$$

represent the *condition equations* imposed on δN_a's, δN_b's and δN_{ab}'s. Again applying Lagrange's method of undetermined multipliers, we multiply Eq. (25) by a constant *a*, and Eq. (26) by a constant *b*, and add to Eq. (24) to get

$$a + \ln\left[\sum_i g_i\, e^{-\epsilon_{ai}/kT}\right] - \ln \overset{\circ}{N}_a = 0\ , \tag{27}$$

$$b + \ln\left[\sum_i g_i\, e^{-\epsilon_{bi}/kT}\right] - \ln \overset{\circ}{N}_b = 0\ , \tag{28}$$

$$a + b + \ln\left[\sum_i g_i\, e^{-\epsilon_{abi}/kT}\right] - \ln \overset{\circ}{N}_{ab} = 0\ . \tag{29}$$

Here $\overset{\circ}{N}_a$, $\overset{\circ}{N}_b$ and $\overset{\circ}{N}_{ab}$ represent the equilibrium values of N_a, N_b, and N_{ab}, respectively. Eliminating *a* and *b* in the preceding equations, we have

$$K_N = \frac{\overset{\circ}{N}_{ab}}{\overset{\circ}{N}_a \overset{\circ}{N}_b} = \frac{\sum g_i\, e^{-\epsilon_{abi}/kT}}{\sum g_i\, e^{-\epsilon_{ai}/kT} . \sum g_i\, e^{-\epsilon_{bi}/kT}}, \tag{30}$$

where K_N is the *equilibrium constant* of the reaction based on the number of molecules present. Introducing the partition function in the form

$$Z = \sum g_i\, e^{-\epsilon_i/kT},$$

as given before, we can also write Eq. (30) as

$$K_N = \frac{\overset{\circ}{N}_{ab}}{\overset{\circ}{N}_a \overset{\circ}{N}_b} = \frac{Z_{ab}}{Z_a Z_b}. \tag{31}$$

Eqs. (30) and (31) are *alternative expressions of the law of chemical equilibrium or the law of mass action,* which forms the basis of computations in chemical equilibrium. It is seen that the equilibrium constant is only a function of the partition functions of the individual species. Once the partition *functions* are evaluated, the equilibrium constant may be determined, and Eq. (31) then provides a relationship between the species concentrations at chemical equilibrium.

Since the Eqs. (11) and (12) must still be satisfied at chemical equilibrium, we have

$$\left.\begin{aligned} N_A &= \overset{\circ}{N}_a + \overset{\circ}{N}_{ab} \\ N_B &= \overset{\circ}{N}_b + \overset{\circ}{N}_{ab} \end{aligned}\right\}. \tag{32}$$

Accordingly, Eqs. (31) and (32) can be regarded as three relations for the three *unknowns* $\overset{\circ}{N}_a$, $\overset{\circ}{N}_b$ and $\overset{\circ}{N}_{ab}$.

It is often convenient to write the equilibrium constant in terms of the *partial pressures* of the constituents. For the reaction considered here,

$$A + B = AB,$$

the equilibrium constant in terms of the partial pressures at equilibrium,

$$\overset{\circ}{P}_a, \quad \overset{\circ}{P}_b, \quad \text{and} \quad \overset{\circ}{P}_{ab},$$

is obtained from Eq. (31), using the equation of state of an ideal gas,

$$K_p = \frac{\overset{\circ}{P}_{ab}}{\overset{\circ}{P}_a \overset{\circ}{P}_b}. \tag{33}$$

Eq. (33) can be simply proved using Dalton's law of partial pressures for ideal gases. Let the total pressure at equilibrium, which is the sum of the partial pressures of the constituents, be P, and N is the total number of molecules of these constituents. Therefore,

$$\overset{\circ}{P}_a + \overset{\circ}{P}_b + \overset{\circ}{P}_{ab} = P.$$

And the partial pressures are given by

$$\frac{\overset{\circ}{P}_a}{P} = \frac{\overset{\circ}{N}_a}{N}, \text{ or } \overset{\circ}{P}_a = \frac{\overset{\circ}{N}_a}{N} P,$$

$$\overset{\circ}{P}_b = \frac{\overset{\circ}{N}_b}{N} P,$$

$$\overset{\circ}{P}_{ab} = \frac{\overset{\circ}{N}_{ab}}{N} P.$$

Hence, substituting in Eq. (33) for the partial pressures in terms of the total pressure P at equilibrium, we have

$$K_P = \frac{\overset{\circ}{P}_{ab}}{\overset{\circ}{P}_a \overset{\circ}{P}_b} = \frac{\frac{\overset{\circ}{N}_{ab}}{N} P}{\frac{\overset{\circ}{N}_a}{N} \frac{\overset{\circ}{N}_b}{N} P2}.$$

$$= \frac{\overset{\circ}{N}_{ab}}{\overset{\circ}{N}_a \overset{\circ}{N}_b} \frac{N}{P} = \frac{Z_{ab}}{Z_a Z_b} \frac{N}{P} = K_N \frac{N}{P} \tag{33$'$}$$

It we now equate $W = \overset{\circ}{W}$ in Eq. (23); that is,

$$\ln \overset{\circ}{W} = \beta U + \overset{\circ}{N}_a \ln \left(\frac{Z_a}{\overset{\circ}{N}_a} \right)^{\overset{\circ}{N}_a} + \overset{\circ}{N}_b \ln \left(\frac{Z_b}{\overset{\circ}{N}_b} \right)^{\overset{\circ}{N}_b}$$

$$+ \overset{\circ}{N}_{ab} \ln \left(\frac{Z_{ab}}{\overset{\circ}{N}_{ab}} \right)^{\overset{\circ}{N}_{ab}} + \overset{\circ}{N}_a + \overset{\circ}{N}_b + \overset{\circ}{N}_{ab}, \tag{23}$$

we can obtain the following expression for the *Gibbs free energy* in terms of N°_{a}, N°_{b} and N°_{ab}

$$G = kT\ln\left[\left(\frac{Z_a}{N^{\circ}_{a}}\right)^{N^{\circ}_{a}}\left(\frac{Z_b}{N^{\circ}_{b}}\right)^{N^{\circ}_{b}}\left(\frac{Z_{ab}}{N^{\circ}_{ab}}\right)^{N^{\circ}_{ab}}\right]. \tag{34}$$

This equation can be derived as follows:

The general expression for *S*, as shown before, is

$$S = k\ln W^{\circ} = Nk\ln\left(\frac{Z}{N}\right) + k\beta U + kN$$

$$= k\left[\ln\left(\frac{Z}{N}\right) + \frac{U}{kT} + N\right].$$

Therefore, the substitution for $\ln W^{\circ}$ in its expanded form shown in Eq. (23)″, gives

$$S = k\left[\frac{U}{kT} + \ln\left(\frac{Z_a}{N^{\circ}_{a}}\right)^{N^{\circ}_{a}} + \ln\left(\frac{Z_b}{N^{\circ}_{b}}\right)^{N^{\circ}_{b}} + \ln\left(\frac{Z_{ab}}{N^{\circ}_{ab}}\right)^{N^{\circ}_{ab}} + N^{\circ}_{a} + N^{\circ}_{b} + N^{\circ}_{ab}\right].$$

For the energy term *TS*, we have from the previous equation,

$$TS = U + kT\ln\left[\left(\frac{Z_a}{N^{\circ}_{a}}\right)^{N^{\circ}_{a}}\left(\frac{Z_a}{N^{\circ}_{b}}\right)^{N^{\circ}_{b}}\left(\frac{Z_a}{N^{\circ}_{ab}}\right)^{N^{\circ}_{ab}}\right] + NkT$$

Hence, substituting in the definition of *G* given by

$$G \equiv U - TS + PV,$$

we obtain

$$G = U - \left[U + kT\ln\left\{\left(\frac{Z_a}{N^{\circ}_{a}}\right)^{N^{\circ}_{a}}\left(\frac{Z_b}{N^{\circ}_{b}}\right)^{N^{\circ}_{b}}\left(\frac{Z_{ab}}{N^{\circ}_{ab}}\right)^{N^{\circ}_{ab}}\right\}\right]$$

$$- NkT + PV.$$

In the case of *an ideal gas*, this equation reduces to

$$G = -kT\ln\left[\left(\frac{Z_a}{N^{\circ}_{a}}\right)^{N^{\circ}_{a}}\left(\frac{Z_b}{N^{\circ}_{b}}\right)^{N^{\circ}_{b}}\left(\frac{Z_{ab}}{N^{\circ}_{ab}}\right)^{N^{\circ}_{ab}}\right], \tag{34}$$

which is the equation we are looking for.

Eqs. (31), (32), and (34) have great theoretical significance. Comparison of Eq. (34) with the equation

$$G_{\text{mixture}} = -kT \ln\left[\left(\frac{Z_1'}{N}\right)^N \left(\frac{Z_2'}{M}\right)^M\right],$$

which we previously* obtained for a mixture of two chemically inert ideal gases, shows that *when equilibrium has been reached,* the Gibbs function for a reactive mixture of A, B, and AB molecules is exactly the same as the Gibbs function which holds for the individual constituents of the mixture; that is, $\overset{\circ}{N}_a$ molecules of gas A, $\overset{\circ}{N}_b$ molecules of gas B, and $\overset{\circ}{N}_{ab}$ molecules of gas AB. Therefore, so long as we know the *equilibrium composition* of a mixture of gases at a given temperature, we can *ignore* the chemical reaction in computing the Gibbs function for the reactive mixture in a state of equilibrium at the same temperature. Furthermore, if we do not know the equilibrium composition, we can determine it from Eq. (31), provided we know the partition functions of the constituent gases.

14.3. Chemical equilibrium

In Section 14.2, the principles of chemical equilibrium were established from statistical concepts. In this section we shall examine more closely the physical significance of the equations derived and establish a clearer link between the statistical approach and the more familiar concepts of chemical thermodynamics.

The pressure of a gas from classical thermodynamics is given by

$$P = -\left(\frac{\partial F}{\partial V}\right)_T,$$

where *F* from statistical thermodynamics is

$$F = -NkT\left[\ln\left(\frac{Z}{N}\right) + 1\right].$$

Thus differentiating this equation at constant temperature, we get

$$\left(\frac{\partial F}{\partial V}\right)_T = -NkT\left(\frac{\partial \ln Z}{\partial V}\right)_T,$$

and through comparison with the pressure equation, we obtain for the pressure the expression,

$$P = NkT\left(\frac{\partial \ln Z}{\partial V}\right)_T.$$

This expression has been derived before, when we considered the thermodynamic properties† of a macroscopic system in terms of the coordinates of the constituent microscopic systems.

* See Section 13.2, Eq. (55).

† See Section 12.6.

The pressure of a mixture of gases using classical thermodynamics is

$$P = -\left(\frac{\partial F}{\partial V}\right)_{N_A, N_B, T} \tag{35}$$

Note that N_a, N_b, and N_{ab} actually change in order to maintain equilibrium when the volume changes.

For our chemical reaction, the Helmholtz function *at equilibrium* is given by

$$F = -kT\left[\ln\left\{\left(\frac{Z_a}{\overset{\circ}{N}_a}\right)^{\overset{\circ}{N}_a}\left(\frac{Z_b}{\overset{\circ}{N}_b}\right)^{\overset{\circ}{N}_b}\left(\frac{Z_{ab}}{\overset{\circ}{N}_{ab}}\right)^{\overset{\circ}{N}_{ab}}\right\} + \overset{\circ}{N}_a + \overset{\circ}{N}_b + \overset{\circ}{N}_{ab}\right]. \tag{36}$$

This equation follows from

$$F = -T\left[k\ln\left\{\left(\frac{Z_1'}{N}\right)^N\left(\frac{Z_2'}{M}\right)^M\right\} + nR\right],$$

or

$$F = -kT\left[\ln\left\{\left(\frac{Z_1'}{N}\right)^N\left(\frac{Z_2'}{M}\right)^M\right\} + (N+M)\right],$$

in the case of two chemically inert ideal gases, as shown before.*

Since the *translational contributions* to the partition functions of the constituents of the mixture are proportional to the volume of an ideal gas, we have

$$\left(\frac{\partial \ln Z_a}{\partial V}\right)_{N_A,\, N_B,\, T} = \left(\frac{\partial \ln V}{\partial V}\right) = \frac{1}{V}, \tag{37}$$

for each constituent. Accordingly, Eq. (35) can first be written as

$$P = kT\frac{\partial}{\partial V}\left[\overset{\circ}{N}_a \ln Z_a + \overset{\circ}{N}_b \ln Z_b + \overset{\circ}{N}_{ab}\ln Z_{ab} - \overset{\circ}{N}_a\left(\ln\overset{\circ}{N}_a - 1\right) - \overset{\circ}{N}_b\left(\ln\overset{\circ}{N}_b - 1\right) - \overset{\circ}{N}_{ab}\left(\ln\overset{\circ}{N}_{ab} - 1\right)\right],$$

* See Section 13.2, Eq. (29).

and then differentiating and using Eq. (37), we obtain

$$P = kT\left[\frac{N^{\circ}_{a}}{V} + \frac{N^{\circ}_{b}}{V} + \frac{N^{\circ}_{ab}}{V} + \frac{\partial N^{\circ}_{a}}{\partial V}\left(\ln Z_a - \ln N^{\circ}_{a}\right)\right.$$

$$\left. + \frac{\partial N^{\circ}_{b}}{\partial V}\left(\ln Z_b - \ln N^{\circ}_{b}\right) + \frac{\partial N^{\circ}_{ab}}{\partial V}\left(\ln Z_{ab} - \ln N^{\circ}_{ab}\right)\right].$$

From Eqs. (11) and (12), we have

$$\left(\frac{\partial N^{\circ}_{a}}{\partial V}\right)_{N_A, N_B, T} = -\left(\frac{\partial N^{\circ}_{ab}}{\partial V}\right)_{N_A, N_B, T},$$

$$\left(\frac{\partial N^{\circ}_{b}}{\partial V}\right)_{N_A, N_B, T} = -\left(\frac{\partial N N^{\circ}_{ab}}{\partial V}\right)_{N_A, N_B, T},$$

or

$$\left(\frac{\partial N^{\circ}_{a}}{\partial V}\right)_{N_A, N_B, T} = -\left(\frac{\partial N^{\circ}_{ab}}{\partial V}\right)_{N_A, N_B, T} = \left(\frac{\partial N^{\circ}_{b}}{\partial V}\right)_{N_A, N_B, T}.$$

Hence,

$$P = \frac{kT}{V}\left(N^{\circ}_{a} + N^{\circ}_{b} + N^{\circ}_{ab}\right) - kT\left(\frac{\partial N^{\circ}_{ab}}{\partial V}\right)_{N_A, N_B, T}$$

$$\times\left(\ln Z_a + \ln Z_b - \ln Z_{ab} - \ln N^{\circ}_{a} - \ln N^{\circ}_{b} + \ln N^{\circ}_{ab}\right).$$

When we use Eq. (31), which is

$$K_N = \frac{N^{\circ}_{ab}}{N^{\circ}_{a} N^{\circ}_{b}} = \frac{Z_{ab}}{Z_a Z_b}, \qquad (31)$$

we find

$$\ln Z_{ab} - \ln Z_a - \ln Z_b = \ln N^{\circ}_{ab} - \ln N^{\circ}_{a} - N^{\circ}_{b},$$

or

$$\ln Z_a + \ln Z_b - \ln Z_{ab} = \ln N^{\circ}_{a} + \ln N^{\circ}_{b} - \ln N^{\circ}_{ab},$$

That is,

$$\ln Z_a + \ln Z_b - \ln Z_{ab} - \ln N^{\circ}_{a} - \ln N^{\circ}_{b} + \ln N^{\circ}_{ab} = 0$$

Hence, the expression in the bracket of the preceding equation which represents *P* disappears, and we obtain*

$$PV = \left(N^{\circ}_{a} + N^{\circ}_{b} + N^{\circ}_{ab}\right) kT. \qquad (38)$$

This equation suggests the *equation of state of an ideal gas.*

* The sum within the parenthesis is not constant, since the total number of molecules changes as the chemical reaction takes place.

14.4. Chemical potentials

We already know that in the general development of thermodynamics, a specification of any two thermodynamic coordinates, such as *temperature and pressure*, are sufficient to determine the state of a system. For a chemically reactive system, two thermodynamic coordinates are sufficient to determine the state of the system only if *the constituents are in chemical equilibrium* so that the chemical composition is given by the law of mass action. If *the constituents are not in chemical equilibrium*, then we must know the composition in addition to the two thermodynamic coordinates, in order to specify the state of the system. Also, since the composition of a chemically reactive system may *vary*, it is useful to introduce the concept of *chemical potentials*, which are defined for the reaction considered here by

$$\mu_a = \left(\frac{\partial F}{\partial N_a}\right)_{N_b, N_{ab}, V, T} \quad ; \quad \mu_b = \left(\frac{\partial F}{\partial N_b}\right)_{N_a, N_{ab}, V, T} \quad ;$$

$$\mu_{ab} = \left(\frac{\partial F}{\partial N_{ab}}\right)_{N_a, N_b, V, T} . \tag{39}$$

It can be proved using classical thermodynamics that the proper or natural independent variables of the thermodynamic coordinates *U*, *H*, *F*, and *G*, as shown by their differential forms,

$$\begin{aligned} dU &= T\,dS - P\,dV, \\ dH &= T\,dS + V\,dP, \\ dF &= -S\,dT - P\,dV, \\ dG &= -S\,dT + V\,dP, \end{aligned}$$

respectively, are

$$U = U(S,V),\ H = H(S,P),\ F = F(T,V) \text{ and } G = G(T,P).$$

Accordingly, the chemical potentials μ_a, μ_b, and μ_{ab} expressed in terms of the thermodynamic coordinate *F* can also be expressed in terms of each of the coordinates U, H, and G. Thus, we consider, for brevity, μ_a:

$$\mu_a = \left(\frac{\partial U}{\partial N_a}\right)_{N_b, N_{ab}, S, V} \quad ; \quad \mu_a = \left(\frac{\partial H}{\partial N_a}\right)_{N_b, N_{ab}, S, P} \quad ;$$

$$\mu_a = \left(\frac{\partial G}{\partial N_a}\right)_{N_b, N_{ab}, T, P} .$$

The other chemical potentials can be expressed in a similar way.

Now returning to Eq. (39), we can show that the substitution of Eq. (36) into this equation gives for the chemical potential of constituent A,

$$\mu_a = -kT\left(\ln Z_a - \ln N^{\circ}_a\right). \tag{40}$$

This equation can be obtained by taking the partial derivative of F with respect to N_a at constant N_b, N_{ab}, V, and T; that is,

$$\mu_a = \left(\frac{\partial F}{\partial N_a}\right)_{N_b, N_{ab}, V, T}$$
$$= -kT\left(\ln Z_a - \ln N^{\circ}_a - 1\right) - kT$$
$$= -kT\left(\ln Z_a - \ln N^{\circ}_a\right). \tag{40}$$

Similarly, the chemical potential of constituent B is

$$\mu_b = -kT\left(\ln Z_b - \ln N^{\circ}_b\right). \tag{41}$$

Adding Eqs. (40) and (41), we obtain

$$\mu_a + \mu_b = -kT\left(\ln Z_a + \ln Z_b - \ln N_a - \ln N_b\right).$$

But according to Eq. (31) we have by taking the logarithm of both sides of the equation.

$$\ln Z_a + \ln Z_b + \ln N^{\circ}_{ab}$$
$$= \ln Z_{ab} + \ln N^{\circ}_a + \ln N^{\circ}_b,$$

that is,

$$\ln Z_a + \ln Z_b - \ln N^{\circ}_a - \ln N^{\circ}_b$$
$$= \ln Z_{ab} - \ln N^{\circ}_{ab}.$$

Hence,

$$\mu_a + \mu_b = -kT\left(\ln Z_{ab} - \ln N^{\circ}_{ab}\right). \tag{42}$$

On the other hand, the chemical potential of constituent AB can be obtained from Eqs. (36) and (39), which is

$$\mu_{ab} = -kT\left(\ln Z_{ab} - \ln N^{\circ}_{ab}\right). \tag{43}$$

Hence, from Eqs. (42) and (43), we get

$$\mu_{ab} = \mu_a + \mu_b. \tag{44}$$

Eq. (44) is *another expression of the law of mass action.*

We want now to relate the preceding results to the partial Gibbs functions. Eq. (36) can be rewritten as

$$F = -kT\Big\{N^{\circ}_a\left(\ln Z_a - \ln N^{\circ}_a\right) + N^{\circ}_b\left(\ln Z_b - \ln N^{\circ}_b\right)$$
$$+ N^{0}_{ab}\left(\ln Z_{ab} - \ln N^{\circ}_{ab}\right)$$
$$+ N^{\circ}_a + N^{\circ}_b + N^{\circ}_{ab}\Big\},$$

then substituting from Eqs. (38), (40), (41), and (43), which are

$$PV = \left(N^\circ_a + N^\circ_b + N^\circ_{ab}\right) kT, \tag{38}$$

$$\mu_a = -kT\left(\ln Z_a - \ln N^\circ_a\right), \tag{40}$$

$$\mu_b = -kT\left(\ln Z_b - \ln N^\circ_b\right), \tag{41}$$

and

$$\mu_{ab} = -kT\left(\ln Z_{ab} - \ln N^\circ_{ab}\right), \tag{43}$$

into this equation, we obtain

$$F = N^\circ_a \mu_a + N^\circ_b \mu_b + N^\circ_{ab}\left(\mu_a + \mu_b\right) - PV.$$

But since by definition

$$G \equiv F + PV,$$

then the substitution for *F* by the preceding equation gives

$$G = N^\circ_a \mu_a + N^\circ_{ab} \mu_a + N^\circ_b \mu_b + N_{ab} \mu_b,$$

$$= \mu_a\left(N^\circ_a + N^\circ_{ab}\right) + \mu_b\left(N^\circ_b + N^\circ_{ab}\right).$$

Now taking $\mu_a = \mu_A$, and $\mu_b = \mu_B$, we have

$$G = \mu_A\left(N^\circ_a + N^\circ_{ab}\right) + \mu_B\left(N^\circ_b + N^\circ_{ab}\right).$$

Substituting for $\left(N^\circ_a + N^\circ_{ab}\right)$ and $\left(N^\circ_b + N^\circ_{ab}\right)$ in this equation by N_A and N_B, respectively, in accordance with Eqs. (11) and (12), we obtain

$$G = N_A \mu_A + N_B \mu_B. \tag{45}$$

It is seen from this equation that

$$\left(\frac{\partial G}{\partial N_A}\right)_{N_B} = \mu_A, \quad \text{and} \quad \left(\frac{\partial G}{\partial N_B}\right)_{N_A} = \mu_B.$$

Hence, the *partial Gibbs functions are identical to the chemical potentials.*

14.5. Equilibrium constants

In the relationships developed thus far, the energy difference between the *arbitrary zero-point energy level* for the reactive system and that of the *actual zero-point energy level* of a particular constituent was implied. Designating *this energy difference* as ϵ_0 for a *particular species*, we have

$$Z = Z_0\, e^{-\epsilon_0/kT}, \tag{46}$$

where *Z* is the partition function for the particular species based on the *arbitrary zero-point energy* level, and Z_0 is the partition function *referred to the actual zero-point energy level of the constituent.* The energy ϵ_0 takes account of the energy gained by the rest of the system when a *chemical bond is broken.* It is clear from our previous analysis that this energy is not included in the translational, rotational, vibrational, electronic, or nuclear contributions which we now associate with the partition function Z_0.

The Gibbs function for the individual constituent considered, based on the *arbitrary zero-point energy land*, is

$$G = -NkT \ln\left(\frac{Z}{N}\right),$$

for an ideal gas, as shown before.* Therefore, using Eq. (46), we have

$$G = -NkT \ln\left(\frac{Z_0 e^{-\epsilon_0/kT}}{N}\right),$$

$$= -NkT \ln\left(\frac{Z_0}{N}\right) + NkT \times \frac{\epsilon_0}{kT}$$

$$= -NkT \ln\left(\frac{Z_0}{N}\right) + N\epsilon_0. \qquad (47)$$

Considering the simple gas reaction

$$A + B \rightarrow AB,$$

we find that the Gibbs function for N molecules of constituent A based on the total pressure $P = NkT/V$

$$G_a = -NkT\ln\left(\frac{Z_{0a}}{N}\right) + N\epsilon_{0a}.$$

Similarly, for the constituents B and AB, we have

$$G_b = -NkT\ln\left(\frac{Z_{0b}}{N}\right) + N\epsilon_{0b},$$

and

$$G_{ab} = -NkT\ln\left(\frac{Z_{0ab}}{N}\right) + N\epsilon_{0ab}.$$

We now define the *change in the specific Gibbs function* at the pressure P and temperature T as

$$\Delta g = g_{ab} - g_a - g_b$$

so that

$$\Delta g = -RT \ln\left(\frac{Z_{0ab} N}{Z_{0a} Z_{0b}}\right) + N\Delta\epsilon_0, \qquad (48)$$

where the zero-point energy difference $\Delta\epsilon_0$ is

$$\Delta\epsilon_0 = \epsilon_{0ab} - \epsilon_{0a} - \epsilon_{0b}.$$

A comparison of Eqs. (46) and (31) shows that

$$K_N = \frac{Z_{ab}}{Z_a Z_b} = \frac{Z_{0ab}\, e^{-\epsilon_{0ab}/kT}}{Z_{0a}\, e^{-\epsilon_{0a}/kT}\, Z_{0b}\, e^{-\epsilon_{0b}/kT}},$$

---

* See Section 12.6, Eq. (30).

Therefore, K_N is given by

$$K_N = \frac{Z_{0ab}}{Z_{0a}\, Z_{0b}} \; e^{-\Delta\epsilon_0/kT}. \tag{49}$$

This is *another expression for the equilibrium constant.* With Eq. (48), Eq. (49) can be written more concisely as

$$NK_N = e^{-\Delta g/RT}. \tag{50}$$

This equation can be arrived at as follows:

Multiplying Eq. (49) by *N*, we get

$$NK_N = \frac{Z_{0ab}\,N}{Z_{0a}\, Z_{0b}} \; e^{-\Delta\epsilon_0/kT},$$

or

$$\ln N k_N = \ln\left(\frac{Z_{0ab}\,N}{Z_{0a}\, Z_{0b}}\right) - \frac{\Delta\epsilon_0}{kT}.$$

Substituting this result into Eq. (48), we obtain

$$\begin{aligned}\Delta g &= -RT \ln NK_N - \frac{\Delta\epsilon_0}{kT} \times RT + N\Delta\epsilon_0, \\ &= -RT \ln N k_N - N\Delta\epsilon_0 + N\Delta\epsilon_0 \\ &= -RT \ln N k_N.\end{aligned}$$

Therefore,

$$-\frac{\Delta g}{RT} = \ln N k_N,$$

or

$$N k_N = e^{-\Delta g/RT},$$

which is Eq. (50).

Eqs. (49) and (50) are important because they establish the relationships among the equilibrium constant, the partition functions, and the changes in the properties of ideal gases undergoing a chemical reaction. Eq. (49) can be generalised in terms of the reaction

$$aA + bB + \ldots = pP + qQ + \ldots$$

$$K_N = \frac{Z_{0p}^{p}\, Z_{0q}^{q} \cdots}{Z_{0a}^{a}\, Z_{0b}^{b} \cdots} \; e^{-\Delta\epsilon_0/kT}.$$

14.6. Computation of equilibrium constants

Before we consider the computation of equilibrium constants, we must first evaluate the Gibbs function, the change in the energy of reactions, and the entropy changes.

It is a standard practice to tabulate the change of the Gibbs function for the formation of each substance at the same temperature and pressure. The free energy change or the Gibbs function change ΔG of a substance is defined as *the increment in the Gibbs function* associated with the reaction that occurs when a given substance is formed from its elements. It is usually given for a standard state, which is considered to be a pressure of one atmosphere and a temperature of 25°C. It is expressed in kilocalories per gram-molecule of the substance. (Note that in the mks system, the term 'mole' implies *kilogram-molecule*, that is, a mass in kilograms numerically equal to the molecular weight.) Also tabulated is the enthalpy change, or heat content change, ΔH, called *heat of formation*, which arises when a compound is formed from its elements. A complete tabulation of the change of the Gibbs function, and the enthalpy change is given in the *National Bureau of standards Circular 500*. Several values from this source are presented in Table 10.

Once a reference value of ΔG_f has been obtained, the zero-point energy difference, $\Delta\epsilon_0$, can be evaluated from Eq. (48), involving specific Gibbs functions, namely

$$\Delta g = -RT \ln\left(\frac{Z_{0ab}\, N}{Z_{0a}\, Z_{0b}}\right) + N\Delta\epsilon_0 . \qquad (48)$$

This equation can then be used to determine ΔG_f as a function of temperature and pressure. As an example, let us consider the dependence of ΔG_f on *pressure*.

The Gibbs function is defined as

$$G \equiv H - TS = U - TS + PV .$$

For an ideal gas, this function can be written as

$$G = U - TS + nRT$$

so that the substitution for the entropy S expressed as

$$S = Nk \ln\left(\frac{Z}{N}\right) + Nk + \frac{U}{T},$$

gives, as shown before

$$G = -NkT \ln\left(\frac{Z}{N}\right),$$

or

$$= -nRT \ln\left(\frac{Z}{N}\right).$$

We know that for an ideal gas only the translational partition function

$$Z_{trans} = \frac{V}{h^3}(2\pi mkT)^{3/2}$$

is affected by a change in the gas pressure at constant temperature. [It is to be noted that this equation (as was shown before) can be derived for a monatomic ideal gas using quantum statistics, and it can also be derived from the Schrödinger equation for a free particle confined to a volume *V*.] Let us express the preceding equation in terms of pressure at constant temperature. That is,

$$Z = \frac{NkT}{P}\frac{1}{h^3}(2\pi mkT)^{3/2}.$$

We see that Z_0, the partition function for an ideal gas at pressure P, is related to Z_0^0; the partition function at the standard pressure P^0 is

$$\left(\frac{Z_0}{Z_0^0}\right)_T = \left(\frac{P^0}{P}\right).$$

From Eq. (48), namely

$$\Delta g = -RT\ln\left(\frac{Z_{0ab}N}{Z_{0a}Z_{0b}}\right) + N\Delta\epsilon_0.$$

The change in the specific Gibbs function at pressure *P*, Δg, is simply related to Δg^0, which is the change in specific Gibbs function at standard pressure P^0, by

$$\left(\Delta g - \Delta g^0\right)_T = -RT\ln\left(\frac{P}{P^0}\right).$$

$$= RT\ln\frac{P^0}{P},$$

the assumption being that the temperature is constant.

On the other hand, the dependence of the equilibrium constant K_N on *pressure* can be determined from the Eq. (33) obtained above, namely

$$K_P = \frac{Z_{ab}}{Z_a Z_P}\frac{N}{P} = k_N\frac{N}{P}. \qquad (33)$$

Thus for the reaction considered, K_N, the equilibrium constant at pressure *P*, is related to K_N^0, which is the equilibrium constant at pressure P^0 by

$$\frac{(NK_N)_P}{(NK_N)_{P^0}} = \left(\frac{P}{P^0}\right),$$

at constant temperature.

However, the dependence of the partition function on temperature for an ideal gas can be quite complex, depending on all the internal states of the constituent molecules. The temperature dependence of the Gibbs function and the enthalpy are often included in thermochemical tables. One of the most complete tabulations of these data are given in the *J A N A F Interim Thermochemical Tables.* Several substances tabulated in the *J A N A F* Tables are included in Appendix C, the enthalpy Table C-1. The values listed are based on a pressure of one atmosphere and are expressed in kilocalories per gram-molecule of the substance considered. Values for O_2, N_2, H_2, F_2, and Cl_2 are not given since these are considered as standard reference substances for which ΔG_f^0 and ΔH_f^0 are defined to be zero. The *J A N A F Tables* were published ten years after the *National Bureau of Standards Circular 500*, and hence some of the values given in Table 1 below may not be in complete agreement with the later values in Table 1 in Appendix C.

Table 1
Changes of Gibb Function and Enthalpy* of Formation at the Standard State
(P = 1 atom, T = 25°C)

Substance	ΔH_f^0	ΔG_f^0	Substance	ΔH_f^0	ΔG_f^0
	kcal mole			Kcal mole	
O	59.159	54.994	H_2S	–4.815	– 7.892
O^+	374.609	—	N	85.565	81.471
O_2	0	0	N^+	422.478	—
O_3	34.0	39.06	N_2	O	O
H	52.089	48.575	NO	21.600	20.719
H^+	367.088	—	NO_2	8.091	12.390
H_2	0	0	NO_3	13	—
OH	10.06	8.93	N_2O	19.49	24.76
H_2O	–57.7979	–54.6357	N_2O_3	20.0	—
H_2O_2	–31.83	—	N_2O_4	2.309	23.491
F	18.3	14.2	N_2O_5	3.6	—
F_2	0	0	NH	59	—
HF	–64.2	–64.7	NH_3	–11.04	–3.976
Cl	29.012	25.192	C	171.698	160.845
Cl_2	0	0	CO	–26.4157	–32.8079
HCl	–22.063	–22.769	CO_2	–94.0518	–94.2598
Br	26.71	19.69	CH	142.1	—
Br_2	7.34	0.751	CH_3OH	–48.08	–38.69
HB_r	–8.66	–12.72	HCN	31.2	28.7
I	25.482	16.766	C_2H_2	54.194	50.000
I_2	14.876	4.63	C_2H_4	12.496	16.282
HI	6.20	0.31	C_2H_6	–20.236	–7.860
S	53.25	43.57	C_2H_6O	–56.24	–40.30
S_2	29.86	—	C_2N_2	73.60	40.81
SO_2	–70.96	–71.79			

* Reference: National Bureau of Standards Circular 500.

CHAPTER 15

Fluctuations

15.1. Density fluctuations in a gas

A necessary consequence of the random motion of the molecules of a gas is the number of molecules per unit volume, in any volume element, is not at every instant equal to the average value but fluctuates about this average. The theory of density fluctuations in a gas was first developed by *Smoluchowski* in 1908, and was followed by another treatment in 1914.

It has been shown before* that when the *Boltzmann* statistics is applied to molecules in a gravitational field in order to obtain the barometric equation, the following equation was first obtained for the partition functions *Z*.

$$Z = \sum g_i \, e^{-\epsilon_i/kT} = \sum g_i \, e^{-\left(mgz + \frac{mv^2}{2}\right)/kT},$$

$$= \frac{m^3}{h^3} \iiiiiint e^{-\frac{mgz}{kT}} \, e^{-\frac{mV^2}{2kT}} \, dx\, dy\, dz\, dv_x\, dv_y\, dv_z,$$

$$= \frac{m^3}{h^3} \iint dx\, dy \int_0^{\infty} e^{-\frac{mgz}{kT}} \, dz \int_{-\infty}^{+\infty}\int_{-\infty}^{+\infty}\int_{-\infty}^{+\infty} e^{-\frac{mV^2}{kT}} \, dv_x\, dv_y\, dv_z,$$

$$= \frac{m^3}{h^3} A \, \frac{kT}{mg} \left(\frac{2\pi kT}{m}\right)^{3/2} = \frac{A}{h^3} \frac{kT}{mg} (2\pi m kT)^{3/2} .$$

where *A* is the horizontal cross section of the column of air.

Subsequently, the number of particles N_i in cell *i* in phase space is

$$N_i = \frac{N}{Z} g_i \, e^{-\left(mgz + \frac{mV^2}{2}\right)/kT},$$

Substituting for g_i and for *Z*, we obtain

$$= N \frac{m^3}{h^3} \, e^{-\left(mgz + \frac{mV^2}{2}\right)/kT} \quad dx\, dy\, dz\, dv_x\, dv_y\, dv_z$$

$$\times \frac{h^3}{A} \frac{mg}{kT} \frac{1}{(2\pi m kT)^{3/2}} .$$

* Sections 9.4 and 9.7.

Using d^6N to denote N_i, we have

$$d^6N = \frac{Nmg}{AkT}\left(\frac{m}{2\pi kT}\right)^{3/2} e^{-\left(mgz + \frac{mV^2}{2}\right)/kT} dx\, dy\, dz\, dv_x\, dv_y\, dv_z. \qquad (1)$$

If we want now to determine the distribution in ordinary space, we integrate Eq. (1) over all values of v_x, v_y, and v_z noting that the triple integral $\int_{-\infty}^{+\infty}\int_{-\infty}^{+\infty}\int_{-\infty}^{+\infty} e^{-m\left(v_x^2 + v_y^2 + v_z^2\right)/2kT} dv_x\, dv_y\, dv_z$ is, as shown before,* equal to $\left(\frac{2\pi kT}{m}\right)^{3/2}$. Thus we get

$$d^3N = \frac{Nmg}{AkT} e^{-(mgz)/kT} dx\, dy\, dz,$$

or

$$\frac{d^3N}{dx\, dy\, dz} = \frac{Nmg}{AkT} e^{-(mgz)/kT}.$$

The left-hand side of this equation is the number of particles per unit volume in ordinary space.

From this equation, the number of particles in a vertical column between the elevations *z* and *z* + dz, denoted by N_z can be obtained by integrating it over *x* and *y*. This gives the horizontal cross section area *A* of the air column so that

$$dN_z = \frac{Nmg}{kT} e^{-(mgz)/kT} dz.$$

Also this equation can be rewritten in the form

$$dN_x = \alpha e^{-\phi/kT} dx, \qquad (2)$$

where ϕ represents the potential energy associated with some coordinate *x*, and α is a constant independent of *x*. The more general statistical theory, instead of considering a large number of *molecules*, considers a large number of *identical systems*, each system containing many molecules. This approach leads to an equation having precisely the form given above except that dN_x represents the *number of systems* each having the potential energy ϕ associated with some coordinate *x*. Each system is assumed to be in contact with a heat reservoir which maintains the system at constant temperature *T*.

To apply the preceding equation to the problem of density fluctuations in a gas, we consider a very large volume of the gas subdivided, in imagination, into a large number *N* of smaller volume elements or systems. The remainder of the gas serves as the *heat reservoir* for any one system. Each system contains a large number of molecules, *n*. (Because of a letter shortage, we now use *n for the number of molecules per system* rather than for the number of molecules per unit volume). For example, we might consider a volume of 10 l or 10,000 cm^3 of a gas and imagine it subdivided into volume elements or systems of 1 mm^3

---

* See Section 9.6. Also see Appendix E, Section 3.

each. Then the number N of volume elements or systems is $10{,}000 \times 10^3/1$ or 10^7. This means that we have in the volume of 10 litres of the gas, a number of systems N equal to 10^7 so that the number of molecules *n per system* at standard conditions is given by

$$n = \frac{6.025 \times 10^{23} \times 10^4}{22400} \times \frac{1}{10^7} \simeq 3 \times 10^{16} \text{ molecules/system.}$$

We now consider each system of molecules enclosed within a perfectly flexible, heat-conducting sac so that there is no exchange of molecules between systems. Then as a result of random molecular motions, the molecules of any one system will at times occupy a volume somewhat smaller than the normal volume V_0, and at other times, they will occupy a volume somewhat greater than the volume V_0. At any one instant, there will be fluctuations in density from system to system, throughout the body of the gas, or, if we follow any one system as time goes on, its density will fluctuate with time.

Since we wish to obtain an expression for the number of systems each of normal volume V_0 which as a consequence of fluctuations have an *actual volume* between V and $V + dV$, we let $x = V$ in Eq. (2) and write

$$dN_V = \alpha\, e^{-\frac{\phi}{kT}} dV. \qquad (3)$$

The next step is to obtain an expression for the potential energy ϕ. We consider the potential energy of a system to be *zero* when it has the volume V_0 and pressure P_0 appropriate to a *perfectly uniform distribution* of molecules throughout the main body of the gas. The potential energy in any other state, of volume V and P, is the *work done on the system* to bring it from the *uniform state* to this state. We assume that during the change, the *remainder of the gas exerts on the system the pressure* P_0. Then if P is the pressure exerted by the molecules of the system at any instant, then the net external pressure is $P_0 - P$, and the work done on the system while its volume changes from V_0 to V, or the potential energy ϕ, is

$$\phi = \int_{V_0}^{V} (P_0 - P)dV = -\int_{V_0}^{V} (P - P_0)dV. \qquad (4)$$

The pressure P is some functions of the volume V, and without making any special assumptions concerning the equation of the state of the gas, we can always use a *Taylor series* to express the pressure P as follows:

$$P = P_0 + \left(\frac{\partial P}{\partial V}\right)_0 (V - V_0) + \left(\frac{\partial^2 P}{\partial V^2}\right)_0 \frac{(V - V_0)^2}{2!} + \left(\frac{\partial^3 P}{\partial V^3}\right)_0 \frac{(V - V_0)^3}{3!} + \ldots, \qquad (5)$$

where the derivatives are to be evaluated at constant temperature and at the volume V_0. The number of terms to be retained in the series depends on the nature of the gas.

We first consider the special case of an ideal gas. Since $PV = nkT$, where n is the number of moles in volume V, that is, the number of molecules in the system, we have

$$\left(\frac{\partial P}{\partial V}\right)_0 = -\frac{1}{V_0^2} nkT,$$

$$\left(\frac{\partial^2 P}{\partial V^2}\right)_0 = \frac{2}{V_0^3} nkT, \text{ etc.}$$

Since from the nature of the problem the *fluctuations* are small

$$(V - V_0) \ll V_0,$$

it follows from Eq. (5) that the terms containing the second and higher derivatives are very small compared with that containing the first derivative so that we can write

$$\begin{aligned} P - P_0 &= -nkT \frac{V - V_0}{V_0^2} \\ &= -\frac{nkT}{V_0} \cdot \frac{V - V_0}{V_0} \\ &= -P_0 \frac{\Delta V}{V_0}. \end{aligned} \quad (6)$$

The fractional change in volume, $\Delta V/V_0$, was called by Smoluchowski the *condensation* δ.

$$\delta = \frac{\Delta V}{V_0} = \frac{V - V_0}{V_0} = \frac{V}{V_0} - 1, \quad (7)$$

$$V - V_0 = V_0 \delta.$$

Hence,

$$d\delta = \frac{1}{V_0} dV. \quad (8)$$

Using Eq. (6), we can write Eq. (4) in the form

$$\phi = \int_{V_0}^{V} P_0 \frac{\Delta V}{V_0} dV = \int_0^{\delta} P_0 \delta . V_0 d\delta$$

$$= \frac{P_0 V_0 \delta^2}{2}.$$

Therefore,

$$\phi = \frac{nkT}{2} \delta^2, \quad (9)$$

and

$$\frac{\phi}{kT} = \frac{n\delta^2}{2}. \quad (9)'$$

Substituting now in Eq. (3) for ϕ/kT by $n\delta^2/2$ from Eq. (9)′, and for dV by $V_0 d\delta$ from Eq. (8), we have for an ideal gas

$$dN_\delta = V_0 \alpha \ e^{-n\delta^2/2} \ d\delta. \quad (10)$$

The constant α can now be determined from the requirement that the integral of dN_δ over all possible values of δ must be equal to the total number of systems N. Now δ is always a small quantity, but let us suppose that it is as large as ± 1 per cent or $\pm 10^{-2}$. Also, suppose that we consider a system so small that it contains only 2×10^6 molecules (in general, n, the number of molecules per system is much larger than this). Then $n\delta^2/2 = 2\times 10^6 \times 10^{-4}/2 = 100$ and $e^{-n\delta^2/2} = e^{-100}$. Hence, under such conditions, the exponential term is already so small when $\delta = \pm 10^{-2}$ that we may without appreciable error take the limits of integration of Eq. (10) from $-\infty$ to $+\infty$. Thus,

$$N = \int dN_\delta = V_0\,\alpha \int_{-\infty}^{\infty} e^{-n\delta^2/2}\, d\delta.$$

But

$$\int_{-\infty}^{\infty} e^{-\frac{n\delta^2}{2}}\, d\delta = \sqrt{\frac{2\pi}{n}}.$$

Therefore,

$$N = V_0\,\alpha \sqrt{\frac{2\pi}{n}}. \tag{11}$$

and

$$V_0\,\alpha = \frac{N}{\sqrt{2\pi/n}} = N\sqrt{n/2\pi}. \tag{12}$$

Then from Eqs. (10) and (12), we obtain

$$\frac{dN_\delta}{N} = \sqrt{\frac{n}{2\pi}}\; e^{-n\delta^2/2}\, d\delta. \tag{13}$$

The ratio dN/N can be considered either as *the fractional number of systems for which the condensation, at any instant, lies between* δ *and* $\delta + d\delta$, or, as *the fraction of time spent by any one system in this state.* Note that the number of systems in which the condensation at any instant has the value δ *decreases* exponentially both with n and δ^2 as implied by Eq. (13).

The mean condensation $\bar{\delta}$ is of course zero, as would be expected, since the distribution is an even function of δ. Let us therefore calculate the root mean square (rms) condensation, $\left(\overline{\delta^2}\right)^{1/2} = \delta_{rms}$. Thus, using a Table of standard integrals,* we obtain

$$\delta_{rms} = \left(\overline{\delta^2}\right)^{1/2} = \left[\frac{\int_{-\infty}^{\infty} \delta^2 e^{-n\delta^2/2}\, d\delta}{\int_{-\infty}^{\infty} e^{-n\delta^2/2}\, d\delta}\right]^{1/2}$$

---

* See Appendix E, Section 3.

$$= \left(\frac{1}{n} \sqrt{\frac{2\pi}{n}} \Big/ \sqrt{\frac{2\pi}{n}} \right)^{\frac{1}{2}} = \frac{1}{\sqrt{n}}. \tag{14}$$

The rms condensation is therefore the reciprocal of the square root of the number of molecules or particles in the system. For example, in a volume element or a system of approximately 1 mm^3 at standard conditions there are, as shown above, 3×10^{16} molecules, so that, δ_{rms} is given by

$$\delta_{rms} = \frac{1}{\sqrt{10^{16}}} = 10^{-8}.$$

Since $\delta = \Delta V/V_0$, thus the rms fluctuation in volume (or *density*), in a volume of this size, is only one part in 100 million. On the other hand, a volume having linear dimensions of the order of a wavelength of light (say 600 mμ $= 6 \times 10^{-7}$ m) contains about 6×10^6 molecules, and in such a volume, $\delta rms = 5 \times 10^{-4}$ or 0.05 per cent. Smoluchowski showed that fluctuations of this order of magnitude resulted in variations of the refractive index throughout a gas, *of sufficient amount* to *cause appreciable scattering of light passing through the gas*. In fact, Smoluchowski derived a formula for light scattering that had been obtained by *Rayleigh* from a consideration of the polarisability of air molecules. The Rayleigh formula predicts a scattering *inversely* proportional to the *fourth power* of the wavelength, and hence the *blue light is scattered more than the red, which accounts for the blue colour of the sky.*

Measurements of the scattering of sunlight by the earth's atmosphere, made at very high altitudes where dust amount is small, have been used to compute the value of Boltzmann's constant *k* and thus, with knowledge of the gas constant *R*, to find Avogadro's number N_a. The precise value of this number is that of *Fowler*, namely 6.05×10^{23} molecules, which he obtained in 1914 from observations at the *Mt. Wilson Observatory* in California.

Using now Eq. (14), we can write Eq. (13) as

$$\frac{dN}{N} = \left(\frac{1}{\delta_{rms}\sqrt{2\pi}} \right) e^{-\frac{\delta^2}{2\delta_{rms}^2}} d\delta. \tag{15}$$

Then if we define a dimensionless quantity *x* by

$$x = \frac{\delta}{\delta_{rms}},$$

so that

$$d\delta = \delta_{rms}\, dx,$$

Eq. (15) can be rewritten in the form

$$\frac{dN}{N} = \frac{1}{\sqrt{2\pi}} e^{-\frac{x^2}{2}} dx. \qquad (16)$$

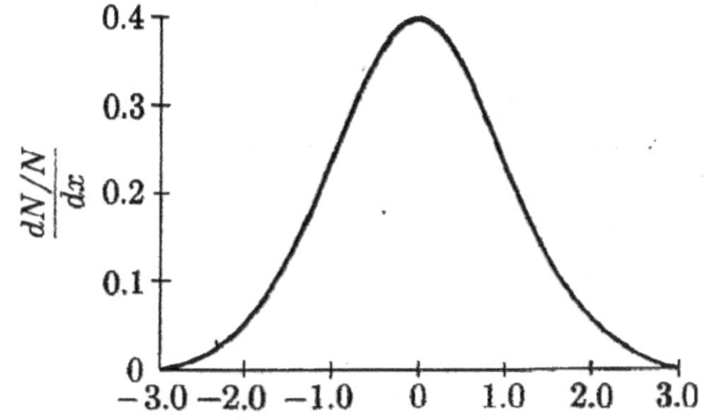

Fig. 1. Normal or Gaussian distribution of random fluctuations.

Fig. 1 is a graph in which the dimensionless ratio $(dN/N)/dx$ is plotted as a function of x. The curve is said to represent a *normal* or *Gaussian* distribution, and it is typical for all types of random fluctuations.

Let us now compute the mean potential energy $\bar{\phi}$ of a system resulting from fluctuations of density. We obtained above the potential energy expression as

$$\phi = \frac{nkT}{2} \delta^2. \qquad (9)$$

Hence, the mean potential energy $\bar{\phi}$ depends upon the mean square condensation. But from Eq. (14) above, $\overline{\delta^2}$ is given by

$$\overline{\delta^2} = \frac{1}{n}. \qquad (17)$$

Therefore,

$$\bar{\phi} = \frac{nkT}{2} \times \frac{1}{n} = \frac{1}{2} kT. \qquad (18)$$

This result is *another illustration of the equipartition principle*. We note from Eq. (9) that ϕ depends on δ^2, and also note from Eq. (18) that the number of particles n per system does not appear in the expression of $\bar{\phi}$. Thus, according to Eq. (18), we can say that the *mean potential energy* of any volume element of an ideal gas due to fluctuations is independent of the number of molecules n in this volume, and is equal to $\frac{1}{2}$ kT.

The *phenomenon of light scattering* is especially pronounced in a substance at or near its critical point, producing *a milky or opalescent appearance*. This effect had been observed for many years but its cause was not known until the theory was formulated by Smoluchowski. For a substance at the critical point, we cannot of course assume that the equation of state of an ideal gas is obeyed, and accordingly, we must return to the general series expansion for the pressure P, which has been given above by Eq. (5). Since at the critical point* we have

$$\frac{\partial P}{\partial V} = \frac{\partial^2 P}{\partial V^2} = 0,$$

the first non-vanishing derivative in Eq. (5) is $\partial^3 P/\partial V^3$, and thus Eq. (4)

$$\phi = -\int_{V_0}^{V} (P - P_0)\, dV \tag{4}$$

becomes

$$\phi = -\int_{V_0}^{V} \frac{\partial^3 P}{d V^3} \frac{(V - V_c)^3}{3!} dV, \tag{19}$$

where we have replaced V_0 by the critical volume V_c. To proceed further, we must know P as a function of V; thus let us assume that *van der Waals* equation is obeyed. Then

$$\left(P + \frac{n^2 a}{V^2}\right)(V - nb) = nkT,$$

$$P = \frac{nkT}{V - nb} - \frac{n^2 a}{V^2}, \tag{20}$$

where n is the number of molecules per system of volume V. It can be proved that the third derivative of P with respect to V at the critical point is

$$\left(\frac{\partial^3 P}{\partial V^3}\right)_c = -\frac{a}{81 b^5}\left(\frac{1}{n}\right)^3 = -\frac{27}{8}\frac{kT_c n}{V_c^4}. \tag{21}$$

Here the terms a and b are constants for any one gas but different for different gases as we already know from van der Waals equation. They are related to V_c, T_c, and P_c by the following equations:

$$V_c = 3nb, \quad T_c = \frac{8a}{27kb}, \quad P_c = \frac{a}{27b^2},$$

respectively, and thus we have for a van der Waals gas

$$\frac{nkT_c}{P_c V_c} = \frac{8}{3}.$$

Now substituting Eq. (21) into Eq. (19), we have

$$\phi = -\int_{V_c}^{V} \frac{27}{8} \frac{nkT_c}{V_c^4} \frac{(V - V_c)^3}{3!} dV, \tag{22}$$

---

* The critical isotherm at this point has not only a horizontal tangent but a point of inflection as well.

and then introducing the condensation δ as given by

$$\delta = \frac{\Delta V}{V_c} = \frac{V - V_c}{V_c} \tag{7}$$

and its differential

$$d\delta = \frac{1}{V_c} dV, \tag{8}$$

which have been given above, into Eq. (22), we obtain

$$\phi = \int_0^\delta \frac{9}{16} nkT_c \, . \, \delta^3 \, d\delta. \tag{23}$$

Therefore,

$$\phi = \frac{9}{16} n\,kT_c \frac{\delta^4}{4} = \frac{9}{64} n\,k\,T_c \, . \, \delta^4, \tag{24}$$

or,

$$\frac{\phi}{kT_c} = \frac{9}{64} n\delta^4. \tag{25}$$

Hence, substituting Eq. (24) into Eq. (3) given above, we obtain

$$d\,N_\delta = \alpha\, e^{-\frac{9\,n\,\delta^4}{64}}\, d\delta, \tag{26}$$

as contrasted with Eq. (10)

$$dN_\delta = V_0\, \alpha\, e^{-\frac{n\,\delta^2}{2}}\, d\delta \tag{10}$$

for an ideal gas. It is evident that according to Eq. (26), the fluctuations now *depend on the fourth power of condensation rather than on the square.*

The constant α can be evaluated using the relation*

$$\int_0^\infty x^n \, e^{-x}\, dx = \Gamma\,(n+1), \tag{27}$$

and from the requirement that $\int dN_\delta = N$, we find that the result expressing the fractional number of systems for which the condensation at any instant lies between δ and $\delta + d\delta$ is

$$\frac{d\,N_\delta}{N} = \frac{\sqrt[4]{9n/64}}{2\,\Gamma(5/4)}\, e^{-\frac{9n\,\delta^4}{64}}\, d\delta. \tag{28}$$

Hence, from Eqs. (26) and (28), the constant α has the value

$$\alpha = N\frac{\sqrt[4]{9n/64}}{2\Gamma\,(5/4)}. \tag{29}$$

Again the mean condensation is zero, while the r m s value evaluated in the usual way, as shown above [see Eq. (14)], is found to be in this case.

$$\delta_{rms} = \frac{0.95}{\sqrt[4]{n}}. \tag{30}$$

---

* A note on the gamma function $\Gamma(n)$ is given in Appendix A.

Since the rms condensation now varies *inversely* with the *fourth root of n*, rather than with the *square root*, they can be relatively large even for relatively large values of *n*. It is *these relatively large density fluctuations which account for the opalescence observed at the critical point.*

15.2. Theory of Brownian motion

The considerations in the preceding section hold also for colloidal suspensions, noting that according to statistical theory a colloidal particle is not fundamentally different from a molecule. Since such particles can be directly observed by a microscope, it is possible by counting to measure fluctuations in their concentration. In such measurements *n* is a very small number, whereas it was a very large number in examples considered in the preceding examples. This requires certain modifications in the Smoluchowski formula, which we need not discuss here since the principles are the same.

If we say that a particle in a suspension undergoes an irregular motion because of collisions by the molecules around it, then this can be considered as another way of expressing the cause of fluctuations in concentration. It is to be pointed out that without such a motion, fluctuations in concentration would be impossible.

The zigzag to-and-fro motion of suspended particles had been known for a long time. It was first observed by the English botanist *Brown* in 1827, and it was thus named after him as *Brownian motion.* The theory of this phenomenon was developed only comparatively recently. The systematic researches of *Wiener* and *Gouy* made it evident that the explanation must be sought in the collisions with the molecules of the surrounding fluid. In 1906, *Einstein* and *Smoluchowski* both derived essentially the same formula from considerations of kinetic theory, although by entirely different methods, and this formula has been verified experimentally by *Perrin, Svedberg, Sedding*, and others.

We now give the derivation of the *Einstein formula* as it was worked by *Langevin* in 1908. This method of attack leads us most quickly to our goal. Langevin assumed that the force on a suspended particle of mass *m* could be considered of two kinds, the first of which is a *'frictional' force f proportional to the velocity*, whereas all other external influences of the surrounding fluid can be combined in the second.

For the motion of a particle in any direction, which we take as the *x*-direction, we have, according to Langevin, an equation of the form

$$m\ddot{x} = -f\dot{x} + X, \tag{1}$$

where $(-f\dot{x})$ denotes the *x*-component of the friction force, and *X* is the combined effects of all other influences. Multiplying Eq. (1) by *x*, we obtain

$$m\ddot{x}x = -f\dot{x}x + Xx. \tag{2}$$

Now

$$\dot{x}x = \frac{1}{2}\frac{d(x^2)}{dt}, \tag{3}$$

and

$$\ddot{x}x = \frac{1}{2}\frac{d}{dt}\left[\frac{d(x^2)}{dt}\right] - \dot{x}^2 . \tag{4}$$

Thus substituting Eqs. (3) and (4) into Eq. (2), we have

$$\frac{m}{2}\frac{d}{dt}\left[\frac{d(x^2)}{dt}\right] - m\dot{x}^2 = -\frac{f}{2}\frac{d(x^2)}{dt} + Xx. \tag{5}$$

We form such an equation for each particle which is suspended in the fluid and take the mean of these expressions for all particles, that is

$$\frac{m}{2}\frac{d}{dt}\overline{\left[\frac{d\left(x^2\right)}{dt}\right]} - m\overline{\dot{x}^2} = -\frac{f}{2}\overline{\frac{d\left(x^2\right)}{dt}} + \overline{Xx}. \tag{6}$$

Let us assume* that the mean value $\overline{Xx}$ vanishes because the force X varies in a completely irregular way. Further, according to the equipartition theorem,

$$m\overline{\dot{x}^2} = kT, \tag{7}$$

so that we obtain finally, since

$$\overline{dx^2/dt} = (d/dt)\overline{x^2},$$

we have

$$\frac{m}{2}\frac{d}{dt}\left[\frac{d\overline{x^2}}{dt}\right] + \frac{1}{2}f\overline{\frac{d\left(x^2\right)}{dt}} = kT. \tag{8}$$

For brevity, let

$$\overline{\frac{d\left(x^2\right)}{dt}} = u,$$

then Eq. (8) can be written as

$$\frac{m}{2}\frac{du}{dt} + \frac{1}{2}fu = kT. \tag{9}$$

Eq. (9) is a differential equation for determining u, the general solution of which is

$$u = kT\frac{2}{f} + Ce^{-\frac{ft}{m}}, \tag{10}$$

where C is an integration constant. Now because of the small value of m, the quotient f/m is a very large number so that the exponential term has no influence after the first extremely small time interval, and thus Eq. (10) reduces to

$$u = \overline{\frac{d\left(x^2\right)}{dt}} = kT\frac{2}{f}. \tag{11}$$

* Strictly a special proof is required.

Integrating Eq. (11) between $t = 0$ and $t = \tau$, we obtain

$$\int_0^\tau d\left(\overline{x^2}\right) = kT\frac{2}{f}\int_0^\tau dt,$$

$$\overline{x^2} - \overline{x_0^2} = kT\frac{2}{f}\tau. \qquad (12)$$

If we now set $x_0 = \text{zero}$ when t is equal to zero, and, because of its small value, write Δx^2 instead of x^2, Eq. (12) can be rewritten as

$$\overline{\Delta x^2} = kT\frac{2}{f}\tau. \qquad (13)$$

The quantity $\overline{\Delta x^2}$ has the following meaning. A particle is observed, at time 0 and time τ. During this time interval, it should have undergone a displacement Δs, whose projection on the x-axis is Δx. The same particle is observed at later times, always separated by the same interval τ, that is, at times 2τ, 3τ, ..., and Δx is determined for each interval. These values are squared and their mean is evaluated, the result being $\overline{\Delta x^2}$. The displacements thus observed are in no sense the actual path of the particle, nor is $\Delta x/\tau$ the x-component of its velocity. For example, in a time interval $\tau = 1$ sec, the particle makes millions of collisions, and what we see are merely its *initial and final positions*, which we connect by a straight line, whereas the actual path is a confused zigzag of linear segments. What we observe is a greatly simplified 'path', an example of which is given in Fig. 2. If one imagines each of the linear segments in this figure to be composed of millions of straight lines, he will begin to approximate the actual path. It is impossible to analyse this complicated motion in all of its details, and we must therefore be satisfied to observe, in a certain time interval τ, the corresponding magnitude of Δx, and then to evaluate $\overline{\Delta x^2}$, which is only loosely related to the actual path.

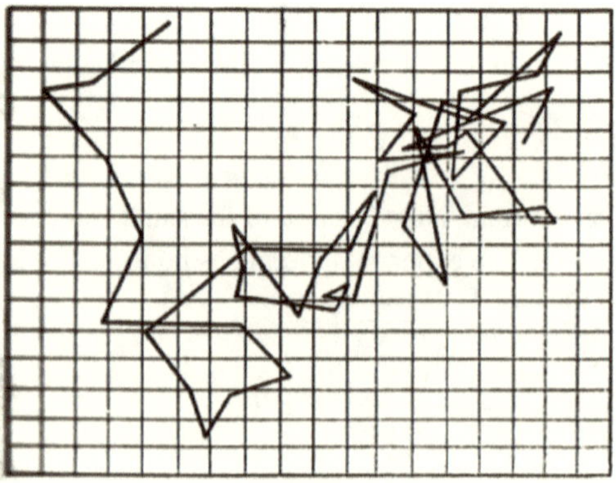

Fig. 2.

An expression for the factor f is now introduced using hydrodynamics. The physical significance of f is a *force per unit of velocity*. But *Stokes'* law gives for the viscous force F

acting on a sphere of radius *R* moving with velocity *v* in a fluid of viscosity η, the expression

$$F = 6\pi \eta R v. \tag{14}$$

Therefore,

$$f = \frac{F}{v} = 6\pi \eta R \tag{15}$$

Substituting for *f* in Eq. (13) by Eq. (15), we have

$$\overline{\Delta x^2} = kT \frac{\tau}{3\pi\eta R},$$

$$\sqrt{\overline{\Delta x^2}} = \sqrt{kT} \frac{\sqrt{\tau}}{\sqrt{3\pi \eta R}}. \tag{16}$$

This is the *Einstein formula* (and, with a slight change in the numerical coefficient, that of *Smoluchowski* also). If the formula is satisfied, we have a convincing proof of the existence of molecules, and since the Brownian motion is brought about by the collisions of molecules, this is of the greatest importance for the verification of the molecular-kinetic theory.

Eq. (16) reveals a very surprising result that $\sqrt{\overline{\Delta x^2}}$ *does not depend on the mass of the particle*. This prediction of the theory was shown by *Perrin* to be satisfied. In his experiments, the masses of the particles observed varied in the ratio 1:15,000; nevertheless all his results, within the limits of experimental error, gave *the same value for the Boltzmann constant k* (or for Avogadro's number N_a which was calculated as $N_a = R/k$). From his best measurements, Perrin found for N_a the value 6.85×10^{26} molecules per kg-molecule, which is undoubtedly too large in the light of later measurements. *Westgren*, using the same methods, obtained 6.04 to 6.05×10^{26} molecules per kg-molecule for N_a, which is probably correct to within 1 per cent.

The results of Perrin, Svedberg, and Westgren all agree in proving that $\sqrt{\overline{\Delta x^2}}$ is proportional to $\sqrt{\tau}$, and *inversely* proportional to $\sqrt{\eta}$. The temperature effect is *not large* since $\sqrt{\overline{\Delta x^2}}$ is proportional to $\sqrt{\tau}$, but since η decreases rapidly with increasing temperature, the pure temperature effect is outweighed by the large decrease in viscosity. The work of Perrin and Svedberg is a brilliant quantitative confirmation of the Einstein–Smoluchowski equation and with it of the molecular-kinetic theory.

It is important to point out here that the diffusion of a non-uniform colloidal suspension is only possible because the individual particles execute Brownian motion. Diffusion, Brownian motion, and fluctuations in concentration are therefore all a *single phenomenon.* In macroscopic observations, we speak of *diffusion*; in substantial ones, of Brownian motion; and in local ones, of *fluctuations in concentration.* Here, the terms 'substantial' and 'local' are used in the same sense as in *hydrodynamics*. In a 'substantial' consideration, we follow a particle of the fluid in its motion; and in a 'local' one, we consider a volume fixed in space, into which particles come and out of which they wander.

CHAPTER 16

Irreversible Processes

16.1. Introduction

One of the characteristics of science is to idealise systems which are very complex in nature to permit a more amenable analysis. Thus the sciences of classical thermodynamics and statistical mechanics are based on the existence of equilibrium systems whose coordinates are *time invariant*. Kinetic theory and the theory of fluctuation do not suffer from this limitation, since microscopic departures from equilibrium are allowed. Fortunately, in many systems of practical interest departures from equilibrium of a microscopic order can be ignored even though from a macroscopic point of view, the system is neither steady nor uniform. However, there are many practical systems in which *microscopic irreversibility* is of greatest importance, and it is these systems which we shall discuss in this chapter.

No completely satisfactory theory of non-equilibrium thermodynamics or irreversible thermodynamics has been developed thus far. Nevertheless, we shall examine some of the general features of irreversible phenomena without restricting ourselves to any one theoretical approach. We shall adopt the *philosophical concepts of the excluded middle and of cause and effect*. Accordingly, we shall assume that a system is either in equilibrium, or it is not. If the system is in equilibrium, the observable effects can be readily related through the science of classical thermodynamics and statistical mechanics. These effects are work, heat, and changes in the properties of the system. If the system is not in equilibrium, we shall seek to define the effects in terms of the causes (or forces) producing the effects.

16.2. Entropy production and continuity equation

Classical thermodynamics is not completely silent on the subject of irreversibility. We know that the entropy of an isolated system in equilibrium is a maximum, and if such a system is not in equilibrium, the entropy will increase but may not decrease. Thus we know that equilibrium lies in the *direction of increasing entropy*. Furthermore, the first law of thermodynamics is valid whether a system is in equilibrium or not. It is to be noted that the term 'isolated system' is quite helpful in expressing the total change in any of the thermodynamic functions accompanying the various processes which occur in any part of this system, whether these processes are irreversible or reversible; this expression or statement is considered as a ruler. Thus, in the case of entropy as a thermodynamic function, we may state the following two rules which were considered earlier:

1. When an *irreversible* change occurs in any part of an isolated system, the total entropy increases.
2. When a *reversible* change occurs in any part of an isolated system, the total entropy remains unchanged.

These two rules can be summarised in one rule, namely

Only those processes can take place in an isolated system for which the total entropy of the system increases, or, in the limit, remains constant.

Let us now consider a system in which the properties vary from point to point and with time. The internal energy of such a system is a function of the space coordinates and time, or

$$u = u\,(x, y, z, t).$$

We can write a *continuity* equation based on the principle of conservation of energy in which heat, work, and entropy are vector quantities in as much as they can have direction and magnitude and are time dependent. Letting $\mathbf{J}_Q$ denote the *heat flow vector,* $\mathbf{J}_W$ the *work flow vector*, and *u the internal energy per unit mass of a differential volume element* of the system, we obtain the continuity equation

$$\rho \frac{\partial}{\partial t}\left(u + \frac{v^2}{2}\right) = -\nabla . \mathbf{J}_Q - \nabla . \mathbf{J}_W . \qquad (1)$$

Or

$$\rho \frac{\partial}{\partial t}\left(u + \frac{v^2}{2}\right) = -\text{ div } \mathbf{J}_Q - \text{div}\,\mathbf{J}_W , \qquad (2)$$

where ∇ is the vector differential operator* del, also known as nabla.

For simplicity, let us assume that the system is a homogeneous and isotropic solid through which heat is conducted and is in contact with a heat reservoir. In this case, there will be no transport of matter in bulk, and if the thermal expansion is neglected, Eq. (2) reduces to

$$\rho \frac{\partial u}{\partial t} = -\text{ div } \mathbf{J}_Q . \qquad (3)$$

Since the work term has been neglected, the internal energy and entropy are related by

$$du = T ds. \qquad (4)$$

It should be noted that Eq. (4) is not restricted to reversible processes. Entropy is a property and is completely determined by the change in the properties *u* and *T*, which are independent of the path. Rearranging Eq. (4) and taking the time derivative, we have

$$\frac{\partial s}{\partial t} = \frac{1}{T}\frac{\partial u}{\partial t},$$

or

$$\rho \frac{\partial s}{\partial t} = \frac{\rho}{T}\frac{\partial u}{\partial t},$$

which when substituted in Eq. (3) gives

$$\rho \frac{\partial s}{\partial t} = -\frac{1}{T}\text{ div } \mathbf{J}_Q . \qquad (5)$$

* See Appendix A, Section 2.

If x is a vector and y is a scalar and both are functions of position, then

$$y \operatorname{div} \mathbf{x} = \operatorname{div}(y\mathbf{x}) - \mathbf{x} \cdot \operatorname{grad} y. \quad (6)$$

Therefore,

$$\frac{1}{T} \operatorname{div} \mathbf{J}_Q = \operatorname{div}\left(\frac{1}{T}\mathbf{J}_Q\right) + \frac{1}{T^2}\left(\mathbf{J}_Q \cdot \operatorname{grad} T\right),$$

and Eq. (5) takes the form

$$\rho \frac{\partial s}{\partial t} = -\operatorname{div}\frac{\mathbf{J}_Q}{T} - \frac{1}{T^2}\left(\mathbf{J}_Q \cdot \operatorname{grad} T\right),$$

or

$$\rho \frac{\partial s}{\partial t} + \operatorname{div}\frac{\mathbf{J}_Q}{T} = -\frac{1}{T^2}\left(\mathbf{J}_Q \cdot \operatorname{grad} T\right). \quad (7)$$

If we let the entropy flow *vector*, or *entropy flux*, be denoted by

$$\mathbf{J}_s = \frac{\mathbf{J}_Q}{T}$$

and

$$\theta = -\frac{1}{T^2}\left(\mathbf{J}_Q \cdot \operatorname{grad} T\right), \quad (8)$$

then Eq. (7) can be expressed more simply as

$$\rho \frac{\partial s}{\partial t} + \operatorname{div}\mathbf{J}_s = \theta. \quad (9)$$

Eqs. (3) and (9) are seen to have the same form except for the term θ in Eq. (9). Hence Eq. (9) can be considered as a *continuity equation for entropy*. The term $(\rho\,\partial s/\partial t)$ in Eq. (9) is the rate of *entropy increase* in the system, $\operatorname{div}\mathbf{J}_s$ is the rate of *entropy outflow*, and θ is the *rate of entropy production for a differential element of volume*. Eq. (9) is more amenable to interpretation if it is converted into volume and surface integrals by means of the Gauss divergence theorem[*+]:

$$\rho \frac{\partial}{\partial t}\int_V s\,dV + \int \mathbf{J}_s \cdot \mathbf{n}\,dA = \int_V \theta\,dV, \quad (10)$$

where dA is an element of area of the system and $\mathbf{n}$ is the outward directed unit vector. The volume integral on the left-hand side of Eq. (10) is the rate of entropy increase within the system. The surface integral is the rate of entropy outflow across the boundary of the system. The integral on the right-hand side of Eq. (10) is the *entropy production of the system.*

If the system is isolated, the surface integral disappears and Eq. (10) reduces to

$$\rho \frac{\partial}{\partial t}\int_V s\,dV = \int_V \theta\,dV, \quad (11)$$

---

[*] See Wilfred Kaplan, *Advanced Calculus*, Addison-Wesley publishing company, Inc., MA, p. 333 (1973).

which states that the rate at which the entropy of the system increases is equal to the volume integral of θ. Therefore, θ is the *rate of entropy production per unit volume*, which is due to irreversible processes occurring within the system. From Eq. (8) it is seen that the entropy of an isolated system remains constant when the temperature gradient is zero. We conclude, therefore, that

$$\theta \geq 0 \tag{12}$$

for an isolated system. This is a more complete statement of the second law than the usual version since it says that *entropy production is positive with time.*

16.3. The Onsager reciprocal relations

In a conducting rod or wire obeying *Ohm's law*, the current *I* in an element of length Δx is proportional to the potential difference ΔV across the element. The proportionality factor is the *reciprocal* of the resistance ΔR of the element:

$$I = \frac{\Delta V}{\Delta R}. \tag{1}$$

If *A* is the cross-sectional area of the element, ρ its electrical resistively, and $\theta = 1/\rho$ its *electrical conductivity*, then

$$\Delta R = \frac{\rho \Delta x}{A} = \frac{\Delta x}{\theta A}. \tag{2}$$

Similarly, the heat current, J_Q, in the element is proportional to the temperature difference ΔT across it, the proportionality factor being the *reciprocal* of the thermal resistance ΔR_{th}:

$$J_Q = \frac{\Delta T}{\Delta R_{th}}. \tag{3}$$

The thermal resistance ΔR_{th} is given by

$$\Delta R_{th} = \frac{\Delta x}{KA}, \tag{4}$$

where *K* is the *thermal conductivity*. Thus the usual expressions for the flow of *charge* given by Eq. (1), and the flow of *heat* given by Eq. (3), have exactly the same form.

Associated with the heat current J_Q is an entropy current J_S given by

$$J_s = J_Q/T. \tag{5}$$

However, the equation for the flow of electricity, that is, Eq. (1), holds only if the temperature difference ΔT across the element is zero, and the equation for the flow of heat, that is, Eq. (3), holds only if $\Delta V = 0$, that is, in the absence of an electric current. Nevertheless, if a temperature difference and an electric current exist *simultaneously*, it is found that the electric current *I* depends not only on the potential difference ΔV but also on the temperature difference ΔT. Similarly, the heat current J_Q and subsequently the entropy current J_S depend on the potential difference ΔV as well as the temperature difference ΔT. Analogous relations exist between many other pairs of the so-called *related flows*. The differences ΔV and ΔT, or other quantities, corresponding to them, are referred to as *driving forces*.

It was first shown by *Onsager* that for a properly selected pair of related flows, also called *conjugated flows*, a simple and useful relation exists between the coefficients of the driving forces. However, there is not a room here to go into the methods by which such a 'properly selected' pair of driving forces can be found. It must suffice to say that if the potential difference ΔV and the temperature difference ΔT are selected as driving forces, the electric current *I* and the entropy current J_S form a pair of conjugated flows, although the electric current *I* and the heat current J_Q do not.

The expressions of these flows can be written as

$$I = L_{11}\,\Delta V + L_{12}\,\Delta T, \tag{6}$$

and

$$J_s = L_{21}\,\Delta V + L_{22}\,\Delta T. \tag{7}$$

From the definition of electrical conductivity θ given by Eq. (2) above, namely,

$$\theta = \frac{\Delta x}{\Delta R} \cdot \frac{1}{A}, \tag{8}$$

and also from the definition of thermal conductivity *K* given by Eq. (4) above, namely,

$$K = \frac{\Delta x}{\Delta R_{th}} \cdot \frac{1}{A}, \tag{9}$$

we can obtain two equations relating the four coefficients L_{11}, L_{12}, L_{21}, and L_{22} to measurable properties of a substance. Onsager's contribution to the theory was to show that for a pair of conjugate flows, the coefficients L_{12} and L_{21} are equal to each other,

$$L_{12} = L_{21}. \tag{10}$$

This equation is known as the *Onsager reciprocal relation.*

16.4. Transport parameters

From the equation

$$J_Q = T\,J_S \tag{5}$$

the following equation

$$J_s = L_{21}\,\Delta V + L_{22}\,\Delta T \tag{7}$$

can be written as

$$J_Q = (T\,L_{21})\,\Delta V + (T L_{22})\,\Delta T. \tag{11}$$

The coefficient of ΔV in this equation is not equal to the coefficient of ΔT in the equation

$$I = L_{11}\,\Delta V + L_{12}\,\Delta T, \tag{6}$$

so that *I* and J_Q are not conjugate flows when ΔV and ΔT are selected us driving forces.

Let us consider a homogeneous rod at a uniform temperature *T* carrying a current *I*. Then $\Delta T = 0$, and from Eq. (6)

$$(I)_{\Delta T = 0} = L_{11}\Delta V.$$

But from Ohm's law

$$(I)_{\Delta T = 0} = \frac{\Delta V}{\Delta R}. \qquad (1)$$

Therefore,

$$L_{11} = \frac{1}{\Delta R}. \qquad (12)$$

Next we take a rod in which there is a heat current J_Q but no electric current. Then *I* = 0 in Eq. (6), and thus we have

$$\left(\Delta V\right)_{I = 0} = -\frac{L_{12}}{L_{11}}\Delta T. \qquad (13)$$

Inserting this expression for ΔV in Eq. (11) and using the Onsager relation $L_{12} = L_{21}$, we obtain

$$\left(J_Q\right)_{I = 0} = T\frac{L_{11}L_{22} - L_{12}^2}{L_{11}}\Delta T. \qquad (14)$$

But the heat current J_Q is given by

$$\left(J_Q\right)_{I = 0} = \frac{\Delta T}{\Delta R_{th}}, \qquad (3)$$

Therefore from Eqs. (3) and (14), we obtain

$$\frac{L_{11}L_{22} - L_{12}^2}{L_{11}} = \frac{1}{T\Delta R_{th}}. \qquad (15)$$

Thus we have two relations between the Ls and properties of the matter, namely Eqs. (13) and (15). A third will be derived in Section 16.5.

On the other hand, the expressions for the flows are

electric current $\quad I = L_{11}\Delta V + L_{12}\Delta T, \qquad (6)$

entropy current $\quad J_s = L_{21}\Delta V + L_{22}\Delta T, \qquad (7)$

heat current $\quad J_Q = (TL_{21})\Delta V + (TL_{22})\Delta T, \qquad (11)$

which have been obtained above, show that both an entropy current J_s and a heat current J_Q exist in a rod carrying an electric current *I*, even though the rod is at uniform temperature and there is no heat flow by *conduction*. That is, the electrons whose motion constitutes the current can be considered to transport *heat* and *entropy*, as well as electric charge. The relation between the electric current *I* and the heat J_Q and entropy J_S currents can be expressed as follows:

Setting $\Delta T = 0$ in the above three equations representing the flows, we have

$$(I)_{\Delta T = 0} = L_{11}\,\Delta V,$$
$$(J_s)_{\Delta T = 0} = L_{21}\,\Delta V,$$
$$(J_Q)_{\Delta T = 0} = T L_{21}\,\Delta V.$$

The ratio of the entropy current J_s to the electric current *I* is represented by S^*, and the ratio of the heat current J_Q to the electric current *I* is represented by Q^*. From the equations above, we have

$$S^* = \left(\frac{J_s}{I}\right)_{\Delta T = 0} = \frac{L_{21}}{L_{11}}, \tag{16}$$

$$Q^* = \left(\frac{J_Q}{I}\right)_{\Delta T = 0} = T\,\frac{L_{21}}{L_{11}}. \tag{17}$$

We see from Eqs. (16) and (17) that

$$Q^* = T\,S^*. \tag{18}$$

The quantities S^* and Q^* are properties of the material of which the rod is composed. In general, they are functions of temperature. Although Eqs. (16) and (17) are obtained by considering the special case where $\Delta T = 0$, they also give the relation between the *entropy* and *heat* currents *transported by an electric current in a flow in which* $\Delta T \neq 0$. In such a case, however, there is an *additional flow* of *entropy* and *heat* as a result of *heat conduction.* This can be brought out more clearly as follows:

We solve the equation

$$I = L_{11}\,\Delta V + L_{12}\,\Delta T \tag{6}$$

for ΔV, that is

$$\Delta V = \frac{I - L_{12}\,\Delta T}{L_{11}}.$$

Then we substitute this equation for ΔV in the equations

$$J_s = L_{21}\,\Delta V + L_{22}\,\Delta L, \tag{7}$$

and

$$J_Q = (T L_{21})\Delta V + (T L_{22})\,\Delta T. \tag{11}$$

Thus, we obtain

$$J_s = L_{21}\left(\frac{I - L_{12}\,\Delta T}{L_{11}}\right) + L_{22}\,\Delta T, \tag{19}$$

and

$$J_Q = (T L_{21})\left(\frac{I - L_{12}\,\Delta T}{L_{11}}\right) + T L_{22}\,\Delta T. \tag{20}$$

Finally, we use the equation

$$L_{12} = L_{21}, \tag{10}$$

together with the following equations:

$$\frac{L_{11} L_{22} - L_{12}^2}{L_{11}} = \frac{1}{T \Delta R_{th}}, \tag{15}$$

$$S^* = \left(\frac{J_s}{I}\right)_{\Delta T = 0} = \frac{L_{21}}{L_{11}}, \tag{16}$$

and

$$Q^* = \left(\frac{J_Q}{I}\right)_{\Delta T = 0} = T \frac{L_{21}}{L_{22}}. \tag{17}$$

In this way, we have for the entropy current J_s using Eq. (19)

$$J_s = \frac{I L_{21} - L_{12}^2 \Delta T}{L_{11}} + L_{22} \Delta T,$$

$$= \frac{L_{21}}{L_{11}} I + \frac{L_{11} L_{22} - L_{12}^2}{L_{11}} \Delta T,$$

$$J_s = S^* I + \frac{1}{T} \frac{\Delta T}{\Delta R_{th}}. \tag{21}$$

In like manner, we obtain for the heat current J_Q using Eq. (20)

$$J_Q = Q^* I + \frac{\Delta T}{\Delta R_{th}}. \tag{22}$$

The terms $\Delta T/\Delta R_{th}$ and $\Delta T/TR_{th}$ represent, respectively, the *heat current* and the *entropy current* that would result from *conduction alone*, in the absence of an electric current *I*. To these must be added the heat and entropy currents transported by the electric current. Thus the quantities S^* and Q^* are called *transport parameters*.

Furthermore, it follows from the equation

$$\left(\Delta V\right)_{I=0} = -\frac{L_{12}}{L_{11}} \Delta T,$$

which is Eq. (13) obtained above, that if a temperature difference ΔT exists across an element of a conducting rod or wire, there is a *potential difference* ΔV *across the element even in the absence of an electric current*. From this equation, and the equation

$$S^* = \frac{L_{21}}{L_{11}} = \frac{L_{12}}{L_{11}},$$

which is Eq. (16) also obtained above, we have

$$-(\Delta V)_{I=0} = S^* \Delta T. \tag{23}$$

This equation is of great importance, since it shall be made use of in discussing the thermoelectric phenomena, as will be seen below.

16.5. Thermoelectric phenomena

Thermoelectric phenomena provide an excellent example of the application of the methods of irreversible thermodynamics. It is useful to consider separately three thermoelectric phenomena, known as the *Seebeck* effect, the *Peltier* effect, and the *Thomson* effect.

1. The Seebeck effect. Fig. 1(a) shows the simplest thermocouple circuit two unlike metals or alloys A and B such as nickel and iron, or platinum and platinum – rhodium, are soldered or welded together to form a closed circuit. If one junction is in contact with a body at a temperature *T* and the other with a body at a lower temperature T_0, it is found that there is an electric current in the circuit. There is also a heat current along the wires.

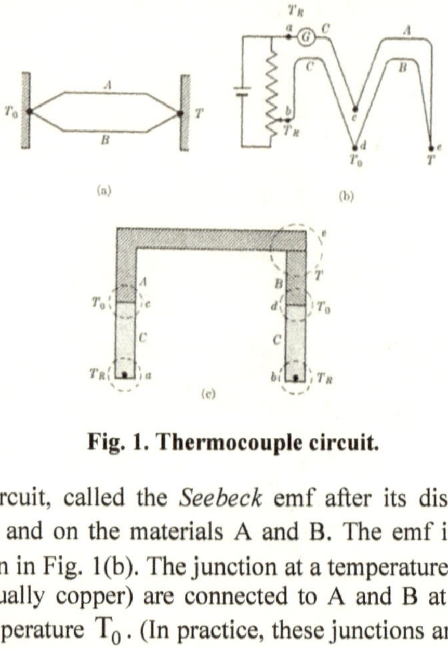

Fig. 1. Thermocouple circuit.

The emf of the circuit, called the *Seebeck* emf after its discoverer, depends on the temperature T_0 and *T* and on the materials A and B. The emf is usually measured by a potentiometer, as shown in Fig. 1(b). The junction at a temperature *T* is denoted by e. *Leads* of a *third* metal C (usually copper) are connected to A and B at junctions c and d. Both junctions are at the temperature T_0. (In practice, these junctions are often inserted in an ice bath.) The copper leads are connected to a potentiometer at points a and b, both of which are the temperature T_R (room temperature). When the potentiometer is balanced, as indicated by zero current in the galvanometer G, the potential difference $V_a - V_b$ equals the Seebeck emf $\in_{A,B}$.

Fig. 1(c), which is lettered to correspond to Fig. 1(b), shows the essential features of the thermocouple alone. In the steady state, temperature differences exist along the wires of the couple, but there is no electric current. We can then apply Eq. (23) in Section 16.4, which is now numbered (1)

$$-(\Delta V)_{I=0} = S^{*}\Delta T, \qquad (1)$$

to each wire of the thermocouple, replacing ΔV and ΔT by dV and dT, respectively, and integrating from one and of each wire to the other. Starting at a point a, we have

$$V_a - V_c = \int_{T_R}^{T_0} S_C^* \, dT, \qquad V_c - V_e = \int_{T_0}^{T} S_A^* \, dT,$$

$$V_e - V_d = \int_{T}^{T_0} S_B^* \, dT, \qquad V_d - V_b = \int_{T_0}^{T_R} S_C^* \, dT.$$

When these equations are added, we obtain

$$V_a - V_c + V_c - V_e + V_e - V_d + V_d - V_b$$

$$= \int_{T_R}^{T_0} S_C^* \, dT + \int_{T_0}^{T} S_A^* \, dT + \int_{T}^{T_0} S_B^* \, dT + \int_{T_0}^{T_R} S_C^* \, dT,$$

indicating that the left-hand sides reduce to $V_a - V_b$, and on the right-hand side, the first and last terms cancel. Consequently, since

$$\epsilon_{A,B} = V_a - V_b,$$

we have

$$\epsilon_{A,B} = \int_{T_0}^{T} \left(S_A^* - S_B^*\right) dT. \tag{2}$$

It is thus evident that the Seebeck emf depends only on the temperatures T and T_0, and on the difference between the entropy transport parameters of the metals A and B. It is independent of metal C, and of the temperature T_R.

If the thermocouple is being used to measure the temperature T, the *rate of change* of emf with temperature is more important than the value of ϵ itself. From Eq. (2), when T_0 is constant, we obtain by differentiation

$$\frac{d\epsilon_{A,B}}{dT} = S_A^* - S_B^*, \tag{3}$$

where the values of S^*_A and S^*_B are those at the temperature T. The derivative $d\epsilon_{A,B}/dT$ is called inappropriately the *thermoelectric power* of the thermocouple.

2. The Peltier effect. Fig. 2 shows the junction between two

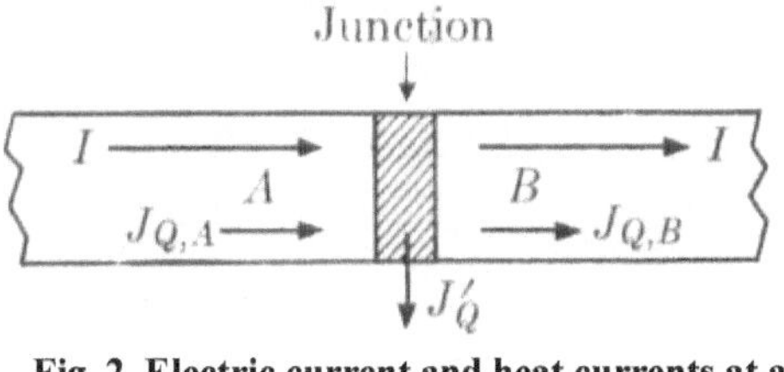

Fig. 2. Electric current and heat currents at a junction of two unlike metals A and B.

unlike metals (or alloys) A and B. The system is at a uniform temperature *T*. If we send an electric current *I* across the junction, it is found that heat is developed but that the rate of development of heat is greater or less than $I^2 R$ heating, the difference depending on the magnitude and direction of the current, on the temperature, and on the materials A and B. This phenomenon is known as the *Peltier effect.*

Let us compute the rate at which heat must be removed from the junction when it is carrying current, in order to maintain the junction at constant temperature *T.* Let V_a and V_b be the potentials of A and B on opposite sides of the junction. The rate at which an electrical work is done on the junction is $I V_{ab} = I^2 R_j$, where R_j is the resistance of the junction. The rate at which heat is transported into the junction by the current *I* at its left face is $J_{a,A} = I Q_A^* = I T S_A^*$, and the rate at which it is transported out of the junction by the electric current *I* across the right face is $J_{a,A} = I Q_A^* = I T S_A^*$. Let J'_Q be the rate at which heat must be removed from the junction to keep its temperature constant. Since the state of the junction does not change, we have from the first law

$$J'_Q = I^2 R_j + I T S_A^* - I T S_B^*$$
$$= I^2 R_j + I \left\{ T \left(S_A^* - S_B^* \right) \right\}. \quad (4)$$

The last term in this equation equals the excess heat that must be removed, over and above the '$I^2 R_j$' heat, and it is called the *Peltier heat.* The quantity $T\left(S_A^* - S_B^*\right)$ is expressed in volts if *I* is in amperes, and is called the *Peltier* emf $\pi_{A,B}$:

$$\pi_{A,B} = T\left(S_A^* - S_B^*\right). \quad (5)$$

Eq. (4) can therefore be written as

$$J'_Q = I^2 R_j + I \pi_{A,B}. \quad (6)$$

The Peltier emf at a junction depends on the temperature, both through the factor *T* and because S_A^* and S_B^* vary with temperature.

3. The Thomson effect. Consider an element of a rod or wire

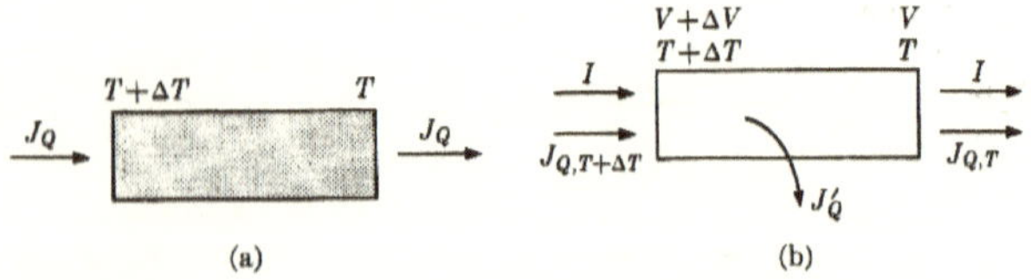

Fig. 3. Electric current and heat current in a homogeneous rod in which there is a temperature gradient.

in the steady state across which there is a temperature difference ΔT. There is then a heat current by conduction J_Q into one face of the element and an equal heat current out of the other face as shown in Fig. 3(a). If we now send an electric current through the element, it is found that heat is developed in the element, but that the rate of development of heat is greater or less than the I^2R heating, the difference depending on the magnitude and direction of the current, on the temperature, and on the material. This phenomenon is known as the *Thomson effect.*

Let us compute the rate at which heat must flow out of the side walls of the element when it is carrying an electric current in order to maintain the temperature at all points the same as it was before the electric current was turned on. The *conduction* heat flows arising from the temperature difference ΔT will then be the same as before, and the heat flowing into the element by conduction across its left face equals that flowing out across the right face. Let J'_Q be the heat current flowing out across the side walls; that is, the rate at which heat must be removed from the element to maintain the temperature gradient the same. [Fig. (3b)]

The rate at which electrical work is done on the element is $I\ \Delta V$. The rate at which heat is transported into the element by the *current I* at its left side is $J_{Q,T+\Delta T} = IQ^*_{T+\Delta T}$, and the rate at which it is transported out across the right face is $J_{Q,T} = IQ^*_T$. (The conduction heat flows are equal and need not be considered). Since the state of the element does not change, we have from the first law

$$\begin{aligned} J'_Q &= I\Delta V + IQ^*_{T+\Delta T} - IQ^*_T \\ &= I\left\{\Delta V + (T+\Delta T)S^*_{T+\Delta T} - TS^*_T\right\}. \end{aligned} \tag{1}$$

But

$$S^*_{T+\Delta T} = S^*_T + \frac{dS^*}{dT}\Delta T, \tag{2}$$

therefore,

$$\begin{aligned} J'_Q &= I\left\{\Delta V + (T+\Delta T)\left(S^*_T + \frac{dS^*}{dT}\Delta T\right) - TS^*_T\right\} \\ &= I\left(\Delta V + T\frac{dS^*}{dT}\Delta T + S^*\Delta T\right), \end{aligned} \tag{3}$$

where we have dropped the subscript *T* on S^* and neglected the small quantity $\frac{dS^*}{dT}\Delta T^2$ or $\Delta S^*\ \Delta T$. From the equation

$$I = L_{11}\Delta V + L_{12}\Delta T, \tag{4}$$

obtained before in Section 16.3, we may express ΔV as

$$\Delta V = \frac{I}{L_{11}} - \frac{L_{12}}{L_{11}}\Delta T. \tag{5}$$

But, again as shown in Section 16.4, we obtained the following equations

$$L_{11} = \frac{1}{\Delta R}, \tag{6}$$

and

$$\frac{L_{12}}{L_{11}} = S^*, \tag{7}$$

so that we have for ΔV the following expression

$$\Delta V = I\Delta R - S^* \Delta T. \tag{8}$$

Hence, substituting for ΔV in Eq. (3) by Eq. (8), we finally have

$$J'_Q = I\left(I\Delta R - S^*\Delta T\right) + I\left(T\frac{dS^*}{dT}\Delta T\right)$$

$$+ I S^* \Delta T,$$

or

$$= I^2 \Delta R + I\left(T\frac{dS^*}{dT}\Delta T\right). \tag{9}$$

The last term in Eq. (9) is equal to the excess heat that must be *removed*, over and above the 'I^2R' heat, and is called the *Thomson heat*. The quantity $\left[T\left(dS^*/dT\right)\Delta T\right]$ is expressed in volts if *I* is in amperes, and is called the *Thomson emf*. The product $T\,dS^*/dT$ is the *Thomson coefficient* σ. However, the usual convention of sign is to consider the Thomson heat as positive if heat must be *added* to keep the temperature the same. We therefore write the Thomson coefficient σ as

$$\sigma = -T\frac{dS^*}{dT}, \tag{10}$$

and subsequently Eq. (9)

$$J'_Q = I^2 \Delta R - I\sigma\Delta T. \tag{11}$$

in order to obtain agreement with the usual sign convention in thermodynamics.

Thus, it is evident from the above that all three thermoelectric effects can be expressed in terms of the entropy transport parameter S^* and the temperature.

The Seebeck emf is

$$\epsilon_{A,B} = \int_{T_1}^{T_2} \left(S_A^* - S_B^*\right) dT,$$

and the thermoelectric power is

$$\frac{d\epsilon_{A,B}}{dT} = S_A^* - S_B^*.$$

The Peltier emf is

$$\pi_{A,B} = T\left(S_A^* - S_B^*\right).$$

The Thomson coefficient is

$$\sigma = -T\frac{dS^*}{dT}.$$

It follows from the second and third of these equation that

$$\pi_{A,B} = T\frac{d\epsilon_{A,B}}{dT},$$

and from the second and fourth that

$$\sigma_A - \sigma_B = -T\frac{d^2\epsilon_{A,B}}{dT^2}.$$

16.6. Thermomechanical phenomena

An important class of phenomena is the coupling of *thermal and pressure* forces in which *matter* flows, or diffuses, in a coupled effect with *heat flow*. Examples of this class are the *fountain effect*, thermo-osmosis, and the *Knudsen effect*. Another important effect is that of *thermo-molecular pressure* which we shall first take up.

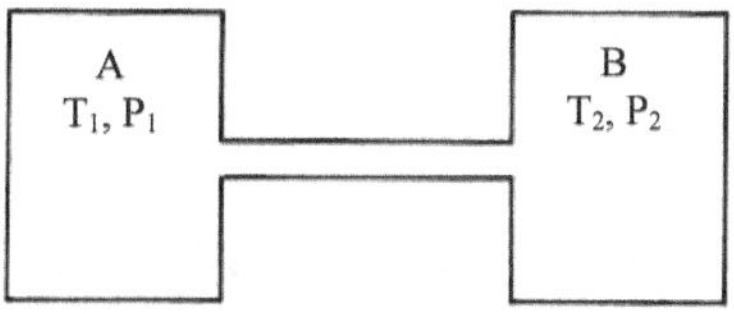

Fig. 1.

1. Thermo-molecular pressure effect

Consider two vessels containing a gas and connected by a capillary tube as shown in Fig. 1. A porous wall or a semipermeable membrane can be used instead of the capillary tube. For simplicity, we shall assume that the volume of the whole system remains constant. Let the absolute temperatures in the two vessels be T_1 and T_2 and the corresponding gas pressures P_1 and P_2. Ordinarily, as a condition of equilibrium, the two pressures must be equal. However, if the diameter of the capillary tube is small compared with the mean free path of the gas molecules, the probability that a molecule will return to vessel A only after a large number of collisions with the molecules in vessel B having attained the temperature T_2 is very high. Hence, one could expect that some other condition of equilibrium other than that of the equality of the two pressures, P_1 and P_2, must be satisfied. According to the *molecular kinetic theory* of an ideal gas, the total number of collisions with the walls of a containing vessel per unit area per unit time made by *n* molecules occupying a unit volume and having all speeds is $\frac{1}{4}n\bar{v}$. Accordingly, the number of gas molecules passing though the capillary tube from vessel A to vessel B is

$$\frac{1}{4}An_A\bar{v}_A, \tag{1}$$

where A is the cross-sectional area of the capillary tube, n_A is the number of molecules per unit volume, and $\overline{v}_A$ is the mean velocity on side A of the capillary tube. The number of gas molecules passing in the opposite direction is

$$\frac{1}{4} A\, n_B\, \overline{v}_B . \tag{2}$$

Equilibrium is attained when the product $n\,\overline{v}$ is equal on both sides of the capillary, that is

$$n_A\, \overline{v}_A \;=\; n_B\, \overline{v}_B . \tag{3}$$

This equation would lead to the result

$$P_1/P_2 \;=\; \sqrt{T_1/T_2}\,, \tag{4}$$

which implies that the temperature difference existing between the two vessels causes the gas to flow and sets up a corresponding pressure difference. The quotient of the corresponding pressure difference and temperature difference, $\frac{\Delta P}{\Delta T}$, is called the *thermo-molecular pressure difference*. It is to be added that the result expressed by Eq. (4) can be simply deduced in another way, as will be seen below when we deal with the *thermomechanical effect*.

2. Thermomechanical effect

If a pressure difference and a constant temperature are maintained in a substance, a flow of *heat* is observed which, as a first approximation, is proportional to the *flow of matter*.

Our first task is to determine the forces relating the two effects. This is accomplished by finding out the appropriate fluxes and forces from the energy source. Let us suppose that the system is contained in two vessels A and B of *equal volume* which are connected by a small hole. The whole system is thermally insulated from the surroundings. When thermodynamic equilibrium is attained, the internal energy *U*, the mass *M*, and the entropy *S* will be the same in the two vessels. Changes in the internal energy and mass in vessel A correspond to changes in the internal energy and mass in vessel B since the system is *insulated.* The change of entropy in vessel A is

$$\Delta S_A \;=\; \left(\frac{\partial S}{\partial U}\right)_M \Delta U_A + \left(\frac{\partial S}{\partial M}\right)_U \Delta M_A + \frac{1}{2}\left(\frac{\partial^2 S}{\partial U^2}\right)_M (\Delta U_A)^2$$
$$+ \frac{\partial^2 S}{\partial U\, \partial M} \Delta U_A\, \Delta M_A + \frac{1}{2}\left(\frac{\partial^2 S}{\partial M^2}\right)_U (\Delta M_A)^2 . \tag{5}$$

The change of entropy in vessel B is given by an equation similar to Eq. (5) but with *negative* changes in internal energy and mass, that is,

$$\Delta S_B = -\left(\frac{\partial S}{\partial U}\right)_M \Delta U_A - \left(\frac{\partial S}{\partial M}\right) \Delta M_A + \frac{1}{2}\left(\frac{\partial^2 S}{\partial U^2}\right)_M (\Delta U_A)^2$$
$$+ \frac{\partial^2 S}{\partial U\, \partial M} \Delta U_A\, \Delta M_A + \frac{1}{2}\left(\frac{\partial^2 S}{\partial M^2}\right)_U (\Delta M)^2 . \tag{6}$$

The average change in entropy for each of the two vessels is

$$\Delta S = \frac{1}{2}(\Delta S_A + \Delta S_B). \tag{7}$$

As a result, we get

$$\Delta S = \frac{1}{2}\left(\frac{\partial^2 S}{\partial U^2}\right)_M (\Delta U_A)^2 + \frac{\partial^2 S}{\partial U\,\partial M}\Delta U_A\,\Delta M_A + \frac{1}{2}\left(\frac{\partial^2 S}{\partial M^2}\right)_U \Delta M_A^2. \tag{8}$$

The time derivative of Eq. (8) is given by

$$\frac{d\Delta S}{dt} = \Delta U\left(\frac{\partial^2 S}{\partial U^2}\right)\frac{d\Delta U}{dt} + \frac{\partial^2 S}{\partial U\,\partial M}\Delta M\,\frac{d\Delta U}{dt} + \Delta M\left(\frac{\partial^2 S}{\partial M^2}\right)\frac{d\Delta M}{dt} + \frac{\partial^2 S}{\partial U\,\partial M}\Delta U\,\frac{d\Delta M}{dt}$$

or

$$= \frac{d\Delta U}{dt}\left[\left(\frac{\partial^2 S}{dU_A^2}\right)\Delta U + \left(\frac{\partial^2 S}{\partial U\partial M}\right)\Delta M\right] + \frac{d\Delta M}{dt}\left[\left(\frac{\partial^2 S}{\partial M^2}\right)\Delta M + \left(\frac{\partial^2 S}{\partial U\partial M}\right)\Delta U\right],$$

or

$$\frac{d\Delta S}{dt} = \frac{d\Delta U}{dt}\Delta\left(\frac{\partial S}{\partial U}\right)_M + \frac{d\Delta M}{dt}\Delta\left(\frac{\partial S}{\partial M}\right)_U. \tag{39}$$

This equation can be put in the form that expresses the sum of the products of the fluxes and forces such as

$$\frac{d\Delta S}{dt} = J_U X_U + J_M X_M, \tag{10}$$

where

$$J_U = \frac{d\Delta U}{dt}, \quad X_U = \Delta\left(\frac{\partial S}{\partial U}\right)_M;$$

$$J_M = \frac{d\Delta M}{dt}, \quad X_M = \Delta\left(\frac{\partial S}{\partial M}\right)_U.$$

The expressions for these *conjugate flows* J_M and J_U are

$$\left.\begin{aligned} J_M &= L_{11} X_M + L_{12} X_U \\ J_U &= L_{21} X_M + L_{22} X_U \end{aligned}\right\}. \tag{11}$$

These two equations relate the flow of matter due to a temperature gradient and the flow of heat due to a concentration gradient.

Now explicit relationships for the forces are required. At constant volume, we have

$$T dS = dU - g\, dM, \tag{12}$$

where g is the specific Gibbs function or chemical potential per unit mass. Therefore,

$$X_U = \Delta \left(\frac{\partial S}{\partial U} \right)_M = \Delta \left(\frac{1}{T} \right) = - \frac{\Delta T}{T^2}. \tag{13}$$

This is because according to Eq. (12),

$$\left(\frac{\partial S}{\partial U} \right)_M = \frac{1}{T}, \text{ and } d\left(\frac{1}{T} \right) = - \frac{1}{T^2} dT,$$

so that

$$\Delta \left(\frac{\partial S}{\partial U} \right)_M = \Delta \left(\frac{1}{T} \right) = - \frac{1}{T^2} \Delta T.$$

Also, since $X_M = \Delta \left(\frac{\partial S}{\partial M} \right)_U$, and $\left(\frac{\partial S}{\partial M} \right)_U = -\left(\frac{g}{T} \right)$.

Therefore,

$$X_M = - \Delta \left(\frac{g}{T} \right),$$

or

$$= \frac{- v\Delta P}{T} + \frac{h \Delta T}{T^2}, \tag{14}$$

Eq. (14) can be obtained in the following way:

$$d\left(\frac{g}{T} \right) = \frac{dg}{T} - \frac{g}{T^2} dT$$

or,

$$\Delta\left(\frac{\mathcal{g}}{T}\right) = \frac{\Delta \mathcal{g}}{T} - \frac{\mathcal{g}}{T^2}\,\Delta T.$$

But

$$\mathcal{g} = h - Ts,$$

$$d\mathcal{g} = -sdT + vdP$$

or,

$$\Delta \mathcal{g} = -s\Delta T + v\Delta P.$$

Therefore,

$$X_M = -\Delta\left(\frac{\mathcal{g}}{T}\right) = -\frac{v\Delta P}{T} + \frac{s\Delta T}{T} - \frac{\mathcal{g}}{T^2}\,\Delta T,$$

$$= -\frac{v\Delta P}{T} + \left(\frac{Ts - g}{T^2}\right)\Delta T,$$

$$= -\frac{v\Delta P}{T} + \frac{h\,\Delta T}{T^2}.$$

Now, when Eq. (13) and (14) are substituted in Eq. (11), we obtain

$$\left.\begin{aligned} J_M &= -\frac{L_{11}v}{T}\,\Delta P + \frac{L_{11}h - L_{12}}{T^2}\,\Delta T \\ J_U &= -\frac{L_{21}v}{T}\,\Delta P + \frac{L_{21}h - L_{22}}{T^2}\,\Delta T \end{aligned}\right\} \qquad (15)$$

In the *special case of uniform temperature, that is,* $\Delta T = 0$, Eq. (15) yields

$$J_U = \frac{L_{21}}{L_{11}}\,J_M = U^*\,J_M, \qquad (16)$$

where the quantity U^*, representing the quotient L_{21}/L_{11}, is the *energy transferred with a unit mass*. This equation describes the thermomechanical effect in which energy is transferred as a consequence of a pressure *difference* when the temperature is *uniform.*

Another special case occurs when there is no mass transfer, that is, $J_M = 0$, but energy transfer takes place due to the non-zero value of J_U. This condition is known as the *stationary state* of the system. From the expression of mass flux J_M, which is given by Eq. (15), we obtain for the stationary state

$$\frac{L_{11} v}{T} \Delta P = \frac{L_{11} h - L_{12}}{T^2} \Delta T,$$

or

$$\frac{\Delta P}{\Delta T} = \frac{h - L_{12}/L_{11}}{vT}. \quad (17)$$

It has been pointed out above in Section 16.1, that the result obtained as expressed in Eq. (4), namely

$$P_1/P_2 = \sqrt{T_1/T_2}, \quad (4)$$

leads to the inference that a temperature difference existing between the two vessels causes matter to flow and sets up a corresponding pressure difference so that the quotient $\Delta P/\Delta T$ is called the *thermo-molecular pressure difference*. This quotient is thus defined now by Eq. (17). On the other hand, the defining relationship for U^* given above by Eq. (16), and the Onsager reciprocal relation $L_{12} = L_{21}$ permits writing Eq. (17) in the form

$$\frac{\Delta P}{\Delta T} = \frac{h - U^*}{vT}, \quad (18)$$

which is an important expression.

It can also be shown that using the relation

$$Q^* = U^* - h, \quad (19)$$

(which can be simply proved) we obtain the alternative form of Eq. (18), namely,

$$\frac{\Delta P}{\Delta T} = -\frac{Q^*}{vT}. \quad (20)$$

Eqs. (18) and (20) are important equations in the thermodynamics of irreversible processes which involve only a single substance. We shall apply these equations to a few interesting phenomena.

(a) Knudsen gas. Let us consider an ideal gas contained in a vessel divided by a partition in which there is a hole whose diameter is small compared with the mean free path of the gas molecules. Every molecule arriving at the hole with pass through it freely. According to the molecular kinetic theory of an ideal gas, the mean energy* of the molecules passing through the hole is

$$\overline{\frac{1}{2} m V^2} = 2kT \tag{21}$$

so that U^*, which is the energy transferred with each unit of mass, is

$$U^* = \frac{2kN_a T}{M} = \frac{2RT}{M}, \tag{22}$$

where *M* now represents the molecular mass. Subsequently, the enthalpy, *h*, per unit mass of a monatomic ideal gas consisting of spherical molecules is given† by

$$h = u + Pv = \frac{3}{2}\frac{RT}{M} + \frac{RT}{M} = \frac{5RT}{2M}. \tag{23}$$

Hence, substituting Eqs. (22) and (23) into Eq. (18), we obtain for an ideal gas the equation

$$\frac{\Delta P}{\Delta T} = \frac{h - U^*}{vT} = \frac{\frac{5}{2}RT/M - 2\,RT/M}{vT} = \frac{1}{2}\frac{R}{Mv} = \frac{1}{2}\frac{P}{T}. \tag{24}$$

The integration of this equation yields Eq. (4) shown above as follows:

$$\int_1^2 \frac{\Delta P}{P} = \frac{1}{2}\int_1^2 \frac{\Delta T}{T}$$

$$\ln\frac{P_1}{P_2} = \ln\left(\frac{T_1}{T_2}\right)^{\frac{1}{2}}.$$

Therefore,

$$P_1/P_2 = \sqrt{T_1/T_2}\,.$$

---

* $v_{rms} = \sqrt{4kT/m}\,.\,(v_{rms})^2 = 4kT/m$, therefore $\overline{\frac{1}{2} m v^2} = 2kT$

† A monatomic gas executing only translational motion has three degrees of freedom so that the specific energy per mole $u = U/n = \frac{3}{2}RT$, and that per unit mass $= \frac{3}{2}\frac{RT}{M}$.

(b) Ordinary ideal gas. For an ordinary ideal gas, the diameter of the hole in the partition is large in comparison with the mean free path. The energy transferred with an element of gas is precisely the whole energy of the gas, or the enthalpy, so that

$$U^* = h, \qquad \text{or} \qquad h - U^* = h - h = 0.$$

Consequently, it is necessary to include in the quantity U^* an internal energy term *u* and the flow energy term *pv* so that

$$U^* = u + pv = h.$$

In this case, we find $\Delta P/\Delta T$ as given by Eq. (24) is equal to zero, and accordingly, we conclude that there is no thermo-mechanical effect for an ideal gas under ordinary conditions.

(c) The fountain effect. It was first observed by *Allen and Jones* in 1938, a fountain effect in their work with liquid helium II. Fig. 2 shows a schematic arrangement of the apparatus with which Allen and Jones first observed the phenomenon. When the helium in vessel A was heated by an electric heater, they observed a flow of helium from vessel B into vessel A as manifested by the rising of the liquid level in vessel A.

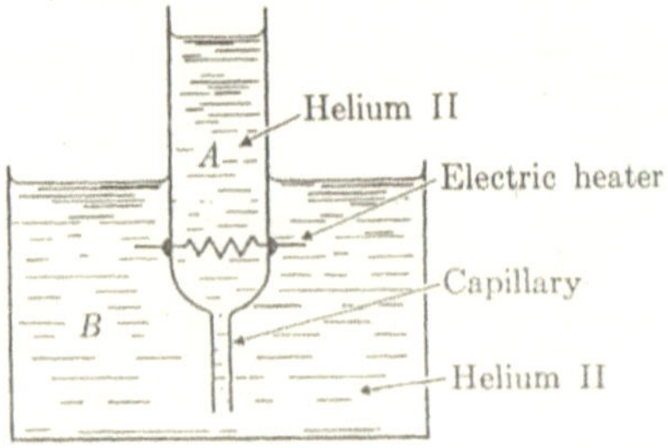

Fig. 2. Allen and Jones apparatus for fountain

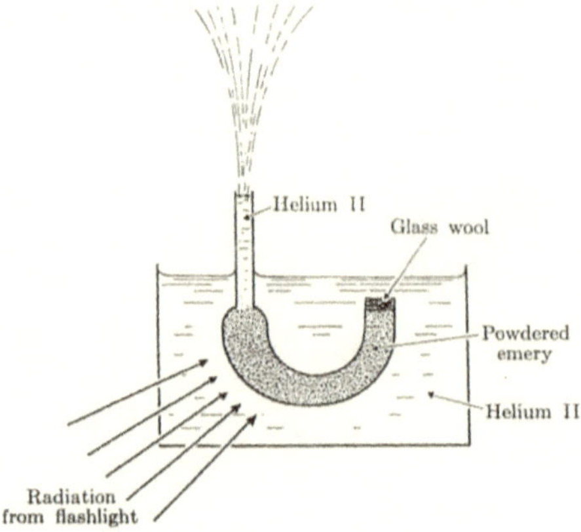

Fig. 3. Allen and Misener apparatus for fountain effect.

Another apparatus designed later by Allen and Misener as shown in Fig. 3 gave a more striking effect. The helium contained in the voids between particles of finely ground emery was heated by radiation from a flashlight. The resulting increase of pressure caused a fountain of liquid helium which reached as high as 30 cm.

It is to be noted that unlike the lighter isotope, ^3He, the common isotope of helium, ^4He, has two liquid phases known as helium I and helium II, the former phase being the high temperature phase* and in all respects representing a normal liquid.

---

* See Section 3.2.

CHAPTER 17

The Fundamental Assumptions of Elementary Statistical Mechanics

17.1. Introduction

The object of statistical mechanics, or the theory of the properties of matter in equilibrium, is to derive the properties of matter in bulk from the known or assumed atomic structure of matter and the laws of interaction between its constituent atoms. The properties of matter in bulk exhibited to our senses are the properties of large collections of individual atoms. The behaviour of an individual atom or molecule is never here in question, only the result of the combined behaviour of the multitude. The establishment of the properties of individual atoms and their laws of interaction is the programme of the quantum theory.

A particular homogenous piece of matter may be a gas, a liquid, a solid, or a surface film. In the case of a gas, we have to deal with a collection of more or less independent atoms or molecules which only occasionally interact sensibly with one another. In the case of a solid, the atoms or molecules are very far from being nearly independent, but we can often analyse the motions of the atoms into nearly independent vibrations of the solid in its normal modes. In other cases such as analysis into nearly independent systems may be less accurate or at present practically impossible. It will then be found that progress in the application of statistical mechanics becomes increasingly difficult. Progress made may be measured by the extent to which it has been possible to analyse the states and motions of the complex material into the states and motions of suitable almost independent systems. For this reason, a convenient habit has grown up of referring to the matter under discussion as an *assembly of systems*. The systems may or may not be atoms, but whatever they are, their properties are determined by quantum theory and are assumed to be completely known. The assembly contains a large number of systems, and the determination of the *observable equilibrium properties* of the assembly, *under given large-scale conditions, is the primary object of statistical mechanics.* In the most recent successful applications, the distinction between the assembly and its constituent systems is sometimes impossible or inconvenient to preserve. The assembly may then be taken to consist of a single system, but in that case, the system itself is *a complicated one* and possesses a very large number of possible states densely distributed in energy.*

---

* The symbol E shall be used here in place of U to denote the energy of a system (or an assembly), when we deal with statistical mechanics, as we did before when we dealt with quantum or wave mechanics.

17.2. The first fundamental assumption of statistical mechanics - the atomic structure of matter

Assumption 1. The atomic constitution of matter. An atom itself is of course not a simple system but consists of a nucleus and a set of electrons in motion about the nucleus. It is often necessary to take explicit account of the electronic structure of the atom (or the molecule) in statistical applications. The electrons of all atoms have identical properties and a negative charge which we denote by $-|e|$, in order not to be in danger of confusion as to whether the symbol e denotes the algebraic charge or the numerical charge on the electron. Each electron has also a spin with spin angular momentum component $\frac{1}{2}(h/2\pi)$, where h is *Planck's* constant, which may be oriented either parallel or antiparallel to any *applied field.* The nucleus has a charge $+Z|e|$, where Z is the atomic number.

The above analysis of atoms into electrons and nuclei may not always be sufficient for the needs of a statistical problem. The nucleus itself (except for hydrogen) is not a simple system but is composed *(probably)* of Z protons and $A–Z$ neutrons, where A is the mass number, that is, the nearest integer to the atomic weight of the species of atom in question on the *chemical scale.* In some applications of statistical mechanics, it is necessary to take explicit account of the detailed structure of the nucleus, but this is seldom or never necessary to applications to terrestrial physics or chemistry. All that is necessary for such problems is to remember that the structure of the nucleus may manifest itself through the *nuclear spin.*

The comments we have just made should make it clear that the *first assumption of statistical mechanics* would be more correctly described as

Assumption 1. Matter is composed of electrons, protons, and neutrons. For ordinary purposes, the protons and neutrons may be regarded as combined into a number of permanent nuclei of definite constitution, internal energy, and spin.

Any assembly is therefore strictly to be regarded as an assembly of a large number of electrons, protons, and neutrons, whose behaviour is controlled by the laws of quantum mechanics.

Terrestrial physicists and chemists requires to know about such an assembly is what will be the ordinarily observable properties of the matter it represents when left to itself to come to equilibrium with its surroundings.

We know, to start with, that the assembly obeys the laws of quantum mechanics, and that, in suitable conditions frequently realised in practice, these laws may be taken to *reduce to their simpler limiting form, classical mechanics.*

17.3. General nature of the states of the assembly

The assembly of electrons, protons, and neutrons, or of electrons and nuclei, is a part of all matter. *All other matter in the universe may naturally be idealised into fields and walls which may be regarded as a conservative field of force* under whose influence the assembly has certain properties which we wish to determine. By conservative, we mean that the field of force is derived from *a potential energy which is a function of position only and is independent of time*. Under the influence of this conservative field of force, the assembly has certain properties which we wish to determine. Under such conditions, the equation of *Schrödinger* for the assembly can be set up and in principle solved. If the assembly is contained in a finite volume with suitable walls, it is known that the quantal equation of Schrödinger admits *discrete eigenvalues* $E_0, E_1, E_2, \ldots$ of the energy, to each of which there correspond *one or more distinct solutions* of the equation which are called eigenfunctions. The *eigenfunctions* themselves are often not of primary importance in statistical mechanics. What is here of primary importance is the mere fact that such solutions exist, *representing possible states of motion of the assembly*. Such states of motion, which are of fundamental importance, are called *stationary states* or simply *states*. By a state, in this sense, we always mean a *state of motion* corresponding to some *one distinct solution of Schrödinger's equation for the assembly* – to each solution one state and *vice versa*. For a large assembly of ideally almost independent systems, there may be a very large number of eigenfunctions and therefore of states corresponding to *any one possible energy value of the assembly*. For other large assemblies, there will be in any case a large number of states corresponding to eigenvalues of the energy lying in the range E, $E + dE$. It is in any event more natural to discuss a *small energy range* rather than an *exact energy*, since no *actual assembly* is really subject to *exactly conservative fields of force*, but is continually losing and gaining small amounts of energy to its surroundings by atomic encounters, even in an equilibrium state.

We, therefore, consider our assembly with a given range or range of energies, to which there corresponds a large number of possible states, and have next to ask how such an assembly will behave according to the laws of quantum mechanics.

If Schrödinger's equation which we have set up to describe the complete assembly were exact, and if it included exactly every interaction between the systems of the assembly and every interaction with walls and outside bodies, then in principle, we could suppose it to be *exactly* solved and every eigenvalue and eigenfunction would then belong to a *true stationary state*, in which the assembly (*genuinely isolated*) could persist for an indefinite time. The properties of the assembly would then be the properties of this *single stationary state*. Such a state of affairs, however, is never actually realisable even to the roughest approximation. In the first place, we have neglected radiation and radiative processes in the formulation of the Schrödinger equation, and we have assumed that all the interactions of the assembly with the external world can be represented by conservative forces. It is in principle impossible to represent the possible interactions of the assembly with radiation by any time-independent conservative fields of force. The result of this is that it is impossible for a strictly *conservative* Schrödinger equation to be set up, and the exact solutions, apparently possible in principle, are actually in practice illusory. The most that is possible is to set up and solve exactly an approximate Schrödinger equation, which must be regarded as subject to time-dependent disturbances, technically known as *perturbations*, inevitably causing transitions from one would-be stationary state to another. Thus, it is never the properties of a single (true) stationary state that concern us, but some sort of average over, or selection form, a large number of stationary states, all perhaps of the same energy or distributed over a small energy range.

Now that we have been forced to admit the principle that it will never be possible for any assembly, with which we are concerned, to be regarded as existing permanently in a true stationary state, no matter how assiduously we set up and solve Schrödinger's equation, we may frankly admit that, in practice, Schrödinger's equation could never be set up and solved to the required accuracy, and that the eigenvalues and stationary states are necessarily only approximate, and transitions between the stationary states are necessarily present. Thus the properties of the assembly can be sought for by averaging in the correct manner over the properties of the right selections of these approximate stationary states.

We are now confronted with the following situation. We have set up and solved the approximate Schrödinger's equation for the assembly, the errors in this equation being represented by perturbation terms, which continually induce transitions among the group of stationary states lying in a defined energy range. The equilibrium properties of the assembly can only be those of the vast majority of stationary states, and we can only hope to calculate the equilibrium properties by taking a suitable overage over all stationary states.

17.4. The second fundamental hypothesis of statistical mechanics - the rule for averaging

Assumption 2. Enumerate all the distinct states of the assembly consistent with a specified energy or energy range and any other given general conditions. The value $\overline{Q}$ of any property Q of the assembly, which will be found to characterise the equilibrium state of the assembly, subject to the specified conditions, may be obtained by averaging over all these states, attaching to each state an equal weight (unity). Formally

$$\overline{Q} = \sum_{\text{states}} Q \Big/ \sum_{\text{states}} 1. \qquad (1)$$

This statement is still vague and must be further refined as will be shown below in Section 17.8.

One may well question whether there are not other functions of the atomic variables of the assembly equally important, whose constancy may form part of the other general conditions mentioned in the enunciation of Assumption 2. A search for such functions will put us in the way of understanding the general concept of *accessibility*. This concept must be introduced before a sharply defined form of Assumption 2 can be given.

17.5. A description of the classical assembly

We have hitherto excluded all reliance on classical mechanics, and almost any reference to it, since we know that classical mechanics is in principle inapplicable to atomic assemblies, and gives correct results only in *suitable limiting cases*. But it will be convenient at this stage to describe a classical assembly and study the *conditions of accessibility in classical mechanics*, since this forms a natural introduction to the conditions of accessibility imposed by the quantum theory.

Any assembly obeying the laws of classical mechanics is fully described when its energy is given as a function of N' generalised positions and the N' generalised momenta corresponding to them. If the external fields of force to which the assembly is subject are *conservative*, the energy of the assembly may be written in *Hamilton's* form $H\left(x, P_x\right)$, where *x* is short for the N' coordinates $x_1, x_2, \ldots, x_{N'}$, and P_x for the corresponding momenta $P_{x_1}, P_{x_2} \ldots P_{x_{N'}}$. In this case (where we neglect the time-dependent forces, remembering, however, that in principle they can never be absent from any actual assembly), no other variables enter *H*. Thus for the idealised assembly in a conservative field of force, the *classical Hamiltonian equations of motion* are

$$\frac{dx}{dt} = \frac{\partial H}{\partial P_x}, \qquad \frac{dP_x}{dt} = -\frac{\partial H}{\partial x}, \tag{1}$$

$2N'$ equations in all. The coordinates $x's$ may be any suitable set, which must just suffice to describe the geometrical configuration of the assembler at any moment. For example, for a set of *N* structureless particles each of mass *m*, a natural set of coordinates would be the 3*N* Cartesian coordinates

$$x_\alpha, \; y_\alpha, \; z_\alpha \qquad (\alpha = 1, 2, \dots, N)$$

of the *N* particles. The corresponding (technically *conjugate*) momenta are then the 3*N* variables

$$m\dot{x}_\alpha, \; m\dot{y}_\alpha, \; m\dot{z}_\alpha \qquad (\alpha = 1, 2, \dots, N).$$

From Eq. (1), one can deduce* that

$$\frac{dH}{dt} = 0, \tag{2}$$

which tells us that the energy does not vary with time. We can write Eq. (2) in the integrated form

$$H(x, P_x) = E,$$

where *E* is a constant. Any such relationship, in which a definite one-valued function of all or any of the 2*N* variables x, P_x is *constant* throughout the motion of the assembly, is called a *uniform integral* of the equations of motion. Thus Eq. (2) tells us that *the energy is a uniform integral of the equations of motion.*

The complete behaviour of a classical assembly is known when all the coordinates *x* and momenta P_x are known as functions of time. This suggests a geometrical representation of the (classical) state of the assembly which is found helpful in any general discussion. The state of the assembly at any instant being fully known when all the *x*'s and P_x's are known, we represent the state of the assembly by point in a *hyperspace* of $2N' \, (= 6N)$ dimensions, the $x's$ and $P_x's$ being Cartesian coordinates of the point. This space, as has been discussed† earlier, it is known as *the phase space of the assembly*, and the path traced out by the point in phase space as its *trajectory*. This trajectory obeys the equations of motion of Eq. (1) of the assembly, and for a conservative assembly, it is fully defined by these equations and its starting point. Two trajectories of a definite conservative assembly can never intersect.

The value of a discussion of classical assembly resides in the *liming principle*, that this assembly can be regarded as a limiting form of the corresponding *quantal assembly*. In order to be able to use classical theorems as an illustration of the corresponding exact quantal ones, it is necessary to know how the discrete eigen states of the quantum theory compare in distribution with the continuously variable states corresponding to points in classical phase space. Fortunately, the complete answer to this question has long been known. It is that for an assembly of N' degrees of freedom each distinct eigenfunction, satisfying the proper boundary conditions and Schrödinger's equation for the assembly,

---

* See Section 4.8.

† See Section 9.1.

corresponds to an extension $h^{N'}$ of classical phase space. This correspondence is suggested in the first instance as a general theorem true for large quantum numbers as part of *Bohr's correspondence principle* and can be checked by direct evaluation in simple cases. Here, it is sufficient to accept as a fact the correspondence between each quantal state and an extension $h^{N'}$ of phase space.

17.6. Accessibility - mainly classical

In classical assemblies of the type just described, it is known that other uniform integrals, besides the energy, may well exist in certain circumstances. The components of linear momentum and angular momentum of the assembly are unaffected by the interactions of the systems of the assembly among themselves, and when *not upset by the outside world are said to be conserved.* They are then such *uniform integrals.* A uniform integral is, as we have already explained, a relationship of the form

$$F(x, P_x) = \text{const.} \tag{1}$$

For example, the constancy of the x-component of linear momentum in an assembly of simple particles of mass m is expressed by the relation

$$\sum_r P_{x\,r} = \text{const.}, \tag{2}$$

and the constancy of the x-component of angular momentum by the equation

$$\sum_r \left(y_r P_{z\,r} - z_r P_{y\,r}\right) = \text{const.} \tag{3}$$

The function F in question is a one-valued (uniform) function of position in phase space. The existence of the uniform integral confines the trajectory to a particular surface in phase space, or, if a small range of the constant in question is allowed, to the region enclosed between two such neighbouring surfaces, exactly as the energy integral confines the trajectory to a single surface or a small range of such surfaces.

The possible existence of such additional uniform integrals besides the energy uniform integral clearly shows the need for caution in our choice of the scope of the averaging or enunciation required by Assumption 2. When such uniform integrals exist, they must be explicitly recognised as part of the general conditions mentioned in the enunciation. We shall be required, when other form integrals exist, to average only over such regions of phase space, or in quantal language over such states of the assembly, as conform to their extra requirements. Such regions or states are called *accessible phase space or accessible states (or complexions)*, and great care must be taken that all the requirements of accessibilities are properly allowed for.

17.7. Accessible states for assemblies of similar systems. The symmetrical and the antisymmetrical group

We shall show here how the symmetry requirements restrict accessibility.

We start with an assembly containing a set of similar systems (e.g. electrons, hydrogen molecules, helium atoms . . .). Let us consider first an assembly containing one such system whose energy in Hamilton's form is $H(x, P_x)$. For such a system, Schrödinger's equation is

$$H\left(x, -\frac{ih}{2\pi}\frac{\partial}{\partial x}\right)\psi = \epsilon\psi. \tag{1}$$

We use ϵ *for the energy of a single system to distinguish it from the energy of the assembly which we have denoted by E.* Possible states of the system correspond to possible *solutions* of this equation, for which ψ satisfies the correct boundary conditions and is bounded and one-valued in the *x*-space *(configuration space)* of the system. For an enclosed system, the permissible values of ϵ can be shown to be discrete, though if the enclosure is a large one, the values of ϵ may lie so close together over parts of their range that they may often be treated by a limiting process as a *classical continuous distribution.* These *permissible values of the energy* ϵ *of a single system, that is, the eigenvalues of Eq.(1) will be supposed to be enumerated by* a subscript σ, that is, ϵ_σ, where σ takes the values 0, 1, 2, . . . , so that no two values of ϵ with different subscripts are equal. *Corresponding to any eigenvalue* ϵ_σ *of* ϵ *there exist one or more* eigenfunctions ψ_σ of Eq. (1). The *number of such eigenfunctions belonging to the eigenvalue* ϵ_σ *is denoted by* ω_σ and is called the *statistical weight or weight* of this state whose energy is ϵ_σ. The value that ω_σ may adopt for any given state σ will be a positive integer (zero is not included). Thus, for a non-degenerate $\omega_\sigma = 1$ while for a degenerate state $\omega_\sigma \neq 1$. If the system is degenerate $(\omega_\sigma \neq 1)$, we can assume that it is reduced to a non-degenerate$(\omega_\sigma = 1)$ one by including suitable conservative perturbing fields, if it is convenient to do so for purposes of discussion.

Let us suppose next that our assembly is built of two such similar systems 1 and 2 with very weak interactions so that to a first approximation the Hamiltonian of the pair is the sum of the two separate Hamiltonians. Then the complete equation of Schrődinger for the assembly is

$$\left[H\left(x_1, i\frac{h}{2\pi}\frac{\partial}{\partial x_1}\right) + H\left(x_2, i\frac{h}{2\pi}\frac{\partial}{\partial x_2}\right) - E\right]\Psi = 0. \tag{2}$$

It can be seen that the equation separates into two parts and that the permissible values of *E* are $\epsilon_\sigma + \epsilon_\tau$, and the corresponding solutions

$$\Psi_{\sigma\tau} = \psi_\sigma(x_1)\,\psi_\tau(x_2), \tag{3}$$

where ϵ_σ and ϵ_τ are eigenvalues of ϵ in Eq. (1), and ψ_σ , ψ_τ are the corresponding eigenfunctions. It is of *utmost importance* to observe that in the limit of vanishing interaction the pair of the systems is essentially *degenerate* except when $\sigma = \tau$ even if the single systems are *not degenerate*. For if $\sigma \neq \tau$, at least two eigenfunctions such as

$$\Psi_{\sigma,\tau} = \psi_\sigma(x_1)\,\psi_\tau(x_2),$$
$$\Psi_{\tau,\sigma} = \psi_\sigma(x_2)\,\psi_\tau(x_1),$$

obtained by permuting the individual systems, correspond to the same eigenvalue of the system $E = \epsilon_\sigma + \epsilon_\tau$. If the single systems are degenerate, i.e. $\omega_\sigma \neq 1$ and $\omega_\tau \neq 1$, then the total number of distinct eigenfunctions corresponding to $E = \epsilon_\sigma + \epsilon_\tau$ is

$$2\,\omega_\sigma\,\omega_\tau \quad (\sigma \neq \tau),$$

or

$$\omega_\sigma^2 \ (\sigma = \tau).$$

The argument is quite general. If the assembly consists of *N* weakly interacting systems, the complete equation of Schrödinger for the assembly is

$$\left[H\left(x_1, -i\frac{h}{2\pi}\frac{\partial}{\partial x_1}\right) + H\left(x_2, -i\frac{h}{2\pi}\frac{\partial}{\partial x_2}\right) + \ldots \right.$$

$$\left. + H\left(x_N, -i\frac{h}{2\pi}\frac{\partial}{\partial x_N}\right) - E\right]\Psi = 0. \quad (4)$$

To the eigenvalue of the energy *E*

$$E = \epsilon_\sigma + \epsilon_\tau + \ldots + \epsilon_\omega, \quad (5)$$

no pair of the subscripts

$$\sigma, \tau, \ldots, \omega$$

being equal, there corresponds a set of N! eigenfunctions obtained by permuting the systems 1, 2, . . . , *N* among the subscripts of the eigenvalues which are $\sigma, \tau, \ldots, \omega$. A simple eigenfunction is

$$\Psi_1 = \psi_\sigma(x_1)\,\psi_\tau(x_2)\ldots\psi_\omega(x_N). \quad (6)$$

For degenerate systems in the assembly, the total number of eigenfunctions corresponding to the eigenvalue given by Eq. (5) is

$$N!\,\omega_\sigma\,\omega_\tau\ldots\omega_\omega. \quad (7)$$

If finally we consider the completely general eigenvalue

$$E = n_\sigma\epsilon_\sigma + n_\tau\epsilon_\tau + \ldots + n_\omega\epsilon_\omega, \quad (8)$$

where the individual eigenvalues are equal in groups of $n_\sigma, n_\tau, \ldots, n_\omega$, so that

$$n_\sigma + n_\tau + \ldots + n_\omega = N, \quad (9)$$

then there are sets of

$$\frac{N!}{n_\sigma!\ n_\tau!\ \ldots n_\omega!} \quad (10)$$

distinct eigenfunctions obtainable by permutation of which

$$\Psi_1 = \psi_\sigma(x_1)\ldots\psi_\sigma(x_{n\sigma})\,\psi_\tau(x_{n\sigma+1})\ldots$$

$$\psi_\tau(x_{n\sigma+n\tau})\ldots\psi_\omega(x_N) \quad (11)$$

is typical. The total number of distinct eigenfunctions[+*] in the degenerate case is

$$\frac{N!\,\omega_\sigma^{n_\sigma}\,\omega_\tau^{n_\tau}\ldots\omega_\omega^{n_\omega}}{n_\sigma!\,n_\tau!\ldots n_\omega!}. \quad (12)$$

If to the assembly thus constructed we add another set of similar *systems* distinct from the set hitherto considered, we obtain a new set of eigenfunctions similar to Eq. (11), each of which can be combined by multiplication with each one of Eq. (11) to give an

---

* Any linear combinations of these eigenfunctions are equally eigenfunctions for the same value of *E*.

independent eigenfunction of the complete assembly. We may not of course permute a pair of distinguishable systems, for in is this case we do not obtain a solution of Schrödinger's equation for the assembly. Thus for an assembly of two distinguishable sets A and B of similar systems corresponding to the eigenvalue

$$E = \left(n_\sigma \epsilon_\sigma + n_\tau \epsilon_\tau + ... + n_\omega \epsilon_\omega\right),$$
$$+ \left(n'_\sigma \epsilon'_\sigma + n'_\tau \epsilon'_\tau + ... + n'_\omega \epsilon'_\omega\right), \qquad (13)$$

where the primed symbols refer to the B systems, the number of complexions is

$$\frac{N_A!\omega_\sigma^{n_\sigma}...\omega_\omega^{n_\omega}}{n_\sigma!\;...\;n_\omega!}\;\frac{N_B!\omega_{\sigma'}^{n'_{\sigma'}}...\omega_{\omega'}^{n'_{\omega'}}}{n'_{\sigma'}!\;...\;n'_{\omega'}!}. \qquad (14)$$

The Eqs. (12) and (14), and similar generalisations, must be used for enumerating the *accessible states* over which we have to average, provided that the introduction of weak interactions between systems of the assembly or between the assembly and the outside world allows the assembly to pass from a state described by one of these eigenfunctions to any other. But this may not be so. It has been shown that the eigenfunctions (Eq. 11) of a set of similar systems, after reorganisation into suitable linear combinations, *necessarily divide into a number of groups*, A, B, . . . , S, defined according to their *symmetry properties.* These groups contain between them all the eigenfunctions belonging to all the eigenvalues of energy, and they possess the extremely important property that *no interaction of whatever type or strength between the systems, or between the systems and the outside world, so long as it is symmetrical in the coordinates of the similar systems, can ever change the assembly from one eigenfunction of one group A to an eigenfunction of any other group B*. Thus if the assembly is originally represented by an eigenfunction of group A, it will for ever be confined to eigenfunctions of group A. Only these states are accessible.

From among the various non-combining groups two groups stand out, conspicuous for the simplicity of their properties and their mathematical form. One group is the group of eigenfunctions which are *symmetrical in all the systems*. This group we shall call simply *the symmetrical group* S. Any eigenfunction of the S group *remains unaltered* when we interchange the coordinates of any two of the component systems. The other group consists of eigenfunctions which are *antisymmetrical in all the systems*. This we shall call the *antisymmetrical group* A. Any eigenfunction of the A group *changes sign* when we interchange the coordinates of any two of the component systems. These groups are unique in that for non-degenerate systems they alone contain at most *one eigenfunction for any given eigenvalue of energy* $E = \sum_\sigma \epsilon_\sigma$, where σ takes the values 0, 1, 2, . . .

17.8. Symmetry type of eigenfunctions for actual assemblies

We know as yet no *a priori* reason why eigenfunctions of only one group, or of one group rather than another, *should be found in nature* for assemblies of particular systems. To determine the appropriate group, we must still appeal to observation, and the appropriate groups will vary from system to system. All material assemblies may, we believe, be correctly analysed into *assemblies of electrons, protons, and neutrons* at least for the ordinary purposes of thermodynamics. For these systems, it is certain that the *appropriate group in each case is the antisymmetrical.* For electrons, this follows from the fact the laws of interaction of electrons must embody *Pauli exclusion principle*, which is fundamental to the interpretation of spectra. According to this principle, we know that two electrons in any atom may never possess the same four quantum numbers, or as we should now say, *may never have the same eigenfunction (including spin).* The *antisymmetrical group A* is the *only group* of assembly eigenfunctions which possesses just this property, that it has no member whenever two systems have the same system eigenfunction. Since the eigenfunctions for the collection of electrons in *any atom* belong to the group A, one must suppose that this is due to the *nature of the electron*, and that the eigenfunctions for the electrons in every assembly belong to group A. For protons, the evidence is less extensive but equally conclusive, depending on the interpretation of the hydrogen band spectrum, and the theory of the rotational heat capacity of hydrogen at low temperatures. For neutrons the evidence is less direct but still adequate, being derived mainly from a study of the properties of the deuteron,* and the rotational heat capacity of deuterium.

Since all assemblies, as we have said, can be analysed into sets of electrons, protons, and neutrons, we therefore enumerate correctly all those distinct states whose eigenfunctions are *antisymmetrical* in the electrons, protons, and neutrons separately. *Each such state will be called a complexion.* So far as is known there are no further absolute restrictions to be imposed on *accessibility*, and in view of the successes of statistical mechanics, it is unlikely that any have been overlooked.

We have now surveyed the field of the general conditions that may or may not limit the accessibility of one quantum state (or region of phase space) from another. Thus this is a convenient moment to summarise the investigation by re-enunciating our adopted version of Assumption 2 in its final form, the most precise available.

Assumption 2. (Adopted version of the fundamental hypothesis.) Enumerate all the distinct accessible states or complexions of the given assembly, each characterised by one linearly independent eigenfunction, consistent with a specified energy or energy range, taking account if necessary of the time scale allowed for the establishment of equilibrium. The value $\overline{Q}$ *of any property* Q *of the assembly, which will be found to characterise the equilibrium state of the assembly subject to the specified conditions, may be obtained by averaging over all these accessible states attaching to each state an equal weight (unity). Formally*

$$\overline{Q} = \sum_{\text{acc. states}} Q \Big/ \sum_{\text{acc. states}} 1. \qquad (1)$$

17.9. Short cuts for enumerating accessible states for ordinary systems

The symmetry requirements which we have already laid down are in theory sufficient to enable us to enumerate correctly the accessible states for any assembly. But it is inconvenient, and gives little physical insight into the nature of an assembly, to regard it as a collection of so many electrons, protons, and neutrons when it is actually an assembly, say, of helium atoms or hydrogen molecules, themselves systems of almost *perfect permanence*. It is wiser to use these units themselves as quasi-*fundamental systems* with secondary derived symmetry rules.

Suppose such a complex system contains l electrons, p protons, and n neutrons. Then, since interchange of a pair of complex systems means interchange of $(l + p + n)$ pairs of electrons, protons, and neutrons, in each of which pairs the eigenfunction of the *assembly* is

* The deuteron is the nucleus of a deuterium atom.

antisymmetrical, then *the eigenfunction of the assembly will be symmetrical in the complex systems if* $(l + p + n)$ *is even, and antisymmetrical if* $(l + p + n)$ *is odd.* This necessary derived symmetry rule has also been proved rigorously to be sufficient* so that we can appeal to the following:

Theorem. It is sufficient in enumerating complexions (accessible states), or rather the eigenfunctions of the assembly representing them, to construct formally and so to enumerate all these linearly independent eigenfunctions which have the correct symmetry properties in the 'permanent' systems (regarded as wholes) of which the assembly is composed. The eigenfunctions of any 'permanent' system must have the correct symmetry properties in the electrons, protons, and neutrons of which that system is composed, but the direct analysis of the complete assembly into electrons, protons, and neutrons may be omitted, for the number of complexions is not thereby affected.

In conformity with this principle, it is possible in most statistical problems to treat atomic nuclei as *permanent complexes*. No matter what the correct analysis of nuclei more fundamental particles may be, we may always proceed by satisfying the symmetry requirements of the assembly merely in all sets of equivalent nuclei and in the extranuclear electrons. The symmetry rule for nuclei which holds universally, so far as is known at present, is that *the eigenfunctions must be antisymmetrical in all similar nuclei when the nuclei have an odd mass number and symmetrical when the mass number is even.* The mass number is equal to the number of protons and neutrons in the nucleus.

We have so far formulated the theorem for the analysis of 'permanent' complex systems into electrons, protons, and neutrons, but there is no need so to restrict it. It applies equally to the analysis of 'permanent ' complex systems into any 'permanent' *secondary systems*. For example, in an assembly of chlorine atoms $^{35}\mathrm{Cl}$, the assembly eigenfunctions must be *antisymmetrical* in the $^{35}\mathrm{Cl}$ nuclei. But the assembly will consist entirely of 'permanent' *molecules* $^{35}\mathrm{Cl}_2$. Applying the principle, therefore, we see that the symmetry requirements of the assembly will be satisfied and that the enumeration of eigenfunctions will be correct if we make the eigenfunction of each molecule *antisymmetrical* in its *nuclei* and then make the assembly eignfunction *symmetrical* in the *molecules*.

The same principle may be extended to an assembly consisting of a crystal in equilibrium with its vapour. The crystal is merely a *supermolecule*, a *system* of which only one is present. We satisfy the symmetry requirements of the assembly by giving its eigenfunctions the *correct symmetry in the free molecules in the vapour phase ignoring those in the crystal*, provided that we correctly enumerate the states of the crystal itself regarded as a 'permanent' system composed of that number of molecules which it happens to contain in any particular example.

17.10. The enumeration of complexions for localised systems

The practical considerations of the preceding section are of value chiefly for *gaseous assemblies*, or at least for assemblies or parts of assemblies in which the *similar systems* have to be treated as on a precisely equal footing, using eigenfunctions from a common store. Examples are *atoms or molecules in the same enclosure in a gas*, or *electrons in a single atom or in a single piece of metal*. The primary rules for electrons, protons, or

* Ehrenfest and Oppenheimer, *Phys. Rev.* **37**, 333 (1931).

neutrons apply of course to all matter and all assemblies without exception, and in the above cases, it reduce to the practical secondary rules. When, however, we need not regard the eigenfunctions to be used are all drawn from a common store, but may discriminate between them, the secondary rules can be still further simplified and limited.

These further limitations are perhaps be approached best by means of the following example. Let us consider an assembly which contains two different enclosures each containing electrons. We may suppose that the electrons can pass from one enclosure to the other, though only with some difficulty. In equilibrium, the electrons will be partitioned between the two enclosures in a proportion which is part of our business to determine. But the enclosures may be supposed to be sufficiently distinct for the description of electronic states as belonging to either one enclosure or the other to be a good approximation. Let the states of the electrons belonging to one enclosure have the eigenfunctions ψ_r and the corresponding energies $\in_r$, and those belonging to the other enclosure ψ_r' and $\in_r'$. Let us suppose that the assembly contains five electrons, and consider the partition in which there are *three* in the first enclosure occupying the states represented by ψ_1, ψ_2, ψ_3, respectively, and two in the second enclosure occupying the states represented by ψ_1', ψ_2'. There is of course just *one accessible state of the assembly* with this specification, with an eigenfunction antisymmetrical in all the electrons, namely that given by the determinant below, in which the subscripts *a*, *b*, *c*, *d*, and *e* of the coordinate *x* specify the electrons.

$$\begin{vmatrix} \psi_1(x_a) & \psi_1(x_b) & \psi_1(x_c) & \psi_1(x_d) & \psi_1(x_e) \\ \psi_2(x_a) & \psi_2(x_b) & \psi_2(x_c) & \psi_2(x_d) & \psi_2(x_e) \\ \psi_3(x_a) & \psi_3(x_b) & \psi_3(x_c) & \psi_3(x_d) & \psi_3(x_e) \\ \psi_1'(x_a) & \psi_1'(x_b) & \psi_1'(x_c) & \psi_1'(x_d) & \psi_1'(x_e) \\ \psi_2'(x_a) & \psi_2'(x_b) & \psi_2'(x_c) & \psi_2'(x_d) & \psi_2'(x_e) \end{vmatrix}, \tag{1}$$

Suppose, however, instead of introducing this perfect antisymmetry we consider that the electrons *a*, *b*, and *c* really belong to the first enclosure and electrons *d* and *e* to the other. If then we construct an eigenfunction for this limitation, in which the factor for the electrons in each enclosure is separately *antisymmetrical* for the electrons of that enclosure, we again get just one eigenfunction namely,

$$\begin{vmatrix} \psi_1(x_a) & \psi_1(x_b) & \psi_1(x_c) \\ \psi_2(x_a) & \psi_2(x_b) & \psi_2(x_c) \\ \psi_3(x_a) & \psi_3(x_b) & \psi_3(x_c) \end{vmatrix} \times \begin{vmatrix} \psi_1'(x_d) & \psi_1'(x_e) \\ \psi_2'(x_d) & \psi_2'(x_e) \end{vmatrix}, \tag{2}$$

and one, as we have seen, is the correct number. In general, if the *assembly* has any number of parts which may be regarded as distinct, each containing a definite number of similar *systems*, then we obtain a correct enumeration of the accessible states of the assembly by making its eigenfunctions show the correct symmetry properties for the exchanges of similar systems in each distinct part of the assembly. It is unnecessary in this enumeration to allow for any exchange of systems between the distinct parts or enclosures.

In certain assemblies, the number of distinct parts or enclosures may be legitimately increased so far that we reach a limit in which every system belongs to its own distinct enclosure or *location*; the *atoms in a crystal lattice provide an example*. Each distinct part of the assembly now has its own *one system*, and if we apply the foregoing rule, we see that we shall enumerate correctly the accessible states of the assembly, *if we keep each system fixed in its own location*, but impose no other symmetry requirement. Such an assembly or part of an assembly is called *an assembly of localised systems*, and if states are enumerated in the way described, no further symmetry restrictions are required.

One important consequence of the lack of symmetry requirements for such assemblies in that the *quantal* enumeration of accessible states reduces to identity with the *classical* enumeration which has long been familiar. Suppose that the *localised systems* of such an assembly have states of energies $\in_1, \in_2, \ldots,$ which are non-degenerate, each possessing *one eigenfunction*. The distribution laws of such an assembly should tell us how many of the systems (N in all) will be found on the average in each state. For this study, we shall find in Chapter 18 that we require to know how many *accessible states* of the assembly correspond to a state of affairs in which n_1 systems have an energy $\in_1$, n_2 an energy $\in_2$ and so on. This number is plainly the number of ways in which the N systems can be parcelled out into groups containing $n_1, n_2, \ldots, n_r$ members, $n_1 + n_2 + \ldots + n_r$ being equal to N. This number is known to be

$$\frac{N!}{n_1!\, n_2! \ldots n_r!}, \tag{3}$$

which is therefore the *required number of accessible states with the given energy distribution.*

The physical difference between an assembly of *localised systems* which requires the classical number given by Eq. (3) of *accessible states* for all the values of the *n*'s, and *an assembly of similar systems in a single enclosure* which takes the quite different number restricted by symmetry requirements, may be illustrated as follows:

In an enclosure containing *similar systems*, there is *no physical sense* in saying that one system has such and such an energy *here* and another one *there*. *Here* and *there* (within the enclosure) have ceased to have any meaning in terms of the *eigenfunctions which describe the states*. But in assemblies of localised systems, *here* and *there* still *have a physical meaning*. The assembly with a system in its *r*th state *here* is physically distinct from another system in its *r*th state *there* and therefore counts as in a distinct state.

CHAPTER 18

The General Theorems for Assemblies of Permanent Systems

18.1. Introduction

The assemblies of permanent systems (*absence of chemical reactions and change of phase*), which are in the highest possible degree independent of one another, are the most amenable to exact treatment. The most natural application is to perfect gases, but the treatment can be extended to crystals.

The highest degree of independence is attained when it is sufficiently accurate to assume throughout the calculations that the energy of the assembly is the sum of the energies of the individual systems and contains no part depending on the coordinates of more than one such system. On this assumption, universally if sometimes tacitly made, some comment is needed. Such an assembly is an *ideal limit* to which an actual assembly may approximate but can never attain. This is because it is essential to the whole idea of an assembly that it should form a *connected dynamical system with a single energy integral, not a number of separate ones.* If, indeed, the energy were really entirely independent of such cross terms, which represent the interactions of the systems, the systems would never interact and the assembly would not be connected. We have therefore to assume that *some such interactions do occur*, but, in this limiting case, so rarely that their contribution to the total energy of the assembly may be *neglected.* They still suffice to preserve connection and ensure that only a *single energy integral exists.* This is an example of the general assertion underlying the whole theory that, while there must exist mechanisms of interaction, their mere existence is sufficient, their nature being *irrelevant* to the laws of equilibrium.

18.2. The assignments of weights

We have stated in Chapter 1 that in averaging to determine the equilibrium state, we shall attach a *weight unity* to every distinct accessible state *(complexion)* of the assembly, *each complexion being defined by an eigenfunction linearly independent of all others so used.* This means that the average value $\overline{Q}$ of any quantity Q is to be calculated by the equation.

$$\overline{Q} = \sum_{\sigma} Q_{\sigma}/C, \tag{1}$$

where Q_{σ} denotes the value of Q for the *particular complexion* σ, the summation $\sum_{\sigma}$ is over *all complexions*, and C denotes *the total number of complexions* so that formally

$$C = \sum_{\sigma} 1. \tag{2}$$

Eqs. (1) and (2) can be rearranged. In the first place, a number Ω_σ of different complexions may for many purposes *possess indistinguishable properties*; for example, the same value of *Q*. We can then regroup the summations of Eqs. (1) and (2) so that they read

$$\overline{Q} = \sum_\sigma \Omega_\sigma Q_\sigma / C, \tag{3}$$

$$C = \sum_\sigma \Omega_\sigma, \tag{4}$$

where the summations are now over the *groups of complexions*. It will often be legitimate for conciseness of expression to speak of each of these groups of complexions, between whose properties we cannot or do not care to distinguish, as a *single state of the assembly of weight* Ω_σ. From Eqs. (3) and (4), we have

$$\overline{Q} = \frac{\sum_\sigma \Omega_\sigma Q_\sigma}{\sum_\sigma \Omega_\sigma}.$$

We next recall that our assemblies are always to be regarded as *collections of practically independent systems, between which, to an approximation, there are no interactions*. It may be shown to follow that *Schrödinger's equation, which determines the eigenfunctions and the energies of the states of the assembly*, immediately separates into equations for the distinct *systems*. The eigenfunction Ψ for the assembly can then be constructed out of products of the eigenfunctions ψ's for these systems. Now to *any one* value of the energy $\in_r$ of an *individual system*, there may belong a number ω_r of *distinct eigenfunctions for the system, between which we need not distinguish*. We speak of ω_r as the *statistical weight or the weight of the system* in the given state. It is convenient to use the term *degenerate* to describe the states for which $\omega_r > 1$, and *non-degenerate* for $\omega_r = 1$. If the assembly consists of *N* such systems, and *in* a *particular state* σ of the assembly, there are n_0, n_1, n_2, ..., n_r, . . . specified systems in states of statistical weight ω_0, ω_1, ω_2, ... ω_r,..., and *if no special limitations of accessibility arise*, then the statistical weight Ω_σ of this group of complexions is given by the equation

$$\Omega_\sigma = \Pi_r \, \omega_r^{n_r}. \tag{5}$$

The number of complexions Ω_σ given in Eq. (5) refers to the complexions provided by *groups of distinct specified systems*, when a *given set* of *systems*, n_0 in number, are in a state of weight ω_0 and so on. The number of complexions is much greater than if any n_0 out of a total *N* systems may be chosen for this state and so on for the groups of complexions in *other states*. The number Ω_σ is then increased by a factor equal to the *number of distinct ways in which groups of* n_0, n_1,... , n_r,... *systems can be chosen from a total of N*. This factor is

$$\frac{N!}{n_0!\, n_1! \ldots n_r! \ldots}, \tag{6}$$

and now

$$\Omega_\sigma = \frac{N!\ \omega_0^{n_0}\ \omega_1^{n_1} \ldots \omega_r^{n_r} \ldots}{n_0!\ n_1! \ldots n_r! \ldots}. \qquad (7)$$

We may again mention that these formulae hold only if there are no special interactions of accessibility.

18.3. Enumeration of accessible states (complexions)

It is necessary next to consider in detail how to enumerate complexions in terms of the states of the component systems of an assembly. It is here and only here that an important divergence is possible between the classical and the quantal enumerations. The number of accessible states (complexions) of the assembly will be different according to whether the assembly is composed of *localised* and *non-localised* systems. We shall now consider a simple example of each type of assembly in which the states of the systems are all *non-degenerate*.

Type 1. Consider first an assembly containing a set of N similar linear oscillators, not necessarily harmonic, which have *fixed positions* in the assembly. They may for example be thought of as *electrons* suitably bound to the atoms of a solid. *Each oscillator* has a similar discrete set of stationary states which may be specified by a quantum number r, having the possible values 0, 1, 2, . . . The eigenfunction of electron a bound to an *atom* α in its rth quantum state may be written *as*

$$\psi_r^\alpha\ (x_a),$$

where x_a is the coordinate specifying the position of electron a. When $n_0,\ n_1,\ \ldots,\ n_t$ are the *numbers of oscillators* in states of quantum numbers 0, 1, . . . , t, we have a *group of states of the assembly*, which may be regarded as a *single state* of weight equal to the number of distinct states in the group. As we saw in Eq. (6) in Section 18.2, we can now construct

$$\frac{N!}{n_0!\, n_1! \ldots n_t!} \qquad (1)$$

distinct eigenfunctions of the type

$$\psi_0^\alpha\ (x_a) \ldots \psi_0^\kappa (x_\kappa) \ldots\ \psi_t^\upsilon\ (x_u), \qquad (2)$$

belonging to the complexion being considered and containing n_0 factors ψ_0 , n_1 factors ψ_1 and so on; the total number of factors is N. Here we have treated the electrons as *permanently* attached to their own atoms.

Type II. Assembly of non-localised systems. Consider next an assembly containing a set of N electrons (or other systems all similar) *free to move* about in the *same* enclosure of given volume. The eigenfunctions now lose their distinctive superscripts

$$\alpha,\ \ldots\ ,\ \kappa,\ \ldots.$$

When now n_0 , n_1 , ... , n_t , are again the numbers of electrons or other systems in states of given quantum number r, we can still make up unsymmetrical eigenfunctions of type given by Eq. (2) to the number given by Eq. (1). But now when we attempt to construct an eigenfunction for the assembly *antisymmetrical* in all the electrons, we can construct only one, if every n is equal to 0 or 1.

We now have the data required for calculating the average properties of an assembly of either type. This we now proceed to do, commencing with an assembly of *localised systems.*

18.4. An assembly of two sets of localised linear oscillators

Let us suppose that an assembly consists of *two large sets of localised linear oscillators A and B* of total numbers N_A and N_B. Each oscillator A has a series of stationary states of weight unity, in which the energy takes the values

$$\epsilon_0, \epsilon_1, \ldots, \epsilon_r, \ldots,$$

and each oscillator B similarly has a series of stationary states of weight unity, with energy values

$$\eta_0, \eta_1 \ldots \eta_s, \ldots.$$

We may emphasise once again that in assigning individual stationary states and energies to the systems separately, we assume that they are practically independent systems, each pursuing its own motion undisturbed for the greater part of time. This is essential to the energy specification and therefore essential to the treatment of assemblies composed of large numbers of practically independent systems. At the same time, we must assume that exchanges of energy between the oscillators are possible and do occasionally take place, otherwise the systems will not form a *connected* assembly and, obviously, cannot possess unique *equilibrium distribution laws*. In the present case, we may think of the exchanges of energy as effected by a few free atoms in an enclosure containing the oscillators so few in number compared with the oscillators that we may ignore their energy altogether.

It is our objective to determine the *distribution laws* of this assembly, that is, the *equilibrium* or *average distribution* of the oscillators among the various states which they can occupy. A specification of this distribution – equilibrium or not – may be referred to as a specification of the *statistical state of the assembly*. This conveys correctly the idea that it is only the *macroscopic state* of the assembly that really interests us, *not the microscopic state*.

An accessible statistical state of the assembly can be specified by choosing any set of *positive integers (zero included)* n_r, n'_s which satisfy the conditions

$$\sum_r n_r = N_A, \tag{1}$$

$$\sum_s n'_s = N_B, \tag{2}$$

$$\sum_r n_r \epsilon_r + \sum_s n'_s \eta_s = E, \tag{3}$$

where E is the energy of the assembly. For quantal assemblies, there is no temptation to use anything but an exact value of E in Eq. (3), but the whole of our preliminary discussion in Chapter 1 referred to a small given energy range E, $E + \mathrm{d}E$ rather than an exact value. At the moment, we must admit that logically Eq. (3) should be replaced by

$$E \leq \sum_r n_r \epsilon_r + \sum_s n'_s \eta_s \leq E + \mathrm{d}E, \tag{3)$'$}$$

where E, $E + \mathrm{d}E$ define the range in which the energy of the assembly may be assumed to lie. For simplicity, we first consider an exact value of E using Eq. (3) rather than Eq. (3)′.

For given N_A, N_B and E any set of positive or zero integers n_r, n'_s, chosen in this way, describes a possible statistical state of the assembly. The energies of the A systems may be referred to an arbitrary zero and likewise the energies of the B systems, these two arbitrary zeros of ϵ_r and η_s determine the *zero* that is to be used for *E*. The conditions in Eqs. (1)–(3) express the facts that the total number of A systems is N_A, that the total number of B systems is N_B, and that the total energy is *E*. Since the assembly is of *Type I* and the system weights are all unity, the number of complexions corresponding to this statistical state is

$$\frac{N_A!}{n_0!\, n_1! \dots n_r! \dots} \; \frac{N_B!}{n'_0!\, n'_1! \dots n'_s! \dots}, \tag{4}$$

by obvious extension of Eq. (6) in Section 18.2. The total number *C* of complexions is given by

$$C = \sum_n \sum_{n'} \frac{N_A!}{n_0!\, n_1! \dots n_r! \dots} \; \frac{N_B!}{n'_0!\, n'_1! \dots n'_s! \dots}, \tag{5}$$

summed for all n, n' subject to Eqs. (1)–(3).

The equilibrium distribution laws for the assembly are obtained by *averaging overall complexions*. We can therefore find at once an expression for the average values $\bar{n}_t$ of n_t, or of any similar quantity, for we have, as in Eq. (3), Section 18.2,

$$C\bar{n}_t = \sum_n \sum_{n'} \frac{n_t\, N_A!}{n_0!\, n_1! \dots n_r! \dots} \; \frac{N_B!}{n'_0!\, n'_1! \dots n'_s! \dots}. \tag{6}$$

One of the most important average quantities is $\bar{E}_A$, the average energy of the systems A. This is given by

$$C\bar{E}_A = \sum_n \sum_{n'} \frac{(\sum_r n_r \epsilon_r)\, N_A!}{n_0!\, n_1! \dots n_r! \dots} \; \frac{N_B!}{n'_0!\, n'_1! \dots n'_s! \dots}. \tag{7}$$

All such summations are of course over the values of n, n' satisfying Eqs. (1)–(3). It can be shown by mathematical operations that the Eqs. (5)–(7) are all *coefficients in certain rather simple power series.**

We commence by constructing the function $f_A(z)$ defined by

$$f_A(z) = z^{\epsilon_0} + z^{\epsilon_1} + z^{\epsilon_2} + \dots = \sum_r z^{\epsilon_r}, \tag{8}$$

called the *partition function* for the systems A, and we consider the product expressed as

$$\{f_A(z)\}^{N_A} = \left\{\sum_r z^{\epsilon_r}\right\}^{N_A}. \tag{9}$$

If this is expanded by the *multinomial theorem* in powers of *z* (which is a simple generalisation of the familiar *binomial theorem*), the general term is

$$\frac{N_A!}{n_0!\, n_1! \dots n_r! \dots} z^{\sum_r n_r \epsilon_r} \qquad \left(\sum_r n_r = N_A\right). \tag{10}$$

Similarly, if we construct $f_B(z)$ defined by

* For details, see R. Fowler and E. A. Guggenheim, *Statistical Thermodynamics*, Chapter 2, Section 206, Cambridge (1965).

$$f_B(z) = z^{\eta_0} + z^{\eta_1} + z^{\eta_2} + \ldots = \Sigma_s z^{\eta_s}, \quad (11)$$

and expand $\{f_B(z)\}^{N_B}$ in powers of *z*, the general term is

$$\frac{N_B!}{n'_0!\, n'_1! \ldots n'_s! \ldots} z^{\Sigma_s n'_s \eta_s} \qquad (\Sigma_s n'_s = N_B). \quad (12)$$

By multiplying these series together, it follows that the *coefficient of* z^E *in the expansion of*

$$\{f_A(z)\}^{N_A} \{f_B(z)\}^{N_B} \quad (13)$$

is

$$\Sigma_n \Sigma_{n'} \frac{N_A!}{n_0!\, n_1! \ldots n_r! \ldots} \frac{N_B!}{n'_0!\, n'_1! \ldots n'_s! \ldots}, \quad (14)$$

summed for all positive (or zero) integral values of n, n' satisfying just the Eqs. (1)–(3). Thus *C*, given by Eq. (5), is the *coefficient* of z^E in the expansion of Eq. (13) in powers of *z*.

A similar expression for $C\overline{E}_A$ can be similarly determined as follows:
We have just seen from Eqs. (9) and (10) that

$$\{f_A(z)\}^{N_A} = \Sigma_n \frac{N_A!}{n_0!\, n_1! \ldots n_r! \ldots} z^{\Sigma_r n_r \epsilon_r}. \quad (15)$$

If we differentiate both sides of this equation with respect to *z* and multiply by *z*, we obtain

$$z \frac{\partial}{\partial z} \{f_A(z)\}^{N_A} = \Sigma_n \frac{N_A!(\Sigma_r n_r \epsilon_r)}{n_0!\, n_1! \ldots n_r! \ldots} z^{\Sigma_r n_r \epsilon_r}. \quad (16)$$

On multiplying both sides of Eq. (16) by $\{f_B(z)\}^{N_B}$,
which is given by

$$\{f_B(z)\}^{N_B} = \Sigma_{n'} \frac{N_B!}{n'_0!\, n'_1! \ldots n'_s! \ldots} \times z^{\Sigma_s n'_s \eta_s},$$

we find

$$\left\{ z \frac{\partial}{\partial z} \left\{ f_A(z)^{N_A} \right\} \right\} \{f_B(z)\}^{N_B} =$$
$$\Sigma_n \Sigma_{n'} \frac{N_A!(\Sigma_n n_r \epsilon_r)}{n_0!\, n_1! \ldots n_r! \ldots} \frac{N_B!}{n'_0!\, n'_1! \ldots n'_s! \ldots} \times z^{\Sigma_r n_r \epsilon_r + \Sigma_s n'_s \eta_s}. \quad (17)$$

These summations are taken over all positive integral values of n and n' satisfying Eqs. (1) and (2). If therefore we select from Eq. (17) the coefficient of z^E, we can see by referring to Eq. (7) that we obtain $C\overline{E}_A$. Thus $C\overline{E}_A$ *is the coefficient of* z^E *in the expansion of*

$$\left[z \frac{\partial}{\partial z} \left\{ f_A(z)^{N_A} \right\} \right] \{f_B(z)\}^{N_B}. \quad (18)$$

It thus appears that the expressions for such quantities as C, $C\overline{E}_A$ are the coefficients in certain power series. A rapid and powerful method of evaluating these is provided by expressing them as *contour integrals* and evaluating the integrals by the method of *steepest descents*. The reader interested in the mathematics will find these given in detail elsewhere.* The more general reader, for whom the present book is intended, will probably be more interested in the results; these are given in Section 18.5.

18.5. Average values and temperature on the statistical scale

We shall slightly generalise the statement of the results obtained in Section 18.4 by removing the trivial restriction that there are only two types of systems A and B.

We consider an assembly consisting of N_A systems of A, N_B systems of B, N_C systems of C, and so on, where N_A, N_B, N_C, ... are large numbers. If the systems A can exist in states 0, 1, 2, . . . , all of unit weight and energies $\epsilon_0, \epsilon_1, \epsilon_2, \ldots$, then the average number $\overline{n}_r$ of systems A in the state r is given by

$$\overline{n}_r = N_A\,\theta^{\epsilon_r}/\Sigma_r\,\theta^{\epsilon_r} = N_A\,\theta^{\epsilon_r}/f_A(\theta), \tag{1}$$

in terms of a *certain parameter* θ, of which the significance will be made clear shortly. The average number $\overline{n'_s}$ of systems B in the state s of energy η_s is similarly

$$\overline{n'_s} = N_B\,\theta^{\eta_s}/\Sigma_s\,\theta^{\eta_s} = N_B\,\theta^{\eta_s}/f_B(\theta). \tag{2}$$

It is evident from Eqs. (1) and (2) that

$$f_A(\theta) = \sum_r \theta^{\epsilon_r},$$

and

$$f_B(\theta) = \sum_s \theta^{\epsilon_s}.$$

There are similar formulae for the systems of other types. According to Eqs. (1) and (2), the essential formulae

$$\Sigma_r\,\overline{n}_r = N_A, \tag{3}$$

$$\Sigma_s\,\overline{n'_s} = N_B, \tag{4}$$

are satisfied automatically.

The average distribution of energy among the various types of systems is the following. The average value of the energy of all the systems A is given† by

$$\overline{E}_A = N_A\,\theta\,\frac{\partial \ln f_A(\theta)}{\partial \theta} = \frac{N_A\,\Sigma_r \epsilon_r\,\theta^{\epsilon_r}}{f_A(\theta)}, \tag{5}$$

which satisfies the essential equation

$$\Sigma_r\,\overline{n}_r\,\epsilon_r = \overline{E}_A. \tag{6}$$

---

* Whittaker and Watson, *Modern Analysis*, Chaps. V–VI (4 ed.) (Cambridge, 1935).

† R. Fowler and E. A. Guggenheim, *Statistical Thermodynamics*, Chapter 2, Sections 204 and 208 (Cambridge, 1965).

Eq. (5) imply that

$$\frac{\partial f(\theta)}{\partial \theta} = \frac{\sum_r \epsilon_r \theta^{\epsilon_r}}{\theta}.$$

The formulae for $\overline{E}_B$, $\overline{E}_C$, and so on are similar. Finally, we have the necessary equality

$$\overline{E}_A + \overline{E}_B + \overline{E}_C + \ldots = \overline{E}. \quad (7)$$

If *E* is given, this fixes θ. Alternatively, if θ is given, this fixes *E*.

It remains to discuss the significance of the parameter θ which, while mathematical in origin, is obviously fundamental in describing the state of the assembly, and should be identifiable with some *physical property* of the assembly. The striking feature of θ is that the average properties of the systems A, those of B, those of C and so on are all determined by *one and the same* θ. Thus, θ is *a parameter helping to define the state of our assembly which must have the same value for all sets of systems in the same assembly.* This is precisely the property which distinguishes the temperature from the other parameters, and we are therefore *forced to the conclusion that* θ *plays the role of temperature.* To be sure, it is not the temperature measured on the usual scale, and we shall therefore call θ the *statistical temperature*, or *the temperature on the statistical scale.* We shall shortly derive the relation between this temperature scale θ and the usual thermodynamic scale *T.* We may *anticipate this derivation* by stating that the relation is

$$\theta = e^{-1/kT}, \quad (8)$$

where *T* is the temperature on the absolute (*Kelvin*) scale, and *k* is universal constant, namely Boltzmann's constant.

18.6. Systems of several degrees of freedom and degenerate systems

Up to the present, we have discussed the average properties of an assembly of *localised linear oscillators*, but we have made no specific assumptions about the oscillators except that they have a discrete series of states each specified by a quantum number, and that *no two of these states have the same energy value.* The formulae obtained will therefore apply to any kind of system with only one degree of freedom. Actually, the most important systems with only one degree of freedom are *linear oscillators.* But although we assumed that the states of our systems were specified by only one quantum number, we never made any use of this assumption. Consequently, our formulae remain true for systems with *p* degrees of freedom if by the state *r* we mean the state defined by the quantum numbers r_1, r_2, ..., r_p and of energy values $\epsilon_{r_1, r_2, \ldots r_p}$, provided no two of these energy values are equal.

Our next step is to *remove* the restriction that no two energy levels are equal and so to extend our formulae to degenerate systems. Suppose that *there are* ω_r *states of equal energy* ϵ_r. We group these states together and call the group a ω_r-fold *degenerate state or a state of weight* ω_r. Now by a suitable small alteration of the conditions to which the systems are each subject, such as the application of external fields, technically known as a *perturbation*, it will generally be possible to separate these ω_r *energy levels* so that the system becomes non-degenerate. We can then apply all our formulae, the partition functions $\sum_r \theta^{\in_r}$ containing *one term* corresponding to *each* of the ω_r energy levels into which the degenerate level has been split. If now we make the perturbation tend to zero, we obtain again our original system with the ω_r-fold degenerate level $\in_r$. The partition function $\sum_r \theta^{\in_r}$ now contains ω_r identical terms, and we therefore write it in the alternative form $\sum_r \omega_r\ \theta^{\in_r}$, the summation now including each degenerate state or each energy level once only.

We shall now restate our results in the form applicable to degenerate as well as non-degenerate systems. Suppose the assembly to consist of N_A systems A, N_B systems B and so on. Suppose the *systems* A have a sequence of stationary states with energies.

$$\in_0, \in_1, \ldots, \in_r, \ldots,$$

and weights

$$\omega_0, \omega_1, \ldots, \omega_r, \ldots.$$

We construct the partition function

$$f(z) = \omega_0 z^{\in_0} + \omega_1 z^{\in_1} + \ldots + \omega_r z^{\in_r} + \ldots. \tag{1}$$

The average number $\bar{n}_r$ of systems A in the *degenerate state of energy* $\in_r$ is then by extension of Eq. (1) in Section 18.5

$$\bar{n}_r = N_A\, \omega_r\, \theta^{\in_r}/f(\theta). \tag{2}$$

The energy associated with this particular group of systems A is therefore given by

$$N_A\, \omega_r\, \in_r\, \theta^{\in_r}/f(\theta), \tag{3}$$

where $f(\theta)$ in Eqs. (2) or (3) is

$$f(\theta) = \sum_r \omega_r\, \theta^{\in_r}.$$

The average energy $\bar{E}_A$ of *all the systems A* is given, by summing Eq. (3) *over all states r*, in the form*

$$\bar{E}_A = N_A \sum_r \omega_r \in_r \theta^{\in_r}/f(\theta) = N_A\, \theta \frac{\partial \ln f(\theta)}{\partial \theta}. \tag{4}$$

Exactly analogous formulae apply to the B and other types of systems, the same θ being used throughout. Finally, we have the necessary equality

$$\bar{E}_A + \bar{E}_B + \ldots = E. \tag{5}$$

If *E* is given, this fixes θ the temperature on the statistical scale. Alternatively, if the temperature θ is given, then the formulae fixes *E*.

The partition function $f(\theta)$ is presumed to refer to the *localised* system. It should be observed that, in the important special case in which the motion splits up into two or more

---

Note that $\frac{\partial \ln f(\theta)}{\partial \theta} = \frac{1}{f(\theta)} \frac{\partial f(\theta)}{\partial \theta} = \frac{1}{f(\theta)} \frac{\sum \in_r \theta^{\in_r}}{\theta}$.

$\frac{\partial \ln f(\theta)}{\partial \ln \theta} = \theta \frac{\partial \ln f(\theta)}{\partial \theta} = \theta \frac{1}{f(\theta)} \frac{\partial f(\theta)}{\partial \theta} = \frac{1}{f(\theta)} \sum \in_r \theta^{\in_r}$, and

* $\frac{\partial f_A(\theta)}{\partial \theta} = \frac{\sum \in_r \theta^{\in_r}}{\theta}$

parts entirely independent of one another, the partition function $f(\theta)$ must *factorise* into functions of the same type, which refer separately to the independent motions. A particular case of this factorisation occurs *approximately for the rotations and vibrations of a diatomic molecule. It occurs accurately for the separation of the translatory motion from the internal motions and rotations of a free molecule*, but in general free molecules are members of a gaseous phase and their complexions require an enumeration of type II.

18.7. Linear harmonic oscillators

It is natural at this stage to consider few examples of special systems and construct their partition functions.

We have already described before* the linear harmonic oscillator; its eigenfunctions ψ_v's are characterised by the single quantum number v which can take the values

$$v = 0, 1, 2, \dots, \tag{1}$$

and the corresponding energy values are

$$\epsilon_v = \left(v + \frac{1}{2}\right) h\nu, \tag{2}$$

relative to an energy zero of rest at the equilibrium position. The partition function, denoted here by $f(\theta)$, is therefore

$$f(\theta) = \Sigma_r \theta^{\epsilon_r} = \Sigma_v \theta^{\left(v+\frac{1}{2}\right)h\nu}$$

$$= \theta^{\frac{1}{2}h\nu}\left(1 - \theta^{h\nu}\right)^{-1}. \tag{3}$$

This equation can be obtained as follows:

$$f(\theta) = \Sigma_v \theta^{\left(v+\frac{1}{2}\right)h\nu}$$

$$= \theta^{\frac{1}{2}h\nu} + \theta^{\frac{3}{2}h\nu} + \theta^{\frac{5}{2}h\nu} + \dots$$

$$= \theta^{\frac{1}{2}h\nu}\left(\theta^0 + \theta^{h\nu} + \theta^{2h\nu} + \dots\right)$$

$$= \theta^{\frac{1}{2}h\nu}\left(1 + \theta^{h\nu} + \theta^{2h\nu} + \dots\right)$$

$$= \theta^{\frac{1}{2}h\nu} \frac{1}{1 - \theta^{h\nu}}$$

$$= \theta^{\frac{1}{2}h\nu}\left(1 - \theta^{h\nu}\right)^{-1}. \tag{3}$$

---

* See Section 8.1.

Hence, in terms of the statistical temperature θ, the average energy of N_A linear harmonic oscillators will be, according to Eq. (4) in Section 18.6, namely,

$$\begin{aligned}\overline{E}_A &= N_A\theta\,\frac{\partial \ln f(\theta)}{\partial\theta}\\ &= N_A\,\theta\,\frac{\partial}{\partial\theta}\ln\left\{\theta^{\frac{1}{2}h\nu}\left(1-\theta^{h\nu}\right)^{-1}\right\}\\ &= N_A\,\theta\,\frac{\partial}{\partial\theta}\left\{\frac{1}{2}h\nu\ln\theta - \ln\left(1-\theta^{h\nu}\right)\right\}\\ &= N_A\left\{\frac{1}{2}h\nu + \frac{h\nu}{\theta^{-h\nu}-1}\right\}.\end{aligned} \tag{4}$$

The average energy of a *single oscillator* is therefore

$$\frac{\overline{E}_A}{N_A} = \frac{1}{2}h\nu + \frac{h\nu}{\theta^{-h\nu}-1}. \tag{5}$$

If we elect to take the *energy zero* as the lowest quantum state ($v = 0$), the energy values become by referring to Eq. (2)

$$\epsilon_v = v\,h\nu, \tag{6}$$

that is, we take $\frac{1}{2}h\nu$ as equal to zero. In this case, the corresponding average energy of an oscillator is

$$\frac{\overline{E}_A}{N_A} = \frac{h\nu}{\theta^{-h\nu}-1}. \tag{7}$$

We shall refer to the energy $\frac{1}{2}h\nu$ of the lowest quantum state of a harmonic oscillator as the *residual energy*; that is,

$$\epsilon_0 = \frac{1}{2}h\nu. \tag{8}$$

This is a grammatically more satisfactory English equivalent of *nullpunktsenergie* than the commonly used 'zero-point energy'.

18.8. An assembly of two sets of non-localised systems

For calculating the equilibrium state of an assembly, we may for simplicity consider an assembly containing systems of two types of *non-localised systems A and B.* They are N_A and N_B in number, and *all their states are first non-degenerate*, with energy values $\epsilon_0, \epsilon_1, \ldots, \epsilon_r, \ldots,$ and $\eta_0, \eta_1, \ldots, \eta_s, \ldots$ for A and B, respectively, all expressible in integers in terms of a suitable unit of energy. The number of *systems* in the states with these energies will be specified as usual by $n_0, n_1, \ldots, n_r, \ldots$ and $n'_0, n'_1, \ldots, n'_s, \ldots$. This set of numbers completely specifies a statistical state of the assembly. We have now to distinguish between two important cases, according as the eigenfunctions are symmetrical or antisymmetrical in all the systems of a given set of numbers.

(i) *The assembly eigenfunctions are symmetrical in all the systems of a given set. Assemblies of such systems are said to obey the Bose-Einstein statistics.*

For such an assembly, there is now one eigenfunction for every set of positive (or zero) integral values of the n_r, n'_s. The total number of complexions C is, therefore, simply equal to the *number of sets* of positive (or zero) integral values of the n_r, n'_s satisfying the necessary equalities

$$\Sigma_r\, n_r = N_A, \quad (1)$$

$$\Sigma_s\, n'_s = N_B, \quad (2)$$

$$\Sigma_r\, n_r \in_r + \Sigma_s\, n'_s \eta_s = E\ . \quad (3)$$

(ii) *The assembly eigenfunctions are antisymmetrical in all the systems of a given set. Assemblies of each systems are said to obey the Fermi-Dirac statistics.*

For such an assembly, there is again one eigenfunction for each set of allowed values of n_r, n'_s, but these allowed values are restricted to 0 and 1, values greater than 1 being restricted in agreement with *Pauli exclusion principle*. The total number of complexions C is therefore equal to the number of sets of values of the n_r, n'_s satisfying the necessary conditions (1), (2), and (3) and subject to the further restriction that each n_r, n'_s must be 0 or 1 but may not exceed 1.

The next step should be the determination of the number C. Since, however, this step is purely mathematical one, we shall dispense with and the reader, who is not particularly interested in the mathematics, can proceed to the next paragraph where the results concerning the average values of the properties of the assembly in the equilibrium state are collected.

18.9. Summary of results

We may summarise in this section the results whose deduction has been excluded in Section 18.8 for it is purely mathematical, and at the same time remove the trivial restriction to assemblies of only two types of systems.

In an assembly containing N_A systems of type A, N_B of type B, and so on, the average number $\overline{n}_r$ of systems of type A, in the non-degenerate state r of energy $\in_r$ is given as shown by Eq. (2) in Section 18.5,

$$\overline{n}_r = N_A\, \theta^{\in_r} \Big/ \textstyle\sum_r \theta^{\in_r} = N_A\, \theta^{\in_r} / f_A(\theta). \quad (1)$$

Without going into mathematical details, it can be proved using *Cauchy's theorem* that

$$\overline{n}_r = \frac{\lambda_A\, \theta^{\in_r}}{1 \pm \lambda_A\, \theta^{\in_r}} = \frac{1}{\lambda_A^{-1}\, \theta^{-\in_r} \pm 1}, \quad (2)$$

where λ_A is a *statistical parameter* whose significance will be made clear shortly. Eq. (2) can also be written in the form

$$\overline{n}_r = \lambda_A \frac{\partial}{\partial \lambda_A} \left\{ \ln \left(1 \pm \lambda_A \theta^{\epsilon_r} \right)^{\pm 1} \right\}, \tag{3}$$

where we have to take the + or − accordingly as the assembly eigenfunctions have to be antisymmetrical or symmetrical in the systems A. The λ_A , λ_B, . . . are parameters of which there is one for each type of system A, B, . . . , but the same parameter λ_A occurs in formulae such as Eq. (2) for all the various states *r* of the A systems. The λ_A , λ_B, . . . are determined by the necessary equalities such as

$$\sum_r \frac{\lambda_A \theta^{\epsilon_r}}{1 \pm \lambda_A \theta^{\epsilon_r}} = \sum_r \overline{n}_r = N_A. \tag{4}$$

The average energy $\overline{E}_A$ of all the N_A systems is given by

$$\overline{E}_A = \sum_r \overline{n}_r \epsilon_r = \sum_r \frac{\epsilon_r \lambda_A \theta^{\epsilon_r}}{1 \pm \lambda_A \theta^{\epsilon_r}}, \tag{5}$$

when Eq. (2) is used. Also Eq. (5) can be obtained from the equation

$$\overline{E}_A = \theta \frac{\partial}{\partial \theta} \sum_r \ln \left(1 \pm \lambda_A \theta^{\epsilon_r} \right)^{\pm 1}. \tag{6}$$

Finally, θ is a parameter with the same value not only for all states *r* of one type of a system, but for all the *various types of systems A, B, . . . The necessary relation*

$$E = \overline{E}_A + \overline{E}_B + \ldots \tag{7}$$

determines θ when *E* is given, but we may equally legitimately regard it as determining *E* when θ is given. For the reasons given in Section 18.5, we may regard θ as the temperature measured on a particular scale which we call the statistical scale. We shall shortly prove that the statistical temperature θ is related to the temperature *T* on the *Kelvin* scale by

$$\theta = e^{-1/kT}, \tag{8}$$

where *k* is Boltzmann's constant.

18.10. Degenerate systems

The equilibrium properties of the assembly have been shown, as implied by Eq. (3) in Section 18.9, to depend only on the functions

$$\ln \left(1 \pm \lambda_A \theta^{\epsilon_r} \right)^{\pm 1},$$

so far as concerns each set of systems. It is therefore now easy to remove the restriction to non-degenerate systems, by allowing the energies to become equal in groups of ω_r. If now the systems A have degenerate states, of energy values ϵ_0 , ϵ_1 , ... ϵ_r, ... and weights

ω_0 , ω_1 , ... ω_r, ..., then the average number of A systems in the degenerate state r can be deduced from either Eq. (3) or Eq. (2) in Section 18.9, that is,

$$\bar{n}_r = \omega_r \lambda_A \frac{\partial}{\partial \lambda_A} \ln \left(1 \pm \lambda_A \theta^{\epsilon_r}\right)^{\pm 1}$$

$$= \frac{\omega_r}{\lambda_A^{-1} \theta^{-\epsilon_r} \pm 1}. \qquad (1)$$

The energy of these particular A systems is $\bar{n}_r \epsilon_r$, is deducible from either this equation, that is,

$$\bar{n}_r \epsilon_r = \omega_r \epsilon_r \lambda_A \frac{\partial}{\partial \lambda_A} \ln \left(1 \pm \lambda_A \theta^{\epsilon_r}\right)^{\pm 1}, \qquad (2)$$

or from Eq. (6) in Section 18.9, that is

$$\bar{n}_r \epsilon_r = \omega_r \theta \frac{\partial}{\partial \theta} \ln \left(1 \pm \lambda_A \theta^{\epsilon_r}\right)^{\pm 1}. \qquad (2)'$$

Therefore, the average energy of *all the A systems* according to Eq. (2)′ is

$$\bar{E}_A = \sum_r \bar{n}_r \epsilon_r = \theta \frac{\partial}{\partial \theta} \sum_r \omega_r \ln \left(1 \pm \lambda_A \theta^{\epsilon_r}\right)^{\pm 1},$$

or,

$$\bar{E}_A = \bar{n}_r \epsilon_r = \sum_r \frac{\omega_r \epsilon_r}{\left(\lambda_A \theta^{\epsilon_r}\right)^{-1} \pm 1}, \qquad (3)$$

according to Eq. (1) . The values of λ_A and θ being fixed by the necessary equations, namely Eqs. (4) and (7) in Section 18.9.

18.11. Classical statistical mechanics

Non-localised systems are free to move in a common enclosure and are distributed among their possible states according to Eq. (1) in Section 18.10, namely,

$$\bar{n}_r = \omega_r \lambda_A \frac{\partial}{\partial \lambda_A} \ln \left(1 \pm \lambda_A \theta^{\epsilon_r}\right)^{\pm 1}$$

$$= \frac{\omega_r}{\lambda_A^{-1} \theta^{-\epsilon_r} \pm 1} = \frac{\omega_r \lambda_A \theta^{\epsilon_r}}{1 \pm \lambda_A \theta^{\epsilon_r}}. \qquad (1)$$

Let us suppose that for all ϵ_r

$$\lambda_A \ll \theta^{-\epsilon_r}, \qquad (2)$$

then we may use the approximation

$$\ln \left(1 + \lambda_A \theta^{\epsilon_r}\right)^{\pm 1} = \lambda_A \theta^{\epsilon_r}, \qquad (3)$$

which is the first term in the expansion of the series . Hence, in this case, both *Fermi-Dirac* and *Bose-Einstein* statistics lead to identical distributions. Using the approximation given by Eq. (3), we can simplify Eq. (1) to

$$\bar{n}_r = \lambda_A \, \omega_r \, \theta^{\epsilon_r} . \quad (4)$$

We also have then

$$N_A = \sum_r \bar{n}_r = \sum_r \lambda_A \, \omega_r \, \theta^{\epsilon_r} = \lambda_A \, f_A(\theta), \quad (5)$$

where $f(\theta)$ is the *partition function* defined above, namely,

$$f_A(\theta) = \sum_r \theta^{\epsilon_r} , \quad \text{(non-degenerate) (6) and}$$

$$f_A(\theta) = \sum_r \omega_r \, \theta^{\epsilon_r} . \quad \text{(degenerate) (7)}$$

If we eliminate λ_A between Eqs. (4) and (5), we obtain

$$\bar{n}_r = N_A \, \omega_r \, \theta^{\epsilon_r} \Big/ \sum_r \omega_r \, \theta^{\epsilon_r} = N_A \, \omega_r \, \theta^{\epsilon_r} \big/ f_A(\theta), \quad (8)$$

which is identical with Eq. (2) in Section 18.6 for the distribution of localised systems.

We have thus obtained the *striking result* that when Eq. (2) above, namely,

$$\lambda_A << \theta^{-\epsilon_r} \quad (2)$$

is obeyed, the Fermi-Dirac statistics, the Bose-Einstein statistics, and the statistical mechanics for *localised systems* all lead to the same distribution laws. *Under these conditions, we do not need to distinguish between the three kinds of statistics, and we say that the assembly obeys classical statistics.*

At the present stage we cannot say, if ever, when the condition (2) will be fulfilled. After we have related θ to T by

$$\theta = e^{-1/kT} ,$$

we shall be able to show that this condition is fulfilled at all temperatures except the very lowest ($T << 1$ K) by gaseous assemblies of any kind of ordinary chemical molecules or atoms, but not by the assembly of 'free' electrons in a metal.

We can anticipate the rigorous discussion of the conditions under which Eq. (2) is obeyed by a semi-quantitative physical discussion. Suppose that the total number N_A of systems A is much *smaller* than the number of possible states of not too high energy, then *the number of complexions with more than one system in any given state will be negligible compared with the number of complexions where each state contains either one system or none*. We may therefore ignore the existence of conceivable complexions containing more than one system in any state. Assemblies obeying the Fermi-Dirac statistics differ from those obeying Bose-Einstein statistics only in that complexions *with more than one system in a given state are forbidden in the former*, but, if the number of such complexions is negligible anyhow, *the distinction becomes unimportant*. We may therefore expect the condition given by Eq. (2) to be *closely related to the condition that the number of systems be small compared with the number of states of not too high energy.*

18.12. Structureless particles moving in a box

As an example of an assembly of non-localised systems, let us consider a number of identical structureless particles in an enclosure. The equation of Schrödinger for a structureless particle of mass m^* in an enclosure of volume *V* and uniform potential energy, which for convenience we take to be zero, is

$$\frac{\partial^2 \psi}{\partial x^2} + \frac{\partial^2 \psi}{\partial y^2} + \frac{\partial^2 \psi}{\partial z^2} + \frac{8\pi^2 m^*}{h^2} \in \psi = 0. \qquad (1)$$

If the enclosure has the form of a rectangular box, the eigenfunctions can be enumerated by three quantum numbers l, m, n each of which can take any *positive* (non-zero) integral value, with the physical meaning that the squares of the momenta parallel to the edges *a*, *b*, and *c* of the box are $\ell^2 h^2/4\pi^2$, $m^2 h^2/4\pi^2$, $n^2 h^2/4\pi^2$, respectively. The momenta themselves have no eigenvalues but an average value zero, corresponding to the fact that a particle with given, l, m, n is moving back and forth between the ends of the box and is as likely to be moving in one direction as in the exactly opposite direction. The energy of the state with given l, m, n is

$$\in_{l,m,n} = \frac{h^2}{8m^*}\left(\frac{l^2}{a^2} + \frac{m^2}{b^2} + \frac{n^2}{c^2}\right). \qquad (2)$$

All such states are non-degenerate. We shall here suppose that for all $\in_r$

$$\lambda << \theta^{-\in_r}, \qquad (3)$$

and shall verify afterwards that this condition is fulfilled. We therefore want to evaluate the partition function

$$f(\theta) = \sum_{l=1}^{\infty} \sum_{m=1}^{\infty} \sum_{n=1}^{\infty} \theta^{\frac{h^2}{8m^*}\left(\frac{l^2}{a^2} + \frac{m^2}{b^2} + \frac{n^2}{c^2}\right)}$$

$$= \sum_{l=1}^{\infty} \theta^{l^2 h^2/8m^* a^2} \sum_{m=1}^{\infty} \theta^{m^2 h^2/8m^* b^2}$$

$$\sum_{n=1}^{\infty} \theta^{n^2 h^2/8m^* c^2}. \qquad (4)$$

We shall further suppose and verify afterwards that

$$h^2 \ln(1/\theta)/8\pi m^* a^2 << 1, \qquad (5)$$

$$h^2 \ln(1/\theta)/8\pi m^* b^2 << 1, \qquad (6)$$

$$h^2 \ln(1/\theta)/8\pi m^* c^2 << 1. \qquad (7)$$

These conditions mean that the separation of the energy states (or the energy levels) is small compared with $1/\ln(1/\theta)$, that is with kT, or $\ln(1/\theta) = 1/kT$, where θ is anticipated to be given by

$$\theta = e^{-1/kT}. \qquad (8)$$

We may then replace the sums in Eq. (4) by integrals and obtain

$$f(\theta) = \int_0^{\infty} \theta^{h^2 l^2/8m^* a^2}\, dl \int_0^{\infty} \theta^{h^2 m^2/8m^* b^2}\, dm$$

$$\times \int_0^{\infty} \theta^{h^2 n^2/8m^* c^2}\, dn.$$

Substituting for θ by $e^{-1/kT}$, we have

$$f(\theta) = \int_0^\infty e^{-\{h^2 \ln(1/\theta)/8m^* a^2\} l^2}\, dl \quad \int_0^\infty e^{-\{h^2 \ln(1/\theta)/8m^* b^2\} m^2}\, dm$$

$$\times \int_0^\infty e^{-\{h^2 \ln(1/\theta)/8m^* c^2\} n^2}\, dn,$$

$$= \left\{\frac{8m^* a^2}{h^2 \ln(1/\theta)} \frac{\pi}{4}\right\}^{\frac{1}{2}} \left\{\frac{8m^* b^2}{h^2 \ln(1/\theta)} \frac{\pi}{4}\right\}^{\frac{1}{2}} \left\{\frac{8m^* c^2}{h^2 \ln(1/\theta)} \frac{\pi}{4}\right\}^{\frac{1}{2}}$$

$$= \left\{\frac{2\pi m^*}{h^2 \ln(1/\theta)}\right\}^{3/2} abc = \left\{\frac{2\pi m^*}{h^2 \ln(1/\theta)}\right\}^{3/2} V, \tag{9}$$

where V being the volume of the box. Anticipating the relation between θ and the Kelvin temperature T, given by Eq. (8), the expression for
$f(\theta)$ becomes

$$f(\theta) = \frac{(2\pi m^* kT)^{3/2} V}{h^3}. \tag{10}$$

Subsequently, according to Eq. (5) in Section 18.11, λ is related to $f(\theta)$ by

$$\lambda = \frac{N}{f(\theta)} = \frac{N}{V} \frac{h^3}{(2\pi m^* kT)^{3/2}}. \tag{11}$$

We must now verify Eqs. (3) and (5)–(7). Anticipating the relation of Eq. (8) between θ and T, and using the numerical values

$$k = 1.37 \times 10^{-16} \text{ ergs/K},$$

$$h = 6.55 \times 10^{-27} \text{ ergs.sec},$$

$$m^* = 1.66 \times 10^{-24} \text{ g} \quad \text{(the mass of a hydrogen atom)},$$

we find for Eq. (5)

$$h^2 \ln(1/\theta)/8m^* a^2 = h^2/8m^* a^2 kT$$

$$= 2.36 \times 10^{-14}/a^2 T. \tag{12}$$

Thus for hydrogen atoms in a box with sides of length 10^{-3} cm at the temperature 0.001 K,

$$h^2 \ln(1/\theta)/8m^* a^2 = 2.36 \times 10^{-5}, \tag{13}$$

and Eqs. (5)–(7) are easily satisfied. They are satisfied *a fortiori* by more massive particles, a larger enclosure, or higher temperatures. Even for electrons, they will usually be comfortably satisfied. We have still to verify the validity of Eq. (3), namely,

$$\lambda << \theta^{-\epsilon_r},$$

or

$$\lambda << e^{\epsilon_r/kT} \quad (\text{all } \epsilon_r). \tag{14}$$

Since kT is positive and all the ϵ_r' s are positive, Eq. (14) will be satisfied, provided we can show that

$$\lambda \ll 1, \tag{15}$$

or, according to Eq. (11), that

$$\frac{N}{V}\frac{h^3}{\left(2\pi m^* kT\right)^{3/2}} \ll 1. \tag{16}$$

Inserting numerical values for k, h, and the value of m^* for a hydrogen atom, the condition given by Eq. (16) becomes

$$5.2 \times 10^{-21}\,\frac{N}{V}\,\frac{1}{T^{3/2}} \ll 1. \tag{17}$$

The molecular density in an ordinary gas at standard temperature and pressure gives $N/V = 2.7 \times 10^{19}$. For the same concentration and T = 9 K, the left-hand side of Eq. (17) becomes 0.005 and the condition is well satisfied. *A fortiori* λ will be much smaller than '1' for more massive particles, higher temperatures, or smaller concentrations. Thus in applications to actual gases, we may assume that λ is small, and we may use *classical statistics*. The only important exception will be an assembly of electrons at the concentrations at which one would expect to find free electrons in metals, about one per atom. On account of the much smaller mass of the electrons, in such assemblies $\lambda \gg 1$ even up to temperatures greater than 2000 K. The formulae of this section are then not applicable.

18.13. Relationship between statistical mechanics and thermodynamics

We have now obtained the distribution laws of the equilibrium (average) state of any assembly of systems, between which the energy of interaction is *negligible*. These laws shall be extended later to apply to assemblies of interacting systems. Before doing so, it is of interest to consider the *relation between the laws already derived and classical thermodynamics*. We have obtained these distribution laws without any reference to thermodynamical ideas, except to point out that there must be a *universal relation between* θ *and the absolute temperature*. The ideas of thermodynamics are entirely foreign to the *foundations* of statistical mechanics, which are mainly *dynamical*. *The proper course is to prove that the laws of thermodynamics are true for the assemblies of statistical mechanics*, if we use suitable analogies to interpret the properties of these assemblies. This was made *abundantly clear by Gibbs*. Such proofs are given in the succeeding sections, and it will be seen that the *direct introduction of the laws of thermodynamics in this way is satisfactorily simple*.

18.14. The laws of thermodynamics

Since the basic laws of thermodynamics can be formulated in several alternative ways, it will be necessary for us to describe shortly the basis which we actually adopt, because the development of the relationship with statistical mechanics is naturally somewhat affected by this choice. *The most logically satisfactory formulation is undoubtedly that of Carathéodory.*[*] We may call this formulation the *classical formulation*, being the one most commonly adopted, although some of the postulates are often assumed without being explicitly stated. This classical formulation is, however, *not the formulation* which is most simply related to the statistical laws which we have obtained. We, therefore, give after the classical formulation *another variant*, which is derivable from the classical formulation and is more related to the statistical laws.

We shall, for the sake of brevity, omit definitions of 'thermodynamic state', 'work', 'thermal insulation', 'adiabatic', and 'thermal contact', but must emphasise that these can be defined without *any reference to temperature*.

The first step is to introduce the *concept of temperature*. As a natural generalisation of experience, we introduce the postulate:

If two assemblies are each in thermal equilibrium with a third assembly, they are in thermal equilibrium with each other.

From this, it may be shown to follow that the condition for *thermal equilibrium* between several assemblies is the equality of a certain single-valued function of the thermodynamic states of the assemblies, which may be called the *temperature* t, any one of the assemblies being used as a 'thermometer' reading the temperature t on a suitable scale. This postulate of the 'existence of temperature' could with advantage be known as *the zeroth law of thermodynamics*. This temperature, whose existence is thus postulated, is measured on a scale which is determined only by the arbitrary choice of the thermometer system, and is called *the empirical temperature* when it is necessary to distinguish it from the *absolute temperature*. For example, the empirical temperature t might be defined[†] to be *the measured volume of a constant quantity of any chosen substance at constant pressure*.

We now formulate the First Law of Thermodynamics as follows:

If a thermally insulated assembly can be taken from a state I to a state II by alternative paths, the work W done on the assembly has the same value for every such (adiabatic) path.

From this, one can deduce that there exists a single-valued function E of the state of an assembly, called its *total energy*, such that for any adiabatic process, the increase ΔE of the energy is equal to the work done W on the assembly. Thus

$$\Delta E = W \quad \text{(adiabatic reversible)}. \tag{1}$$

In particular, in an assembly mechanically as well as *thermally isolated the energy remains constant, thus*

$$\Delta E = 0 \quad \text{(isolated assembly)}. \tag{2}$$

It follows that in any interaction between two parts of an *isolated assembly* gain by energy in the one part is equal to the loss of energy in the other. It is to be noticed that for this definition of energy it is necessary and sufficient that it be possible by an adiabatic process to change the assembly either from state I to state II or from state II to state I.

We now define the heat q *absorbed* by an assembly as the increase in total energy of the assembly less the work *done on the assembly*, thus

$$q = \Delta E - W \qquad \text{(all processes)}. \tag{3}$$

---

* Carathéodory, *Math. Ann.* **67**, 355 (1909). For a simple account, see Born, *Phys. Zeit.* **22**, 218, 249, 282 (1921).

† See Section 5.1, pp. 222–223.

We now formulate the Second Law of Thermodynamics as follows; it is the classical formulation.

There exist single-valued functions of state, T, called the absolute temperature, and S, called the entropy, such that

(1) T is a function of t only.

(2) The entropy of any assembly is equal to the sum of the entropies of its parts.

(3) For any infinitesimal change in any completely homogeneous assembly.

$$d'q \le T\,dS; \tag{4}$$

the equality sign holds for quasi-static processes, the inequality for natural (irreversible) processes. If an assembly is not completely homogeneous, the Eq. (4) applies to each of its homogeneous parts.

These postulates *form a sufficient basis of thermodynamics*. From them one can, for instance, derive *Carnot's formula* for the efficiency of a reversible cycle.* Actually, we shall have more use for the branch of thermodynamics which gives us the conditions for complete equilibrium. We shall briefly outline the procedure for deriving these conditions from the postulates.

If an assembly is in *complete equilibrium*, then every conceivable infinitesimal change must be *quasi-static*. The condition for complete equilibrium is therefore by Eq. (4) that for every infinitesimal process

$$d'q_{rev} = TdS \qquad \text{(equilibrium).} \tag{5}$$

If we combine this with Eq. (3) for an infinitesimal process, the condition becomes

$$d'q_{rev} = TdS = dE - d'W \qquad \text{(equilibrium).} \tag{6}$$

Certain special forms of Eq. (6) are of particular importance. We quote two conditions, namely,

(a) For an assembly *isolated thermally and mechanically*, that is,

$$dS = 0, \quad d'W = 0,$$

we have

$$dE = 0 \qquad \text{(equilibrium).} \tag{7}$$

(b) For an assembly *mechanically isolated in a thermostat*,

$$d'W = 0, \quad dT = 0,$$

we have

$$d(E - TS) = 0 \qquad \text{(equilibrium).} \tag{8}$$

The quantity (*E–TS*) occurring in Eq. (8), as we already know, is called the *Helmholtz free energy, and is denoted by F*. It plays a fundamental part in the alternative formulation given below, and in our derivation of the laws of classical thermodynamics from those of statistical thermodynamics as previously shown in Section 9.6.

This completes the classical formulation. To make use of it one proceeds as follows. One assumes that any assembly, if not completely homogenous, can be divided into a number of completely homogeneous parts called *phases* each with its own *energy*, *entropy*, and *temperature*. The energy and entropy of the whole assembly are then equal to the sum of those of the constituent phases. A complete description of the state of the whole system

---

* See Section 1.5.

involves a complete description of each phase. To define completely a given phase, we first define its *composition*, that is, the number N_A, N_B,... of the various types of systems (molecules) that it contains. Next, we specify the values of all geometrical parameters *x*. [Usually the only important one in practice is the volume *V*.] Having done this, the phase still has one degree of freedom, and we are required to specify *one further quantity to complete the description of its state*. In the classical formulation sketched above, the quantity chosen is the *entropy*. We thus have for the given phase the independent variables $S_1, x_1, \ldots, N_A, N_B, \ldots$ or for brevity *S*, *x*, *N*. Any other property of the phase may then be regarded as a function of *S*, *x*, *N*, and in particular the total energy *E* may be so regarded. We may therefore write for any variation in the state or nature of the phase

$$E = E(S, x, N)$$

so that

$$dE = \frac{\partial E}{\partial S} dS + \sum \frac{\partial E}{\partial x} dx + \sum \frac{\partial E}{\partial N} dN \; . \qquad (9)$$

If in particular we consider a variation at constant configuration and constant composition, Eq. (9) reduces to

$$dE = \left(\frac{\partial E}{\partial S}\right)_{x,N} dS, \qquad (x, N \text{ constant}), \qquad (10)$$

and *E* becomes *a single-valued function of S*. The only process that can take place at constant configuration and constant composition is the absorption of heat, for at constant configuration there can be no work. In this case, Eq. (3) reduces to

$$d'q = dE. \qquad (11)$$

It must be remembered that constant configuration includes constancy of any long-range fields produced by bodies outside the assembly. Substituting Eq. (11) into Eq. (10), we obtain

$$d'q = \left(\frac{\partial E}{\partial S}\right)_{x,N} dS. \qquad (12)$$

If we arrange that the absorption of heat be quasi-static, that is, reversible, we have according to Eq. (4)

$$d'q = T\,dS, \qquad \text{(equilibrium)} \qquad (13)$$

and so by comparison of Eqs. (12) and (13)

$$\left(\frac{\partial E}{\partial S}\right)_{x,S} = T. \qquad (14)$$

Since for given *x* and *N* we know that the total (internal) energy *E* is a *single-valued function of S, the Eq. (14) must be independent of the process used in its derivation.* Therefore, Eq. (9) can be rewritten as

$$dE = T\,dS + \sum \frac{\partial E}{\partial x} dx + \sum \frac{\partial E}{\partial N} dN \, . \qquad (15)$$

Provided the phase is in mechanical equilibrium with its surroundings, $-(\partial E/\partial x)$ is *simply the generalised mechanical force exerted by the phase tending to increase x. This we denote by X.* The coefficients $\partial E/\partial N$ are the quantities introduced by *Gibbs* and denoted by μ. They are called[*] the *partial potentials.* Introducing these symbols, we rewrite Eq. (15) as

$$dE = TdS - \sum X dx + \sum \mu dN . \qquad (16)$$

This formula shows the dependence of E on the independent variables S, x , N ; E, thus regarded as a function of these variables, is called the *thermodynamic potential* for the variables S, x, and N. Eq. (16) is called the *fundamental formula* for this set of independent variables. These are one such formula for each phase in the assembly. These formulae form the usual *starting point* for determining equilibrium properties after the *manner of Gibbs.*

By using fundamental formulae of the type in Eq. (16), one such for each phase, one can deduce the following:

(*a*) That heat flows always from a *higher* temperature to a *lower* one.

(*b*) That any *geometrical boundary* moves from a higher pressure to a lower one.

(*c*) That chemical changes always proceed in such a direction that certain linear combinations of the *partial potentials decrease.*

As a corollary of (*a*), we have the fact that there is thermal equilibrium between two phases of equal temperature, which is already known from the zeroth law. Since we are mainly interested in the equilibrium properties of assemblies, we can dispense with deduction (*a*) and confine ourselves to assemblies of uniform temperature. We can then transform the formulation of the second law to a more convenient form, which is *less general than the classical formulation,* but only to the trivial extent that one postulates equality of temperature throughout the assembly.

Adopting this standpoint, we are able in our choice of independent variables to *replace the entropies of each of the phases by the temperature of the whole assembly.* The whole formulation can thereby be modified. To achieve this, we define the *Helmholtz free energy* F by

$$F \equiv E - TS, \qquad (17)$$

and differentiating Eq. (17) and combining the result with Eq. (16), we obtain

$$dF = -SdT - \sum X dx + \sum \mu dN. \qquad (18)$$

Eq. (18) is the *fundamental formula* for the variables T, x, and N; F is the *thermodynamic potential* for these variables. One could, if one wished, use Eq. (18) as the *basis of the second law.* The variables T, x, and N are indeed, from the statistical point of view, more 'natural' than S, x, and N, but we can improve the naturalness still further by getting rid of all need to refer to S. This we now proceed to do.

According to Eq. (18), we have

$$\frac{\partial F}{\partial T} = -S, \qquad (19)$$

and consequently, by using Eq. (17) with Eq. (19), we obtain

$$\frac{\partial (F/T)}{\partial T} = \frac{1}{T}\frac{\partial F}{\partial T} - \frac{F}{T^2} = -\frac{(TS + F)}{T^2} = -\frac{E}{T^2}. \qquad (20)$$

---

[*] See R. Fowler and E.A. Guggenheim, *Statistical Thermodynamics*, Chapter 2, Section 202, (Cambridge, 1965).

Then by using Eq. (20), we can eliminate S from Eq. (18) and obtain

$$d\left(\frac{F}{T}\right) = -\frac{E}{T^2}dT - \frac{1}{T}\sum X dx + \frac{1}{T}\sum \mu dN. \qquad (21)$$

It is this formula which we choose to adopt as our basis for the *alternative version of the second law*. We now proceed to give our alternative formulation.

The zeroth law, the first law, and the definition of heat are preserved in the form already given. It is only in the formulation of the second law that there is a departure from the classical formulation.

The formulation of the second law, which we now adopt or the adopted version of the second law, is the following:

Second law (alternative formulation). There exist single-valued functions of state, T, called the absolute temperature, and F, called the Helmholtz free energy, such that

1. *T is a function of t only.*
2. *The Helmholtz free energy of an assembly is equal to the sum of the free energy of its parts.*
3. *When work is done on the assembly isothermally*
   a. $dF \le d'W \quad (dT = 0);$ (22)
   b. *the equality sign refers to quasi-static processes, and the inequality sign to actual (natural) processes.*
4. *When F is regarded as a function of T, x, and N, its dependence on the temperature is given by*

$$\frac{\partial (F/T)}{\partial T} = -\frac{E}{T^2}. \qquad (23)$$

Applying this *statistical formulation* we need to indicate, only quite briefly, how it leads to the same equilibrium laws as the *classical formulation.* Choosing the independent variables T, x, and N, we write

$$d\left(\frac{F}{T}\right) = \frac{\partial (F/T)}{\partial T} dT + \frac{1}{T}\sum\left(\frac{\partial F}{\partial x}\right)dx + \frac{1}{T}\sum\left(\frac{\partial F}{\partial N}\right)dN. \qquad (24)$$

The coefficient of dT is determined by Eq. (23). Eq. (22), when applied to quasi-static variations of the geometry of the assembly at *fixed composition,* tells us that $-\partial F/\partial x = X$. Alternatively, we may take this as the *definition of the generalised forces* X. Finally, we define the *partial potentials* μ by the relation

$$\mu_i = \left(\frac{\partial F}{\partial N_i}\right)_{T,x,n_j\,(j\neq i)}. \qquad (25)$$

We thus have the thermodynamic formula, Eq. (24), rewritten as

$$d\left(\frac{F}{T}\right) = -\frac{E}{T^2}dT - \frac{1}{T}\sum X dx + \frac{1}{T}\sum \mu dN, \qquad (26)$$

or alternatively

$$T\,d\left(\frac{F}{T}\right) = -\frac{E}{T}dT - \sum X\,dx + \sum \mu\,dN, \qquad (27)$$

in agreement with Eq. (21) obtained by the *classical derivation.*

The *condition for complete equilibrium of an assembly* is that any change whatever should be quasi-static, and according to Eq. (22), this can be formulated as

$$dT = 0, \qquad dF = d'W \qquad \text{(equilibrium)}, \qquad (28)$$

and in particular, considering *chemical changes*, keeping the geometry of the assembly unaltered, we obtain

$$dT = 0,\ dx = 0, \qquad dF = 0 \ \text{(equilibrium)}. \qquad (29)$$

Eq. (29) is the *most convenient general form for the conditions of chemical equilibrium.*

It is not essential to introduce other thermodynamic functions than E, F, T, and μ, which are now defined. For completeness, however, we can *define the entropy S* by

$$S \equiv (E - F)/T. \qquad (30)$$

Using Eqs. (20) and (23), we then have

$$\frac{\partial F}{\partial T} = T\frac{\partial (F/T)}{\partial T} + \frac{F}{T}$$

$$= -\frac{E}{T} + \frac{F}{T} = -S, \qquad (31)$$

Alternatively, we may define S by Eq. (31) and deduce Eq. (30).

18.15. Derivation of thermodynamics from statistical mechanics

We have already found the correct analogy to the zeroth law, since θ has the required properties of temperature. The first law is also an obvious deduction from our premises, if we define the total energy E of thermodynamics by the equation

$$E = \bar{E} = \sum_A \bar{E}_A = \sum_A \sum_r \bar{n}_r \epsilon_r\,. \qquad (1)$$

To deduce the second law in the required form, we have to define suitable functions to act as T and F. We shall show that this can be done. Since θ can be identified as the *empirical temperature*, we may *tentatively define T* by

$$kT = \frac{1}{\ln(1/\theta)},$$

or

$$\ln\theta = -1/kT, \qquad (2)$$

where k is any constant, merely fixing the size of the degree; it is the Boltzmann constant. We then *define F* by the equations

$$F = \sum_A F_A, \qquad (3)$$

$$F_A \ln\theta = -\frac{F_A}{kT}$$

$$= \sum_r \omega_r \ln\left(1 \pm \lambda_A\,\theta^{\epsilon_r}\right)^{\pm 1} - N_A \ln \lambda_A\,. \qquad (4)$$

We have now to show that F and T have the *properties of free energy (Helmholtz) and absolute temperature, respectively,* required by the *adopted version of the second law* in Section 18.14.

The additive property of F is already catered for by Eq. (3). In order to derive the other required properties of F, we have to treat F as a function of θ (or T), x, and N_A. We therefore note that λ_A *depends on* θ, x, and N_A, *while* $\in_r$ *depends on* x *only*. The complete differential of $F_A \ln \theta$ is then

$$d(F_A \ln\theta) = \frac{\partial(F_A \ln\theta)}{\partial \ln\theta} d\ln\theta + \ln\theta \sum_r \frac{\partial F_A}{\partial \in_r} \frac{\partial \in_r}{\partial x} dx + \ln\theta \frac{\partial F_A}{\partial N_A} dN_A$$
$$+ \ln\theta \frac{\partial F_A}{\partial \lambda_A}\left\{\frac{\partial \lambda_A}{\partial \ln\theta} d\ln\theta + \frac{\partial \lambda_A}{\partial x} dx + \frac{\partial \lambda_A}{\partial N_A} dN_A\right\}. \tag{5}$$

We require next the value of $\partial F_A/\partial \lambda_A$ to insert into Eq. (5). From Eq. (4), we deduce

$$\lambda_A \ln\theta \frac{\partial F_A}{\partial \lambda_A} = \sum_r \frac{\omega_r}{\lambda_A^{-1}\theta^{-\in_r} \pm 1} - N_A.$$

But $\bar{n}_r$ is given as shown above in Section 18.10 by Eq. (1), namely,

$$\bar{n}_r = \frac{\omega_r}{\lambda_A^{-1}\theta^{-\in_r} \pm 1}.$$

Therefore,

$$\lambda_A \ln\theta \frac{\partial F_A}{\partial \lambda_A} = \sum_r \bar{n}_r - N_A = 0. \tag{6}$$

Consequently, in Eq. (5), we may omit all terms containing $\partial F_A/\partial \lambda_A$ as a factor, and treat λ_A as though it were not variable so that Eq. (5) simplifies to

$$d(F_A \ln\theta) = \frac{\partial(F_A \ln\theta)}{\partial \ln\theta} d\ln\theta + \ln\theta \sum_r \frac{\partial F_A}{\partial \in_r} \frac{\partial \in_r}{\partial x} dx$$
$$+ \ln\theta \sum_A \frac{\partial F_A}{\partial N_A} dN_A, \tag{7}$$

all differentiations being performed treating λ_A *as constant.* If we now compare Eq. (4) with the equation expressing $\bar{E}$, which we obtained before in Section 18.10, Eq. (3), namely

$$\bar{E}_A = \sum_r \bar{n}_r \in_r = \theta \frac{\partial}{\partial \theta} \sum_r \omega_r \ln\left(1 \pm \lambda_A \theta^{\in_r}\right)^{\pm 1},$$

we obtain

$$\left(\frac{\partial(F_A \ln\theta)}{\partial \ln\theta}\right)_{x, N_A} = \left(\frac{\partial(F_A \ln\theta)}{\partial \ln\theta}\right)_{x, N_A, \lambda_A}$$
$$= \theta \frac{\partial}{\partial \theta} \sum_r \omega_r \ln\left(1 \pm \lambda \theta^{u_r}\right)^{\pm 1} = \bar{E}_A. \tag{8}$$

Similarly, by comparing Eq. (4) with the equation

$$\overline{X} = -\sum_r \frac{\partial \epsilon_r}{\partial x} \frac{\omega_r \lambda \theta^{\epsilon_r}}{1 \pm \lambda \theta^{\epsilon_r}},$$

(that might be derived in the following way

$$\overline{X} \equiv -\frac{\partial \overline{E}_A}{\partial x},$$

$$= -\sum_r \frac{\partial (\epsilon_r \overline{n}_r)}{\partial x} = -\sum \frac{\partial \epsilon_r}{\partial x} \overline{n}_r$$

$$= -\sum_r \frac{\partial \epsilon_r}{\partial x} \times \frac{\omega_r}{\lambda^{-1} \theta^{-\epsilon_r} \pm 1},$$

$$= -\sum_r \frac{\partial \epsilon_r}{\partial x} \frac{\omega_r \lambda \theta^{\epsilon_r}}{1 \pm \lambda \theta^{\epsilon_r}}),$$

we obtain

$$\left(\frac{\partial F_A}{\partial x}\right)_{\theta, N} = \sum_r \left(\frac{\partial F_A}{\partial \epsilon_r}\right)_{\theta, N_A, \lambda_A} \frac{\partial \epsilon_r}{\partial x} = \sum_r \frac{\partial \epsilon_r}{\partial x} \frac{\omega_r \lambda_r \theta^{\epsilon_r}}{1 \pm \lambda_A \theta^{\epsilon_r}} = -\overline{X}. \tag{9}$$

Further from Eq. (4), we derive

$$\left(\frac{\partial F_A}{\partial N_A}\right)_{\theta, x} = \left(\frac{\partial F_A}{\partial N_A}\right)_{\theta, x, \lambda_A} = -\frac{\ln \lambda_A}{\ln \theta}. \tag{10}$$

Hence, substituting Eqs. (8)–(10) into Eq. (7) and summing for all species, we obtain using Eq. (3)

$$d(F \ln \theta) = \overline{E}\, d \ln \theta - \ln \theta \sum \overline{X} dx - \sum_A \ln \lambda_A\, dN_A. \tag{11}$$

We now substitute in this equation from Eq. (2), which *tentatively defined T* as a certain function of θ, and obtain for $\partial(F/T)$, the expression

$$\partial\left(\frac{F}{T}\right) = -\frac{\overline{E}}{T^2} dT - \frac{1}{T} \sum \overline{X} dx + k \sum_A \ln \lambda_A\, dN_A. \tag{12}$$

If we now compare Eq. (12) *obtained statistically* by using the tentative definitions given by Eqs. (2)–(4) of *T* and *F*, with the *thermodynamic formula*, Eq. (26), derived in Section 18.14, namely

$$d\left(\frac{F}{T}\right) = -\frac{E}{T^2} dT - \frac{1}{T} \sum X dx + \frac{1}{T} \sum \mu dN, \tag{26}$$

we find that the dependence of *F* on *T* and *x* is identical in the two formulae, provided we identify *E* with $\overline{E}$ and *X* with $\overline{X}$. This partly justifies our identification of *T* and *F* defined according to Eqs. (2)–(4) with the thermodynamic temperature and free energy, respectively. To complete the justification for this identification, we must show that the terms of Eqs. (12) and (26), of Section 18.14, in dN_A are also equivalent to each other. We can tentatively make these terms equivalent by assuming that *each λ is related to a partial potential μ according to a relation of the form*

$$\mu_A = kT \ln \lambda_A, \tag{13}$$

or

$$\lambda_A = e^{\mu_A/kT}. \tag{14}$$

We shall show that these relations between the *statistical* λ's and the *thermodynamic* μ's are correct, and thus complete the justification for our assumption that *F* defined by Eqs. (3) and (4) is identical with the *thermodynamic free energy of Helmholtz.*

The criterion for the correctness of the tentative identification of Eq. (13) or (14) is that the same conditions for complete equilibrium should be obtained by the *statistical* method as by the *thermodynamic*. We cannot at this stage give the most general proof of this, but *in effect*, there are only two independent types of equilibria of importance, namely the *equilibrium for a particular kind of system partitioned between several parts (phases) of the assembly (physical phase equilibrium)* and that for *the interconversion of different types of systems (chemical equilibrium).* We shall now show that Eq. (14) leads correctly to the conditions for physical phase equilibrium; we postpone consideration of chemical equilibrium to Chapter 20.

Let us examine first the statistical condition for equilibrium in a two-phase assembly. In all our derivations the subscripts A, B, . . . each referred to a set of *identical* systems, that is systems which have the *same accessible states*. This means that *all the* A's *must be not only chemically identical but also in the same enclosure*. If we have *two sets of systems of identical chemical nature but confined to different enclosures (phases)* in thermal contact, then all our formulae are applicable to the assembly consisting of the *two enclosures*, provided we denote the *two separate sets of chemically identical systems* by different symbols A, A′ ; B , B′; and so on. The two parts of the assembly will have a common temperature (θ or T), but there will in general be different values of the λ's in the separate enclosures, say λ_A , λ_B ... in the one and λ'_A , λ'_B ,... in the other. Suppose now that we replace the *barrier* by a *membrane permeable only to the systems A (including, of course,* A′). The A's and A′'s redistribute themselves until equilibrium is reached. Only in the special case that the two enclosures (phases) were already in equilibrium *will no change take place*. But when all the states primed and unprimed are accessible to all the systems of type A, statistical theory *requires that there must be a common value of* λ_A *for all the systems A whichever enclosure they happen to be in*. Hence, in the special case, that no change takes place when the barrier is removed, we know that λ_A must have already had the same value in both enclosures before the barrier was removed. Thus *the condition for equilibrium between the two enclosures (phases) with respect to the system A is*

$$\lambda'_A = \lambda_A. \tag{15}$$

There will be a similar equilibrium condition for each other chemical species B, C, ... We shall in Chapter 20 give somewhat different derivation of the equilibrium condition given by Eq. (15).

We can now show that the *statistical condition* of equilibrium expressed by Eq. (15) is equivalent to the *thermodynamic condition*. We saw in Section 18.14 that the *thermodynamic condition for complete equilibrium* could be expressed in the form

$$dT = 0, \quad dx = 0, \quad dF = 0 \text{ (equilibrium)}. \tag{16}$$

In the present example, the process to be considered is the transfer, at constant temperature and constant geometry, of some of the single species A from one of the two phases to the other. For this simple process, the conditions in Eq. (16) reduce to

$$\mu_A \, dN_A + \mu'_A \, dN'_A = 0, \tag{17}$$

subject to

$$dN_A + dN'_A = 0, \tag{18}$$

the primed symbols referring to the second phase. We can eliminate dN_A, dN'_A from Eqs. (17) and (18) and so obtain as *the thermodynamic condition for equilibrium between two phases the form*

$$\mu'_A = \mu_A. \tag{19}$$

If now the λ'_A and μ's are related according to Eq. (13), we may replace Eq. (19) by

$$\lambda'_A = \lambda_A, \tag{20}$$

in agreement with the *statistical condition* given by Eq. (15). We have thus shown that the same condition in Eq. (15) or (20) for the (physical) equilibrium of a given type of system between two phases is obtained both from *statistical laws* and from the *thermodynamic law* of *making F minimum for given T and x*. We may thus be assured that *F* defined by Eqs. (3) and (4) has all the essential properties of the *thermodynamic free energy*, and so comparison of Eqs. (12) and (26) of the Section 15.14 shows that the *statistical* and *thermodynamic* temperature *scales* are correctly related by Eq. (2).

18.16. Thermodynamic transcription

Having now established the complete equivalence of the equilibrium laws derivable from statistical mechanics and from thermodynamics, we are justified in making all use of thermodynamic formulae wherever convenient. It is therefore expedient to recapitulate the relations between the *statistical and thermodynamic functions*. First, we have the relation between the two temperature scales.

$$\theta = e^{-1/kT} \tag{1}$$

or

$$T = \frac{1}{k \ln(1/\theta)}. \tag{2}$$

The *thermodynamic* total energy *E* is simply equal to the *statistical average* energy $\overline{E}$, while the Helmholtz free emerge F can, according to the following two equations

$$\overline{X} = -\sum_r \frac{\partial \epsilon_r}{\partial x} \omega_r \lambda \frac{\partial}{\partial \lambda} \ln\left(1 \pm \lambda \theta^{\epsilon_r}\right)^{\pm 1} \tag{3}$$

$$= -\sum_r \frac{\partial \epsilon_r}{\partial x} \frac{\omega_r \lambda \theta^{\epsilon_r}}{1 \pm \lambda \theta^{\epsilon_r}},$$

$$d\left(\frac{F}{T}\right) = -\frac{\overline{E}}{T^2} dT - \frac{1}{T} \sum \overline{X} dx + k \sum_A \ln \lambda_A \, dN_A, \tag{4}$$

be regarded as the potential of the average force *X* for isothermal changes. Thus we have the strikingly simple relations

$$E = \overline{E}, \tag{5}$$

$$\frac{\partial F}{\partial x} = \frac{\partial \overline{E}}{\partial x} = -\overline{X}. \tag{6}$$

If we substitute Eq. (1) into

$$F_A \ln \theta = -\frac{F_A}{kT} = \sum_r \omega_r \ln\left(1 \pm \lambda_A \theta^{\epsilon_r}\right)^{\pm 1} - N_A \ln \lambda_A , \tag{7}$$

we obtain

$$F_A = -kT \sum_r \omega_r \ln\left(1 \pm \lambda_A \theta^{-\epsilon_r/kT}\right)^{\pm 1} + N_A kT \ln \lambda_A, \tag{8}$$

while *F* is given by

$$F = \sum_A F_A . \tag{9}$$

Similarly, substituting Eqs. (1) and (5) into the equation

$$\overline{E}_A = \sum_r \overline{n}_r \epsilon_r = \theta \frac{\partial}{\partial \theta} \sum_r \omega_r \ln\left(1 \pm \lambda_A \theta^{\epsilon_r}\right)^{\pm 1} = \sum_r \frac{\omega_r \epsilon_r}{\left(\lambda_A \theta^{\epsilon_r}\right)^{-1} \pm 1}, \tag{10}$$

and also using Eq. (20) in Section 18.14, we obtain

$$E_A = -T^2 \frac{\partial\left(F_A/T\right)}{\partial T} = \sum_r \frac{\omega_r \epsilon_r}{e^{(\epsilon_r/kT)}/\lambda_A \pm 1} \tag{11}$$

while *E* is given by

$$E = \sum_A E_A . \tag{12}$$

We also have the relation between the statistical parameters λ_A and the partial potentials μ_A

$$\mu_A = kT \ln \lambda_A . \tag{13}$$

An appropriate thermodynamic name for λ_A is the *absolute activity* of A, to distinguish it form the *relative activity** a_A defined by

a_A,

$$\mu_A = \mu_A(T) + kT \ln a_A \tag{14}$$

* Lewis and Randall, *Thermodynamics*, p. 255 (McGraw-Hill, 1923).

where $\mu_A^0(T)$ is arbitrarily fixed at each temperature. The condition for heterogeneous equilibrium of systems of type A between two parts I and II of an assembly can then be expressed in the equivalent forms

$$\mu_A^I = \mu_A^{II}, \tag{15}$$

$$\lambda_A^I = \lambda_A^{II}. \tag{16}$$

These relations between the statistical quantities and the thermodynamic quantities lead, as we have already shown, to the thermodynamic formula

$$d\left(\frac{F_A}{T}\right) = -\frac{E_A}{T^2}dT - \frac{1}{T}\sum X_A dx + \frac{1}{T}\mu_A dN_A, \tag{17}$$

or by summation over all types of systems

$$d\left(\frac{F}{T}\right) = -\frac{E}{T^2}dT - \frac{1}{T}\sum X dx + \frac{1}{T}\sum \mu dN. \tag{18}$$

As already described in Section 18.14, we can define the entropy *S* by

$$S = \sum S_A, \tag{19}$$

$$S_A = (E_A - F_A)/T, \tag{20}$$

and deduce that

$$\left(\frac{\partial F_A}{\partial T}\right)_{x,N_A} = -S_A. \tag{21}$$

By combining Eq. (20) with Eqs. (8) and (11), we obtain the statistical formula for the entropy of the systems A

$$S_A = k\sum_r \omega_r \ln\left(1 \pm \lambda_A e^{-\epsilon_r/kT}\right)^{\pm 1} + \frac{1}{T}\sum_r \frac{\omega_r \epsilon_r}{e^{\epsilon_r/kT}/\lambda_A \pm 1}$$
$$- N_A k \ln \lambda_A. \tag{22}$$

18.17. Thermodynamic formulae of classical statistics

We have shown above that both Fermi-Dirac and Bose-Einstein statistics lead to the same thermodynamic laws. Since the statistical formulae for localised systems can be derived formally as a limiting case of either the Fermi-Dirac or the Bose-Einstein formulae by assuming $\lambda \ll 1$, it is evident that this third form of statistics also leads to the thermodynamic laws. We have already mentioned that for almost all assemblies, the condition $\lambda \ll 1$ is satisfied, and the three forms of statistics can take the common form of classical statistics. We then have with sufficient accuracy

$$\ln\left(1 \pm \lambda_A e^{-\epsilon_r/kT}\right)^{\pm 1} = \lambda_A e^{-\epsilon_r/kT} \tag{1}$$

and Eq. (8), in Section 18.16, reduces to

$$F_A = -kT\lambda_A \sum_r \omega_r e^{-\epsilon_r/kT} + N_A kT \ln \lambda_A,$$

or

$$= -kT\lambda_A f_A(T) + N_A kT \ln \lambda_A. \tag{2}$$

But, as shown before by Eq. (5) in Section 18.11, we have

$$\lambda_A = N_A / f_A(T), \tag{3}$$

so that by substituting Eq. (3) into Eq. (2), we obtain

$$F_A = -N_A kT \{\ln f_A(T) - \ln N_A + 1\}. \tag{4}$$

Differentiating Eq. (4) with respect to N_A and using the definition of μ_A, given before by Eq. (25) in Section 18.14, namely,

$$\mu_A = \left(\frac{\partial F}{\partial N_A}\right), \tag{5}$$

we obtain

$$\mu_A = kT \ln \frac{N_A}{f_A(T)}, \tag{6}$$

which could alternatively be obtained by substituting Eq. (3) into the equation

$$\mu_A = kT \ln \lambda_A.$$

given in Section 18.16.

It is important to point out that Eqs. (4) and (6) are commonly made use of owing to the fact that once the partition functions $f_A(T)$ for various types of assembly are known, the Helmholtz free energy *F* of the assembly can be constructed by means of Eq. (4), and then any required equilibrium properties can be derived by the usual thermodynamic formulae. For instance, for the energy E_A, we obtain from Eq. (4) using Eq. (20) in Section 18.14, the equations

$$E_A = -T^2 . \frac{\partial (F_A/T)}{\partial T},$$

$$= -T^2 . \left\{- N_A k \frac{\partial \ln f_A(T)}{\partial T}\right\},$$

$$= N_A kT^2 \frac{\partial \ln f_A(T)}{\partial T}. \tag{7}$$

Also, using Eq. (4), the generalised force *X* corresponding to the coordinate *x* takes the form

$$X = -\frac{\partial F_A}{\partial x} = N_A kT \frac{\partial \ln f_A(T)}{\partial x}. \tag{8}$$

The contribution $C_A^{(x)}$ of the systems A to the *heat capacity of the assembly, for constant geometrical parameters x*, is *defined* by

$$C_A^{(x)} = \frac{\partial E_A}{\partial T}, \tag{9}$$

and thus substituting from Eq. (7) into Eq. (9), we obtain*

$$C_A^{(x)} = N_A k \frac{\partial}{\partial T}\left\{T^2 \frac{\partial \ln f_A(T)}{\partial T}\right\}$$

$$= N_A k \left(\frac{1}{T}\right)^2 \frac{\partial^2 \ln f_A(T)}{\partial (1/T)^2}. \qquad (10)$$

18.18. Boltzmann constant and the gas constant

The constant k merely fixes the size and sign of the degree on the T scale. We shall discuss first the sign. If we expand $f_A(T)$ in Eq. (10) in Section 18.17 and perform the differentiation, we obtain

$$\frac{C_A^{(x)}}{N_A k} = \left(\frac{1}{T}\right)^2 \frac{\partial^2}{\partial (1/T)^2}\left\{\ln \sum\nolimits_r \omega_r\, e^{-\epsilon_r/kT}\right\}$$

$$= \frac{1}{k^2 T^2}\left[\frac{\sum_r \omega_r \epsilon_r^2\, e^{-\epsilon_r/kT}}{\sum_r \omega_r\, e^{-\epsilon_r/kT}} - \left(\frac{\sum_r \omega_r \epsilon_r\, e^{-\epsilon_r/kT}}{\sum_r \omega_r\, e^{-\epsilon_r/kT}}\right)^2\right] >> 0. \qquad (1)$$

The universal convention for the absolute temperature is that *E increases* as *T increases*; consequently, we must choose *k positive.*

The Kelvin scale of temperature T is *obtained by adjusting the value of k so that 100 degrees separate the ice point and the steam point (at one atmosphere) of water*. The required value of *k* is 1.371×10^{-16} erg/degree. With this assigned value, *k* is a *universal constant* known as *Boltzmann's constant.*

According to the definition of the degree,† the ice point lies within a few hundredths of a degree of 273.15 K, but its exact value will always be a matter of some uncertainty. This small uncertainty is of trivial importance at ordinary temperatures, but it can cause serious ambiguity when one expresses temperatures below 1 K on the *Celsius scale* (commonly called the *centigrade scale*), which has its zero at the ice point. Giauque‡ has therefore made the useful suggestion that the degree should be redefined by fixing *exactly* the ice point. Whether this is fixed at 273.1 K or 273.2 K, or at any intermediate value, the difference between the steam point and the ice point will *still* be 100 degrees within the present experimental uncertainty.

The average value of any extensive property *per mole* is equal to that *per molecule* multiplied by Avogadro's number $N_a = 6.06_4 \times 10^{23}$. When energies are expressed *per mole, then k has to be replaced by*

$$N_a k = R. \qquad (2)$$

---

* Note that $\frac{\partial(1/T)}{\partial T} = -\frac{1}{T^2}$, so that $\partial(1/T) = -\frac{1}{T^2}\partial T$ or $\partial T = -T^2\,\partial(1/T)$

† See Section 5.1.

‡ Ciauque, Nature **143**, 623 (1939).

This quantity is called the gas constant and has a value equal to 8.314×10^7 ergs/deg, or 8.314 joules/deg, or 1.986 cal./deg.

18.19. Distribution laws in terms of the Kelvin temperature *T*

Now that we have established the relation given by Eq. (1) in Section 18.16 between θ and T, it is convenient to reformulate the *distribution law* in terms of T. Thus by substituting this equation, namely,

$$\theta = e^{-1/kT}, \tag{1}$$

into the equation

$$\begin{aligned} \bar{n}_r &= \omega_r \lambda_A \frac{\partial}{\partial \lambda_A} \ln \left(1 \pm \lambda_A \theta^{\epsilon_r}\right)^{\pm 1} \\ &= \frac{\omega_r}{\lambda_A^{-1} \theta^{-\epsilon_r} \pm 1}, \end{aligned} \tag{2}$$

which is Eq. (1) in Section 18.10, we obtain

$$\bar{n}_r = \frac{\omega_r}{e^{\epsilon_r/kT}/\lambda_A \pm 1}, \tag{3}$$

with λ_A determined by the necessary equality

$$\sum_r n_r = N_A, \tag{4}$$

is expressed as, shown before in Section 18.17, by

$$\lambda_A = N_A/f(t). \tag{5}$$

For classical statistics $(\lambda_A << 1)$, Eq. (3) reduces to

$$\bar{n}_r = \lambda_A \omega_r e^{-\epsilon_r/kT}, \tag{6}$$

and we have with the help of Eq. (5) the following relations:

$$\begin{aligned} \bar{n}_r &= \lambda_A \omega_r e^{-\epsilon_r/kT} = N_A \frac{\omega_r e^{-\epsilon_r/kT}}{\sum_r \omega_r e^{-\epsilon_r/kT}} \\ &= N_A \frac{\omega_r e^{-\epsilon_r/kT}}{f_A(T)}. \end{aligned} \tag{7}$$

Eqs. (6) and (7) could equally have been obtained by substituting Eq. (1) into

$$\bar{n}_r = \lambda_A \omega_r \theta^{\epsilon_r}, \tag{8}$$

or

$$\begin{aligned} \bar{n}_r &= N_A \omega_r \theta^{\epsilon_r} \Big/ \sum_r \omega_r \theta^{\epsilon_r} \\ &= N_A \omega_r \theta^{\epsilon_r} / f_A(\theta). \end{aligned} \tag{9}$$

These two equations have been derived in Section 18.11 and denoted by Eqs. (4) and (8), respectively.

CHAPTER 19

Permanent Perfect Gases

19.1. Nature of perfect gases

We may define a perfect gas as an assembly of systems between which the mutual energy of interactions is negligible. We may then regard the energy of the whole assembly as the sum of the kinetic energy of translation of the individual systems, and the rotational, vibrational, and electronic energies of the individual systems, referred to axes passing through the centre of mass of each system. For convenience, we shall refer to the sum of the *rotational, vibrational, and electronic energies as the internal energy*. This being so, the partition function, *which we shall denote from now onwards by Z* [instead of $f(\theta)$ or $f(T)$ used before in Chapter 18], *splits into two factors*, one Z_{trans} for the translational energy, and the other Z_{int} for the internal energy as we have done before in Chapter 11. If the rotational and the vibrational motions may be treated as independent, then Z_{int} factorises into Z_{rot} for rotational energy, Z_{vib} for vibrational energy, and Z_{elec} for electronic energy. At present, we are unjustifiably ignoring the nuclear structure, and treating the nucleus as a massive point. We shall correct this inaccuracy in Section 19.11.

We shall begin by considering the translational motion and deduce the laws of perfect gases. Since we shall see that *monatomic molecules have no variable internal energy*, many of their properties are completely determined by the *translational partition function* Z_{trans}. We shall afterwards consider the rotational and vibration motions so as to deduce the properties of gases having *diatomic and polyatomic molecules*.

We already know from Chapter 18 that the Helmholtz free energy *F* and the total energy *E* were each expressed as the sum of contributions.

$$F_A, F_B, \ldots$$

and

$$E_A, E_B, \ldots$$

of sets of systems A, B, . . . *The condition for F and E to be expressible in this form is the absence of any appreciable interaction between the systems A and B* . . . This requirement is satisfied if either (1) the several sets A, B, . . . are in *different enclosures*, or (2) the several sets are *perfect gases* in the *same enclosure*. The former condition is fulfilled when $F_A, F_B, \ldots$ and $E_A, E_B, \ldots$ denote the contributions of *several homogeneous phases*; in the assemblies under discussion in this chapter, it is the *second condition which is fulfilled.*

19.2. Quantised translational motion

We have already derived in Chapter 18 the formulae for the translational motion of an assembly of particles in an enclosure. We are interested only in the case where *classical statistics* is valid. If the mass of the molecules of type A is m_A, then by Eq. (10) in Section 18.12, the *translational partition function* $Z_{A\,\mathrm{trans}}$ for these molecules is

$$Z_{A\,\mathrm{trans}} = (2\pi\, m_A\, kT)^{3/2}\, \frac{V}{h^3}. \tag{1}$$

Consequently, by the equation

$$F_A = N_A\, kT\, (\ln Z_A - \ln N_A + 1),$$

which has also been derived before in Section 18.17, Eq. (4), the Helmholtz free energy F_A is given by

$$\begin{aligned} F &= -\,kT\,\Sigma_A N_A \left\{\ln Z_{A\,\mathrm{trans}}\, Z_{A\,\mathrm{int}} - \ln N_A + 1\right\} \\ &= -kT\,\Sigma_A N_A \left\{\ln Z_{A\,\mathrm{trans}} - \ln N_A + 1\right\} \\ &\quad -kT\,\Sigma_A N_A \ln Z_{A\,\mathrm{int}}, \\ &= -kT\,\Sigma_A N_A \left\{\ln \frac{(2\pi\, m_A\, kT)^{3/2}\, V}{h^3} - \ln N_A + 1\right\} \\ &\quad -kT\,\Sigma_A N_A \ln Z_{A\,\mathrm{int}}, \\ &= -\,kT\,\Sigma_A N_A \left\{\ln \frac{(2\pi m_A\, kT)^{3/2}\, V}{h^3} - \ln N_A + 1\right\} \\ &\quad + F^{(Z_{\mathrm{int}})} \; . \end{aligned} \tag{2}$$

Here $F^{(Z_{\mathrm{int}})}$ is the contribution of the internal (including rotational) degrees of freedom, and from their nature, we may expect $Z_{A\,\mathrm{int}}$ and so $F^{(Z_{\mathrm{int}})}$ to be independent of *V*. If we now regard *V* as a parameter or *generalised coordinate*, then the *conjugate generalised force* is *the pressure P*, and consequently,

$$\begin{aligned} P &= -\partial F_A/\partial V \\ &= \Sigma_A N_A\, kT/V. \end{aligned} \tag{3}$$

This is the *familiar equation of state for a mixture of perfect gases* and includes the *laws of Boyle, Charles, Avogadro, and Dalton.* For an assembly of systems of a *single type*, Eq. (3) reduces to

$$P = NkT/V. \tag{4}$$

It is hardly necessary to discuss the validity of these formulae. Since actual gases are not perfect, the properties of perfect gases cannot strictly be said to be observed. They must be obtained by *extrapolation to zero concentration* from the actual observations at ordinary concentrations. This presents no serious difficulty and introduces little uncertainty into the results. It is a *commonplace* that Eq. (3) is accurately the limit of the actual equation of state for all permanent gases or gas mixtures at all temperatures, except very near to the absolute zero, when *Bose-Einstein* or *Fermi-Dirac statistics* must be used. For most of the simpler gases, the equation of state is already very near to its limiting form at normal pressures of the order of one atmosphere, even if the temperature is low.

We shall require to use two alternative forms of Eq. (3).

(i) If we define the *partial pressure* p_A of systems A in a gas, *whether perfect or not*, by

$$p_A = \frac{N_A}{\Sigma_A N_A} P, \quad \Sigma_A p_A = P, \tag{5}$$

then for a *perfect gas* we have, according to Eq. (3),

$$p_A = N_A kT/V. \tag{6}$$

(ii) For any *phase whatever* the *partial molecular volume* V_A of a system *A is defined thermodynamically by*

$$V_A = \left(\frac{\partial V}{\partial N_A}\right)_{T,P}, \tag{7}$$

so that for a perfect gas mixture we obtain from Eq. (3) by differentiation with respect to N_A, we obtain

$$V_A = kT/P. \tag{8}$$

We also deduce from Eq. (2) that the *total energy* of the gas is

$$E = -T^2\left(\frac{\partial F/T}{\partial T}\right)_V = \Sigma_A \frac{3}{2} N_A kT + E^{(Z_{int})}, \tag{9}$$

where $E^{(Z_{int})}$ is the energy due to the internal degrees of freedom (rotational and vibrational) and is independent of *V* (or *P*).

The partial *molecular energy* E_A of a system A in any phase is *defined thermodynamically by*

$$E_A = (\partial E/\partial N_A)_{T,P}. \tag{10}$$

When E_A is correctly defined by Eq. (10), then for any kind of assembly

$$E = \Sigma_A N_A E_A. \tag{11}$$

According to Eqs. (9) and (10), we thus obtain for a perfect gas

$$E_A = \frac{3}{2} kT + E^{(Z_{int})}, \tag{12}$$

where $E^{(Z_{int})}$ is independent of *V* (or *P*).

Incidentally, since *E* for a perfect gas is independent of *V* (or *P*) at constant temperature, we have from Eq. (10)

$$(\partial E/\partial N_A)_{T,V} = E_A = (\partial E/\partial N_A)_{T,P}, \tag{13}$$

but it is important to remember that the equality in this equation is peculiar to perfect gases, while the relations in Eqs. (10) and (11) are true for all homogeneous phases. For free electrons* in a metal, Eq. (11) is true but not Eq. (13).

19.3. Classical and unexcited degrees of freedom

For any classical degree of freedom, the partition function is given according to the limiting process by *replacing the sum by an integral.* Thus, for a single classical degree of freedom

$$Z = \frac{1}{h} \int e^{-\epsilon/kT} dx\, dp_x . \tag{1}$$

As we have seen before, the condition for a particular degree of freedom to be *classical is that all the separations between successive energy levels are much smaller than kT.* The opposite extreme case occurs when

$$\epsilon_1 - \epsilon_0 >> kT , \tag{2}$$

where ϵ_0 and ϵ_1 denote the energies of the lowest and next to the lowest energy levels. In this case, the partition function reduces to its *first term*,

$$Z = \omega_0 \, e^{-\epsilon_0/kT} , \tag{3}$$

and we refer to such a degree of freedom as an *unexcited degree of freedom.* It will usually be convenient to define *the zero of energy* for such a degree of freedom as that of the lowest state so that

$$\epsilon_0 = 0 , \tag{4}$$

and then Z reduces to

$$Z = \omega_0 . \tag{5}$$

19.4. Electronic degrees of freedom

One of the factors of the partition function Z_{int} for internal energy is that due to the electronic degrees of freedom, of which there are *four (three translational and one of spin)* for each electron. Under ordinary conditions, the separation between the lowest energy level ϵ_0 and the next lowest ϵ_1 of most, but not all, normal molecules greatly *exceeds kT.* The electronic degrees of freedom are then *unexcited*, and the electronic partition function Z_{elec}

$$Z_{elec} = \Sigma_r \, \upsilon_r \, e^{-\epsilon_r/kT} , \tag{1}$$

is reduced to its first term

$$Z_{elec} = \upsilon_0 \, e^{-\epsilon_0/kT} , \tag{2}$$

---

* See R. Fowler and E.A. Guggenheim, *Statistical Thermodynamics*, Chapter 3, Section 301, (Cambridge, 1965).

where υ_r denotes the weight of the rth electronic level. If we choose as *zero the normal energy level of the electron*, that is, $\epsilon_0 = 0$, then the electronic partition function Z_{elec} becomes simply

$$Z_{elec} = \upsilon_0 . \tag{3}$$

We have therefore merely to insert the factor υ_0 into the partition function, and may otherwise *ignore the electronic degrees of freedom.* In the rather exceptional case, where the *second electronic level may not be ignored*, we have merely to substitute

$$\upsilon_0 + \upsilon_1 \, e^{-\epsilon_1/kT}$$

for υ_0 in Eq. (3), that is,

$$Z_{elec} = \upsilon_0 + \upsilon_1 \, e^{-\epsilon_1/kT} . \tag{4}$$

This is the case for NO. A discussion of the values of υ_0 shall be considered for various molecules in Chapter 20.

19.5. Heat capacities of perfect gases

According to Eq. (12) in Section 19.2, the energy *per molecule* of a *perfect gas of one component is of the form*

$$E = \frac{3}{2} kT + E^{(Z_{int})}, \tag{1}$$

where $E^{(Z_{int})}$ is the energy due to the internal degrees of freedom (rotational and vibrational) and by its nature is independent of V (or P).

By definition, C_V, the molecular heat capacity at constant volume, is given by

$$C_V = \left(\frac{\partial E}{\partial T} \right)_V , \tag{2}$$

and therefore by Eq. (1)

$$C_V = \frac{3}{2} k + \frac{d E^{(Z_{int})}}{dT} . \tag{3}$$

Similarly by the definition of C_P, the molecular heat capacity at constant pressure, we have

$$C_P = \left\{ \frac{\partial (E + PV)}{\partial T} \right\}_P$$

$$= \left(\frac{\partial E}{\partial T} \right)_P + P \left(\frac{\partial V}{\partial T} \right)_P . \tag{4}$$

Since for a perfect gas at constant temperature *E* does not depend on *P* or *V*, we have, using Eq. (8) in Section 19.2, which is

$$V_A = \frac{kT}{P}, \tag{5}$$

$$C_P = C_V + P\left(\frac{\partial V}{\partial T}\right)_P,$$

$$= \frac{3}{2}k + \frac{dE^{(Z_{int})}}{dT} + k. \tag{6}$$

Comparing Eqs. (3) and (6), we see that

$$C_p - C_V = k. \tag{7}$$

The relation in Eq. (7) is well known to be obeyed accurately by gases so that the values of C_V and C_P are both determined when we know either separately, or their ratio C_P/C_V, usually denoted by γ.

In order to progress further, we have to know the form of $E^{(Z_{int})}$ and this requires a discussion of the internal degrees of freedom. For this purpose, it is most convenient to deal separately with gases having monatomic, diatomic, and polyatomic molecules. It is sufficient here to consider, for simplicity, monatomic molecules.

19.6. Monatomic molecules

Apart from the electronic degrees of freedom, which may be assumed *unexcited*, a monatomic molecule has no internal degrees of freedom. But, in fact, this statement is not exactly true and will require to be revised when we consider nuclear spin. The internal partition function Z_{int} thus reduces to υ_0 when the lowest electronic level is chosen *as energy zero* as implied by Eq. (3) in Section 19.4. Hence for the *resultant partition function of a monatomic molecule*, we have

$$Z_{trans} = \frac{(2\pi m k T)^{3/2} V}{h^3} \upsilon_0. \tag{2}$$

According to the *Maxwell-Boltzmann* expression for the Helmholtz free energy *F* of an assembly of *N* molecules, which is

$$F = -NkT\left\{\ln\left(\frac{Z}{N}\right) + 1\right\},$$

it is therefore given by

$$F = -NkT \ln\left\{\frac{(2\pi m k T)^{3/2} V \upsilon_0}{h^3}\right\} + NkT(\ln N - 1),$$

$$= -NkT\left\{\ln\frac{(2\pi mk)^{3/2}\upsilon_0}{h^3} + \frac{3}{2}\ln T + \ln V\right\} - NkT$$

$$+ NkT\ln N$$

$$= -NkT\left\{\ln\frac{(2\pi mk)^{3/2}\upsilon_0}{h^3}\right\} - \frac{3}{2}NkT\ln T$$

$$+ NkT\ln N - NkT\ln V - NkT,$$

$$= NkT\left\{\ln\frac{N}{V} - 1\right\} - \frac{3}{2}NkT\ln T$$

$$-NkT\left\{\ln\frac{(2\pi mk)^{3/2}\upsilon_0}{h^3}\right\}. \quad (3)$$

For the energy *E*, we have

$$E = NkT^2\,\frac{d(\ln Z)}{dT}, \quad (4)$$

where $\ln Z$ is given by

$$\ln Z_{trans} = \frac{3}{2}\ln(2\pi mkT) + \ln V - 3\ln h + \ln\upsilon_0,$$

and,

$$\frac{\partial \ln Z}{\partial T} = \frac{3}{2}\,\frac{1}{T},$$

so that the substitution in Eq. (4) gives

$$E = \frac{3}{2}NkT. \quad (5)$$

Also *E* can be obtained from the relation

$$E = -T^2\left(\frac{\partial F/T}{\partial T}\right)_{V,N}, \quad (6)$$

using Eq. (3). Thus, we first have

$$F/T = -Nk\ln\left\{\frac{(2\pi mk)^{3/2}\upsilon_0}{h^3}\right\} +$$

$$Nk\left\{\ln\left(\frac{N}{V}\right) - 1\right\} - \frac{3}{2}Nk\ln T,$$

then

$$\left(\frac{\partial(F/T)}{\partial T}\right)_{V,N} = -\frac{3}{2}Nk\,\frac{1}{T}. \quad (7)$$

Hence, the substitution of this equation into Eq. (6) gives

$$E = -T^2 \times -\frac{3}{2} N k \frac{1}{T}$$

$$= \frac{3}{2} N k T,$$

which is Eq. (5).

For the pressure P, we have

$$P = -\left(\frac{\partial F}{\partial V}\right)_{T,N}, \tag{8}$$

where *F* is given by Eq. (3). Therefore,

$$P = -\left(\frac{\partial F}{\partial V}\right)_{T,V} = \frac{N k T}{V}. \tag{9}$$

For the partial potential μ, we have

$$\mu = \left(\frac{\partial F}{\partial N}\right)_{T,V}. \tag{10}$$

Therefore, using Eq. (3) we obtain

$$\mu = \left(\frac{\partial F}{\partial N}\right)_{T,V} = N k T \times \frac{1}{N} - k T \ln V + k T \ln N - k T$$

$$-k T \left\{\frac{(2\pi m k)^{3/2} \upsilon_0}{h^3}\right\} - \frac{3}{2} k T \ln T,$$

$$= k T \ln\left(\frac{N}{V}\right) - \frac{3}{2} k T \ln T$$

$$-k T \ln \left\{\frac{(2\pi m k)^{3/2} \upsilon_0}{h^3}\right\}. \tag{11}$$

or, using Eq. (6), derived above in Section 19.2, namely,

$$p_A = N_A \frac{k T}{V}, \text{ (for a perfect gas)} \tag{12}$$

we obtain

$$\mu = k T \ln p - \frac{5}{2} k T \ln T - k T \ln \left\{\frac{(2\pi m)^{3/2} k^{5/2} \upsilon_0}{h^3}\right\}. \tag{13}$$

19.7. Heat capacities of monatomic molecules

According to Eq. (5) in Section 19.6, we have for the energy *per molecule* of a monatomic gas, *referred to its lowest state as energy zero*, the equation

$$E = \frac{3}{2} k T. \tag{1}$$

Also, according to the formulae

$$C_v = \frac{3}{2}k + \frac{dE^{(Z_{int})}}{dT}, \qquad (2)$$

and

$$C_P = \frac{3}{2}k + \frac{dE^{(Z_{int})}}{dT} + k, \qquad (3)$$

which have been derived in Section 19.5, we have therefore

$$C_V = \frac{3}{2}k, \qquad (4)$$

$$C_P = \frac{5}{2}k, \qquad (5)$$

$$\gamma = C_P / C_V = \frac{5}{3} \qquad (6)$$

Of the three quantities, C_V , C_P, and γ, the first is extremely difficult to measure and it is usual to determine either C_P by a flow method or γ by measuring the velocity of sound. The experimental data, extrapolated to zero pressure, are then compared with theory as was done* by *Kundt* and *Warburg*; *Rayleigh* and *Ramsay*; *Ramsay*; *Heuse*; *Robitzsch*, and *others*. The theoretical values of C_P and γ for monatomic gases such as He, Ne, A, Kr, Xe, Na, K, Hg, which were investigated, are $C_P/k = 2.5$; $\gamma = C_P/C_V = 1.667$. The agreement with the experimental values is good.

We have assumed above that the electronic degrees of freedom are *unexcited*, and we should perhaps verify this. For the species ordinarily capable of existing as atoms in the free state, the energy ϵ_1 required for excitation from the *normal* to the *first excited state* varies from 4 to 20 eV, while *kT* in electronvolt has the value $8.60 \times 10^{-5}\,T$. Accordingly, ϵ_1/kT has a value which is at least $0.5 \times 10^5/T$, and so in the electronic partition function

$$Z_{elec} = \Sigma_r \upsilon_r e^{-\epsilon_r/kT}, \qquad (4)$$

the first excited term, $\upsilon_1 e^{-\epsilon_1/kT}$, and *a fortiori* all subsequent excited terms are *negligible* compared with the normal term υ_0 unless *T* is at least 10.000 K.

19.8. Diatomic molecules of gases at moderate temperatures

In addition to the types of motion and energy content which they share with free atoms, diatomic molecules possess further types of motion. The atomic nuclei can rotate about their centre of mass, to a first approximation like *a rigid body*, and can vibrate along the line joining them, to a first approximation like *a simple harmonic oscillator*. If the molecule is nearly rigid, so that the frequency of these vibrations is high, the *rotations and vibrations*

---

* See R. Fowler and E. A. Guggenheim, *Statistical Thermodynamics*, Chapter 3, Section 311, (Cambridge, 1965).

are nearly independent of each other. Moreover, at fairly low temperatures, the nuclei vibrations will be *unexcited*, and the whole extra motion reduces to the rotations of a rigid boy. The non-vibrating molecule must indeed *stretch under the centrifugal forces*, but for *stiff* molecules of *high* vibrational frequency, this effect will be small for moderate rotations, that is, at low temperatures.

Just as for monatomic molecules, the electronic motions are *usually completely unexcited*, and the electronic partition function reduces to its first term

$$Z_{elec} = \upsilon_0 \, e^{-\epsilon_0/kT}, \tag{1}$$

or, if the energy zero is chosen as the normal state, the partition function reduces to

$$Z_{elec} = \upsilon_0 . \tag{2}$$

It may, however, happen that there is *another electronic state low enough* to contribute appreciably to the partition function, in which case we have to replace υ_0 by $\upsilon_0 + \upsilon_1 \, e^{-\epsilon_1/kT}$ so that

$$Z_{elec} = \upsilon_0 + \upsilon_1 \, e^{-\epsilon_1/kT} . \tag{3}$$

Important examples actually occur among simple permanent gases (e.g. NO). To account for the properties of diatomic gases, we are therefore chiefly need to consider in some detail the rotational partition functions.

19.9. Rotational degrees of freedom

Apart from the electronic motions (and nuclear spin that will be discussed in Section 19.11), a *diatomic molecule* has zero moment of inertia I_a about the internuclear axis, which we denote by a . Also, the component $\mathbf{P}_a$ of the total angular momentum $\mathbf{P}$ of the molecule along the a axis is zero ($\mathbf{P}_a = I_a \, \omega_a$, where ω_a is the angular velocity). The other principal axes of inertia denoted by I_b and I_c are equal and mutually perpendicular to the internuclear axis a. This is also the case for polyatomic linear molecules. Using classical mechanics, the energy of a rotating linear molecule is

$$P^2/2I_b \tag{1}$$

In Section 8.3, the quantised rotation of molecules had been discussed in some detail. It was shown there that for quantised rotation of a molecule, the letter **J** is used to denote the total angular momentum vector in place of the symbol **P**, and that the permitted values of **J** are

$$\mathbf{J} = \sqrt{J(J+1)} \, h/2\pi, \quad J = 0, 1, 2, \ldots, \tag{2}$$

where J is a quantum number that may be any positive integer, zero included. Also that in the case of a linear molecule $\mathbf{P}^2$ in the classical expression, given by Eq. (1), is replaced by $\mathbf{J}^2$, and then the quantisation condition given by Eq. (2) is introduced to obtain the rotational energy ϵ_{rot} of the molecule; that is,

$$\epsilon_{rot} = J(J+1) \, h^2/8\pi^2 I_b . \tag{3}$$

Further, consideration of the wave functions for rotation of a linear molecule shows that the energy levels are $(2J+1)$-fold degenerate; that is, there are $2J+1$ coincident sub-levels for each value of J.

Based on the above, the rotational partition function Z_{rot} is

$$Z_{rot} = \sum_{J=0}^{\infty} (2J + 1)\, e^{-\epsilon_r/kt}, \tag{4}$$

$$= \sum_{J=0}^{\infty} (2J + 1)\, e^{-hJ(J+1)B/kT}, \tag{5}$$

where B is a constant known as the *rotational constant* of the molecule and equal to $h/8\pi^2 I_b$. Thus, Eq. (5) takes the form

$$Z_{rot} = \sum_{J=0}^{\infty} (2J + 1)\, \theta^{J(J+1)h^2/8\pi^2 I_b kT}. \tag{6}$$

Now if we substitute for

$$h^2/8\pi^2 I_b k$$

by Θ_r in Eq. (6), we obtain

$$Z_{rot} = \sum_{J=0}^{\infty} (2J + 1)\, e^{-J(J+1)\Theta_r/T}. \tag{7}$$

When $\Theta_r \ll T$, as is the case at ordinary temperatures for all diatomic molecules, and even at low temperatures for diatomic molecules not containing a hydrogen atom, we may use the *approximation of Mulholland,* since Eq. (7) representing the rotational partition function cannot be further simplified without making approximations*. Thus putting Θ/T in Eq. (7) equal to r, we have

$$Z_{rot} = \sum_{J+0}^{\infty} (2J + 1)\, e^{-J(J+1)r}. \tag{8}$$

Under the condition that $\theta_r \ll T$ or $r \ll 1$, we can replace the sum in Eq. (8) by an integral; that is,

$$Z_{rot} = \int_0^{\infty} (2J + 1)\, e^{-J(J+1)r}\, dJ. \tag{9}$$

Then using the substitution

$$J(J+1)r = \xi,$$

$$(2J + 1)\, r\, dJ = d\xi, \tag{10}$$

we obtain

$$Z_{rot} = \frac{1}{r}\int_0^{\infty} e^{-\xi}\, d\xi = \frac{1}{r} = \frac{T}{\Theta}. \tag{11}$$

Now a more exact evaluation has been obtained by Mulholland who has shown that

$$Z_{rot} = \frac{1}{r}\left\{1 + \frac{1}{3}r + \frac{1}{15}r^2 + o\, r^3\right\}, \tag{12}$$

---

* Mulholland, *Proc. Camb. Phil. Soc.* **24**, 280 (1928).

or,

$$Z_{rot} = \frac{T}{\Theta_r}\left\{1 + \frac{1}{3}\frac{\Theta_r}{T} + \frac{1}{15}\frac{\Theta_r^{\ 2}}{T^2} + o\left(\frac{\Theta_r}{T}\right)^3\right\}. \quad (13)$$

From Eq. (13), we derive for F_{rot}, the *contribution of the rotational degree of freedom to the Helmholtz free energy*,

$$F_{rot} = -NkT \ln Z_{rot}$$

$$= NkT\left\{\ln\frac{\Theta_r}{T} - \frac{1}{3}\frac{\Theta_r}{T} - \frac{1}{90}\left(\frac{\Theta_r}{T}\right)^2 + o\left(\frac{\Theta_r}{T}\right)^3\right\}. \quad (14)$$

For E_{rot}, the contribution of the rotational degrees of freedom to the total energy

$$E_{rot} = -T^2\frac{\partial(F_{rot}/T)}{\partial T} = NkT\left\{1 - \frac{1}{3}\frac{\Theta_r}{T} - \frac{1}{45}\left(\frac{\Theta_r}{T}\right)^2 + o\left(\frac{\Theta_r}{T}\right)^3\right\}. \quad (15)$$

Finally for C_{rot}, the contribution of the rotational degrees of freedom to the molecular heat capacity, we obtain by differentiating Eq. (15) with respect to *T* and dividing by *N*.

$$C_{rot} = k\left\{1 + \frac{1}{45}\left(\frac{\Theta_r}{T}\right)^2 + o\left(\frac{\Theta_r}{T}\right)^3\right\}. \quad (16)$$

19.10. Classical treatment of rotation of diatomic molecules

We have seen that, at ordinary temperatures, $\Theta_r \ll T$ at least for all molecules not containing hydrogen. This means that the rotational degrees of freedom are classical, and we may therefore make use of the limiting principle to replace the *partition function* by the corresponding *phase integral.** The value of the *phase integral* which gives the limiting form of the partition function Z_{rot} for a rigid rotator is

$$\frac{8\pi^2 I_b kT}{h^2}, \quad (1)$$

in agreement with the leading term of Eq. (13) in Section 19.9, for $\Theta \ll T$, this term being T/Θ

$$\Theta = h^2/8\pi^2 I_b k.$$

* See, for example, Lamb, *Hydrodynamics*. Chapter V (Cambridge, 1924).

It may be briefly noted here that in the derivation of the phase integral for a rigid diatomic molecule, the internuclear axis has the usual spherical polar coordinates θ, ϕ, then the conjugate momenta P_θ, P_ϕ are given by

$$P_\theta = I_b \dot{\theta}, \quad P_\phi = I_b \sin^2\theta . \phi, \qquad (2)$$

the limits of θ being $0, \pi$; those of ϕ $0, 2\pi$; and those of P_θ, P_ϕ each $-\infty, +\infty$. The classical energy equation

$$\frac{1}{2I_b}\left(P_\theta^2 + \frac{1}{\sin^2\theta} P_\phi^2\right) \qquad (3)$$

is used for solving the corresponding Schrődinger equation.

19.11. Nuclear symmetry

In evaluating the phase integral that gives Eq. (1) in Section 19.10 was paid no attention to *symmetry with respect to the two nuclei.* In molecules with two *different* nuclei (heteronuclear molecules), there is no need to do so. In the case of molecules containing two *identical* nuclei (homonuclear molecules) on the other hand, each *half rotation* brings the molecule back to a position *indistinguishable* from that from which it started. If then we use the phase integral discussed in brief in Section 19.11 with the limits there given, we are counting *twice over every physically distinguishable orientation.* The question arises whether one should not therefore correct for this by dividing the result by two. Actually, this procedure leads to correct results. Hence, according to the limiting principle, we must expect a similar correction to arise in a strict *quantal treatment.* We shall now consider how *this arises,* and shall show that it comes from the *necessity of distinguishing between eigenfunctions symmetrical and antisymmetrical in the nuclei.*

Up to the present, we have assumed that the motion of a molecule can be decomposed into translational motion of the centre of mass, rotational motion, vibrational motion, and electronic motion, with a corresponding *factorisation of the eigenfunction.* We have tacitly ignored any possible nuclear structure. Strictly, we must allow for this by introducing nuclear factors, but the relevant factor at terrestrial temperature is the factor *due to the spin of the lowest nuclear state since all other states have energies exceeding that of the lowest state by amounts which are exceedingly large compared with kT.*

As shown before in Section 8.3, a nucleus in capable of rotation, or spin, about an axis. This spin generates a quantised angular momentum vector **I** known as *nuclear spin angular momentum* whose magnitude is given by

$$\mathbf{I} = \sqrt{I(I+1)}\, h/2\pi. \qquad (1)$$

Here *I*, the quantum number for nuclear spin, can assume integral, half-integral, or zero values. Eq. (1) is a typical wave mechanical equation for the quantisation of spin angular momentum vector. The angular momentum **I** can be thought of as the result of the angular momenta of the protons and neutrons of which the nucleus is composed. Each of these fundamental particles is considered to have a spin quantum number $I = \frac{1}{2}$, the value of the proton being experimentally observable. As the deuterium nucleus, or deuteron, possesses one proton and one neutron the resultant spin must be 1 or 0, depending on whether the individual spins are aligned parallel or antiparallel. Experiment shows that the deuteron in the ground state has I = 1 corresponding to parallel alignment.

Owing to this nuclear spin, there will be *orientational quantisation in an applied external magnetic field, with an eigenfunction for each such orientation.* In the absence of an external field, *these orientated states become indistinguishable, but their number* ρ *remains unaltered, and the nucleus is therefore usefully described as being in a state of spin weight* ρ.

Let us first consider a *heteronuclear diatomic molecule* made up of a nucleus a with ρ_a orientational eigenfunctions $\psi_1, \psi_2, \ldots, \psi_{\rho_a}$, and a nucleus b with ρ_b orientational eigenfunctions $\psi'_1, \psi'_2 \ldots, \psi'\rho_b$. Then for the molecule, we have *nuclear eigenfunctions of the type*

$$\psi_r(a)\,\psi'_s(b).$$

Thus the consideration of nuclear spin orientation introduces into the partition function of *the heteronuclear diatomic molecule the extra nuclear weight factor* $\rho_a\,\rho_b$.

Now consider by contrast a *homonuclear diatomic molecule* made up of two identical nuclei a and b *each having* ρ *orientational eigenfunctions*

$$\psi_1, \psi_2, \ldots, \psi_\rho.$$

When the two identical nuclei form part of a single system, such as a diatomic molecule, we can obtain nuclear spin eigenfunctions for the system by *combination of these two sets for the single nuclei.* In particular, we can form

$$\frac{1}{2}\rho(\rho-1)$$

eigenfunctions of the type

$$\psi_r(a)\;\psi_s(b) - \psi_s(a)\,\psi_r(b),$$

antisymmetrical in the nuclei. We can similarly form

$$\frac{1}{2}\rho(\rho-1)$$

eigenfunctions of the type

$$\psi_r(a)\,\psi_s(b) + \psi_s(a)\,\psi_r(b)$$

symmetrical in the nuclei; and also ρ eigenfunctions of the type

$$\psi_r(\mathrm{a})\,\psi_s(\mathrm{b}),$$

also symmetrical in the nuclei. Thus, *in all* we can form $\frac{1}{2}\rho(\rho-1)$ *antisymmetrical* and $\frac{1}{2}\rho(\rho-1)+\rho=\frac{1}{2}\rho(\rho+1)$ *symmetrical* orientational eigenfunctions for the nuclear spins of the molecule. It will be noticed that *the number of antisymmetrical and symmetrical eigenfunctions* is ρ^2, the same as the number of functions for a *heteronuclear diatomic molecule, containing two nuclei with equal* ρ *values.*

We have to next consider the *symmetry properties with respect* to *the nuclei of the eigenfunctions for the translations, vibrations, and rotations of the molecule.* The translational motion, being that *of the centre of mass of the molecule, is independent of the relative positions of the nuclei; the eigenfunction does not contain the coordinates of the nuclei and so is symmetrical in the nuclei.* The vibrational eigenfunction is a *function of the distance apart of the two nuclei. Since this distance is unaffected by an interchange of the two nuclei, the vibrational eigenfunction is also symmetrical in the two nuclei.* The rotational eigenfunctions are the *spherical harmonics*. Those with $J = 0, 2, 4, 6, \ldots$, that is, even rotational quantum number, are *symmetrical* in the nuclei, those with $J = 1, 3, 5, \ldots$, that is, odd number, are *antisymmetrical* in the nuclei.

Now for any given type of nuclei, only those states are accessible which are *antisymmetrical in the nuclei*, or only those which are *symmetrical*. So far as is known, *the antisymmetrical states occur for nuclei with odd* mass numbers, and the symmetrical for those with even mass numbers.* If the nuclei may be regarded as composed of protons and neutrons only, this conforms to the rule that *the eigenfunction for every accessible state is antisymmetrical in electrons, in protons, and in neutrons.*

Let us now consider the application of these considerations to *diatomic molecules*. If the molecule is *heteronuclear*, we have already seen that consideration of nuclear spin merely introduces an extra weight factor $\rho_a \rho_b$. For *homonuclear molecules* the situation is the following.

If the nuclei are of *odd* mass number, then to obtain *resultant* eigenfunctions *antisymmetrical* in the nuclei, the even rotational eigenfunctions (J = 0, 2, 4, . . .), being symmetrical in the nuclei, have to be combined with one of $\frac{1}{2}\rho(\rho - 1)$ *antisymmetrical nuclear functions*, and the odd rotational functions ($J = 1, 3, \ldots$) being antisymmetrical in the nuclei, have to be combined with one of the $\frac{1}{2}\rho(\rho + 1)$ *symmetrical nuclear functions*. Thus the *even* rotational states have an extra weight factor $\frac{1}{2}\rho(\rho - 1)$ and the *odd* rotational states an extra factor $\frac{1}{2}\rho(\rho + 1)$.

On the other hand, when the mass number is *even*, then to obtain resultant eigenfunctions *symmetrical* in the nuclei, the rotational eigenfunctions have to be combined with nuclear eigenfunctions in the reverse manner. The result is that *even* rotational states have an extra weight factor $\frac{1}{2}\rho(\rho + 1)$ and the *odd* rotational states have an extra factor $\frac{1}{2}\rho(\rho - 1)$. We thus obtain the following partition functions $Z_{r.n}$ for the *combined rotations and nuclear orientations.*

Heteronuclear:

$$Z_{r.n} = \rho_a \rho_b \sum_{J=0}^{\infty} (2J + 1)\, e^{-J(J+1)\Theta_r/T}\,, \tag{1}$$

---

* Nuclei with odd mass number must have half-integral values of the spin quantum number I, whereas nuclei with even mass number (hence with even total number of protons and neutrons) must have zero or integral values of I.

Homonuclear, odd mass number, or antisymmetrical states:

$$Z_{r.n} = \frac{1}{2}\rho(\rho-1) \sum_{J=0,2\ldots}^{\infty} (2J+1)\, e^{-J(J+1)\Theta_r/T}$$

$$+ \frac{1}{2}\rho(\rho+1) \sum_{J=1,3}^{\infty} (2J+1)\, e^{-J(J+1)\Theta_r/T}. \tag{2}$$

Homonuclear, even mass number, or symmetrical states:

$$Z_{r.n} = \frac{1}{2}\rho(\rho+1) \sum_{J=0,2\ldots}^{\infty} (2J+1)\, e^{-J(J+1)\Theta_r/T}$$

$$+ \frac{1}{2}\rho(\rho-1) \sum_{J=1,3\ldots}^{\infty} (2J+1)\, e^{-J(J+1)\Theta_r/T}. \tag{3}$$

In the simplest case of nuclei without spin, $\rho = 1$, and we notice that the requirements of *nuclear symmetry* result in the complete disappearance of *alternate* rotational states of a homonuclear molecule. This has been completely verified spectroscopically for the molecules of oxygen of ^{16}O.

For all diatomic molecules, we may use the approximation $\Theta \ll T$ at *ordinary temperatures*. Now it can be shown that under this condition we have with a high degree of accuracy

$$\sum_{J=0,2,\ldots}^{\infty} (2J+1)\, e^{-J(J+1)\Theta_r/T}$$

$$= \sum_{J=1,3,\ldots}^{\infty} (2J+1)\, e^{-J(J+1)\Theta_r/T}$$

$$= \frac{1}{2} \sum_{J=0,1,\ldots}^{\infty} (2J+1)\, e^{-J(J+1)\Theta_r/T}. \tag{4}$$

and consequently the partition functions $Z_{r.n}$ for combined rotations and nuclear spins have the form $Z_{r.n}$

Heteronuclear: $$Z_{r.n} = Z_{rot}\, \rho_a \rho_b, \tag{5}$$

Homonuclear: $$Z_{r.n} = \frac{1}{2} Z_{rot}\, \rho^2, \tag{6}$$

where Z_{rot} is given by

$$Z_{rot} = \sum_{J=0,1,\ldots}^{\infty} (2J+1)\, e^{-J(J+1)\Theta_r/T} \tag{7}$$

This analysis has required a broadening of the classification of the states of a rotating homonuclear molecule. We have hitherto treated atomic nuclei as structureless points, a treatment which has proved inadequate. When a *nucleus* has ρ possible orientations, it must be *assigned a weight* ρ times as great as we have hitherto used for *structureless points*. The diatomic molecule therefore should have a weight $\rho_a\,\rho_b$ due to the orientations of the nuclei. The extra factor ρ^2 in homonuclear molecules reduced alternately to $\frac{1}{2}\rho(\rho-1)$ and $\frac{1}{2}\rho(\rho+1)$ by the requirements of symmetry in the nuclei, and therefore effectively to $\frac{1}{2}\rho^2$, except at very low temperatures. This reduction from ρ^2 to $\frac{1}{2}\rho^2$ is the quantal language of the introduction of the *symmetry number* σ of classical statistics.*

We accordingly define a symmetry number σ equal to 2 if the two nuclei in the molecule are identical and equal to 1 otherwise. We can combine Eqs. (5) and (6) into the single formula

$$Z_{r.n} = Z_{rot}\,\rho_a\,\rho_b/\sigma, \tag{8}$$

when Z_{rot} is defined by Eq. (7). Using Eq. (13) given above in Section 19.9, Eq. (18) can be put in the form

$$Z_{r.n} = \frac{T}{\Theta_r}\left\{1 + \frac{1}{3}\frac{\Theta_r}{T} + \frac{1}{15}\frac{\Theta_r^{\,2}}{T^2} + o\left(\frac{\Theta_r}{T}\right)^3\right\}\frac{\rho_a\,\rho_b}{\sigma} \quad (\Theta \ll T). \tag{9}$$

In most cases, it will be sufficient to retain only the first term so that

$$Z_{r.\,n} = \frac{T}{\Theta_r}\frac{\rho_a\,\rho_b}{\sigma} = \frac{8\,\pi^2\,I_b\,k\,T}{h^2}\frac{\rho_a\,\rho_b}{\sigma}. \tag{10}$$

It has been tacitly assumed that the electronic eigenfunction, which is necessarily *antisymmetrical with respect to any pair of electrons, is symmetrical with respect to the two identical nuclei of the diatomic molecule.* This is in fact the case for the ground electronic state of most symmetrical diatomic molecules.

19.12. Amended formulae for monatomic molecules

If we wish to bring our formulae for monatomic molecules, which we have obtained above, namely Eqs. (1)–(13) of Section 19.6, into line with those for diatomic molecules, we have merely to introduce the *extra factor* ρ *for the number of possible nuclear orientational eigenfunctions.* We have then for the complete partition function of a monatomic molecule

$$Z_{trans} = \frac{(2\pi m k T)^{3/2}}{h^3}\,\upsilon_0\,\rho, \tag{1}$$

* Ehrenfest and Trkal, *Proc. Sec. Sci.*, Amsterdam **23**, 162 (1920).

and so for the Helmholtz free energy *F* of a gas with monatomic molecules, using Eq. (3) of Section 19.6,

$$F = NkT\left\{\ln\left(\frac{N}{V}\right) - 1\right\} - \frac{3}{2}NkT\ln T -$$

$$NkT\ln\left\{\frac{(2\pi mk)^{3/2}}{h^3}\upsilon_0\rho\right\}. \tag{2}$$

For its partial potential $\mu = \left(\frac{\partial F}{\partial N}\right)_{T,V}$, we have

$$\mu = kT\ln\left(\frac{N}{V}\right) - \frac{3}{2}kT\ln T - kT\ln\left\{\frac{(2\pi mk)^{3/2}}{h^3}\upsilon_0\rho\right\}, \tag{3}$$

or

$$\mu = kT\ln - \frac{5}{2}kT\ln T - kT\ln\left\{\frac{(2\pi m)^{3/2}k^{5/2}}{h^3}\upsilon_0\rho\right\}. \tag{4}$$

For the absolute activity

$$\lambda = PT^{-5/2}\frac{h^3}{(2\pi m)^{3/2}k^{5/2}\upsilon_0\rho}. \tag{5}$$

This equation can be derived as follows:

With sufficient accuracy we have shown before in Section 18.17 that

$$\ln\left(1 \pm \lambda_A e^{-\epsilon_r/kT}\right)^{\pm 1} = \lambda_A e^{-\epsilon_r/kT}, \tag{6}$$

and

$$F_A = -N_A kT\left(\ln Z_{A\,\mathrm{trans}} - \ln N_A + 1\right). \tag{7}$$

Differentiating Eq. (7) with respect to N_A and using the definition of the partial potential μ by the relation

$$\mu_A = \left(\frac{\partial F}{\partial N_A}\right)_{T,P}, \tag{8}$$

we obtain

$$\mu_A = kT\ln\frac{N_A}{Z_A}. \tag{9}$$

But

$$\lambda_A = N_A/Z_A, \tag{10}$$

as also shown in Section 18.11, therefore

$$\mu_A = kT\ln\lambda_A. \tag{11}$$

Now substituting for μ in Eq. (4) by Eq. (11), we obtain

$$kT \ln \lambda = kT \ln P - \frac{5}{2} kT \ln T$$
$$-kT \ln \left\{ \frac{(2\pi m)^{3/2}\, k^{5/2}}{h^3} \upsilon_0\, \rho \right\}.$$

Hence,

$$\ln \lambda = \ln P - \ln T^{5/2}$$
$$- \ln \left\{ \frac{(2\pi m)^{3/2}\, k^{5/2}}{h^3} \upsilon_0\, \rho \right\},$$

that is,

$$\lambda = P T^{-5/2} \frac{h^3}{(2\pi m)^{3/2}\, \upsilon_0\, \rho},$$

which is the equation we are looking for.

CHAPTER 20

Statistical Approach to Chemical Equilibria and Evaporation. The Third Law of Thermodynamics and Nernst's Heat Theorem

20.1. Chemical constants

We shall translate in this section the general laws of homogeneous gaseous chemical equilibria into a more familiar thermodynamic form. We consider the general gaseous reactions.

$$aA + bB + \ldots = l\,L + mM + \ldots, \tag{1}$$

where A, B, . . ., L, M, . . . denote molecular species and *a*, *b*, . . ., l, *m*, . . . are small positive integers. We can abbreviate Eq. (1) to

$$\sum aA = \sum l\,L. \tag{2}$$

The condition for chemical equilibrium is

$$\sum a\mu_A = \sum l\,\mu_L, \tag{3}$$

where $\mu\ (= \partial F/\partial N)\ T, P$ is the quantity introduced by *Gibbs* and denoted by μ. It is called partial potential as pointed out before. Now we have for each partial potential μ a formula of definite type, according to whether the species is monatomic, diatomic, and polyatomic linear or non-linear. We have already obtained these formulae in Chapter 19 on the assumption that the rotational degrees of freedom are classical. They were there expressed as functions of the concentrations *N*/*V*, and, alternatively, as functions of the partial pressures. The two forms are of course simply correlated by using the equation of state for a perfect gas mixture in the form.

$$p_A = N_A kT/V. \tag{4}$$

It is convenient for our purpose to regard μ_A as a function of the partial pressure p_A. In Chapter 19, we chose as the internal energy zero for each molecular species the state of lowest internal energy. This choice is the natural one as long as each molecular species is regarded as a permanent system. But we are now allowing molecules to react with one another chemically, and it is therefore necessary to revise the energy zeros so that they become consistent with the energy changes accompanying chemical reaction. We therefore denote the energy of the lowest internal state of the molecule A by $\in_A^0$. *We may, if we wish, set* $\in_A^0$ *zero for the usual molecular form of each element, but* $\in_A^0$ *will still differ from zero for compounds. The partial potential*

$$\mu_A = \left(\frac{\partial F}{\partial N_A}\right)_{T,P} \tag{5}$$

will now *contain an extra term* $\in_A^0$ as compared with the formulae used in Chapter 19. Thus instead of

$$\mu = kT\ln P - \frac{5}{2}kT\ln T - kT\ln\left\{\frac{(2\pi m)^{3/2}\,k^{5/2}}{h^3}\times \upsilon_0\,\rho\right\}, \tag{6}$$

for monatomic molecules, we have the amended formula

$$\ln\lambda = \frac{\mu}{kT} = \frac{\in^0}{kT} + \ln p - \frac{5}{2}\ln T - \ln\left\{\frac{(2\pi m)^{3/2}\,k^{5/2}}{h^3}\upsilon_0\,\rho\right\}. \tag{7}$$

The formulae for different types of molecules (*diatomic* and *polyatomic*) are conveniently summarised in the form

$$\ln\lambda = \frac{\mu}{kT} = \frac{\in^0}{kT} + \ln p - \frac{C_p^0}{k}\ln T$$
$$- \sum_v \ln Z_{vib} - j - \sum \ln\rho, \tag{8}$$

where C_p^0, j are *molecular constants* (independent of *T*, *P*), Z_{vib} is the partition function for a single mode of vibration, υ_0 is the *electronic weight of the normal state*, and σ is the symmetry number of the molecule. The values for monatomic molecules are as follows:

$$Z_{vib} = 1,\quad C_p^0/k = \frac{5}{2},\quad j = \ln\left\{\frac{(2\pi m)^{3/2}\,k^{5/2}}{h^3}\upsilon_0\right\}. \tag{9}$$

The reason for separating the term $\sum \ln\rho$ in Eq. (8) (where ρ represents the number of possible orientations of the nucleus in a molecule known as the *nuclear spin weight*), instead of including it in *j* is that, in applying Eq. (8) to any chemical equilibrium, we will find that all the terms involving nuclear spin weights cancel, and there is therefore no need to evaluate them. The term $\sum \ln\rho$ is thus chemically unimportant, and could be omitted without affecting any chemical deductions. It is not even necessary to know the nuclear spin weights, and actually they are not yet all known. This cancellation of the nuclear spin weights is dependent on the rotational degrees of freedom of the gaseous molecules being at least approximately classical. Strictly speaking, the rotational-nuclear partition function $Z_{r.n}$ is the sum of terms each of which is the product of a nuclear spin weight and a rotational factor. These terms can be grouped according to the symmetry properties of the rotational eigenfunctions, and the nuclear spin weight factor will have the same value for each of these groups, but will vary from one group to another. A rotational-nuclear partition function denoted by $Z_{r.n}$ cannot strictly be factorised into a rotational and a nuclear factor. But if the rotational degrees of freedom are classical or nearly classical, then it can be shown that (except for H_2 , D_2 at low temperatures).

$$Z_{r.n} = Z_{rot}\,\frac{\Pi\rho}{\sigma}, \tag{10}$$

where Z_{rot} is the *complete rotational partition function containing terms for all rotational states whatever their symmetry properties*, $\Pi \rho$ is the *product of the nuclear spin weights of the atoms in the molecule*, and σ is the symmetry number of the molecule defined as the *number of indistinguishable orientations of the molecule.*

Let us now transform the equilibrium condition expressed in Eq. (3) for the chemical reaction Eq. (1) or Eq. (2) into the familiar law of *mass action* in terms of partial vapour pressures, namely

$$\frac{\Pi_L (P_L)^l}{\Pi_A (P_A)^a} = K_P(T). \tag{11}$$

By using Eq. (8) for the partial potential of each species, it can be proved that the equilibrium constant K_P is given by

$$\begin{aligned} \ln K_P = & -\frac{1}{kT}\left(\Sigma_L\, l \in_L^0 - \Sigma_A\, a \in_A^0\right) \\ & + \left\{ \Sigma_L\, l \frac{\left(C_P^0\right)_L}{k} - \Sigma_A\, a \frac{\left(C_P^0\right)_A}{k} \right\} \ln T \\ & + \left\{ \Sigma_L\, l\, \Sigma_v \ln Z_{vib\,L} - \Sigma_A\, a\, \Sigma_v \ln Z_{vib\,A} \right\} \\ & + \left\{ \Sigma_L\, l\, j_L - \Sigma_A\, a\, j_A \right\}. \end{aligned} \tag{12}$$

The values of C_P^0/k and *j* for monatomic molecules are given by Eq. (9); those for diatomic molecules and polyatomic *linear* molecules are

$$C_P^0/k = \frac{7}{2}, \quad j = \ln\left\{\frac{(2\pi m)^{3/2}\, k^{5/2}}{h^3}\right.$$

$$\left.\times \frac{8\pi^2 I_b k}{h^2} \frac{\upsilon_0}{\sigma}\right\}, \qquad (13)$$

where $I_b\,(= I_c)$ is the moment of inertia of the linear molecule, the moment of inertia I_a about the internuclear axis I_a being equal to zero. It is now evident from Eq. (12) that the nuclear spin weights of the atoms present in the molecule cancel. Eq. (12) can also be written* (without going into mathematical details) as

$$\ln K_P = -\frac{\Delta H^0}{kT} + \frac{\Delta \sum C_P^0}{k} \ln T +$$

$$\int_0^T \frac{dT_1}{T_1^2} \int_0^{T_1} \frac{\Delta \Sigma C'(T_2)}{k}\, dT_2 + \Delta \Sigma j, \qquad (14)$$

where the operator Δ denotes the increase in the value of any quantity when the chemical reaction takes place to the extent represented by $\Sigma a A \rightarrow \Sigma l L$. ΔH^0 is the limiting value of ΔH or ΔE as $T \rightarrow 0$. Thus written Eq. (14) is purely thermodynamic and is in principle due to *Nernst*. Thermodynamics can predict nothing concerning the values of the chemical or molecular constant *j*, but asserts only that there is one chemical constant for each gaseous molecular species, and that this same constant will occur in all chemical equilibria in which that particular species takes part. This is the reason for the name *chemical constant*. Statistical mechanics gives us the values of these constants.

20.2. Equilibrium between a crystal and its vapour

We turn now to the equilibrium between a crystal and its vapour. The thermodynamic condition for equilibrium is

$$\mu^G = \mu^K, \qquad (1)$$

where the superscript G refers to the gaseous phase, and the superscript K to the crystal phase. If then we can by *statistical methods* evaluate μ^G and μ^K, the equilibrium is statistically determined.

---

* See R. Fowler and E.A. Guggenheim, *Statistical Thermodynamics*, Chapter 5, Section 510, (Cambridge, 1965).

We can obtain an explicit formula for the vapour pressure of a crystal by substituting into the thermodynamic equilibrium condition given by Eq. (1) the values for μ^G and μ^K obtained by statistical methods.

For μ^G we have by Eq. (8) derived in Section 20.1, namely

$$\ln\lambda = \frac{\mu^G}{kT} = \frac{\epsilon_0}{kT} + \ln p - \frac{C_p^0}{k}\ln T$$

$$-\Sigma_v \ln Z_{vib} - j - \Sigma \ln\rho, \tag{2}$$

while for μ^K we have by definition

$$\mu^K = -kT\ln Z_{crys} + PV, \tag{3}$$

where Z_{crys} denotes the partition function per atom in the crystal; it is a function of temperature and volume so that

$$\frac{\mu^K}{kT} = -\ln Z_{crys} + \frac{PV}{kT}. \tag{4}$$

We must now consider the form of Z_{crys}. In all formulae where evaporation and condensation are permitted, it is essential that the gaseous and the crystal partition functions should be referred to the same energy zero. We, therefore, have for the crystal the partition function as

$$Z_{crys} = e^{-\left(\epsilon_0 - \chi\right)/kT}\,\alpha Z_{int}, \tag{5}$$

where ϵ_0 denotes the energy of the lowest quantum state of a gaseous molecule, and χ denotes the excess of this energy over that of a molecule in the crystal in its lowest quantum state; α is the partition function per molecule for the *acoustical modes*; Z_{int} is the partition function for the internal degrees freedom of the molecule in the crystal. We can further usually assume that, in the accessible temperature range, the only factors of Z_{int}, which are temperature dependent, are those from the *internal vibrational modes*. We may then write

$$Z_{int} = Q\,o\,\upsilon_0^K\,\Pi\,\rho, \tag{6}$$

where Q is the *partition function for all internal vibrational modes*, υ_0^K is the weight of the lowest electronic state, and $\pi\rho$ is the product of nuclear spin weights. Finally, o is *an orientational factor*, usually unity; it can, however, differ from unity if there are several distinguishable orientations of the molecule in the crystal with effectively equal energies. We shall have much to say about o, but at this stage all we need emphasise is that σ, like υ_0^K and ρ's, is temperature independent.

If we now substitute Eq. (6) into Eq. (5) and then Eq. (5) into Eq. (4), we obtain

$$\frac{\mu^{K}}{kT} = \frac{\epsilon_0 - \chi}{kT} + \frac{PV}{kT} - \ln\{\alpha Q\}$$

$$- \ln\left\{ o\, \upsilon_0{}^{K}\, \Pi\, \rho \right\}. \qquad (7)$$

If we now substitute Eqs. (2) and (7) into Eq. (1), we obtain for the vapour pressure over the crystal.

$$\ln P = -\chi/kT + PV/kT + \left\{\left(C_p^0/k\right)\ln T + \Sigma \ln Z_{vib}\right\}^G$$
$$-\{\ln\alpha + \ln Q\}^K + j^G - \ln\left\{o\, \upsilon_0{}^K\right\}. \qquad (8)$$

According to Eq. (8), the nuclear spin weights are again cancelled. This result is, as it must be, independent of the common energy zero used for the energy states in the crystal and in the gas.

The statistical form of the vapour pressure over the crystal given by Eq. (8) can be translated into the thermodynamic form where the quantity,

$$j - \ln o\, \upsilon_0{}^K, \qquad (9)$$

that is appearing as the last two terms in Eq. (8) is denoted by a constant i called the *vapour pressure constant.*

20.3. Relation between vapour pressure constants and chemical constants

Both Eq. (8) in Section 20.2 for the vapour pressure of a crystal, and Eq. (12) or Eq. (14) in Section 20.1 for the equilibrium constant of a gaseous equilibrium are obtainable by pure thermodynamics, and were effectively first so obtained by *Nernst.* But classical thermodynamics can predict nothing about the values of the two integration constants, namely the vapour pressure constant i in the former case, and the chemical constant j in the latter. Statistical mechanics, however, predicts the values of both, and, in particular, the relation between them,

$$i = j - \ln o\, \upsilon_0{}^K. \qquad (1)$$

For the majority of crystals o and $\upsilon_0{}^K$ are both unity, in which case i and j are identical. The two constants are, however, in principle different* and must not be confused.

We shall see that one version of Nernst's heat theorem is equivalent to the assertion that

$$o\, \upsilon_0{}^K = 1, \qquad (2)$$

which leads to

$$i = j \qquad (3)$$

---

* Eucken and Fried, *Zeit. Phys.* **29**, 36 (1924).

If this were valid for all crystals, there would be no necessity to continue to distinguish between the chemical constants j and the vapour pressure constants i as they would be equal. We shall discuss later the conditions for the validity of Eq. (2) when we deal below with the *third law using statistical mechanics*. Meanwhile, we shall continue to distinguish between i and j.

20.4. Effects of ortho-para separations

Molecular hydrogen (and deuterium) must frequently be treated as two separate substances, para-hydrogen* and ortho-hydrogen,† owing to the slow rate of interchange between the para and ortho states in the molecular form. There are special reasons why this division becomes prominent and important for hydrogen; the first is that its rotational degrees of freedom in the gas become non-classical at easily accessible temperatures; the second we shall examine in Section 20.12. The division, however, must undoubtedly occur for all other molecules of type X_2, and similar more complicated divisions which occur for polyatomic molecules of this or higher symmetry. In calculating chemical constants and vapour pressure constants for such substances, we commonly ignore these divisions, just as we ignore the separation of an element into isotopes. It is important to see why this neglect is in general justified. We shall be content to examine only the simplest case of a molecule X_2 in which the spin weight of each atomic nucleus is ρ.

(1) *The symmetrical molecule* X_2 *with equilibrium maintained between the para and ortho states*. We know that the nuclear symmetry requirements in the eigenfunctions can be taken care of in the gas phase by introducing the symmetry factor $1/\sigma = \frac{1}{2}$, and setting

$$Z_{r.n} = Z_{\text{rot}} \frac{1}{2} \rho^2, \tag{1}$$

provided the rotational degrees of freedom are classical or nearly classical. We shall now show that there is a similar justification in the solid phase. In general, and certainly for all molecules X_2 no longer rotating in the crystal, each molecule X_2 will have, apart from nuclear spins, one para and one ortho state of least energy, these energies being effectively equal. These two states are the correct stationary states, which correspond to the two classical states provided by a single natural direction of equilibrium, and the possibility of turning the molecule end for end. *The lowest energy terms in the partition function for the vibrations and nuclear orientations of each molecule in the crystal are therefore*

$$\frac{1}{2} \rho(\rho - 1) + \frac{1}{2}\rho (\rho + 1)\, e^{\delta/kT}, \tag{2}$$

where δ is the (negligible) energy difference between the para and ortho vibrational states. These two terms therefore reduce to the factor ρ^2, and the same factor occurs in all higher vibrational states when these need to be retained. Thus the nuclear spins in the crystal still

---

* The molecule with antisymmetrical nuclear spin eigenfunctions is called para-hydrogen.

† The molecule with symmetrical nuclear spin eigenfunctions is called ortho-hydrogen.

make the standard contribution to $\Pi\rho$, which could with equal propriety be regarded as the simple factor ρ per atom in the crystal.

(2) *The symmetrical molecule* X_2 *when the para and ortho states must be regarded as effectively non-combining.* The above arguments apply only when there is full equilibrium between the para and ortho states of molecules of each type. This equilibrium in general is not preserved, and the gas and the crystal must each be treated as a perfect mixture of distinct sets of para and ortho molecules, exactly equivalent to an isotopic mixture. The rotational partition functions $Z_{rot}(o)$ and $Z_{rot}(p)$ for the two types of gaseous molecules consist one of the even and the other of the odd terms of Z_{rot}. But provided the rotational degrees of freedom are classical or approximately classical, we can assume that with a high degree of accuracy

$$Z_{rot}(o) = Z_{rot}(p) = \frac{1}{2} Z_{rot}. \tag{3}$$

The two forms have therefore the same rotational partition function.

In the crystal phase, the two forms have sets of states which *differ insignificantly* in energy and lead therefore to *equal partition functions.* If effectively rotating, the same arguments apply as in the gas phase so that in all cases and in both phases the rotational and vibrational partition functions of para and ortho molecules may be taken to be equal. It follows from the equality that the para and ortho forms occur in effectively the same proportions in the crystal and the gas. The para and ortho forms have nuclear spin weights $\frac{1}{2}\rho(\rho-1)$ and $\frac{1}{2}\rho(\rho+1)$, respectively, but these make equal contributions to the partition functions of each type in each phase and therefore also for the mixture. The entropy of mixing terms are also equal in both phases. We thus find again that the existence of para and ortho forms may be *ignored.*

This reasoning does not apply to H_2 and D_2 because *their rotational degrees of freedom are far from classical.* These molecules are also exceptional in another respect, and they are discussed separately in Section 20.12 when we deal with the special cases of hydrogen and deuterium.

20.5. Molecular entropies of gas and crystal

Many authors prefer to express their thermodynamic results in terms of entropy. We shall therefore derive the relevant formulae. For the partial potential of any single gas if we do not assume the rotational and vibrational motions to be necessarily independent, we can write instead of Eqs. (8) and (9) shown above in Section 20.1 the equation

$$\begin{aligned}\mu^G(T,P) &= \epsilon^0 + kT\ln P - \frac{5}{2}kT\ln T \\ &\quad - kT\ln Z_{rot}Z_{vib} - kT\ln\left\{\frac{(2\pi m)^{3/2}k^{5/2}}{h^3}\right. \\ &\qquad \left.\times\frac{\upsilon_0\Pi\rho}{\sigma}\right\},\end{aligned} \tag{1}$$

where $Z_{rot}\,Z_{vib}$ denotes the rotational-vibrational partition function, including all terms regardless of their symmetry characters; these are taken care of the by the symmetry number σ. The corresponding formula for the *molecular entropy* is

$$S^G\left(T,P\right) = -\left(\frac{\partial \mu^G}{\partial T}\right)_P$$

$$= -k\ln P + \frac{5}{2}k\ln T + k\ln Z_{rot}\,Z_{vib}$$

$$+kT\,\frac{\partial \ln Z_{rot}\,Z_{vib}}{\partial T}$$

$$+k\left[\ln\left\{\frac{(2\pi m)^{3/2}\,k^{5/2}}{h^3}\;\frac{\upsilon_0\,\Pi\,\rho}{\sigma}\right\}+\frac{5}{2}\right]. \tag{2}$$

For the crystal, we have by the equations

$$\mu^K = -k\,T\,\ln Z_{crys} + PV, \tag{3}$$

and

$$Z_{crys} = e^{-(\epsilon_0-\chi)/kT}\;\alpha\,Z_{int}\,, \tag{4}$$

which were shown before in Section 20.2 as Eqs. (4) and (5), respectively, the partial potential μ^K expressed as

$$\mu^K\,(T,P) = \epsilon_0 - \chi + PV - kT\,\ln(\alpha Z_{int}), \tag{5}$$

and for ordinary pressures where $PV \ll kT$, this equation reduces to

$$\mu^K\,(T,0) = \epsilon_0 - \chi - kT\,\ln(\alpha Z_{int}). \tag{6}$$

Hence, the molecular entropy at ordinary pressure for the crystal is given by

$$S^K\,(T,0) = -\left(\frac{\partial \mu^K}{\partial T}\right)_{P\simeq 0} = k\ln(\alpha Z_{int})$$

$$+kT\,\frac{\partial\{\ln\alpha\,Z_{int}\}}{\partial T}. \tag{7}$$

If we now make $T \rightarrow 0$, we have

$$\alpha \rightarrow 1,\ \text{and}\ j \rightarrow o\,\upsilon_0{}^K\,\Pi\,\rho. \tag{8}$$

Eq. (8) is in effect the *definition of the orientational weight* o. We thus obtain*

$$S^K(0,0) = k \ln\left(o\upsilon_0^K \Pi\rho\right). \tag{9}$$

If we now subtract Eq. (9) from Eq. (2) we obtain

$$S^G(T,P) - S^K(0,0) = -k\ln P + \frac{5}{2}k\ln T + k\ln Z_{rot} Z_{vib}$$

$$+kT\frac{\partial \ln Z_{rot} Z_{vib}}{\partial T}$$

$$+k\left[\ln\left\{\frac{(2\pi m)^{3/2} k^{5/2}}{h^3} \frac{\upsilon_0^G}{\sigma o \upsilon_0^K}\right\} + \frac{5}{2}\right], \tag{10}$$

the nuclear weights cancelling as usual.

Now the left side of Eq. (10) is directly determinable by calorimetric measurements apart from an *inevitable* extrapolation to zero temperature. For we have

$$S^G(T,P) - S^K(0,0) = \underset{T^0\to 0}{\mathrm{Lt}} \int_{T^0}^{T} \frac{d'q_{rev}}{T}, \tag{11}$$

where $d'q_{rev}$ is the element of heat absorbed, and the path of integration is any reversible one leading from the crystal at temperature T^0, and negligible pressure to the gas at temperature *T* and pressure *P*. This path may, and usually will, pass through the liquid state. There may be one or more phase transitions on the way, and each will make its contribution to the integral. If now the energies of the molecule have been determined spectroscopically, then $Z_{rot}\ Z_{vib}$ and its temperature coefficient can be computed for each *T*, while υ_0^G and σ are also determinable spectroscopically, or otherwise. Thus everything on the right of Eq. (10) is known except $o\upsilon_0^K$. By comparison of the calorimetric and spectroscopic data, we can determine the value of $o\upsilon_0^K$. If Nernst's heat theorem is obeyed, the value unity should be found.

20.6. Absolute entropy

It is to be known that there is *no physical significance in an absolute value of an entropy*. It is only entropy differences between interconvertible states of an assembly that are physically significant. We use a *convention* such that, when every system is in its *lowest quantum state*, the *contribution of each system to the entropy is*

$$k \ln \omega_0, \tag{1}$$

where ω_0 is the product of the weight of the lowest translational–rotational–vibrational–electronic state and the product of the nuclear spin weight $\Pi\rho$. But, if *we cared to do so, we might replace Eq. (1) by*

$$k \ln \omega_0 + k \ln \Pi a, \tag{2}$$

* Eq. (9) will be made use of later in Section 20.8.

where a is an *arbitrary constant characteristic of each atomic species*, and the product is taken over all atoms in the system, these arbitrary factors being a_e for the electron, a_p for the proton, and a_n for the neutron. Since we are still too ignorant of the details of the nuclear structure to be able to specify what the correct values of the a's would be for the electron, proton, and neutron, we see that there is nothing absolute about Eq. (1). Thus the simplest and most convenient convention allowable in the absence of nuclear transformations is to put $a = 1$ for all nuclei.

Returning to practical thermodynamics, we may of course assign, if we please, the value zero to the entropy of all perfect crystals of a single element in its idealised state at the absolute zero of temperature, but even this has no theoretical significance on account of nuclear spin weights. For the purpose of calculating experimental results some conventional zero must be chosen, and the above choice or a similar one is thus often convenient. But its conventional character will no longer be so likely to be overlooked that any importance will in future be attached to absolute entropy, an idea which has caused much confusion and been of very little assistance in developing the subject.

To prevent all possibility of misunderstanding, it should be pointed out that in naming the λ's 'absolute activities' we are using the epithet 'absolute' in an entirely different sense from that under discussion here, the purpose being to contrast the λ's with the relative activities a's used by the school of G. N. Lewis. Thus as shown before in Section 18.16, the statistical parameters λ_A and partial potentials μ_A are related by

$$\mu_A = kT \ln \lambda_A, \tag{13}$$

whereas the *relative* activity a_A is defined by

$$\mu_A = \mu_A^0(T) + kT \ln a_A, \tag{14}$$

where μ_A^0 is *arbitrarily fixed at each temperature*. Subsequently, if we compare two thermodynamic states denoted by single and double primes, respectively, then

$$\mu' - \mu'' = kT \ln \lambda'/\lambda''$$

is identically true, whereas the analogous

$$\mu' - \mu'' = kT \ln a'/a''$$

is valid only if the two thermodynamic states have the same temperature.

20.7. Extrapolation to absolute zero

In deriving all our formulae, we have assumed that we could select a temperature T^0 at which in the gas the vibrational degrees of freedom are *unexcited*, and the rotational degrees of freedom, except for monatomic molecules, are effectively classical. The *formula* so obtained are then valid for all temperatures *greater* than T^0. We choose for T^0 a temperature, in a range where the molecular heat capacity C^K is *tending rapidly to zero* according to some simple theoretical or semi-theoretical formula, such as that due to *Debye* noted before. We then extrapolate to $T = 0$, using the convention that C^K continues to obey this law. In other words, C^K is defined as follows:

For $T > T^0$, C^K is the observed molecular heat capacity of the crystal.

For $T < T^0$, C^K is defined by extrapolation according to the simplest formula with a theoretical basis that is verified to hold for C^K when $T \approx T^0$.

In the lower ranges of temperature, there is no *experimental value* for the heat capacity. When temperatures much lower than T^0 become experimentally accessible, we may expect to find two alternative kinds of behaviour.

(i) In some cases, we shall find striking abnormalities in C^K and, with a new much lower T^0, the new value of $o\upsilon_0{}^K$ *will become unity.* The important case of H_2 comes into this class.

(ii) In other cases, we shall find that the actual value of C^K agrees closely with the extrapolated value here defined, and $o\upsilon_0{}^K$ is unaffected, remaining *greater than unity*. The measurements then in fact refer to a state of the solid which is strictly only metastable. There is then always a possibility that by performing the experiments *extremely slowly* the true equilibrium state may be attained. Different values of C^K and $o\upsilon_0{}^K$ will then be found. But there are many cases in which the time required to reach true equilibrium is *impossibly long at these low temperatures.*

20.8. Basis of comparison between theory and experiment

There are several alternative methods of comparing the theoretical formulae with experiment. The method to be used in each case must depend not only on the kind of the experimental data available, but also on the knowledge of the values of the quantities occurring in the theoretical formulae. These methods of comparison shall be discussed in turn.

The molecular mass *m* is always known, and the moment of inertia I_b is known from spectroscopic data for most diatomic molecules. The moments of inertia I_a, I_b, I_c for many simple polyatomic molecules are also known from spectroscopic data. The *symmetry number* σ is usually known from the chemical constitution, if not from spectroscopic data. The *electronic weight* $\upsilon_0{}^G$ *of the normal state* of the gaseous molecule is also determined spectroscopically for all simple molecules. It is unity for the great majority of the molecules, and may safely be assumed unity for all molecules ordinarily called 'saturated'.

It follows that, at least for simple molecules, one usually has all the data for computing a theoretical value of the *chemical constants j*. But a single chemical constant cannot be compared with experiment; it is only linear combinations of the form

$$\Delta \Sigma j = \Sigma_L l\, j_L - \Sigma_A a\, j_A, \qquad (1)$$

occurring in formulae such as Eq. (12) in Section 20.1, for $\ln K_P$, that can be checked. If the theoretical and experimental values of $\Delta \Sigma j$ agree for a number of reactions involving certain molecular species A, L, . . . this agreement affords indirect confirmation of the correctness of the theoretical values of the individual chemical constants *j*.

Alternatively, one might compare theoretical and experimental values of the vapour pressure constants *i*. This, however, requires the knowledge of the theoretical values of $o\upsilon_0{}^K$ as well as the quantities mentioned above. For this it is necessary to know how the crystals concerned are constructed – that is, whether they should be regarded as built up of atomic, molecular (diatomic), or even of multimolecular units. It is safe to assume that *when* $\upsilon_0{}^G$ *is unity, as it is for the majority of stable molecules, then* $\upsilon_0{}^K$ *is also unity*. When $\upsilon_0{}^G$ exceeds unity, we may not assume that $\upsilon_0{}^K$ is equal to $\upsilon_0{}^G$. Often the molecules will combine or interact in the crystal to make $\upsilon_0{}^K$ less than $\upsilon_0{}^G$, and often equal to unity. To determine the value of o it is further necessary to know whether the molecules can *still rotate freely in the crystal at the low-limit temperature* T^0. If they cannot rotate freely, it is necessary to know how many distinct orientations are possible with almost equal energy. *A combination of purely calorimetric measurements with a determination of the transition temperature, for a process involving only crystal phases, gives an experimental value for linear combinations of the* $\ln o\,\upsilon_0^K$ *of the following type*

$$\Delta \sum \ln o\upsilon_0{}^K = \Delta \sum (j - i). \qquad (3)$$

We shall enumerate in Section 20.10 a number of processes of this type, for which accurate measurements are available. In all these cases, the experimental value found for Eq. (3) is zero. This *can hardly be fortuitous*, and suggests that for all the crystals involved $o\upsilon_0{}^K$ is unity. If $o\upsilon_0{}^K$ were always unity, there would be as we have already said no need to distinguish between the vapour pressure constants *i* and the chemical constants *j*. One would then have all the required theoretical data for calculating the vapour pressure constants *i*, and those calculated values could be compared with the experimental values. Actually, we shall see later that the value of $o\upsilon_0{}^K$ associated with extrapolation from the values of T^0 ordinarily used is probably unity for *most, but not all, crystals*. We therefore proceed by computing a *theoretical* value of *j* and compare this with the *experimental* value of *i*. If in most cases we find that these values are equal within the experimental accuracy, this confirms the supposition that for these substances

$$i = j, \quad o\upsilon_0{}^K = 1. \qquad (4)$$

In the remaining cases, where the calculated value of *j* and the *experimental* value of *i* disagree, we can compute a value of $o\upsilon_0{}^K$ from Eq. (1) given above in Section 20.3, namely

$$i = j - \ln o\upsilon_0{}^K, \qquad (5)$$

and then consider whether, from the chemical structure of the molecule, this value of $o\upsilon_0{}^K$ is explicable by reasonable assumptions about the structure of the crystal.

This method of comparison is convenient for monatomic molecules and also for diatomic molecules, if at ordinary temperatures the rotational degrees of freedom are classical, and the vibrational degrees of freedom either unexcited or at least separable from the rotational. When this is not the case, an *alternative procedure is adopted.* From spectroscopic data, one calculates the quantity

$$S^G(T,P) - k \ln \Pi\rho, \tag{6}$$

which is obtained by *merely omitting the nuclear spin weights from* Eq. (2) *in Section* 20.5. This quantity is then compared with the experimental value of

$$S^G(T,P) - S^K(0,0), \tag{7}$$

measured as

$$\underset{P^0,T^0 \to 0}{\mathrm{Lt}} \int_{T^0,P^0}^{T,P} d'q_{rev}/T, \tag{8}$$

the extrapolation from $T = T^0$ to $T^0 = 0$ being performed as described above in Section 20.7. It should be recalled here that according to Eq. (9) in Section 20.5, $S^K(0,0)$ is given by

$$S^K(0,0) = k \ln \{o\upsilon_0{}^K \,\Pi\rho\}, \tag{9}$$

so that the subtraction of Eq. (7) from Eq. (6) will produce the quantity

$$k \ln o\upsilon_0{}^K, \tag{10}$$

which we are looking for. It has been found that the value obtained for $\ln o\upsilon_0{}^K$ *in this way is often but not always zero.*

The procedure just described is that adopted by *Giauque* and *his collaborators*. Giauque called the quantity defined by Eq. (6) the *spectroscopic entropy* (*nuclear spin omitted*), and the quantity defined by Eq. (7) the *calorimetric entropy*, of the gaseous molecule at the temperature *T* and pressure *P*. The calculated spectroscopic entropy, using Eq. (6), is the *entropy excess* over that of an *idealised state*, in which all the molecules have the same energy, a translational–rotational–vibrational–electronic weight 1, and a nuclear spin weight $\Pi\rho$ equal to the product of the spin weights ρ of the constituent atoms. The measured calorimetric entropy defined by Eq. (7) is the entropy excess over the value obtained by extrapolation to $T = 0$ for the crystal. The use of a smooth extrapolation presupposes, among other things, that separation of isotopes does not occur. Correspondingly, one ignores the entropy contribution due to the mixing of isotopes in computing the *spectroscopic entropy.* The conventional nature of an entropy zero used in such calculations has been emphasised in Section 20.6.

If equality is found between the spectroscopic entropy and the calorimetric entropy, that is, the difference given by Eq. (10) is equal to zero, it means that smooth extrapolation of the heat capacity *leads* to a crystal at the absolute zero in which the molecules have a translational–rotational–vibrational–electronic weight unity; in other words,

$$o\upsilon_0{}^K = 1. \tag{11}$$

20.9. Chemical constants and vapour pressure constants in practical units

In practice, it is customary to use logarithms to base 10, and to measure vapour pressure in atmospheres rather than in dynes/cm^2. If we rewrite our formulae in this manner, we obtain new integration constants j', i' defined by

$$j' = \frac{j - \ln P^+}{\ln 10}, \tag{1}$$

$$i' = \frac{i - \ln P^+}{\ln 10}, \tag{2}$$

where P^+ is the value of an atmosphere in dynes/cm^2. These new constants j', i' are conveniently called the *practical chemical constant* and the *practical vapour pressure constant*, respectively.

The corresponding formulae for i' are given immediately from Eqs. (1) and (2) by

$$i' = j' - \log_{10} o\,\upsilon_0{}^{K}. \tag{3}$$

The numerical value of j' for *monatomic molecules*, for which

$$C^0{}_P/k = \frac{5}{2},$$

and

$$j = \ln\left\{\frac{(2\pi m)^{3/2}\,k^{5/2}}{h^3}\,\upsilon_0\right\},$$

as shown by Eq. (9) in Section 20.1, is given as

$$j' = -1.587 + \frac{3}{2}\log_{10} M + \log_{10} \upsilon_0{}^{G}. \tag{4}$$

Here *M* is the conventional chemical molecular weight.

For linear molecules with classical relations:

$$C_P^0 / k = \frac{7}{2},$$

and

$$j = \ln\left\{\frac{(2\pi m)^{3/2}\,k^{5/2}}{h^3}\,\frac{8\pi^2 I_b\,k}{h^2}\,\frac{\upsilon_0}{\sigma}\right\},$$

as shown by Eq. (13) in Section 20.1, the numerical value of j' is

$$j' = -3.85 + \frac{3}{2}\log_{10} M + \log_{10}\left(10^{40} I_b\right) + \log_{10}\frac{\upsilon_0{}^{G}}{\sigma}.$$

However, it may be noted that the use of chemical constants is rapidly dying, their place being taken by numerical values of the thermodynamic functions* μ , H , and S.

20.10. Experimental data for reactions between crystals

We now consider a physical or chemical process between crystalline phases only. If the heat capacity of each phase has been measured down to temperatures sufficiently low to extrapolate smoothly to $T = 0$ by a formula such as that of Debye, and the heat of reaction has been measured at any one temperature, these data are sufficient to determine the transition temperature, that is the temperature at which the process becomes reversed, provided one knows the value of

$$\Delta \left\{ \Sigma \ln o\upsilon_0{}^{K} \right\}. \tag{1}$$

Alternatively, from the knowledge of the transition temperature, one can calculate the value of Eq. (1).

The simplest processes of this type are transitions between two crystalline forms α and β of a single substance,

$$\alpha = \beta. \tag{2}$$

Experimental study of these tell us the value of

$$\ln \left\{ o\upsilon_0{}^{K} \right\}^{\alpha} - \ln \left\{ o\upsilon_0{}^{K} \right\}^{\beta} . \tag{3}$$

The earlier determinations carried out by various authors included the transitions rhombic sulphur to monoclinic sulphur, white tin to grey tin, quartz to cristobalite, and calcite to aragonite. The accuracy was not high, except for the tin transition, but in all the cases mentioned it is sufficient to exclude a value of Eq. (3) numerically as great as $\pm \ln 2$. The experimental evidence is therefore in all cases consistent with the supposition $o\upsilon_0{}^{K} = 1$ for both crystalline forms. The value zero for Eq. (3) for the rhombic sulphur to monoclinic sulphur has been accurately confirmed by a recent redetermination of the heat capacities of the two forms.

For the following chemical reactions between crystalline phases

$$Pb + 2I = PbI_2$$

$$Ag + I = AgI$$

$$Hg + AgCl = Ag + \frac{1}{2} Hg_2 Cl_2$$

$$Pb + Hg_2 Cl = 2Hg + PbCl_2,$$

* For reference see R. Fowler and E, A. Guggenheim, *Statistical Thermodynamics*, Appendix, Section A3. p. 687 (Cambridge, 1965).

there are calorimetric data and electromotive force measurements, from which to evaluate the integration constant. The accuracy is not very high except for the reaction between silver and iodine. In no case does the absolute magnitude found for the integration constant j' given by Eq. (4) *exceed one-third* the value of ln 2, and the data are consistent, within the experimental accuracy, with the supposition that $o\,\upsilon_0^{\,K}$ is unity for each crystalline phase concerned.

20.11. Experimental vapour pressure data for monatomic molecules

For *monatomic* molecules, we have always

$$C_P = C^0{}_P = \frac{5}{2}k, \qquad (1)$$

and the practical vapour pressure constant i' (according to Eqs. (3) and (4) obtained above in Section 20.9, respectively) is given by

$$i' = -1.587 + \frac{3}{2}\log_{10} M + \log_{10}\, \upsilon_0^{\,G}/\upsilon_0^{\,K}. \qquad (2)$$

For a monatomic molecule, an orientational factor o *other than unity cannot occur*. That o is also unity for the halogens is explained below. Experimental data for such molecules were obtained.*

The atoms He, Ne, A, Kr, Xe, Mg, Zn, Cd, and Hg in the vapour are in ^1S states (the normal ground states) with an electronic spin weight $\upsilon_0^{\,G} = 1$, while Na and K are in ^2S states with a weight $\upsilon_0^{\,G} = 2$. All other electronic states for these atoms lie so high that they do not contribute to the partition function at the relevant temperatures. In Tl vapour, the atoms are in a $^2P_{1/2}$ state of weight 2, the upper state $^2P_{3/2}$ of the doublet lying too high to contribute to the partition function. Similarly, in Pb vapour, the atoms are in a $^3P_0$ state of weight unity, the other states $^3P_1$, $^3P_2$ of the triplet lying too high to contribute to the partition function. For a free halogen atom, the *normal state* is $^2P_{3/2}$ of weight 4, *but in this case, the other state* $^2P_{1/2}$ *of weight 2 of the doublet does not lie high enough to be entirely negligible*. Hence, in place of $\upsilon_0^{\,G} = 4$ for the normal state, we must use

$$Z_{elec} = \upsilon_0^{\,G} + \upsilon_1\, e^{-\epsilon_1/kT}$$

$$= 4 + 2\, e^{-\epsilon_1/kT}, \qquad (3)$$

---

* See R. Fowler and E. Guggenheim, *Statistical Thermodynamics*, Chapter 5, Section 530, (Cambridge, 1965).

where $\in_1$ is the energy difference between the two terms forming the doublet. For the three halogen atoms Cl, Br, I their chemical constants were measured at 1,000 and 1,700 K, 1,350 K, and 1,200 K, respectively. It was found that at these temperatures it is only for Cl that the upper level of the doublet is significant. The calculated j' value for Cl was obtained by using

$$\log_{10} Z_{elec} = 0.70, \tag{4}$$

in place of

$$\log_{10} \upsilon_0{}^G = 0.60,$$

which in turn corresponds to

$$\log_{10} Z_{elec} = \log_{10} \upsilon_0{}^G = \log_{10} 4,$$

as shown in Eq. (3).
For Br and I, the value

$$\log_{10} Z_{elec} = \log_{10} \upsilon_0{}^K \tag{5}$$
$$= \log_{10} 4$$
$$= 0.60 \tag{6}$$

was used.

The fact that the values of j' calculated from the known values of $\upsilon_0{}^G$ were found to *agree in all cases for monatomic molecules* within the experimented error with the *observed values of* i', had led to the conclusion that in every case

$$\upsilon_0{}^K = 1 \tag{7}$$

It is to be noted that the equation

$$j' = i' + \log_{10} \upsilon_0{}^K$$

was used for the calculation of the integration constant j', this equation being Eq. (3) derived above in Section 20.9, where the orientational weight o for each molecule is set unity.

We must now consider the physical significance of $\upsilon_0{}^K = 1$. For Ne, A, Kr, and Xe, since the crystal is an atomic one, one naturally expects the electronic state to be ^1S as in the vapour. For the halogens, the crystal structure is one of the molecules Cl_2, Br_2, or I_2 in $^1\Sigma$ states. The electronic weight of the normal state of the *molecule is therefore unity*. The weight of the lowest vibrational state of the *crystal regarded as formed of molecules* is also unity, and there is no orientational weight o other than unity, when the requirement of antisymmetry in the nuclei is satisfied. Now it remains for us to explain why $\upsilon_0{}^K$ is *unity* for the metals including those such as Na, K, and Tl for which $\upsilon_0{}^G$ is *not unity*. The explanation depends on the electron theory of metals, according to which we may regard the *atoms in the metal* as dissociated into electrons and positive ions, e.g.

Na^+, Mg^{++}, $(Tl)^+$ (or Tl^{+++}); the ions are in a ^1S state of weight unity (nuclear weights being as always neglected), and therefore contribute a factor unity to $\upsilon_0{}^K$. The electrons may be considered to form an electron gas obeying the *Fermi-Dirac statistics* in an enclosure at nearly constant potential. According to this form of quantum statistics, it has been concluded that at all ordinary temperatures, and *a fortiori* at low temperatures, the *N* electrons occupy the *N* states of lowest energy; this distribution corresponds to a single eigenfunction for the assembly. Thus $\upsilon_0{}^K$ remains unity in agreement with experiment.

20.12. The special cases of hydrogen and deuterium

We have still to consider hydrogen and deuterium, as these molecules present special features meriting careful examination. The moment of inertia I_b of H_2 is only 0.463×10^{-40} gm.cm^2 so that

$$\Theta_r = \frac{h^2}{8\pi^2 I_b k} = \frac{39.6}{10^{40} I_b} \tag{1}$$

$$= 85.4 \text{ deg}.$$

Thus the deviation of $\ln Z_{rot}$ from its classical value $\ln T/\Theta_r$ at 300 K will be $\frac{1}{3}\Theta_r/300 = 85.4/900 = 0.095$. The corresponding correction in *j* is 0.095 and that in j' is 0.041, which is comparable with the experimental uncertainty. Thus in the case of H_2, the classical approximation for Z_{rot}, may be just allowable at *temperatures above 300 K*, but is certainly not allowable at lower temperatures. It may be emphasised that the deviation of $\ln Z_{rot}$ and so of the effective chemical constant *j*, from its classical value, in the case of H_2 molecule above 300 K, may well be *appreciable* even though deviations of C_P/k from its classical value $\frac{7}{2}$ are *inappreciable*. The *rotational contribution* to the heat capacity of a *diatomic molecule* according to the equation

$$C^{rot} = k\left\{1 + \frac{1}{45}\frac{\Theta_r^2}{T^2} + o\left(\frac{\Theta_r}{T}\right)^3\right\}, \tag{2}$$

which is Eq. (16) obtained before in Section 19.9, is given by

$$\frac{C^{rot}}{k} = 1 + \frac{1}{45}\left(\frac{\Theta_r}{T}\right)^2 = 1 + \frac{1}{45}\left(\frac{h^2}{8\pi^2 I_b k T}\right)^2$$

$$= 1 + \frac{1}{45}\left(\frac{39.6}{10^{40} I_b T}\right)^2, \tag{3}$$

where the term $o\left(\frac{\Theta_r}{T}\right)^3$ is omitted for brevity. For H_2 at 300 K this gives $C^{rot}/k = 1.002$, thus differing insensibly from unity. On the other hand, the error in j', computed above as 0.041, corresponds to an error of no less than 10 per cent in the partial pressure of the gas. For deuterium D_2 the error in j' is about half as great as for hydrogen H_2 at the same temperature.

For the molecules H_2, HD, and D_2, we may therefore use the usual formulae for the chemical constant *j* of a diatomic molecule only at temperatures above 300 K. We may further continue to use such formulae at slightly lower temperatures if we add a correction term $\frac{1}{3}\Theta_r/T \ln 10$ to the value of j' calculated by assuming classical rotations. All temperatures at which this procedure is allowable are considerably higher than the critical temperature 33 K, and one has then no use for a vapour pressure constant *i*.

At *lower temperatures*, when the condition $T \gg \Theta_r$ *fails*, we have to treat H_2 as a mixture of two non-interconvertible species $p-H_2$ and $o-H_2$. This also applies to D_2. Not only does the classical condition $T \gg \Theta_r$ fails for H_2, except for temperatures above 300 K, but it is comparatively easy to reach temperatures at which the opposite condition $T \ll \Theta_r$ holds. At such temperatures, the rotations will be effectively unexcited. In the case of $p-H_2$, the normal rotational state, of quantum number 0, has weight 1, and the first excited state of quantum number 2, has weight* 5 and energy $6k\Theta_r$. The ratio† of the number of molecules of $p-H_2$ in the rotational state 2 to the number in the rotational state 0 is therefore $5e^{-6\Theta_r/T}:1$. At the temperature *T* = 43 K, corresponding to $\Theta_r/T = 2$, this fraction is only 3×10^{-5}. For $o-H_2$ the ratio of the number of molecules in the state of rotational quantum number 3 to the number in the ground state of rotational quantum number 1 is similarly $7e^{-12\Theta_r/T}:3e^{-2\Theta_r/T}$. At $T = 43\,K$ this fraction is only 5×10^{-9}. Thus, at this temperature, the fraction of H_2 molecules, whether para or ortho, *not in the lowest accessible rotational state is entirely negligible.* This is equally true for D_2 at temperatures half as great as for H_2.

Owing to their small moments of inertia and their nearly spherical external field of force, the molecules H_2, D_2 and HD continue to rotate in the crystal down to temperatures as low as 12 K. These rotations are of course not classical, and below 40 K only the rotational states of least quantum number consistent with the symmetry of the para or ortho form are occupied. At still lower temperatures, an orientational interaction between the molecules will remove the *triple degeneracy* of the lowest rotational state of rotational quantum number 1 of ortho-hydrogen. This should show itself by an *extra contribution* to the heat capacity superposed on the contribution of the lattice vibrations. This has in fact

---

* See Section 9.4.

† See Section 8.2.

been observed at temperatures below 12 K by Simon* et al in 1931. It should be noticed that the extra heat capacity increases with the relative amount of ortho-hydrogen, and vanishes for pure para-hydrogen, because the *lowest* rotational state of the latter corresponding to a rotational quantum number 0 is not degenerate.

For H_2 the para molecules have a least rotational quantum number 0 and weight 1, and the ortho molecules have a least rotational quantum number 1 and weight 3. We also recall that ordinary H_2 is a mixture of one part para to three parts ortho; this proportion being equal to the ratio of the nuclear spin weights, namely 1:3. We thus have for the value of the orientational factor o in the crystal

$$\ln o = \ln 1 = 0 \qquad (p-H_2), \tag{4}$$

$$\ln o = \ln 3 \qquad (o-H_2). \tag{5}$$

The nuclear spin weights to be used with those values of o are 1 for the para and 3 for the ortho. To derive the value of o for the ordinary metastable mixture of para- and ortho-hydrogen, we make use of the principle that for the ordinary mixture we obtain *correct results* by ignoring differences between the nuclear spin weights of the para and ortho forms, provided we also ignore the factor due to mixing. We thus obtain

$$\ln o = \frac{1}{4}\ln 1 + \frac{3}{4}\ln 3 = \frac{3}{4}\ln 3 \text{ (ordinary } H_2\text{)}, \tag{6}$$

the *associated nuclear spin weight* being 2^2 for the molecule corresponding to a spin quantum number $= \frac{1}{2}$ for each nucleus. For the more general example of a mixture of $1-x$ parts of para- and x parts of ortho-hydrogen, we cannot use any such short cut. For such a mixture we obtain† for the limiting of the partition function $\underset{T=0}{Z}$ per molecule in the crystal, taking account of the mixing factor as well as the nuclear and rotational factors, and assuming the electronic spin weight to be unity,

$$\begin{aligned}\ln \underset{T=0}{Z} &= \{(1-x)\ln 1 + x\ln 9\} \\ &\quad - \{(1-x)\ln(1-x) + x\ln x\} \\ &= (1-x)\ln\left(\frac{1}{1-x}\right) + x\ln\left(\frac{9}{x}\right)\end{aligned} \tag{7}$$

For the special case $x = \frac{3}{4}$ for ortho-hydrogen, Eq. (7) reduces to

$$\ln \underset{T=0}{Z} = \frac{3}{4}\ln 3 + \ln 4 \tag{8}$$

* Mendelssohn, Ruhemann, and Simon, *Zeit. Plys. Chem.* B **15**, 121 (1931).

† See R. Fowler and E. Guggenheim, *Statistical Thermodynamics*, Chapter 5, Section 531 (Cambridge, 1965).

Now if we substitute in this equation for $\frac{3}{4}\ln 3$ by $\ln o$ from Eq. (6), and for $\ln 4$ by $\ln \rho^2$ since the associated nuclear spin weight is 2^2, we obtain

$$\ln \underset{T=0}{Z} = \ln o + \ln \rho^2. \tag{9}$$

For D_2 it is the ortho molecules (with nuclear spin weight 6) that have a *lowest rotational state* of quantum number 0 and thus weight 1, and the para molecules (with nuclear spin weight 3) that have a *lowest rotational state* of quantum number 1 and thus weight 3. *The proportion of ortho to para molecules in the ordinary metastable mixture of* D_2 *is also equal to the ratio of the nuclear spin weights, namely 6:3.* Hence, the values of o are given by

$$\ln o = \ln 1 = 0 \qquad (o-D_2), \tag{10}$$

$$\ln o = \ln 3 \qquad (p-D_2), \tag{11}$$

$$\ln o = \frac{6}{9}\ln 1 + \frac{3}{9}\ln 3 = \frac{1}{3}\ln 3 \qquad (\text{ordinary } D_2). \tag{12}$$

For HD, since there are no para and ortho forms, all the molecules at low temperatures will be in the lowest rotational state of quantum number 0 and weight 1, and we have simply $o = 1$.

The experimental data for HD are incomplete, since the heat capacities of crystalline and liquid HD have not been measured.

20.13. Comparison of calorimetric and spectroscopic data for diatomic molecules

When the available calorimetric and spectroscopic data are sufficiently accurate, it may be convenient to compare, as described in Sections 20.5 and 20.8,

$$S^G(T,P) - S^K(0,0), \tag{1}$$

determined calorimetrically with

$$S^G(T,P) - k\Sigma \ln \rho, \tag{2}$$

determined spectroscopically. When Eq. (1) is subtracted from Eq. (2), one obtains

$$S^K(0,0) - k\Sigma \ln \rho, \tag{3}$$

but $S^K(0,0)$ is related to the orientational weight o, as shown above, by

$$S^K(0,0) = k \ln \left\{o\, \upsilon_0{}^K\, \Pi\rho\right\} \tag{4}$$

which is Eq. (9) in Section 20.5, therefore Eq. (3) reduces to

$$k \ln \left\{o\, \upsilon_0{}^K\, \Pi\rho\right\} - \Sigma \ln \rho,$$

that is,

$$k \ln o\, \upsilon_0{}^{K}. \tag{5}$$

This is the procedure adopted by *Giauque* and his collaborators, who call Eq. (1) the 'calorimetric entropy' and Eq. (2) the 'spectroscopic entropy (without nuclear spin)' per molecule of gas at the temperature *T* and pressure *P*.

To evaluate the 'spectroscopic entropy' for one mole, we insert numerical values into Eq. (2) in Section 20.5, and obtain

$$N_a S^G(T,P) - R\Sigma \ln \rho = -R\ln\frac{P}{P^+} + \frac{5}{2}R\ln T + R\ln Z_{rot} Z_{vib}$$

$$+ R\frac{\partial \ln Z_{rot} Z_{vib}}{\partial T} + \frac{3}{2} R \ln M$$

$$-R \ln\sigma - 2.300 \text{ cal/deg.mole}, \tag{6}$$

where P^+ is one atmosphere measured in the same units as $P(P/P^+$ is thus the value of the pressure in atmospheres, and *M* is the conventional chemical molecular weight. When it is allowable to separate the rotational degrees of freedom from the vibrational, and to treat the former as classical, we may replace Eq. (6) by

$$N_a S^G(T,P) - R\Sigma \ln \rho = -R\ln\frac{P}{P^+} + \frac{7}{2}R\ln T$$

$$+R \ln Z_{vib} + R\frac{\partial \ln Z_{vib}}{\partial T}$$

$$+\frac{3}{2} R \ln M + R \ln\left(10^{40}\, I_b\right)$$

$$-R \ln \sigma - 7.62 \text{ cal./deg. mole}, \tag{7}$$

where I_b is the principal moment of inertia in $g{\cdot}cm^2$.

Giauque did not use the approximation contained in Eq. (7) but calculates the 'spectroscopic entropy' from Eq. (6).

20.14. Comparison of calorimetric and spectroscopic data for polyatomic molecules

We turn now to polyatomic molecules. Eqs. (1)–(12) in Section 20.12 and Eqs. (1)–(7) in Section 20.13 apply to polyatomic as well as to diatomic molecules. When it is allowable to separate the rotational degrees of freedom from the vibrational, and to treat the former as classical, we may replace Eq. (6) in Section 20.13 by

$$N_a S^G(T,P) - R\Sigma \ln \rho = -R\ln\frac{P}{P^+} + 4R\ln T$$

$$+R\Sigma_v \ln Z_{vib} + \frac{3}{2} R \ln M$$

$$+R\Sigma_v \frac{\partial \ln Z_{vib}}{\partial T} - R \ln \sigma$$

$$+\frac{1}{2} \ln \left(10^{120} I_a I_b I_c\right) - 9.14 \text{ cal./deg. mole}, \qquad (1)$$

which is valid for non-linear molecules. For linear polyatomic molecules, we have merely to use Eq. (7) in Section 20.13 with each term in Z_{vib} replaced by a *sum of several similar terms in* Z_{vib} *for the several normal modes.*

We shall now discuss those polyatomic molecules for which both accurate calorimetric data and reliable spectroscopic data are available.* The data are given in Table 1, where we have used Giauque's nomenclature, namely 'calorimetric entropy' for

$$S^G (T,P) - S^K (0,0),$$

and 'spectroscopic entropy' for

$$S^G (T,P) - k\Sigma \ln \rho,$$

these two equations being Eqs. (1) and (2), respectively, in Section 20.15. The values of the 'calorimetric entropy' are corrected from actual gas to ideal gas using the thermodynamic formula

$$S^{ideal} - S^{actual} = \int_{P^*}^{P} dP \left\{ \left(\frac{\partial S}{\partial P}\right)_T^{ideal} - \left(\frac{\partial S}{\partial P}\right)_T^{actual} \right\}$$

$$= \int_{P^*}^{P} dP \left\{ \left(\frac{\partial V}{\partial T}\right)_P^{actual} - \left(\frac{\partial V}{\partial T}\right)_P^{ideal} \right\}, \qquad (2)$$

where P^* is any pressure sufficiently low for the gas to be ideal. If the (P, V, T) relation for the gas is known, this correction is easily calculated. It is always small, usually not exceeding 0.3 cal./deg. mole.

In the *sixth column* of Table 1 are given the values of $o\, \upsilon_0{}^K$ that *have to be assumed to give agreement*; when these values are assumed for $o\, \upsilon_0{}^K$, and $R \ln o\, \upsilon_0{}^K$ is subtracted from the spectroscopic entropies given in the *fifth column*, we obtain the values given in the *seventh column.*

* A useful survey of the data previous to 1936 were compiled by Kassel, *Chem. Rev.* **18**, 277 (1936).

Table (1)

Molar entropies of polyatomic gases of their boiling-points at 1 atmosphere

Substance	Symmetry number σ	Boiling-point K	"Calorimetric entropy" observed cal./deg. mole	"Spectroscopic entropy" cal./deg. mole	$ov_0^K = o$ assumed	"Calorimetric entropy" calculated cal./deg. mole	References
OH_2	2	*	44·28	45·10	$\frac{3}{2}$	44·29	(1)
OD_2	2	*	45·89	46·66	$\frac{3}{2}$	45·85	(2)
SH_2	2	212·77	46·38	46·44	1	46·44	(3)
		212·8	46·33	46·42	1	46·42	(4)
NH_3	3	239·68	44·13	44·10	1	44·10	(5)
PH_3	3	185·38	46·39	46·5 ± 1·0	1	46·5 ± 1·0	(6)
		185·7	46·39	46·4	1	46·4	(7)
CH_4	12	111·5	36·53	36·61	1	36·62	(8)
CH_3D	3	99·7	36·72	39·49	4	36·73	(9)
CO_2	2	194·67	47·59	47·55	1	47·55	(10)
NNO	1	184·59	47·36	48·50	2	47·12	(11)
CS_2	2	318·39	57·48	57·60	1	57·60	(12)
SCO	1	222·87	52·56	52·66	1	52·66	(13)
SO_2	2	263·08	58·07	58·23 ± 0·15	1	58·23 ± 0·15	(14)
C_2H_4	4	169·4	47·36	47·35	1	47·35	(15)
CH_3Br	3	276·66	57·86	57·99	1	57·99	(16)

* The values given for OH_2 and OD_2 are for 298·1 K., and not for the boiling-point.

References to Table 6

(1) Giauque and Stout, *J. Am. Chem. Soc.* **58**, 1144 (1936).
(2) Long and Kemp, *J. Am. Chem. Soc.* **58**, 1829 (1936).
(3) Giauque and Blue, *J. Am. Chem. Soc.* **58**, 831 (1936).
(4) Clusius and Frank, *Zeit. Physikal. Chem.* B, **34**, 420 (1936).
(5) Overstreet and Giauque, *J. Am. Chem. Soc.* **59**, 254 (1937).
(6) Stephenson and Giauque, *J. Chem. Phys.* **5**, 149 (1937).
(7) Clusius and Frank, *Zeit. Physikal. Chem.* B, **34**, 405 (1936).
(8) Frank and Clusius, *Zeit. Physikal. Chem.* B, **36**, 291 (1937).
(9) Clusius, Popp and Frank, *Physica*, **4** (no. 10), 1105 (1937).
(10) Giauque and Egan, *J. Chem. Phys.* **5**, 45 (1937).
(11) Blue and Giauque, *J. Am. Chem. Soc.* **57**, 991 (1935).
(12) Brown and Manov, *J. Am. Chem. Soc.* **59**, 500 (1937).
(13) Kemp and Giauque, *J. Am. Chem. Soc.* **59**, 79 (1937).
(14) Giauque and Stephenson, *J. Am. Chem. Soc.* **60**, 1389 (1938)
(15) Egan and Kemp, *J. Am. Chem. Soc.* **59**, 1264 (1937).
(16) Egan and Kemp, *J. Am. Chem. Soc.* **60**, 2097 (1938).

It will be observed that in most, but not all, cases this value of $o\ \upsilon_0^{K}$ is unity. We need to discuss only the exceptions. All the molecules are chemically saturated so that $\upsilon_0^{K} = \upsilon_0^{G} = 1$. Any deviation of $o\ \upsilon_0^{K}$ from unity must therefore be due to $o \neq 1$.

The value 2 for o in the case of N_2O may be explained by supposing that the linear NNO molecule, owing to the high degree of symmetry, can, like the CO molecule, be *reversed end for end* without appreciably altering the energy of the crystal. The equilibrium distribution of directions will remain random down to temperatures for which *kT* is comparable with the energy difference in the two orientations. If by this temperature the chances of *reversal* have become *negligible*, the crystal will remain for all lower temperatures with random orientations of the molecules $(o = 2)$. It is interesting to observe that the SCO in *not sufficiently symmetrical* to behave in this way. For the molecule CH_3D, the value 4 for o takes account of the *four effectively equivalent orientations* of the molecule due to the D atom being isotopic with the three H atoms. We notice that the product σo is $3 \times 4 = 12$, equal to the value for CH_4 and CD_4. Thus the observed behaviour of CH_3D provides *experimental confirmation of the rule that isotopic molecules have equal values of* σo, even when the separate values of σ and o differ. We may note that for CH_2D_2 we should have $\sigma = 2$ and $o = 6$ so that we again recover $\sigma o = 12$.

The *discrepancy* between the calorimetric and spectroscopic entropies of H_2O and D_2O was interpreted by Pauling[13] as due to an *indefiniteness* in the position of the *hydrogen atoms*, or more strictly the *hydrogen nuclei*, in the crystal. Pauling made the following assumptions:

1. In ice, each oxygen atom has two hydrogens attached to it at distances about 0.95 Å, forming a molecule, the HOH angle being about 105° as in the *gas molecule*.
2. Each HOH molecule is oriented so that its two H atoms are directed approximately towards two of the four O atoms which surround it tetrahedrally.
3. The orientations of the adjacent HOH molecules are such that only one H atom lies approximately along each O-O axis.
4. Under ordinary conditions, the interaction of non-adjacent molecules is not such as to stabilise appreciably any one of the many configurations satisfying the preceding conditions relative to the others.

Pauling then calculated o by two alternative methods which lead to the same conclusion. We quote both.

A given molecule can orient itself in *six* ways satisfying condition 2, but the chance that the adjacent molecules will permit a given orientation is $\left(\frac{1}{2}\right)^2$, since each adjacent molecule has two tetrahedral directions occupied and two unoccupied by H atoms; thus the change of given direction being available for the original molecule is $\frac{1}{2}$. We thus obtain

$$o = 6\left(\frac{1}{2}\right)^2 = \frac{3}{2}.$$

---

[13] Linus Pauling, J. Am. Chem. Soc. **57**, 2680 (1935).

The same result is given by the following equivalent argument. If there are N molecules, and if we ignore condition 1, there are 2^{2N} configurations satisfying conditions 2 and 3, each of the $2N$ H nuclei having the choice of two positions, one near one O atom and the other near the other. Some of these are ruled out by condition 1. Let us now consider a given O atom and four surrounding H atoms. There are $2^4 = 16$ arrangements of the 4H atoms, but of these only $(4\times3)/2 = 6$ satisfy condition 1. We thus have

$$o^N = 2^{2N}(6/16)^N = \left(\frac{3}{2}\right)^N,$$

or

$$o = \frac{3}{2}.$$

This calculated value of o is in excellent agreement with the experimental value for both H_2O and D_2O. It is of interest to note that H_2S shows no analogous effect.

There are also experimental data for BF_3, CF_4, and SF_6, but the heats of evaporation have not been measured directly and so have to be computed from the dependence of the vapour pressure on the temperature. This procedure introduces an experimental uncertainty greater than that of the data in Table 1, probably about 0.4 cal./deg. mole. Within this degree of accuracy, there is equality between the calorimetric entropy and the spectroscopic entropy.

20.15. The third law of thermodynamics.
(i) The experimental value of O

We are now due to undertake a discussion of the *third law of thermodynamics* and *Nernst's heat theorem.* The law and the theorem have been in the past subject to much controversy, but one may perhaps claim that *substantial agreement has recently been achieved.*

In view of recent controversies, we shall start with a discussion of the meaning of the experimental values of $o\,\upsilon_0^K$, before we give the statement of the third law and develop its consequences.

In discussing Nernst's heat theorem, we are concerned with the behaviour of the entropy change in an isothermal reaction when $T \to 0$. The discussions of the earlier part of this chapter have shown us that

For any isothermal reactions between crystals

$$\underset{T\to0}{\mathrm{Lt}}\ \Delta S = k\,\Delta\sum \ln o\,\upsilon_0^K. \qquad (1)$$

These discussions have also shown us how in practice this limit is evaluated, determinations of $o\,\upsilon_0^K$ for various crystals having been made.

We have seen that usually $o\,\upsilon_0^K = 1$. The occasional values of $o\,\upsilon_0^K > 1$ can arise in two distinct ways, which we consider in turn.

(a) Extrapolation from too high a value of T^0. We emphasise again that measurements at the absolute zero cannot be made, and the only physical meaning for a quantity such as ΔS at the absolute zero is the limit of this quantity as $T \to 0$. The evaluation of this limit will always involve an extrapolation to $T = 0$ from the lowest convenient temperature T^0 for which measurements of heat capacities have been made. In our theoretical formulae, various sets of states are treated as *degenerate*, but none of these sets need be *strictly degenerate;* so long as their energy differences $\Delta\epsilon$ are small compared with kT, it is correct to ignore these differences and treat the set simply as a *degenerate state* with the corresponding extra weight factor. If the observations stop at a temperature T^0 sufficiently high, no effect of the ignored separations will be seen in the heat capacities, and one must include the *extra weight factors in the partition function.* If, however, observations are pushed lower $(kT^0 \approx \Delta\epsilon)$, we reach temperatures at which, for equilibrium to be maintained, the upper states of the set must be gradually emptied. Finally, we may reach still lower temperatures $(kT^0 \ll \Delta\epsilon)$ where the upper states can be ignored, and the effective states are no longer degenerate. During this change, the heat capacity will show temporarily *exceptionally large values.*

We have already seen in Section 20.12 that ortho-hydrogen, and to a less extent the ordinary metastable mixture of ortho- and para-hydrogen, shows a rise in the heat capacity when the temperature is reduced below 12 K. For H_2 undoubtedly $\upsilon_0{}^K = \upsilon_0{}^G = 1$, and so we need consider only the orientational factor o. The values $o = 3$ for ortho-hydrogen and $o = 3^{3/4}$ for the ordinary metastable mixture are evidently due to too high a value of T^0. If the heat capacity measurements could be continued to still lower temperatures below 12 K, the value of $k \ln o\nu^k$ between the limits 0.3 K and 12 K would be $k \ln 3$ for ortho-hydrogen or $\frac{3}{4} k \ln 3$ for ordinary hydrogen. One would then be able to choose $T^0 \approx 0.3\,K$ and would find $o = 1$. Similarly, the weight factors $o = 3$ for para-deuterium and $o = 3^{1/3}$ for the ordinary mixture of ortho- and para-deuterium would be reduced to unity by using a sufficiently low T^0. Thus for reactions involving H_2 and D_2 extrapolated to zero from $T^0 \approx 0.3\,K$, *no contribution would be expected towards a non-zero value of* o *by the* H_2 *or* D_2 *involved.*

(b) Frozen-in varieties of orientation. There are a number of molecules such as CO, NO, N_2O, H_2O, and D_2O whose crystals have non-unit values of o and so of $o\,\upsilon_0{}^K$. *These extra weight factors* we have attributed to extra orientations of one type or another for CO and NNO to the possibility of turning an almost symmetrical linear molecule *end for end*; for NO to the possibility of two distinct orientations for the molecule N_2O_2; for H_2O and D_2O to the possible different arrangements of the H or D (Section 20.14). *The energy differences of these orientations are small*, and for any example it may be that *the various orientations form a still effectively degenerate but unfrozen state at* T^0 so that the extra weight factor would be removed by choice of a lower T^0, an example of case (a). This,

however, is unlikely. It is much more likely that we have a case of *metastable equilibrium with the random orientations frozen in.* In general, we should perhaps *expect such metastable equilibria,* whenever we have a crystal built of molecules with their two ends *physically distinguishable* but extremely similar in size and force field. For we may then expect the molecule to have two or more equilibrium orientations in the crystal, only one of which is *completely stable,* the remainder being *metastable,* leading to a value of $o > 1$. The metastable orientations lie above the stable orientations by an energy difference $\Delta\epsilon$, which is much smaller than kT_K where T_K is the minimum temperature required for reorientation at an appreciable speed. In other words, when we reach temperatures so low that it matters to the molecule which way it points, it is already *practically impossible for it to turn round.* The situation is exactly the same for orientations here as it is for spatial arrangements and rearrangements in a glass or in a chilled metallic alloy, and we might describe such phases as 'orientationally amorphous'. *The molecules will remain frozen at all lower temperatures, being distributed perfectly at random between their two or more orientations.* The frozen crystal with o orientations occupied at random will have an entropy $k \ln o$ per molecule *greater than that of an ideal crystal* with regular orientations, and this is the entropy which one would naturally regard as *the true limiting value of the entropy at the absolute zero.* It is interesting to note that for an almost symmetrical molecule of the type discussed here, we obtain the correct value of its vapour pressure constant by treating it as truly symmetrical, for then we replace the true value $o = 2$, $\sigma = 1$ by the *fictitious values* $o = 1$, $\sigma = 2$, and the product $o\sigma$ is unaltered.

20.16. The third law of thermodynamics.
(ii) Historical sketch

After this introductory survey of the statistical interpretation of the experimental facts in the form of the observed values of $o\,\upsilon_0^{K}$, we pass to a brief historical sketch of the third law of thermodynamics and Nernst's heat theorem had been discussed earlier in Chapter 7 using classical thermodynamics.

One of the earliest enunciations of Nernst's heat theorem took the form:

> *In any isothermal process between condensed systems (including glasses, supercooled liquids, and solutions)*

$$\underset{T\to 0}{\mathrm{Lt}}\ \Delta S = 0 \quad . \qquad (1)$$

Provided we assign to $\underset{T\to 0}{\mathrm{Lt}}$ the usual meaning corresponding to a smooth extrapolation, it was found that Eq. (1) was not true *either for glasses or for solutions.*

It was therefore necessary to revise the enunciation, and restricted enunciations of the theorem were later proposed in the forms:

(a) *For any isothermal process involving only pure crystals,*

$$\underset{T\to 0}{\mathrm{Lt}}\ \Delta S = 0 \quad . \qquad (2)$$

(b) *If the entropy of each element in the crystalline state stable at $T = 0$ be taken as zero, every substance has then a finite positive entropy, but at $T = 0$ the entropy may become zero, and does become zero for all perfect crystalline substances including compounds.*

Formulation (b) is equivalent to one first given by *Lewis* and *Gibson.*[14] It will be observed that they are careful so to formulate the theorem that the idea of *absolute entropy* is not introduced.

These enunciations avoid the difficulty created by phases such as glasses *ordinarily recognised as amorphous*, but the empirical foundations even of these enunciations soon become none too secure as the facts became better known.

Further progress in the understanding of the theorem has been greatly helped by a better appreciation of its *statistical interpretation through studies of the formula*

$$\mathop{\mathrm{Lt}}_{T \to 0} \Delta S = k \, \Delta \, (\ln o \upsilon_0{}^K), \tag{3}$$

and especially of the conditions under which we may expect $o = 1$ for all substances concerned in any reaction. But, however wide or narrow these conditions turn out to be, it is already clear that the theorem

$$\Delta S \to 0, \tag{4}$$

is not universally valid according to its normal interpretation. It was therefore natural to search for a suitable *principle of universal validity which might rank as a third law of thermodynamics and from which Nernst's heat theorem when true would follow.*

Such a principle, fulfilling all requirements, is available – *The principle of the unattainability of the absolute zero*, first enunciated by Nernst.[15] We shall adopt this principle as the *third law*.

20.17. The third law of thermodynamics.
(iii) Its thermodynamic formulation and consequences

The principle of the *unattainability of the absolute zero* on the available evidence is of completely general validity. This principle, which we call the *third law of thermodynamics*, may be more precisely enunciated as follows:

It is impossible by any procedure, no matter how idealised, to reduce any assembly to the absolute zero in a finite number of operations.

The type of evidence that suggests this principle most strongly is that derived from experiments on the adiabatic demagnetisation of paramagnetic salts, the discussion of which was given[16] earlier. Accepting this principle, we proceed to examine its consequences. It should be noticed that the assumption we make here is that as $T \to 0$, the heat capacities all tend to zero *sufficiently fast* for the entropies to remain finite.

Let us consider any process (e.g. change of volume, change of external field, chemical reaction) denoted formally by

$$\alpha \to \beta; \tag{1}$$

[14] Lewis and Gibson. *J. Am. Chem. Soc*, 42, 1529 (1920). See also Lewis and Randall, *Thermodynamics*, p. 448 (McGraw-Hill, 1923).

[15] Nernst, *Berlin, Sitzungsber*, p. 134 (1912).

[16] See Section 16.13.

we shall use the subscripts α and β to denote properties of the assembly in the states α and β, respectively. Then the entropies of the assembly in these two states depend on the temperature according to the formulae.

$$S_\alpha = S_\alpha^0 + \int_0^T \frac{C_\alpha}{T} dT, \tag{2}$$

$$S_\beta = S_\beta^0 + \int_0^T \frac{C_\beta}{T} dT, \tag{3}$$

where S_α^0 , S_β^0 are the limiting values of S_α , S_β for $T \to 0$. We know from quantum theory that both the integrals *converge*. Suppose now that we start with the assembly in the state α at the temperature T' and that we can make the *process* $\alpha \to \beta$ *take place adiabatically*. Let the final temperature after the assembly has reached the state β be T''. We are now going to consider the possibility or impossibility of T'' being zero. From the second law we know that, for an adiabatic process defined by its initial and final states, the change of entropy is zero if the process takes place quasi-statistically (i.e. reversibly) but positive otherwise. It is therefore clear that the chances of attaining as low a final T as possible are most *favourable* when the process takes place quasi-statistically. *We need therefore consider only such a quasi-static path.* For the quasi-static adiabatic process being considered, we have then by Eqs. (2) and (3)

$$S_\alpha^0 + \int_0^{T'} \frac{C_\alpha}{T} dT = S_\beta^0 + \int_0^{T''} \frac{C_\beta}{T} dT. \tag{4}$$

If T'' *is to be zero*, we must then have

$$S_\beta^0 - S_\alpha^0 = \int_0^{T'} \frac{C_\alpha}{T} dT. \tag{5}$$

Now if $S_\beta^0 - S_\alpha^0 > 0$, it will always be possible to choose an initial T' satisfying Eq. (5), and by making the process $\alpha \to \beta$ take place from the initial T' it will be possible to reach $T'' = 0$. From the *premise of the unattainability of* $T = 0$, we can therefore conclude that

$$S_\beta^0 \le S_\alpha^0. \tag{6}$$

Similarly, we can show that if we can make the reverse process $\beta \to \alpha$ take place quasi-statistically and adiabatically, then we can reach $T = 0$ from an initial temperature T' satisfying

$$S_\alpha^0 - S_\beta^0 = \int_0^{T'} \frac{C_\beta}{T} dT. \tag{7}$$

Further, if $S_\alpha^0 - S_A^0 > 0$, we can always choose an initial T' satisfying Eq. (7). From the *unattainability* of $T = 0$, we can therefore conclude that

$$S_\alpha^0 \le S_\beta^0. \tag{8}$$

From Eqs. (6) and (8), we deduce

$$S_\alpha^0 = S_\beta^0 . \qquad (9)$$

This is precisely *Nernst's heat theorem.*

We can also show *conversely* that, given Eq. (9), neither the *process* $\alpha \to \beta$ nor the *reverse process* $\beta \to \alpha$ can be used to reach $T = 0$. This is because if we assume Eq. (9) to be true, we now have, according to Eq. (4), for the adiabatic process $\alpha \to \beta$ the initial temperature T' and the final temperature T'' related by

$$\int_0^{T'} \frac{C_\alpha}{T} dT = \int_0^{T''} \frac{C_\beta}{T} dT. \qquad (10)$$

To reach $T'' = 0$ we should require

$$\int_0^{T'} \frac{C_\alpha}{T} dT = 0. \qquad (11)$$

But since $C_\alpha > 0$ always for any finite T, it is *impossible to satisfy* Eq. (11). Hence the process $\alpha \to \beta$ *cannot be used to reach* $T = 0$. The proof for the *reverse process* is exactly similar.

From the derivation of Nernst's heat theorem in Eq. (9), we see that *its deduction from the third law of thermodynamics* (*the unattainability of* $T = 0$) involves the assumption that the states α and β can be *connected by a reversible path.*

If all the phases concerned are phases in complete internal equilibrium, the reactions concerned must presumably by regarded as *ideally reversible*. If any phase is naturally in *metastable internal equilibrium*, a reaction or process affecting it may or may not disturb the *frozen metastable equilibrium.* If it does not disturb this equilibrium, then the reaction or process must still be regarded as *reversible*, but otherwise it will be *irreversible*. We must therefore enunciate *Nernst's heat theorem* in the following accurate form:

For any isothermal process involving only phases in internal equilibrium or, alternatively, if any phase is in frozen metastable equilibrium, provided the process does not disturb this frozen equilibrium,

$$\operatorname*{Lt}_{T \to 0} \Delta S = 0. \qquad (12)$$

This includes all physical changes which depend on the variation of some continuous external parameter such as pressure or magnetic field strength. It also covers all chemical reactions between phases in *internal equilibrium*, but does not include chemical reactions involving phases in metastable equilibrium.

Any conceivable *isothermal process* involving a frozen metastable equilibrium, which does disturb the metastability, can obviously proceed only in the direction *which decreases this metastability*. It follows at once from the third law that for such a reaction proceeding in the only possible direction

$$\operatorname*{Lt}_{T \to 0} \Delta S < 0, \qquad (13)$$

and such a process is even less efficient[17] for reaching $T = 0$ than one for which $\mathop{\mathrm{Lt}}_{T\to 0} \Delta S = 0$.

20.18. The third law of thermodynamics. (iv) Statistical interpretation

We are now in a position to give a proper statistical interpretation of Nernst's heat theorem. We have seen in the earlier sections of this chapter whose results are collected and discussed in Section 20.15 that, for all reactions between crystals,

$$\mathop{\mathrm{Lt}}_{T\to 0} \Delta S = k\, \Delta \sum \ln o\, \upsilon_0^{\,K}, \qquad (1)$$

where all contributions of nuclear spins, isotopic mixtures, and ortho-para separations can be ignored. Further, provided that extrapolations are made to zero from a suitably chosen T^0, which is of the order commonly used in practice (except that, for H_2 and for D_2, T^0 should be about 0.3 K), we have found that $o\,\upsilon_0^{\,K}$ is *different from unity only for orientationally amorphous phases in frozen metastable equilibrium*. For these phases $o > 1$ and so $o\,\upsilon_0^{\,K} > 1$. Hence, for any process which *thaws or removes the metastability*.

$$\mathop{\mathrm{Lt}}_{T\to 0} \Delta S < 0. \qquad (2)$$

For any process, on the other hand, which does not concern systems for which $o\,\upsilon_0^{\,K} > 1$, or in which values of $o\,\upsilon_0^{\,K}$ if greater than unity occur but are not affected,

$$\mathop{\mathrm{Lt}}_{T\to 0} \Delta S = 0. \qquad (3)$$

These statistical convulsions are therefore in full agreement with the third law and our deductions from it.

It has been argued[18] that *irregular solid phases* must be excluded from the field of the theorems *a priori*, because, since they are phases in *metastable equilibrium* and not in *true internal equilibrium*, the theorems of thermodynamics, including Nernst's heat theorem, which *state equalities*, cannot be applied to them. This point of view, however, proves too drastic and cannot be reasonably maintained.[19] In its extreme form, this view implies that *a phase in frozen equilibrium has no definite value of the entropy*. In the first place, we observe that there is no need to hold this view in order to be able to give a perfectly general and acceptable form of the third law of thermodynamics and Nernst's heat theorem such as we have given above. Again an entropy difference between a metastable solid phase and the liquid phase can always be established and measured, because by carrying out the melting process at the proper speed it can be made to go reversibly, or practically reversibly. In the particular case of NNO, it has been shown[20] that the *thermal properties* of the crystal at the

---

[17] See Simon, *Zeit. Phys.* **41**, 806 (1927).
[18] Simon, *Zeit. Anorg. Chem.* **203**, 226 (1931).
[19] Compare Eastman and Milner, *J. Chem. Phys.* **1**, 451 (1933).
[20] Eucken and Veith, *Zeit. Phys. Chem.* B, **35**, 463 (1937).

lowest temperatures are entirely independent of *the rate of cooling to these temperatures.* It follows that the changes occurring during heating and cooling are *reversible*, although the crystal, being orientationally amorphous, *is unquestionably metastable at the lowest temperatures.* Thus the *metastable solid phase must be held to have a definite entropy, just, as much as any other phase in which the substance can exist.*

It is argued again that a glass cannot have a definite vapour pressure. This may well be true. For if the process of evaporation and re-condensation removes the metastability, the vapour pressure would progressively change over to the vapour pressure characteristic of the regular crystalline solid. But even if it is true that a particular glass has no definite vapour pressure, this does not mean that the glass has no definite entropy, but merely, as is obvious, *the process of evaporation is irreversible.* On the other hand, it is conceivable that the process of evaporation from an irregular solid like a glass is, in the absence of the crystalline form, actually reversible and that the re-condensation on the amorphous substance leaves the solid amorphous. In this case, the glass would resemble a supersaturated solution in having a definite vapour pressure in spite of its metastability.

A well-known assembly, to which no one hesitates to apply the theorems of thermodynamics, is a gaseous mixture of oxygen and hydrogen. *This is only in metastable equilibrium*, since in the *true equilibrium state* there should be a large conversion of the oxygen and hydrogen into water or steam. The difference of entropy (or of free energy) between such a metastable assembly and the stable assembly into which it might be changed irreversibly has a *well-defined and measurable value.*

Careful consideration of all such cases leads to the conclusion that all *thermodynamic systems are liable to be only metastable*, and all applications of the equalities stated by the theorems of thermodynamics to be of only *relative validity.* Actually we hardly ever, if ever at all, deal with complete equilibrium in the strict sense of the word, as quoted by Giauque[21] and Johnston, for even among elementary atoms probably some are unstable relative to others, but, except for the recognisably radioactive elements, the transformations are negligibly slow. The fact that there are such slow transformations occurring does not, however, matter to thermodynamics. In order to be able to use the equalities stated by the theorems, we must be able to classify all processes of change into two classes, those that are *very fast* and those that are *very slow* compared with the changes that we wish to impose experimentally. *The processes of the slow group we can entirely ignore. Those of the fast group will maintain complete equilibrium among the states or phases that they connect, and the imposed changes will be perfectly reversible.* Processes of intermediate speed which are neither fast nor slow are, however, fatal, and inevitably make an imposed change *irreversible.* The situation in pure thermodynamics is in fact the same as we found it in discussing the underlying principles of accessibility in statistical mechanics in Chapter 17. There seems no good reason for supposing that processes and phenomena at very low temperatures are any different in this respect from those at ordinary temperatures, to which we are better accustomed, and there is no need to make such artificial distinctions in order to preserve a general form of Nernst's heat theorem.

---

[21] Giauque and Johnston, *Phys. Rev.* **36**, 1592 (1930).

BIBLIOGRAPHY

Selected Bibliography

The fundamental investigators in their gigantic power grapple with and subjugate the subject of their inquiry and imprint on it the forms of conceptual thought. Those who know the course of the development of science will judge more freely and more correctly the significance of any present scientific movement than those who, limited in their views to the age in which their own lives have been spent, contemplate merely the trend of intellectual events at the present moment.

History and Biography

Histories

H. Crew, *The Rise of Modern Physics* (New York: Williams and Wilkins, 1928, 1935). Modern and readable. The best book of its sort so far published in English.

W. C. D. Dampier Whetham and M. D. Dampier Whetham, *Cambridge Readings in the Literature of Sciences* (London: Cambridge University Press, 1924). A book of extracts from the writing of men of science chosen to illustrate the development of definite subjects in thought of succeeding ages.

I. B. Hart, *Makers of Science: Mathematics, Physics, Astronomy* (Oxford: Oxford University Press, 1923). Brief, interesting accounts of the lives and works of *Aristotle Archimedes, Roger Bacon, Copernicus, Kepler, Galileo, Descartes, Newton, Boyle, Davy, Faraday, Kelvin, etc.* and their relations to each other and to scientific progress.

D. Mckkie and N. H. De V. Heathcote, *The Discovery of Specific and Latent Heats* (London: Arnold, 1935). A scholarly study of the foundations of the modern science of heat.

A. WOLF, *A History of Science, Technology, and Philosophy in the Sixteenth and Seventeenth Centuries* (London: Allen & Unwin, 1935). This interesting volume treats in detail and with profuse illustrations one of the most fruitful periods of scientific development.

Biographies

Thomas Andrews (1813–1885)

P. G. Tait and A. Crum Brown, 'Memoir' in *The Scientific Papers of Thomas Andrews* (London: Macmillan, 1889).

Archimedes (*c*. 287–212 BC)

T. L. Heath, *Archimedes* (London: Macmillan, 1920). Also, *The Works of Archimedes* (London: Cambridge University Press, 1897). Much of our information regarding the life of Archimedes comes from Plutarch's *Life of Marcellus*.

Amedeo Avogadro (1776–1856)

I. Guareschi, *Amedeo Avogadro e la Teoria moleculare* (1901); German translation by O. Merchens, *Amedio Avogadro und die Moleculartheorie* (London: Barth, 1903).

Ludwig Boltzmann (1844–1906)

P. Lenard, *Great Man of Science* (London: Macmillan, 1933). p. 350.
C. H. Bryan, *Nature* **74** (1906). p. 569.

Robert Boyle (1627–1691)

T. Birch, The *Works of the Honorable Robert Boyle* (London, 1744, 1772).
F. Mason, *Robert Boyle, a Biography* (London: Constable, 1914).

Nicolas Léonard Sadi Carnot (1796–1832)

M. H. CARNOT, 'Life of Sadi Carnot', *Reflections on the Motive Power of Heat* R. H. Thurston (ed.), (New York: Wiley, 1897). *Nature* **130** (1932). p. 266.
E. H. Johnson, *Scientific Monthly* **36** (1933). p. 131.

Henry Cavendish (1731–1810)

G. Wilson, *The Life of the Hon. Henry Cavendish* (London: Cavendish Society, 1851). The standard life of Cavendish.
Introduction by J. C. MAXWELL and E. THORPE to Vols. I and II of *The Scientific Papers of the Honorable Henry Cavendish*, F. R. S. (London: Cambridge University Press, 1921).

Rudolf Julius Emmanuel Clausius (1822–1888)

J. W. Gibbs, 'Rudolf Julius Emmanuel Clausius', *Proceedings of the American Academy*, New Series **16**, 458 (1889). Also *The Collected Works of J. Willard Gibbs* (London: Longmans, Green, 1828). Vol. II, pp. 261–267.
Proceedings of the Royal Society **48,** i (1890).

Charles Auguistin Coulomb (1736–1806)

T. Young, 'Life of Coulomb' *Miscellaneous Works of Thomas Young* (London: Murray, 1855). Vol. II, pp. 527–541.
P. Lenard, *Great Men of Science* (London: Macmillan, 1933). pp. 149–158.

John Dalton (1766–1844)

W. C. Henry, *Memoirs of the Life and Scientific Researches of John Dalton* (London: Cavendish Society, 1854). The standard work on Dalton.

H. Rosooe, *John Dalton and the Rise of Modern Chemistry* (London: Macmillan, 1895). An especially interesting biography.
R. A. Smith, *Memoir of John Dalton, and History of the Atomic Theory* (London: Bailliere, 1856). Emphasises Dalton's contributions to atomic theory.

Leonhard Euler (1707–1783)

R. E. Langer, *Scripta Mathematica*, **3**, (1935). p. 61, 131.
M. J. N. C. De Condorcet, 'Eloge d'Euler', *Lettres de L. Euler à une princesse d' Allemagne* (London: Hachette, 1842).

Michael Faraday (1791–1867)

R. Appleyard, *A Tribute to Michael Faraday* (London: Constable, 1931). An eulogy and appreciation of Faraday as a man as well as a scientist.
J. H. Gladstone, *Michael Faraday* (New York: Harper, 1872). Interesting personal reminiscences.
B. Jones, *The Life and Letters of Faraday*, 2 vols. (London: Longmans, Green, 1870). The standard biography of Faraday by one of his intimate friends.
S. P. Thompson, *Michael Faraday, His Life and Work* (London: Cassell, 1901). The most satisfactory biography of Faraday as a scientist.
J. Tyndall, *Faraday as a Discoverer* (New York: Appleton, 1880). An excellent summary of Faraday's work by a great scientist and expositor.

Joseph Fourier (1768–1830)

F. Arago, *Biographies of Distinguished Scientific Men*, W. H. Smyth, B. Powell, and R. Grant (Trans.), (London, 1957). The memoir on Fourier is reproduced in the *Smithsonian Institution Reports* (1871).
Nature **125**, (1930), p. 710.

Galileo Galilei (1564–1642)

W. W. Bryant, *Galileo* (Washington, DC: Sheldon Press, 1925).
J. J. Fahie, *Galileo, His life and Work* (London: Murray, 1903). The best biography of Galileo in English.

Louis Joseph Gay-Lussac (1778–1850)

F. Arago, 'Êtage de Gay-Lussac'. *Proceedings of the Royal Society* **5,** 1013 (1943–1850).

Josiah Willard Gibbs (1839–1903)

H. A. Bumstead, *American Journal of Science* **16** (4) (1903). Also *The Collected Words of J. Willard Gibbs* (London: Longmans, Green, 1928, Vol. 1, pp. xiii–xxvii.
J. Johnston, *Journal of Chemical Education* **5** (1928). p. 207.
E. E. Slosson, *Leading American Man of Science*, D. S. Jordon (ed.), (London: Holt, 1910), pp. 311–362.
E. B. WILSON, *Scientific Monthly* **32** (1931). p. 211.

Hermann Von Helmholtz (1821–1894)

L. Koenigsberger, *Hermann von Helmholtz* (London: Wieweg, 1902), F. A. Welby (Trans.), (Oxford: Oxford University Press, 1906).

Journal of the Optical Society of America **6** (1922), p. 312, 327, 336.

A. W. Rucker, *Fortnightly Review* (1894); repr., *Smithsonian Institution Report* (1994).

Proceedings of the Royal Society **59,** xvii (1895–1896).

Robert Hooke (1635–1703)

R. T. Gunther, *Early Science in Oxford* (Privately printed, Oxford, 1930), VI. This is the life written by R. Waller as an introduction to *the Posthumous Works of Robert Hooke* (1705). To it extracts have been added from J. Ward's *Lives of the Gresham Professors* (1740), and J. Aubrey's *Short Lives.*

H. W. Robinson and W Adams (ed.), *The Diary of Robert Hooke* (London: Taylor & Francis 1935).

James Prescott Joule (1818–1889)

O. Reynolds, *Memoirs and Proceedings of the Manchester Literary and Philosophical Society* (4) **(6)** (1892). 'The best biography of a scientist in the English language'.

J. Dewar, *Proceedings of the Royal Institution* **13** (1890). p. 1.

J. T. Bottomley, *Nature* **26** (1882). p. 617.

Lord Kelvin (William Thomson) (1824–1907)

A. Gray, *Lord Kelvin, an Account of His Scientific Life and Work* (London: Dent, 1908).

S. P. Thompson, *The Life of William Thomson, Baron Kelvin of Largs*, 2 vols. (London: Macmillan, 1910). The standard biography of Kelvin, with numerous quotations from original documents and letters which speak for themselves.

E. King, *Lord Kelvin's Early Home* (London: Macmillan, 1909). An intimate account by his sister.

A. G. King, *Kelvin the Man* (London: Holder & Stoughton, 1925). An intimate picture of Kelvin's personality by his niece.

Pierre Simon de Laplace (1749–1827)

F. Arago, *Biographies of Distinguished Scientific Men*, W. H. Smyth, B. Powell, and R. Grant (Trans.), (London, 1857). The memoir on Laplace is reproduced in the *Smithsonian Institution Report* (1874).

W. W. Rouse Ball, *A Short Account of the History of Mathematics* (London: Macmillan, 1927). pp. 412–421.

James Clerk Maxwell (1831–1879)

L. Campbell and W. Garnett, *The Life of James Clerk Maxwell* (London: Macmillan, 1882). The standard biography of Maxwell. Lewis Campbell was a school-fellow and lifelong friend of Maxwell; Garnett was his demonstrator at Cambridge.

R. T. Glazebrook, *James Clerk Maxwell and Modern Physics* (London: Macmillan, 1896). Briefer than the preceding biography.

James Clerk Maxwell, *A Communication Volume, 1831–1931* (London: Macmillan, 1931).
Brief essays by J. J. Thomson, Planck, Einstein, Larmor, Jeans, Lodge, Glazebrook, and others on various phases of Maxwell's life and accomplishments.

Isaac Newton (1642–1727)

D. Brewster, *Memoirs of the Life, Writing and Discoveries of Sir Isaac Newton* (London: Edmonton & Douglas, 1860). The standard life of Newton to which frequent reference was made.

S.Brodetsky, *Sir Isaac Newton* (London: Methuen, 1927). Brief and well written; intended for the reader who possesses only a moderate grounding in the elements of science.

W. J. Greenstreet, *Isaac Newton* (London: Bell, 1927). A collection of rather technical essays on little-known parts of Newton's work.

I. T. More, *Isaac Newton, a Biography* (New York: Scribner, 1934). Easily the best biography of Newton so far written and the only really adequate one.

Heike Kamerlinch Onnes (1853–1926)

F.A. Freeth, *Nature* **117** (1926). p. 350. *Smithsonian Institution Report* (1926), p. 533.

Blaisf Pascal (1623–1662)

L. G. J. Chevalier, *Blaise Pascal* (London: Longmans, Green, 1930).

W. W. Rousf Ball, *A Short Account of the History of Mathematics* (London: Macmillan, 1927). pp. 281–288.

P. Lenard, *Great Man of Science* (London: Macmillan, 1933). pp. 49–50.

Lord Rayleigh (John William Strutt) (1842–1919)

R. J. Strutt, *John William Strutt, Third Baron Rayleigh* (London: Arnold, 1924). An interesting biography, by his son.

A. Schuister, *Proceedings of the Royal Society* **98**, i (1920–1921).

Johannes Oiderik Van Der Waals (1837–1923)

H. Kamerlingh Onnes, *Nature* **111** 1923, p. 209.

James Watt (1736–1819)

H. W. Dickinson and Rhys Jenkins, *James Watt and the Steam Engine* (Oxford: Oxford University Press, 1927). A memorial volume prepared for the Committee of the Watt Centenary Commemoration at Birmingham in 1919.

H. W. Dickinson, *James Watt: Craftsman and Engineer* (London: Cambridge University press, 1936).

F.Bramwell, *James Watt* (London 1899). Also in the *Dictionary of National Biography*, Vol 60, pp. 51–62. 'The best short biography extant, by an engineer of ripe experience' .

T. H. Marshall, *James Watt* (London: London, Boston, 1925).

Otto Von Guericke (1602–1686)

Experimental Nora (ut vocantur) Magdeburgica de Vacuo Spatio (New York: Amsterdam, 1672). This work contains descriptions of the first air pump, many interesting experiments on air pressure, the barometer, the Magdeburg hemispheres, the manometer, the lever, thermoscope, and some pioneer experiments on electricity.

Thomas Young (1773–1829)

G. Peacock, *Life of Thomas Young* (London: Murray, 1855).
F. Oldham, *Thomas Young, F. R. S.* (London: Arnold, 1933).
H. B. Williams, *Journal of the Optical Society of America* **20** 1930, p. 35.

Appendices

APPENDIX A

Contents

I. GREEK ALPHABET USED AS SYMBOLS

Greek letter	Greek name	Greek letter	Greek name
Α α	Alpha	Ν ν	Nu
Β β	Beta	Ξ ξ	Xi
Γ γ	Gamma	Ο ο	Omicron
Δ δ	Delta	Π π	Pi
Ε ε	Epsilon	Ρ ρ	Rho
Ζ ζ	Zeta	Σ σ	Sigma
Η η	Eta	Τ τ	Tau
Θ θ	Theta	Υ υ	Upsilon
Ι ι	Iota	Φ φ	Phi
Κ κ	Kappa	Χ χ	Chi
Λ λ	Lambda	Ψ ψ	Psi
Μ μ	Mu	Ω ω	Omega

II. METRIC PREFIXES

	Abbreviation or Symbol	Factor		Abbreviation or Symbol	Factor
deca-	da	10	deci-	d	10^{-1}
hecto-	h	10^{2}	centi-	c	10^{-2}
kilo-	k	10^{3}	milli-	m	10^{-3}
mega-	M	10^{6}	micro-	μ	10^{6}
giga-	G	10^{9}	nano-	n	10^{-9}
tera-	T	10^{12}	pico-	p	10^{-12}
pecta-	P	10^{15}	femto-	f	10^{15}
exa-	E	10^{18}	atto-	a	10^{-18}

III. SI UNITS

1. Base units

Physical quantity	Name	Abbreviation or symbol
Length	metre	m
Mass	kilogram	kg
Time	second	s
Electric current	ampere	A
Temperature	kelvin	K
Amount of substance	mole	mol or mole
Luminous intensity	candela	cd
Electric charge	coulomb	C

2. Derived units with special names

Physical quantity	Name	Abbreviation or symbol
Frequency	hertz	Hz
Energy	joule	J
Force	newton	N
Power	watt	W
Pressure	pascal	Pa
Electromotive	volt	V
Electric resistance	ohm	–
Electric conductance	siemens	S
Electric capacitance	farad	F

S I units (continued)

Physical quantity	**Name**	**Abbreviation or symbol**
Magnetic flux	weber	Wb
Inductance	henry	H
Magnetic flux density	tesla	T
Luminous flux	lumen	lm
Illumination	lux	lx

3. Supplementary units

Physical quantity	**Name**	**Abbreviation or symbol**
phase angle	radian	rad
solid angle	steradian	sr

SECTION 1

Infinite Series and Their Uses

Introduction

There is a well-defined distinction between the approximate values of a physical constant, which are seldom known to more than three or four significant figures, and the approximate value of the *incommensurables* π, e, $\sqrt{2}$, ... which can be calculated to any desired degree of accuracy. In scientific work, we are rarely concerned with absolute errors.

I. Maclaurin's theorem

Maclaurin's theorem determines the law for the expansion of a function of a single variable in a series of ascending powers of that variable.

Let the variable be denoted by *x*, then,

$$u = f(x).$$

Assume that *u* can be developed in ascending powers of *x*, namely,

$$u = f(x) = A + Bx + Cx^2 + Dx^3 \ldots, \tag{1}$$

where A, B, C, D, . . . , are constants independent of *x*, but depend on the original function. It is required to determine the values of these constants, in order that the above assumption may be true for all values of *x*.

By successive differentiation of Eq. (1),

$$\frac{du}{dx} = \frac{df(x)}{dx} = B + 2Cx + 3Dx^2 + \ldots; \tag{2}$$

$$\frac{d^2u}{dx^2} = \frac{d f'(x)}{dx} = 2C + 2.3.\,Dx + \ldots; \tag{3}$$

$$\frac{d^3u}{dx^3} = \frac{d f''(x)}{dx} = 2.3.D + \ldots. \tag{4}$$

By hypothesis, Eq. (1) is true whatever the value of *x*, and, therefore, the constants A, B, C, D . . . are the same whatever value be assigned to *x*. Now substitute $x = 0$ in Eqs. (2)–(4). *Let V be the value assumed by u when x = 0.*

Hence, from Eq. (1),

$$\left.\begin{array}{lll} & V = f(0) = A, & \therefore A = V; \\ \text{from (2)} & \dfrac{dV}{dx} = f'(0) = 1.B, & \therefore B = \dfrac{dV}{dx}; \\ \text{from (3)} & \dfrac{d^2V}{dx^2} = f''(0) = 1.2C, & \therefore C = \dfrac{1}{2!}\dfrac{d^2V}{dx^2}; \\ \text{from (4)} & \dfrac{d^3V}{dx^3} = f'''(0) = 1.2.3\,D, & \therefore D = \dfrac{1}{3!}\dfrac{d^3V}{dx^3} \end{array}\right\}. \tag{5}$$

Substituting the above values of A, B, C, . . . , in Eq. (1), we get

$$u = V + \frac{dV}{dx}\frac{x}{1} + \frac{d^2V}{dx^2}\frac{x^2}{2!} + \frac{d^3V}{dx^3}\frac{x^3}{3!} + \dots . \qquad (6)$$

The series on the right-hand side is known as *Maclaurin's Series*. The first term is what the series becomes when $x = 0$; the second term is what the first derivative of the function becomes when $x = 0$, multiplied by x; the third is the product of the second derivative of the function when $x = 0$, into x^2 divided by factorial 21 . . .

The notation "$f^n(0)$" means that $f(x)$ is to be differentiated n times, and x equated to zero in the resulting expression. Therefore using this notation, Eq. (6) assumes the form

$$u = f(0) + f'(0)\frac{x}{1} + f''(0)\frac{x^2}{1.2} + f'''(0)\frac{x^3}{1.2.3} + \dots . \qquad (7)$$

II. Useful deductions from Maclaurin's theorem

Let us first rewrite *Maclaurin's series* in its two forms expressed above by Eqs. (6) and (7), namely,

$$u = V + \frac{dV}{dx}\frac{x}{1} + \frac{d^2V}{dx^2}\frac{x^2}{2!} + \frac{d^3V}{dx^3}\frac{x^3}{3!} + \dots , \qquad (6)$$

and

$$u = f(0) + f'(0)\frac{x}{1} + f''(0)\frac{x^2}{2!} + f'''(0)\frac{x^3}{3!} + \dots , \qquad (7)$$

which is obtained on the basis of the assumption that $u = f(x)$ can be developed in ascending powers of x; that is,

$$u = f(x) = A + Bx + Cx^2 + Dx^3 \dots . \qquad (8)$$

The following are some examples of the use of Maclaurin's formula.

(a) Binomial series

In order to expand any function by Maclaurin's theorem, we first obtain the successive differential coefficients of u, and x is then equated to zero. This fixes the values of the different constants.

Let

$$u = (a + x)^m,$$

$$\frac{du}{dx} = n(a + x)^{n-1}, \qquad \therefore f'(0) = n a^{n-1};$$

$$\frac{d^2u}{dx^2} = n(n-1)(a + x)^{n-2}, \qquad \therefore f''(0) = n(n-1)a^{n-2}$$

$$\frac{d^3u}{dx^3} = n(n-1)(n-2)(a + x)^{n-3}, \qquad \therefore f'''(0) = n(n-1)(n-2)a^{n-3}$$

and so on.

Now substitute these values in Maclaurin's series Eq. (7),

$$(a + x)^n = a^n + \frac{n}{1}a^{n-1}x + \frac{n(n-1)}{1.2}a^{n-2}x^2 + \ldots,, \tag{9}$$

a result known as the *binomial series*, true for positive, negative, or fractional values of *x*.

(b) Trigonometric series

Suppose

$$u = f(x) = \sin x.$$

Note that $\frac{du}{dx} = \frac{d(\sin x)}{dx} = \cos x; \frac{d^2u}{dx^2} = \frac{d^2(\sin x)}{dx^2} = \frac{d(\cos x)}{dx}$ $= -\sin x$, etc.; and that $\sin 0 = 0$, $-\sin 0 = 0$, $\cos 0 = 1$, $-\cos\theta = -1$.

Hence, we get

$$u = f(x) = \sin x$$

$$f(0) = 0$$

$$\frac{du}{dx} = \cos x, \qquad \therefore f'(0) = 1;$$

$$\frac{d^2u}{dx^2} = \frac{d\cos x}{dx} = -\sin x, \qquad \therefore f''(0) = 0;$$

$$\frac{d^3u}{dx^3} = -\frac{d\sin}{dx} = -\cos x, \qquad \therefore f'''(0) = -1;$$

and so on.

Now, we substitute these values in Maclaurin's series Eq. (7),

$$\sin x = 0 + \frac{x}{1} + 0 - \frac{x^3}{3!} + \ldots$$

$$= \frac{x}{1} - \frac{x^3}{3!} + \frac{x^5}{5!} - \frac{x^7}{7!} \ldots \tag{10}$$

This is the *sine series*.

In the same manner, we find the *cosine series*

$$\cos x = 1 - \frac{x^2}{2!} + \frac{x^4}{4!} - \frac{x^6}{6!} + \ldots.. \tag{11}$$

These series are employed for calculating the numerical values of angles

$$\sin\left(\frac{1}{2}\pi - x\right) = \cos x, \tag{12}$$

and

$$\cos\left(\frac{1}{2}\pi - x\right) = \sin x. \tag{13}$$

In like manner, the *tangent series* is given by

$$\tan x = \frac{x}{1} + \frac{2x^3}{3!} + \frac{16x^5}{5!} + \ldots;$$

or

$$\tan x = \frac{x}{1} + \frac{x^3}{3} + \frac{2x^5}{15} + \ldots. \tag{14}$$

(c) Inverse trigonometric series

Let $\theta = \tan^{-1} x$, that is, $\tan \theta = x$. It can be shown that

$$\tan^{-1} x = x - \frac{x^3}{3} + \frac{x^5}{5} - \ldots, \tag{15}$$

or,

$$\theta = \tan \theta - \frac{1}{3}\tan^3 \theta + \frac{1}{5}\tan^5 \theta - \ldots. \tag{16}$$

which is known as *Gregory's series*. This series is known to be *converging* when θ lies between $-\frac{1}{4}\pi$ and $\frac{1}{4}\pi$; and it has been employed for calculating the numerical value of π.

Let $\theta = 45° = \frac{1}{4}\pi,$

$\therefore \quad x = \tan 45° = 1, \text{ or } \frac{\pi}{4} = \tan^{-1} x.$

Substitution for $\tan^{-1} x$ by $\frac{\pi}{4}$, and for *x* by 1 in Eq. (15) gives

$$\frac{\pi}{4} = 1 - \frac{1}{3} + \frac{1}{5} - \frac{1}{7} + \frac{1}{9} - \frac{1}{11} \cdots \quad (17)$$

This is the so-called *Leibnitz series.*

(d) The numerical value of π

Eq. (17) can be rewritten in the following form:

$$\frac{\pi}{4} = \left(1 - \frac{1}{3}\right) + \left(\frac{1}{5} - \frac{1}{7}\right) + \left(\frac{1}{9} - \frac{1}{11}\right) + \ldots;$$

or

$$\frac{\pi}{4} = \frac{2}{1 \times 3} + \frac{2}{5 \times 7} + \frac{2}{9 \times 11} + \ldots;$$

$$\therefore \quad \frac{\pi}{8} = \frac{1}{1 \times 3} + \frac{1}{5 \times 7} + \frac{1}{9 \times 11} + \ldots.$$

It will be observed that $\tan^{-1} x \; (= \theta)$ is not to be referred to the *degree-minute-second system* of units, but to the unit of *circular system*, namely, the *radian.* Suppose $x = \sqrt{\frac{1}{3}}$, then $\tan^{-1}\sqrt{\frac{1}{3}} = 30^\circ = \frac{\pi}{6}$ radians. Substituting this value in Eq. (5), namely,

$$\tan^{-1} x = x - \frac{x^3}{3} + \frac{x^5}{5} - \ldots, \quad (15)$$

and collecting the positive and negative terms in separate brackets, we get.

$$\frac{\pi}{6} = \left(\frac{1}{\sqrt{3}} + \frac{1}{5\sqrt{3^5}} + \ldots\right) - \left(\frac{1}{3\sqrt{3^3}} + \frac{1}{7\sqrt{3^7}} + \ldots\right). \quad (18)$$

To further illustrate, we shall compute the numerical value of π to five correct decimal places. We shall find out the summation of six terms enclosed in each of the two brackets. The result is

$$\frac{\pi}{6} = 0.591039 - 0.067440 = 0.523599$$

$$\therefore \quad \pi = 0.523599 \times 6 \qquad = 3.14159. \quad (19)$$

The correct value of π to *seven decimal places* is 3.1415926.

(e) Exponential series

Let

$$u = f(x) = e^x,$$

and assume that $f(x)$ can be developed in an ascending powers of x given by

$$u = f(x) = e^x = A + Bx + Cx^2 + Dx^3 + \ldots, \quad (1)$$

where A, B, C, D . . . , are constants independent of x. It is required to determine the values of these constants, in order that the above assumption may be true for all values of x.

The *successive differentiation* of Eq. (1) yields

$$\frac{du}{dx} = \frac{d\,f(x)}{dx} = f'(x) = B + 2Cx + 3Dx^2 + \ldots; \quad (2)$$

$$\frac{d^2u}{dx^2} = \frac{d\,f'(x)}{dx} = 2C + 2.3.Dx + \ldots; \quad (3)$$

$$\frac{d^3u}{dx^3} = \frac{d\,f''(x)}{dx} = 2.3.D + \ldots. \quad (4)$$

By hypothesis, Eq. (1) is true whatever the value of x, and, therefore, the constants A, B, C, D, . . . are the same whatever value is assigned to x. Now substitute $x = 0$ in Eqs. (2)–(4). Let V denote the value assumed by u when $x = 0$. Hence,

$$\left.\begin{array}{ll}
\text{from Eq. (1)} & V = f(0) = A, \qquad \therefore A = V; \\
\text{from Eq. (2),} & \dfrac{dV}{dx} = \dfrac{d\,f(0)}{dx} = f'(0) = 1.B, \qquad \therefore B = \dfrac{dV}{dx}, \\
& \qquad = f'(0); \\
\text{from Eq. (3),} & \dfrac{d^2V}{dx^2} = \dfrac{d\,f'(0)}{dx} = f''(0) = 1.2.C, \quad \therefore C = \dfrac{1}{2!}\dfrac{d^2V}{dx^2} \\
& \qquad = \dfrac{1}{2!} f''(0); \\
\text{from Eq. (4),} & \dfrac{d^3V}{dx^3} = \dfrac{d\,f''(0)}{dx} = f'''(0) = 1.2.3.D, \quad \therefore D = \dfrac{1}{3!}\dfrac{d^3V}{dx^3} \\
& \qquad = \dfrac{1}{3!} f'''(0).
\end{array}\right\} \quad (5)$$

Substituting the above values of A, B, C, D, . . . , in Eq. (1), we obtain

$$u = V + \frac{dV}{dx}\frac{x}{1} + \frac{d^2V}{dx^2}\frac{x^2}{2!} + \frac{d^3V}{dx^3}\frac{x^3}{3!} + ..., \qquad (6)$$

or

$$u = f(0) + f'(0)\frac{x}{1} + f''(0)\frac{x^2}{2!} + f'''(0)\frac{x^3}{3!} + ... \qquad (7)$$

Since the exponential e^x is put as a function of x and denoted by u,

$$u = f(x) = e^x,$$

then it follows that Eq. (7) on substitution takes the form

$$e^x = 1 + \frac{x}{1} + \frac{x^2}{2!} + \frac{x^3}{3!} + ... \qquad (8)$$

This result is known as the *exponential series*, true for positive, negative, or fractional values of x.

If $x = 1$, we have

$$e = 1 + 1 + \frac{1}{2!} + \frac{1}{3!} + ... \qquad (9)$$

The numerical value of e to four decimal places is thus equal to 2.7169.

In general, an *exponential series* expresses the development of e^x, a^x, or any other exponential function in a series of ascending powers of x and coefficients independent of x. For example, a^x, where $a = e^k$, i.e, $k = \ln a$, is given by

$$a^x = 1 + kx + \frac{k^2x^2}{2!} + \frac{k^3x^3}{3!} + ... \qquad (10)$$

Also, e^{-x} expansion can be shown to be

$$e^{-x} = 1 - \frac{x}{1} + \frac{x^2}{2!} + \frac{x^3}{3!} + \frac{x^4}{4!} + ... \qquad (11)$$

(f) Euler's sine and cosine series

If we substitute $\sqrt{-1}\,.\,x$, or, what is the same thing, ix, in place of x, in the series of e^x, namely,

$$e^x = 1 + \frac{x}{1} + \frac{x^2}{2!} + \frac{x^3}{3!} + ..., \qquad (8)$$

we obtain,

$$e^{ix} = 1 + \frac{ix}{1} - \frac{x^2}{2!} - \frac{ix^3}{3!} + \frac{x^4}{4!} + \frac{ix^5}{5!} - \ldots,$$

$$\therefore \quad e^{ix} = \left(1 - \frac{x^2}{2!} + \frac{x^4}{4!} - \ldots\right) + i\left(\frac{x}{1} - \frac{x^3}{3!} + \frac{x^5}{5!} \ldots\right).. \qquad (12)$$

The first expression in brackets is the cosine series, and the second, the sine series, has been shown above. Hence,

$$e^{ix} = \cos x + i \sin x. \qquad (13)$$

In the same way, it can be shown that

$$e^{-ix} = 1 - \frac{ix}{1} - \frac{x^2}{2!} + \frac{ix^3}{3!} + \frac{x^4}{4!} - \frac{ix^5}{5!} - \ldots;$$

$$\therefore \quad e^{-ix} = \left(1 - \frac{x^2}{2!} + \frac{x^4}{4!} - \ldots\right) - i\left(\frac{x}{1} - \frac{x^3}{3!} + \frac{x^5}{5!} - \ldots\right). \qquad (14)$$

or,

$$e^{-ix} = \cos x - i \sin x. \qquad (15)$$

Combining Eqs. (13) and (15), we get

$$\cos x = \frac{1}{2}\left(e^{ix} + e^{-ix}\right);$$

$$i \sin x = \frac{1}{2}\left(e^{ix} - e^{-ix}\right), \qquad (16)$$

or

$$\sin x = \frac{1}{2i}\left(e^{ix} - e^{-ix}\right).$$

It is important to point out that the development by Maclaurin's series cannot be used if the function or any of its derivatives becomes infinite or discontinuous when x is equated to zero. For example, the first differential coefficient of $f(x) = \sqrt{x}$ is $\frac{1}{2} \cdot \frac{1}{x^{1/2}}$, which is infinite for $x = 0$; in other words, the series is no longer convergent. The same thing will be with the functions

$$\log x, \ \cot x, \ 1/x, \ a^{1/x}, \text{ and } \sec^{-1} x.$$

3. Plane trigonometry formulae

A brief outline on *plane trigonometry* will be given below which may perhaps be of some assistance.

Trigonometry deals with the relations between the *sides and angles of triangles*; if the triangle is drawn on a plane surface, we have plane trigonometry; if the triangle is drawn on the surface of a sphere, we have spherical trigonometry.

(a) The measurement of angles

An angle is the intersection of two lines. The magnitude of an angle depends only on the relative directions, or slopes of the lines, and is independent of their length. In *practical work*, angles are usually measured in degrees, minutes, and seconds. These units are the subdivisions of a right angle *defined* as

1 right angle	=	90 degrees, written 90°;
1 degree	=	60 minutes, written 60′;
1 minute	=	60 seconds, written 60″.

In *theoretical calculations*, however, this system of degree-minute-second is replaced by another as can be seen from the following:

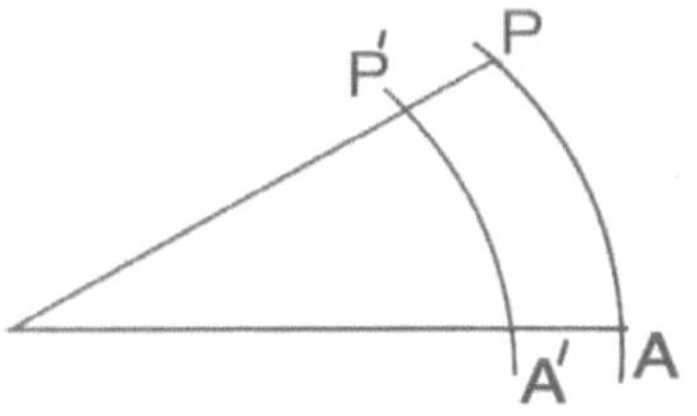

Fig. 1

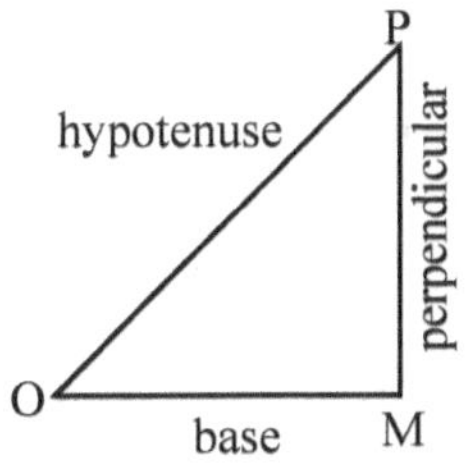

Fig. 2

In Fig.1., the length of the circular arcs $\mathrm{P'A'}$, PA, drawn from the centre O, are proportional to the lengths of the radii $\mathrm{OA'}\ (= \mathrm{OP'})$ and $\mathrm{OA}\ (= \mathrm{OP})$,

or

$$\frac{\text{arc P}'\text{A}'}{\text{radius OA}'} = \frac{\text{arc PA}}{\text{radius OA}}.$$

If the angle at the centre O is constant, the ratio, arc/radius, is also constant. *This ratio therefore furnishes a method for measuring the magnitude of an angle*. The ratio

$$\frac{\text{arc}}{\text{radius}} = 1,$$

is called *a radian.*

Two right angles = 180° = π radians, where π = 180° has the numerical value 3.14159 estimated above. The ratio, arc/radius, is called the *circular or radian measure of an angle*, and the *unit angle* in this system is the *radian.*

(b) The relation between the degrees and radians

Consider a circle of radius *r* or circumference $2\pi r$ so that if *r* is unity, the circumference is 2π. The angle 360 corresponds to the arc whose length is 2π. Therefore, 1 radian corresponds to an angle equal to $360/2\pi \quad = 180/\pi$

$$= \frac{180}{3.14159} = 57^\circ\ 17'\ 44.8'';$$

that is,

$$1 \text{ radian} = 57^\circ\ 17'\ 44.8'' \qquad (1)$$

$$= 57.2958^\circ$$

or, conversely,

$$1^\circ = 0.0175 \text{ radians.}$$

(c) Trigonometric ratios of an angle as functions of the sides of a triangle

In the triangle OPM, Fig. 2,

(i) The ratio $\frac{MP}{OM} = \frac{\text{perpendicular}}{\text{base}}$ is the *tangent* of the angle POM, and written, tan POM.

(ii) The ratio $\frac{OM}{MP} = \frac{\text{base}}{\text{perpendicular}}$ is the *cotangent* of the angle POM, and written, cot POM. Note that *cotangent* of an angle is the reciprocal of its *tangent*.

(iii) The ratio $\frac{MP}{OP} = \frac{\text{perpendicular}}{\text{hypotenuse}}$ is the sine of the angle POM, and written, sin POM.

(iv) The ratio $\frac{OP}{MP} = \frac{\text{hypotenuse}}{\text{perpendicular}}$ is the cosecant of the angle POM, and written, cosec POM. The cosecant of an angle is the reciprocal of its sine.

(v) The ratio $\frac{OM}{OP} = \frac{\text{base}}{\text{hypotenuse}}$ is the cosine of the angle POM, and written cos POM.

(vi) The ratio $\frac{OP}{OM} = \frac{\text{hypotenuse}}{\text{base}}$ is the secant of the angle POM, and written, sec POM. The secant of an angle is the reciprocal of its cosine.

Summary of trigonometric formulae.

Note: $\pi = 180°$; or , 3.14159 ;

one radian = 57.2958°
= 57° 17′ 44.8″.

$$\left.\begin{aligned} &\sin\left(\frac{\pi}{2}-x\right)=\cos x; \quad \cos\left(\frac{\pi}{2}-x\right)=\sin x;\\ &\operatorname{cosec}\left(\frac{\pi}{2}-x\right)=\sec x; \quad \sec\left(\frac{\pi}{2}-x\right)=\operatorname{cosec} x;\\ &\tan\left(\frac{\pi}{2}-x\right)=\cot x; \quad \cot\left(\frac{\pi}{2}-x\right)=\tan x. \end{aligned}\right\} \qquad (1)$$

$$\frac{1}{\sin x}=\operatorname{cosec} x; \quad \frac{1}{\cos x}=\sec x; \quad \frac{1}{\tan x}=\cot x.$$

$$\left.\begin{aligned} \sin(180°-x) &= \sin(\pi-x)=\sin x\\ \cos(180°-x) &= \cos(\pi-x)=-\cos x\\ \tan(180°-x) &= \tan(\pi-x)=-\tan x\\ \cot(180°-x) &= \cot(\pi-x)=-\cot x \end{aligned}\right\}. \qquad (2)$$

The angle $180° - x$, or $\pi - x$, is called the *supplement* of the angle *x*.

The angle $90° - x$, or $\frac{\pi}{2} - x$ is called the *complement* of the angle *x*.

$$\left.\begin{aligned} &\sin\left(\frac{\pi}{2}+x\right)=\cos x; \quad \cos\left(\frac{\pi}{2}+x\right)=-\sin x;\\ &\tan\left(\frac{\pi}{2}+x\right)=-\cot x; \quad \cot\left(\frac{\pi}{2}+x\right)=-\tan x. \end{aligned}\right\}. \qquad (3)$$

$$\left.\begin{aligned} &\sin(\pi+x)=-\sin x; \quad \cos(\pi+x)=-\cos x;\\ &\tan(\pi+x)=\tan x; \quad \cot(\pi+x)=\cot x. \end{aligned}\right\}. \qquad (4)$$

$$\left.\begin{aligned} &\sin(-x)=-\sin x; \quad \cos(-x)=\cos x;\\ &\tan(-x)=-\tan x. \end{aligned}\right\}. \qquad (5)$$

$$\underset{x\to 0}{\mathrm{Lt}}\ \frac{\sin x}{x}=\frac{\tan x}{x}=\cos x=1; \quad \frac{\sin^{-1}x}{x}=\frac{\tan^{-1}x}{x}=1 \qquad (6)$$

$$\sin^2 x + \cos^2 x = 1. \qquad (7)$$

IV. The hyperbolic functions

We shall now explain the origin of a new class of functions, which are used as tools in mathematical reasoning.

In plane trigonometry, an angle is conveniently measured as a function of the arc of a circle. Thus, if l' denotes the length of an arc subtending an angle θ at the centre, and r' the radius of the circle, then

$$\theta = \frac{\text{length of arc}}{\text{length of radius}} = \frac{l'}{r'}.$$

This is called the *circular measure of an angle* and, for this reason trigonometrical functions are sometimes called *circular functions*. This property is possessed by no plane curve *other than the circle*. For instance, the hyperbola, though symmetrically placed with respect to its centre, is not at all points equidistant from it. The same thing is true of the ellipse. The parabola has no centre.

If l denotes the length of the arc of any hyperbola which cuts the x-axis at a distance r ($= a$) from the centre, as shown in Fig. 3,

the *ratio*

$$u = \frac{l}{r},$$

is called a *hyperbolic function of u*, just as the ratio l'/r' is called a *circular function of θ.*

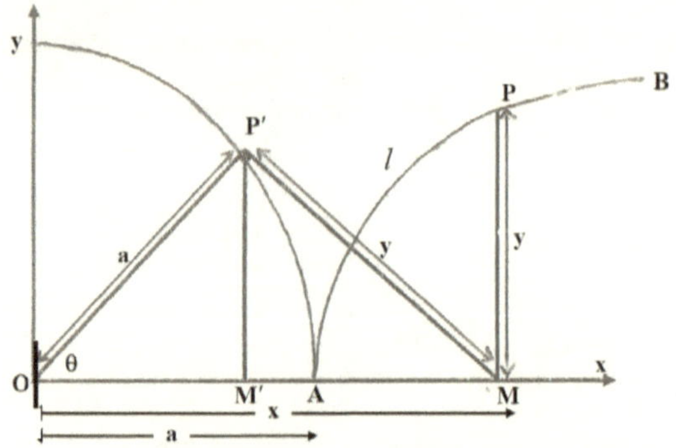

Fig. 3. The correlation between the circular angle *θ* and the hyperbolic functions of $u = \frac{x}{a}$ and $\frac{y}{a}$.

It is found that if l denotes the length of the *arc of a rectangular hyperbola*,

$$x^2 - y^2 = a^2. \tag{1}$$

Between the coordinates having *abscissa a* and *x* the ratio $\frac{x}{a}$ is given by

$$\frac{x}{a} = \frac{1}{2}\left(e^{u} + e^{-u}\right). \tag{2}$$

But this relation is previously developed for cos *x*

$$\cos x = \frac{1}{2}\left(e^{ix} + e^{-ix}\right),$$

where i*x* stands for *u*. Thus the ratio $\frac{x}{a}$ expressed by Eq. (2) is defined as the *hyperbolic cosine of u*. It is usually written cosh *u*, or hy cos *u*, and pronounced 'cosh *u*', or 'h cosine *u*'.

Accordingly, Eq. (2) can be rewritten in the following form:

$$\frac{x}{a} = \cosh u = \frac{1}{2}\left(e^{u} + e^{-u}\right)$$

$$= 1 + \frac{u^2}{2!} + \frac{u^4}{4!} + \ldots\ .. \tag{3}$$

In the same way, proceeding from Eq. (1), it can be shown that

$$\frac{y}{a} = \sqrt{\frac{x^2}{a^2} - 1} = \sqrt{\frac{e^{2u} + 2 + e^{-2u}}{4} - 1}$$

$$= \sqrt{\frac{e^{2u} - 2 + e^{-2u}}{4}}$$

$$= \frac{1}{2}\left(e^{u} - e^{-u}\right). \tag{4}$$

This relation is previously developed for i sin *x*

$$i \sin x = \frac{1}{2}\left(e^{ix} - e^{-ix}\right),$$

where i*x* stands for *u*. The ratio $\frac{y}{a}$ is defined as the *hyperbolic sine of u*. It is written sinh *u* or hy sin *u*.

Accordingly, Eq. (4) can be expressed in the form

$$\frac{y}{a} = \sinh u = \frac{1}{2}\left(e^{u} - e^{-u}\right)$$

$$= u + \frac{u^3}{3!} + \frac{u^5}{5!} + \ldots\ . \tag{5}$$

The remaining four hyperbolic functions, analogous to the remaining four trigonometrical functions, are

tanh *u*, cosech *u*, sech *u*, and coth *u*.

Values of these functions may be deduced from their relations with sinh *u* and cosh *u*.

Thus,

$$\left.\begin{aligned} \tanh u &= \frac{\sinh u}{\cosh u}; \\ \coth u &= \frac{1}{\tanh u}; \\ \operatorname{sech} u &= \frac{1}{\cosh u}; \\ \operatorname{cosech} u &= \frac{1}{\sinh u}. \end{aligned}\right\} \qquad (6).$$

Unlike the circular functions, the ratios $\frac{x}{a}$ and $\frac{y}{a}$, when referred to the hyperbola, *do not represent angles. Any hyperbolic function expresses a certain relation between the coordinates of a given portion on the arc of a rectangular hyperbola.*

We want now to find the correlations between the circular trigonometric functions and the hyperbolic functions. For this purpose, we reconsider Fig. (3) shown above.

Let O be the centre of the hyperbola APB, described about the coordinate axes Ox and Oy. From any point P (x, y) drop a perpendicular PM on to the x-axis. Let OM = x, MP = y, OA = a.

For the rectangular hyperbola,

$$x^2 - y^2 = a^2, \qquad (1)$$

as pointed out above by Eq. (1) . Since the ratio $\frac{x}{a}$ is defined as the hyperbolic cosine of u, and the ratio $\frac{y}{a}$ the hyperbolic sine of u as shown above, therefore by substitution in Eq. (1), we obtain

$$a^2 \cosh^2 u - a^2 \sinh^2 u = a^2,$$

or

$$\cosh^2 u - \sinh^2 u = 1. \qquad (7)$$

This formula thus resembles the well-known trigonometrical relation.

$$\cos^2 x + \sin^2 x = 1.$$

Now draw $P'M$ a tangent to the *circle* AP' at P'. Drop a perpendicular $P'M'$ on to the *x*-axis. Let the angle $M'OP' = \theta$.
Therefore, in the right-angle triangle $MP'O$, we have

$$\sec\theta = \frac{x}{a},$$

but

$$\frac{x}{a} = \cosh u.$$

Therefore,

$$\sec\theta = \cosh u,$$

or

$$\frac{x}{a} = \cosh u = \sec\theta. \tag{8}$$

Also we have

$$\tan\theta = \frac{MP'}{a},$$

but

$$(MP')^2 + a^2 = x^2,$$

and for the rectangular hyperbola

$$x^2 - y^2 = a^2,$$

or

$$y^2 + a^2 = x^2.$$

Therefore

$$(MP')^2 + a^2 = y^2 + a^2,$$

or

$$MP' = y.$$

Hence,

$$\tan\theta = \frac{y}{a}.$$

But again

$$\frac{y}{a} = \sinh u.$$

Therefore,

$$\sinh u = \tan\theta. \tag{9}$$

(a) Conversion formulae

Corresponding with the trigonometric formulae, there are a great number of relations among the hyperbolic functions such as Eq. (7) above.
Thus,

$$\cosh 2x = 1 + 2\sinh^2 x = 2\cosh^2 x - 1, \tag{10}$$

$$\sinh x - \sinh y = 2\cosh\frac{1}{2}(x+y).\ \sinh\frac{1}{2}(x-y), \tag{11}$$

and so on.

(b) Graphic representation of hyperbolic functions

We know that the trigonometric sine, cosine, etc. are *periodic* functions, whereas the hyperbolic functions are *exponential*, not periodic. This will be evident if we plot the six hyperbolic functions on squared papers. For example, the plots for $y = \cosh x$, and $y = \operatorname{sech} x$ are shown in Fig. (4).

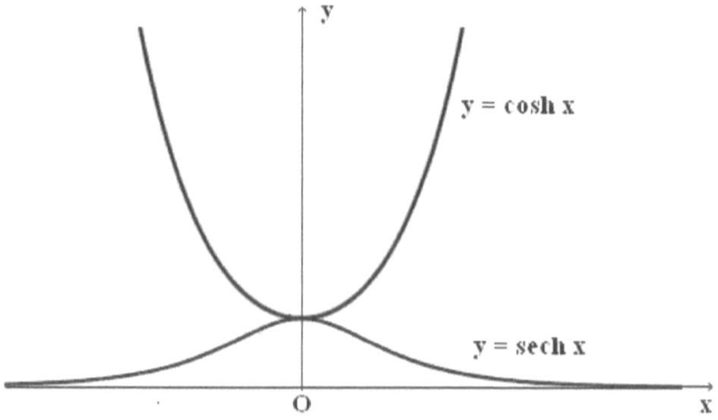

Fig. 4. Graphs of cosh *x* and sech *x*.

The graph of $y = \cosh x$ is known in statistics as the 'catenary'.

Important relations between trigonometric and hyperbolic functions:

$$e^{x} = \cos ix - i \sin ix \tag{1}$$

$$e^{-x} = \cos ix + i \sin ix \tag{2}$$

$$\therefore \quad \cos ix = \frac{e^{x} + e^{-x}}{2} \tag{3}$$

and

$$\sin ix = -\frac{e^{x} - e^{-x}}{2i} \tag{4}$$

$$e^{ix} = \cos x + i \sin x \tag{5}$$

$$e^{-ix} = \cos x - i \sin x \tag{6}$$

$$\therefore \quad \cos x = \frac{e^{ix} + e^{-ix}}{2} \tag{7}$$

and

$$\sin x = \frac{e^{ix} - e^{-ix}}{2i} \tag{8}$$

$$\cosh x = \frac{e^{x} + e^{-x}}{2} \tag{9}$$

$$\sinh x = \frac{e^{x} - e^{-x}}{2} \tag{10}$$

$$\cosh ix = \frac{e^{ix} + e^{-ix}}{2} \tag{11}$$

$$\sinh ix = \frac{e^{ix} - e^{-ix}}{2}. \tag{12}$$

Based on the above equations from Eqs. (1)–(12), it will be evident that

(i) The relation between $\cos ix$ and $\cosh x$ is

$$\cos ix = \cosh x; \tag{13}$$

(ii) The relation between $\cos x$ and $\cosh ix$ is

$$\cos x = \cosh ix; \tag{14}$$

(iii) The relation between sin ix and sinh x is that between

$$-\frac{e^{x} - e^{-x}}{2i} \quad \text{and} \quad \frac{e^{x} - e}{2},$$

which is

$$\sin ix = -\frac{1}{i} \sinh x,$$

that is,

$$\sin ix = i \sinh x. \tag{15}$$

The relation between sin x and sinh ix is that between

$$\frac{e^{x} - e^{-ix}}{2i} \quad \text{and} \quad \frac{e^{-ix} - e^{ix}}{2},$$

which is

$$\sin x = \frac{\sinh ix}{i}. \tag{16}$$

Making use of these relations, the exponentials e^{x} and e^{x} given by Eqs. (1) and (2) can be expressed as

$$e^{x} = \cosh x + \sinh x, \tag{17}$$

and

$$e^{-x} = \cosh x - \sinh x, \tag{18}$$

respectively.

V. Inverse hyperbolic functions

As an example, let us determine the function $\sinh^{-1} y$. Put this function equal to the variable x, i.e.,

$$\sinh^{-1} y = x. \tag{1}$$

then using Eq. (5) in Section 4 rewritten in the form

$$\sinh x = \frac{1}{2}\left(e^{x} - e^{-x}\right), \tag{2}$$

we obtain,

$$y = \frac{1}{2}\left(e^{x} - e^{-x}\right). \tag{3}$$

Therefore,

$$e^{2x} - 2ye^{x} - 1 = 0. \tag{4}$$

The solution of this equation is

$$e^{x} = \frac{+2y \pm \sqrt{4y^2 + 4}}{2}$$

$$= y \pm \sqrt{y^2 + 1}. \tag{5}$$

Taking the logarithm of both sides, we obtain

$$\ln e^{x} = \ln\left(y \pm \sqrt{y^2 + 1}\right). \tag{6}$$

Therefore,

$$x = \sinh^{-1} y = \ln\left(y + \sqrt{y^2 + 1}\right), \tag{7}$$

where the negative sign is excluded from the logarithmic term for real values of *x*.

In the same fashion

$$\cosh^{-1} y = \ln\left(y + \sqrt{y^2 - 1}\right), \tag{8}$$

$$\operatorname{sech}^{-1} y = \ln\left(y + \sqrt{1 - y^2}\right). \tag{9}$$

It is to be remembered that

$$\left\{\begin{array}{lll} \sinh x, & \operatorname{cosech} x & = \dfrac{1}{\sinh x} \\ \cosh x, & \operatorname{sech} x & = \dfrac{1}{\cosh x} \\ \tanh x, & \coth x & = \dfrac{1}{\tanh x} \end{array}\right\}. \tag{10}$$

As an example of the application of the relation between the inverse hyperbolic functions, we might consider the *re-expression of the equation*

$$\operatorname{sech}^{-1} \frac{\in''}{\in''} = \beta \ln \frac{\omega}{\omega_m},$$

where $\in''$ is the dielectric loss of a substance at an angular frequency ω, and $\in''_m$ is its maximum dielectric loss at an angular frequency ω_m, *in terms of the inverse hyperbolic function* $\cosh^{-1} \in''_m/\in''$.

Since

$$\operatorname{sech}^{-1} \frac{\epsilon''}{\epsilon''_m} = \beta \ln \frac{\omega}{\omega_m}.$$

Therefore,

$$\frac{\epsilon''}{\epsilon''_m} = \operatorname{sech}\left(\beta \ln \frac{\omega}{\omega_m}\right). \tag{1}$$

But

$$\operatorname{sech} x = \frac{1}{\cosh x}.$$

Therefore,

$$\operatorname{sech}\left(\beta \ln \frac{\omega}{\omega_m}\right) = \frac{1}{\cosh\left(\beta \ln \frac{\omega}{\omega_m}\right)},$$

or

$$\cosh\left(\beta \ln \frac{\omega}{\omega_m}\right) = \frac{1}{\operatorname{sech}\left(\beta \ln \frac{\omega}{\omega_m}\right)}.$$

Using Eq. (1), we have

$$\cosh\left(\beta \ln \frac{\omega}{\omega_m}\right) = \frac{1}{\epsilon''/\epsilon''_m},$$

$$= \frac{\epsilon''_m}{\epsilon''}. \tag{2}$$

Hence,

$$\cosh^{-1} \frac{\epsilon''_m}{\epsilon''} = \beta \ln \frac{\omega}{\omega_m}. \tag{3}$$

Eq. (3) is known as *Fuoss–Kirkwood equation*; its graphical representation can be achieved as follows.
Since

$$\cosh^{-1} y = \ln\left(y + \sqrt{y^2 - 1}\right),$$

as shown by Eq. (8) above,

therefore,

$$\cosh^{-1}\frac{\epsilon''_m}{\epsilon''} = \ln\left(\frac{\epsilon''_m}{\epsilon''} + \sqrt{\left(\frac{\epsilon''_m}{\epsilon''}\right)^2 - 1}\right). \qquad (4)$$

An estimation of the left-hand side of Eq. (3), made by computing the numerical value of right side of Eq. (4), is then plotted against $\ln\frac{\omega}{\omega_m}$, a linear relationship might be obtained so that β could be determined from the slope of the line.

SECTION 2

I. Vectors and Scalars

A *vector* is a quantity having both magnitude and direction, such as displacement, velocity, force, and acceleration.

Graphically, a vector is represented by an arrow OP (Fig. 1) defining the direction, the magnitude of the vector being indicated by the length of the arrow. The tail end O of the arrow is called the *origin* or *initial point* of the vector, *and the head P* is called the *terminal point or terminus.*

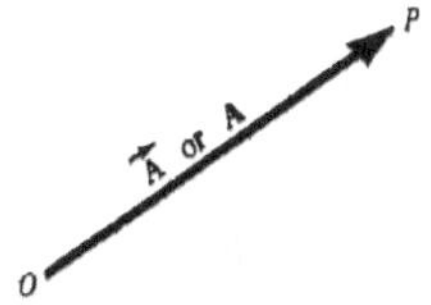

Fig. 1.

Analytically, a vector is represented by a letter with an arrow over it, as $\vec{A}$ in Fig. 1, and the magnitude is denoted by $|\vec{\mathbf{A}}|$ or A. In printed works, *bold-faced type*, such as **A**, is used to indicate the vector $\vec{\mathbf{A}}$ while $|\mathbf{A}|$ or A indicates its magnitude. We shall use this bold-faced notation. The vector OP is indicated as $\overrightarrow{OP}$ or ***OP***, and the magnitude is denoted by $|\overrightarrow{OP}|$, $\overline{OP}$, or $|\mathbf{OP}|$.

A *scalar* is a quantity having magnitude but no direction, such as mass, length, time, temperature, and any real number. Scalars are indicated by letters in ordinary type as in elementary algebra. Operations with scalars follow the same rules as in elementary algebra.

Vector algebra.

(1) The *sum* or *resultant* of vectors **A** and **B** is a vector **C** formed by placing the initial point of **B** on the terminal point of **A** and then joining the initial point of **A** to the terminal point of **B** (Fig. 2).

This sum is written **A** + **B**; that is, **C** = **A** + **B**

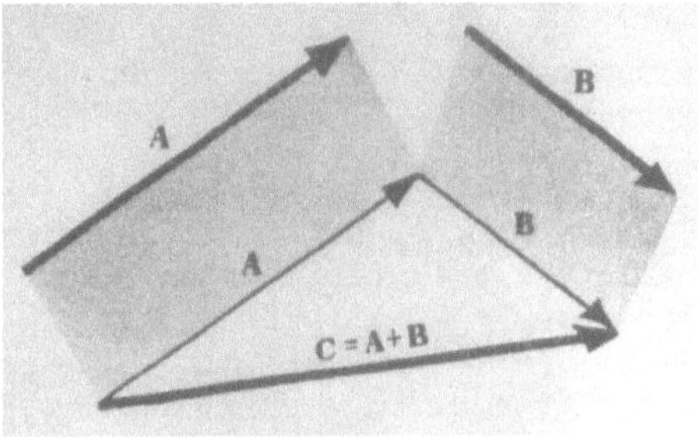

Fig. 2.

(2) The *differences* of vectors **A** and **B**, represented by **A. B**, is that vector **C** which added to **B** yields vector **A**. Equivalently,

A – **B** can be defined as the sum **A** + (–**B**).

If **A** = **B**, then **A** – **B** can be defined as the *null* or *zero vector* and is represented by the symbol **0** or simply 0. It has zero magnitude and no specific direction. A vector which is not null is a *proper vector*.

(3) The *product* of a vector **A** by a scalar *m* is a vector *m* **A** with magnitude $|m|$ times the magnitude of **A** and with direction the same as or apposite to that of **A**, according as *m* is positive or negative. If *m* = 0, *m* **A** is the *null vector*.

Laws of vector algebra. If **A**, **B**, and **C** are vectors and *m* and *n* are scalars, then

1. **A** + **B** = **B** + **A** Commutative law for addition.
2. **A** + (**B** + **C**) = (**A** + **B**) + **C** Associative law for addition.
3. *m* **A** = **A** *m* Commutative law for multiplication.
4. *m* (*n* **A**) = (*mn*) **A** Associative law for multiplication.
5. (*m* + *n*) **A** = *m* **A** + *n* **A** Distributive law.
6. *m* (**A** + **B**) = *m* **A** + *m* **B** Distributive law.

These laws enable us to treat vector equations in the same way as ordinary algebraic equations. For example, if **A** + **B** = **C**, then by transposing **A** = **C** – **B**.

A unit vector is a vector having unit magnitude. If **A** is a vector with magnitude $A \neq 0$, then $\mathbf{A}/A$ is a unit vector having the same direction as **A**.

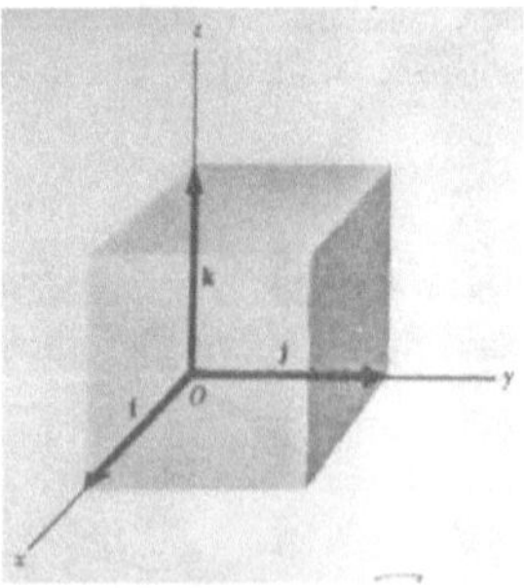

Fig. 3.

Any vector **A** can be represented by a unit vector **a** in the direction of **A** multiplied by the magnitude of **A**. In symbols,

A = A **a**.

The rectangular unit vectors **i, j, k**. An important set of unit vectors, are those having the directions of the positive *x*, *y*, and *z* of a three-dimensional rectangular coordinate system, and are denoted respectively by **i, j**, and **k** (Fig. 3).

We shall use *right-handed rectangular coordinate systems unless otherwise stated.* Such a system derives its name from the fact that a right threaded screw rotating through 90° from Ox to Oy will advance in the positive *z* direction, as in Fig. 3 above.

In general, three vectors **A**, **B**, and **C** which have coincident initial points and are not *coplanar* are said to form a *right-handed system* or *dextral system* if a *right threaded screw* rotated through an angle less than 180° from **A** to **B** will advance in the direction **C** as shown in Fig. 4.

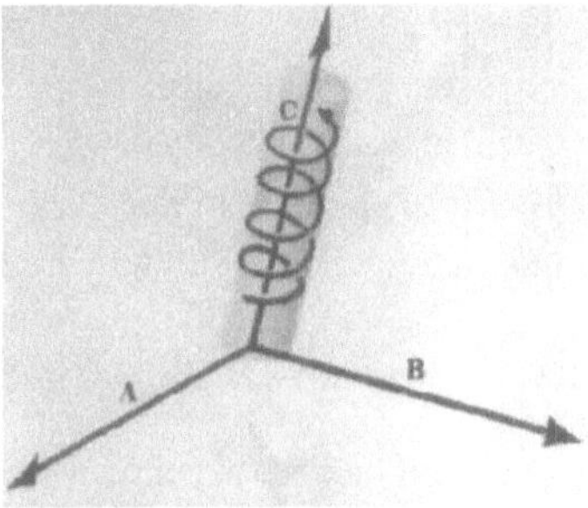

Fig. 4

Components of a vector. Any vector **A** in three dimensions can be represented with initial point at the origin O of a rectangular coordinate system (Fig. 5). Let (A, A_2, A_3) be the rectangular coordinates of the terminal point of vector **A** with initial point at O. The vectors $A_1\mathbf{i}$, $A_2\mathbf{j}$ and $A_3\mathbf{k}$ are called the *rectangular component vectors of A* in the *x*, *y*, and *z* directions, respectively. A_1, A_2 and A_3 are called the rectangular components of **A** in the *x*, *y*, and *z* directions, respectively.

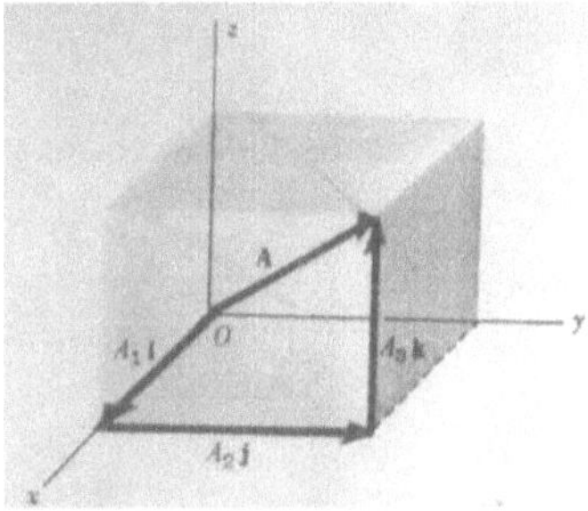

Fig. 5.

The sum or resultant of A_1i, A_2j, and A_3k is the vector **A** so that we can write

$$\mathbf{A} = A_1\mathbf{i} + A_2\mathbf{j} + A_3\mathbf{k}.$$

The magnitude of **A** is

$$A = |\mathbf{A}| = \sqrt{A_1^2 + A_2^2 + A_3^2}.$$

In particular, the *position vector* or *radius vector* **r** from O to the point (x, y, z) is written

$$\mathbf{r} = x\mathbf{i} + y\mathbf{j} + z\mathbf{k}$$

and the magnitude

$$r = |\mathbf{r}| = \sqrt{x^2 + y^2 + z^2}.$$

Scalar field. If to each point (x, y, z) of a region R in space there corresponds a number or scalar $\phi(x, y, z)$, then ϕ is called a *scalar function of position* or *scalar point function,* and we say that a *scalar field* ϕ has been defined in R.

Examples. (1) The temperature at any point within or in the earth's surface at a certain time defines a scalar field.

(2) $\phi(x, y, z) = x^3 y - z^2$ defines a scalar field.

A scalar field which is independent of time is called a *stationary* or *steady-state scalar field.*

Vector field. If to each point (x, y, z) of a region R in space there corresponds a vector **V** (x, y, z), then **V** is called a *vector function of position* or *vector point function,* and we say that *a vector field* **V** has been defined in R.

Examples. (1) If the velocity at any point (x, y, z) within a moving field is known at a certain time, then a *vector field* is defined.

(2) $\mathbf{V}(x, y, z) = xy^2\mathbf{i} + 2yz^3\mathbf{j} + z^2\mathbf{k}$ defines a *vector field.*

A vector field which is independent of time is called a *stationary* or *steady-state vector field.*

Solved Problems

1. State which of the following are vectors and which are scalars.

(a) weight	(c) specific heat	(e) density	(g) volume
(b) calorie	(d) momentum	(f) energy	(h) distance
(i) speed	(j) magnetic field intensity		

Ans.

(a) scalar	(c) scalar	(e) scalar	(g) scalar
(b) scalar	(d) vector	(f) scalar	(h) scalar
(i) scalar	(j) vector		

An automobile travels three miles due north, then five miles north-east. Represent these displacements graphically and determine the resultant displacement (a) graphically And (b) analytically.

Ans. Vector **OP** or **A** represents displacement of three miles due north. Vector **PQ** or **B** represents displacement of five miles north-east. Vector **OQ** or **C** represents the resultant displacement or sum of vectors **A** and **B**, that is, **C** = **A** + **B**. This is the *triangle law of vector addition.*

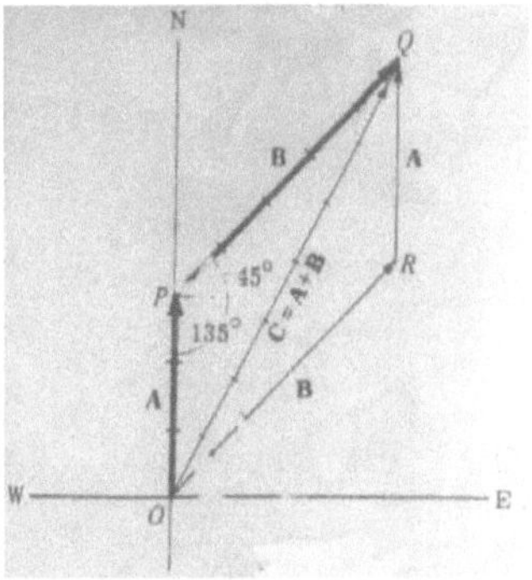

Fig. 6.

The resultant vector **OQ** can also be obtained by constructing the diagonal of the parallelogram OPQR having vectors **OP** = **A** and **OR** equal to vector **PQ** or **B** as sides. This is the *parallelogram law of vector addition*

(a) *Graphical determination of resultant.* Lay off the one-mile unit on vector **OQ** to find the magnitude of this vector. It is found to be 7.4 miles approximately. Angle EOQ = 61.5° using *a protractor*. Then vector **OQ** has magnitude 7.4 miles and direction 61.5° north of east. Also the angle OQP = 16.5°.

(b) *Analytical determination of resultant.* From triangle OPQ, denoting the magnitudes of **A, B, C** by A, B, C, we have by the law of cosines.

$$C^2 = A^2 + B^2 - 2AB\cos\angle OPQ$$

$$= 3^2 + 5^2 - 2(3 \times 5)\cos 135°$$

$$= 34 - 30\cos(180° - 45°)$$

$$= 34 + 30\cos 45$$

$$= 34 + 15\sqrt{2} = 55.21.$$

Therefore,

$$C = 7.43 \quad \text{(approximately).}$$

By the law of sines,

$$\frac{A}{\sin \angle OQP} = \frac{C}{\sin \angle OPQ}.$$

Then

$$\sin \angle OQP = \frac{A}{C} \sin \angle OPQ$$

$$= \frac{3}{7.43} \sin 135^\circ$$

$$= 0.2855,$$

and

$$\angle OQP = 16^\circ\ 35',$$

Hence, the vector **OQ** has magnitude 7.43 miles and direction $(45^\circ + 16^\circ\ 35') = 61^\circ\ 35'$ north of east.

3. Given three non-coplanar vectors **a**, **b**, and **c**. Find an expression for any vector **r** in three-dimensional space.

Ans. Non-coplanar vector are vectors which are not parallel to the same plane. Hence, when their initial points coincide, they do not lie in the same plane.

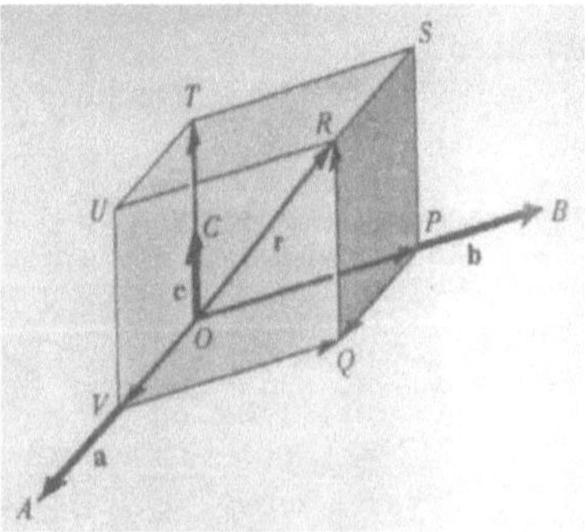

Fig. 6.

Let **r** be any vector in space having its initial point coincident with the initial points of **a**, **b**, and **c** at O. Through the terminal point of **r** pass planes parallel respectively to the planes determined by **a** and **b**, **b** and **c**, and **c** and **a**; and complete the parallelepiped PQRSTUV by extension of the lines of action of **a**, **b**, and **c**. From the adjoining figure, we have for the vectors **OV**, **OP** and **OT**:

$\mathbf{OV} = x\,(\mathbf{OA}) = x\mathbf{a}$, where x is a scalar

$\mathbf{OP} = y\,(\mathbf{OB}) = y\mathbf{b}$, where y is a scalar

$\mathbf{OT} = z\,(\mathbf{OC}) = z\mathbf{c}$, where z is a scalar.

But

$$\mathbf{OR} = \mathbf{OV} + \mathbf{VQ} + \mathbf{QR}$$
$$= \mathbf{OV} + \mathbf{OP} + \mathbf{OT}.$$

Therefore

$$\mathbf{r} = x\mathbf{a} + y\mathbf{b} + z\mathbf{c}.$$

The vectors $x\mathbf{a}$, $y\mathbf{b}$, and $z\mathbf{c}$ are called *component vectors* of $\mathbf{r}$ in directions $\mathbf{a}$, $\mathbf{b}$, and $\mathbf{c}$, respectively. The vectors $\mathbf{a}$, $\mathbf{b}$, and $\mathbf{c}$ are called *base vectors* in three dimensions.

As a special case, if $\mathbf{a}$, $\mathbf{b}$, and $\mathbf{c}$ are the unit vectors $\mathbf{i}$, $\mathbf{j}$, and $\mathbf{k}$, which are mutually perpendicular, we see that any vector $\mathbf{r}$ can be expressed uniquely in terms of $\mathbf{i}$, $\mathbf{j}$, and $\mathbf{k}$ by the expression

$$\mathbf{r} = x\mathbf{i} + y\mathbf{j} + z\mathbf{k}.$$

Also, if $\mathbf{c} = 0$, then $\mathbf{r}$ must lie in the plane of $\mathbf{a}$ and $\mathbf{b}$.

4. Prove that if $\mathbf{a}$ and $\mathbf{b}$ are non-collinear,* then

$$x\mathbf{a} + y\mathbf{b} = 0$$

implies

$$x = y = 0.$$

Ans. Suppose that $x \neq 0$, then $x\mathbf{a} + y\mathbf{b} = 0$ implies $x\mathbf{a} = -y\mathbf{b}$ or $\mathbf{a} = -(y/x)\,\mathbf{b}$,; that is, $\mathbf{a}$ and $\mathbf{b}$ must be parallel to the same line (collinear) contrary to hypothesis. Thus if $x = O$, then $y\mathbf{b} = 0$, from which $y = 0$.

5. Prove that the magnitude A of the vector $\mathbf{A}$.

$$A_1\mathbf{i} + A_2\mathbf{j} + A_3\mathbf{k} \text{ is } A = \sqrt{A_1^2 + A_2^2 + A_3^2}.$$

Ans. By the parallelogram theorem as shown in Fig. 8, we have

$$(\overline{OP})^2 = (\overline{OQ})^2 + (\overline{QP})^2,$$

where $(\overline{OP})^2$ denotes the magnitude of vector **OP**.
Similarly,

$$(\overline{OQ})^2 = (\overline{OR})^2 + (\overline{RQ})^2.$$

Then

$$(\overline{OP})^2 = (\overline{OR})^2 + (\overline{RQ})^2 + (\overline{QP})^2,$$

* Non-collinear vectors are vectors which are not parallel to the same line so that when their initial points coincide, they determine a plane.

or

$$A^2 = A_1^2 + A_2^2 + A_3^2, \text{ i.e. } A^2 = A_1^2 + A_2^2 + A_3^2.$$

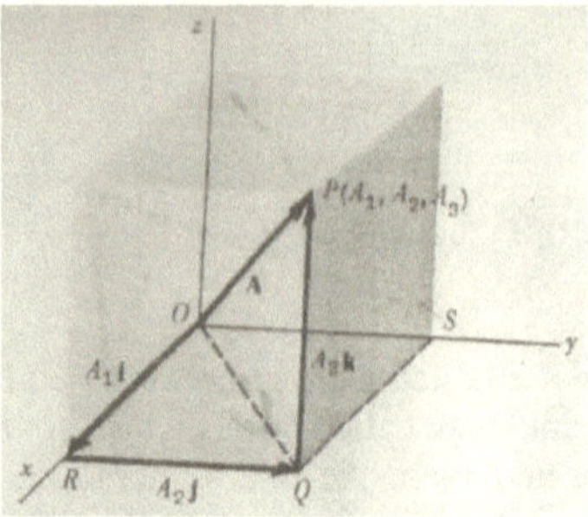

Fig. 8.

6. Determine the angles α, β and γ which the vector $\mathbf{r} = x\mathbf{i} + y\mathbf{j} + z\mathbf{k}$ makes with the positive directions of the coordinate axes, and show that

$$\cos^2\alpha + \cos^2\beta + \cos^2\gamma = 1.$$

Ans. Referring to Fig. 9, triangle OAP is a right triangle with right angle at A, then,

$$\cos\alpha = \frac{x}{|\mathbf{r}|}.$$

Similarly, from right triangles OBP and OCP,

$$\cos\beta = \frac{y}{|\mathbf{r}|},$$

and

$$\cos\gamma = \frac{z}{|\mathbf{r}|}.$$

Also

$$|\mathbf{r}| = r = \sqrt{x^2 + y^2 + z^2}.$$

Thus

$$\cos\alpha = \frac{x}{r},$$

$$\cos\beta = \frac{y}{r},$$

$$\cos\gamma = \frac{z}{r}$$

from which α, β, and γ can be obtained. From these, it follows that

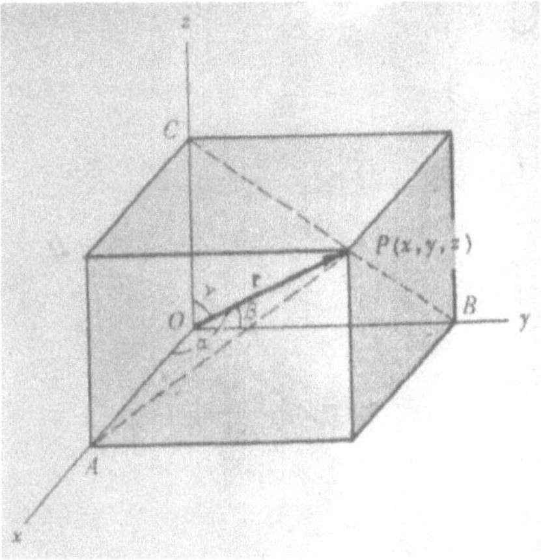

Fig. 9

$$\cos^2\alpha + \cos^2\beta + \cos^2\gamma = \frac{x^2 + y^2 + z^2}{r^2} = 1$$

The numbers $\cos\alpha$, $\cos\beta$, and $\cos\gamma$ are called the *direction (or directional) cosines* of the vector **OP**.

II. The Dot and Cross Product

The dot or scalar product of two vectors **A** and **B**, denoted by **A. B** (read **A** dot **B**), is defined as the product of the magnitudes of **A** and **B** and the cosine of the angle θ between them. In symbols

$$\mathbf{A.\,B} = AB\cos\theta, \qquad 0 \le \theta \le \pi.$$

Note that **A. B** is a *scalar* and not a vector.
The following laws are valid:

1. **A. B** = **B.A** Commutative law for dot products
2. **A. (B + C) = A. B + A. C** Distribution law
3. *m* **(A. B)** = (*m* **A**). **B** = A. (*m* **B**) = **(A. B)** *m*, where *m* is a scalar
4. **i. i = j. j = k. k** = 1, **i. j = j. k = k. i** = 0
5. If $\mathbf{A} = A_1\mathbf{i} + A_2\mathbf{j} + A_3\mathbf{k}$

And

$$\mathbf{B} = B_1\mathbf{i} + B_2\mathbf{j} + B_3\mathbf{k},$$

Then

$$\mathbf{A.\,B} = A_1B_1 + A_2B_2 + A_3B_3$$

$$\mathbf{A.A} = A^2 = A_1^2 + A_2^2 + A_3^2$$

$$\mathbf{B.B} = B^2 = B_1^2 + B_2^2 + B_3^2$$

6. If **A.B** = 0 and **A** and **B** are not null vectors, then **A** and **B** are perpendicular.

The cross or vector product of **A** and **B** is a vector $\mathbf{C} = \mathbf{A} \times \mathbf{B}$ (read **A** cross **B**). The magnitude of $\mathbf{A} \times \mathbf{B}$ is defined as the product of the magnitudes of **A** and **B** and the sine of the angle θ between them. The direction of the vector $\mathbf{C} = \mathbf{A} \times \mathbf{B}$ is perpendicular to the plane of **A** and **B** such that **A**, **B**, and **C** form a right-handed system. In symbols,

$$\mathbf{A} \times \mathbf{B} = AB\sin\theta\,\mathbf{u}, \qquad 0 \le \theta \le \pi,$$

where **u** is a unit vector indicating the direction of $\mathbf{A} \times \mathbf{B}$. If **A** = **B**, or if **A** is parallel to **B**, then $\sin\theta = 0$, and w0e define $\mathbf{A} \times \mathbf{B} = \mathbf{0}$.

The following laws are valid.

1. $\mathbf{A} \times \mathbf{B} = -\mathbf{B} \times \mathbf{A}$ (Commutative law for cross products fails)
2. $\mathbf{A} \times (\mathbf{B} + \mathbf{C}) = \mathbf{A} \times \mathbf{B} + \mathbf{A} \times \mathbf{C}$ Distributive law
3. $m(\mathbf{A} \times \mathbf{B}) = (m\mathbf{A}) \times \mathbf{B} = \mathbf{A} \times (m\mathbf{B}) = (\mathbf{A} \times \mathbf{B})m$, where *m* is a scalar
4. $\mathbf{i} \times \mathbf{i} = \mathbf{j} \times \mathbf{j} = \mathbf{k} \times \mathbf{k} = \mathbf{0}$; $\mathbf{i} \times \mathbf{j} = \mathbf{k}$, $\mathbf{j} \times \mathbf{k} = \mathbf{i}$, $\mathbf{k} \times \mathbf{i} = \mathbf{j}$
5. If $\mathbf{A} = A_1\mathbf{i} + A_2\mathbf{j} + A_3\mathbf{k}$ and

   $\mathbf{B} = B_1\mathbf{i} + B_2\mathbf{j} + B_3\mathbf{k}$, then

$$\mathbf{A} \times \mathbf{B} = \begin{vmatrix} \mathbf{i} & \mathbf{j} & \mathbf{k} \\ A_1 & A_2 & A_3 \\ B_1 & B_2 & B_3 \end{vmatrix}.$$

6. The magnitude of $\mathbf{A} \times \mathbf{B}$ is the same as the area of a parallelogram with sides **A** and **B**.
7. If $\mathbf{A} \times \mathbf{B} = \mathbf{0}$, and **A** and **B** are not null vectors, then **A** and **B** are parallel.

Triple Products. Dot and cross multiplication of three vectors **A, B,** and **C** may produce meaningful products of the form $(\mathbf{A.B})\mathbf{C}$, $\mathbf{A}.(\mathbf{B}\times\mathbf{C})$ and $\mathbf{A}\times(\mathbf{B}\times\mathbf{C})$. The following laws are valid.

1. $(\mathbf{A.B})\,\mathbf{C} \neq \mathbf{A}\,(\mathbf{B}\,.\,\mathbf{C})$
2. $\mathbf{A}.(\mathbf{B}\times\mathbf{C}) = \mathbf{B}.(\mathbf{C}\times\mathbf{A}) = \mathbf{C}.(\mathbf{A}\times\mathbf{B}) = \text{volume}$ of a parallelepiped having **A**, **B**, and **C** as edges, or the negative of this volume, according as **A**, **B**, and **C** do or do not form a right-handed system. (The parallelepiped is a solid body of which each face is a parallelogram, the latter being a plane four-sided figure with its opposite sides are parallel to each other).* If $\mathbf{A} = A_1\mathbf{i} + A_2\mathbf{j} + A_3\mathbf{k}$, then

$$\mathbf{A}.(\mathbf{B}\times\mathbf{C}) = \begin{vmatrix} A_1 & A_2 & A_3 \\ B_1 & B_2 & B_3 \\ C_1 & C_2 & C_3 \end{vmatrix}.$$

3. $\mathbf{A}\times(\mathbf{B}\times\mathbf{C}) \neq (\mathbf{A}\times\mathbf{B})\times\mathbf{C}$ (Associative law for cross products fails).
4. $\mathbf{A}\times(\mathbf{B}\times\mathbf{C}) = (\mathbf{A.C})\,\mathbf{B} - (\mathbf{A.B})\mathbf{C}$

$$(\mathbf{A}\times\mathbf{B})\times\mathbf{C} = (\mathbf{A.C})\,\mathbf{B} - (\mathbf{B.C})\,\mathbf{A}$$

The product $\mathbf{A}.(\mathbf{B}\times\mathbf{C})$ is sometimes called the *scalar triple product* or *box product* and may be denoted by **[ABC]**.

The product $\mathbf{A}\times(\mathbf{B}\times\mathbf{C})$ is called the *vector triple product.*

In $\mathbf{A}.(\mathbf{B}\times\mathbf{C})$ parentheses are sometimes omitted and we write $\mathbf{A}.\,\mathbf{B}\times\mathbf{C}$. However, parentheses must be used in $\mathbf{A}\times(\mathbf{B}\times\mathbf{C})$.

III. The Gradient, Divergence, and Curl

The vector differential operator del, written ∇, is defined by

$$\nabla = \frac{\partial}{\partial x}\mathbf{i} + \frac{\partial}{\partial y}\mathbf{j} + \frac{\partial}{\partial z}\mathbf{k}$$

$$= \mathbf{i}\frac{\partial}{\partial x} + \mathbf{j}\frac{\partial}{\partial y} + \mathbf{k}\frac{\partial}{\partial z}.$$

This *vector operator* possesses properties analogous to those of ordinary vectors. It is useful in defining three quantities which arise in practical applications and are known as the *gradient*, the *divergence*, and the *curl*. The operator ∇ is also known as *nabla.*

---

* The cube is a solid body with six equal square faces.

The gradient. Let $\phi(x, y, z)$ be defined and differentiable at each point (x, y, z) in a certain region of space (i.e. ϕ defines a differentiable *scalar* field). Then the *gradient* of $\boldsymbol{\phi}$, written $\nabla\phi$ or grad ϕ, is defined by

$$\nabla\phi = \left(\frac{\partial}{\partial x}\mathbf{i} + \frac{\partial}{\partial y}\mathbf{j} + \frac{\partial}{\partial z}\mathbf{k}\right)\phi$$

$$= \frac{\partial\phi}{\partial x}\mathbf{i} + \frac{\partial\phi}{\partial y}\mathbf{j} + \frac{\partial\phi}{\partial z}\mathbf{k}.$$

Note that $\nabla\phi$ defines a vector field.

The component of $\nabla\phi$ in the direction of a unit vector **a** is given by $\nabla\phi.\mathbf{a}$ and is called the *directional derivative* of ϕ in the *direction* **a**. Physically, this is the rate of change of ϕ at (x, y, z) in the direction **a**.

The divergence. Let $\mathbf{V}(x, y, z) = V_1\mathbf{i} + V_2\mathbf{j} + V_3\mathbf{k}$ be defined and differentiable at each point (x, y, z) in a certain region of space (i.e. **V** defines a differentiable *vector* field).

Then the divergence of **V**, written $\nabla.\mathbf{V}$ or div **V**, is defined by

$$\nabla.\mathbf{V} = \left(\frac{\partial}{\partial x}\mathbf{i} + \frac{\partial}{\partial y}\mathbf{j} + \frac{\partial}{\partial z}\mathbf{k}\right).$$

$$(V_1\mathbf{i} + V_1\,\mathrm{j} + V_1\mathbf{k})$$

$$= \frac{\partial V_1}{\partial x} + \frac{\partial V_2}{\partial y} + \frac{\partial V_3}{\partial z}.$$

Note the analogy with $\mathbf{A.B} = A_1B_1 + A_2B_2 + A_3B_3$.
Also note that $\nabla.\mathbf{V} \neq \mathbf{V}.\nabla$.

The curl. If **V** (x, y, z) is a differentiable vector field, then the *curl* or *rotation* of **V**, written **V**, or curl **V** or rot V, is defined by

$$\nabla\times\mathbf{V} = \left(\frac{\partial}{\partial x}\mathbf{i} + \frac{\partial}{\partial y}\mathbf{j} + \frac{\partial}{\partial z}\mathbf{k}\right)\times$$

$$(V_1\mathbf{i} + V_2\,\mathrm{j} + V_3\mathbf{k})$$

$$= \begin{vmatrix} \mathbf{i} & \mathbf{j} & \mathbf{k} \\ \frac{\partial}{\partial x} & \frac{\partial}{\partial y} & \frac{\partial}{\partial z} \\ V_1 & V_2 & V_3 \end{vmatrix}$$

$$= \begin{vmatrix} \frac{\partial}{\partial y} & \frac{\partial}{\partial z} \\ V_2 & V_3 \end{vmatrix} \mathbf{i} - \begin{vmatrix} \frac{\partial}{\partial x} & \frac{\partial}{\partial z} \\ V_1 & V_3 \end{vmatrix} \mathbf{j}$$

$$+ \begin{vmatrix} \frac{\partial}{\partial x} & \frac{\partial}{\partial y} \\ V_1 & V_2 \end{vmatrix} \mathbf{k}$$

$$= \left(\frac{\partial V_3}{\partial y} - \frac{\partial V_2}{\partial z}\right)\mathbf{i} + \left(\frac{\partial V_1}{\partial z} - \frac{\partial V_3}{\partial x}\right)\mathbf{j} + \left(\frac{\partial V_2}{\partial x} - \frac{\partial V_1}{\partial y}\right)\mathbf{k}.$$

Note that in the expansion of the determinant the operators $\frac{\partial}{\partial x}, \frac{\partial}{\partial y}, \frac{\partial}{\partial z}$ must *precede* V_1, V_2, V_3.

Formulae involving ∇. If **A** and **B** are differentiable *vector* functions and ϕ and ψ are differentiable *scalar* functions of position (*x*, *y*, *z*), then

1. $\nabla(\phi + \psi) = \nabla\phi + \nabla\psi$ or grad $(\phi + \psi)$ = grad ϕ + grad ψ

2. $\nabla.(\mathbf{A} + \mathbf{B}) = \nabla.\mathbf{A} + \nabla.\mathbf{B}$ or div (**A** + **B**) = div **A** + div **B**

3. $\nabla\times(\mathbf{A}+\mathbf{B}) = \nabla\times\mathbf{A} + \nabla\times\mathbf{B}$ or curl (**A** + **B**) = curl **A** + curl **B**

4. $\nabla.(\phi\mathbf{A}) = (\nabla\phi).\mathbf{A} + \phi(\nabla.\mathbf{A})$

5. $\nabla\times(\phi\mathbf{A}) = (\nabla\phi)\times\mathbf{A} + \phi(\nabla\times\mathbf{A})$

6. $\nabla.(\mathbf{A}+\mathbf{B}) = \mathbf{B}.(\nabla\times\mathbf{A}) - \mathbf{A}.(\nabla\times\mathbf{B})$

7. $\nabla\times(\mathbf{A}\times\mathbf{B}) = (\mathbf{B}.\nabla)\mathbf{A} - \mathbf{B}(\nabla.\mathbf{A}) - (\mathbf{A}.\nabla)\mathbf{B} + \mathbf{A}(\nabla.\mathbf{B})$

8. $\nabla(\mathbf{A}.\mathbf{B}) = (\mathbf{B}.\nabla)\mathbf{A} + (\mathbf{A}.\nabla)\mathbf{B} + \mathbf{B}\times(\nabla\times\mathbf{A}) + \mathbf{A}\times(\nabla\times\mathbf{B})$

9. $\nabla.(\nabla\phi) = \nabla^2\phi = \frac{\partial^2\phi}{\partial x^2} + \frac{\partial^2\phi}{\partial y^2} + \frac{\partial^2\phi}{\partial z^2}$

   where $\nabla^2 = \frac{\partial^2}{\partial x^2} + \frac{\partial^2}{\partial y^2} + \frac{\partial^2}{\partial z^2}$ is called

   the *Laplacian operator*.

10. $\nabla\times(\nabla\phi) = \mathbf{0}$. The curl of the gradient of ϕ is zero.

11. $\nabla.(\nabla\times\mathbf{A}) = 0$. The divergence of the curl of **A** is zero.

SECTION 3

The Gamma Function

Gamma Function

The gamma function denoted by $\Gamma(n)$ is defined by

$$\Gamma(n) = \int_0^\infty x^{n-1} e^{-x}\, dx \qquad (1)$$

which is convergent for $n > 0$.

A recurrence formula for the gamma function is

$$\Gamma(n+1) = n\,\Gamma(n) \qquad (2)$$

where $\Gamma(1)=1$ From Eq. (2), $\Gamma(n)$ can be determined for all $n > 0$ when the values for $1 \le n < 2$ (or any other interval of unit length) are known (see Table below).

In particular, if n is a positive integer, then

$$\Gamma(n+1) = n! \qquad n = 1, 2, 3, \ldots \;. \qquad (3)$$

For this reason $\Gamma(n)$ is sometimes called the *factorial function.*

Examples: $\Gamma(2) = 1! = 1, \qquad \Gamma(6) = 5! = 120,$

$$\frac{\Gamma(5)}{\Gamma(3)} = \frac{4!}{2!} = 12.$$

It can be shown that

$$\Gamma\left(\frac{1}{2}\right) = \sqrt{\pi}\,. \qquad (4)$$

The recurrence relation

$$\Gamma(n+1) = n\Gamma(n) \qquad (2)$$

is a difference equation which has Eq. (1)

$$\Gamma(n) = \int_0^\infty x^{n-1} e^{-x}\, dx \qquad (1)$$

as a *solution.* By taking Eq. (1) as the definition of $\Gamma(n)$ for $n > 0$, we can generalise the gamma function to $n < 0$ by use of Eq. (2) in the form

$$\Gamma(n) = \frac{\Gamma(n+1)}{n} \qquad (5)$$

The process is called *analytic continuation.*

Table of values of the gamma function

N	$\Gamma(n)$
1.00	1.0000
1.10	0.9514
1.20	0.9182
1.30	0.8975
1.40	0.8873
1.50	0.8862
2.00	1.0000

Solved Problems

1. Prove (a) $\Gamma(n+1) = n\Gamma(n), \; n > 0$

(b) $\Gamma(n+1) = n!, \; n = 1, 2, 3, \ldots$

Ans.

(a) $$\Gamma(n+1) = \int_0^\infty x^n e^{-x} dx = \underset{M\to\infty}{\text{Lt}} \int_0^M x^n e^{-x} dx$$

$$= \underset{M\to\infty}{\text{Lt}} \left\{ (x^n)(-e^{-x}) \Big|_0^M - \int_0^M (-e^{-x})(n x^{n-1}) dx \right\}$$

$$= \underset{M\to\infty}{\text{Lt}} \left\{ -M^n e^{-M} + n \int_0^M x^{n-1} e^{-x} dx \right\}$$

$$= \Delta H - T\Delta S \qquad \text{if } n > 0.$$

This is the recurrence formula for the gamma function, which is Eq. (2) above.

(b) $$\Gamma(1) = \int_0^\infty x^0 e^{-x} dx = \int_0^\infty e^{-x} dx$$

$$= \underset{M\to\infty}{\text{Lt}} \int_0^M e^{-x} dx$$

$$= \underset{M\to\infty}{\text{Lt}} \; -e^{-x} \Big|_0^M$$

$$= \underset{M\to\infty}{\text{Lt}} \left(-e^{-M} + e^0\right) = \underset{M\to\infty}{\text{Lt}} \left(1 - e^{-M}\right)$$

$$= 1$$

Put $n = 1, 2, 3, \ldots$ in $\Gamma(n+1) = n\Gamma(n)$. Then

$$\Gamma(2) = 1\Gamma(1) = 1$$

$$\Gamma(3) = 2\Gamma(2) = 2 \times 1 = 2!$$

$$\Gamma(4) = 3\Gamma(3) = 3 \times 2! = 3 \times 2 \times 1 = 3!$$

In general,

$$\Gamma(n+1) = n!$$

if n is a positive integer, that is, $n = 1, 2, 3, \ldots$.

As a miscellaneous example involving gamma function is the result

$$\Gamma(x)\,\Gamma(1-x) = \frac{\pi}{\sin x\pi}, \qquad 0 < x < 1.$$

Thus in particular if $x = \frac{1}{2}$, we have

$$\Gamma\left(\frac{1}{2}\right) . \Gamma\left(\frac{1}{2}\right) = \frac{\pi}{\sin 90^\circ} = \pi$$

$$\Gamma\left(\frac{1}{2}\right) = \sqrt{\pi}.$$

SECTION 4

1. Alternating Current

Introduction

The alternating current technique is chosen for the detailed treatment of conduction in polymers because it provides some information not readily available by direct current methods, and also enables one to avoid some of the problems which direct current leads to such as polarisation by space charge built up.

(1) Direct current technique. In semiconductors, non-ohmic processes play important role. This has led to the belief that the deviation from *Ohm's* law is an evidence that the conduction mechanism is *electronic.* However, this evidence is not true because there are many factors that can lead to non-ohmic behaviour in polymers. Generally, the possible sources of non-ohmic conduction can be summarised as ionic, ionic and electrolytic, and electronic.

(2) Alternating current technique. It is known in servomechanism and in circuit design engineering that almost any relatively smooth frequency response curve can be matched by some combination of resistors and capacitors that is, by an RC network an example of which is

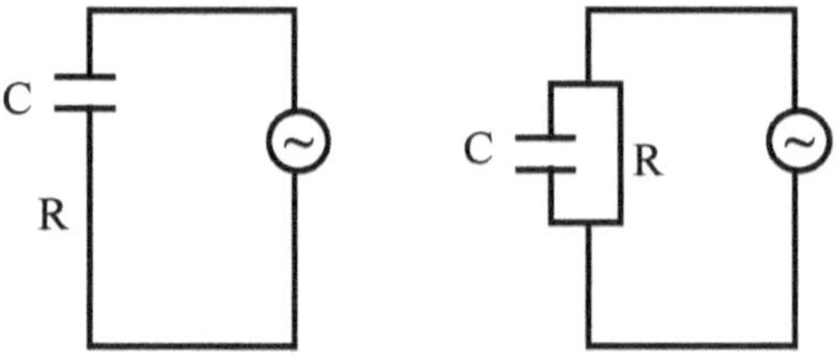

It is to be noted that

(i) The current in the capacitor or C *leads* the sinusoidal voltage $V\left(=V_{\circ}\cos\omega t\right)$ by 90° in phase. This is because

$$I_C = \frac{dQ}{dt}, \qquad Q = CV,$$

and

$$I_C = C\frac{dV}{dt}, \tag{1}$$

so that

$$= -CV_{\circ}\left(\sin\omega t\right)\times\omega$$

$$= -C\omega V_{\circ}\sin\omega t$$

$$= C\omega V_{\circ}\cos\left(\omega t + \frac{\pi}{2}\right); \tag{2}$$

(ii) The current and voltage are in phase in an ideal resistor R; and

(iii) If the capacitor C and the resistance R are in series combination, the impedances add, that is,

$$Z_S = R_S + \frac{1}{i\omega C} = R_S - \frac{i}{\omega C}, \qquad (3)$$

whereas if they are in parallel combination, the admittances add, that is,

$$y = \frac{1}{R_P} + \frac{1}{1/i\omega C_P} = \frac{1}{R_P} + i\omega C_P.$$

(3) Alternating current theory. Fig. 1 represents the forced oscillations in a circuit containing an inductance L, a capacitor C, and a resistance R in series.

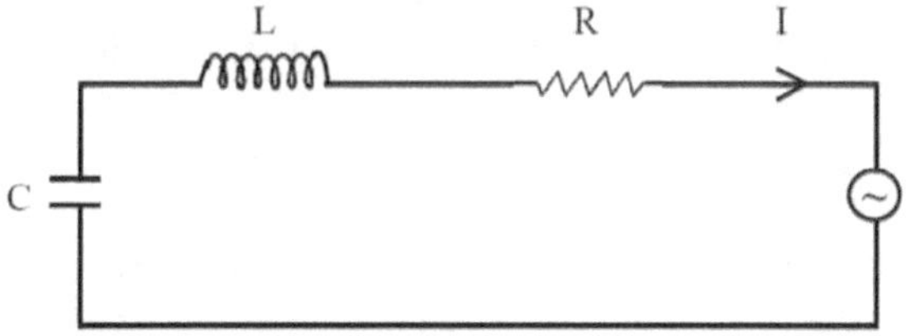

Fig. 1.

If *I* is the current flowing at any instant, the voltage set in L is

$$L\frac{dI}{dT},$$

and the voltage drop across C and R are respectively

$$\frac{Q}{C} \text{ and } RI.$$

Therefore, for this circuit, we have

$$L\frac{dI}{dT} + \frac{Q}{C} + RI = V = V_{\circ} \cos \omega t \qquad (1)$$

It is evident that for this *series circuit the voltages add.*

If we want to evaluate the current *I*, we determine the rate of increase of charge on the capacitor; that is, $\frac{dQ}{dt}$. Thus differentiating Eq. (1) with respect to *t*, we obtain

$$L\frac{d^2 I}{dt^2} + \frac{I}{C} + R\frac{dI}{dt} = -V_{\circ}\,\omega \sin \omega t.. \qquad (2)$$

This is a differential equation whose solution consists of two parts:

The first part, which is known as the *complementary part*, can be obtained by solving the equation

$$L\frac{d^2 I}{dt^2} + R\frac{dI}{dt} + \frac{I}{C} = 0. \qquad (3)$$

The solution of Eq. (3) is

$$I = e^{-iR/2L}\left(A\,e^{nt} + B\,e^{-nt}\right),$$

where *n* is given by

$$n^2 = \left(\frac{R}{2L}\right)^2 - \frac{1}{LC},$$

and A and B are constants determined by the initial conditions. That is, at $t = 0$, $I = 0$ so that

$$A\,e^{nt} + Be^{-nt} \rightarrow A + B = 0,$$

or

$$A = -B.$$

This solution represents a *transient flow* of current. This current is produced by the application of the emf, $V = V_\circ \cos \omega t$, with the assumption that it starts to act at the instant $t = 0$.

In all practical applications, the transient current decays rapidly in magnitude owing to the exponential term $e^{-iR/2L}$ and becomes negligible within few seconds of the circuit being closed. If the conditions are such that the transient current is oscillatory, its frequency will be the natural frequency determined by the values of L, C, and R and not that of the applied emf.

The second part of the solution of Eq. (2) is known as the *particular integral*; it may be expressed

$$I = \frac{V_\circ}{Z}\cos\left(\omega t - \phi\right), \tag{4}$$

where *Z* is known as the impedance* of the circuit and ϕ, the phase angle, is given by its tangent as

$$\tan\phi = \left(\omega L - \frac{1}{\omega C}\right)\Big/R. \tag{5}$$

$$\tan\phi = \left(\omega L - \frac{1}{\omega C}\right)\Big/R.$$

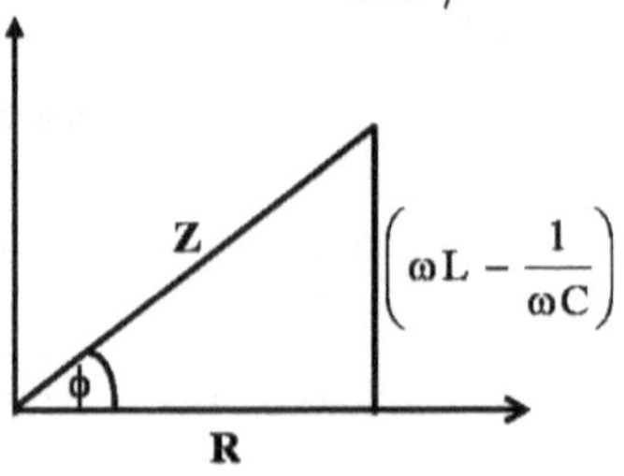

Fig. 2.

* The total resistance of an electric circuit to the flow of alternating current is called impedance, and is denoted by Z.

The angle ϕ appears here mathematically as

$$\phi = \tan^{-1} \frac{\omega L - \dfrac{1}{\omega C}}{R},$$

but it does not express any physical meaning except that *I* and *V* are not of the same phase. The impedance *Z* is given by

$$Z = \sqrt{R^2 + (\omega L - \frac{1}{\omega C})^2}. \qquad (6)$$

This solution is also known as the *steady-state* solution, owing to the fact that Eq. (4) gives the current at any time after the transient current has become negligible. It is to be noted that the *frequency* of the current is the same as that of the applied emf; that is, $\omega/2\pi$, so that the current can be considered to be in *forced* oscillation. But, on the other hand, Eq. (4) as a solution of Eq. (2) says that the flowing current is not in phase with the applied voltage; this difference in phase angle ϕ will disappear if it happens that the applied frequency ω satisfies the condition

$$\omega L - \frac{1}{\omega C} = 0,$$

that is,

$$\omega = \frac{1}{\sqrt{LC}} = \omega_o.$$

Under such conditions, the current has not any *damping resistance*. The circuit is thus in *resonance*, and the current amplitude is a maximum. At any instant, *I* is given by

$$I = \frac{V_o}{Z} \cos \omega t,$$

where $Z = R$.

It is important to point out that at frequencies other than the resonance frequency, the value of the impedance Z of the circuit is related to the quantities R, L, and C by Eq. (6), and not by simple additive relation that holds for the resistances in series. The reason for this is that the voltages across the different elements are not in phase, as a consequence of which the total voltage amplitude is not the sum of the individual amplitudes. This difficulty as would be seen below can be overcome by the introduction of *complex numbers* to express the impedances and to represent them graphically on the *Argand diagram* where it is usual to represent real quantities by *vectors* drawn parallel to the *x*-axis, and imaginary quantities by *vectors* drawn parallel to the *y*-axis.

The use of complex impedances enables us to apply *Kirchhoff's* laws to alternating current networks. In this way, we can find the values of the current and voltage in any branch with out having to solve a differential equation as we have done above.

2. Use of Vectors

Assume that a current *I* flows through R, L, and C in *series*, where *I is equal to* $I_o \cos \omega t$.

The voltages across the three elements of the circuit are, respectively,

$$V_R = IR = RI_o \cos \omega t \tag{1}$$

$$V_L = L\frac{dI}{dT} = -\omega L I_o \sin \omega t$$

$$= \omega L I_o \cos\left(\omega t + \frac{\pi}{2}\right), \tag{2}$$

$$V_C = \frac{Q}{C} = \frac{\int I_C\, dt}{C}$$

$$= \frac{\int I_o \cos \omega t\ dt}{C} = \frac{I_o \sin \omega t}{\omega C}$$

$$= \frac{I_o}{\omega c} \cos\left(\omega t - \frac{\pi}{2}\right). \tag{3}$$

It is evident from these equations that (i) the *voltage* across the resistance R is *in phase* with the applied *current*, (ii) the voltage across the inductance L leads the current in phase by $\frac{\pi}{2}$, and (iii) the voltage across the capacitor lags behind the current in phase by $\frac{\pi}{2}$.

In order to visualise the physical meaning of leading or lagging in phase between the current and the voltages across the inductance L and the capacitor C, respectively, we may represent these voltages, as well as the voltage across R, by *vectors*. This representation can be achieved in such a way that if the voltage across the resistance R is represented by a vector *parallel* to the *x*-axis, the voltage across the inductance L can be represented by a vector *parallel* to the *y*-axis, and that across the capacitor by vector also *parallel* to the *y*-axis but in opposite direction. The lengths of the vectors are taken to be proportional to the magnitudes of R, ωL, and $\frac{1}{\omega C}$, respectively.

One may have an objection to this *vector representation of scalar quantities* such as voltage, resistance, and reactance. However, this objection may be considered *invalid* on the basis that if the current is represented by a *unit vector* along the resistance R or parallel to the direction of R, then the impedance *Z* multiplied by unity for the current gives the amplitude of the voltage. In other words, on this basis the voltage amplitude across the resistance R is represented by the resistance R itself. Similarly, the voltage amplitude across the reactance of the capacitor $\frac{1}{\omega C}$ is represented by the reactance $\frac{1}{\omega C}$ itself, and the voltage amplitude across the reactance of the inductance ωL is represented by the reactance ωL itself.

Suppose now that we like to determine the voltages across two of the elements, say, R and L. This may be obtained by adding the two individual voltage vectors derived above

$$V_R = R I_\circ \cos\omega t,$$

$$V_L = \omega L I_\circ \cos\left(\omega t + \frac{\pi}{2}\right),$$

as shown graphically in Fig. 2, where the voltage V_R is represented by a vector parallel to the *x*-axis and V_L by a vector parallel to the *y*-axis. The lengths of these vectors are taken proportional to the magnitudes of R and ωL, respectively so that Fig. 2 may be considered as the graphical representation of impedances.

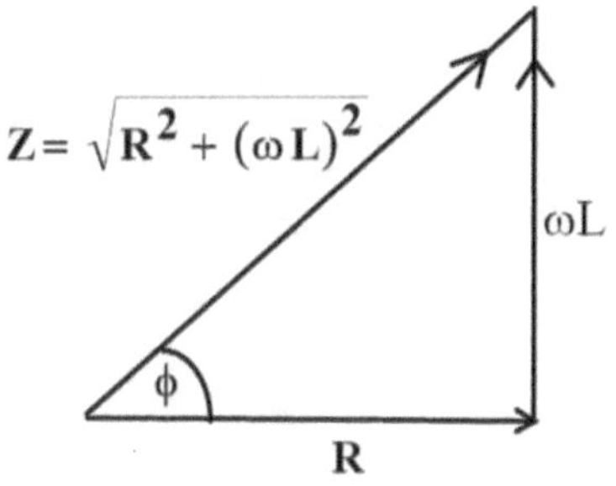

Fig. 2

Algebraically, the resultant voltage is

$$V = V_R + V_L = R I_\circ \cos\omega t + \omega L I_\circ \cos\left(\omega t + \frac{\pi}{2}\right)$$

$$= R I_\circ \cos\omega t - \omega L I_\circ \sin\omega t$$

$$= I_\circ (R\cos\omega t - \omega L \sin\omega t).$$

On the other hand, making use of the preceding graphical representation that

$$\omega L = R \tan\phi,$$

we have

$$V = I_\circ \left(R\cos\omega t - R\frac{\sin\phi}{\cos\phi}\sin\omega t \right)$$

$$= \frac{I_\circ R}{\cos\phi}(\cos\omega t \cos\phi - \sin\omega t \sin\phi)$$

$$= \frac{I_\circ R}{R/Z}(\cos\omega t \cos\phi - \sin\omega t \sin\phi)$$

$$= I_\circ Z \cos(\omega t + \phi).$$

The total impedance Z of the circuit is given by the length of the hypotenuse of the triangle (Fig. 2), that is,

$$Z = \sqrt{R^2 + (\omega L)^2}.$$

Let us now consider *another example* for determining the voltages across a resistance R is series with a capacitance C.
The voltage across R is

$$V_R = R I_\circ \cos \omega t.$$

The voltage across C is

$$V_C = \frac{I_\circ}{\omega C} \cos\left(\omega t - \frac{\pi}{2}\right).$$

Based on the above, we consider these voltages as vectors so that by graphical representation the vector V_R is taken parallel to the *x*-axis and the vector V_C antiparallel to the *y*-axis, the lengths of these two vectors being taken propositional to R and $\frac{1}{\omega C}$, respectively as shown in Fig. 3.

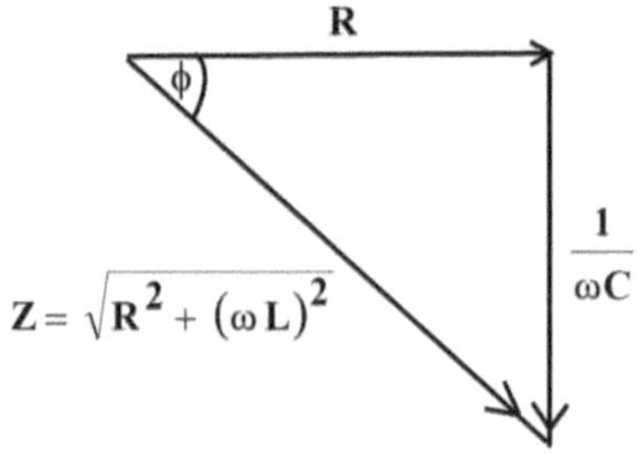

Fig. 3.

This figure may be regarded as the graphical representation of impedances so that

$$Z = \sqrt{R^2 + \left(\frac{1}{\omega C}\right)^2}.$$

The total voltage is given by

$$V = R I_\circ \cos \omega t + \frac{I_\circ}{\omega C} \sin \omega t = R I_\circ \cos \omega t$$

$$+ R I_\circ \tan \phi \sin \omega t$$

$$= I_\circ Z \cos(\omega t - \phi).$$

For all components in *series* the vector diagram for the representation of impedances is shown in Fig. 4.

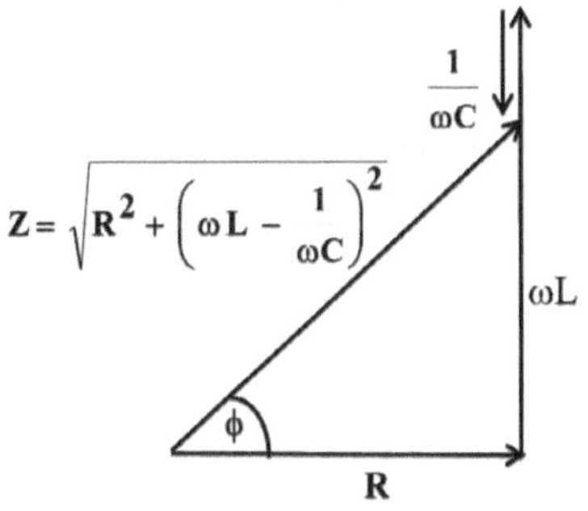

Fig. 4.

The total impedance Z is given by the length of the hypotenuse of the triangle, that is,

$$Z = \sqrt{R^2 + \left(\omega L - \frac{1}{\omega C}\right)^2}.$$

The total voltage V is obtained algebraically as

$$V = I_o R \cos \omega t + \frac{I_o}{\omega C} \sin \omega t - \omega L I_o \sin \omega t$$

$$= I_o R \cos \omega t + I_o \sin \omega t \left(\frac{1}{\omega C} - \omega L\right).$$

But

$$\omega L - \frac{1}{\omega C} = R \tan \phi,.$$

Therefore,

$$V = I_o R \left(\cos \omega t \cos \phi - \sin \omega t \sin \phi\right)$$
$$= I_o Z \left(\cos \omega t \cos \phi - \sin \omega t \sin \phi\right)$$

$$= I_o Z \cos(\omega t + \phi).$$

3. Use of Complex Numbers

The phase of voltage which leads* the current by $\pi/2$ in the inductance or lags+ behind it by $\pi/2$ in the capacitance is taken into account by the introduction of complex numbers

---

+ $V_L = \omega L I_o \cos\left(\omega t + \frac{\pi}{2}\right)$, $V_C = \frac{I_o}{\omega C} \cos\left(\omega t - \frac{\pi}{2}\right)$

* as shown above.

to represent the impedances using the complex exponential $e^{i\omega t}$ which is the sum of sine and cosine functions

$$e^{i\omega t} = \cos\omega t + i\sin\omega t.$$

Here we are interested in the real and imaginary parts. For the graphical representation of the impedances, we use the *Argand* diagram as noted above, where it is usual to represent the real quantities by vectors drawn parallel to the *x*-axis, and the imaginary quantities by vectors parallel to the *y*-axis.

If we now write the current

$$I = I_{\circ} \cos\omega t$$

as

$$I = I_{\circ} \operatorname{Re} e^{i\omega t},$$

we have

$$I = I_{\circ} \cos\omega t,$$

and

$$RI = RI_{\circ} \cos\omega t. \tag{1}$$

Also quantities as $\frac{dI}{dt}$,

or

$$Q = \int I\,dt$$

become

$$\frac{dI}{dt} = I_{\circ} \operatorname{Re} \frac{d\,e^{i\omega t}}{dt},$$

$$= I_{\circ} \operatorname{Re} \frac{d}{dt}(\cos\omega t + i\sin\omega t),$$

$$= I_{\circ} \frac{d\cos\omega t}{dt} = -I_{\circ}\,\omega\sin\omega t.$$

Therefore,

$$L\frac{dI}{dT} = -I_{\circ}\,\omega L \sin\omega t. \tag{2}$$

Also, we have

$$Q = \int I\,dt$$

$$= I_{\circ} \int \cos\omega t\,dt,$$

that is,

$$= I_\circ \text{ Re} \int e^{i\omega t}\, dt$$

$$= I_\circ \text{ Re}\left(\frac{e^{i\omega t}}{i\,\omega}\right) = I_\circ \text{ Re}\left(\frac{\cos\omega t + i\sin\omega t}{i\,\omega}\right)$$

$$= I_\circ \text{ Re}\left(\frac{\cos\omega t}{i\,\omega} + \frac{\sin\omega t}{i\,\omega}\right)$$

$$= \frac{I_\circ}{\omega}\sin\,\omega t.$$

Therefore,

$$\frac{Q}{C} = \frac{I_\circ}{\omega C}\sin\omega t. \tag{3}$$

Accordingly, the summation of Eqs. (1)–(3) gives the total *V*, that is,

$$V = I_\circ\left(R\cos\omega t - \omega L\,\sin\omega t + \frac{1}{\omega C}\sin\omega t\right), \tag{4}$$

This equation is the same as that obtained directly from a relation given incidentally above, namely,

$$V = RI + L\frac{dI}{dT} + \frac{Q}{C}, \tag{5}$$

where

$$I = I_\circ \cos\omega t,$$

$$V_R = R\,I_\circ \cos\omega t,$$

$$V_L = L\frac{dI}{dT} = -\,\omega L\,I_\circ \sin\omega t,$$

$$V_C = \frac{Q}{C} = \frac{I_\circ}{\omega C}\sin\,\omega t,$$

so that by the summation of V_R , V_L and V_C we obtain in Eq. (4) *without the introduction of i in the exponential*; that is, without the use of the complex exponential $e^{i\omega t}$.

On the other hand, instead of using each term in Eq. (4) separately, we follow the procedure.

$$V = R\,I_\circ e^{i\omega t} + L\,I_\circ \frac{d\,e^{i\omega t}}{d\,t} + \frac{I_\circ}{C}\int e^{i\omega t}$$

$$= R\,I_\circ\, e^{i\omega t} + L\,I_\circ\, e^{i\omega t} \times i\omega + \frac{I_\circ}{C}\frac{e^{i\omega t}}{i\omega}$$

$$= I_\circ\, e^{i\omega t}\left(R + i\omega L + \frac{1}{i\omega C}\right). \qquad (7)$$

Since the current I is given by the real part of $e^{i\omega t}$; that is,

$$I = I_\circ \cos u t,$$

then the voltage V in Eq. (7) may be found by taking the real part of this equation; that is,

$$V = \text{Re}\left[I_\circ (\cos\omega t + i\sin\omega t)\left(R + i\omega L + \frac{1}{i\omega C}\right)\right]$$

$$= I_\circ\left(R\cos\omega t - \omega L\sin\omega t + \frac{1}{\omega C}\sin\omega t\right).$$

This is the same equation obtained above by taking each term separately; it is Eq. (4). Also it is the same equation obtained directly from the relation

$$V = RI + L\frac{dI}{dt} + \frac{Q}{C}. \qquad (5)$$

Without of introduction of complex exponentials, but algebraically *using vectors*.

Now, the importance of Eq. (7) is due to the fact that the inductance L and the capacitance C can be represented by *complex impedance operators* $i\omega L$ and $\frac{1}{i\omega C}$, respectively, which may be *added* to one other if the elements are in *series* as in the case of resistances. Thus, using the *Argand* diagram in which real quantities are represented by vectors drawn parallel to the *x*-axis and the imaginary quantities by vectors parallel to the *y*-axis, the complex impedance operator *vector Z* of a circuit consisting of R, L, and C in *series* can be represented *in magnitude* by the same diagram shown above in Fig. 4 (Section 2) which was obtained for the vectorial representation of the impedances RI, ωL, and $\frac{1}{\omega C}$.

The modulus of **Z**, or its magnitude *Z*, will therefore be given by

$$|\mathbf{Z}| = Z = \sqrt{R^2 + \left(\omega L - \frac{1}{\omega C}\right)^2}, \qquad (8)$$

where

$$\mathbf{Z} = R + i\left(\omega L - \frac{1}{\omega C}\right).$$

Also, **Z** can be written as

$$\mathbf{Z} = R + iX, \tag{9}$$

where

$$X = \omega L - \frac{1}{\omega C}.$$

or, it can be written as

$$\mathbf{Z} = Ze^{i\phi} \tag{10}$$

Accordingly, from Eqs. (9) and (10) we have

$$R + iX = Z(\cos\phi + i\sin\phi).$$

Therefore,

$$R = Z\cos\phi.$$

$$X = Z\sin\phi,$$

or,

$$\omega L - \frac{1}{\omega C} = Z\sin\phi.$$

Again this can be represented graphically by the diagram shown in Fig. 4, where the magnitude *Z* is given by

$$Z = \sqrt{R^2 + \left(\omega L - \frac{1}{\omega C}\right)^2}.$$

represents the length of the hypotenuse of the triangle.

It is to be noted that the above arguments are based on the assumption that the current flowing in a circuit of R, L, and C is

$$I = I_\circ \cos\omega t.$$

A question might arise here saying what would be the situation if *I* is

$$I = I_\circ \sin\omega t\ ? \tag{11}$$

The reply will be that in this case we consider the imaginary of the complex exponential $e^{i\omega t}$; that is,

$$I = I_\circ \sin\omega t \tag{11}$$

will be written as

$$I = I_\circ \text{ Im.}\left(e^{i\omega t}\right)$$

$$= I_\circ \text{ Im.}\left(\cos\omega t + i\sin\omega t\right)$$

$$= I_\circ \sin u\, t \tag{11}$$

which is the original equation.

Subsequently, a quantity such as

$$\frac{dQ}{dt} = I,$$

or

$$Q = \int I\,dt$$

becomes

$$Q = \int I_\circ \, \text{Im.} \left(e^{i\omega t}\, dt\right)$$

$$= I_\circ \, \text{Im.} \int e^{i\omega t}\, dt)$$

$$= I_\circ \, \text{Im.} \left(\frac{e^{i\omega t}}{i\omega}\right)$$

$$= I_\circ \, \text{Im.} \left(\frac{\cos\omega t}{i\omega} + \frac{i\sin\omega t}{i\omega}\right)$$

$$= -\frac{I_\circ}{\omega}\cos\omega t.$$

Therefore, $\frac{Q}{C}$ is given by

$$\frac{Q}{C} = -\frac{I_\circ}{\omega C}\cos\omega t. \qquad (12)$$

Also the quantity $\frac{dI}{dt}$ becomes

$$\frac{dI}{dt} = \frac{d\left(I_\circ \, \text{Im.}\, e^{i\omega t}\right)}{dt} = I_\circ \, \text{Im.} \frac{de^{i\omega t}}{dt}$$

$$= I_\circ \, \text{Im.} \left(e^{i\omega t} \times i\omega\right)$$

$$= I_\circ \, \text{Im.} \left(i\omega\cos\omega t + i^2\omega\sin\omega t\right)$$

$$= I_\circ \, \omega\cos\omega t.$$

Therefore,

$$L\frac{dI}{dt} = I_\circ \, \omega L\cos\omega t. \qquad (13)$$

Accordingly, the total voltage *V* which is given by

$$V = RI + L\frac{dI}{dt} + \frac{Q}{C},$$

is the summation of Eqs. (11)–(13), that is,

$$V = I_{\circ}\left(R\sin\omega t + \omega L\cos\omega t - \frac{1}{\omega C}\cos\omega t\right) \tag{14}$$

in this case when $I = I_{\circ}\sin\omega t$, whereas in the previous case when $I = I_{\circ}\cos\omega t$, the total voltage *V* is given by

$$V = I_{\circ}\left(R\cos\omega t - \omega L\sin\omega t + \frac{1}{\omega C}\sin\omega t\right)$$

as shown above, that is, Eq. (4).

4. The Rate of Doing Work (or Power)

For a circuit having R, L, and C in series, as shown in Fig. 1 above, where the applied voltage *V* is

$$V = V_{\circ}\cos\omega t, \tag{1}$$

the current flowing in the circuit is given by

$$I = \frac{V_{\circ}}{Z}\cos(\omega t - \phi), \tag{2}$$

or

$$I = I_{\circ}\cos(\omega t - \phi) \tag{3}$$

where ϕ is the phase angle

$$\phi = \tan^{-1}\frac{\omega L - \dfrac{1}{\omega C}}{R}, \tag{4}$$

and *Z* is the total impedance of the circuit

$$Z = \sqrt{R^2 + \left(\omega L - \frac{1}{\omega C}\right)^2}. \tag{5}$$

Since in this circuit the voltages across the three elements add, then the total voltage *V* which is given by

$$V = RI + L\frac{dI}{dT} + \frac{Q}{C} \tag{6}$$

can be proved (as has been shown above) to be

$$V = I_{\circ}Z(\cos\omega t\cos\phi - \sin\omega t\sin\phi), \tag{7}$$

or,

$$= I_\circ Z \cos(\omega t + \phi). \tag{8}$$

Therefore, the rate of doing work in the circuit, that is, the electric power drawn from a generator having such a circuit, is

$$V \times I = I_\circ Z (\cos \omega t \cos \phi - \sin \omega t \sin \phi)$$

$$\times \frac{V_\circ}{Z} \cos(\omega t - \phi), \tag{9}$$

where *I* is substituted for the right-hand side of Eq. (2).

Since

$$I_\circ = \frac{V_\circ}{Z}. \tag{10}$$

Therefore,

$$V \times I = V_\circ \times I_\circ (\cos \omega t \cos \phi - \sin \omega t \sin \phi)$$

$$\times \cos(\omega t - \phi),$$

or

$$V \times I = \frac{V_\circ^2}{Z} (\cos \omega t \cos \phi - \sin \omega t \sin \phi)$$

$$\times \cos(\omega t - \phi). \tag{11}$$

Since the current *I* flowing in the resistance R is *in phase* with the applied emf or voltage, then the preceding equation may be reduced to

$$V \times I = \frac{V_\circ^2}{Z} \left(\cos^2 \omega t \cos \phi - \cos \omega t \sin \omega \sin \phi\right). \tag{12}$$

We can now find the mean rate of doing work, by averaging the right-hand side of the last equation over one or more periods of oscillation. The mean value of $\cos^2 \omega t$ is $\frac{1}{2}$, whereas that of $\cos \omega t \sin \omega t$ is zero. Hence, the mean power drawn from a generator of such a circuit is given by

$$\overline{W} = \frac{1}{2} \frac{V_\circ^2}{Z} \cos \phi, \tag{13}$$

or, since

$$V_\circ = I_\circ Z,$$

we have

$$\overline{W} = \frac{1}{2} I_0^2 Z \cos \phi. \tag{14}$$

Furthermore, since $\cos\phi$ is given by

$$\cos\phi = \frac{R}{Z},$$

then Eq. (14) can be put in the form

$$\overline{W} = \frac{1}{2} I_\circ^2 R \tag{15}$$

This means that the power delivered by the generator averaged over a period (or more) is *dissipated as heat* in the resistance R of the circuit.

Also, since

$$V = V_\circ \cos\omega t,$$

or

$$V^2 = V_\circ^2 \cos^2 \omega t,$$

and the average of $\cos^2 \omega t = \frac{1}{2}$, then

$$\overline{V^2} = \frac{1}{2} V_\circ^2,$$

where $\overline{V^2}$ is the mean square voltage, and Eq. (14) can be written as

$$\overline{W} = \frac{1}{2} \frac{\overline{V^2}}{Z} \frac{1}{\cos^2 \omega t} \cos\phi,$$

$$= \frac{V^2}{Z} \cos\phi,$$

so that by substituting for $\frac{V}{Z}$ in this equation by

$$\frac{V}{Z} = I,$$

we obtain

$$\overline{W} = \overline{V}.\overline{I} \cos\phi. \tag{16}$$

This is an important equation, according to which the ratio

$$\frac{\overline{W}}{\overline{V}.\overline{I}} = \cos\phi \tag{17}$$

represents the fraction of the product $\overline{V}.\overline{I}$ which is dissipated in the material as *Joule's heat*. It is known as the *power factor* of the circuit (or the generator), and according to Eq. (17), it is denoted by $\cos\phi$.

If the circuit behaves as a *pure resistance* as occurs when the *resonance condition*

$$\omega L = \frac{1}{\omega C},$$

or

$$\omega = \omega_\circ = \frac{1}{\sqrt{LC}} \tag{18}$$

is satisfied in the circuit, the *current has not any damping resistance*; that is, the phase angle ϕ will be equal to zero and, hence, the power factor $\cos\phi$ *will be equal to unity*. If, however, the *circuit contains no resistance, the power factor is zero*.

5. Kirchhoff's Laws

These laws are rules which enable us to solve complicated problems of networks by a systematic process.

First law. *The algebraic sum of the currents at a junction is zero.*

This law states that a junction cannot accumulate charge. The algebraic sum of the currents entering the junction shown in Fig. 1 with four arms is $i_1 - i_2 - i_3 - i_4$, and by the law this is equal to zero.

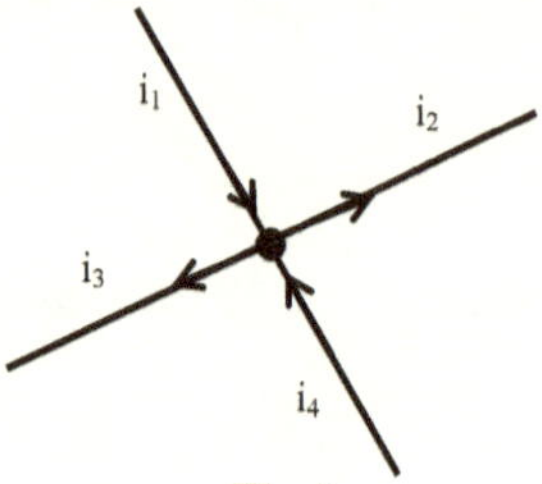

Fig. 1

Second law. *The total emf in a circuit is equal to the voltage drop round it.*

This law states that the potential of a point in a circuit depends only on the position of that point so that the work done on a charge in being taken around a closed circuit is zero.

Fig. 2 below shows a closed circuit ABCDEFA which is a part of larger circuit. In going from A to B, there is a voltage drop $i_1 R_1$, from B to C an emf rise of V_1, from C to D a voltage drop $i_2 R_2$, from D to E an emf drop V_2, from E to F a voltage drop $- i_3 R_3$, and from F to A, a voltage drop $- i_4 R_4$.

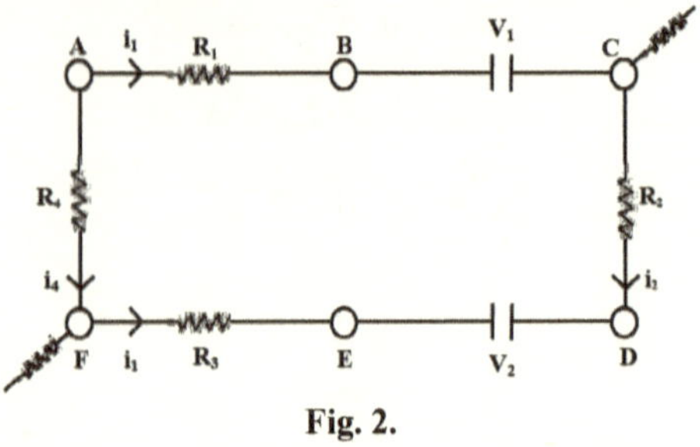

Fig. 2.

According to the second law of *Kirchhoff* and taking the emfs and voltage drops in the clockwise direction, we have

$$V_1 - V_2 = i_1 R_1 + i_2 R_2 - i_3 R_3 - i_4 R_4.$$

If we want now to obtain such a relation, we just calculate the work done on a unit charge as it is taken around the circuit. It is easily seen that

$$i_3 R_3 - V_2 - i_2 R_2 + V_1 - i_1 R_1 + i_4 R_4$$

is equal to zero. This is evidently the same result obtained above by applying Kirchoff's second law of equating the emf and voltage drops around the circuit.

APPENDIX B

1. Definitions of Important Units with Special Names

(1) The *ampere* is the amount of current in each of two long straight parallel wires, one metre apart, which causes a force of 2×10^{-7} newtons on each metre of wire. It is named after the French physicist A.-M. Ampère (1775–1836).

(2) The *coulomb* is the amount of charge that flows in one second when the current is one *ampere*. It is named after the French engineer C. A. de Coulomb (1736–1806). (Conversely, the current in amperes is defined as the number of coulombs flowing in one second). This unit shall be reconsidered below.

(3) The *volt* is the potential difference between two points such that one joule of work is done in moving one coulomb of charge between these points. It is named after the Italian physicist count Alessandro Volta (1754–1824). The definition of the joule as an electrical energy unit given later by the English physicist J. P. Joule (1818–1889), namely, it is the electrical energy of an electric current flowing through a potential difference of on volt for one second, as has been shown before, is evidently based *a priori* on the above definition of the volt.

(4) The *watt* is the amount of electrical energy flow per second; that is, electrical work done per second known as *power* which corresponds to one joule per second. Thus a current of one ampere to a potential difference of one volt corresponds to one watt of power. Put another way, one watt is equal to one ampere multiplied by one volt,

$$1 \text{ watt} = 1 \text{ ampere} \times 1 \text{ volt}.$$

Therefore,

$$1 \text{ watt} = 1 \text{ coulomb} \times 1 \text{ volt} / \text{sec}$$

or

$$1 \text{ watt} = 1 \text{ joule per sec ond},$$

that is,

$$1 \text{ joule} = 1 \text{ watt} . \text{sec}.$$

(Also, commercially, we speak of the kilowatt which is equal to 1000 watt, the plural of watt being watt.) The watt was named after James Watt who was a Scottish engineer; he greatly improved the steam engine.

(5) The *ohm* is the resistance of a material which allows a current of one ampere to flow if the potential difference is one volt. It is named after the German physicist G. S. Ohm (1789–1854).

(6) The *siemens* is the unit of electrical conductance; it is the reciprocal of the ohm, named after E. W. von Siemens who was German electrical engineer (1816–1892).

(7) The *farad* is the passage of one coulomb of charge between two plates of potential difference of one volt. It is named after Faraday Michael (1791–1867), an English physicist and chemist; discoverer of electromagnetic induction.

(8) The *electronvolt* is the quantity of electrical energy acquired by an electron in falling through a potential difference of one volt; it is denoted by eV. According to this definition,

$$1\,\text{ev} = \frac{4803 \times 10^{-10}}{299.79} \times 1 \quad = 1.602 \times 10^{-19} \text{ (coul)} \times \text{(volt)}$$

$$= 1.602 \times 10^{-19} \text{ volt . caulomb}$$

$$= 1.602 \times 10^{-19} \text{ joule,}$$

where the magnitude of the charge of e is 1.602×10^{-19}coulomb in derived units, the electronic charge in esu $= 4.803 \times 10^{-10}$, and the volt is $1/299.79$ e s u of potential.

Chemists are habituated to expressing energy quantities in calories or kilocalories per mole, and on this basis, we have by definition

$$1\,\text{ev} = \frac{1.602 \times 10^{-19} \text{ (joule)} \times 6.023 \times 10^{23} \text{ (mole)}}{4.18 \qquad\qquad \times 10^{3}}$$

$$= 23.06 \text{ kcal / mole}.$$

Further, it is sometimes convenient to relate the quantum of energy to a *characteristic* temperature: for example,

$$1\,\text{ev} = \frac{1.602 \times 10^{-12} \text{ (erg/molecule)}}{1.3804 \times 10^{-16} \text{ (erg/molecule/K)}}$$

$$= 11.605 \text{ K}.$$

(9) The *weber* is the derived unit of magnetic flux; it is named after the German physicist W. E. Weber (1891).

(10) The *pascal* is the derived unit of pressure, named after Blaise Pascal who was French mathematician, physicist, and religious philosopher (1623–1662).

(11) The *kelvin* is a degree of temperature on the absolute scale of temperature which was realised by Lord Kelvin with zero at $-273.15°\text{C}$. It was thus named after him; his name before was William Thomson (1824–1907). It is abbreviated by K.

Lord Kelvin was one of the great English physicists of the nineteenth century, and was noted for his work in thermodynamics, introduction of the absolute temperature scale, and involvement in laying the first transatlantic telephone cable. His first ideas regarding the absolute scale of temperature (on which the kelvin or the degree is equivalent to the Celsius degree) was published in 1848, but was vastly improved in 1851.

Another great English physicist, who was counted together with some other remarkable physicists such as J. Clark Maxwell, Lord Rayleigh, Lord Rutherford, is Sir J. J. Thomson (1856–1940); he discovered the electron. These great physicists were in succession the directors of the Cavendish Laboratory at Cambridge University, England, within the period 1871–1937; this laboratory was among the most productive laboratories in the world at that time.

(12) The *gauss* is the electromagnetic unit of magnetic induction. It is named after K. E. Gauss (1777–1855), who was German mathematician, astronomer, and physicist. (The plural of gauss is gauss).

(13) The *hertz* is the derived unit of frequency of electromagnetic waves; it is equal to one cycle per second, and symbolised by Hz. It is named after H. W. Hertz (1857–1894) who was a German physicist; a pioneer of radio communication.

(14) The cm^{-1} is an energy unit equivalent to 1.9861 erg in the cgs system; by the basic equation of quantum theory the passage of energy between matter and radiation occurs in quanta of magnitude ΔE, where $\Delta E = h\nu$. when ΔE is expressed in terms of the erg, and ν as reciprocal seconds, *Planck's constant h* has the value 6.625×10^{-27} erg. sec. in the cgs system. With frequency in wave numbers denoted by $\sigma \left(= \frac{1}{\lambda} \text{ cm}^{-1} \right)$, ΔE is given by

$$\Delta E = h c \sigma,$$

where *c*, the velocity of light, is the product of the wavelength λ and frequency ν.. Thus, substantiating for *h* by 6.625×10^{-27} erg. sec. and for *c* by 2.9979×10^{-10} cm/sec in the last equation, we obtain

$$\Delta E(\text{erg}) = 6.625 \times 10^{-27} (\text{erg.sec}) \times 2.9979$$

$$\times 10^{10} (\text{cm/sec}) \times \sigma(\text{cm}^{-1})$$

According to this relation, 1 cm^{-1} is given by

$$1\,\text{cm}^{-1} = 6.625 \times 2.9979 \times 10^{-17}$$

$$= 1.9861 \times 10^{-16} \text{ erg.}$$

Spectroscopists often express their energy terms in wave numbers, that is, in cm^{-1}, by means of this relation.

2. Conversion of Energy Units

With the help of Table 1, the interconversion of 1 calorie, 1 joule, and 1 litre-atmosphere into one another can be made through conversion factors.

Table 1

Conversion factors

	Calorie	Joule	Litre-atmosphere
1 calorie	1	4.1840	0.041291
1 joule	0.23901	1	0.0098689
1 litre-atmosphere	24.218	101.328	1

Also with the help of another table, Table 2, the energy quantities can be interconverted from one unit into another unit with the help of conversion factors.

Table 2

Interconversion of energy quantities expressed in different units

erg	eV	cm^{-1}	K	Kcal/mole
1	6.242×10^{11}	5.035×10^{15}	7.244×10^{15}	1.439×10^{13}
1.602×10^{-12}	1	8.066×10^{3}	1.1605×10^{4}	2.306×10^{1}
1.986×10^{-16}	1.240×10^{-4}	1	1.439×10^{0}	2.859×10^{-3}
1.380×10^{-16}	8.617×10^{-5}	6.950×10^{-1}	1	1.987×10^{-2}
6.949×10^{-14}	4.338×10^{-2}	3.499×10^{3}	5.034×10^{2}	1

This table should be explanatory. Along any row, all the entries correspond to the *same quantity* of energy expressed in the units given at the top of each column.

3. The Unit of Electric Charge

Coulomb provided direct experimental confirmation of the inverse square law for electric charges suggested by *Priestly* before. He summarised his results in a single equation which describes the force that two small charged spheres A and B at a distance *R* from one another exert on each other

$$F_{\text{electric}} = k_c \frac{Q_A Q_B}{R^2}, \tag{1}$$

where k_c is a constant whose value depends on the units of charge and distance that are being used. This law is known as *Coulomb's law* and the constant k_c *Coulomb's constant.*

We could use Eq. (1) to define a unit of charge. For example, we could arbitrarily let the magnitude of k_c be exactly 1 and define a unit charge so that two unit charges separated by a unit distance exert a unit force on each other. There is a set of units based on this choice. However, in the system of electrical units, we shall find more conversance to use the mks system and to derive the unit of charge from the unit of current, which is the ampere. The unit of charge is called the coulomb, and is defined as shown before as the amount of charge that flows past a point in a wire in one second when the current is equal to one ampere. The ampere is a familiar unit because it is frequently used to measure the current drawn by electrical appliances. By experiment, the constant k_c in Eq. (1) is found to be equal to about nine billion newton. m^2 per $coul^2$, that is,

$$k_c = 9 \times 10^9 \frac{\text{newton . m}^2}{(\text{coulomb})^2}. \qquad (2)$$

This means that two objects, each with a net charge of one coulomb, separated by a distance of one metre, could exert forces on each other of nine billion newtons. This force is roughly the same as a weight of one million tons. We never observe such large forces, because we cannot actually collect that much excess charge in one place, or exert enough force to bring two such charges so close together. The mutual repulsion of like charges is so strong that it is difficult to keep a charge, of more than a thousandth of a coulomb on an object of ordinary size.

Eq. (2) expressing the experimental value of Coulomb's constant as

$$k_c = 9 \times 10^9 \frac{\text{newton . m}^2}{\text{coul}^2},$$

or

$$= 9 \times 10^9 \frac{\text{kg.m}^3}{\text{coul}^2\ \text{.sec}^2},$$

can be written in form

$$k_c = \frac{1}{4\pi \in_\circ} \frac{\text{kg . m}^3}{\text{coul}^2\ \text{. sec}^2}. \qquad (3)$$

This is because the value of $\in_\circ$, the permittivity of vacuum, as found experimentally is

$$\in_\circ = \frac{1}{36\pi} \times 10^{-9} \text{ fasad/m}, \qquad (4)$$

so that the substitution for $\in_\circ$ in Eq. (3) from Eq. (4) gives the magnitude of k_c as

$$k_c = \frac{1}{4\pi} \times 36\pi \times 10^9 = 9 \times 10^9. \qquad (5)$$

4. The Unit of Electronic Charge

Based on the specifications of the fundamental primary unit of charge in the rationalised mks system, which is the coulomb indicated in Table 1(a), it seems instructive to determine the value of the electronic charge *e* in esu.

Since in the electrolysis process electrons are supplied to the cathode and removed from the anode; that is, it is a process of electron transfer between the electrodes and surrounding ions, then the laws of *Faraday* can be regarded as the result of counting off one electron from each univalent atom and two electrons from each divalent ion, etc. Thus, since each gram-atom of a univalent element, that is, one mole contains 6.0238×10^{23} electrons are involved in the passage of 96,494 coulombs. Similarly, for a divalent element $2 \times 96,494$, coulombs are involved in the passage of $2 \times 96,494$ coulombs.

Therefore, the charge of the electron is given by

$$e = \frac{96,494}{6.0238 \times 10^{23}}$$

$$= 1.602 \times 10^{-19} \text{ coulomb.}$$

Since as shown in Table 1(a)

$$1 \text{ coulomb} = 10^{-1} \text{ emu of charge,}$$

and

$$1 \text{ emu of charge} = \text{velocity of light} \times 1 \text{ esu of charge}$$

$$= 2.998 \times 10^{10} \text{ esu of charge,}$$

therefore the electronic charge *e* is given by

$$e = 1.602 \times 10^{-19} \times \frac{1}{10} \times 2.998 \times 10^{10}$$

$$= 4.8028 \times 10^{-10} \text{ esu.}$$

Or, since as shown in Table 1

$$1 \text{ coulomb} = 3 \text{ (or } 2.998) \times 10^{10} \text{ esu of charge,}$$

we directly have

$$e = 1.602 \times 10^{-19} \times 2.998 \times 10^{9}$$

$$= 4.8028 \times 10^{-10} \text{ e.s.u.}$$

It may be added here that an electron charge determined by X-ray and other methods is 4.80251 esu. The value obtained above from the experimental value of the faraday is thus in close agreement with the latter more precise value.

5. Dimensions and Units in mks and cgs Systems

Introduction

There probably never was a student who at one time or another was not confused by the many different systems of units used in various fields of physics, chemistry, and engineering. Such perplexity disappears once the concepts behind dimensions and unit systems are properly understood.

The word *dimension* is used to mean a name given to *any measurable quantity*. Length, mass, time, velocity, and area are dimensions.

The former three quantities, namely length, mass, and time are generally used as the *three fundamental dimensions* for the description of the mechanical world.

The word *unit* is a quantity chosen as a standard in terms of which other quantities can be expressed.

Electricity and *magnetism* represent two new phenomena not contained in the frame work of mechanical concepts. They are linked with each other through *Ampere's circuited law*, *Faraday's induction law,* or summarily through *Maxwell's equations.* Hence, it is logical to add one new *fundamental quantity* to the above three dimensions of the mechanical system.

It was the tendency in earlier times to make the three fundamental dimensions in mechanics suffice by prescribing that either the dielectric constant (electrostatic system) or the permeability (electromagnetic system) be a plain number. Such a supposition *eliminates a fourth dimension.* This is because the dielectric constant and permeability are interlinked through relations, such as the phase velocity, which contains only mechanical dimensions.

However, the result of this artificial reduction is that *fractional exponents* appear in many of the dimensional equations for these fractions have no physical sense. This difficulty can be avoided, and all concepts simplified by referring to *four fundamental quantities.* The *electric charge* as the *fourth dimension* was adopted, and thus the *four fundamental dimensions* are *length, mass, time, and electric charge.*

The units of these four dimensions are *metre, kilogram, second, and coulomb,* respectively.

And accordingly, the *corresponding dimensional system* is known as the *mks system.* The abbreviations of these units are, respectively,

m, kg, s, and C.

The first three fundamental dimensions can be expressed in other units, namely

centimetre, gram, and second,

respectively. Thus the *corresponding dimensional system* is known as the *c gs system.* The abbreviations of these units are, respectively,

m, g, and s.

The fourth fundamental dimension, namely, the electric charge whose unit in the m k s system is the *coulomb*, is expressed in the cgs dimensional system in *electrostatic units* (esu) as well as in *electromagnetic units* (emu) by using a *multiplication factor.* This will be clearer in Table 1.

It is to be noted that the electrostatic and electromagnetic units belong to the cgs dimensional system, which comprises Tables 1(a)–1(f).

In Table 1, all quantities *except* the four quantities having the fundamental dimensions – length, mass, time, and electric charge – are characterised by possessing certain equations known as *defining equations.* These equations, which are also known as *dimensional equations,* are used to derive for any quantity the *appropriate primary units* which, of course, should depend upon the dimensional system selected. Up to this stage we speak of *three*-dimensional systems, namely (a) the *mks system* as exemplified in Table 1, (b) the *absolute flt engineering system,* and (c) the *common fmlt engineering system,* the illustrations for the latter two engineering systems being shown in Tables 2 and 3, respectively.

The primary units resulting in this way can therefore be considered as *rationalised primary units* in the mks system or in the engineering systems.

Furthermore, to shorten the *defining (or dimensional)* equations, we can obtain the so-called *derived units* by making use of the *Ampere's circuited law.* In this law, the derived units for the currents, potential (or electromotive force), and resistance are ampere, volt, and ohm, respectively. Added to these units is the farad unit as a derived unit for the capacitance in the mks and flt and fmlt systems. The derived units are also considered as *rationalised derived units* in these systems; they are named after their *discoverers.*

It is thus evident that the m k s system (as well as other dimensional systems) comprises rationalised *primary* and *derived units.*

Although the esu and emu belong to the cgs system, it has become convenient to speak of esu system, as well as, emu system, but bearing in mind that they originate from the cgs system.

The *unrationalised esu system* postulates that the *permittivity of free space* $\in_o$ is unity, and the *unrationalised e mu system* also postulates that the *permeability of free space* μ_o *is unity.* The unrationalised emu system is known as the *Gaussian system.*

Finally, it is to be pointed out that the recent system of units known as *System International SI* is *practically* the rationalised mks units.

Examples of defining or dimensional equations, primary units and derived units in the m k s dimensional system, multiplication factors for converting the derived units in the mks system into esu and emu in the egs dimensional system, and finally the ratio of esu to e mu in terms of the velocity *c* of light are given for various quantities in the following series of tables designed by von Hippel*; Tables 1(a)–1(f).

* Von Hippel, Massachusetts Institute of Technology, Boston, USA, Laboratory of Insulators, 1969.

Table 1(a)

Quantity	Symbol	Defining equation	Units in the m k s dimensional system		Multiplication factor for converting derived units in the m k s system into e s u and e m u in the c g s dimensional system		Ratio of e s u to e m u
			Primary units	Derived units	e s u	e m u	
Electric charge	Q	It has no defining or dimensional equation being selected as one of the fundamental dimensions in the mechanical world.	coulomb	coulomb	3×10^9	1×10^{-1}	$c(= 3\times10^{10})$
Electro-motive force	e m f	$\int_1^2 E.dl.$ E is the electric field strength.	$\frac{kg\,m/sec^2}{coul}m = \frac{kg\,m^2}{sec^2\,coul}$	volt	$\frac{1}{300}$	1×10^8	$\frac{1}{c}$
Electric field strength.	E	$E = \frac{F}{Q}$ F is the force.	$\frac{kg\ m/sec^2}{coul} = \frac{kg\,m}{sec^2\,coul}$	volt/m	$\frac{1}{3} \times 10^{-4}$	1×10^6	$\frac{1}{c}$

Table 1(b)

Quantity	Symbol	Defining equation	Units in the m k s dimensional system		Multiplication factor for converting derived units in the m k s system into e s u and e m u in the c g s dimensional system		Ratio of e s u to e m u
			Primary units	Derived units	e s u	e m u	
Direct current resistively	ρ	$\rho = \frac{1}{\sigma}$	$\frac{kg\,m^3}{sec\ coul^3}$	ohm m	$\frac{1}{9} \times 10^{-9}$	1×10^{11}	$\frac{1}{c^2}$
Los tangent (dissipation factor)	$\tan\delta$	$\tan\delta = \frac{\epsilon''}{\epsilon'}$					
Quality factor of a dielectric	Q	$Q = \frac{1}{\tan\delta}$					
Electric displacement	D	$D = \int D.n\,dA$ n is a unit vector perpendicular to the plates of the condenser	$coul/m^2$	Farad $volt/m^2$ $\left(\frac{sec^2\,coul^2}{kg\,m^2} \times \frac{kg\,m^2}{sec^2\,coul} \times \frac{1}{m^2}\right)$	$12\pi \times 10^5$ $\left(4\pi \times 9 \times 10^{11} \times \frac{1}{300} \times 10^{-4} = 12\pi \times 10^5\right)$	$4\pi \times 10^{-5}$ $\left(4\pi \times 1 \times 10^{-9} \times 10^8 \times 10^{-4} = 4\pi \times 10^{-5}\right)$	c
Potential (see e m f)	ϕ	$\phi_1 = \int_{\infty}^{1} E.dl$ E is the electric field strength	$\frac{kg\,m^2}{sec^2\ coul}$	volt	$\frac{1}{300}$	1×10^8	$\frac{1}{c}$

Table 1(c)

Quantity	Symbol	Defining equation	Units in the m k s dimensional system		Multiplication factor for converting derived units in the m k s system into e s u and e m u in the c g s dimensional system		Ratio of e s u to e m u
			Primary units	Derived units	e s u	e m u	
Capacitance	C	$C = Q/V$	$\sec^2 \text{coul}^2/\text{kg m}^2$	farad	9×10^{11}	1×10^{-9}	$\frac{1}{c^2}$
Complex dielectric constant	ϵ°	$\epsilon^{\circ} = \epsilon' - i\epsilon''$	$\sec^2 \text{coul}^2/\text{kg m}^3$	farad/m	$36\pi \times 10^9$	$4\pi \times 10^9$	c^2
Direct current conductance	G	$G = 1/R$ $= \frac{I}{V}$	$\frac{\text{coul/sec}}{\text{kg m}^2/\sec^2 \text{coul}}$ $= \sec \text{coul}^2/\text{kg m}^2$	Ohm^{-1} or mho or $(\text{siemens})^{-1}$	9×10^{11}	1×10^{-9}	c^2
Direct current specific conductivity	σ	$\sigma = \frac{J}{F}$ $= \frac{\text{current density}}{\text{electric field strength}}$	$\frac{\text{ampere/m}^2}{\text{volt/m}}$ $= \frac{\frac{\text{coul}}{\sec}\frac{1}{\text{m}^2}}{\text{kg m}/\sec^2 \text{coul}}$ $= \sec \text{coul}^2/\text{kg m}^2/\text{m}$	$\text{ohm}^{-1}, \text{m}^{-1}$ or, mho/m or, $(\text{siemens})^{-1}.\ \text{m}^{-1}$	9×10^{9}	1×10^{11}	c^2

Table 1(d)

Quantity	Symbol	Defining equation	Units in the m k s dimensional system		Multiplication factor for converting derived units in the m k s system into e s u and e m u in the c g s dimensional system		Ratio of e s u to e m u
			Primary units	Derived units	e s u	e m u	
Current	I	$I = \frac{dQ}{dt}$	coul/sec	ampère	3×10^9	1×10^{-1}	c
Current density	J	$J = \frac{dI}{dA}$	coul/sec m^2	ampere/m^2			
Dielectric constant (Permittivity)	$\in$ or $\in_\circ \in_r$	$\in = \frac{D}{E}$ D is the displacement, E is the field strength	$\frac{\text{coul/m}^2}{\frac{\text{kg m}}{\text{sec}^2\text{ coul}}} = \frac{\text{sec}^2\text{ coul}^2}{\text{kg m}^3}$	farad/m	$(4\pi)(9\times10^{11}\times10^{-2}) = 36\pi \times 10^9$	$(4\pi)(1\times10^{-9}\times10^{-2}) = 4\pi \times 10$	c
Dielectric specific conductivity	σ	$\sigma = \in'' \omega$, (where $\sigma = J/E$, so that $\in'' = \frac{J_{loss}}{\omega E}$)	$\frac{\text{sec}^2\text{ coul}^2}{\text{kg m}^3} \times \text{sec}^{-1} = \frac{\text{sec coul}^2}{\text{kg m}^2\text{.m}}$	$\text{ohm}^{-1}\text{ m}^{-1}$ or mho/m or $(\text{siemens})^{-1}.\text{m}^{-1}$	9×10^9	1×10^{-11}	c^2
Direct current resistance	R	$R = \frac{V}{I}$	$\frac{\text{kg m}^2}{\text{sec}^2\text{ coul}} \times \frac{\text{sec}}{\text{coul}} = \frac{\text{kg m}^2}{\text{sec coul}^2}$	ohm	$\frac{1}{9} \times 10^{-11}$	1×10^9	$\frac{1}{c^2}$

N.B. $\in_\circ$ is the permittivity of a vacuum, $\in_r$ is the permittivity relative which is practically measured and denoted by $\in$ dropping the subscript for convenience or brevity, and putting the superscript (′) for its frequency dependence.

Table 1(e)

Quantity	Symbol	Defining equation	Units in the m k s dimensional system		Multiplication factor for converting derived units in the m k s system into e s u and e m u in the c g s dimensional system		Ratio of e s u to e m u
			Primary units	Derived units	e s u	e m u	
Force	F or f	Mass ×acceleration	kgm/sec^2	newton	1×10^5	1×10^{-9}	1
Electrostatic energy	U	$U = Q \times \int_{\infty} E\,dl$	$coul \times \frac{kg\ m^2}{sec^2\ coul} = \frac{kg\ m^2}{sec^2}$	joule	1×10^7	1×10^7	1
Magnetic field strength	H	$\int H\,dl$	coul/m sec	ampere/m	3×10^7 $(= 3 \times 10^9/10^2)$	1×10^{-3}	c
Electric dipole moment	μ	charge×distance	coul m	ampere sec m	3×10^{11}	1×10^1	c
Permeability	μ	$\mu = \frac{B}{H}$ B is the magnetic induction H is the magnetic field strength					

Table 1(f)

Quantity	Symbol	Defining equation	Units in the m k s dimensional system		Multiplication factor for converting derived units in the m k s system into e s u and e m u in the c g s dimensional system		Ratio of e s u to emu
			Primary units	Derived units	e s u	e m u	
Potential difference (see emf)	$\phi_2 - \phi_1$	$\phi_2 - \phi_1 = -\text{emf}$	$\frac{\text{kg m}^2}{\text{sec}^2\text{ coul}}$	volt	$\frac{1}{300}$	1×10^8	$\frac{1}{c}$
Voltage (see emf)	V	$V = -\text{emf}$	$\frac{\text{kg m}^2}{\text{sec}^2\text{ coul}}$	volt	$\frac{1}{300}$	1×10^8	$\frac{1}{c}$
Power	P	$P = \frac{dW}{dt}$	$\frac{\text{kg m}^2}{\text{sec}^3}$	joule/sec or watt	1×10^7	1×10^7	1
Torque	T	$T = F \times d$	$\frac{\text{kg m}}{\text{sec}^2}\text{ m} = \frac{\text{kg m}^2}{\text{sec}^2}$	newton m	1×10^7	1×10^7	1
Inductance	L	$L = \frac{V}{dI/dt}$	$\frac{\frac{\text{kg m}^2}{\text{sec}^2\text{ coul}}}{\frac{\text{coul}}{\text{sec}^2}} = \frac{\text{kg m}^2}{\text{coul}^2}$	henry	$\frac{1}{36\pi}\times10^{-11}$	$\frac{1}{4\pi}\times10^9$	$\frac{1}{c^2}$
Impedance (intrinsic)	Z	$Z = \frac{E}{H}$ E is the electric field strength H is the magnetic strength	$\frac{\frac{\text{kg m}}{\text{sec}^2\text{ coul}}}{\text{coul/m sec}} = \frac{\text{kg m}^2}{\text{sec coul}}$	ohm	$\frac{1}{9}\times10^{-11}$	1×10^9	$\frac{1}{c^2}$

6. Engineering Dimensional Systems

Introduction

Recalling that the word *dimension* is a name given to any measurable quantity such as length, mass, time, velocity, area, etc. We now consider the concept called *primary quantities* for a particular dimensional system. They may be defined as the quantities for which arbitrary scales of measure are set up. On the other hand, we consider the concept of the so-called *secondary quantities* which are defined as those quantities whose dimensions are expressed in terms of the dimensions of the primary quantities. The essential point to grasp is that the quantities for which arbitrary scales of measure are set up can be picked in many ways, and it just happens that the early scientists picked certain sets.

In the dimensional systems customarily employed, the length and time are *primary* quantities, and velocity and area are *secondary* quantities.

The primary scales of measure are expressed in terms of units. For example, foot, inch, and metre are all different units with the common dimension of length. Each of the primary scales is based on a carefully chosen standard. The international standard of length used to be the *distance between two marks on a platinum–iridium bar,* but in 1960, an atomic standard based on the *wavelength of the orange-red line in the spectrum of krypton 86* was adopted by international agreement. The metre is now defined as *1,650,763.73 times this wavelength.* The present standard of time is the *second.* Until 1960 it was defined as $1/86,400$ *of a mean solar day.*

Length and *time* are *primary quantities* in all dimensional systems in common use. In some systems, *mass* is also taken as a *primary quantity. A standard 1 kilogram mass of platinum–iridium is kept at the International Bureau of Weights and Measures in France.* Also, in other dimensional systems, *force* is chosen as a *primary* quantity; the *standard* of force can be taken as the *weight of the platinum–iridium mass at some prescribed point on the earth.* Nature has provided an excellent standard of charge in the electron so that it can be used as a *basis of charge scale.*

The dimensions of *secondary quantities* are fixed by the equations which relate them to the primary quantities. For example, in the mks system, the length (metre), mass (kilogram), and time (second) are taken as the primary quantities. Force is a secondary quantity in the system. Thus the dimensions and magnitude of force in this system results from the selection of *unity* for the value of the constant K_N in *Newton's second law*, which then becomes for a particle

$$\text{force} = \text{mass} \times \text{acceleration}.$$

The *acceleration,* which is a secondary quantity in this system, has the dimensions of $\text{length}/(\text{time})^2$. Accordingly, the force has the dimensions of $\text{mass} \times \text{length}/(\text{time})^2$, and the units $\text{kg.m}/\text{sec}^2$ in the mks system.

In engineering, the first dimensional system is the *absolute engineering flt system* as has been noted above in which force, length, and time are the *primary quantities*. Here the *mass* is considered as a *secondary quantity* and accordingly its dimensions and units result from arbitrarily setting Newton's constant $K_N = 1$. Newton's law then is again

$$\text{force} = \text{mass} \times \text{acceleration},$$

and hence in this absolute engineering flt system, the mass has the dimensions of $\text{force}/(\text{length})/(\text{time})^2$ and the units of $\text{lbf.sec}^2/\text{ft}$; it is a secondary unit. It may be noted that the symbol 'lb' is the abbreviation from the *Latin libra taken as the unit of weight and known as the pound.* Also that the unit of force in this system as well as in the second dimensional engineering system is denoted by the abbreviation lb f, that is, pound force, where the unit of mass in the two engineering systems is lb m, that is, pound mass.

It is important to realise that the constants in physical laws are often picked to have the value equal to one in certain dimensional systems or given the experimentally determined value in other systems.

In the second dimensional engineering system known as the common f m lt engineering system, as noted above, both force and mass plus length and time are chosen as the *primary quantities.* The scales of force and mass are chosen such that the weight of an object at sea level is numerically equal to its mass. The force is assigned the units of lb f (pound force) and the mass is assigned the units of lb m (pound mass). In this fmlt engineering unit system, Newton's constant k_N is not equal to 1; it is considered as having the experimentally determined value of

$$k_N = \frac{1}{32.1739}\ \text{lbf.sec}^2/\text{lbm.ft}.$$

This is because expressing Newton's equation

$$\text{force} = k_N \times \text{mass} \times \text{acceleration}$$

in units, we have

$$\text{lbf} = k_N\ \text{lbm} \times \text{ft}/\text{sec}^2$$

so that k_N has the units

$$\frac{\text{lbf.sec}^2}{\text{lbm.ft}}.$$

It is customary to denote the reciprocal of K_N by g_c in the fmlt engineering system, that is,

$$g_c = 32.1739\ \text{lbm.ft}/\text{lbf.sec}^2.$$

The constant g_c is a *universal constant* which emerges in Newton's law as a result of the arbitrary choices for standard of measure. It has the value of unity and is dimensionless in the mks, cgs, and absolute engineering flt systems.

A tabulation of the two engineering flt and fmlt systems is presented in Tables (2) and (3), respectively.

Table 2

	mks system	cgs system	*Absolute* flt *engineering* system	common fmlt *engineering* system+
Primary quantities				
Length	meter,	centimeter, m	foot, ft	foot, ft
Mass	kilogram, kg	gram, g	. . .	pound mass, lbm
Time	second, sec	second, sec	. . .	second, sec
Force	. . .	. . .	pound force, lbf	pound force, lbf
Newton's second law, $F = \frac{m\,dV}{g_c\,dt}$	$g_c \equiv 1$ (selected)	$g_c \equiv 1$ (selected)	$g_c \equiv 1$ (selected)	$g_c = 32.17$ ft-lbm/lbf-sec^2 (experimental)
Secondary quantities				
Force	kg-m/sec^2	g-cm/sec^2	. . .	. .
Mass	. . .	. . .	lbf-sec^2/ft	. . .
Energy, $dW = F\,dX$	kg-m^2/sec^2	g-cm^2/sec^2	ft-lbf	ft-lbf
Power, $\dot{W}$	kg-m^2/sec^3	g-cm^2/sec^3	ft-lbf/sec	ft-lbf/sec
Aliases*				
Force	1 newton $\equiv$ 1 kg-m/sec^2	1 dyne $\equiv$ 1 g-cm/sec^2	. . .	. . .
Mass	. . .	. . .	1 slug $\equiv$ 1 lbf-sec^2/ft	. . .
Energy	1 joule $\equiv$ 1 kg-m^2/sec^2 = 1 newton-m	1 erg $\equiv$ 1 g-cm^2/sec^2 = 1 dyne-cm	. . .	. . .
Power	1 watt $\equiv$ 1 kg-m^2/sec^3 = 1 joule/sec	. . .	. . .	. . .

+A body having a weight of 1 lbf on the surface of the earth will have a mass of approximately 1 lbm.

*Alias $\equiv$ otherwise or plural aliases

Table 3

	mks system	Absolute cgs *electrostatic (esu)* system	Absolute cgs *magnetostatic (emu)* system
Primary quantities			
Length	meter, m	Centimeter, cm	Centimeter, cm
Mass	kilogram, kg	gram, g	gram, g
Time	second, sec	second, sec	second, sec
Charge	Coulomb, coul	. . .	. . .
Coulomb's law,	$k_C = \frac{1}{4\pi\epsilon_0}$	$k_C \equiv 1$ (selected)	$k_C = 0.8992 \times 10^{17}$ cm^2-sec^{-2} (experimental)
$\mathsf{F}_{12} = k_C \frac{Q_1Q_2}{R_{12}^2}$	$\epsilon_0 = 8.854 \times 10^{-12}$ coul^2-sec^2-kg^{-1}-m^{-3} (experimental)		
Biot-Savart law, $\mathsf{F}_{12} = k_B \frac{Q_1Q_2\mathbf{V}_1 \times (\mathbf{V}_2 \times \mathbf{R})}{\lvert\mathbf{R}\rvert^3}$	$k_B = \frac{\mu_0}{4\pi}$	$k_B = 1.112 \times 10^{-17}$ sec^2-cm^{-2} (experimental)	$k_B \equiv 1$ (selected)
	$\mu_0 = 1.256 \times 10^{-6}$ kg-m-coul^{-2} (experimental)		
Secondary quantities			
Charge	. . .	g$^{1/2}$-cm$^{3/2}$-sec^{-1}	g$^{1/2}$-cm$^{1/2}$
Current density, $J = \dot{Q}/A$	coul-m^{-2} sec^{-1}	g$^{1/2}$-cm$^{-1/2}$-sec^{-2}	g$^{1/2}$-cm$^{-3/2}$-sec^{-1}
Current, $I = \int \mathbf{J} \cdot d\mathbf{A}$	coul-sec^{-1}	g$^{1/2}$-cm$^{3/2}$-sec^{-2}	g$^{1/2}$-cm$^{1/2}$-sec^{-1}
Electric and magnetic fields, $\mathbf{F} = Q(\mathbf{E} + \mathbf{V} \times \mathbf{B})$			
Electric field strength **E**	kg-m-coul^{-1}-sec^{-2}	g$^{1/2}$-cm$^{-1/2}$-sec^{-1}	g$^{1/2}$-cm$^{-1/2}$-sec^{-2}
Magnetic induction **B**	kg-sec^{-1}-coul^{-1}	g$^{1/2}$-cm$^{-3/2}$	g$^{1/2}$-cm$^{-1/2}$-sec^{-1}

Table 3 (continued)

	mks system	*Absolute cgs–esu* system	*Absolute cgs–emu* system
Electrical potential $\mathcal{E}$, $d\mathcal{E} = \mathbf{E} \cdot d\mathbf{L}$	kg-m^2-coul^{-1}-sec^{-2}	g$^{1/2}$-cm$^{1/2}$-sec^{-1}	g$^{1/2}$-cm$^{3/2}$-sec^{-2}
Electric displacement, polarization:	$\mathbf{D} = \epsilon_0\mathbf{E} + \mathbf{P}$	$\mathbf{D} = \mathbf{E} + 4\pi\mathbf{P}$	
Displacement $\mathbf{D}$	coul-m^{-2}	g$^{1/2}$-cm$^{-1/2}$-sec^{-1}	
Polarization $\mathbf{P}$	coul-m^{-2}	g$^{1/2}$-cm$^{-1/2}$-sec^{-1}	
Magnetic field strength, magnetization:	$\mathbf{B} = \mu_0(\mathbf{H} + \mathbf{M})$		$\mathbf{B} = \mathbf{H} + 4\pi\mathbf{M}$
Magnetic field strength $\mathbf{H}$	coul-m^{-1}-sec^{-1}		g$^{1/2}$-cm$^{-1/2}$-sec^{-1}
Magnetization $\mathbf{M}$	coul-m^{-1}-sec^{-1}		g$^{1/2}$-cm$^{-1/2}$-sec^{-1}
Electric flux, $\Phi_E = \int \mathbf{E} \cdot d\mathbf{A}$	kg-m^3-coul^{-1}-sec^{-2}	g$^{1/2}$-cm$^{3/2}$-sec^{-1}	
Magnetic flux, $\Phi_B = \int \mathbf{B} \cdot d\mathbf{A}$	kg-m^2-coul^{-1}-sec^{-1}		g$^{1/2}$-cm$^{3/2}$-sec^{-1}
Electrical conductivity, J/E	coul^2-sec-kg^{-1}-m^{-3}	sec^{-1}	sec-cm^{-2}
Resistance, $R = \Delta\mathcal{E}/I$	kg-m^2-coul^{-2}-sec^{-1}	sec-cm^{-1}	cm-sec^{-1}
Capacitance, $C = I/(d\mathcal{E}/dt)$	coul^2-sec^2-kg^{-1}-m^{-2}	cm	sec^2-cm^{-1}
Inductance, $L = \Delta\mathcal{E}/(dI/dt)$	kg-m^2-coul^{-2}	sec^2-cm^{-1}	cm.
Aliases			
Charge		1 statcoul$^+$ ≡ 1 g$^{1/2}$-cm$^{3/2}$-sec^{-1}	1 abcoul* ≡ 1 g$^{1/2}$-cm$^{1/2}$
Current	1 amp ≡ 1 coul-sec^{-1}	1 statamp ≡ 1 g$^{1/2}$-cm$^{3/2}$-sec^{-2}	1 abamp ≡ 1 g$^{1/2}$-cm$^{1/2}$-sec^{-1}
Magnetic induction	1 tesla ≡ 1 kg-sec^{-1}-coul^{-1}		1 gauss ≡ 1 g$^{1/2}$-cm$^{-1/2}$-sec^{-1}
Electric potential	1 volt ≡ 1 kg-m^2-coul^{-1}-sec^{-2}	1 statvolt ≡ 1 g$^{1/2}$-cm$^{1/2}$-sec^{-1}	1 abvolt ≡ 1 g$^{1/2}$-cm$^{3/2}$-sec^{-2}
Magnetic flux	1 weber ≡ 1 kg-m^2-coul^{-1}-sec^{-1}		
Resistance	1 ohm ≡ 1 kg-m^2-coul^{-2}-sec^{-1}	1 statohm ≡ 1 sec-cm^{-1}	1 abohm ≡ 1 cm-sec^{-1}
Capacitance	1 farad ≡ 1 coul^2-sec^2-kg^{-1}-m^{-2}	1 statfarad ≡ 1 cm	1 abfarad ≡ 1 sec^2-cm^{-1}
Inductance	1 henry ≡ 1 kg-m^2-coul^{-2}	1 stathenry ≡ 1 sec^2-cm^{-1}	1 abhenry ≡ 1 cm

+ statcoul: $\text{force} = \dfrac{(\text{charge})^2}{(\text{distance})^2}$, **therefore** $(\text{charge})^2 = \text{force} \times (\text{distcence})^2 = \text{g} \times \dfrac{\text{cm}}{\text{sec}^2}\ \text{cm}^2$, **or** $\text{charge} = \text{g}^{1/2}\ \text{cm}^{3/2}\ \text{sec}^{-1}$.

This is the alias of charge denoted by statcoul in the absolute cgs electrostatic (esu) system.

*** abcoul : statcoul / c units** $= \dfrac{\text{g}^{1/2}\ \text{cm}^{3/2}\ \text{sec}^{-1}}{\text{cm}\ \text{sec}^{-1}} = \text{g}^{1/2}\ \text{cm}^{1/2}$.

7. Fundamental Physical Constants

Table 4

Symbol	Name	Value
c	Velocity of light	2.9979×10^{10} cm sec^{-1}
e	Electronic charge*	1.6021×10^{-19} coul
Na	Avogadro's number	6.0225×10^{23} molecules
h	Planck's constant	6.6256×10^{-27} erg. s.
F	Faraday constant	96,487 coul eq^{-1}
R	Gas constant	82.056 c. c. atm mole^{-1} deg^{-1} 1.9872 cal mole^{-1} deg^{-1} 8.3143 joule mole^{-1} deg^{-1}
k	Boltzmann's constant	1.3805×10^{-16} erg deg^{-1}
m_e	Mass of electron	9.107×10^{-31} kg
m_p	Mass of proton	1.6725×10^{-27} kg
μ_B	Bohr magneton†	9.27×10^{-24} joule weber^{-1}
μ_n	Nuclear magneton‡	5.0504×10^{-3} joule gauss^{-1}
$\epsilon_\circ$	Permittivity of vacuum	$\frac{1}{36\pi} \times 10^{-9}$ farad . m^{-1} $= 8.854 \times 10^{-12}$ farad m^{-1}
$\mu_\circ$	Permeability of vacuum	$4\pi \times 10^{-7}$ henry m^{-1}

* The electronic charge e in esu is 4.8029×10^{-10}.

† The Bohr magneton μ_B is the group of the constants $\frac{e}{2 m_e c} \frac{h}{2\pi}$.

‡ The nuclear mugneton μ_n is the group of the constants $\frac{e}{2 m_p c} \frac{h}{2\pi}$.

$\frac{1}{N_\circ}$	Mass of unit of atomic weight	1.6603×10^{-27} kg
1 eV	Electron volt	1.602×10^{-19} joule

$$k_c \text{, Coulomb constant} \quad = \frac{1}{4\pi\epsilon_\circ} = 8.987 \times 10^9 \ \frac{kg.m^3}{coul^2.sec^2}$$

$$= 9 \times 10^9 \ \frac{newton.m^2}{coul^2}.$$

$$k_B \text{, Biot–Savart constant} \quad = \frac{\mu_\circ}{4\pi} = 1.0000 \times 10^{-7} \ \frac{kg.m}{coul^2}.$$

$$k_N \text{, Newton constant} \quad = \frac{1}{37.174} \ \frac{lbf.sec^2}{ft.lbm}.$$

$$g_c \text{, Reciprocal of Newton constant} \quad = 32.174 \ \frac{ft.lbm}{lbf.sec^2}.$$

It is to be pointed out that the substitution for $\epsilon_\circ$, the permittivity of vacuum, in the expression $\frac{1}{4\pi\epsilon_\circ}$ for the coulomb constant, k_c, by its value

$$\epsilon_\circ = \frac{1}{36\pi} \times 10^{-9} \ \text{farad/m},$$

gives

$$\frac{1}{4\pi\epsilon_\circ} = \frac{1}{4\pi} \times 36\pi \times 10^9 = 9 \times 10^9 \ \text{m/farad},$$

and since the primary units of the farad as shown above in Table 1(c) are $sec^2\,coul^2/kg\,m^2$, we have

$$k_c = \frac{1}{4\pi\epsilon_\circ} = 9 \times 10^9 \ \frac{kg.m^2.m}{sec^2\,coul^2} = 9 \times 10^9 \ \frac{newton.m^2}{coul^2}.$$

8. Selected Dimensional Equivalents

Table 5

Length	1 m = 3.280 ft = 39.37 in. 1 cm ≡ 10^{-2} m = 0.394 in. = 0.0328 ft 1 mm ≡ 10^{-3} m 1 micron (μ) ≡ 10^{-6} m 1 angstrom (Å) ≡ 10^{-10} m
Time	1 hr ≡ 3600 sec = 60 min 1 millisec ≡ 10^{-3} sec 1 microsec (μsec) ≡ 10^{-6} sec 1 nanosec (nsec) = 10^{-9} sec
Mass	1 kg ≡ 1000 g = 2.2046 lbm = 6.8521 × 10^{-2} slugs 1 slug ≡ 1 lbf·sec²/ft = 32.174 lbm
Force	1 newton ≡ 1 kg·m/sec² = 0.2248 lbf 1 dyne ≡ 1 g·cm/sec² 1 lbf = 4.448 × 10^5 dynes = 4.448 newtons
Energy	1 joule ≡ 1 kg·m²/sec² 1 Btu ≡ 778.16 ft·lbf = 1.055 × 10^{10} ergs = 252 cal 1 cal ≡ 4.186 joules 1 kcal ≡ 4186 joules = 1000 cal 1 erg ≡ 1 g-cm²/sec² 1 ev ≡ 1.602 × 10^{-19} joules
Power	1 watt ≡ 1 kg·m²/sec³ = 1 joule/sec 1 hp ≡ 550 ft·lbf/sec 1 hp = 2545 Btu/hr = 746 watts 1 kw ≡ 1000 watts = 3413 Btu/hr
Pressure	1 atm ≡ 14.696 lbf/in.² = 760 torr 1 mmHg = 0.01934 lbf/in.² ≡ 1 torr 1 dyne/cm² = 145.04 × 10^{-7} lbf/in.² 1 bar = 14.504 lbf/in.² ≡ 10^6 dynes/cm² 1 micron (μ) ≡ 10^{-6} mHg = 10^{-3} mmHg
Volume	1 gal ≡ 0.13368 ft³ 1 liter ≡ 1000.028 cm³
Temperature	1 K = 1 C° = 1.8 F° = 1.8 R° 0°C corresponds to 32°F, 273.16 K, and 491.69°R
Magnetic quantities	1 gauss ≡ 1 $g^{1/2}/cm^{1/2}$ sec 1 gauss = 10^3 coul/m-sec for M⁺ 1 gauss = $(1/4\pi) \times 10^3$ coul/m-sec for H• 1 gauss = 10^{-4} tesla for B* 1 tesla ≡ 1 kg/coul sec

⁺ **M = magnetization, magnetic dipole moment per unit volume.**

• **H = magnetic field strength;**

* **B = magnetic induction.**

9. Physical-Chemical Constants

Table 6

The following tables give the recommended values of the fundamental constants for physical chemistry as of July 1, 1951.* They are based upon the reanalysis and re-evaluation of experimental values by DuMond and Cohen.†

Values of the Basic Constants

Velocity of light	c	2.997902×10^{10} cm sec^{-1}
Planck constant	h	6.6238×10^{-27} erg sec molecule^{-1}
Avogadro constant	N	6.0238×10^{23} molecules mole^{-1}
Faraday constant	F	96,493 coulombs equivalent^{-1} 23,062 cal (volt equiv)$^{-1}$
Absolute temperature of the "ice" point, 0°C	$T_{0°C}$	273.16 K.
Pressure-volume product for 1 mole of a gas at 0° and zero pressure	$(pV)_{T_{0°}}^{p=0}$	22,414.6 cm^3 atm mole^{-1} 22.4140 liter atm mole^{-1} 2271.16 joules mole^{-1}

Values of the Derived Constants

Electronic charge	$e = \frac{F}{N}$	1.60186×10^{-19} coulomb 1.60186×10^{-20} emu 4.8022×10^{-10} esu
Gas constant	$R = \frac{(pV)_{T_{0°}}^{p=0}}{T_{0°}}$	1.9872 cal deg^{-1} mole^{-1} 82.057 cm^3 atm deg^{-1} mole^{-1} 0.082054 l atm deg^{-1} mole^{-1} 8.3144 joules deg^{-1} mole^{-1}
Boltzmann constant	$k = R/N$	1.38026×10^{-16} erg deg^{-1} molecule^{-1}

Values of the Defined Constants

Standard gravity	980.665 cm sec^{-2}
Standard atmosphere	1,013,250 dynes cm^{-2}
Standard millimeter of mercury pressure	$\frac{1}{760}$ atm
Calorie (thermochemical)	4.1840 joules 4.18331 international joules 41.2929 cm^3 atm 0.0412917 l atm

Values of Certain Auxiliary Relations

1 second (mean solar) = 1.00273791 sidereal seconds
1 joule = 0.999835 international joule
1 ohm = 0.999505 international ohm
1 ampere = 1.000165 international amperes
1 volt = 0.999670 international volt
1 coulomb = 1.000165 international coulombs
1 watt = 0.999835 international watt
1 liter = 1,000.028 cm^3

* Rossini, Gucker, Johnston, Rauling and Vinal, J. A.m. Chem. Sec., **74**, 2699 (1952).

† Du Mond and Cohen, phys. Rev., **82**, 555 (1951).

We are now at a stage to understand how the table of *selected dimensional equivalents, namely* Table 5, shown above, has been achieved.

Dimensional equivalents are easily obtained by considering a given magnitude of any quantity one may select. Examples can make clear the procedure that should be followed. Thus two examples of this procedure follow.

(1) We find the dimensional equivalent of 1 newton of force in the fmlt engineering unit system.

To determine this dimensional equivalent, consider a body having a mass of 1 kg being accelerated by a force at the rate of $1\,\mathrm{m/sec^2}$. The force acting on the body in the mks system is given by

$$\text{force} = \text{mass} \times \text{acceleration}$$

$$= 1\,\mathrm{kg\,.\,m/sec^2}\,.$$

The mass of the body in the fmlt engineering system is 2.2046 lb m, and its acceleration is

$$1\,\mathrm{m} \times \left(3.280\ \mathrm{ft/m/sec^2}\right) = 3.280\ \mathrm{ft/sec^2}\,.$$

The force in this engineering system in which k_N is not set to zero is given by

$$\text{force} = \frac{1}{g_c} \times \text{mass} \times \text{acceleration}\,,$$

where

$$g_c = 32.1739\ \mathrm{lb\,m\,ft/lbf.sec^2}\,.$$

Therefore, the force which is 1 newton in the mks system is *equivalent* in the fmlt engineering system to

$$\frac{2.2046\ \mathrm{lbm} \times 3.280\ \mathrm{ft/sec^2}}{32.179\ \mathrm{lbm\,ft/lbf.sec^2}}$$

$$= 0.2248\ \mathrm{lbf}\,.$$

Hence,

$$1\,\text{newton} = 1\,\mathrm{kg.m/sec^2} = 0.2248\ \mathrm{lbf}\,.$$

Or, conversely the dimensional equivalent of 1 lb f in the mks system is

$$1\,\mathrm{lbf} = 4.448 \times 10^5\ \text{dynes} = 4.448\ \text{newtons}.$$

(2) As another example, we determine the dimensional equivalents of the magnetic induction **B** and the magnetisation **M** in the mks system, and in the absolute cgs engineering magnetostatic (emu) system . This calculation is complicated by the difference in definition.

We have

$$\mathbf{B} = \mathbf{H} + 4\pi\,\mathbf{M}$$

in the magnetostatic cgs system, **H** being the magnetic field strength. In the mks system, **B** is defined by

$$\mathbf{B} = \mu_\circ \, (\mathbf{H} + \mathbf{M}),$$

where $\mu_\circ$ is the permeability of vacuum.

Let us consider two charges, each of 1 C, separated by a distance of 1 m, moving in opposite directions perpendicular to the line between them at $1\,\text{m/sec}$ as shown in the figure below.

Experimentally it is found, as shown before, that the repulsive force between two *stationary* point charges Q_1 and Q_2 is proportional to the product of their charges and inversely proportional to the square of the distance *R* between

$$F = k_c \frac{Q_1 Q_2}{R^2}.$$

This is *Coulomb's law*, the constant k_c being dependent on the arbitrary choices of unit forces and charges.

Moving charges exert additional forces upon one another. If the charges are not moving too rapidly, the additional forces are given by the *Biot–Savart law.** The constant of proportionality k_B in this expression similarly depends upon the arbitrary choices of forces and charges scales; it is given in the mks system by

$$k_B = \frac{\mu_\circ}{4\pi} = 1.0000 \times 10^{-7}\ \text{kg.m/coul}^2,$$

where $\mu_\circ$ the permeability of free space or vacuum is equal to $1.256 \times 10^{-6}\ \text{kg.m/coul}^2$ (experimentally).

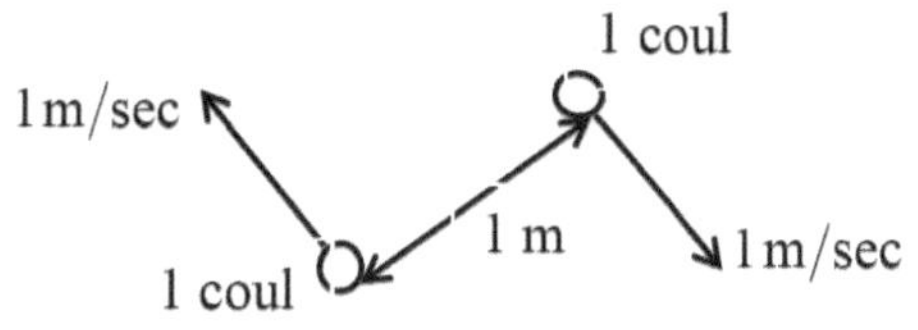

Fig. 1.

Accordingly, using the Biot–Savart law in our problem the force **F** in the *mks system* is given by

$$F = \frac{1.256 \times 10^{-6}}{4\pi}\left(\text{kg.m/coul}^2\right) \times 1\,\text{coul}^2 \times 1\,\text{m}^2/\text{sec}^2 \times \frac{1\,\text{m}}{1\,\text{m}^3}$$

$$= \frac{1.256 \times 10^{-6}}{4\pi}\ \text{kg.m/sec}^2 = 10^{-7}\ \text{kg.m/sec}^2. \qquad (1)$$

---

* See Table 3 above for the Biot–Savart law, for the expression of $\Delta S_2 = 100 \times 1 \int_{273}^{323} \frac{dT}{T} = 100 \ln \frac{323}{273}$, and for the value of $\Delta S_{total} = 100\left(\ln \frac{323}{373} + \ln \frac{323}{273}\right)$.

In the absolute cgs magnetostatic (emu) unit system, the charge, distance, and speed have the following magnitudes:

$$Q = 0.1\,g^{1/2}.cm^{1/2}\ ,\ R = 100\ cm\ ,\ V = 100 cm/sec,$$

respectively, so that the force in this system according to Biot–Savart law, where $k_B = 1$ (as shown in Table 2) is given by

$$F = 0.1 g.cm \times 100\,cm^2/sec^2 \times \frac{100\,cm}{100^3\,cm^3} = 0.01 g.cm/sec^2. \qquad (2)$$

The magnetic induction **B** in *either system* is expressed as

$$\mathbf{B} = \frac{\mathbf{F}}{QV}.$$

Thus in the rationalised mks system, **B** is given by

$$B = \frac{10^{-7}\,kg.m/sec^2}{1\,coul \times 1\,m/sec}$$

$$= 10^{-7}\,kg/coul.sec. \qquad (3)$$

It is evident from Eqs. (1) and (2) that using the dimensional equivalents, the calculation of the force **F** in the mks system or in the cgs magnetostatic (emu) system has given the same result. That is, **F** in the formers system is equal to $10^{-7}\ kg.m/sec^2$ and in the latter system is 0.01 $g.cm/sec^2$ which on using the dimensional equivalents is equal to $10^{-2}/10^3 \times 10^{-2} = 10^{-7}\ kg.cm/sec^2$. The same can be said with respect to the magnetic induction **B** whose value in the mks system as calculated above is

$$10^{-7}\,kg/coul.sec, \qquad (3)$$

and its value in the cgs magnetostatic (emu) system is

$$B = \frac{F}{QV} = \frac{0.01\,g.cm/sec^2}{0.1\,g^{1/2}\,cm^{1/2} \times 100\,cm/sec} \quad \left(for\,Q = 0.1 g^{1/2}.cm^{1/2}\right)$$

$$= 10^{-3}\,g^{1/2}\,cm^{1/2}\,sec^{-1}. \qquad (4)$$

The former expression of **B**, which is given by Eq. (3), can be simply re-expressed as

$$\mathbf{B} = 10^{-7} \times 10^3\,g/0.1 g^{1/2}\,cm^{1/2}.sec$$

$$= 10^{-3}\,g^{1/2}\,cm^{-1/2}\,sec^{-1}.$$

It is thus dimensionally equivalent to that obtained in the cgs magnetostatic (emu) system given by Eq. (4), where 1 coulomb in the mks system $= \frac{1}{10}$ emu of charge, so the Q in Eq.

(4) is substituted for by $0.1\,g^{1/2}\,cm^{1/2}$ in terms of the aliases shown in Table 3 above. Again, in terms of other aliases in the same table, the *tesla*, which is the unit of magnetic induction **B**, is given in the mks system by

$$1\,\text{tesla} = 1\,\text{kg}\,.\,\text{coul}^{-1}\ \text{sec}^{-1}. \tag{5}$$

On the other hand, *the gauss which is also the unit of **B** is given in* the absolute cgs magnetostatic (emu) system by

$$1\,\text{gauss} = 1\,g^{1/2}\ \text{cm}^{-1/2}\,\text{sec}^{-1}\ , \tag{6}$$

so that the tesla is related to the gauss by the following relation

$$1\,\text{tesla} = 10^3\,g \times 10 \times g^{-1/2}\,\text{cm}^{-1/2}\ \text{sec}^{-1}$$

$$= 10^4\,g^{1/2}\,\text{cm}^{-1/2}\,\text{sec}^{-1}. \tag{7}$$

Hence from Eqs. (6) and (7), we obtain

$$1\,\text{tesla} = 10^4\,\text{gauss} \qquad \text{for } \mathbf{B}.$$

It is to be pointed out that the correspondence between tesla and gauss for the magnetic induction **B**, which by definition is given as shown above by

$$\mathbf{B} = \mathbf{H} + 4\pi\ \mathbf{M},$$

and

$$\mathbf{B} = \mu_\circ \mathbf{H} + \mathbf{M}\ ,$$

in the cgs magnetostatic (emu) system and in the mks system, respectively, does not apply to **H** and **M.**

In order to find the dimensional equivalents for **H** as we have done for **B,** we consider a field with magnetic induction **B** = 1 tesla and **M** = 0. Then in the rationalised practical mks system, **H** is given by

$$\mathbf{H} = \frac{\mathbf{B}}{\mu_\circ} \tag{8}$$

$$= \frac{1\,\text{kg/coul.sec}}{1.256 \times 10^{-6}\ \text{kg.m/coul}^2}.$$

$$\mathbf{H} = \frac{1}{1.256 \times 10^{-6}}\ \text{coul/m. sec}. \tag{9}$$

In the absolute cgs magnetostatic (emu) system we have for

$$\mathbf{M} = 0$$

$$\mathbf{B} = \mathbf{H} = 10^4\ \text{gauss}\,. \tag{10}$$

Equating the two values for **H** given by Eqs. (9) and (10), we find

$$10^4 \text{ gauss} = \frac{1}{1.256 \times 10^{-6}} \text{ coul/m.sec},$$

or,

$$1 \text{ gauss} = \frac{1}{1.256 \times 10^{-2}} \text{ coul/m.sec.} \qquad (11)$$

Since numerically $1.256 \times 10^{-1/2}$ is equal to $4\pi \times 10^{-3}$, Eq. (11) can be put in the form

$$1 \text{ gauss} = (1/4\pi) \times 10^3 \text{ coul/m.sec for } \mathbf{H},$$

as shown in Table 5 above.

Now to obtain the dimensional equivalents for **M**, we imagine a field of magnetic induction **B** of 1 tesla with **H** = 0. Then, in the rationalised practical mks unit system, where

$$\mathbf{B} = \mu_\circ (\mathbf{H} + \mathbf{M}),$$

we have for $H = 0$,

$$\mathbf{M} = \frac{\mathbf{B}}{\mu_\circ} = \frac{1 \text{ kg/coul.sec}}{1.256 \times 10^{-6} \text{ kg . m/coul}^2}.$$

Therefore,

$$\mathbf{M} = \frac{1}{1.256 \times 10^{-6}} \text{ coul/m.sec.} \qquad (12)$$

In the absolute cgs magnetostatic (emu) unit system, we have for $H = 0$ using the corresponding equation

$$\mathbf{B} = \mathbf{H} + 4\pi.\mathbf{M},$$

$$\mathbf{M} = \frac{\mathbf{B}}{4\pi}.$$

Accordingly, when $\mathbf{B} = 1 \text{ tesla} = 10^4 \text{ gauss}$, we have

$$\mathbf{M} = \frac{10^4}{4\pi} \text{ gauss.} \qquad (13)$$

Equating the two values for **M** from Eqs. (12) and (13), we have

$$\frac{10^4}{4\pi} \text{ gauss} = \frac{1}{1.256 \times 10^{-6}} \text{ coul/m.sec}$$

or,

$$1 \text{ gauss of } \mathbf{M} = 10^3 \text{ coul/m.sec.} \qquad (14)$$

Since 1 tesla of $\mathbf{B} = 10^4$ gauss, as has been obtained above, the dimensional equivalence for **B** might be put in the form

$$1\,\text{gauss of } \mathbf{B} = 10^{-4}\ \text{kg/coul. sec}, \qquad (15)$$

1 tesla being equal to $1\,\text{kg/coul.sec}$ as shown by Eq. (5) above.

This example illustrates the importance of keeping straight just what the dimensional equivalent is for. One gauss is not always equivalent to the same number of coul/m.sec as implied by Eq. (14); sometimes it is equivalent to a number of kg/coul . sec as shown by Eq. (15).

Converting units from one system to another is as confusing task, but it can be made simple by noting that multiplication of anything by unity leaves it unchanged. The number '1' can be written in many useful ways, two of which are

$$1 = \frac{12\,\text{in.}}{1\,\text{ft.}}$$

$$1 = \frac{778.16\ \text{ft.lbf}}{1\,\text{Btu}}.$$

Then, to express an energy density of $50\,\text{Btu/in}^3$ in terms of of ft.lbf/ft^2,

$$\frac{E}{V} = 50\,\text{Btu/in}^3 \times (778.16\,\text{ft . lbf/Btu}) \times (12\,\text{in/ft})^3$$

$$= 6.72 \times 10^7\ \text{ft. lbf/ft}^3.$$

It is to be noted that dimensions provide an easy check on equations and are an essential part of the answer to any engineering problem as well as to any other dimensional problem.

At the end it seems useful to quote from the table of the 'selected dimensional equivalents', which is Table 5 given above, the following equivalents for the magnetic properties:

$$1\,\text{gauss} = 1\,\text{g}^{1/2}/\text{cm}^{1/2}\,\text{.sec.}$$

$$1\,\text{tesla} = 1\,\text{kg/coul.sec.}$$

$$1\,\text{gauss} = 10^3\ \text{coul/m.sec for } \mathbf{M}.$$

$$1\,\text{gauss} = (1/4\pi) \times 10^3\ \text{coul/m.sec} \quad \text{for } \mathbf{H}.$$

$$1\,\text{gauss} = 10^{-4}\ \text{tesla for } \mathbf{B}.$$

APPENDIX C

Table C – 1

CHANGE OF GIBBS FUNCTION AND ENTHALPY DURING FORMATION AS A FUNCTION OF TEMPERATURE AT STANDARD PRESSURE ($P = 1$ ATM)*†
(Expressed in kcal/gm-mole)

Substance / T, K	CH_4		CO		CO_2		H		OH	
	ΔH_f	ΔG_f	ΔH_f	ΔG_f	ΔH_f	ΔG_f	ΔH_f	ΔG_f	ΔH_f	ΔG_f
0	−15.991	−15.991	−27.200	−27.200	−93.965	−93.965	51.632	51.632	9.273	9.273
100	−16.728	−15.400	−26.890	−28.745	−93.925	−94.080	51.750	50.773	9.202	8.906
200	−17.216	−13.909	−26.601	−30.719	−94.019	−94.190	51.945	49.715	9.296	8.567
298	−17.895	−12.145	−26.417	−32.783	−94.054	−94.265	52.102	48.587	9.330	8.194
300	−17.909	−12.110	−26.414	−32.823	−94.055	−94.267	52.104	48.565	9.330	8.187
400	−18.636	−10.066	−26.318	−34.975	−94.070	−94.335	52.254	47.363	9.339	7.805
500	−19.316	−7.845	−26.296	−37.144	−94.091	−94.399	52.402	46.123	9.330	7.422
600	−19.916	−5.493	−26.332	−39.311	−94.124	−94.458	52.548	44.853	9.306	7.042
700	−20.429	−3.046	−26.409	−41.468	−94.169	−94.510	52.694	43.559	9.274	6.668
800	−20.857	−.533	−26.514	−43.612	−94.218	−94.556	52.838	42.244	9.233	6.298
900	−21.207	2.029	−26.637	−45.744	−94.270	−94.596	52.979	40.912	9.189	5.934
1000	−21.482	4.625	−26.771	−47.859	−94.321	−94.628	53.117	39.563	9.144	5.574
1100	−21.696	7.247	−26.914	−49.962	−94.371	−94.658	53.250	38.201	9.100	5.220
1200	−21.854	9.887	−27.062	−52.049	−94.419	−94.681	53.380	36.828	9.058	4.869
1300	−21.971	12.535	−27.218	−54.126	−94.469	−94.701	53.505	35.443	9.018	4.521
1400	−22.050	15.195	−27.376	−56.189	−94.515	−94.716	53.625	34.049	8.980	4.177
1500	−22.104	17.859	−27.537	−58.241	−94.562	−94.728	53.739	32.647	8.943	3.835

* All substances are taken to be ideal gases; O_2, N_2, H_2 are not included since they are reference quantities for which $\Delta H = \Delta G = 0$.

+ Reference, *JANAF Interim Thermochemical Tables*, The Dow Chemical Company, Midland, Michigan, Dec. 31, 1961.

Table C – 1 (continued)

Substance / T, K	H_2O		N		NO		O		O_3	
	ΔH_f	ΔG_f	ΔH_f	ΔG_f	ΔH_f	ΔG_f	ΔH_f	ΔG_f	ΔH_f	ΔG_f
0	−57.103	−57.103	112.520	112.520	21.528	21.528	58.989	58.989	34.739	34.739
100	−57.433	−56.557	112.670	111.459	21.581	21.330	59.169	57.993	36.141	37.573
200	−57.579	−55.635	112.819	110.191	21.631	21.056	59.379	56.733	34.254	37.411
298	−57.798	−54.636	112.965	108.870	21.652	20.769	59.559	55.396	34.100	38.997
300	−57.803	−54.617	112.968	108.845	21.652	20.763	59.562	55.370	34.098	39.028
400	−58.042	−53.519	113.116	107.448	21.662	20.466	59.725	53.948	34.026	40.684
500	−58.277	−52.361	113.261	106.014	21.667	20.166	59.870	52.486	34.019	42.351
600	−58.500	−51.156	113.402	104.551	21.670	19.865	59.998	50.997	34.048	44.015
700	−58.710	−49.915	113.535	103.066	21.674	19.565	60.114	49.488	34.097	45.672
800	−58.905	−48.646	113.660	101.562	21.678	19.263	60.217	47.963	34.154	47.321
900	−59.084	−47.352	113.778	100.042	21.682	18.961	60.311	46.425	34.217	48.963
1000	−59.246	−46.040	113.887	98.510	21.687	18.658	60.397	44.877	34.282	50.599
1100	−59.391	−44.712	113.990	96.967	21.692	18.355	60.478	43.322	34.347	52.227
1200	−59.519	−43.371	114.087	95.415	21.696	18.051	60.553	41.759	34.411	53.850
1300	−59.634	−42.022	114.178	93.856	21.701	17.747	60.623	40.189	34.474	55.468
1400	−59.734	−40.663	114.264	92.289	21.704	17.443	60.690	38.616	34.535	57.080
1500	−59.824	−39.297	114.346	90.716	21.706	17.138	60.753	37.036	34.593	58.688
1600	−59.906	−37.927	114.425	89.139	21.707	16.834	60.812	35.453	34.649	60.293
1700	−59.977	−36.549	114.501	87.556	21.707	16.530	60.869	33.866	34.703	61.895
1800	−60.041	−35.170	114.573	85.968	21.706	16.224	60.922	32.276	34.751	63.492
1900	−60.099	−33.786	114.644	84.378	21.703	15.921	60.973	30.684	34.795	65.088
2000	−60.150	−32.401	114.712	82.784	21.699	15.618	61.021	29.089	34.835	66.680

Table C – 2

Values of ΔH_f°, ΔG_f°, ΔS_f° and ΔC_P°

Substance	ΔH_f°, kcal/mole	ΔG_f°, kcal/mole	ΔS_f° cal/mole-deg	ΔC_P, cal/mole-deg
$Ag(s)$	0.00	0.00	10.21	6.09
$Ag^+(aq)$	25.31	18.43	17.67	
$AgCl(s)$	−30.36	−26.22	22.97	12.14
$Ba(s)$	0.00	0.00	15.1	6.30
$Ba^{2+}(aq)$	−128.67		3.0	
$Br(g)$	26.71	19.69	41.81	4.97
$Br^-(aq)$	−28.90	−24.57	19.29	
$Br_2(g)$	7.34	0.75	58.64	8.60
$Br_2(l)$	0.00	0.00	36.4	
$C(g)$	171.70	160.84	37.76	4.98
C(diamond)	0.45	0.68	0.58	1.45
C(graphite)	0.00	0.00	1.36	2.07
$CCl_4(g)$	−25.5	−15.3	73.95	19.96
$CCl_4(l)$	−33.3	−16.4	51.25	31.49
$CHCl_3(l)$	−31.5	−17.1	48.5	27.8
$CH_4(g)$	−17.89	−12.14	44.50	8.54
$CO(g)$	−26.42	−32.81	47.30	6.96
$CO_2(g)$	−94.05	−94.26	51.06	8.87
$CO_3^{2-}(aq)$	−161.63	−126.22	−12.7	
$H_2CO_3(aq)$	−167.0	−149.0	45.7	
$HCO_3^-(aq)$	−165.18	−140.31	22.7	
$C_2H_2(g)$	54.19	50.00	48.00	10.50
$C_2H_4(g)$	12.50	16.28	52.45	10.41
$C_2H_6(g)$	−20.24	−7.86	54.85	12.58
$C_3H_8(g)$	−24.82	−5.61	64.5	
$CH_3OH(l)$	−57.02	−39.73	30.3	19.5
$C_2H_5OH(l)$	−66.36	−41.77	38.4	26.64

Table C – 2 (continued)

Substance	ΔH_f°, kcal/mole	ΔG_f°, kcal/mole	ΔS_f° cal/mole-deg	ΔC_P, cal/mole-deg
$H_2O_2(aq)$	−45.68	−31.47		
$H_2S(g)$	−4.82	−7.89	49.15	8.12
$HS^-(aq)$	−4.22	3.01	14.6	
$S^{2-}(aq)$	10.0	20.0	5.3	
$Hg(g)$	14.54	7.59	41.80	4.97
$Hg(l)$	0.00	0.00	18.5	6.65
$Hg_2Cl_2(s)$	−63.32	−50.35	46.8	24.3
$I(g)$	25.48	16.77	43.18	4.97
$I^-(aq)$	−13.37	−12.35	26.14	
$I_2(g)$	14.88	4.63	62.28	8.81
$I_2(s)$	0.00	0.00	27.9	13.14
$K(s)$	0.00	0.00	15.2	6.97
$K^+(aq)$	−60.04	−67.47	24.5	
$KCl(s)$	−104.18	−97.59	19.76	12.31
$KNO_3(s)$	−117.76	−93.96	68.87	23.01
$Mg(s)$	0.00	0.00	7.77	5.71
$Mg^{2+}(aq)$	−110.41	−108.99	−28.2	
$MgCl_2(s)$	−153.40	−141.57	21.4	17.04
$Mn(s)$	0.00	0.00	7.59	6.29
$Mn^{2+}(aq)$	−52.3	−53.4	−20	
$MnO_2(s)$	−124.2	−111.1	12.7	12.91
$MnO_4^-(aq)$	−129.7	−107.4	45.4	
$N(g)$	112.98	108.88	36.62	4.97
$NH_3(g)$	−11.04	−3.98	46.01	8.52
$NH_4^+(aq)$	−31.74	−19.00	26.97	
$NO(g)$	21.60	20.72	50.34	7.14
$NO_2(g)$	8.09	12.39	57.47	9.06
$N_2(g)$	0.00	0.00	45.77	6.96
$N_2O(g)$	19.49	24.76	52.58	9.25
$N_2O_4(g)$	2.31	23.49	72.73	18.90
$Na(s)$	0.00	0.00	12.2	6.79
$Na^+(aq)$	−57.28	−62.59	14.4	11.88
$NaCl(s)$	−98.23	−91.78	17.30	11.88
$NaHCO_3(s)$	−226.5	−203.6	24.4	20.94

Table C-2 (Continued)

Substance	ΔH_f°, kcal/mole	ΔG_f°, kcal/mole	ΔS_f° cal/mole-deg	ΔC_P, cal/mole-deg
$HCOOH(aq)$	−98.0	−85.1	39.1	
$HCOO^-(aq)$	−98.0	−80.0	21.9	
$CH_3COOH(l)$	−116.4	−93.8	38.2	29.5
$C_6H_6(l)$	11.72	29.76	41.3	32.53
$C_6H_6(g)$	19.82	30.99	64.34	19.52
$Ca(s)$	0.00	0.00	9.95	6.28
$Ca^{2+}(aq)$	−129.77	−132.18	−13.2	
$CaCO_3$(calcite)	−288.45	−269.78	22.2	19.57
$CaCO_3$(aragonite)	−288.49	−269.53	21.2	19.42
$CaO(s)$	−151.9	−144.4	9.5	10.23
$Ca(OH)_2(s)$	−235.80	−214.33	18.2	20.2
$Cl(g)$	29.01	25.19	39.46	5.22
$Cl^-(aq)$	−40.02	−31.35	13.17	
$Cl_2(g)$	0.00	0.00	53.29	8.11
$Cr(s)$	0.00	0.00	5.68	5.58
$Cr^{3+}(aq)$	−61.2	−51.5	−73.5	
$Cr_2O^{2-}(aq)$	−364.0	−315.4	51.5	
$Cu(s)$	0.00	0.00	7.96	5.85
$Cu^{2+}(aq)$	15.39	15.53	−23.6	
$CuCl(s)$	−32.2	−28.4	21.9	
$Fe(s)$	0.00	0.00	6.49	6.03
$Fe^{2+}(aq)$	−21.0	−20.30	−27.1	
$Fe^{3+}(aq)$	−11.4	−2.52	−70.1	
$Fe_2O_3(s)$	−196.5	−177.1	21.5	25.0
$Fe_3O_4(s)$	−267.0	−242.4	35.0	34.3
$H(g)$	52.09	48.58	27.39	4.97
$H^+(aq)$	0.00	0.00	0.00	0.00
$HBr(g)$	−8.66	−12.72	47.44	6.96
$HCl(g)$	−22.06	−22.77	44.62	6.96
$HI(g)$	6.20	0.31	49.31	6.97
$H_2(g)$	0.00	0.00	31.21	6.89
$H_2O(g)$	−57.80	−54.64	45.11	8.02
$H_2O(l)$	−68.32	−56.69	16.72	18.00
$H_2O_2(l)$	−44.84	−27.24		

Table C-2 (Continued)

Substance	ΔH_f°, kcal/mole	ΔG_f°, kcal/mole	ΔS_f° cal/mole-deg	ΔC_P, cal/mole-deg
$Na_2CO_3(s)$	−270.3	−250.4	32.5	26.41
$ONCl(g)$	12.57	15.86	63.0	9.37
$O(g)$	59.16	55.00	38.47	5.24
$OH^-(aq)$	−54.96	−37.60	−2.52	
$O_2(g)$	0.00	0.00	49.00	7.02
$O_3(g)$	34.0	39.06	56.8	9.37
$Pb(s)$	0.00	0.00	15.51	6.41
$PbCl_2(s)$	−85.85	−75.04	32.6	18.4
$S(s$, rhombic)	0.00	0.00	7.62	5.40
$S(s$, monoclinic)	0.071	0.023	7.78	5.65
$SO_2(g)$	−70.96	−71.79	59.40	9.51
$SO_3(g)$	−94.45	−88.52	61.24	12.10
$SO_4^{2-}(aq)$	−216.90	−177.34	4.1	
$Zn(s)$	0.00	0.00	9.95	5.99
$Zn^{2+}(aq)$	−36.43	−35.18	−25.4	
$ZnCl_2(s)$	−99.40	−88.26	25.9	18.3
$ZnO(s)$	−83.17	−76.05	10.5	9.62

APPENDIX D

(1) Table of the Natural Elements

Substance	*Symbol*	*Atomic number*	$\hat{M}$†	*Substance*	*Symbol*	*Atomic number*	$\hat{M}$
Aluminum	Al	13	26.97	Molybdenum	Mo	42	95.95
Antimony	Sb	51	121.76	Neodymium	Nd	60	144.27
Argon	A	18	39.944	Neon	Ne	10	20.183
Arsenic	As	33	74.91	Nickel	Ni	28	58.69
Barium	Ba	56	137.36	Nitrogen	N	7	14.008
Beryllium	Be	4	9.02	Osmium	Os	76	190.2
Bismuth	Bi	83	209.00	Oxygen	O	8	16.0000
Boron	B	5	10.82	Palladium	Pd	46	106.7
Bromine	Br	35	79.916	Phosphorus	P	15	30.98
Cadmium	Cd	48	112.41	Platinum	Pt	78	195.23
Calcium	Ca	20	40.08	Potassium	K	19	39.096
Carbon	C	6	12.010	Praseodymium	Pr	59	140.92
Cerium	Ce	58	140.13	Proctactinium	Pa	91	231
Cesium	Cs	55	132.91	Radium	Ra	88	226.05
Chlorine	Cl	17	35.457	Radon	Rn	86	222
Chromium	Cr	24	52.01	Rhenium	Re	75	186.31
Cobalt	Co	27	58.94	Rhodium	Rh	45	102.91
Columbium	Cb	41	92.91	Rubidium	Rb	37	85.48
Copper	Cu	29	63.54	Ruthenium	Ru	44	101.7
Dysprosium	Dy	66	162.46	Samarium	Sm	62	150.43
Erbium	Er	68	167.2	Scandium	Sc	21	45.10
Europium	Eu	63	152.0	Selenium	Se	34	78.96
Fluorine	F	9	19.00	Silicon	Si	14	28.06
Gadolinium	Gd	64	156.9	Silver	Ag	47	107.880
Gallium	Ga	31	69.72	Sodium	Na	11	22.997
Germanium	Ge	32	72.60	Strontium	Sr	38	87.63
Gold	Au	79	197.2	Sulfur	S	16	32.066
Hafnium	Hf	72	178.6	Tantalum	Ta	73	180.88
Helium	He	2	4.003	Tellurium	Te	52	127.61
Holmium	Ho	67	164.94	Terbium	Tb	65	159.2
Hydrogen	H	1	1.0080	Thallium	Tl	81	204.39
Indium	In	49	114.76	Thorium	Th	90	232.12
Iodine	I	53	126.92	Thulium	Tm	69	169.4
Iridium	Ir	77	193.1	Tin	Sn	50	118.70
Iron	Fe	26	55.85	Titanium	Ti	22	47.90
Krypton	Kr	36	83.7	Tungsten	W	74	183.92
Lanthanum	La	57	138.92	Uranium	U	92	238.07
Lead	Pb	82	207.21	Vanadium	V	23	50.95
Lithium	Li	3	6.940	Xenon	Xe	54	131.3
Lutecium	Lu	71	174.99	Ytterbium	Yb	70	173.04
Magnesium	Mg	12	24.32	Yttrium	Y	39	88.92
Manganese	Mn	25	54.93	Zinc	Zn	30	65.38
Mercury	Hg	80	200.61	Zirconium	Zr	40	91.22

Based on a tabulation published by the *Journal of the American Chemical Society*.

† $\hat{M}$ in g/gmole, kg/kg-mole, or lbm/lbmole.

(2) Data[+] Concerning The Elementary Particles That Combine To Build Up All The Atoms Of The Periodic System Of Elements

NAME AND SYMBOL	CHARGE	MASS	ENERGY	ATOMIC MASS	NUMBER	DIAMETER
Positive + Negative − Neutral 0	absolute e. s. u.	gram	mc^2 ergs	relative $^{12}C = 12$	in one gram	cm
NEUTRON, n	0	1.67482×10^{-24}	1.5053×10^{-3}	1.0086654	5.9708×10^{23}	1.4×10^{-13}
PROTON, p	$+4.80298 \times 10^{-10}$	1.67252×10^{-24}	1.5032×10^{-3}	1.0072766	5.9790×10^{23}	1.4×10^{-13}
ELECTRON, β^-	-4.80298×10^{-10}	9.1091×10^{-28}	8.1869×10^{-7}	5.48597×10^{-4}	1.0978×10^{27}	unknown
POSITRON, β^+	$+4.80298 \times 10^{-10}$	9.1091×10^{-28}	8.1869×10^{-7}	5.48597×10^{-4}	1.0978×10^{27}	unknown

[+] From Key to the Periodic Chart of the Atoms 1969, Edition Sargent-Welch Scientific Company 7300 N. Linder Avenue. Skokie; Illinois 60076

(3) Periodic Table of The Elements

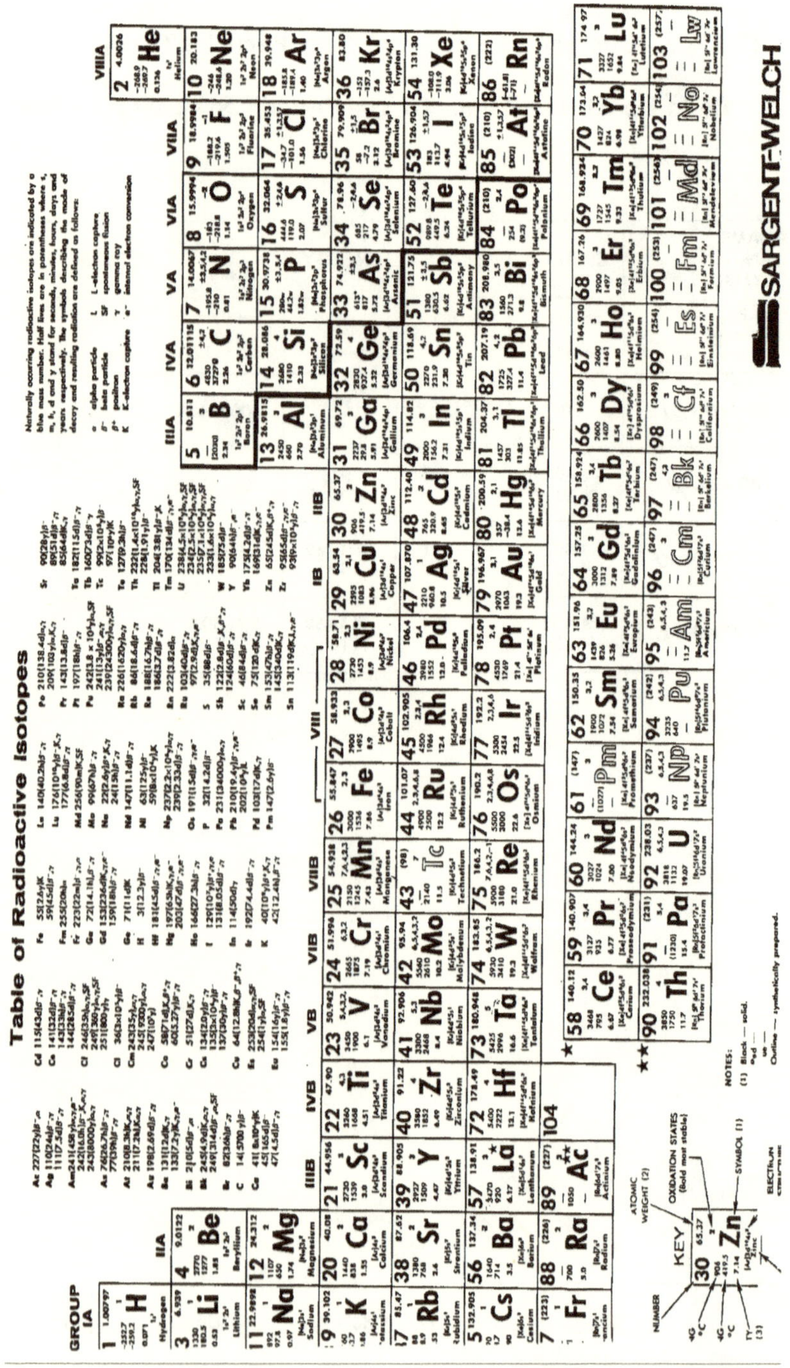

(4)

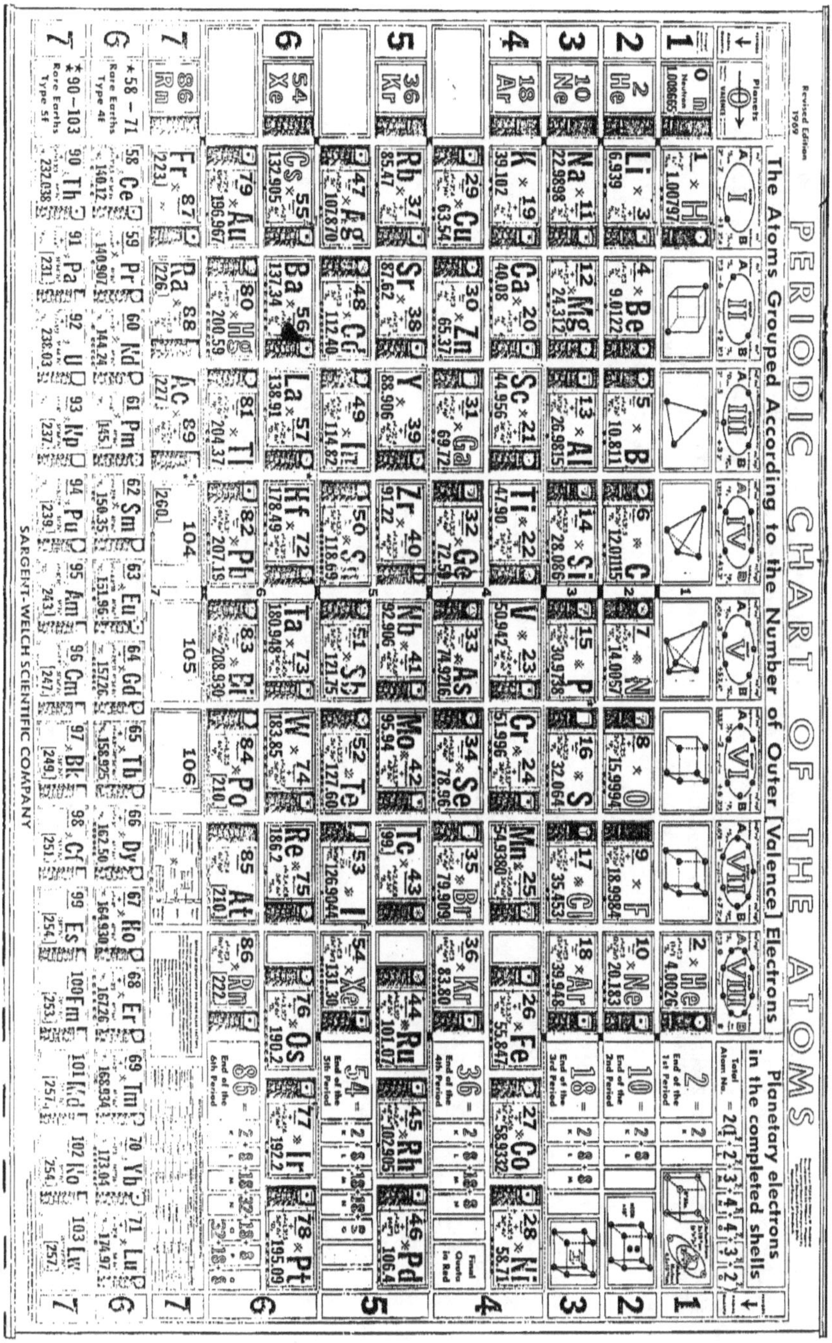

Revised Edition 1969
PERIODIC CHART OF THE ATOMS
The Atoms Grouped According to the Number of Outer [Valence] Electrons
Planetary electrons in the completed shells
Rare Earths Type 4f
Rare Earths Type 5f
SARGENT-WELCH SCIENTIFIC COMPANY

(5) Table of Radioactive Isotopes

Ac 227(22y)β^-,α

Ag 110(24s)β^-,γ; 111(7.5d)β^-,γ

Am 241(458y)α,γ,e^-; 242(16.0h)β^-,K,α,γ; 243(8000y)α,γ

As 76(26.7h)β^-,γ; 77(39h)β^-,γ

At 210(8.3h)K,α,γ; 211(7.2h),Kα,γ

Au 198(2.69d)β^-,γ

Ba 131(12d)K,γ; 133(7.2y)K,γ,e^-

Bi 210(5d)β^-,α

Bk 245(4.9d)K,α,γ; 249(314d)β^-,α,SF

Br 82(36h)β^-,γ

C 14(5700y)β^-

Ca 41(8x10^4y)K; 45(165d)β^-; 47(4.5d)β^-,γ

Cd 115(43d)β^-,γ

Ce 141(32d)β^-,γ; 143(33h)β^-,γ; 144(285d)β^-,γ

Cf 246(35h)α,γ,SF; 249(360y)α,γ,SF; 251(800y)γ

Cl 36(3x10^5y)β^-

Cm 243(35y),α,γ; 245(9300y),α,γ; 247(10^7y)

Co 58(71d),K,β^+,γ; 60(5.27y)β^-,γ

Cr 51(27d),K,γ

Cs 134(2.0y)β^-,γ; 135(3x10^6y)β^-; 137(30y)β^-,γ

Cu 64(12.8h)K,β^-,β^+,γ

Es 253(20d)α,γ,SF; 254(1y)α,SF

Eu 154(16y)β^-,γ; 155(1.8y)β^-,γ

Fe 55(2.6y)K; 59(45d)β^-,γ

Fm 255(20h)α

Fr 223(22m)β^-,γ,α

Ga 72(14.1h),β^-,γ

Gd 153(236d)K,γ,e^-; 159(18h)β^-,γ

Ge 71(11d)K

H 3(12.3y)β^-

Hf 181(45d)β^-,γ,e^-

Hg 197(65h)K,γ,e^-; 203(47d)β^-,γ,e^-

Ho 166(27.3h)β^-,γ

I 129(10^7y)β^+,γ,e; 131(8.05d)β^-,γ

In 114(50d)γ

Ir 192(74.4d)β^-,γ

K 40(10^9y)β^+,K,γ; 42(12.4h),β^-,γ

La 140(40.2h)β^-,γ

Lu 176(10^{10}y)β^-,K,γ; 177(6.8d)β^-,γ

Md 256(90m)K,SF

Mo 99(67h)β^-,γ

Na 22(2.6y)β^+,K,γ; 24(15h)β^-,γ

Nd 147(11.1d)β^-,γ

Ni 63(125y)β^-; 59(8x10^4y),K

Np 237(2.2x10^6y)α,γ; 239(2.33d)β^-,γ

Os 191(15d)β^-,γ,e^-

P 32(14.2d)β^-

Pa 231(34000y)α,γ

Pb 210(19.4y)β^-,γ,e^-; 202(10^5y)L

Pd 103(17d)K,γ

Pm 147(2.6y)β^-

Po 210(138.4d)α,γ; 209(103y)α,K,γ

Pr 143(13.8d)β^-

Pt 197(18h)β^-,γ

Pu 242(3.8 × 10^5y)α,SF; 241(13y)β^-,α,γ; 239(24300y)α,γ,SF

Ra 226(1620y)α,γ

Rb 86(18.6d)β^-,γ

Re 188(16.7h)β^-,γ; 186(3.7d)β^-,γ

Rn 222(3.82d)α

Ru 103(40d)β^-,γ; 97(2.9d),K,γ,e^-

S 35(88d)β^-

Sb 122(2.8d)β^-,K,β^+,γ; 124(60d)β^-,γ

Sc 46(84d)β^-,γ

Se 75(120d)K,γ

Sm 153(47h)β^-,γ; 145(340d)K,γ

Sn 113(119d)K,L,γ,e^-

Sr 90(28y)β^-; 89(51d)β^-,γ; 85(64d)K,γ

Ta 182(115d)β^-,γ

Tb 160(73d)$\beta^-$$\gamma$

Tc 99(2x10^5y)β^-; 97(10^6y)K

Te 127(9.3h)β^-

Th 232(1.4x10^{10}y)α,γ,SF; 228(1.91y)β^-

Tl 204(3.81y)β^-,K

Tm 170(134d)β^-,γ,e^-

U 238(4.5x10^9y)α,γ,SF; 234(2.5x10^5y)α,γ,SF; 235(7.1x10^8y)α,γ,SF; 233(1.6x10^5y)α,γ

W 185(75d)β^-

Y 90(64h)β^-,e^-

Yb 175(4.2d)β^-,γ; 169(31d)K,γ,e^-

Zn 65(245d)K,β^+,γ

Zr 95(65d)β^-,γ,e^-; 93(9x10^5y)β^-,γ

Naturally occurring radioactive isotopes are indicated by a blue mass number. Half lives are in parentheses where s, m, h, d and y stand for seconds, minutes, hours, days and years respectively. The symbols describing the mode of decay and resulting radiation are defined as follows:

α	alpha particle	L	L-electron capture
β^-	beta particle	SF	spontaneous fission
β^+	positron	γ	gamma ray
K	K-electron capture	e^-	internal electron conversion

(6) Percent Ionic Character of a Single Chemical Bond

Difference in electronegativity	0.1	0.2	0.3	0.4	0.5	0.6	0.7	0.8	0.9	1.0	1.1	1.2	1.3	1.4	1.5	1.6
Percent ionic character %	0.5	1	2	4	6	9	12	15	19	22	26	30	34	39	43	47

Difference in electronegativity	1.7	1.8	1.9	2.0	2.1	2.2	2.3	2.4	2.5	2.6	2.7	2.8	2.9	3.0	3.1	3.2
Percent ionic character %	51	55	59	63	67	70	74	76	79	82	84	86	88	89	91	92

(7) Sub – Atomic Particles

	Electron	Positron	Proton	Neutron	Photon	Neutrino	Meson					Hyperon				
Symbol	e^-, β^-	e^+, β^+	†p	†n	γ	†ν	$\mu^\pm$	$\pi^\pm$	†π^0	†K^+	†K^0	†Λ^0	†Σ^+	†Σ^-	†Ξ^-	Ω^-
Mass*	1	1	1836.12	1838.65	0	~0	206.84	273.23	264.4	966.6	974.4	2181.4	2327.7	2343.2	2584	1686
Charge**	−1	+1	+1	0	0	0	±1	±1	0	+1	0	0	+1	−1	−1	−1
Spin	½	½	½	½	1	½	½	0	0	0	0	½	½-integral	½-integral	½-integral	
Magnetic Moment	1.00 B.m.	1.00 B.m.	2.793 n.m.	−1.913 n.m.	0	~0		0	0							
Mean Life (sec.)	stable	stable	stable	1.11×10^3	stable	stable	2.22×10^{-6}	2.54×10^{-8}	$\sim10^{-15}$	$\sim10^{-8}$	$\sim10^{-10}$–10^{-7}	2.8×10^{-10}	$\sim5\times10^{-11}$	$\sim10^{-10}$	$\sim10^{-10}$	$\sim10^{-9}$
Decay Modes				$\rightarrow p+e^-+\bar{\nu}$			$\rightarrow e^\pm+\nu+\bar{\nu}$	$\rightarrow \mu^\pm+\nu$	$\rightarrow \gamma+\gamma$ $\rightarrow e^++e^-+\gamma$	complex	complex	$\rightarrow p+\pi^-$ $\rightarrow n+\pi^0$	$\rightarrow p+\pi^0$ $\rightarrow n+\pi^+$	$\rightarrow n+\pi^-$	$\rightarrow \Lambda^0+\pi^-$	$\pi^-+\Xi^-$

B.m. = Bohr magneton   n.m. = Nuclear magneton   *In units of 9.1083×10^{-31} kg.   **In units of 4.80286×10^{-10} esu.   †Exists as an antiparticle not listed.

APPENDIX E

1. Significant Figures and Notations of Powers of Ten

The idea of significant figures provide a useful method of indicating the accuracy of a numerical result and at the same time of minimising the labour of computations. A *significant figure* is a digit that is believed to be nearer the actual value than any other. Zeros are significant if any other digits precede them in the number, otherwise not. Thus there are three significant figures in 204, 340, 400, and 0.00540. If, in measuring the length of an object with a measuring rod capable of being read to a tenth of a millimetre, the result is two metres, thirty-four centimetres, and no millimetres, this is to be recorded as 2.3400 m, 234.00 cm, or 2340.0 mm but not as 2.34 m, 234 cm, or 2340 mm.

If it is desired to indicate that a value lies between 2,400 and 2,600 and therefore has *two* significant figures, one obviously cannot write this as 2,500, for 2500 has *four* significant figures and indicates that the value lies between 2,499 and 2,501. To avoid this ambiguity, 2500 should be written as 2.5×10^3 when it has only *two* significant figures or as 2.5×10^3 if it has *three*.

Much unnecessary work of calculation can be avoided by employing the following rules for dropping meaningless or non-significant figures:

1. *In easting off non-significant figures*, if the value of the rejected figures is greater than a half unit in the last place retained, increase the last digit retained by 1; if it is less than half, leave this digit unchanged; if it equals half a unit, increase the digit by 1 half the time only – for example, when the last retained digit is even.
2. *In sums and differences*, drop energy digit that falls under a non-significant digit in any of the quantities to be added or subtracted. Thus the sum of 216.526, 16.5, and 2.054 is 235.1, not 235.080.
3. *In products or quotients*, retain the number of significant figures that appear in the least accurately known quantity involved. Thus the product of 314.428 and 11.0 is 3.46×10^3, not 3458.708.
4. *In computing with logarithms*, when any of the quantities which are to be multiplied or divided can be trusted no closer than 0.01 per cent, use a five-place table; if no closer than 0.1 per cent, use a four-place table; if no closer than 1 per cent, use a slide rule.
5. *Where angles are involved*, distances expressed to 2, 3, 4, or 5 significant figures call for angles expressed to the nearest 30 min, 5 min, 1 min, or 0.1 min, respectively, and vice versa.

In writing a number as a power of ten, the number is written as the product of two factors: the first factor contains as many digits as there are significant figures, the decimal point always being made to appear at the right of the first digit; the second factor is a power of ten. Thus the statement that the speed of light in vacuum is 2.99796×10^{10} cm.sec^{-1} implies that this speed has been determined to six significant figures.

2. Common Approximations

In order to avoid laborious computations, approximation formulae should be used whenever possible. Whenever an approximation suggests itself, however, the error introduced by using it should be investigated and the approximation not made unless this error is small enough to leave unaffected any figure that otherwise could be trusted in the result. Examples are shown in Table 1 below

Table 1

True value	Approximate value	When applicable	Approximate error introduced
$1+a+a^2$ . . .	$1+a$	a small	$-a^2$ *
$(1+a)(1+b)$ . .	$1+a+b$	a and b small	$-ab$
$(1\pm a)^m$	$1\pm ma$ †	a small	$-\frac{1}{2}m(m-1)a^2$
$\sqrt{ab}$	$\frac{1}{2}(a+b)$	b nearly equal to a	$+(b-a)^2/8a$
$\sin a$	a radians	a small	$+a^3/6$
$\cos a$	1	a small	$+a^2/2$
$\tan a$	a radians	a small	$-a^3/3$
$\tan a$	$\sin a$	a small	$-a^3/2$

* For example, when $a=0.1$, the error is 1 percent; when $a=0.01$, the error is 0.01 percent.

† m may be either a positive or a negative integer or a fraction.

3. Some Standard Integrals

$$f(n) = \int_0^{\infty} x^n e^{-ax^2}\, dx$$

n	$f(n)$	n	$f(n)$
0	$\frac{1}{2}\sqrt{\pi/a}$	1	$1/2a$
2	$\frac{1}{4}\sqrt{\pi/a^3}$	3	$1/2a^2$
4	$\frac{3}{8}\sqrt{\pi/a^5}$	5	$1/a^3$
6	$\frac{15}{16}\sqrt{\pi/a^7}$	7	$3/a^4$

If n is even, $$\int_{-\infty}^{+\infty} x^n e^{-ax^2}\, dx = 2f(n).$$

If n is odd, $$\int_{-\infty}^{+\infty} x^n e^{-ax^2}\, dx = 0.$$

NAME INDEX

V

W

Y

SUBJECT INDEX

F

G

H

I

J

K

L

M

N

V

W

www.ingramcontent.com/pod-product-compliance
Lightning Source LLC
Chambersburg PA
CBHW031809170526
45157CB00001B/10
9781465373182